Knowledge and Space

Volume 12

Series editor
Peter Meusburger, Department of Geography, Heidelberg University, Heidelberg, Germany

Knowledge and Space

This book series entitled "Knowledge and Space" is dedicated to topics dealing with the production, dissemination, spatial distribution, and application of knowledge. Recent work on the spatial dimension of knowledge, education, and science; learning organizations; and creative milieus has underlined the importance of spatial disparities and local contexts in the creation, legitimation, diffusion, and application of new knowledge. These studies have shown that spatial disparities in knowledge and creativity are not short-term transitional events but rather a fundamental structural element of society and the economy.

The volumes in the series on Knowledge and Space cover a broad range of topics relevant to all disciplines in the humanities and social sciences focusing on knowledge, intellectual capital, and human capital: clashes of knowledge; milieus of creativity; geographies of science; cultural memories; knowledge and the economy; learning organizations; knowledge and power; ethnic and cultural dimensions of knowledge; knowledge and action; and mobilities of knowledge. These topics are analyzed and discussed by scholars from a range of disciplines, schools of thought, and academic cultures.

Knowledge and Space is the outcome of an agreement concluded by the Klaus Tschira Foundation and Springer in 2006.

More information about this series at http://www.springer.com/series/7568

Peter Meusburger • Michael Heffernan
Laura Suarsana
Editors

Geographies of the University

Springer Open

Klaus Tschira Stiftung
Gemeinnützige GmbH

KTS

Editors
Peter Meusburger
Department of Geography
Heidelberg University
Heidelberg, Germany

Michael Heffernan
School of Geography
University of Nottingham
Nottingham, UK

Laura Suarsana
Department of Geography
Heidelberg University
Heidelberg, Germany

ISSN 1877-9220
Knowledge and Space
ISBN 978-3-030-09272-6 ISBN 978-3-319-75593-9 (eBook)
https://doi.org/10.1007/978-3-319-75593-9

Printed on acid-free paper

This Springer imprint is published by the registered company Springer International Publishing AG, part of Springer Nature.
The registered company address is: Gewerbestrasse 11, 6330 Cham, Switzerland

This book is dedicated to Peter Meusburger (1942–2017), the spiritus rector of the Knowledge & Space symposia and the founding editor of this book series. Drawing on his own pioneering research on the productive capacity of local milieux to generate new ideas and fresh thinking, Peter worked consistently to bring together scholars from across the globe and from different disciplines and intellectual traditions to share their ideas and theories about the relationship between human knowledge and geography in the many unique spaces he created in Heidelberg. Sadly, Peter passed away just as this volume went into production in December 2017. We cherish his inspirational leadership as an original researcher, a passionate teacher, and a much-loved colleague.

Johannes Glückler, Mike Heffernan, and Laura Suarsana

Acknowledgements

The editors thank the Klaus Tschira Stiftung for funding the symposia and book series on Knowledge and Space. The team of the Klaus Tschira Stiftung and the Studio Villa Bosch have always been contributing greatly to the success of the symposia. Together with all the authors in this volume, we are especially grateful to David Antal and James Bell for their tireless dedication to quality as technical editors of the chapters and as translators for some of them. Volker Schniepp at the Department of Geography at Heidelberg University has generously helped get figures and maps into shape for publication. We also thank all student assistants and colleagues from the Department of Geography who have helped accomplish the symposium as well as this twelfth volume in the Klaus Tschira Symposia book series. We are especially grateful to Katrin Janzen, Andreas Kalström, Laura Krauß, Pia Liepe, Leslie Ludwig, Anna Mateja Schmidt, Florence Wieder, and Marius Zipf.

Contents

Contributors

Hanne Kirstine Adriansen Danish School of Education, Aarhus University, Copenhagen, Denmark

Inge Adriansen Department of Political Science and Public Management, University of Southern Denmark, Odense, Denmark

Julia Boger World University Service, Deutsches Komitee e.V., Wiesbaden, Germany

Helmut Bott Institute of Urban Planning and Design, University of Stuttgart, Stuttgart, Germany

Allan Cochrane Faculty of Arts & Social Sciences, The Open University, Milton Keynes, UK

Jennifer L. Croissant Department of Gender and Women's Studies, University of Arizona, Tucson, AZ, USA

Flavia T. J. Curvelo Magdaniel Faculty of Architecture and the Built Environment, Delft University of Technology, Delft, The Netherlands

Alexandra C. den Heijer Faculty of Architecture and the Built Environment, Delft University of Technology, Delft, The Netherlands

Henry Etzkowitz International Triple Helix Institute (ITHI), Palo Alto, CA, USA

Johannes Glückler Department of Geography, Heidelberg University, Heidelberg, Germany

John Goddard Centre for Urban and Regional Development Studies (CURDS), Claremont Bridge, Newcastle University, Newcastle upon Tyne, UK

Michael Heffernan School of Geography, University of Nottingham, Nottingham, UK

Howard Hotson Faculty of History, St. Anne's College, University of Oxford, Oxford, UK

Heike Jöns Department of Geography, Loughborough University, Loughborough, UK

Jane Kenway Faculty of Education, Monash University, Victoria, Australia

Jane Knight Department of Leadership, Higher and Adult Education, OISE, University of Toronto, Toronto, Canada; Ali Mazrui Centre for Higher Education Studies, University of Johannesburg, Johannesburg, South Africa

Peter Meusburger Department of Geography, Heidelberg University, Heidelberg, Germany

Robert Panitz Department of Geography, Heidelberg University, Heidelberg, Germany

Ferenc Probáld Department of Regional Geography, Eötvös Loránd University, Budapest, Hungary

Christine Sattler Institute of Psychology, Heidelberg University, Heidelberg, Germany

Eike W. Schamp Department of Human Geography, Goethe University Frankfurt, Frankfurt am Main, Germany

Rainer Christoph Schwinges University of Bern, Institute of History, Bern, Switzerland

Karlheinz Sonntag Institute of Psychology, Heidelberg University, Heidelberg, Germany

Laura Suarsana Department of Geography, Heidelberg University, Heidelberg, Germany

Anthony Welch Department of Education and Social Work, The University of Sydney, Sydney, Australia

Christian Wuttke Department of Geography, Heidelberg University, Heidelberg, Germany

Carl Zillich International Building Exhibition, IBA Heidelberg GmbH, Heidelberg, Germany

Chapter 1
Geographies of the University: An Introduction

Michael Heffernan, Laura Suarsana, and Peter Meusburger

This volume analyzes the history and character of the modern university from a variety of disciplinary perspectives, with particular emphasis on the constitutive significance of geography as a factor shaping the internal and external dynamics of universities and the national and international systems of higher education in which they have operated. In considering the geographies of the university, the essays in this volume deploy two interlinked conceptual approaches derived from Manuel Castells's (1996) formulations of the spatial logics that he claims constitute the essential characteristics of past and present societies: the *space of places* and the *space of flows*. The first approach adopts a placed-based perspective and focuses on the spatial organization of the settings, practices, and ideologies that constitute the key functions of contemporary universities at different geographical scales, including their research, teaching, and learning, as well as their administration, enterprise, and public engagement. The second, flow-based approach addresses the wider networks that constitute universities as seats of research, learning, and expertise. It encompasses, for example, their recruitment of students, academic staff, and other employees; their outgoing and incoming mobilities of people, resources, and knowledges; their conferment of degrees and awards; their formal and informal collaboration in research, teaching, management, enterprise, and public engagement; and their local, regional, national, and international impacts.

Research on universities has a long tradition in several disciplines. Geographical research about universities came to prominence in the 1950s and 1960s as higher

M. Heffernan (✉)
School of Geography, University of Nottingham, Nottingham, UK
e-mail: mike.heffernan@nottingham.ac.uk

L. Suarsana · P. Meusburger
Department of Geography, Heidelberg University, Heidelberg, Germany
e-mail: suarsana@uni-heidelberg.de

P. Meusburger et al. (eds.), *Geographies of the University*, Knowledge and Space 12,
https://doi.org/10.1007/978-3-319-75593-9_1

education rapidly began to expand across the globe. In developed countries university expansion was motivated by a desire to increase the number of graduates in the workforce, especially in the sciences, and to enhance access to higher education for students from disadvantaged backgrounds (Anderson, 2006, pp. 131–133). In developing countries of South America and newly independent former colonies in Africa and Asia, an expanded system of higher education was usually a major priority for economic development and nation-state building (Jöns, 2016, pp. 322–324). Early explicitly geographical investigations of this process of university expansion during the 1950s and 1960s include James W. Harvey's study on the economic relationships between the University of California and the city of Berkeley in the United States (Harvey, 1958); William Balchin's work on the actual and potential university locations in the United Kingdom (Balchin, 1959); Serge Vassal's research on the impact of new university campuses on the integration and fragmentation of urban space in France (Vassal, 1969); and Alois Mayr's comparative investigations of old and new universities in Germany (Mayr, 1979; see chapter by Heffernan & Jöns in this volume, pp. 247–248).

During the 1970s, a wave of geographical research on higher education emerged, notably in Germany, where Robert Geipel, Professor of Applied Geography at the Technical University in Munich, was especially influential. This innovative work examined the locations of universities within cities, regions, and nation states; catchment areas for student recruitment; the economic impacts of universities on local and regional communities; the academic performance of universities, measured by a range of variables now widely deployed as part of the governance of higher education; and the mobility and career trajectories of students and academics (Geipel, 1968, 1971; Giese, 1987; Meusburger, 1976, 1990; for an overview, see Meusburger, 1998, pp. 438–460).

Over the subsequent decades, studies on the geographies of the university have proliferated thematically and in terms of geographical scales to include the regional and social origin of university students (e.g., Giese, 1982; Hoare, 1991; Holdsworth, 2009a; Nutz, 1991) and professors (e.g., Meusburger, 1986, 1990; Meusburger & Schuch, 2010; Weick, 1995); the nature and outcomes of international student mobility (e.g., Brooks & Waters, 2011; Findlay, King, Smith, Geddes, & Skeldon, 2012; Li, Findlay, Jowett, & Skeldon, 1996) and academic travel (e.g., Heffernan & Jöns, 2013; Jöns, 2003, 2008, 2015); the transformation of towns and cities through students (e.g., Chatterton, 2000; Holdsworth, 2009b; Smith, 2008); the development of university-business-government relations and wider regional impacts of universities (e.g., Harrison & Turok, 2017; Lawton Smith, 2007; Lawton Smith, Glasson, Romeo, Waters, & Chadwick, 2013; Lawton Smith, Keeble, Lawson, Morre, & Wilkinson, 2001); the emergence of ethnic inequalities in higher education through different degrees of cultural conformity (Freytag, 2003, 2016); the role of geography in the new plateglass universities of the 1960s (Johnston, 2004); the politics of honorary degree conferment (Heffernan & Jöns, 2007); the internationalization of higher education through the development of knowledge and education hubs (Knight, 2013; Olds, 2007) and branch campuses (Geddie, 2012); the international mobility of degree programs (Waters & Leung, 2013); and critical perspectives on world university rankings (Jöns & Hoyler, 2013;

see also the relevant contributions in Mayr & Nutz, 2002, and the recent review by Freytag, Jahnke, and Kramer, 2015, pp. 20–26).

The most ambitious historical geography of a single university is probably the interdisciplinary *Wissenschaftsatlas* of Heidelberg University, which affords a long-term historical and geographical analysis of the wider spatial relations of this globally important center of higher learning and research (Heffernan, 2013; Meusburger & Schuch, 2012). The *Wissenschaftsatlas* exemplifies how the four key processes that Jöns (2016) identified as pivotal for the development of modern schools and universities from about 1450 to 1970 unfolded in one institution—Heidelberg University—because this university was profoundly affected by "cycles of expansion and contraction"; the emergence of a "core set of common practices," such as doctoral and scientific laboratory research; the "professionalization of learning, teaching, and research"; and a "complex transition from a humanistic to a scientific paradigm" of knowledge production and exchange (p. 310).

This introduction aims to contextualize the 20 peer-reviewed chapters of this book within existing academic literature on five main themes that reflect the book's structure—historical perspectives; the university, knowledge, and governance; the university and the city; the university and the region; and the international university.

Historical Perspectives

Traditions of higher learning were well developed in the ancient civilizations of China, India, and the Islamic world. Some Medieval Islamic universities, such as al-Qarawiyyin in the Moroccan city of Fez (founded 859) and al Azhar in the Egyptian capital of Cairo (970), survive to the present day. The first European universities emerged from informal gatherings of students and scholars in Bologna, Paris, and Oxford from the eleventh century onward. Three distinctive features characterized the medieval European foundings: first, their corporate organization as *universitates magistrorum et scholarium*, involving rights and privileges as well as rules and obligations; second, the award of academic degrees; and third, the organization into the four medieval faculties of philosophy, law, medicine, and theology (Rüegg, 1992a, pp. xix–xx; 1992b, pp. 3–8; Shils & Roberts, 2004).

Universities were founded by Europeans in colonized territories since the early modern period, early examples including the foundings in the Caribbean city of Santo Domingo (1538), Lima (1571), and Mexico City (1595) on the mainland of South America, and in Cambridge, Massachusetts, in North America (Harvard College, 1636; see Roberts, Rodriguez Cruz, & Herbst, 1996). Out of about 1,500 universities in operation globally by 1970, only 131 (9%) were founded before the nineteenth century. Subsequently, the global expansion of higher education increased with relatively equal shares of new universities being founded at ever shorter intervals: 1801–1900: 25%; 1901–1945: 21%; 1946–1960: 25%; and 1961–1970: 20% (Jöns, 2016, p. 324). Since the 1970s, higher education has expanded by approximately 18,500 universities worldwide (92% of those existing in 2012; Krull, 2012, p. 120), growth that has raised participation rates of young people in higher education considerably, albeit very unevenly on a global scale. In the United Kingdom, age-cohort specific participation rates increased from 14% in 1970 to 49% in 2015, whereas participation rates of all postsecondary students varied between different world regions—in 2007—from only 4% in sub-Saharan Africa and 35% in Latin America and the Caribbean to over 70% in North America and Western Europe (Altbach, Reisberg, & Rumbley, 2009, p. vii; Department for Education, 2017, p. 3; Robertson, 2010, p. 19).

The contributors of the first five chapters of this book discuss historical geographies of the European university from the Middle Ages up to the twentieth century, including methodological considerations about the historiographic mapping of landscapes of higher education; the analysis of international academic mobility of students and academics; and facets of the politics of university expansion through the founding of new universities. Rainer Christoph Schwinges presents the Repertorium Academicum Germanicum (RAG), a database that facilitates intensive research on the geographies and social influence of universities. The RAG research project aims to create a digital database containing the biographical data of all graduate scholars who worked within the Holy Roman Empire between 1250 and 1550. Furthermore, the RAG implements the *Gelehrtenatlas* (Atlas of Scholars), a web-based geographic information system that allows for the analysis and visualization of academic mobility, a university's catchment area, and subsequent career trajectories. Beyond the tracking of individuals, the database also assists in the tracking of cohorts, such as scholars from specific areas, in order to compare the catchment areas of different universities. Schwinges demonstrates the capacities and research opportunities of the RAG by using the example of the influential German scholar Winand von Steeg and several depictions of the catchment areas of universities such as Prague and Erfurt.

Peter Meusburger and Ferenc Probáld analyze the scientific and cultural relations between Heidelberg University and Hungary. For various reasons, Hungarian students have been among the most mobile in Europe since the Middle Ages. Reviewing five centuries, the authors focus on the historical periods of 1595 to 1621 and 1789 to 1919, when the relations between Heidelberg University and the Carpathian Basin were especially close. The chapter elucidates to what extent international universities such as Heidelberg are influenced by national and international political developments, power relations, and interests, and how these lead to results ranging from outstanding scientific achievements to academic mediocracy and even irrelevance. The share of Hungarian students attending Heidelberg University was influenced by political, social, religious, intellectual, and economic developments and varied widely over time. The authors examine the causes of fluctuating student mobility from Hungary and Transylvania to Heidelberg in the context of a wider discussion of the relationship between Hungary and Germany. Their analysis yields insights into the regional and social backgrounds of Hungarian students as well as the faculties they visited, their later professions, and the influence of Heidelberg professors on the cultural, academic, and political development in Hungary. The chapter also provides details on biographies and networks of selected outstanding individuals.

Howard Hotson illustrates the value of digital, highly granular data for intellectual histories of Europe. Drawing on matriculation registers from universities in the Holy Roman Empire during the Thirty Years' War, he shows how student migration functioned as a form of intellectual exchange and how the war transformed the academic geography of the Empire and surrounding regions in ways that reflected and sustained existing denominational differences. Methodologically, Hotson points the way toward an intriguing intellectual historical geography of seventeenth-century Europe and calls for more coherent, instantly navigable data sets that would allow further analysis and visualization.

Hanne Kirstine Adriansen and Inge Adriansen adopt a geohistorical perspective to explore the political geography and discourses related to four university foundings in the Danish monarchy from the fourteenth to the twentieth century. They show how the founding of universities was a means for manifesting political independence and supporting the creation of a nation-state. In addition, they reflect more generally on universities as national symbols and institutions, arguing that universities play important roles in the preservation of national language and unofficial national symbols, and may even serve as national symbols, granting countries control over education and knowledge production. The authors conclude that a geographical approach to university history is very valuable and explore the role of internationalization for universities in relation to local and national interests because international exchanges might contribute to the decolonization of knowledge.

Michael Heffernan and Heike Jöns reconstruct the decision-making processes that led to the founding of a new Scottish university in the county town of Stirling. Their analysis of the practices and deliberations of the University Grants Committee (UGC) is set within the historical context of the postwar expansion of British higher education during the late 1950s and early 1960s. Clearly marked by the publication of the Robbins Report on Higher Education in 1963, the Robbins committee recommended that the existing higher education system be extended by a number of new university institutions, including the establishment of a fifth new university in Scotland. The authors turn to previously unexamined documents from the UK National Archives to analyze the debates and decisions on the question of which of the seven competing locations the government was to choose as the location for the new Scottish university. Heffernan and Jöns argue that the absence of geographers in the decision-making contributed to the UGC members' lack of discussion about the long-term economic, social, and cultural consequences of a new university in each of the seven competing places. This absence paved the way for intense lobbying and counterlobbying practices involving different alumni networks that favored Stirling and reflected wider policy cultures in the United Kingdom at the time.

The University, Knowledge, and Governance

Universities have been studied in a range of disciplines because they are important creative environments that generate research and innovation across the sciences and the humanities and fuel economic growth through technological and cultural innovation. Universities educate knowledge workers and future decision-makers in economy and society, and contribute to local, regional, national, and supranational economic and sociocultural development (e.g., Cochrane & Williams, 2013; Goddard & Vallance, 2013; Lawton Smith, 2007; Saxenian, 1994, 2006). Considering this multiscalar nature of universities and their wider impacts as well as the historical origin of universities as institutions chartered initially by both the pope and the king and later also by other agents such as municipal councils, national governments, industrial philanthropists, and other wealthy people (Anderson, 2004, 2006; Rüegg, 1992b, 2004), university governance is situated, as conceptualized by Clark (1983), within the interplay of academic oligarchy, the state, and the market. This raises the key question about the strategies that can be pursued in relation to each of these three spheres in order to enhance creativity and innovation.

In the late nineteenth century a key competitive advantage of the hegemonic German research universities was the government support they received. From 1870 to 1914, affluent states such as Baden invested up to 4.7% of their budgets in universities and other scientific institutions. This support resulted in more generous professorial salaries and better equipped laboratories than was possible at universities that relied on endowments (Meusburger & Schuch, 2010, p. 62). Technological innovation in the industrial society was often dominated by dyadic collaborative relationships between industry and government (Etzkowitz & Leydesdorff, 2000). However, initiatives such as those of Vannevar Bush, Professor of Electrical Engineering at the Massachusetts Institute of Technology (M.I.T.), to develop the well-known vacuum tube firm Raytheon in Cambridge, Massachusetts, in the 1920s, and his subsequent role as the director of the Office of Scientific Research and Development (OSRD) from 1941 to 1947, increasingly entangled university research in both industry and government (Saxenian, 1994, pp. 13–14). The subsequent development of flourishing high-tech regions around M.I.T. (Route 128) and Stanford University (Silicon Valley) inspired the triple-helix model of university-industry-government relations that Etzkowitz (1993) and Etzkowitz and Leydesdorff (1995, 1997) identified not only as the cornerstone of the technoscientific complex underpinning American hegemony in the twentieth century, or what Senator J. William Fulbright famously called the "military-industrial-academic complex" in a Senate speech of 1967 (Shapin, 2012, p. 16), but, more generally, as the basis of productive technological innovation in the knowledge society.

The authors in the second section of this book discuss different aspects of university governance that contribute to innovation, creativity, and quality standards, yet they also reflect on the opposite phenomena of ignorance and absence, which also need to be regarded as a constituent part of settings in which knowledge is produced and debated. Peter Meusburger delineates theoretical and methodological issues of knowledge environments in universities. Until the early 1980s, most research on scientific creativity centered on personal attributes of scholars. Few authors found it necessary to take the social, cultural, and scientific environment into account. Yet, creativity is never the result of individual action alone. A stimulating environment and a talented individual must come together and interact before a creative process can occur. Several milieu factors can promote or hinder scientific creativity and academic careers as well. This chapter addresses the following questions: What is meant by the term *knowledge environment*? Which components make up a local knowledge environment? In what way can a local knowledge environment affect goals, decisions, learning, research processes, and careers of academics? How can one verify the consequences of a knowledge environment?

Henry Etzkowitz reflects on the role of the government in university-based innovation. He discusses the development of a triple-helix system of innovation as a basis of innovation policy under laissez-faire conditions in the United States, which he compares to the development of more statist regimes elsewhere. He describes how, during World War II, direct links were established between government, industry, and the university in the United States. These relations led to a shift in attitude among the scientists involved, whose prewar opposition to government funding was reversed. Universities increasingly sought government funding for research, and a new organizational model transferred large part of decision-making on innovative processes and products to scientists. After World War II, initially linear models of innovation were adopted, relying on the government to supply funding for research in expectance of outcomes such as innovation, technical solutions, and new ideas. Evaluations in the 1960s, however, showed that little research had been turned into innovation, a realization that led to a more structured approach and an enhanced role of government. The changing role that government in the United States and other countries has in innovation is normatively and analytically conceptualized through the triple helix of industry, science, and government, which Etzkowitz describes as the key relationship for innovation in a knowledge-based society.

Christine Sattler and Karl-Heinz Sonntag present the theoretical background and selected evidence from the project "heiQUALITY Cultures." The main objective of this project is to create an empirically based instrument to operationalize quality cultures within higher education institutions. The project has led to the development of the "Quality Culture Inventory (QCI)," which enables organizations to evaluate their current quality-culture empirically and to analyze quality-oriented leadership and strength as well as weaknesses of the organization's quality culture profile. The authors highlight that this process depends on the acceptance and openness of the participants to reflect on the organization's quality culture. The use of both an organizational-psychological survey and a structural-formal questionnaire generates results that enable reflection on quality cultures in higher education institutions and on ways to improve these through targeted-oriented intervention measures.

Jennifer L. Croissant contributes to the emerging debate on ignorance as an "ethnographic object." She adds to the development of theory and lays the foundation for cross-case comparisons in studies on ignorance by identifying points at which the study of ignorance and the study of absence as a broader concept intersect and diverge. She first explores the properties of ignorance and analyzes agnotology and other concepts of ignorance for their different disciplinary origins, discussing the term *ignorance* in relation to synonyms and contrasting terms and proposing five attributes that can be applied to case studies of ignorance and nonknowledge: ontology and epistemology, chronicity, granularity, scale, and intentionality. In a second step, she relates studies of agnotology to general concepts of absences, such as privatives, silences and invisibilities, and symmetry and stupidity.

The University and the City

According to the cultural historian Peter Burke (2000), "the rise of cities and the rise of universities occurred together in Europe from the twelfth century onwards" (p. 33). This great importance of both the local environment for the flourishing of universities and of universities for the economic and sociocultural prosperity of towns and cities has been analyzed for some of the most renowned universities in Europe and the United States from a long-term historical perspective (Bender, 1986). In 1252, for example, King Konrad IV (1228–1254) promoted his newly founded university in Salerno by referring to the beautiful location of the city. Similarly, the founding of other medieval universities, such as Prague, Erfurt, Heidelberg, Regensburg, Ingolstadt, and Tübingen, were justified by geographical arguments, including population size, the healthy location of the city, the beauty of the city, and the security of the food supply (Lorenz, 1999, p. 9).

The important role of universities for the economic prosperity of towns and cities was confirmed in a study of English cities by Parkinson et al. (2006), who argued that universities are "the key to innovation in the city" if they "successfully recruit and retain university graduates" and "encourage sustainable links between the city, the university and local businesses" (p. 104). According to Huber (2012), this twofold strategy of attracting and retaining firms and R&D workers can be effectively supported by cluster policies focusing on labor market initiatives and brand management rather than merely local networking strategies. Goddard and Vallance's (2013) book *The University and the City* explicitly aims to widen research "from the previously dominant focus in this field on universities as agents of knowledge-based development in the economic and political spaces of regions" (p. 1). They seek to do so by discussing research on student life and processes of studentification in combination with studies on the role that university campuses have played in urban development, as well as other impacts of the university on diverse social, cultural, economic, and sustainable features of the city (e.g., Armstrong, Darrall, & Grove-White, 1997; Benneworth, Charles, & Madanipour, 2010; Chatterton 1999, 2000; Gumprecht, 2008; Smith, 2008; Smith & Hubbard, 2014).

Such multidimensional interrelations between universities and cities are discussed in the third section of this book. John Goddard explores the changing nature of links between the university and the city from both a theoretical and practice-oriented perspective. He claims that universities are "key institutions in society" (p. 356), for which a relationship with the surrounding actors of market, government, and civil society is inevitable. He regards civic universities as "urban anchor institutions" (p. 356) that represent possible sources of stability in local economies because of their low susceptibility to economic downturns. Goddard also inquires into the active contributions of universities to place-making, innovation, and social and economic development and develops the approach of a quadruple helix. This latter concept extends the triple-helix model (see chapter by Etzkowitz in this volume) by including civil society and social innovation in the conceptualization of external collaborations of universities. With universities facing growing expectations to contribute to social challenges, as expressed in the European Commission's concept of responsible research and innovation, he argues that the performance of civic roles implies tensions between the university, its local surroundings, and internal structures. Based on the examples of universities working with their four English host cities—Newcastle, Manchester, Sheffield, and Bristol—Goddard's proposed model of the civic university "integrates teaching, research, and engagement with the outside world such that each enhances the other" (p. 362).

Helmut Bott illustrates the change in architectural concepts of university buildings and in the spatial relationship between university, city, and landscape from a long-term historical perspective. He gives an overview on the early European universities, which were similar to urban monasteries and were integrated into the power structures and interests of feudal clerical and secular powers. He shows how the establishment of applied and experimental sciences and the turn toward research universities since the Renaissance has led to new architectural requirements, such as laboratories, observatories, and botanical gardens, and how the invention of letterpress printing created a demand for large libraries as a new building typology. Bott describes how, beginning in the eighteenth century, triple-wing palace universities became a new paradigm that opened them to public space. In the nineteenth century radical reform and liberalization of universities created new faculties and modern research universities such as the University of Berlin. European universities became integrated into urban patterns, a change that combined internal, semipublic, and public space. Bott regards the development of the Anglo-American universities since the seventeenth century as a different type of design because they often consist of ensembles of detached buildings within rural landscapes outside cities, a characteristic that has led to many of today's picturesque university campuses. A current trend he identifies in university design is reurbanization through a growing reintegration of campuses into urban development and structures.

Alexandra C. den Heijer and Flavia T. J. Curvelo Magdaniel probe the relation between universities and cities for the physical setting and functional mix of campuses in the past, present, and future. They argue that universities and the cities they are located in have the same goal of attracting talent and stimulating innovation, the attainment of which depends greatly on the local environment. The authors present empirical evidence from a comparative exploratory study of an international sample of 39 campuses. With regard to physical campus-city relations and functions, they find an "enduring shift in campus development from peripheral to inner-city locations" and a "shift from monofunctional to multifunctional campuses" (p. 451).

Den Heijer and Curvelo Magdaniel conclude that contemporary universities increasingly become part of the city, with which they share ever more physical and functional resources. They argue that this integration holds growing potential for closer campus-city cooperation and increasing awareness among diverse interest groups about how the multiple physical and functional campus-city relations could improve decision-making in both spheres.

Carl Zillich provides insights into the International Building Exhibition (IBA) in Heidelberg. Reviewing the history of IBAs in Germany, he states that the strategies of the IBA were transformed according to the societal system, change that ushered in the rather soft criteria of IBAs today to allow the IBA to adapt to locality and context. IBAs emphasize not only architecture but also societal evolution and the "urban realm and its underlying governance" (p. 464). The IBA in Heidelberg started in late 2012 and will function as an urban laboratory until 2022, aiming to create spatial potentials for innovation. As a platform, network, and development agency, the IBA brings together actors from the fields of education, science and research, and other realms related to the knowledge-based society, including actors from the private, public, and other sectors of society. The IBA in Heidelberg thus integrates a variety of public and private interests in a combination of top-down and bottom-up strategies. Although not funding construction itself, the event helps to create ideas for new practices of urban development that will address future challenges of Heidelberg City as a so-called "knowledge pearl."

The University and the Region

Since the mid-twentieth century, leading American research universities, including M.I.T. and Stanford University, have had enormous impact on their regional economies by generating a number of start-up and spin-off companies in knowledge-intensive industries such as personal computers and semiconductors—and, more recently, biotechnology—in Silicon Valley and along Route 128. However, research on the most visible British high-tech clusters in Cambridgeshire and Oxfordshire has raised questions about the value of spatial proximity for high-tech innovation, for their firms' most important sources of innovation have been clients and customers located on the national and international scales (Lawton Smith, 2007; Lawton Smith et al., 2001). In order to evaluate the role of physical proximity in a cluster for innovation, it seems to be important to consider variations between high-tech sectors (e.g., information technology [IT] or biotech), the stage of the industry cycle, and the job roles of research participants. In Cambridgeshire, close university-business interactions were important for creating the cluster in the 1970s, but they have decreased with the maturity of the IT industry cycle and the related shift from a product-based to a producer-service-based high-tech system with technical consultancies (Lawton Smith et al., 2001, 2013). Moreover, Huber's (2012) research on the Cambridge IT cluster has demonstrated that the perceived benefits from local knowledge spillovers and networks vary between R&D managers, who value the access to *business knowledge* through local personal networks, and engineers, who see less need to interact locally for accessing new *technological knowledge*.

Based on a decentralized but cooperative industrial system, informal business culture, and support of entrepreneurial risk-taking, Silicon Valley has become a model for regional clusters of high-tech innovation around the world (Saxenian, 1994, 2006). Cook and Joseph (2001) have argued that the global transfer of Silicon Valley's business culture has been difficult because of the historical and geographical specificity in which this leading cluster of technological innovation emerged. Yet, Saxenian (2006) has demonstrated that U.S.-educated engineers who had immigrated from India, China, and Taiwan, and who had returned there after working in Silicon Valley for some time, eventually founded their own companies in their home countries, thereby contributing to the emergence of cross-regional transnational communities of high-tech innovation. This brain circulation, combining return migration with transnationalism, has fostered the emergence of successful high-tech regions in Asia and has thus proven successful at transferring Silicon Valley's business culture internationally through embodied personal experience and face-to-face interactions (Saxenian, 2006).

Drawing on the insight that "a cluster's economic prospect depends on its internal interactions and its ability to identify and access external knowledge sources located far away" (Maskell, Bathelt, & Malmberg, 2006, p. 998), the essays in this third section of the book contribute to a wider research agenda by diversifying the existing research focus on Anglo-American high-tech regions geographically and thematically in two ways. First, they highlight the economic and labor-market impacts of universities in Africa and China as well as Europe; second, they call attention to effects that universities have on local communities and sociocultural relationships. Johannes Glückler, Robert Panitz, and Christian Wuttke examine the impact of universities on the economy of the German federal state of Baden-Württemberg. They argue that universities have qualitative and complex regional impacts that are impossible to quantify fully in monetary terms over long periods and from geographical perspectives. They therefore look closely at short-term, periodic monetary effects by calculating the regional expenditures of the university and its members for goods and services, which increase autonomous demand and lever production and employment within a region and beyond. The authors analyze the gross expenditures of the universities and their members as a direct effect on the regional economy and compute the corresponding rise in production and associated supplies in related sectors of the economy as an indirect effect. This rise triggers increased employment in the upstream sectors, which leads to additional income that the authors compute as induced effect. According to their results, the attraction of students and external funding has a stronger influence on the regional impact than public expenditures do.

In a case study of three universities in Cameroon, Eike W. Schamp argues that regional knowledge spillovers in Africa, especially in the nonmetropolitan context, differ from those on other continents in that there is an absence of spillovers through codified, formalized communication, such as licenses, contract research, and collaborative publication. He attributes this lack to the "particular historical, societal, economic, and political context" (p. 533) because the nonmetropolitan type of university is peripheral in terms of academic performance, resources, and political influence, which he considers typical for many young public African universities. The author argues that regional engagement and spillovers of knowledge to local societies still occur through several forms of less visible, possibly even "invisible" communications. These are relevant for knowledge spillovers between the universities and their region and are crucial for communicating tacit knowledge to local informal and nonprofit stakeholders as relevant agents of development in peripheral regions.

Julia Boger analyzes the reintegration of academics into regional labor markets in sub-Saharan Africa. Specifically, she studies the experience of graduate students who pursued their education abroad and who have been employed in academia since returning to Ghana and Cameroon, two countries intensely affected by outmigration. Boger points out that sub-Saharan universities lack financial resources, infrastructure, and qualified personnel, and often cannot meet the demand for higher education. Highly educated return migrants are expected to function as agents of change that will spark development through the transfer of knowledge. On the basis of empirical data on the labor market entries of the returning graduates and their professional reintegration, she describes their entry into the labor market as long and difficult. She concludes that the work of returned graduates in the higher education sector can stimulate or cultivate development processes, for example, through capacity building, community services, and innovation at the institutional level. She states, though, that the universities do not fully realize the potential expertise of the returning academics.

Anthony Welch then critically examines conventional views on global regionalism through his mapping of connectivity and cross-border-relations between China's southern borderlands and ASEAN member states in higher education. He gives an overview of relations between China and ASEAN countries and discusses the character, qualities, and limits of China-ASEAN regionalism. He states that ASEAN regionalism is arguably at a very low level because of national resistance to supranational regional initiatives and because of gaps between aspirations and their actual implementations. He then examines regional relations in higher education networks and includes examples from the case of the "Borderland" university from China's southern borderlands, which plays an important role in higher education relations in the ASEAN region. Welch argues that cross-border flows are very intense, though irregular and often illegal, which he considers challenging for the conceptualization of global regions and regionalism, especially because he expects further intensification of these relations in the future.

The International University

Universities are sites of cultural encounter and exchange through diverse international linkages among their students, researchers, and academics (e.g., Gunter & Raghuram, 2017; Jöns, 2015; Madge, Raghuram, & Noxolo, 2015; Meusburger & Schuch, 2012; Tournès & Scott-Smith, 2017). They have been key knowledge hubs in recent globalization processes "shaped by an increasingly integrated world economy, new information and communications technology (ICT), the emergence of an international knowledge network, the role of the English language, and other forces beyond the control of academic institutions" (Altbach et al., 2009, p. iv). The internationalization of higher education—defined by Knight (2003) as "the process of integrating an international, intercultural or global dimension into the purpose, functions, and delivery of postsecondary education" (p. 2)—has ranked highly on policy agendas of governments and universities in order "to respond to the many demands placed upon them by globalization and as a way for higher education to prepare individuals for engagement in a globalized world" (Altbach et al., 2009, pp. 23–24). After an initial focus on mobility of students, researchers, and academics (Altbach, 1989; Tournés & Scott-Smith, 2017) and curriculum development (OECD, 1996), internationalization strategies have proliferated since the 1990s. They include interinstitutional partnerships, such as joint and mobile degree programs, and the creation of international branch campuses (Knight, 2003; Olds, 2007).

Contemporary global geographies of higher education are highly uneven. This asymmetry is largely the result of long-term historical path dependencies linked to shifts in economic growth that either preceded or coincided with changing academic mobilities and knowledge centers (Taylor, Hoyler, & Evans, 2008). Uneven global power relations were also reinforced through academic travels, collaborations, and appointment practices in the context of European imperialism (Ellis, 2017; Jöns, 2017; Pietsch, 2017). With these historical experiences in mind, Jöns (2015, pp. 385–386) has argued that increasing flows of students, scientists, and scholars from and to China since the mid-1990s indicate an ongoing global shift of knowledge centers that will most likely redirect emphasis from transatlantic to transpacific scientific interactions and knowledge networks. These wider geographical changes are shaped by both a growing commercialization of higher education and the proliferation of neoliberal Anglophone audit cultures in the form of university rankings, which have reinforced the development of different tiers of global higher education (Findlay et al., 2012; Jöns & Hoyler, 2013; Marginson, 2016).

Whereas transnational education programs contribute to epistemological globalization in some ways, Waters and Leung (2017) have shown that these programs can produce highly ambiguous results for the immobile international students because they lack authentic experiences and language skills acquired in the country that exports its degree programs. By comparison, the increasing importance of technologies, especially virtual communication, in university research, teaching, and learning has quite different impacts on the geographies of the university. Rye's (2014) research, for example, shows that global online education can offer access to more cost effective and democratic forms of education for students from developing countries. By contrast, Storme, Faulconbridge, Beaverstock, Derudder, and Witlox (2016) have shown that virtual mobility cannot substitute for physical mobility of researchers and academics, for face-to-face contacts remain important for the exchange of tacit knowledge and the creation of social network ties.

By asking how globalization has affected universities and how universities have contributed to globalization, the authors of the final three chapters of this book investigate ways in which universities are integrated into international networks and developments (Robertson Olds, Dale, & Dang, 2016). Allan Cochrane explores the relationship between universities as place-based institutions and wider globalization processes. His central argument is that even though universities are in many ways connected and active at a global level, they are still locally fixed and embedded within their regions, on which they have significant impacts. He considers the changing conceptualizations of the geographies of higher education and explores the concept of globally integrated, but regionalized, universities that are "placed as development nodes and transmission belts and as active partners in communities" (p. 606). Cochrane examines universities for their institutional and discursive practices, presenting four case studies on different relations between universities and their regions, which are all linked to geographical reimaginations of the universities in their specific places and wider networks. He stresses that the strategic place-based operations and business practices of universities (as employers) have significant local impacts, including local partnerships, property development, and unplanned or unintended consequences, such as changing demographics or a change in the reputation of the city or area.

Jane Kenway examines the geography of the contemporary university with regard to international student mobility and associated university practices. She proposes an understanding of universities as being not only territorially rooted, national, and subnational institutions but also places of regional and transnational routes. She argues that universities have become unbound and examines how "roots and routes" of students and universities conflict and intersect. Kenway discusses student mobility at the global level and highlights the asymmetrical distribution of international student enrollment. By studying mobile Asian students, she illustrates how flows of people, knowledge, and emotion were shaped by responses to Kuan-Hsing Chen's (2010) *Asia as a Method: Toward Deimperialization*, which partly conflicted with the dominant roots of the grounded university.

Jane Knight analyzes international education hubs framed as a new development and a third generation of cross-border higher education, which builds on and includes student mobility as well as the mobility of degree programs and education providers. She defines international education hubs as "a planned effort to build a critical mass of local and international actors strategically engaged in cross-border education, training, knowledge-production and innovation initiatives" (p. 644). Knight offers a typology consisting of three major types: student hubs, which are mainly engaged in education and training; talent hubs, which are geared primarily to generating a skilled workforce; and knowledge-innovation hubs, which produce and distribute knowledge and innovation. She then applies this typology to those six international education hubs that exist at the international level. Lastly, she analyzes international education hubs in their relation to the previous generations of cross-border higher education with regard to geographic outreach and impact.

In conclusion, the chapters in this volume present empirically grounded and theoretically informed research perspectives on the multiscalar geographies of the university as they are practiced in different disciplinary and linguistic epistemic communities. The authors illuminate the great value of both historical and contemporary research perspectives on place-based and flow-based aspects of university life in order to improve understanding of the nature of the university and to inform policies that help shape its multiple and, one hopes, sustainable futures. In that sense this book relaunches an interdisciplinary research agenda on geographies of the university that engages with the spatial dimensions of all the functions of the university and the practices of its members from both historical and contemporary perspectives. Additional inquiry is needed on the topics addressed in this book—and on the many topics that were not addressed—both to diversify research perspectives geographically and thematically and to produce new comparative insights.

References

Altbach, P. G. (1989). The new internationalism: Foreign students and scholars. *Studies in Higher Education, 14*(2), 125–136. doi:https://doi.org/10.1080/03075078912331377446

Altbach, P. G., Reisberg, L., & Rumbley, L. E. (2009). *Trends in global higher education: Tracking an academic revolution*. Paris: UNESCO. Retrieved from http://www.cep.edu.rs/public/Altbach,_Reisberg,_Rumbley_Tracking_an_Academic_Revolution,_UNESCO_2009.pdf

Anderson, R. D. (2004). *European universities from the enlightenment to 1914*. Oxford, UK: Oxford University Press.

Anderson, R. D. (2006). *British universities: Past and present*. London: Hambledon Continuum.

Armstrong, H. W., Darrall, J., & Grove-White, R. (1997). Maximising the local economic, environmental and social benefits of a university: Lancaster University. *GeoJournal, 41*, 339–350. doi:https://doi.org/10.1023/A:1006849629834

Balchin, W. G. V. (1959). *Universities in Great Britain: A geographical conspectus*. Swansea: University College. (Typescript)

Bender, T. (1986). *The university and the city*. Oxford, UK: Oxford University Press.

Benneworth, P., Charles, D., & Madanipour, A. (2010). Building localized interactions between universities and cities through university spatial development. *European Planning Studies, 18*, 1611–1629. doi:https://doi.org/10.1080/09654313.2010.504345

Brooks, R., & Waters, J. (2011). *Student mobilities, migration and the internationalization of higher education*. Basingstoke: Palgrave Macmillan.

Burke, P. (2000). *A social history of knowledge: From Gutenberg to Diderot*. Cambridge, MA: Polity.

Castells, M. (1996). *The rise of the network society*. The Information Age: Economy, Society, and Culture: Vol. 1. Oxford, UK: Blackwell.

Chatterton, P. (1999). University students and city centres: The formation of exclusive geographies: The case of Bristol, UK. *Geoforum*, 30, 117–133. doi:https://doi.org/10.1016/S0016-7185(98)00028-1

Chatterton, P. (2000). The cultural role of universities in the community: Revisiting the university: Community debate. *Environment and Planning A, 32*, 165–181. doi:https://doi.org/10.1068/a3243

Chen, K.-H. (2010). *Asia as method: Toward deimperialization*. Durham: Duke University Press.

Clark, B. R. (1983). *The higher education system: Academic organization in cross-national perspective*. Berkeley: University of California Press.

Cochrane, A., & Williams, R. (2013). Putting higher education in its place: The socio-political geographies of English universities. *Policy & Politics, 41*, 43–58. doi:https://doi.org/10.1332/030557312X645775

Cook, I., & Joseph, R. (2001). Rethinking Silicon Valley: New perspectives on regional development. *Prometheus: Critical Studies in Innovation*, 19, 377–393. doi:https://doi.org/10.1080/08109020110091431

Department for Education. (2017). Participation rates in higher education: Academic years 2006/2007–2015/2016 (Provisional). Retrieved from https://www.gov.uk/government/uploads/system/uploads/attachment_data/file/648165/HEIPR_PUBLICATION_2015-16.pdf

Ellis, H. (2017). Collaboration and knowledge exchange between scholars in Britain and the Empire, 1830–1914. In H. Jöns, P. Meusburger, & M. Heffernan (Eds.), *Mobilities of knowledge* (pp. 141–155). Knowledge and Space: Vol. 10. Cham, Switzerland: Springer. doi:https://doi.org/10.1007/978-3-319-44654-7_8

Etzkowitz, H. (1993). Enterprises from science: The origins of science-based regional economic development. *Minerva, 31*, 326–360. doi:https://doi.org/10.1007/BF01098626

Etzkowitz, H., & Leydesdorff, L. (1995). The triple helix—University-industry-government relations: A laboratory for knowledge based economic development. *EASST Review, 14,* 14–19. Retrieved from https://ssrn.com/abstract=2480085

Etzkowitz, H., & Leydesdorff, L. (Eds.). (1997). *Universities and the global knowledge economy: A triple helix of university-industry-government relations.* London: Cassell.

Etzkowitz, H., & Leydesdorff, L. (2000). The dynamics of innovation: From national systems and "Mode 2" to a triple helix of university-industrygovernment relations. *Research Policy, 29*, 109–123. doi:https://doi.org/10.1016/S0048-7333(99)00055-4

Findlay, A. M., King, R., Smith, F. M., Geddes, A., & Skeldon, R. (2012). World class? An investigation of globalisation, difference and international student mobility. *Transactions of the Institute of British Geographers, New Series*, 37, 118–131. doi:https://doi.org/10.1111/j.1475-5661.2011.00454.x

Freytag, T. (2003). *Bildungswesen, Bildungsverhalten und kulturelle Identität: Ursachen für das unterdurchschnittliche Ausbildungsniveau der hispanischen Bevölkerung in New Mexico* [Education, educational behavior, and cultural identity: Causes of the below-average level of education in the Hispanic population in New Mexico]. Heidelberger Geographische Arbeiten: Vol. 118. Heidelberg: Department of Geography.

Freytag, T. (2016). Educational inequalities reflecting sociocultural and geographical embeddedness? Exploring the place of Hispanics and Hispanic Cultures in higher education and research institutions in New Mexico. In P. Meusburger, T. Freytag, & L. Suarsana (Eds.), *Ethnic and cultural dimensions of knowledge* (pp. 93–108). Knowledge and Space: Vol. 8. Dordrecht, The Netherlands: Springer.

Freytag, T., Jahnke, H., & Kramer, C. (2015). *Bildungsgeographie* [Geographies of education]. Darmstadt: Wissenschaftliche Buchgesellschaft.

Geddie, K. (2012). Constructing transnational higher education spaces: International branch campus developments in the United Arab Emirates. In R. Brooks, A. Fuller, & J. Waters (Eds.), *Changing spaces of education: New perspectives on the nature of learning* (pp. 39–58). Abingdon, UK: Routledge.

Geipel, R. (1968). Der Standort der Geographie des Bildungswesens innerhalb der Sozialgeographie [The position of the geography of education within social geography]. In K. Ruppert (Ed.), *Zum Standort der Sozialgeographie: Wolfgang Hartke zum 60. Geburtstag* (pp. 155–161). Münchener Studien zur Wirtschafts- und Sozialgeographie: Vol. 4. Kallmünz: Lassleben.

Geipel, R. (1971). Die Universität als Gegenstand sozialgeographischer Forschung [The university as an object of social geographic research]. *Mitteilungen der Geographischen Gesellschaft in München, 56,* 17–31.

Giese, E. (1982). Die Anziehungskraft wissenschaftlicher Hochschulen in der Bundesrepublik Deutschland für Studenten [The attractiveness of scientific universities in the Federal Republic of Germany for students]. *Die Erde, 113,* 115–132.

Giese, E. (1987). Hochschulen als Gegenstand regionalwissenschaftlicher Forschung [Universities as a subject of regional studies research]. In E. Giese (Ed.), *Aktuelle Beiträge zur Hochschulforschung* (pp. 1–24). Gießener Geographische Schriften: Vol. 62. Giessen: Justus Liebig University.

Goddard, J., & Vallance, P. (2013). *The university and the city*. London: Routledge.

Gumprecht, B. (2008). *The American college town.* Amherst: University of Massachusetts Press.

Gunter, A., & Raghuram, P. (2017). International study in the global south: Linking institutional, staff, student, and knowledge mobilities. *Globalisation, Societies and Education*. Advance online publication. doi:https://doi.org/10.1080/14767724.2017.1401453

Harrison, J., & Turok, I. (2017). Universities, knowledge and regional development. *Regional Studies, 51*, 977–981. doi:https://doi.org/10.1080/00343404.2017.1328189

Harvey, J. W. (1958). *The university and the city: A study of economic relationships between the University of California and the city of Berkeley*. Unpublished report, Bureau of Public Administration, University of California, Berkeley.

Heffernan, M. (2013). Review: Wissenschaftsatlas of Heidelberg University: Spatio-temporal relations of academic knowledge production. *Journal of Historical Geography*, *40*, 113–114. doi:https://doi.org/10.1016/j.jhg.2013.01.008

Heffernan, M., & Jöns, H. (2007). Degrees of influence: The politics of honorary degrees in the universities of Oxford and Cambridge, 1900–2000. *Minerva*, *45*, 389–416. doi:https://doi.org/10.1007/s11024-007-9065-8

Heffernan, M., & Jöns, H. (2013). Research travel and disciplinary identities in the University of Cambridge, 1885–1955. *The British Journal for the History of Science*, *46*, 255–286. doi: https://doi.org/10.1017/S000708741200074X

Hoare, A. (1991). Bias in university enrolments: A regional analysis. *Regional Studies*, *25*, 459–470. doi:https://doi.org/10.1080/00343409112331346637

Holdsworth, C. (2009a). 'Going away to uni': Mobility, modernity, and independence of English higher education students. *Environment and Planning A*, *41*, 1849–1864. doi:https://doi.org/10.1068/a41177

Holdsworth, C. (2009b). Between two worlds: Local students in higher education and 'scouse'/student identities. *Population, Space and Place*, *15*, 225–237. doi:https://doi.org/10.1002/psp.511

Huber, F. (2012). Do clusters really matter for innovation practices in information technology? Questioning the significance of technological knowledge spillovers. *Journal of Economic Geography*, *12*, 107–126. doi:https://doi.org/10.1093/jeg/lbq058

Johnston, R. (2004). Institutions and disciplinary fortunes: Two moments in the history of UK geography in the 1960s I: Geography in the 'plateglass universities'. *Progress in Human Geography, 28*, 57–77. doi:https://doi.org/10.1191/0309132504ph470oa

Jöns, H. (2003). *Grenzüberschreitende Mobilität und Kooperation in den Wissenschaften: Deutschlandaufenthalte US-amerikanischer Humboldt-Forschungspreisträger aus einer erweiterten Akteursnetzwerkperspektive* [Boundary-crossing academic mobility and collaboration: U.S. Humboldt Award Winners in Germany from an extended actor-network perspective]. Heidelberger Geographische Arbeiten: Vol. 116. Heidelberg: Selbstverlag des Geographischen Instituts der Universität Heidelberg.

Jöns, H. (2008). Academic travel from Cambridge University and the formation of centres of knowledge, 1885–1954 *Journal of Historical Geography, 34*, 338–362. doi:https://doi.org/10.1016/j.jhg.2007.11.006

Jöns, H. (2015). Talent mobility and the shifting geographies of Latourian knowledge hubs. *Population, Space and Place*, 21, 372–389. doi:https://doi.org/10.1002/psp.1878

Jöns, H. (2016). Modern school and university. In B. Lightman (Ed.), *A companion to the history of science* (pp. 310–327). Oxford, UK: Wiley Blackwell.

Jöns, H. (2017). The University of Cambridge, academic expertise, and the British Empire, 1885–1962. In H. Jöns, P. Meusburger, & M. Heffernan (Eds.), *Mobilities of knowledge* (pp. 185–210). Knowledge and Space: Vol. 10. Cham, Switzerland: Springer. doi:https://doi.org/10.1007/978-3-319-44654-7_10

Jöns, H., & Hoyler, M. (2013). Global geographies of higher education: The perspective of world university rankings. *Geoforum, 46*, 45–59. doi:https://doi.org/10.1016/j.geoforum.2012.12.014

Knight, J. (2003). *Internationalization of higher education practices and priorities: 2003 IAU survey report*. Paris: International Association of Universities. Retrieved from https://www.iau-aiu.net/IMG/pdf/internationalisation_2003_0.pdf

Knight, J. (Ed.). (2013). *International education hubs: Student, talent, knowledge-innovation models*. Dordrecht, The Netherlands: Springer.

Krull, W. (2012). Who is leading whom, where to, what for: And how? Governance and empowerment in the university of the twenty-first century. In A. Altmann & B. Ebersberger (Eds.), *Universities in change: Managing higher education institutions in the age of globalization* (pp. 117−133). New York: Springer. doi:https://doi.org/10.1007/978-1-4614-4590-6_7

Lawton Smith, H. (2007). Universities, innovation, and territorial development: A review of the evidence. *Environment and Planning C: Government and Policy, 25*, 98−114. doi:https://doi.org/10.1068/c0561

Lawton Smith, H., Glasson, J., Romeo, S., Waters, R., & Chadwick, A. (2013). Entrepreneurial regions: Evidence from Oxfordshire and Cambridgeshire. *Social Science Information, 52*, 653−673. doi:https://doi.org/10.1177/0539018413499978

Lawton Smith, H., Keeble, D., Lawson, C., Morre, B., & Wilkinson, F. (2001). University-business interaction in the Oxford and Cambridge regions. *Tijdschrift voor economische en sociale geografie, 92*, 88−99. doi:https://doi.org/10.1111/1467-9663.00141

Li, F. L. N, Findlay, A. M., Jowett, A. J., & Skeldon, R. (1996). Migrating to learn and learning to migrate: A study of the experiences and intentions of international student migrants. *International Journal of Population Geography, 2*, 51–67. doi:https://doi.org/10.1002/(SICI)1099-1220(199603)2:1<51::AID-IJPG17>3.0.CO;2-B

Lorenz, S. (1999). Fehlgeschlagen, gescheitert, erfolglos: Vergebliche Versuche von Universitätsgründungen in Regensburg, Lüneburg, Breslau und Pforzheim [Abortive, futile, unsuccessful: Vain attempts to found universities in Regensburg, Lüneburg, Wroclaw, and Pforzheim]. In S.Lorenz, O. Auge, & N. Becker (Eds.), *Attempto—oder wie stiftet man eine Universität? Die Universitätsgründungen der sogenannten zweiten Gründungswelle im Vergleich* (pp. 7–18). Contubernium: Vol. 50. Stuttgart: Steiner.

Madge, C., Raghuram, P., & Noxolo, P. (2015). Conceptualizing international education: From international student to international study. *Progress in Human Geography, 39*, 681–701. doi:https://doi.org/10.1177/0309132514526442

Marginson, S. (2016). Global stratification in higher education. In S. Slaughter & B. Taylor (Eds.), *Higher education, stratification, and workforce development* (pp. 13−34). Higher Education Dynamics: Vol. 45. Cham, Switzerland: Springer.

Maskell, P., Bathelt, H., & Malmberg, A. (2006). Building global knowledge pipelines: The role of temporary clusters. *European Planning Studies, 14*, 997−1013. doi:https://doi.org/10.1080/09654310600852332

Mayr, A. (1979). *Universität und Stadt: Ein stadt-, wirtschafts- und sozialgeographischer Vergleich alter und neuer Hochschulstandorte in der Bundesrepublik Deutschland* [University and city: Old and new campuses in the Federal Republic of Germany from the perspective of urban, economic, and social geography]. Münstersche Geographische Arbeiten: Vol. 1. Paderborn: Schöningh.

Mayr, A., & Nutz, M. (Eds.). (2002). *Nationalatlas Bundesrepublik Deutschland: Bildung und Kultur* [Nationalatlas of the Federal Republic of Germany: Education and Culture]. Nationalatlas Bundesrepublik Deutschland: Vol. 6. Heidelberg: Spektrum Akademischer Verlag.

Meusburger, P. (1976). Entwicklung, Stellung und Aufgaben einer Geographie des Bildungswesens: Eine Zwischenbilanz [Development, standing, and tasks of a geography of the educational system: An interim review]. *Mitteilungen der Österreichischen Geographischen Gesellschaft*, 118, 9−54.

Meusburger, P. (1986). Die Heidelberger Professoren im Jahre 1984 mit besonderer Berücksichtigung ihrer regionalen und sozialen Herkunft [The Heidelberg professors in 1984, with special consideration of their regional and social origins]. *Beiträge zur Hochschulforschung, 1/2,* 63–106.

Meusburger, P. (1990). Die regionale und soziale Herkunft der Heidelberger Professoren zwischen 1850 und 1932 [The regional and social origin of the Heidelberg professors between 1850 and 1932]. In P. Meusburger & J. Schmude (Eds.), *Bildungsgeographische Arbeiten über Baden-Württemberg* (pp. 187–239). Heidelberger Geographische Arbeiten: Vol. 88. Heidelberg: Selbstverlag des Geographischen Instituts.

Meusburger, P. (1998). *Bildungsgeographie: Wissen und Ausbildung in der räumlichen Dimension* [Educational geography: Knowledge and education in the spatial dimension]. Heidelberg: Spektrum Akademischer Verlag.

Meusburger, P., & Schuch, T. (2010). From mediocrity and existential crisis to scientific excellence: Heidelberg University between 1803 and 1932. In P. Meusburger, D. N. Livingstone, & H. Jöns (Eds.), *Geographies of science* (pp. 57–93). Knowledge and Space: Vol. 3. Dordrecht, The Netherlands: Springer. doi:https://doi.org/10.1007/978-90-481-8611-2_4

Meusburger, P., & Schuch, T. (Eds.). (2012). *Wissenschaftsatlas of Heidelberg University: Spatio-temporal relations of academic knowledge production.* Knittlingen: Bibliotheca Palatina.

Nutz, M. (1991). *Räumliche Mobilität der Studierenden und Struktur des Hochschulwesens in der Bundesrepublik Deutschland: Eine Analyse des Entscheidungsverhaltens bei der Studienortwahl und der Einzugsgebiete der Universitäten.* [Regional mobility of students and the structure of higher education in the Federal Republic of Germany: An analysis of the decision-making on university choice and the regional catchment areas of universities]. Kölner Geographische Arbeiten: Vol. 54. Köln: Geographisches Institut der Universität zu Köln.

OECD. (Ed.). (1996, October). *Internationalisation of higher education.* OECD Documents, Centre for Educational Research and Innovation, OECD, Paris.

Olds, K. (2007). Global assemblage: Singapore, foreign universities, and the construction of a "global education hub." *World Development, 35,* 959–975. doi:https://doi.org/10.1016/j.worlddev.2006.05.014

Parkinson, M., Champion, T., Evans, R., Simmie, J., Turok, I., Crookston, M., ... Wood, P. (2006). State of the English cities: A research study: Vol. 1. Retrieved from https://www.liverpool.ac.uk/media/livacuk/publicpolicypractice/State,of,Cities,V1.pdf

Pietsch, T. (2017). Geographies of selection: Academic appointments in the British academic world, 1850–1939. In H. Jöns, P. Meusburger, & M. Heffernan, (Eds.), *Mobilities of knowledge* (pp. 157–183). Knowledge and Space: Vol. 10. Cham, Switzerland: Springer. doi:https://doi.org/10.1007/978-3-319-44654-7_9

Roberts, J., Rodriguez Cruz, A. M., & Herbst, J. (1996). Exporting models. In H. de Ridder-Symoens (Ed.), *Universities in early modern Europe (1500–1800)* (pp. 256–282). A History of the University in Europe: Vol. 2. Cambridge, UK: Cambridge University Press.

Robertson, S. L. (2010). Globalising UK higher education (LLAKES Research Paper 16). Retrieved from https://susanleerobertson.files.wordpress.com/2012/07/2010-robertson-globalising-higher-ed.pdf

Robertson, S. L., Olds, K., Dale, R., & Dang, Q. A. (Eds.). (2016). *Global regionalisms and higher education: Projects, processes, politics.* Cheltenham: Edward Elgar.

Rüegg, W. (1992a). Foreword. In H. de Ridder-Symoens (Ed.), *Universities in the Middle Ages* (pp. xix–xxvii). A history of the university in Europe: Vol. I. Cambridge, UK: Cambridge University Press.

Rüegg, W. (1992b). Themes. In H. de Ridder-Symoens (Ed.), *Universities in the Middle Ages* (pp. 3–34). A history of the university in Europe: Vol. I: Cambridge, UK: Cambridge University Press.

Rüegg, W. (2004). Themes. In W. Rüegg (Ed.), *Universities in the nineteenth and early twentieth centuries (1800–1945)* (pp. 3–31). A History of the University in Europe: Vol. 3. Cambridge, UK: Cambridge University Press.

Rye, S. A. (2014). The educational space of global online higher education. *Geoforum, 51,* 6–14. doi:https://doi.org/10.1016/j.geoforum.2013.09.011

Saxenian, A. (1994). *Regional advantage: Culture and competition in Silicon Valley and Route 128.* Cambridge, MA: Harvard University Press.

Saxenian A. (2006). *The new argonauts: Regional advantage in a global economy*. Cambridge, MA: Harvard University Press.
Shapin, S. (2012). The Ivory Tower: The history of a figure of speech and its cultural uses. *The British Journal for the History of Science 45* (01), 1-27. doi:https://doi.org/10.1017/S0007087412000118
Shils, E., & Roberts, J. (2004). The diffusion of European models outside Europe. In W. Rüegg (Ed.), *Universities in the nineteenth and early twentieth centuries (1800–1945)* (pp. 163–230). A History of the University in Europe: Vol. 3. Cambridge, UK: Cambridge University Press.
Smith, D. P. (2008). The politics of studentification and '(un)balanced' urban populations: Lessons for gentrification and sustainable communities? *Urban Studies, 45*, 2541–2564. doi:https://doi.org/10.1177/0042098008097108
Smith, D. P., & Hubbard, P. (2014). The segregation of educated youth and dynamic geographies of studentification. *Area, 46*, 92–100. doi:https://doi.org/10.1111/area.12054
Storme, T., Faulconbridge, J. R., Beaverstock, J. V., Derudder, B., & Witlox, F. (2016). Mobility and professional networks in academia: An exploration of the obligations of presence. *Mobilities, 12*, 405–424. doi:https://doi.org/10.1080/17450101.2015.1116884
Taylor, P. J., Hoyler, M., & Evans, D. M. (2008). A geohistorical study of 'the rise of modern science': Mapping scientific practice through urban networks: 1500–1900. *Minerva, 46*, 391–410. doi:https://doi.org/10.1007/s11024-008-9109-8
Tournès, L., & Scott-Smith, G. (Ed.). (2017). *Global Exchanges: Scholarships and transnational circulations in the modern world*. New York: Berghahn Books.
Vassal, S. (1969). Les nouveaux ensembles universitaires français: Éléments de géographie urbaine [The new French university groups: Elements of urban geography]. *Annales de Géographie, 78* (426), 131–157. Retrieved from http://www.persee.fr/doc/geo_0003-4010_1969_num_78_426_15836
Waters, J., & Leung, M. (2013). Immobile transnationalisms? Young people and their *in situ* experiences of 'international' education in Hong Kong. *Urban Studies, 50,* 606–620. doi: https://doi.org/10.1177/0042098012468902
Waters, J., & Leung, M. (2017). Trans-knowledge? Geography, mobility, and knowledge in transnational education. In H. Jöns, P. Meusburger, & M. Heffernan, (Eds.), *Mobilities of knowledge* (pp. 269–285). Knowledge and Space: Vol. 10. Cham, Switzerland: Springer. doi: https://doi.org/10.1007/978-3-319-44654-7_14
Weick, C. 1995. *Räumliche Mobilität und Karriere: Eine individualstatistische Analyse der baden-württembergischen Universitätsprofessoren unter besonderer Berücksichtigung demographischer Strukuren* [Geographical mobility and career: An individual statistical analysis of the university professors of Baden-Württemberg under particular consideration of demographic structures]. Heidelberger Geographische Arbeiten Vol. 101. Heidelberg: Selbstverlag des Geographischen Instituts der Universität Heidelberg.

Part I
Historical Perspectives

Chapter 2
The Repertorium Academicum Germanicum (RAG) and the Geography of German Universities and Academics (1350–1550)

Rainer Christoph Schwinges

In the Holy Roman Empire universities were not founded by universal rulers such as kings, emperors, or popes but rather by local sovereigns or municipal authorities. Founding universities was considered a sovereign right like the right to build castles, found cities, or endow churches and monasteries. The three great royal dynasties of late medieval Germany led the way. The House of Luxembourg created a university in Prague; the House of Habsburg, in Vienna; and the house of Wittelsbach, in Heidelberg. Other princes of the Empire followed suit and established their own state universities, as was the case with the Houses of Saxony (the Universities of Leipzig and Wittenberg), Brabant (Louvain), Bavaria (Ingolstadt), Württemberg (Tübingen), and Brandenburg (Frankfurt on the Oder). Major cities such as Cologne, Erfurt, and Basle did so as well.

As important as the universal powers were for the legitimation of the universities—and as important as they remained under the ancien régime—it was the regional powers that sustained these new institutions through the centuries. The Empire was a loosely federated territorial entity, a political structure that explains the rapid proliferation of the universities (Fig. 2.1). At the turn of the sixteenth century, there were already 17 of them in the German Empire, more than anywhere else in Europe. For their founders—whether sovereigns or cities—universities mostly meant prestige and profit for the dynasty, dominion, or location: In Germany they stood in the service of their lands or rulers (see Schwinges, 1998a, 1998b; Verger, 1994).

However, what was more of a marginal benefit to the universities' founders—academic studies, science, and scholarship—ended up initiating one of the most momentous developments in the Empire. After the founding of Prague University in 1348 and of the subsequent German universities, it was possible to acquire scholarly

R. C. Schwinges (✉)
University of Bern, Institute of History, Bern, Switzerland
e-mail: rainer.schwinges@hist.unibe.ch

P. Meusburger et al. (eds.), *Geographies of the University*, Knowledge and Space 12,
https://doi.org/10.1007/978-3-319-75593-9_2

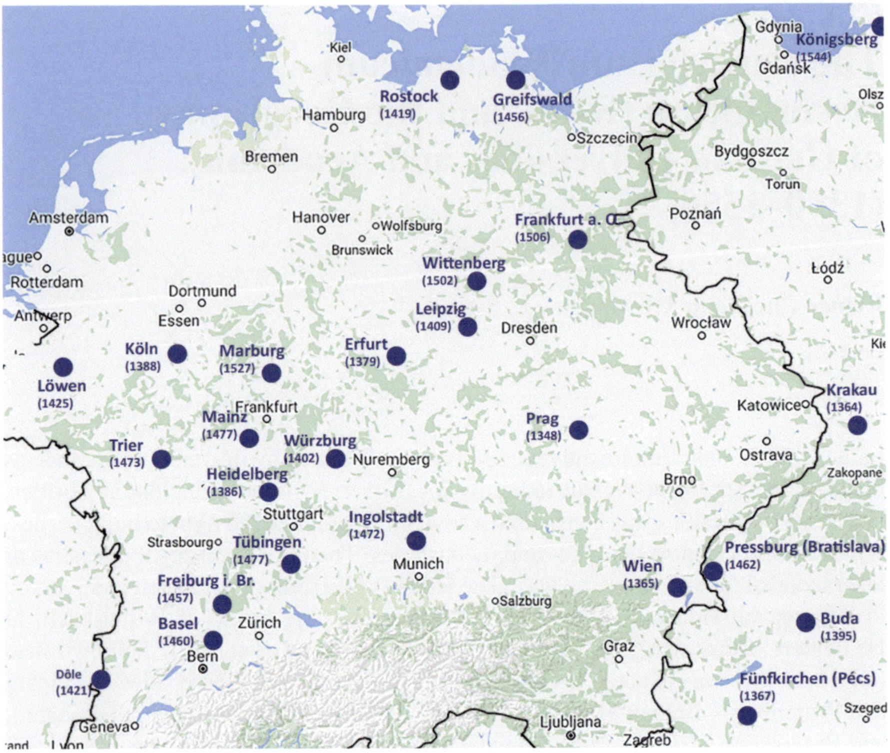

Fig. 2.1 The RAG website showing the German Universities founded in the Middle Ages. Source and copyright: RAG. Reprinted with permission.

knowledge within the borders of the Empire north of the Alps for the first time. The numbers of students and graduates continued to increase, and the supply of academics of varying degrees and faculties steadily rose. These scholars came from the older universities as well as from the new ones, the founding of which increased the recruitment pool in their surrounding areas. With a growth rate of 1.75% per year since 1400 and a total of six thousand people attending annually, the German universities generated an unprecedented volume of academics in the last two decades of the fifteenth century (Schwinges, 1984, pp. 14–16; Schwinges, 1986a, pp. 23–37). By 1550, these numbers had summed to about 300,000 persons. The local sovereigns and cities gradually learned to use this supply of students, scholars, and academic experts.

A purely chronological examination of the history of medieval education and science and the pertinent institutions would, of course, inevitably begin with France and Italy—Europe's centers of theology and philosophy in Paris and of law and medicine in Bologna, Padua, Pavia, Siena, Florence, and Montpellier. Their equivalents founded north of the Alps 150 years later do not seem to compare favorably to these universities. Quantitatively and qualitatively, it took quite a long time for the younger, northern institutions of learning to catch up with those of western and southern Europe (see Schwinges, 1994, pp. 187–191).

However, considering the Empire's graduates and their achievements in theology, canon and civil law, medicine, and the liberal arts,[1] and given their accomplishments regarding the nation, the cities, the church, and society in general, the universities of the Empire well deserve focused attention. For nowhere else in Europe is it possible to learn so much about scholars offering themselves and their knowledge to the world. This opportunity constitutes the initial setting for a comprehensive endeavor, the Repertorium Academicum Germanicum (RAG) and the *Gelehrtenatlas* (Atlas of Scholars).

Repertorium Academicum Germanicum

The RAG is a study in German of the personal histories, life paths, knowledge, and social impact of the academically educated scholars of the Old Empire[2] and of the specific culture that developed around these persons (see Andresen, 2011; Hesse, 2007, 2016; Maurer, 2011; Schwinges, 2008a, 2013; Wagner, 2010). Several other experts and I have been working on this project for more than a decade and have presented results to the public from the outset at www.rag-online.org[3]

The goal of the RAG is (a) to encompass all of the approximately 60,000 scholars who graduated with a *magister artium* (master of liberal arts) or higher degree between 1250 and 1550 and who were active within the Empire and (b) to provide exhaustive curriculum vitae based on a wide selection of sources. So far, all of our approximately 60,000 known figures can be accessed in our online database. Ultimately, we want to understand how the medieval and premodern foundations of a knowledge economy came to exist via these people.

The RAG classifies as scholars all persons who acquired the title of *magister artium* or *baccalareus*, *licentiatus*, or *doctor* of one of the higher faculties (law, medicine, or theology). It includes all those who demonstrably completed study in one of these higher faculties without taking the exams necessary for graduation. That practice was common among the higher nobility, who usually did not want to forgo academic education but whose status was still above university degrees, for they considered it demeaning to be examined by their social inferiors. These approximately 60,000 graduates spearheaded the 300,000 above-mentioned persons

[1]Grammar, logic, rhetoric, arithmetic, geometry, music, and astronomy.

[2]The term Old Empire is equivalent to the Holy Roman Empire from the Middle Ages until its dissolution in 1806.

[3]Under the auspices of the Historical Commission at the Bavarian Academy of Sciences (Munich), the RAG has long been funded by the Deutsche Forschungsgemeinschaft (DFG) and the Schweizerischer Nationalfonds (SNF). The Fritz Thyssen Foundation, too, supported the project from 2001 to 2006. In 2007 the German Union of Academies agreed to contribute to our funding, as did the Swiss Academy of Humanities and Social Sciences in 2008.

originating in the Empire and studying at European universities before 1550. Our ability to state those numbers is due directly to the excellent source material for the Old Empire. The relevant registries and other records of the universities and faculties, including those of the German nations in France and Italy, are preserved in unmatched quantity and are partially accessible digitally or in print (see Giessler-Wirsig & Böhm-Klein, 2006; Paquet, 1992). In addition, the Vatican Archives and the Apostolic Penitentiary contain a wealth of sources from which we have most fruitfully drawn information on the regions of the Old Empire (see Esch, 2010, pp. 52–58; Hörnschemeyer, 2009; Schmugge, 2003; Schmugge, 1998–2014). Only our primary sources lie in these repositories, however. For details on the lives of scholars beyond academia, we comb through virtually the entire historical record.

Arising from the intent to provide a who's who database of the scholars in the Old Empire, the RAG offers broad perspectives. It facilitates qualitative and quantitative conclusions about the Empire's intellectual elite as a whole, about their European connections, and about types and ways of knowledge transfer and personal relations, and it makes institutional and territorial comparisons possible. Processes of academization and professionalization within the German realm, which so far have only been assumed, can now be confirmed empirically.[4]

An Initial Example—A Database Query

A user's query prompts the first search in a search box. With a theoretically unlimited number of fields, the search can be narrowed down by further queries. The period can be specified on the time line running from 1250 to 1550. To find Winand von Steeg, for example (see Fig. 2.2), the user clicks to call up a short biography of him (see Table 2.1). The information is organized thematically, with the first bloc giving his personal data; the second bloc, the data on his education; the third bloc, the year-by-year information on his secular and spiritual activities, offices, positions, and functions; and the fourth bloc (not shown in Table 2.1), sources, literature, and current Internet links to other online data sources concerning the searched person.

To visualize the paths of a scholar's life—at least from the geographical perspective—one clicks to call up a map of Europe showing all the places where the person stayed or left his marks. In Winand's case the image shows the stages of his education (the yellow symbols) in the Netherlands, perhaps at Deventer with the famous Brethren of the Common Life, and at the Universities of Heidelberg and Würzburg. The violet symbols represent his spiritual career stations in Rome, at the Councils of Constance and Basle, and as a churchman in Cologne and Bacharach; his secular activities in the service of King Sigismund in Austria and Hungary; and as a legal advisor to the cities of Nuremberg and Augsburg. Clicking on the life

[4]For basic ideas and relevant literature see Hammerstein (2003) and Schwinges (2008a, 2008b).

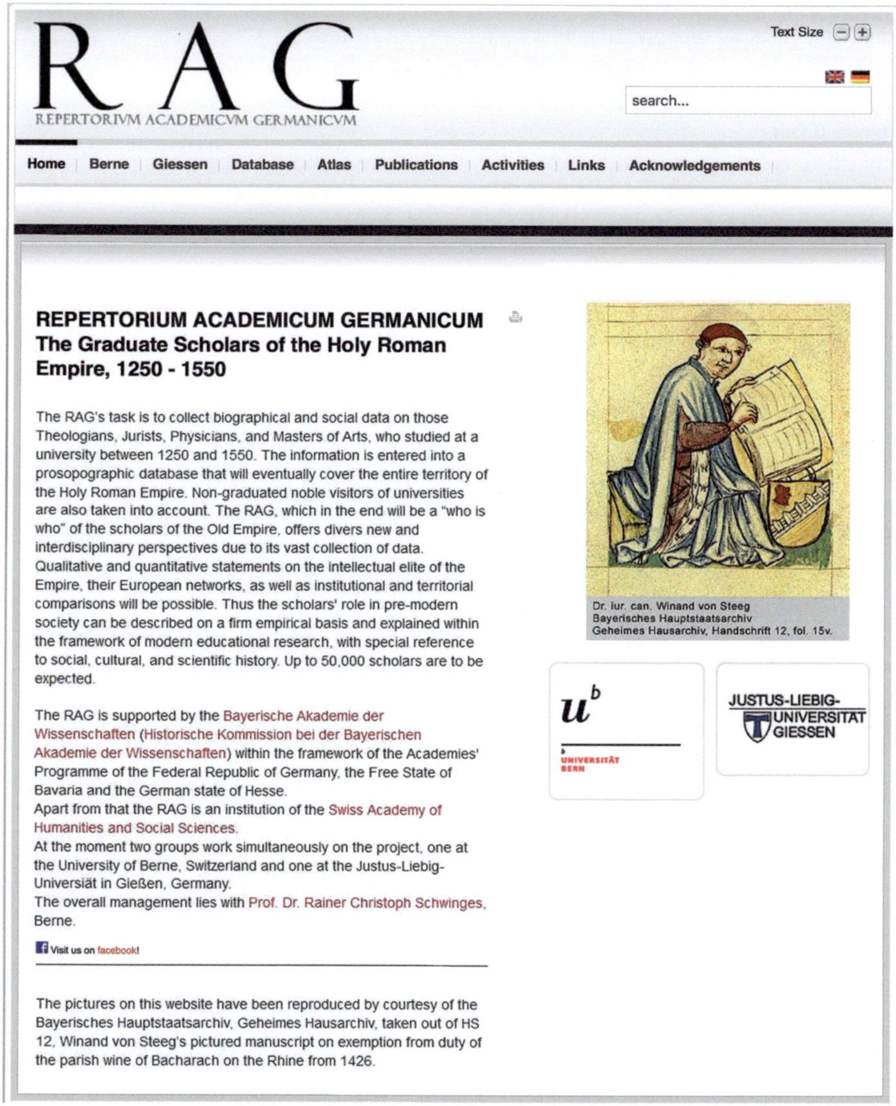

Fig. 2.2 The RAG website, showing a self-portrait of the scholar Winand von Steeg writing an approval certificate. Source and copyright: RAG. Reprinted with permission.

stations (e.g., in Heidelberg, Figs. 2.3 & 2.4) displays each local event along with the corresponding observations. Of course, one can browse for additional scholars and contemporaries and then compare their careers with those of Winand (for his biography see also Schmidt & Heimpel, 1977, pp. 9–31).

Table 2.1 Career steps, activities, and whereabouts of Winand von Steeg (Winand Ort von Steeg; Winandus de Stega; Winandus Ort; Winandus Ort de Stega; Wynand der Stalberger; Wynandus de Stega; Wynandus Ort; Wynandus Ort de Stega) according to a query in the Repertorium Academicum Germanicum (RAG)

Personal Data				
	Origin, geographic		1394	Steeg, diocese: Trier
	Origin, social		1371-05-01	Upper class Steeg
	Birth		1371-05-01	Steeg
Possible	Death		1453-01-19	Koblenz
Possible	Death		1453-07-09	
Education				
Possible	School attendance	before	1394	Deventer, School of the Brethren of the Common Life
	Registration		1394 summer	Heidelberg, group registration: yes
	Graduation		1396-07	Heidelberg, baccalaureus artium
	Graduation		1401-01	Heidelberg, baccalaureus iuris
	Graduation	before	1404	Würzburg, Dr. decretorum (doctor of canon law)
Activities				
	Stay	between	1391 and 1392	Rome, Curia, benefices
	Minister	start	1392-12-16	Weisel, St. Andreas
	Cleric		1392-12-16	Diocese: Trier
	Stay	from	1397 to 1398	Rome, Curia, benefices
	Professor	start	1403	Würzburg, law faculty, subject: Clementines, subject: Liber Sextus
	Professor	end	1411	
	Assessor		1405	
		around	1409–1410	
	Assessor		1404	Würzburg, manorial court
	Canon	start	1405-05-09	Würzburg, St. Johannes in Haug
	Canon	end	1453	
	Canon	after	1405-05-09	Limburg, St. George
Possible	Rector		1407	Würzburg, University
	Preacher		1407	Würzburg
	Preacher		1409	
			1410	
	Speaker		1407	Würzburg, graduation
	General vicar		1409–1410	Diocese: Würzburg
	Arbitrator	before	1409	Würzburg, client: Carmelites
	Lawyer	around	1409	Würzburg, client: Scots' monastery
	Envoy		1410-10-28	Rothenburg, client: German Order
	Legal consultant	start	1411	Nuremberg
	Preacher		1412-12-18	Passau
	Preacher		1413	
			1414	
	Envoy		1415	Konstanz, council, client: Nuremberg, city
	Witness		1416-10-03	Regensburg
	Stay	from	1416 to 1417	Brandenburg, court
	Witness	before	1417-10-22	Amberg, court, marriage contract
	Attendance		1417-11	Konstanz, council
	Trial	start	1417	Passau, cathedral, subject: benefices
	Trial	end	1421	
	Attendance		1418-05	Konstanz, council

(continued)

Table 2.1 (continued)

Activities			
Stay		1418 and 1419	Passau, Vienna, Preßburg, Linz, Stuhlweissenburg, royal court
Secretary		1419-04-14	Buda
Legal consultant		1419-11	Augsburg
Legal consultant		1421-03	
Stay		1419-04-14	Buda, royal court
Stay		1419-08	Bacharach
Legal consultant	start before	1419-07-11	Nuremberg
Legal consultant	end after	1422-03-23	
Scholar		1420-10-02	Würzburg, St. Johannes in Haug, resignation
Scholar	end	1423	
Priest	start before	1420-10-02	Resignation
Priest	end	1421-07-08	Hartkirchen, St. Peter
Altarist		1420-10-02	
Altarist	further evidence	1431-12-17	Würzburg, St. Maria
Altarist		1420-10-02	
Altarist	further evidence	1431-12-17	Röttingen, St. Peter
Canon	start	1421-03-21	Cologne, St. Andreas
Canon	end	1453	
Priest	start	1421-07-16	Bacharach, St. Peter
Priest	end	1453	
Envoy		1421 and 1422	Nuremberg, Imperial Diet
Artist		1426	Bacharach, book illumination, mural painting
Artist	around	1426	Steeg, St. Anna, mural painting
Secretary		1426	
Consultant		1426	
Artist	around	1430	Oberdiebach, St. Mauritius, mural painting
Chaplain	around	1430	
Canon	start	1431-12-17	Koblenz, St. Kastor
Canon	further evidence	1451	
Cantor		1431-12-17	Koblenz, St. Kastor
Dean	start	1439-09	Koblenz, St. Kastor, resignation
Dean	end before	1447-08-14	
Supplicant		1439-10-23	Koblenz, St. Kastor, benefices: Dean
Priest	start	1441-12	Ostheim, St. George
Priest	end	1453	
Judge		1441	Diocese: Trier, ecclesiastical court(s)
Judge		1443	

Note. Retrieved April 20, 2017, from http://www.rag-online.org/pnd/118633562 (translated by the author). The content and data of RAG are in German.

Fig. 2.3 The RAG website showing Winand von Steeg's whereabouts in life
Source and copyright: RAG. Reprinted with permission.

The *Gelehrtenatlas* (Atlas of Scholars)[5]

As the examples illustrate, we have begun to build an atlas of scholars, parallel to the RAG and based on its data. This *Gelehrtenatlas* is a web-based geographical information system, a tool for analyzing and visualizing the geographical mobility of our scholars. It will eventually also show the transfer of knowledge within the Old Empire. In this series entitled Knowledge and Space, I hardly need to justify why the distribution of scholars of every discipline since the late medieval period is one of European history's most fascinating topics, along with the application of their knowledge in their areas of origin and occupation. Students and scholars were a group in transit from their place of origin, the school town, and the place where they attended university—not to mention several benefices and occupations during and

[5]The digital *Gelehrtenatlas* described in this chapter is still a prototype, a work in progress. A new application, the Repertorium Academicum Germanicum, is scheduled for late 2017. By applying the principles of "incoming" and "outgoing," one may include not only the universities but all institutions (e.g., churches, towns, and courts) where scholars have worked during their lifetimes. For the work in progress, consult www.rag-online.org.

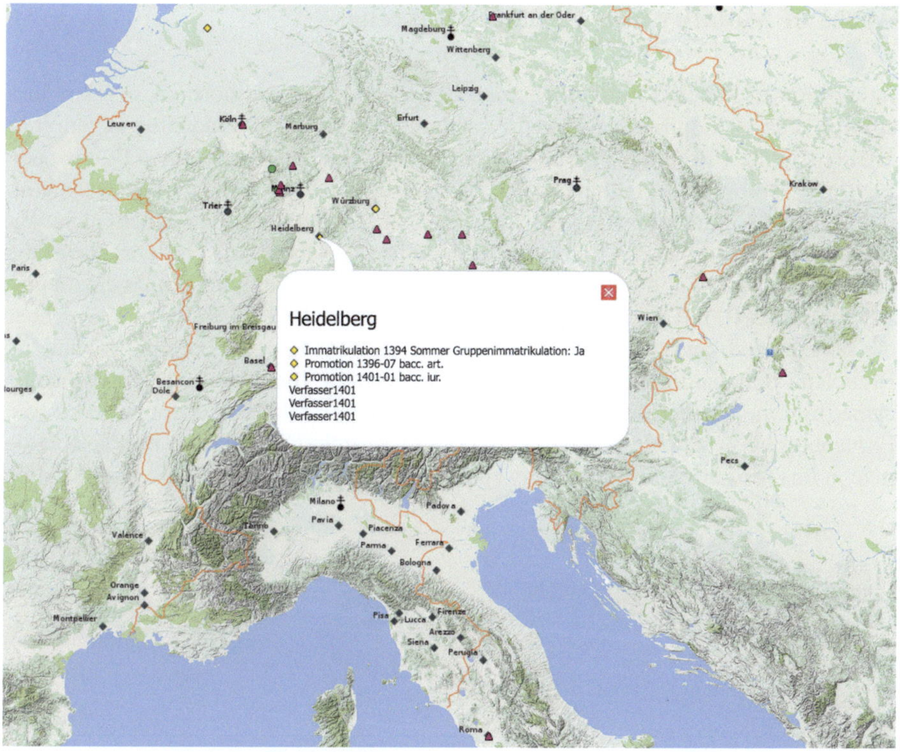

Fig. 2.4 The RAG website showing the dates of Winand von Steeg at the University of Heidelberg. Clicking on a place name (e.g., Heidelberg) calls up a list of Winand von Steeg's activities there Source and copyright: RAG. Reprinted with permission.

after their studies—to the places where their careers took them or their masters sent them. The masters themselves were similarly diverse, ranging from churches and monasteries, emperors and kings, and local sovereigns to cities, schools, and universities. The scholars traveled and operated alone or in groups. They mingled with people, forged links, and left traces both social and intellectual.

During their lives, scholars crafted conference spaces, which functioned both physically and virtually through personal correspondence and dissemination of their writings, for example, or through testaments and foundings. Such geographical data is available along with prosopographic data on our online database (www.rag-online.org), making it possible to examine personal records for a scholar's name, place of origin, universities attended, graduate degrees, registration dates, and sometimes other facts of education and career moves. The details can be linked to geographic or topographic references throughout the Empire and the rest of Europe. We thereby hope to build a fundamental work for addressing both current and future research questions ranging from those about the regional peculiarities of professionalization and transfer to those about the functioning and collaboration of whole regions (for past experiments see, for example, Schwinges, 1986b, 1988, 2003 with maps).

Areas of Mobility and Catchment Areas

In addition to tracking individuals, one can try to present cohorts of scholars in a specific area of mobility, such as the catchment areas of the graduates in canon or civil law at the University of Bologna or Padua within a period defined by the user. Figure 2.5 shows that graduates came from all over the Empire, but primarily from the Rhineland, extending from the Netherlands to Switzerland. The latter catchment area may not be surprising, for during the Middle Ages the Rhine area was one of the Empire's leaders in political, economic, ecclesiastical, and cultural management (see Schwinges, 2001). Adding in the law graduates from the University of Padua over the same period, one can easily compare both catchment areas. Even for Padua, the Rhineland remained the center for recruiting lawyers, but, obviously, more people came from the southeastern parts of the Empire, ranging from Franconia to Austria. Explaining why this pattern emerged requires research. Note that the bubbles on the map in this resolution represent only spatial clusters but that zooming in may enable

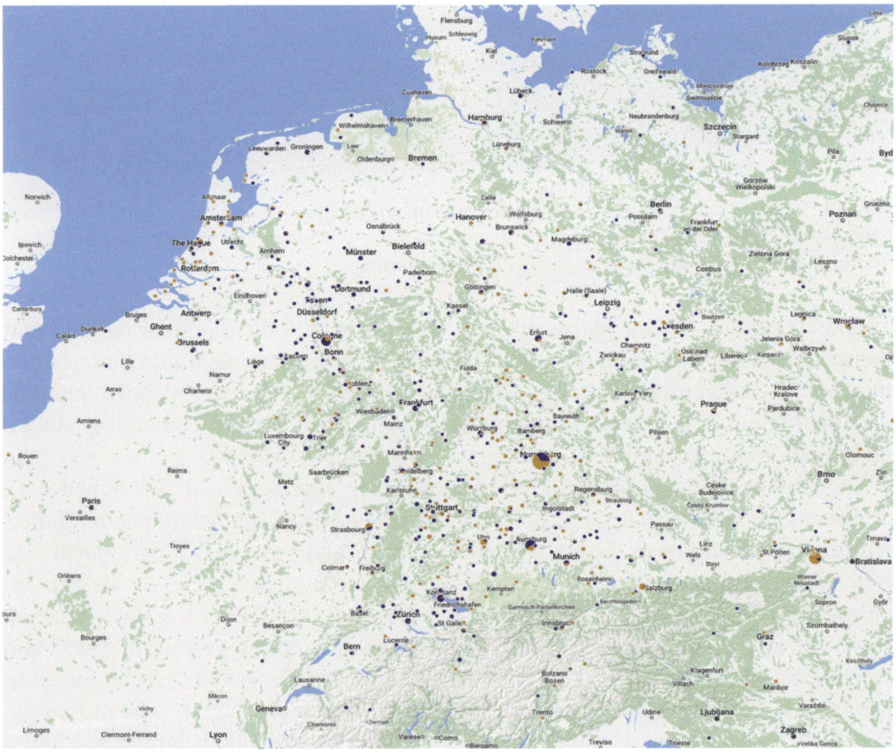

Fig. 2.5 The RAG website showing the catchment areas of Bologna (violet) and Padua (orange) in the Old Empire, 1250–1550.
Source and copyright: RAG. Reprinted with permission.

one to find smaller places of scholars' origins. In the final display of the map, all of these locations can be placed on a list of the persons in question.

Another way to visualize a catchment query is through comparative depiction and analysis of the catchment areas of the German and other European universities, considering either all students or only the graduates. This approach will enable users to draw connections that have been observed only inadequately, if at all. There are maps for individual universities at different points in time, as in the *Wissenschaftsatlas of Heidelberg University* (Meusburger & Schuch, 2012), but there has been no comprehensive, comparative survey of the university spaces, no *Atlas of German Universities and/or Scholars* from their origins until 1550 or later. Those waters are uncharted, a blind spot that requires us to identify and accurately locate all places of origin, profession, and life of the scholars included in the RAG. At present, our target accuracy is more than 80%.

The catchment area of each university has a what is known as a core region, the area in which the university and its scholars were socially rooted at first. The German universities required the protection of a local or municipal authority to ensure the continuity that a foreign body such as a university always initially represented (see Schwinges, 1986a, pp. 229–260; 2000, pp. 36–41). If the founding had no support from a sovereign, it would fail, as happened numerous times (e.g., Cracow in 1364; Vienna in 1365; and Würzburg, Chełmno, Buda, and Geneva). In this spirit prospering universities had a wide, secure regional base that they gradually expanded, forging a relatively stable network that linked the university, the sovereign, and the city with their environment. This framework used to be mistaken for insularity and attributed to recruitment customs supposedly different from those of the nineteenth and twentieth centuries. In fact, however, that insularity should be seen as regional *proximity*, which was ordinarily the basis for successful recruitment of scholars in the late Middle Ages. Anything beyond that scope was an exception calling for a specific explanation.

The universities in the Empire up to the time of the Reformation about the 1520s generally developed their catchment areas in three steps. In an early phase a tight network of relationships was forged between the university, the court, the city, and the core region. This period was followed by one of development and strengthening, during which the core region was expanded, new areas for recruitment found, and the potential of the existing ones boosted. Lastly, during a phase of expansion, the university sometimes attracted as many or even more students from outside its core region than from within it, owing to the extensive geographic mobility of the persons attending. In the more densely populated regions of the Empire—the Rhine area, southern Germany, and the Danube region—the core regions were obviously more compact. Toward the east, with its decreased population density in the low mountain ranges, and in the Baltic, the core regions looked different but generally had the same function.

The maps available by clicking on the localities in Figures 2.6 through 2.11 compare the emergence of core regions as well as their expansion and full range for several universities: Erfurt, Prague, Heidelberg, Vienna, Cologne, Rostock, Greifswald, Wittenberg, and Leipzig. Core regions of the University of Erfurt

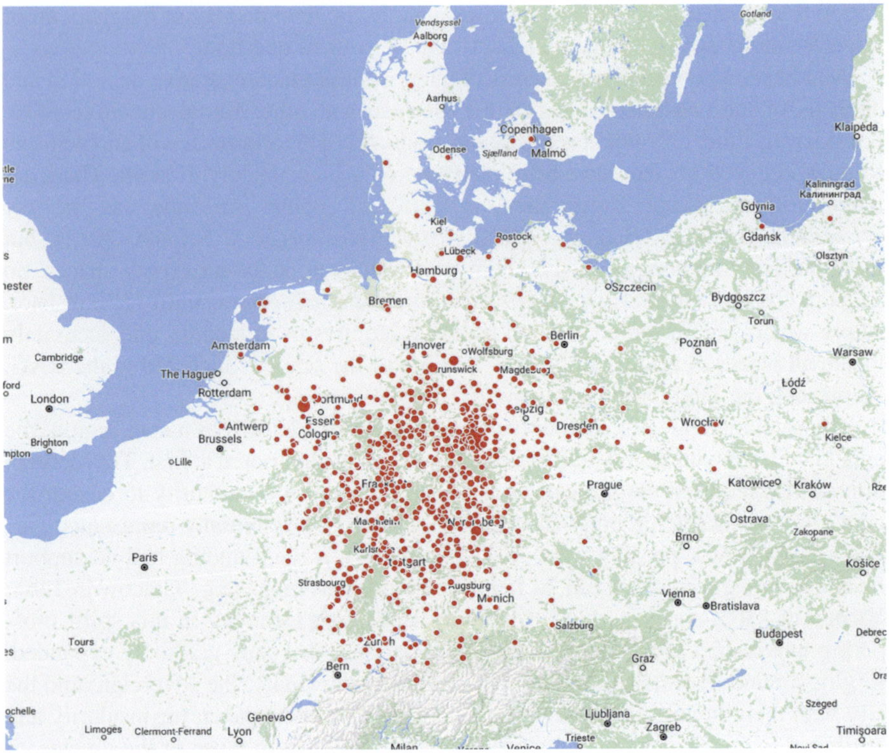

Fig. 2.6 The RAG website showing the core regions and catchment areas of the University of Erfurt, 1440–1550.
Source and copyright: RAG. Reprinted with permission.

(Fig. 2.6) were Thuringia, Hesse, and the southern parts of Lower Saxony. The orientation of recruitment was clearly to the West, nearly between the Elbe and Rhine rivers, territory that also constituted the main sphere of influence of the Prince Elector Archbishop of Mainz (see Schwinges, 1995, pp. 213–222). Core regions usually formed very quickly because they were not randomly composed. An individual's choice of a university was usually not an abstract decision; it was not a rational choice for or against a college town. Apart from political and economic factors, personal connections and bonds of patronage had the greatest influence on the decision (see chapter by Meusburger & Probáld in this volume).

Between students and professors alike, the most apparent connections were those between fellow countrymen. Wherever the students or scholars hailed from, the most common trait between them was simply that the new university was close to their place of origin. Local provenance therefore always meant provenance from a certain social area (see Fig. 2.7, for example). Everyone who left Prague for Heidelberg in

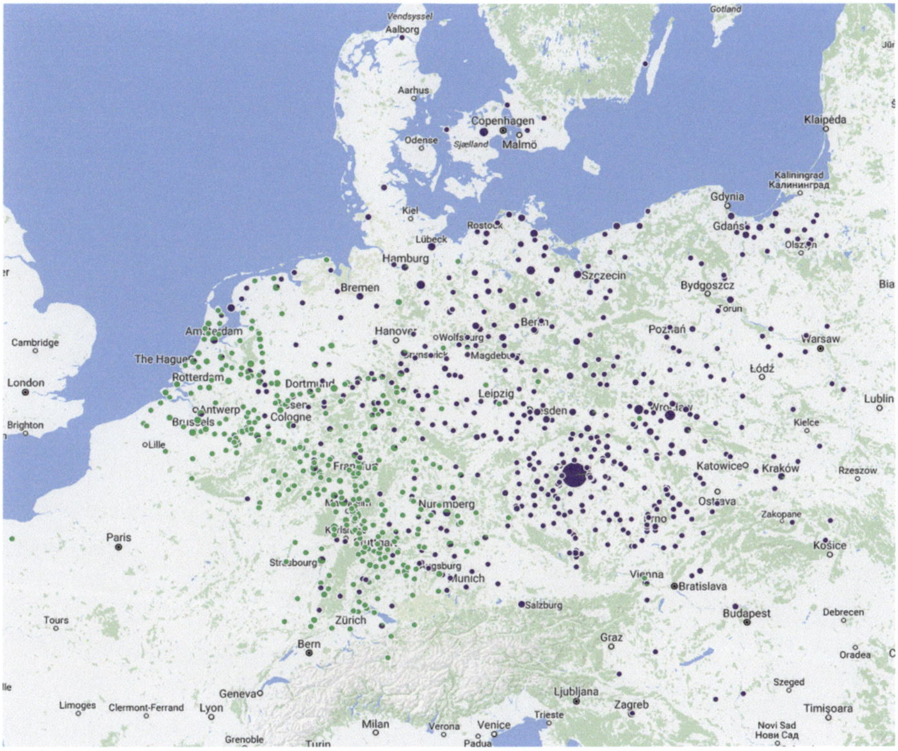

Fig. 2.7 The RAG website showing the catchment areas of Prague (purple) and Heidelberg (green), 1386–1420.
Source and copyright: RAG. Reprinted with permission.

and after 1386 came from the Bishopric of Worms, the Electoral Palatinate, or the Rhineland (including the Netherlands). Everyone else would have been as foreign in Heidelberg as they were in Prague, so they just remained in their home area. The renegades had good reason to expect better social opportunities at the new university founded closer to home than in the surroundings of foreign universities. The determining factors for a successful career were bonds to sovereigns, families, friendships, and relationships between fellow countrymen—in short, the system of patronage and social networks. The universities themselves were embedded in those networks from the very beginning and became ever more deeply enmeshed therein from generation to generation. Where this system failed to develop, prosperity was short-lived.

Cities located along key communication routes had the best prospects for attracting students from outside their immediate surroundings. Such places included Vienna, Cologne, and Erfurt (Fig. 2.8). The same advantage was enjoyed by smaller,

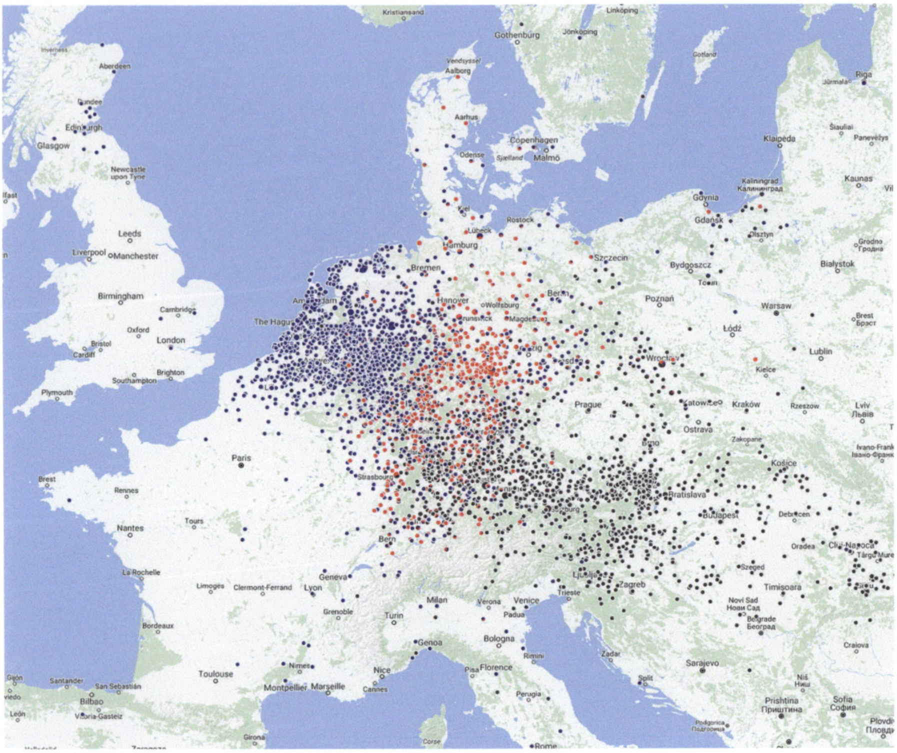

Fig. 2.8 The RAG website showing the catchment areas of the Universities of Vienna (gray), Cologne (blue) and Erfurt (red), 1395–1520.
Source and copyright: RAG. Reprinted with permission.

regional universities managed in collaboration with the town in which they were sited. Examples were Rostock and Greifswald (Fig. 2.9), which were Hanseatic universities as it were. They were centers of attraction far beyond their regions of Mecklenburg and Pomerania and drew students from the entire coastal Hansa region from Flanders to the Baltic. But Greifswald was a very small university of no more than 60 students a year and, accordingly, few graduates (Asche, 2010; Link, 2000, with maps; Pluns, 2007).

The University of Cologne was especially successful in attracting students. Starting with a rather extensive core region consisting of the Bishoprics of Cologne, Utrecht, and Liège, it was recruiting from all over the northwestern parts of the Empire by the end of the fifteenth century, with the external recruits even competing with those from the areas of other universities in the south (Fig. 2.8). Sometimes, old social and economic ties (even long-distance ones) and more recent compatriot links between students and teachers proved stronger than the pull of their more closely

Fig. 2.9 The RAG website showing the catchment areas of the Universities of Rostock (chartreuse) and Greifswald (blue), 1425–1550.
Source and copyright: RAG. Reprinted with permission.

located universities (Meuthen, 1988, pp. 80–85; Schwinges, 1986a, pp. 244–260). This observation also applies to Vienna (Fig. 2.8), which recruited heavily from southern Germany, where most of the imperial cities were naturally aligned with Vienna as one of the imperial centers. Another example is Wittenberg, with its great catchment area after the beginning of the Reformation (Fig. 2.10).

What was true even for Wittenberg was also clearly demonstrated by the Universities of Erfurt, Heidelberg, Leipzig, Vienna, and Cologne up to 1550: The three phases distinguished in this chapter made for great variety and enabled each of these centers of higher learning to cultivate a typical local identity (Fig. 2.11). Regional universities founded by sovereigns or cities recruited their students and graduates mainly from their own area or region; the tendencies of both students and scholars to migrate were aligned with demand.

Because of the importance of regionalism, switching of catchment areas from one university to the next was not an issue. There was no noteworthy interchange between the universities. Although large universities such as Cologne, Erfurt, and Leipzig (Fig. 2.11) expanded into foreign spaces, their encroachment essentially did

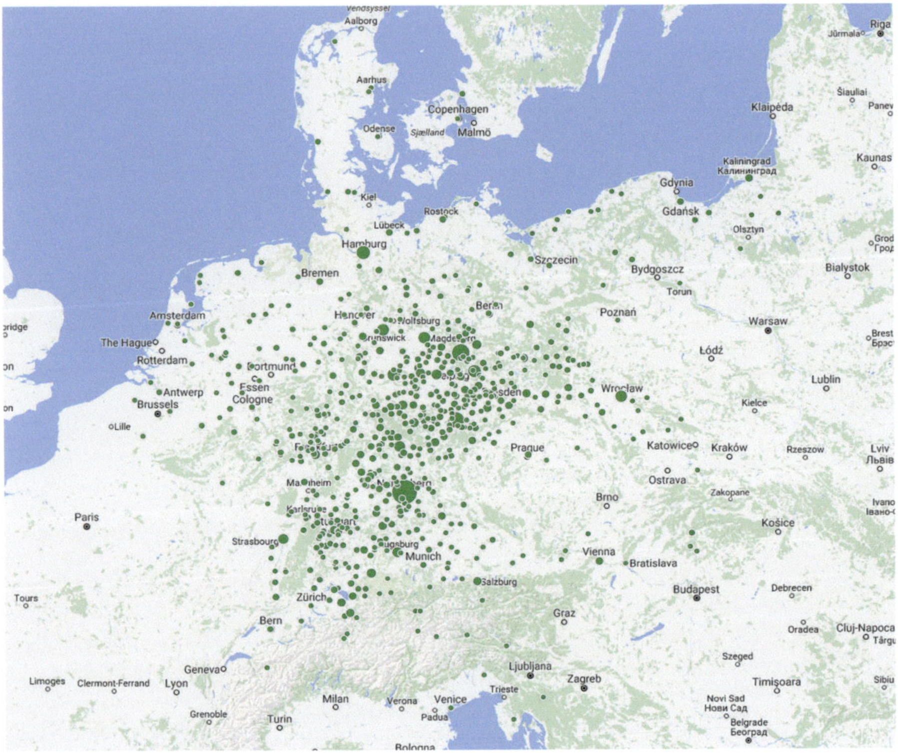

Fig. 2.10 The RAG website showing the catchment area of the University of Wittenberg, 1502–1550.
Source and copyright: RAG. Reprinted with permission.

not compromise others. The mobilization of university regions should not to be mistaken for the mobility of university students; the two processes are distinct. Relocation from one university to another was a primarily social act that must be seen in relation to the social step ladder. That is, mobility of students or academics traveling beyond a university can be interpreted as aristocratic conduct. The nobility, the cities' patricians, and the senior church dignitaries switched universities more frequently than other groups did. It was mainly the German upper class that undertook the great educational journeys to France and Italy. Over 70% of the German law students in Bologna were nobles, as currently verifiable up to 1500 (see Schmid, 2006; Schmutz, 2000, pp. 77–84). Generally, the switchers' network of local and social relationships remained intact. It was common to move within the same greater area—from Cologne to Heidelberg or Louvain, for example—with only a minority choosing more distant universities such as Erfurt, Leipzig, or Basle.

The university towns, however, were not the only important places for recruiting students or academics during the three phases in which university catchment areas developed. Every place of origin played a crucial role, too. Many factors could have a massive influence on an individual's access to a university: the developmental

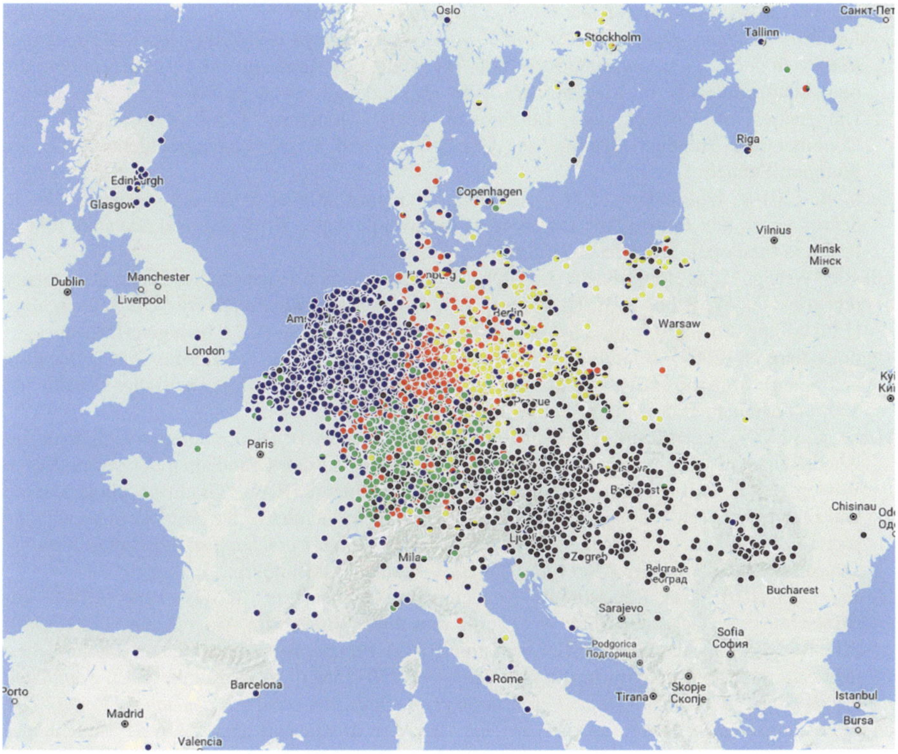

Fig. 2.11 The RAG website showing the catchment areas of the Universities of Erfurt (red), Heidelberg (green), Leipzig (yellow), Vienna (gray) and Cologne (blue), 1400–1550.
Source and copyright: RAG. Reprinted with permission.

stage of his place of origin; that place's status as a city, market town, or village; its population; parochial and academic circumstances; the economy and communications; and the complicated issues of authority and property, including the matter of the local ruler's stability and continuity. But these aspects go beyond the scope of this chapter, for proximity and regionalism are not just geographical matters but rather also matters of social relationships.

References

Andresen, S. (2011). Der Blick hinter die Kulissen des Repertorium Academicum Germanicum (RAG) [A look behind the scenes of the Repertorium Academicum Germanicum (RAG)]. In S. Andresen & R. C. Schwinges (Eds.), *Über Mobilität von Studenten und Gelehrten zwischen dem Reich und Italien 1400–1600 / Della mobilità degli studiosi e eruditi fra l'Impero e l'Italia 1400–1600* (pp. 1–18). Repertorium Academicum Germanicum—Forschungen: Vol. 1. Zurich: vdf Hochschulverlag.

Asche, M. (2010). *Von der reichen hansischen Bürgeruniversität zur armen mecklenburgischen Landeshochschule: Das regionale und soziale Besucherprofil der Universitäten Rostock und Bützow in der Frühen Neuzeit (1500–1800)* [From the rich Hanseatic citizen's university to the poor state university of Mecklenburg: The regional and social profile of students of the Universities of Rostock and Bützow in Early Modernity (1500–1800)] (2nd ed.). Contubernium–Tübinger Beiträge zur Universitäts- und Wissenschaftsgeschichte: Vol. 70. Stuttgart: Steiner.

Esch, A. (2010). *Wahre Geschichten aus dem Mittelalter. Kleine Schicksale selbst erzählt in Schreiben an den Papst* [True stories from the Middle Ages: Brief personal accounts told in letters to the Pope]. Munich: C. H. Beck.

Giessler-Wirsig, E., & Böhm-Klein, J. (2006). Universitäts- und Hochschulmatrikeln [University registries]. In W. Ribbe & E. Henning (Eds.), *Taschenbuch für Familiengeschichtsforschung* (13th ed., pp. 167–170, with a list of enrollments on CD, pp. 96–132). Insingen: Degener.

Hammerstein, N. (2003). *Bildung und Wissenschaft vom 15. bis zum 17. Jahrhundert* [Education and science from the fifteenth to the seventeenth century]. Enzyklopädie Deutscher Geschichte: Vol. 64. Munich: Oldenbourg.

Hesse, C. (2007). Repertorium Academicum Germanicum (RAG): Die graduierten Gelehrten des Alten Reiches zwischen 1250 und 1550. Auf dem Weg zu den Grundlagen der vormodernen Wissensgesellschaft [Repertorium Academicum Germanicum (RAG): Graduate scholars of the Old Empire between 1250 and 1550. On the way to the principles of the premodern knowledge society]. In H. Zedelmaier (Ed.), *Jahrbuch der historischen Forschung in der Bundesrepublik Deutschland, Berichtsjahr 2006* (pp. 105–108). Munich: Oldenbourg.

Hesse, C. (2016). Das Repertorium Academicum Germanicum (RAG). Perspektiven zur Erforschung der Gelehrten, ihrer Netzwerke und ihres Wirkens im Alten Reich (1250–1550) [The Repertorium Academicum Germanicum (RAG): Perspectives of research on scholars, their networks, and their activity in the Old Empire (1250–1550)]. In C. Reinle (Ed.), *Stand und Perspektiven der Sozial- und Verfassungsgeschichte zum römisch-deutschen Reich. Der Forschungseinfluss Peter Moraws auf die deutsche Mediävistik* (pp. 53–64). Studien und Texte zur Geistes- und Sozialgeschichte des Mittelalters: Vol. 10. Affalterbach: Didymos.

Hörnschemeyer, J. (2009). DENQ [Digital editions of postmedieval sources]. In M. Matheus & H. Wolf (Eds.), *Bleibt im Vatikanischen Geheimarchiv vieles zu geheim? Historische Grundlagenforschung in Mittelalter und Neuzeit* (pp. 13–18). Rome: Deutsches Historisches Institut. Retrieved from http://dhi-roma.it/fileadmin/user_upload/pdf-dateien/Online-Publikationen/Dresden_Histtag/Hist_Grundlagenforschung_Mittelalter_Neuzeit.pdf

Link, A. (2000). *Auf dem Weg zur Landesuniversität. Studien zur Herkunft spätmittelalterlicher Studenten am Beispiel Greifswald (1456–1524)* [On the way to the state university: Studies on the origin of late-medieval students, Greifswald (1456–1524)]. Beiträge zur Geschichte der Universität Greifswald: Vol. 1. Stuttgart: Steiner.

Maurer, T. (2011). La mobilité des savants germaniques au Moyen Âge et le résultat du Repertorium Academicum Germanicum [The mobility of German scholars in the Middle Ages and the result of the Repertorium Academicum Germanicum]. In J. Hiernard, D. Turrel, & Y. Delmas-Rigoutsos (Eds.), *Les routes européennes du savoir. Vita peregrinatio. Fin du Moyen Âge–XVII*[e] *siècle* (pp. 39–47). Paris: Les Indes savantes.

Meusburger, P., & Schuch, T. (Eds.). (2012). *Wissenschaftsatlas of Heidelberg University: Spatio-temporal relations of academic knowledge production*. Knittlingen: Bibliotheca Palatina.

Meuthen, E. (1988). *Die alte Universität* [The old university]. Kölner Universitätsgeschichte: Vol. 1. Cologne: Böhlau.

Paquet, J. (1992). *Les matricules universitaires* [University registries]. Typologie des sources du moyen âge occidental: Vol. 65. Turnhout: Brepols.

Pluns, M. A. (2007). *Die Universität Rostock 1418–1563: Eine Hochschule im Spannungsfeld zwischen Stadt, Landesherren und wendischen Hansestädten* [Rostock University, 1418–1563: A university poised between city, sovereign prince, and Wendish Hansa towns]. Quellen und Darstellungen zur Hansischen Geschichte: Vol. 58. Cologne: Böhlau.

Schmid, T. (2006). Iusticie cultores: Die deutschen Rechtsstudenten an der Universität Bologna 1426–1525 [The harvesters of law: German students of law at the University of Bologna 1426–

1525]. (Unpublished master's thesis). University of Bern, Philosophical Historical Faculty. Berne, Switzerland.

Schmidt, A., & Heimpel, H. (1977). *Winand von Steeg (1371–1453), ein mittelrheinischer Gelehrter und Künstler und die Bilderhandschrift über Zollfreiheit des Bacharacher Pfarrweins auf dem Rhein aus dem Jahr 1426* (Handschrift 12 des Bayerischen Geheimen Hausarchivs zu München) [Winand von Steeg (1371–1453), a scholar and artist from the Middle-Rhine and the 1426 illuminated manuscript about exemption from import duties on wine from the Bacharach parish on the Rhine from the year 1426 (Manuscript 12 of the Bavarian State Secret Archives in Munich)]. Abhandlungen. Bayerische Akademie der Wissenschaften, Philosophisch-Historische Klasse; N.F.: Vol. 81. Munich: Verlag der Bayerischen Akademie der Wissenschaften.

Schmugge, L. (2003). Über die Pönitentiarie zur Universität [From the Vatican penitentiary to the university]. In C. Hesse, B. Immenhauser, O. Landolt, & B. Studer (Eds.), *Personen der Geschichte—Geschichte der Personen. Studien zur Kreuzungs-, Sozial- und Bildungsgeschichte. Festschrift für Rainer Christoph Schwinges* (pp. 255–268). Basle: Schwabe.

Schmugge, L. (Series Ed.). (1998–2014). *Repertorium Poenitentiariae Germanicum*. 9 Vols. Tübingen: Max Niemeyer. Berlin: De Gruyter.

Schmutz, J. (2000). *Juristen für das Reich. Die deutschen Rechtsstudenten an der Universität Bologna 1265–1425* [Lawyers for the Empire: German students of law at Bologna University, 1265–1425]. Veröffentlichungen der Gesellschaft für Universitäts- und Wissenschaftsgeschichte: Vol. 2, parts 1–2. Basle: Schwabe.

Schwinges, R. C. (1984). Universitätsbesuch im Reich vom 14. zum 16. Jahrhundert: Wachstum und Konjunkturen [University attendance in the Empire from the fourteenth to the sixteenth century: Growth and fluctuation]. *Geschichte und Gesellschaft. Zeitschrift für Historische Sozialwissenschaft, 10,* 5–30.

Schwinges, R. C. (1986a). *Deutsche Universitätsbesucher im 14. und 15. Jahrhundert. Studien zur Sozialgeschichte des Alten Reiches* [Germans university attendance in the fourteenth and fifteenth centuries: Studies in social history of the Old Empire]. Veröffentlichungen des Instituts für Europäische Geschichte Mainz, Abteilung Universalgeschichte: Vol. 123, Beiträge zur Sozial- und Verfassungsgeschichte des Alten Reiches: Vol. 6. Stuttgart: Steiner.

Schwinges, R. C. (1986b). Zur Prosopographie studentischer Reisegruppen im 15. Jahrhundert [On the prosopography of traveling groups of students in the fifteenth century]. In N. Bulst & J.-P. Genet (Eds.), *Medieval lives and the historian: Studies in medieval prosopography* (pp. 333–341). Kalamazoo: Medieval Institute Publications.

Schwinges, R. C. (1988). Migration und Austausch: Studentenwanderungen im Deutschen Reich des späten Mittelalters [Migration and exchange: Mobility of students in the German Empire of the late Middle Ages]. In G. Jaritz & A. Müller (Eds.), *Migration in der Feudalgesellschaft* (pp. 141–155). Studien zur Historischen Sozialwissenschaft: Vol. 8. Frankfurt am Main & New York: Campus.

Schwinges, R. C. (1994). Admission. In H. de Ridder-Symoens (Ed.), *Universities in the Middle Ages* (2nd ed., pp. 171–194). A history of the university in Europe: Vol. 1. Cambridge, UK: Cambridge University Press.

Schwinges, R. C. (1995). Erfurts Universitätsbesucher im 15. Jahrhundert. Frequenz und räumliche Herkunft [Attendance at the University of Erfurt in the fifteenth century: Rate and regional origin]. In U. Weiß (Ed.), *Erfurt. Geschichte und Gegenwart* (pp. 207–222). Schriften des Vereins für die Geschichte und Altertumskunde von Erfurt: Vol. 2. Weimar: Hermann Böhlaus Nachfolger.

Schwinges, R. C. (1998a). The medieval German university: Transformation and innovation. *Paedagogica Historica, 34*, 375–388. doi:https://doi.org/10.1080/0030923980340204

Schwinges, R. C. (1998b). Prestige und gemeiner Nutzen. Universitätsgründungen im deutschen Spätmittelalter [Prestige and common benefit: University foundations in Germany's late Middle Ages]. *Berichte zur Wissenschaftsgeschichte, 21,* 5–17. doi:https://doi.org/10.1002/bewi.19980210104

Schwinges, R. C. (2000). On recruitment in German universities from the fourteenth to the sixteenth century. In W. J. Courtenay & J. Miethke (Eds.), *Universities and schooling in medieval society* (pp. 32–48). Education and Society in the Middle Ages and Renaissance: Vol. 10. Leiden: Brill.

Schwinges, R. C. (2001). Innovationsräume und Universitäten in der älteren deutschen Vormoderne [Spaces of innovation and universities in early German premodernity]. In R. C. Schwinges, P. Messerli, & T. Münger (Eds.), *Innovationsräume. Woher das Neue kommt—in Vergangenheit und Gegenwart* (pp. 31–44). Zurich: vdf Hochschulverlag an der ETH Zürich.

Schwinges, R. C. (2003). Entre régionalité et mobilité: Les effectifs des universités dans l'Empire romain germanique aux XVe et XVIe siècles [Between regionality and mobility: Attendance rates at the universities of the Roman-German Empire in the fifteenth and sixteenth centuries]. In M. Bideaux & M.-M. Fragonard (Eds.), *Les échanges entre les universités Européennes à la Renaissance* (pp. 359–373). *Colloque international organisé par la Société Francaise d'Etude du XVIe Siècle . . . Valence, 15–18 mai 2002.* Travaux d'humanisme et renaissance: Vol. 384. Geneva: Droz.

Schwinges, R. C. (2008a). Repertorium Academicum Germanicum: Ein Who's Who der graduierten Gelehrten des Alten Reiches (1250–1550) [Repertorium Academicum Germanicum: A who's who of graduate scholars of the Old Empire (1250–1550). In P. Moraw (Ed.), *Gesammelte Beiträge zur Deutschen und Europäischen Universitätsgeschichte. Strukturen, Personen, Entwicklungen* (pp. 577–602). Education and Society in the Middle Ages and Renaissance: Vol. 31. Leiden: Brill

Schwinges, R. C. (2008b). *Studenten und Gelehrte: Studien zur Sozial- und Kulturgeschichte deutscher Universitäten im Mittelalter* [Students and scholars: A social and cultural history of German medieval universities]. Education and Society in the Middle Ages and Renaissance: Vol.32. Leiden: Brill

Schwinges, R. C. (2013). Das Repertorium Academicum Germanicum (RAG): Ein digitales Forschungsvorhaben zur Geschichte der Gelehrten des Alten Reiches (1250–1550) [The Repertorium Academicum Germanicum (RAG): A digital research project about the history of scholars in the Old Empire (1250–1550)]. *Jahrbuch für Universitätsgeschichte, 16*, 215–232.

Verger, J. (1994). Patterns. In H. de Ridder-Symoens (Ed.), *Universities in the Middle Ages* (2nd ed., pp. 45–65). A history of the university in Europe: Vol. 1. Cambridge, UK: Cambridge University Press.

Wagner, F. (2010). Das Repertorium Academicum Germanicum: Mehr als ein Who's Who des gelehrten Deutschland vom 13. bis zum 16. Jahrhundert [The Repertorium Academicum Germanicum: More than a who's who of scholarly Germany from the thirteenth to the sixteenth century]. In U. Morgenstern & T. Riechert (Eds.), *Catalogus Professorum Lipsiensis. Konzeption, technische Umsetzung und Anwendungen für Professorenkataloge im Semantic Web* (pp. 145–150). Leipziger Beiträge zur Informatik: Vol. 21. Leipzig: Leipziger Informatikverbund.

Chapter 3
Scientific and Cultural Relations between Heidelberg University and Hungary over Five Centuries

Peter Meusburger and Ferenc Probáld

Mobility of Students, Scholars, and Knowledge

The intellectual acclaim and attractiveness of a university, the scientific and cultural relations between regions, and the spatial diffusion of scientific knowledge and intellectual currents are affected by a host of factors. A main one is the mobility of students and scholars, which has been exceptionally evident since the Middle Ages. Erasmus of Rotterdam (born in Rotterdam between 1464–1469, died in Basle in 1536), Michel de Montaigne (1533–1592), and other humanists held that a person could be truly educated only through travel. They regarded learning through experience to be superior to learning through books and pointed out that travel can hone judgment, broaden a person's educational horizon, and facilitate the acquisition of foreign languages (de Ridder-Symoens, 1996, pp. 416–419; Stagl, 1995, pp. 72, 78–79). For hundreds of years, studying at several foreign universities and going on the *peregrinatio animi causa*—the Grand Tour, Cavalier's Tour, or study tour—across Europe were vital to the cultivation of the whole person among the elites and figured as the crowning glory of humanistic erudition (see Almási, 2014; Black, 1983, 1992; Chaney, 1998; Heiss, 2005; Horn, 2014; Kühnel, 1964; Miethke, 1995, 2004; Paravicini, 2005). "Between the seventeenth and the early nineteenth centuries the Grand Tour provided the canonical model for aristocratic and bourgeois travel within Europe" (Gregory, 2000, p. 314). In those times, such travel was

Some parts of this chapter have already appeared in German (Meusburger, 2010), Hungarian (Meusburger & Probáld, 2010) and English (Meusburger & Probáld, 2012). Unless otherwise stated, all translations of originally non-English passages are our own with David Antal.

P. Meusburger
Department of Geography, Heidelberg University, Heidelberg, Germany

F. Probáld (✉)
Department of Regional Geography, Eötvös Loránd University, Budapest, Hungary
e-mail: probald.ferenc@gmail.com

P. Meusburger et al. (eds.), *Geographies of the University*, Knowledge and Space 12,
https://doi.org/10.1007/978-3-319-75593-9_3

instrumental in the spread of knowledge also because books circulated predominantly by means of regionally mobile students (Sienerth, 2007, pp. 286–289).

Geography, history, science studies, and other disciplines have long reflected an interest in connections between travel and the production of knowledge (Almási, 2014; Clifford, 1992; Driver, 2001, 2017; Duncan & Gregory, 1999; Gregory, 1999, 2000; Horn, 2014; Jöns, Heffernan, & Meusburger, 2017; Naylor & Ryan, 2010; Thomas, 2015). Most studies on this subject tend to fall into one of two categories: (a) knowledge acquired on expeditions by researchers and explorers in Africa, South America, and Asia and on travel commissioned by colonial authorities, and (b) knowledge gained through Grand Tours, study tours, and formal learning outside one's home country.

Building on this observation, we discuss some of the salient structural factors that can affect a university's appeal and bear on interregional scientific relations, as exemplified by Heidelberg University and Hungary. Hungary lends itself to this topic because its students were among Europe's most mobile for several centuries. The above-average mobility of Hungarian students resulted in great part from the fact that the universities of Pécs (founded in 1367), Buda (founded in 1395), and Pozsony (Bratislava; founded in 1465) were closed for lack of royal support shortly after they had opened and that other attempts to create universities in Hungary failed as well (see Ábel, 1881; Altmann, 1997; Csizmadia, 1965; Diener, 1963; Domonkos, 1997; Schönebaum, 1926; Szögi, 1995, 2008; Zhang, 2013).[1] Moreover, the history of studying abroad and knowledge transfer is particularly well documented and researched in Hungary—as shown in the following sections of this chapter.

For lack of universities in their own land, students from Hungary and Transylvania had little choice but to study in another country, making the *peregrinatio hungarica* the chief vehicle for knowledge transfer between western Europe and the Carpathian Basin for several centuries. From the 1100s through the 1400s, most of the students from Hungary flocked to the universities of Paris, Bologna, Padua, Vienna, Prague, and Cracow. Before the Reformation approximately 66% of the students from Transylvania studied in Vienna, about 30% in Cracow, and 3% to 4% were distributed across 17 other universities (Offner, 2006, p. 292). In the wake of the Reformation as of the sixteenth century, universities in Germany and Switzerland became the favored destination of students from Hungary and Transylvania.[2] The educational migration of Hungarian students shifted again in the 1620s—this time to the Netherlands—because some German universities had been hard hit during the Thirty Years' War. Some of them had had to interrupt or cease teaching, and a number of famous Protestant professors of German universities had fled to the Netherlands during the Thirty Years' War (Gömöri, 1985; Ladányi, 1997; Murdock, 2000). From the mid-nineteenth century on, students from Hungary once more

---

[1]A Jesuit university was opened in Nagyszombat (today's Trnava in Slovakia) in 1635, but it had only two faculties and was not meant for Protestant students.

[2]Transylvania was ruled by Hungarian princes from the mid-sixteenth century to 1690, then passed to the Habsburg crown. In 1848 and from 1867 through 1918, the region again formed part of the Kingdom of Hungary. For linguistic reasons, the terms *Hungary* and *Hungarian* are sometimes used in this chapter for both parts of the country.

turned to German and Austrian universities, which had regained their leading position.

The ebb and flow of relations between German universities and Hungarian students was not merely a function of the academic reputation of the professors, the changing intellectual standard associated with the universities' locations, or the usually confessional nature of the networks and communicational spaces of German professors. The choice of places at which students from Hungary and Transylvania studied outside their country depended additionally on the political interests of the Hungarian rulers and nobles as well as on military conflicts, confessional disputes, incipient intellectual currents, and worldviews. Among Hungarian students from families of modest means, the choice of where to study hinged also on scholarships conferred by Hungarian nobles and cities; the organizations (*coetus*) of the Hungarian students themselves at the foreign universities; and a range of free board, grants, and student dormitories or hostels, which varied considerably from one university to the next.[3] In summary, changes in student migration resulted from the concurrence of spatial rearrangements of intellectual authority and political power, financial and emotional support by communities, and personal relations and networks of scholars.

Of course, students from other countries, too, moved about widely and frequently. The literature on this mobility and the regions from which the universities drew their students (catchment areas) is almost overwhelming (e.g., Birkenmaier, 2012; Courtenay, 2000; de Ridder-Symoens, 1996; Dotzauer, 1977; Fata, 2006, 2014; Fata & Schindling, 2006; Hebeisen & Schmid, 2003; Honeck & Meusburger, 2012; Irrgang, 2003; Maleczek, 1995; Miethke, 1985; Moraw, 1994; Schindling, 2006; Schwinges, 1982, 1984, 1986, 1988, 1994, 1998, 2000, 2001, 2006; Siebe, 2006; Szabó, 1999b, 2006; Szögi, 2001, 2006; Verger, 1970). In relation to population size, however, the intensity and internationality of student mobility from Hungary was more pronounced and enduring than from any other country.

Szögi (2006, p. 388) has stated that approximately 74,000 students from Hungary and Transylvania were enrolled at foreign universities between 1526 and 1919. Some 39,000 of them were at universities in the Habsburg empire (primarily Vienna, Graz, Prague, and Cracow), around 22,000 in Germany, and about 13,000 in other European countries. The number of those registered is not to be equated with the number of students, however, for most students enrolled at more than one university in the course of their studies. Taking these multiple enrollments into account, Szögi (2001, 2006, p. 388) has assumed that at least 15,000 students from Hungary studied at German universities in these four centuries.

Heidelberg University (founded in 1386) was selected for this study for several reasons. First, it served for many centuries as a magnet for students from the Carpathian Basin. Second, this university's peak intellectual periods repeatedly fell

[3]Compared to students from relatively poor social strata in other countries (see Hebeisen & Schmidt, 2003; Irrgang, 2003), those from Hungary seemed to have received fairly adequate scholarships and free board, particularly if they were studying theology abroad.

prey to wars, political catastrophes, confessional politics, and existential crises and had to start from scratch several times. Third, the institution's registries contain relatively detailed personal information about the students. Fourth, Heidelberg University played a prominent part in the spread of Protestantism in the Carpathian Basin.

This chapter addresses the questions of which political, social, religious, intellectual, and economic influences helped or hindered the educational migration of Hungarian students to Heidelberg, which regions and social strata the students from Hungary and Transylvania came from, which faculties they enrolled in at Heidelberg University, which leadership functions these students later assumed in their home country, and what sway individual Heidelberg professors had over the cultural, academic, and political development in Hungary and Transylvania. The study covers nearly five centuries but focuses analytically on two historical eras—1595 to 1621 and 1789 to 1919—during which especially close relations existed for various reasons between Heidelberg University and the countries in the Carpathian Basin. We chose these two spans for conceptual reasons but also because of the availability of sources (Heltai, 1982, 1994, 1999, 2006; Szabó, 2005–2007; Szögi, 2001, 2006; Teutsch, 1872; Toepke, 1886; Toepke & Hintzelmann, 1903, 1904, 1906; Tonk, 1979).

Methodologically, we do not aim to produce a "linearized" or "verticalized" history of the university (Moraw, 1982, p. 2), which is apt to impart a teleological slant to the historical development. Instead, we take a "horizontal approach," as described by the university historian Moraw (1982, p. 2; 1983, p. 525) and recommended by other historians as well (e.g., Schwinges, 2005). Geographers call it context-related research. It directs attention to the complex mutual relations a university has with its social, economic, and political environment (see also the chapter by Meusburger in this volume). The intention is to show the dependence of scientific relations on certain basic conditions; to shed light on the knowledge milieu at university sites; to reveal the intellectual impact that universities, faculties, and individual professors had on cultural, scientific, and political developments in certain regions; and to identify other factors that bear on the academic attraction and growth of a university as well as on the recruitment and mobility of scholars and students.

The First Gilded Age of the Relations between Heidelberg University and Hungary and Reasons for its Decline

The first mention of a student from Hungary in the registries of Heidelberg University appears in 1502. For the next six decades, no other student from that country studied at Heidelberg University.[4] Not until the 1560s did Heidelberg experience a relatively large influx of students from Hungary and Transylvania. This steep rise in

[4]Heidelberg University's catchment area in the first half of the sixteenth century was much smaller than that of the universities of Rostock and Greifswald, for example (Schwinges, 1986, pp. 186–187).

Heidelberg University's significance among students from the Carpathian Basin resulted from the confluence of six partly intertwined historical events.

- The reformation ignited by Martin Luther in 1517 and the rapid spread of Protestantism in Hungary.
- The publication of the Heidelberg Catechism in 1563, which quickly reached many areas of Hungary (Kohnle & Wolgast, 2012).
- The reduction in the importance of the University of Wittenberg because of confessional disputes. Until 1590 Wittenberg had drawn far more students from Hungary and Transylvania than any other German university had, but then it lost many of its students to Heidelberg because of internal conflicts between Lutherans and Calvinists.[5]
- The 150-year Ottoman occupation of central and southern Hungary after the Battle of Mohács in 1526. The partition of Hungary enabled Transylvania to become relatively independent, to resist the Catholic Counter-Reformation, and to become a Calvinist center seeking coalitions with other mainstays of Protestantism in Europe.
- Heidelberg University's close link with the sovereign (Moraw, 1983).[6] Heidelberg enjoyed status as a Calvinist center of political and intellectual power and as one of the most notable cultural hubs of the late Renaissance. It was a town where famous humanists from all over Europe were active, including many Protestant religious refugees from France and the Netherlands. To Protestants from Hungary and Transylvania in the late sixteenth and early seventeenth centuries, Heidelberg was, in a sense, the "Gateway to Europe" (Heltai, 1982, p. 347).
- The founding of several academically outstanding Protestant academies[7] in Hungary and Transylvania during the sixteenth and seventeenth centuries (e.g.,

[5]Juhász (2010, p. 71), Fata and Millisits (2010, p. 533), and others point out that the terms *Calvinist* and *Calvinism* were not used by adherents of the Reformed church to refer to themselves or to their faith until the rebirth of interest in Calvinism in the late nineteenth century. The Reformed population of Transylvania and Hungary declined to be identified solely with Calvin.

[6]Moraw (1983) meant that the Count Palatine of the Rhine (the Heidelberg Prince Elector) closely supervised and controlled Heidelberg University. In particular, the university reform of 1452 was a "clear expression of the sovereign's rule over the university" (p. 526). It was manifested partly in the fact that professors were charged with key responsibilities in service at court and that sons of senior officials at court were appointed as professors (p. 527). Unlike the case in Cracow, for example, where there was a close exchange between the merchant elite and the university (Bernhard, 2015, p. 75), the situation in Heidelberg never allowed for the emergence of a self-confident mercantile community that had, or even fostered, an interest in the university.

[7]According to Sipos (2010, p. 267), nearly all the professors at the Nagyenyed Academy, a Calvinist school of higher learning, had graduated from foreign universities. The academy also had German professors such as Johann H. Bisterfeld (1605–1665) and Johann H. Alsted (1588–1638), who declined several calls to German universities and other institutions of higher learning (p. 278). Both Bisterfeld and Alsted had studied in Heidelberg.

Pápa, Sárospatak, Debrecen, Kolozsvár, Nagyenyed, Nagyvárad, and Gyulafehérvár.[8] They prepared their graduates superbly for successful study at foreign universities. "Students sent by the Reformed [i.e., Calvinist] academies to foreign universities constituted the new Protestant educated elite, who brought Calvinist doctrine and its educational ideals to Hungary and Transylvania" (Fata & Schindling, 2010, p. x)

In the following sections we first explain some of these influences and contexts at a structural level and then illustrate them further through the biographies of selected scholars, intellectuals, and politicians. The examples are also intended to show how sensitive a university system is to the world around it, how the centers of scientific knowledge production repeatedly move from one city to another,[9] and how much impact the social networks of professors and students can have on the knowledge transfer between elites of different countries, on the cultural development of regions, and on the diplomatic relations between countries.

Heidelberg as a Political and Intellectual Center of Calvinism

Like the creation of many other universities in the late fourteenth and early fifteenth century, the founding of Heidelberg University in 1386 did not serve the local population but rather the interests of its founder—the most powerful of the Holy Roman Empire's seven Prince Electors[10]—aggrandizing his power by increasing his prestige and the resources of his dominion. In 1559, under the rule of the first Calvinist Prince Elector, Frederick III (ruled from 1559 to 1576), Heidelberg became a ranking European spiritual and political center of Calvinism and a fulcrum of anti-Habsburg diplomacy, having large sway over imperial policy.[11] As a thriving hub of the late Renaissance and Humanism in the sixteenth century, Heidelberg University's intellectual beacon radiated throughout Europe. Humanism did much to spread the Reformation in Hungary, with many Hungarian humanists embracing the new religious movement (Bernhard, 2015, p. 227).

As Kohnle (2001), Seidel (2004), Wolgast (1998), Zepf (2001), and others have written in their detailed accounts, Heidelberg University's professors have never been

[8]The German place name of Gyulafehérvár was Weißenburg. Today it belongs to Romania and is called Alba Julia. For the sake of consistency and readability in this chapter, we generally use the Hungarian names to refer to Hungarian places.

[9]On the shift of the European research centers in the natural sciences between 1500 and 1900, see, for instance, Hoyler and Taylor (2012) and Taylor, Hoyler, and Evans (2010).

[10]The seven Prince Electors held the privilege of electing the King of the Germans and future Holy Roman emperor. Such elections took place when a dynasty changed, but they were also needed by the son of the reigning king in order to be approved by the Prince Electors as his father's rightful successor. The most important of the seven Prince Electors resided in Heidelberg.

[11]Heidelberg at that time was called "the center of Calvinist-Reformed academia" (Wolgast, 1986, p. 40) and the "international Mecca of the anti-Spanish, anti-Roman intelligentsia" (Wolgast, 1996, p. 293).

as scientifically, culturally, and politically influential and as internationally well networked as in the years from 1560 to 1622. The electoral court in Heidelberg and the university attracted officials, scholars, and students from many parts of Europe, particularly from territories with populations having a high percentage of Calvinists or in which Calvinists were politically persecuted. Many humanists were also scholars, poets, ambassadors, and policy advisors at court. Heidelberg University's faculties all boasted internationally renowned and prolific scholars whose thoughts and works disseminated in all of Europe by means of literary Latin. The Heidelberg humanists corresponded with scholars, diplomats, and princes in many areas of Europe, were embedded in international networks, and were academically highly mobile (see Baar-Cantoni & Wolgast, 2012; Kühlmann, Hartmann, & El Kholi, 2012; Neumaier, 2012b; Strohm & Becker, 2012; Strohm & Hofmann, 2012; Wolgast, 1998). The internationality of Heidelberg University, which was astonishing for that period, was apparent partly in the fact that 11 of the 16 professors teaching in the Faculty of Theology between 1559 and 1622 were foreigners. Four more hailed from other territories of Germany, and only one had been born in the Electoral Palatinate (Wolgast, 2015, p. 92). Of these 16 professors, 11 had been persecuted for their Protestant faith at the places where they used to work (in the Spanish Low Countries, France, Italy, and eastern central Europe) (p. 92). The 17 rectors[12] elected between December 1559 and December 1575 included 3 from Italy, 1 from France, 3 from the Low Countries, 1 from Switzerland, and 3 from the Holy Roman Empire outside the Palatinate (Kohnle, 2001, p. 145). The students were of similarly international provenance. In 1563, 40.3% of them were foreigners; between 1564 and 1609, the share of foreigners in the student body usually exceeded 50% (Kohnle, 2001, p. 153).

Conditions and Factors that Promoted the Spread of Protestantism in Hungary

After the capitulation of Buda in 1541, southern and central Hungary belonged to the Ottoman empire. The west and north of Hungary constituted the Kingdom of Hungary and stood under Habsburg rule, and the principality of Transylvania in the east enjoyed relatively great political independence under Ottoman suzerainty.[13] To protect this independence from the House of Habsburg, which was Catholic, the

[12] In much of continental Europe, the rector is traditionally a university's highest academic official.

[13] Transylvania was obliged to consult the Ottomans only on important matters of foreign policy and paid annual tribute to the sultan.

Fig. 3.1 The partition of the Kingdom of Hungary after the Treaty of Speyer, 1570. Sources: Fata (2000, p. 17), Glatz (1995, p. 61). Design by the authors. Cartography: U. Selgert, V. Schniepp.

princes of Transylvania were interested in cultivating close cultural, scientific, and political relations with Europe's Protestant centers of power.[14]

The tripartite political division of Hungary (Fig. 3.1) had an abiding effect on the conditions framing ecclesiastical policy in the Reformation and Counter-Reformation (or re-Catholization) and led to dissimilar confessional structures and student mobility in the three regions. According to Benda (1976), between 80% and 85% of the population in Hungary was Catholic in 1525. The rest consisted of Orthodox Christians, Jews, heretics, and other groups. Between 1570 and 1620 some 75% to 80% of Hungary's population turned to embrace Protestantism, with only about 10% remaining Catholic (Benda, 1976; Murdock, 2000; Szabó, 2010; Zach, 2004, pp. 159–160). Backed by the aristocracy as of the 1540s, the Reformation surged through the Habsburg part of Hungary (western and northern Hungary), with most of the Germans and Slovaks being Lutheran and most of the ethnic Hungarians being Calvinist (Asche, 2004). At that time the Protestants dominated the educational system and letterpress printing. Of the 168 grammar schools that existed in Hungary

[14]Calvinism in Transylvania and eastern Hungary also greatly contributed to the rise of a political identity among ethnic Hungarians.

in the sixteenth century, 134 were Protestant (Bucsay, 1977, p. 161), and Protestants wrote an estimated 90% of the more than 500 books printed in Hungarian in the sixteenth century (p. 159).

In the Ottoman part of Hungary, Lutheranism and the Reformed church spread unhindered, whereas the organized Catholic church steadily weakened. Islamization of the native population did not occur in central Hungary (Asche, 2004, pp. 38–39; Bernhard, 2015, p. 447; Zach, 2004, p. 160). Transylvania became the center of ascendant Calvinism in the Carpathian region of the sixteenth century, and in the seventeenth century it long resisted the Counter-Reformation led from the west by Peter Pázmány (1570–1637), who became archbishop of Esztergom in 1616 (Höchsmann, 1894–1895, 1896–1897) and cardinal in 1629 (on Pázmány's role see Bucsay, 1977, pp. 164–167).

The University of Wittenberg and Heidelberg University became focal points for students from Hungary and Transylvania after the Reformation (Szabó, 1992, 1999b, 2006) because these two institutions were linked with four of the most eminent authorities of Protestantism: Martin Luther (1483–1546), Philipp Melanchthon (1497–1560), Zacharias Ursinus (1534–1584), and David Pareus (1548–1622). Not only were the first two men towering figures in Wittenberg, their names became closely associated with Heidelberg as well. With Luther, the latter connection took root just months after his 95 theses appeared in Wittenberg, for he received the opportunity to air and defend his ideas on reform publicly in a disputation, or scholarly debate at Heidelberg University (Baar-Cantoni, 2012; Mühlen, 1985; Scheible, 1983; Seebaß, 1983). This encounter, during which he presented his doctrine of justification in a series of assertions, or premises, took place on April 26, 1518, in the lecture hall of the Faculty of Philosophy (corresponding to the Arts and Humanities in a modern U.S. context). A good dozen of the students attending the event went on to become notable reformers in southern Germany (Baar-Cantoni, 2012, p. 57).

Philipp Melanchthon

Heidelberg's second connection to Protestantism came through Philipp Melanchthon (Fig. 3.2), an intimate friend of Martin Luther. Melanchthon was a remarkable, immensely authoritative reformer, humanist, church policy-maker, university reformer, and educator. Without his organizational skill as a Protestant negotiator in imperial diets and religious discussions, the Reformation might well have taken a less successful course than it did.[15] Melanchthon "began his path to Calvinist theology . . . in Heidelberg, continued it in Tübingen, and completed it in Wittenberg" (Rieger, 2010, p. 153). He was born in Bretten, about 30 miles south of Heidelberg. Extraordinarily gifted, he was permitted to enroll at Heidelberg University on October 14, 1509, when he was just 12½ years old. He earned a baccalaureate

[15]This summary of Melanchthon's biography and magnitude is based on Greschat (2010), Jesse (2005), Lorenz (2010), Köpf (2010a, 2010b, 2010c), Rieger (2010), Schweitzer, Lorenz, & Seidl (2010), Stempel (1993), and Stupperich (1990, 1996).

Fig. 3.2 Philipp Melanchthon. Source: Heidelberg University Archives. Reprinted with permission.

by the age of 14. In September 1512, Melanchthon transferred to the University of Tübingen, where in January 1514, not yet 17 years old, he completed his *Magister artium* (the academic degree common at that time for philosophical, linguistic, cultural, or historical subjects in the arts and humanities).[16]

At the age of 21, Melanchthon was appointed to the newly created chair of Greek at the University of Wittenberg, where he deeply impressed Martin Luther and other scholars with his brilliant inaugural lecture on the need for extensive academic reform. In the winter semester of the academic year beginning in 1523, he became the rector of the University of Wittenberg for the first time. He formulated new university statutes, reformed the curriculum, proved to be an excellent organizer, wrote commentaries on ancient Greek authors, and composed noteworthy textbooks

[16]On Melanchthon's activity at the University of Tübingen, see Köpf (2010a, 2010b, 2010c) and Lorenz (2010).

on Greek,[17] rhetoric, ethics, physics,[18] history, and geography. His analytical reports and recommendations helped shape the university reforms in Frankfurt on the Oder, Marburg, Tübingen, Greifswald, Rostock, Leipzig, Jena, and Heidelberg (Asche, 2010, p. 82; Köpf, 2010b, p. 25). Melanchthon was also probably one of the first people to recognize the unmistakable import that the school system has for a country's economic development (for details see Meusburger, 2013, p. 22). A paramount figure of German intellectual life, he was acclaimed by his contemporaries as *Praeceptor Germaniae*—the teacher of Germany (for details see Frank & Treu, 2001; Gebhardt, 2008; Rieger, 2010). Perhaps no other scholar of his day was as communicative and persuasive as he, or as well connected with so many leading educational institutions, learned societies, humanist circles, cities, church administrators, and princely and royal courts.

Melanchthon is also an example of how much networks, webs of relationships, and knowledge milieus can mean for the academic socialization and later careers of students. He illustrates how much the intellectual appeal of universities is marked by the people working there and by their range of contacts. Melanchthon's strongest ally was his great uncle, Johannes Reuchlin (1455–1522), the humanist, diplomat, distinguished Hebraic language scholar, and pivotal man in a key European network. When Prince Elector of Saxony Frederick the Wise (1463–1525)—who had founded the University of Wittenberg in 1502 (see Stievermann, 1999)—asked Reuchlin to recommend a suitable candidate for the recently created chair of Greek in Wittenberg, Reuchlin named his student and grandnephew, Melanchthon. Reuchlin had already paved the way for him in Heidelberg and Tübingen (for details see Köpf, 2010b, pp. 29, 37). Like his great uncle, Melanchthon himself later became the heart of a European-wide personal network from which students from Hungary and Transylvania especially benefited.

Owing to his personal contacts and extensive correspondence, Melanchthon was exceptionally well informed about Hungary and Transylvania (see Scheible, 1985). By 1521 his friend and former fellow pupil from grammar school days in Pforzheim, Simon Grynaeus (1493–1541), had become the school director of St. George's Chapel at the royal castle in Buda and librarian of the Corvinus library, one of Europe's best at that time.[19] Grynaeus became a professor at Heidelberg University in 1524 and at the University of Basle in 1529 (Scheible, 1985, pp. 37–38).

[17]Melanchthon's book on Greek grammar appeared in 1518 and underwent its 17th printing in 1544. It became the most widely used schoolbook in Europe's Reformed regions (Rieger, 2010, p. 153).

[18]Melanchthon rejected the heliocentric concept of the world, however (Reich, 2010, p. 140).

[19]The library's destruction after the Ottoman conquest of Buda (1541) was an incalculable loss of culture and knowledge and ultimately hindered the exchange of knowledge between Buda and other European countries.

Melanchthon was in touch with the most prominent exponents of the early Protestant movement in Buda.[20] One of the first people to spread Melanchthon's ideas in the Carpathian Basin was Mátyás Dévai Biró (?–1547), Hungary's celebrated reformer, who was the second Hungarian to enroll at the University of Wittenberg (1529). He was a reformer in various locations of western and northern Hungary after 1538, later in Transylvania as well (Bernhard, 2010, pp. 37, 47). Other major theologians of the Hungarian and Transylvanian Reformed church, such as István Szegedi Kis (1505–1572) and Péter Méliusz Juhász (1532–1572), had likewise studied under Melanchthon (Juhász, 2010, p. 65).

While in Wittenberg, Melanchthon attracted students from Hungary and Transylvania and was probably one of the main reasons why Hungarian students found Wittenberg to be the most desirable university in Germany for several decades. Wittenberg in the sixteenth century accounted for 68.4% of all academics among the Transylvanian "Saxons";[21] in the seventeenth century, 41.3% (Offner, 2006, p. 292). A few of the Hungarian students—such as György Albani Csirke, who later became the Holy Roman Emperor's ambassador to the Ottoman court—lived in Melanchthon's house, and he held lectures and devotions in Latin for Hungarians who had not yet mastered German (Bernhard, 2010, p. 49; Szabó, 2010, p. 86).

Melanchthon's publications circulated widely in Hungary and Transylvania. As an undisputed authority on theology and education, Melanchthon wrote numerous expert reports—recommendations for Hungarian and Transylvanian clergy, teachers, city councilors, and officials. Melanchthon's *Confessio Augustana variata* (Altered Augsburg Confession) of 1540 also served as the model for the rules governing church life, belief, and worship in the royal free cities and the mining towns of upper Hungary, the German municipalities in the Zips region (today a region in northeastern Slovakia; Slovakian *Spiš*, Hungarian *Szepes*), and the Transylvanian towns (see Asche, 2004, pp. 36–37). Because Melanchthon exerted considerable influence on Hungary's cultural elite, he was (and is) occasionally also known as *Praeceptor Hungariae*—the teacher of Hungary (Beck, 2016).

For students from Hungary and Transylvania, Wittenberg was undeniably the premier German university in the sixteenth century (Fig. 3.3). More than half of the approximately 4,300 Hungarian students who attended German universities in the sixteenth and seventeenth centuries were enrolled at Wittenberg, followed by Frankfurt on the Oder, Leipzig, Jena, Heidelberg, and Tübingen (Tar, 2007, p. 19).

Wittenberg did not lose this position until the reign of Saxon Prince Elector August (1553–1586), when orthodox Lutherans and Melanchthon's supporters clashed over the theological question about the nature of the Lord's Supper. Melanchthon's students were belittled as Crypto-Calvinists or Philippists,

[20]Many of the Protestant humanists fled Buda for Germany after the city's capitulation to the Ottomans. Hungary's manifold close economic relations with the rest of Europe were thenceforth heavily restricted.

[21]In Transylvania all German-speakers were called Saxons. The term has nothing to do with the territory of Saxony.

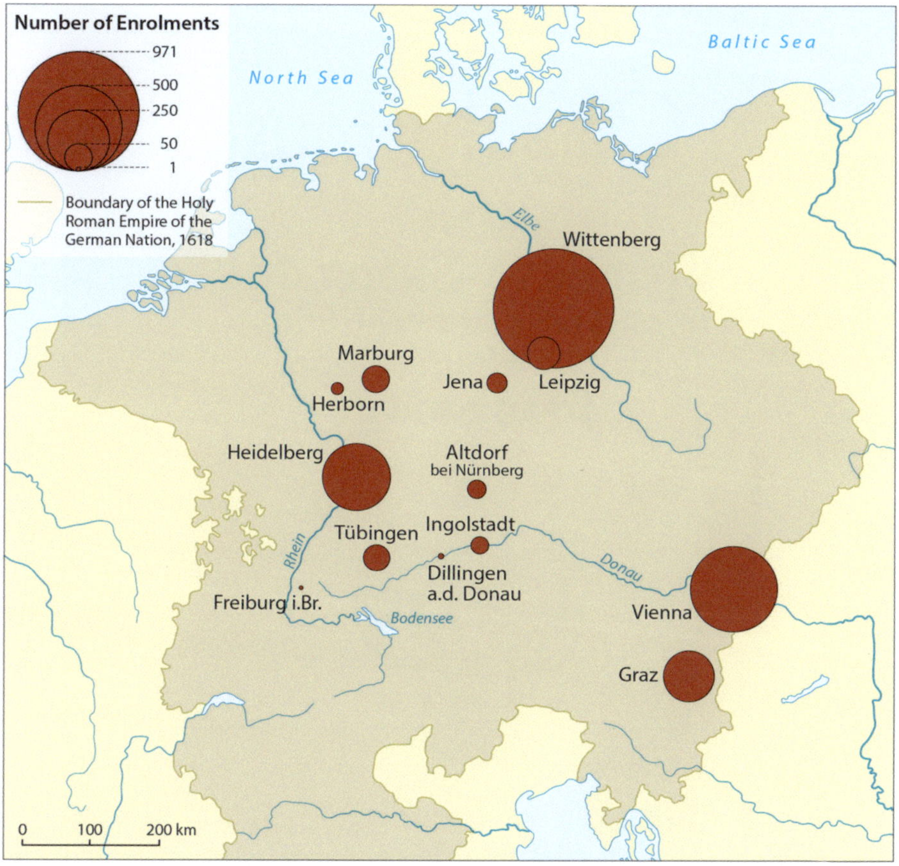

Fig. 3.3 Hungarian students at German universities, 1560–1620.
Data sources: Heltai (1982, 1999); Teutsch (1872). Basic map: Leibniz-Institut für Länderkunde, Leipzig (IfL). Cartography: V. Schniepp.

systematically persecuted, and finally driven from the university once and for all in 1592. The waves of purges and subsequent unrest eventually eroded the intellectual level and reputation of the University of Wittenberg (Ludwig, 2009; Szabó, 1993). The Calvinist students from Hungary and Transylvania thereafter looked to Heidelberg, with Wittenberg remaining acceptable only to Lutheran students of theology.

Zacharias Ursinus and the Significance of the Heidelberg Catechism

The Heidelberg Catechism (Fig. 3.4), published by Zacharias Ursinus (Fig. 3.5) by commission of Prince Elector Friedrich III in 1563, Count Palatine of the Rhine,

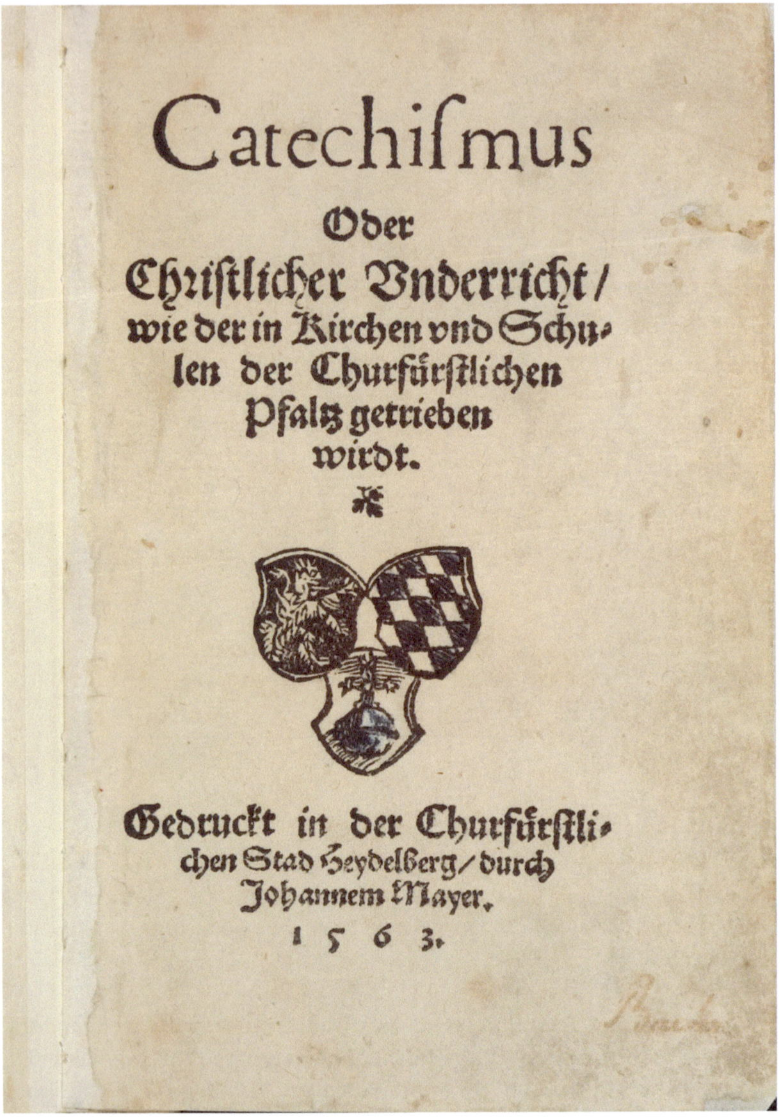

Catechiſmus

Oder

Chriſtlicher Vnderricht/
wie der in Kirchen vnd Schu-
len der Churfürſtlichen
Pfaltz getrieben
wirdt.

Gedruckt in der Churfürſtli-
chen Stad Heydelberg/ durch
Johannem Mayer.
1 5 6 3.

Fig. 3.4 Title page of the first edition of the Heidelberg Catechism in 1563.
Source: Heidelberg University Archives. Reprinted with permission.

galvanized the cultural relations between Heidelberg and Hungary.[22] The document not only incorporated Calvin's teachings but also articulated its own theological position, which assimilated seminal theological ideas of Calvin, Melanchthon, and Protestants of northern Switzerland. The aim was to harmonize and intelligibly

[22]Presumably, other Heidelberg University theologians also worked on the Catechism (see Wolgast, 2015, p. 89).

Fig. 3.5 Zacharias Ursinus, key author of the Heidelberg Catechism.
Source: Heidelberg University Archives.
Reprinted with permission.

reconcile the increasingly doctrinaire and divergent views held by orthodox Lutherans and Calvinists (for details see Bierma, 2015; Henss, 1983; Millisits, 2012; Módis, 1967; Nagy, 1967; Schilling, 2010; Zach, 2004).

In 1564 a disagreement over the Lord's Supper arose in Transylvania between the preachers of the primarily German-speaking Lutherans and the ethnic Hungarians, who espoused Calvin's teachings. The clerics eventually asked the famous Heidelberg theologians for their opinion on the thorny theological concern. The Heidelberg professors responded with a letter in which they set forth their standpoint and referred the Hungarian fellow Christians to the new Heidelberg Catechism.[23] The letter was printed in Kolozsvár (today's Cluj-Napoca) in 1565 and published under the title *Epistola Professorum Theologiae Academiae Heydelbergensis* (Epistle by the Professors of Theology at Heidelberg University). The Heidelberg Catechism itself was initially printed in Kolozsvár the next year—an edition that did not yet contain certain critical passages. The complete Hungarian translation appeared in 1577 (Bartha, 1967; Millisits, 2012; Nagy, 1967, p. 41). The Heidelberg Catechism was so successful that it soon supplanted the other Protestant catechisms existing in Hungary at the time.

Another explanation for this swift acceptance is surely that the Heidelberg Catechism offered the analytical culmination (Zach, 2004, p. 162) of what some Hungarians had tentatively already begun trying to formulate—a comprehensible compromise between the disparate Protestant doctrines. Nagy (1967, p. 32) pointed

[23]Bernhard (2015, p. 438) assumes that the original text of the Heidelberg Catechism was known in the eastern parts of Hungary by late 1564.

out that the German translation of the briefer confession written by the French reformer Theodore Beza (also called Théodore Bèze, 1519–1605) appeared in Heidelberg in 1562 and verifiably informed many passages of the Heidelberg Catechism. Hence, the Heidelberg Catechism was inspired by the same Beza whose longer statement of confession had been adopted by the Hungarians as their own at the Synods of Tarcal (1562) and Torda (1563). This circumstance certainly helps account for the ready reception of the Heidelberg Catechism in Hungary. The national synod of Szatmárnémeti in 1646 recognized the Heidelberg Catechism along with the fundamental Second Helvetic Confession as the second authentic doctrinal tract of the Reformed church of Hungary and Transylvania. These synods mandated that theology students swear an oath before and after studying outside Hungary that they would always conform to the doctrine laid out in the Heidelberg Catechism.

The Heidelberg Catechism was translated into English, French, Hungarian, Czech, and Romanian as early as the seventeenth century (Ehrenpreis, 2015, p. 312). Before long it was obtainable in 40 languages and was eventually reissued 150 times in Hungary alone (Baar-Cantoni & Wolgast, 2012, pp. 67–69; Henss, 1983; Zach, 2004, p. 163).[24] Such wide dissemination owed not least to the tract's availability in the native languages of several Protestant countries, where they were used to teach reading in schools in the sixteenth and seventeenth centuries (Ehrenpreis, 2015, p. 315). In that instructional capacity the Heidelberg Catechism decisively shaped the educational systems of Hungary, and other Protestant areas for centuries (Nagy, 1967, p. 48). Under Habsburg absolutism, however, some of its paragraphs were censored, so several Hungarian publications of it could appear only as abridged editions.

David Pareus as a Main Authority on Calvinism and an Instrumental Councilor to the Ruler of Transylvania

For more than two decades the greatest intellectual magnet drawing Calvinist students from Hungary and Transylvania was the Heidelberg professor David Pareus (Fig. 3.6), a student of Zacharias Ursinus. Pareus held a chair at Heidelberg University from 1598 to 1621 and was regarded in Hungary and Transylvania as the supreme authority on Calvinist orthodoxy. The magnitude of Pareus's impact on the spiritual, cultural, and political development in Hungary and Transylvania is partly evident from the fact that Pareus chaired 293 disputations[25] by Hungarian

[24]The most recent Hungarian translation of the Catechism was published in 2013 in commemoration of the Confession's 450th anniversary.

[25]From the Middle Ages until after the dawn of modern times, the *disputatio* was one approach to scientific exchange and the foremost method of resolving scientific disputes. In this regard it differs from the defense of a dissertation today—the oral examination in partial fulfillment of the requirements for a doctorate. Hence, some of the 173 students from Hungary mentioned by Heltai (1982, 1999) faced more than one *disputatio*.

Fig. 3.6 Portrait of David Pareus. Source: Heidelberg University Archives. Reprinted with permission.

students—92% of all the disputations by Hungarian students in Heidelberg at that time (Heltai, 2006, p. 68; Ulrichs, 1993), and many of his former students took over important functions in Hungary and Transylvania. In Pareus's summer house, the *Pareanum* at the foot of the Schlossberg, the Hungarian students would meet with their professors and fellow students for convivial conversation (Röhrs, 1971, p. 7), not only discussing theological topics but also making personal friends and building academic networks that were later cultivated in lively exchanges of letters (Seidel, 2004, p. 246). Pareus's students from Hungary and Transylvania obviously had fond memories of this time. Writing to Pareus in 1609, the nobleman Újfalvi Katona (1572–1610) described the four months he had spent in Heidelberg as a scientifically stimulating, all-too brief interlude compared to his more than two years in Wittenberg, which he characterized as having been nothing but a waste of time and money (Seidel, 2004, p. 232).

After returning to Hungary or Transylvania and advancing to high positions and offices, they continued to regard Pareus as the highest authority and to seek his counsel on daily matters of church practice. Pareus kept himself thoroughly informed about events in Hungary and Transylvania and maintained a lively correspondence with the reigning prince in Transylvania, Gábor (Gabriel) Bethlen (1580–1629), who ruled from 1613 to 1629[26] and held Pareus to be the preeminent interlocutor in the Reformed world. Pareus enthusiastically welcomed and endorsed the prince's project of founding a Calvinist academy in Gyulafehérvár/Weissenburg. It is partly due to Pareus's advice that this school's regulations were modeled on educational institutions in the Electoral Palatinate and Hesse. One of the people Bethlen brought to Transylvania to help develop the academy in Gyulafehérvár/Weissenburg was the German scholar Johann Heinrich Bisterfeld (1605–1655), a student of Pareus. Bisterfeld remained there for 25 years as the leading professor, built a dense European network of personal contacts with scholars and princes with whom he corresponded frequently, and took part in crucial diplomatic negotiations on several occasions (Viskolcz, 2010).

Gábor Bethlen assured Pareus in a letter that he was very impressed by *Irenicum* (1614, Pareus's chief work) and "holds it in his hands day and night" (Seidel, 2004, p. 247). The irenic theology that Pareus authentically practiced, which may be seen as a forerunner of today's ecumenical undertakings, emphasized the shared foundation of the Christian faith and rejected both religiously rooted violence and harsh theological polemics (Hotson, 1995, 2004). This notion was embraced by most of Pareus's students, too, and was fully consistent with the interests of the prince, whose reign was beset by foreign and domestic tensions. Although Gábor Bethlen favored the Calvinist (Reformed) church above all, he did set store in avoiding an escalation of religious conflict. He had to keep in mind Transylvania's exceedingly complicated confessional circumstances, which had moved the provincial diet to adopt a law of unprecedented tolerance guaranteeing religious freedom as early as 1568.[27]

It would be wrong, however, to reduce Pareus's significance solely to Hungary and Transylvania. Between 1599 and 1619 he conducted over 400 disputations and profoundly influenced students from many European countries, such as the Prague scholar Comenius,[28] who studied for a year at Heidelberg University in 1613–1614 and engaged in a formal scholarly debate (*disputatio*) with Pareus in the Faculty of Theology on March 19, 1614 (Hotson, 1995, pp. 438, 450–453; Röhrs, 1971, p. 5). Despite the brevity of his study in Heidelberg, the impression on him was lasting (on the nature of the teaching that Comenius received in Heidelberg and on the content of his theses, see Hotson, 1995). Pareus and his *Irenicum* underpinned the thoughts behind Comenius's efforts to end to the naval war between England and Holland and behind his general exhortations in *Panorthosia* (Universal Reform) for

[26]Bethlen had acquired his title of prince in 1613 with the active support of the Sublime Porte (the Ottoman feudal government, with the sultan and his court at its center) but against the resistance of the court in Vienna. He was thereafter able to consolidate Transylvania's independence. In 1620 he was elected as king of Hungary but refused to be crowned. In 1621 he had to renounce the Hungarian crown forever in the Peace of Nikolsburg.

[27]There were three recognized Protestant confessions in Transylvania in the early seventeenth century. Some of the nobility and the Szekler were Catholic, and the Romanians have always remained Greek Orthodox despite repeated Protestant attempts to convert them (Leppin & Wien, 2005).

[28]Comenius registered under the name Johannes Amos Niuanos Moravus, perhaps one reason why he has received relatively little attention in the history of Heidelberg University thus far (Röhrs, 1971, pp. 2–5).

princes to strive for peace and security (p. 8). Comenius dedicated the foreword of his *Unum Necessarium* (The One Thing Necessary, 1668) to Friedrich V's son, Ruprecht, who was living in exile in London (p. 11). Comenius traveled to Hungary in 1650 and taught there for four years at the Calvinist academy in Sárospatak, where he wrote his famous works *Orbis sensualium pictus* (The Visible World in Pictures; see Comenius, 1705) and *Schola ludus* (School by Play).

Fluctuation in the Number of Students

Journeying constantly from one place of learning to another (*peregrinatio academica*), the students from Hungary and Transylvania were well networked with each other. They therefore usually responded quickly to scientific disputes, religious conflicts, crises at universities, and turns in the political and social conditions in Germany and Hungary. When Heidelberg University reverted to Lutheranism under Prince Elector Ludwig VI (ruled 1576–1583), many professors had to leave Heidelberg because the ruler's religion dictated that of the subjects (the principle of *cuius regio, ejus religio*). Not only did this exodus lower the university's intellectual caliber, the confessional shift also meant that no students from Hungary or Transylvania enrolled there between 1578 and 1582. Not until 1583, under Calvinist Prince Elector Friedrich IV (ruled 1583–1610), did the numbers of students from Hungary, Transylvania, and other parts of Europe soar again (Kohnle, 2001, p. 147). For about 30 years after the expulsion of the Crypto-Calvinists from the University of Wittenberg in 1592 (see section on Philipp Melanchthon, above), Heidelberg University became the key German university for students from Hungary and Transylvania (Fig. 3.7). A consequence of the Bocskay uprising[29] against the Habsburgs (1605–1606) and the clashes with the Ottomans was that the number of Hungarian students temporarily plummeted again in the early seventeenth century (Fig. 3.7). It then recovered, peaking between 1608 and 1621. As Heltai (2006, pp. 68, 72) has reported, the flood of Hungarian and Transylvanian students to Heidelberg University after 1614 also stemmed from the fact that contours of a European alliance of Protestant princes began to emerge after the marriage of Prince Elector Friedrich V, who resided in Heidelberg, to Elizabeth Stuart, the daughter of James I, King of England (James VI, King of Scotland) in 1612–1613 (see Marshall, 2003). Gábor Bethlen wanted to join this coalition of Protestant rulers, so he promoted the tide of Transylvanian students to Heidelberg for political reasons. Assisted by his students, Bethlen succeeded in intensifying his contacts to the court of the Heidelberg Prince Elector to "break out of political isolation" (Heltai, 2006,

[29]István Bocskay was elected prince of Transylvania (1605) and led an uprising against the Habsburg emperor Rudolf II., king of Hungary. In the Treaty of Vienna (1606) various constitutional and religious rights and privileges were granted to the Hungarians in both Transylvania and Royal Hungary. The accord also recognized Bocskay as the Prince of Transylvania and guaranteed the right of Transylvanians to elect their own independent princes in the future.

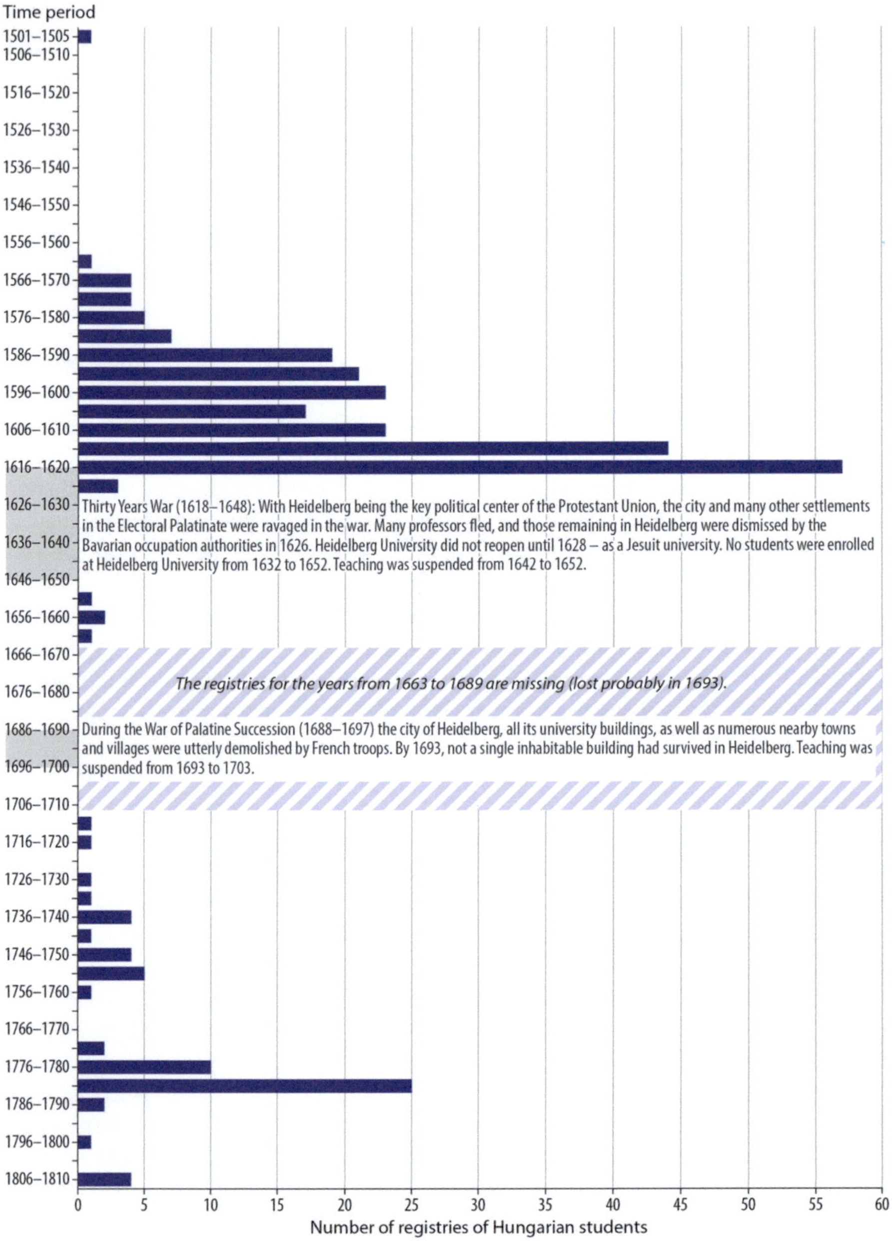

Fig. 3.7 Students from Hungary and Transylvania studying at Heidelberg University, 1500–1810. Data sources: Heltai (1982, 1999); Teutsch (1872). Design by authors.

p. 73). The Heidelberg peregrination vitally contributed to bringing Transylvanians into the Thirty Years' War on the side of the Protestant princes (p. 73).

Social Origins, Choice of a Place to Study, and Choice of Occupation of the Students from Hungary and Transylvania, 1595–1621

Hungary's tripartite division affected not only the confessional profile of the Hungarian and Transylvanian students abroad but also their political interests, diplomatic relations, and choice of where to study. The Catholics preferred the Catholic universities of Graz, Vienna, and Cracow. The Lutherans, who accounted for most of the German-speaking population in Hungary and Transylvania, usually chose to study at German universities at which Lutheran professors predominated. The Calvinists gravitated mostly to Heidelberg, Geneva, and, after the outbreak of the Thirty Years' War, to universities in the Low Countries (see Hotson, 2008; Murdock, 2000; and the chapter by Hotson in this volume).

Heltai (1982, 1999) has reported that 175 students from Transylvania and Hungary enrolled at Heidelberg University between 1595 and 1621, almost all of them Calvinist Hungarians. The vast majority of the students came from the urban middle class, which was still relatively small then. In the sixteenth and seventeenth centuries, an era in which both society and politics revolved around issues of religion, clerics, especially the famous preachers, could achieve high standing. A church career at that time offered very attractive paths of upward social mobility for the sons of artisans, merchants, free peasants, and even a few serfs.[30] Students whose families could not afford to have them study outside Hungary could apply for a scholarship from the magistrate of the area in which they lived, from aristocratic patrons, or from the prince himself. Preachers' sons who intended to follow their fathers into the ministry were particularly well represented among Heidelberg students. Only about 8% of the students from Hungary and Transylvania were nobles.

Reconstructing the careers of 109 of the 175 Hungarian students who studied in Heidelberg between 1595 and 1621, Heltai (1982, 1999) found that 10 graduates had later pursued a secular occupation, 11 had gone on to employment as teachers in the church schools, and 88 had become preachers (pastors) in the Reformed church in Hungary or Transylvania. Of these clerics, 29 rose to the rank of deacons, and 10 became superintendents (comparable to a Catholic bishop)—key positions in the Calvinist church hierarchy, which was rapidly expanding in that era (for details see Heltai, 2006).[31]

---

[30] In Transylvania the freedom of learning was legally guaranteed for serfs as well.

[31] Unless otherwise indicated, the following information on students of Heidelberg University is from Heltai (2006).

For example, János Keserűi Dajka (1580–1633), who enrolled at Heidelberg University in 1609 and who was a student of Pareus, became a superintendent (bishop) of the Calvinist church of Transylvania and court chaplain to the ruling prince, Gábor Bethlen. Dajka recommended that his prince send students to Heidelberg University each year. The personal ties between Pareus and Dajka were so close that Pareus dedicated to him the volume *Collegiorum theologicorum pars altera* (1620), a collection of disputations that he had edited (Heltai, 2006, pp. 65–66). Péter Alvinczi (1570–1634) served as court chaplain, confidant, and political counselor to the princes István Bocskay and Gábor Bethlen (Heltai, 1994). Caspar Boithius (1595–died after 1640) studied in Heidelberg with Gábor Bethlen's support from 1617 to 1620 and later became a linchpin in the relations between the Palatinate and Hungary. Between 1615 and 1660 an unbroken succession of men from Heidelberg served at the court of Gyulafehérvár/Weissenburg and as superintendent (bishop) of Transylvania: János Keserűi Dajka, István Milotai Nyilas, István Geleji Katona, and György Csulai (Heltai, 2006, p. 72).

István Milotai Nyilas (1571–1623) was another court chaplain to Prince Gábor Bethlen for a time and did much to shape the Transylvanian school system. István Geleji Katona (1589–1649), who enrolled at Wittenberg in 1596 and at Heidelberg in 1598, became famous because he was court chaplain to three Transylvanian princes, tutor to István Bethlen (1606–1632)—a nephew of Prince Gábor Bethlen and potential heir to the throne—and superintendent of Transylvania. His *Magyar Grammatikatska* (Little Hungarian Grammar Book, published in Gyulafehévár in 1645) greatly nurtured the development of Hungary's linguistic culture (see Nagy, 1967, p. 44; Seidel, 2004, pp. 233–236).

Albert Szenczi Molnár (1574–1634), one of his time's most versatile and best known Hungarian scholars, offers an impressive example of high academic mobility and the seminal role of networks, therefore merits somewhat more detailed consideration in this overview (for additional information see Dézsi, 1897; Giebermann, 2005; Szabó, 1999a, 2003; Vásárhelyi, 1985, 1999, 2006, 2014). Molnár left the Calvinist academy of Debrecen at the age of 16 and enrolled at several German universities in the course of time, including Wittenberg (1590), Heidelberg (1592 and 1596), Herborn, Strasbourg, and Altdorf. At the University of Altdorf, he edited the *Elementa Grammatica Latinae* (Elements of Latin Grammar) and a Hungarian-Latin-Greek dictionary that was printed in Nuremberg in 1604. The dictionary's high quality and the fact that Latin was the official language in Hungary until the mid-nineteenth century ensured that work's use in Hungary for two-hundred years.

In Altdorf Molnár also worked on his main opus, the *Psalterium Hungaricum*, the Hungarian translation of the Psalms of David. From 1607 to 1611, he resided in Marburg, where the Calvinist landgrave Moritz von Hessen-Kassel aided him financially. It was there that Molnár once again revised the Hungarian translation of the Bible, published a new translation of the Heidelberg Catechism, and wrote the first solid grammar of the Hungarian language. In Geneva, the bastion of Calvinism, he consulted with the acclaimed Calvinist theologian Theodore Beza. After moving back to Heidelberg from Geneva, he worked primarily with the poet Martin Opitz, whom Gábor Bethlen later called to the academy in Gyulafehérvár.

In 1614 Molnár went to Hanau briefly, soon thereafter to Amberg, and a year later to Oppenheim. There he worked as the precentor (*Kantor*) of the grammar school and, from 1617 to 1619, as its rector. In 1619 he again took up residence in Heidelberg, where in 1622 he was brutalized by troops and lost much of his library when the city fell to troops commanded by Count Tilly (1659–1632), a field marshal commanding forces of the Catholic League during the Thirty Years' War. After fleeing Heidelberg, he worked at the University of Leiden and then made his way back to Hungary. Within a year he went to Hanau via Heidelberg. The counts of Hanau granted the Calvinist refugees asylum at that time. After two years in the town, Molnár resettled in Hungary for good, this time in Transylvania, where Prince Gábor Bethlen maintained him.

Many students from Hungary and Transylvania received scholarships for their studies in other countries. Their patrons attached certain conditions to this help, so the choice of what and where to study did not always lie with the students alone. The patrons of 81 students are known for the years from 1595 to 1621 (Heltai, 2006, p. 69). Gábor Bethlen, for example, sent 17 students to Heidelberg at his own cost beginning in 1614; the Rákóczi family, 8 students. Grants were provided by many other aristocratic families, too—such as Lórántffy, Thököly, Dobó, Bocskay, and Mágocsy (for details see Bernhard, 2015, pp. 529–532)—as well as the minor nobility, middle-class, and towns. The overriding aim of these scholarships was to build an efficient Calvinist ecclesiastical, educational, and administrative structure in Transylvania (Asche, 2004, p. 44). The eminent value attached to educationally motivated migration abroad is apparent from a law forbidding authorities, even the prince himself, to thwart a student's study tour by denying a travel permit (Tonk & Szabó, 1993).

Unlike Silesian students, a substantial percentage of whom remained in the Electoral Palatinate after completing their studies in Heidelberg, nearly all Hungarian students went back to their homeland as they were obliged to do by their patrons. The students were expected to make the knowledge gained abroad available to the principality, church, or benefactor. These expectations were explicit, as shown by a letter written by Prince Gábor Bethlen to Boithius:

> [I]f you are thinking of continuing your studies, we inform you that we are prepared to assume their cost for four years. We desire you to go from Heidelberg to Padua for a year or half a year and from there to Paris, where you will stay half a year and come back to us after your study tour. . . . In order for you to understand our intentions in this context, we wish to advise that you diligently lay the foundations not only of your theological but also of your philosophical knowledge so that we can use your expertise both in God's House and in secular matters as well as in foreign policy or wherever we wish after your return. (Quoted in Ladányi, 1999, p. 216; 2001, p. 136).

This letter documents that the decisions on the location(s), subject(s), duration, and academic degree relating to the recipient's study in a foreign country were heavily swayed by the institutions and aristocratic families financing the venture. Church politics and foreign policy thus also influenced the choice of university at which one was to study.

Pareus's former students constituted a close network that played a role for a long time not only in the development of the Calvinist church and Hungarian intellectual life but also in the cultural and foreign policy of the Transylvanian princes. Heidelberg University's faculty of theology at that time figured decisively in the training of the Hungarian and particularly the Transylvanian educational elite for more than 20 years. The Hungarian students who had enrolled with the faculty of theology did not confine themselves to studying theology. The core training of future preachers also involved knowledge of philosophy, skill in rhetoric and grammar, and the reading of ancient texts. The students in Heidelberg also came into contact with late-humanist art and literature, tracts on theories of the state, and writings in the natural sciences. They could also supplement these academic studies by learning to dance, fence, and ride.

After resettling back home, the Heidelberg alumni constituted an exceptional stratum of the Hungarian and Transylvanian intellectual elite, one characterized by notable upward social mobility and excellent training (Heltai, 1994, 2006). Many of them kept up written correspondence with thinkers at the centers of the Reformation in the Holy Roman Empire, the Low Countries, England, and Switzerland, cultivating the contact essential for bringing new knowledge into Transylvania (see Seidel, 2004, pp. 227–228). Those who extended their time abroad, such as Albert Szenczi Molnár, endeavored to pave the way for the students newly arriving from Hungary and Transylvania.

Cavalier's Tours (Grand Tours)[32] *by Aristocratic Students*

In Europe of the early seventeenth century, the educational ideal of the nobility altered fundamentally. Chivalric upbringing was thenceforth supplemented by training in literature, natural science, and especially law so as to prepare the nobility for service at court, in the state administration, and in diplomacy (de Ridder-Symoens, 1996, p. 432). As a result, the percentage of aristocrats among the students increased markedly. The largest contingents of aristocrats among students in seventeenth-century southern Germany were at the University of Ingolstadt and Heidelberg University (Müller, 1984, pp. 38–39). Prominent Hungarian aristocrats who studied in Heidelberg included István Bethlen, who enrolled at Heidelberg University for the 1619–1620 academic year together with three other grandees. Another estimable aristocrat was Miklós (Nikolaus) Bethlen (1642–1716; see below, this section).

[32]Cavalier's Tour and Grand Tour are not equivalent. The term *Cavalier's Tour* is generally used for high-ranking nobility. The term *Grand Tour* has a broader definition, notably in Anglo-Saxon research, and also includes students of the middle classes. Cavalier's Tours may differ from Grand Tours in the duration of the journey, the assigned tasks that have to be fulfilled during the journey (e.g., diplomatic missions), the means of transportation, accommodations at the destinations (high-ranking nobles were often invited to stay at courts), and other criteria of social distinction (for details see Leibetseder, 2004, 2013).

Barons Stephan and Nikolaus Thököly of Késmárk, who were from one of Hungary's wealthiest aristocratic families, enrolled at Heidelberg University in 1596 and 1603, respectively. Nobles from the Bornemisza, Barcsay, Újfalvi Katona, und Decius families enrolled there in this period, too.

Unlike the students from the middle and lower strata of society, most young nobles did not intend to acquire occupational training or an academic title during their study outside their country. For Hungary's higher nobility and the wealthiest landowners, studying abroad was often linked with a Cavalier's Tour, which had been very popular throughout Europe since the late sixteenth century. Its purpose was to culminate aristocratic, courtly education; prepare the social elites for their future responsibilities; raise the political status of aristocratic families; and pursue political objectives (for details see Almási, 2014; Black, 1983, 1992; Chaney, 1998; de Ridder-Symoens, 1996; Freller, 2007; Heiss, 2005; Kühnel, 1964; Leibetseder, 2004, 2013; Pánek & Polívka, 2005; Paravicini, 2005; Schwinges, 2005). Aristocratic students used the Cavalier's Tours to visit various courts, participate in ceremonies there; acquire genteel etiquette, court ritual, and language skills; make political contacts; and conduct confidential, diplomatic missions. The Cavalier's Tour thus served not only the appropriation of knowledge and competence but also "the necessary symbolic distinction" (Paravicini, 2005, p. 667). In the sixteenth and seventeenth centuries the Cavalier's Tour was a must in the education of the higher nobility, tantamount to an initiation into the world of the social and political elite and a staple of cultural transfer (see Heiss, 2005, pp. 218–219; Pánek & Polívka, 2005, pp. 68–69; Paravicini, 2005, pp. 658–660). The students from the higher nobility were usually accompanied by an experienced tutor (*ephorus*), who was responsible for the study and all organizational and financial aspects of the journey. This person had to be fluent in foreign languages and, of course, had to enjoy the family's full trust (for details see Garms-Cornides, 2005; Ugry, 2014).

The Cavalier's Tour by Hungarian nobles, depending on their confession, took routes similar to those of the educational migrations embarked on by nonaristocratic students. The Lutherans and Calvinists headed above all to the Protestant regions of the Holy Roman Empire, with Heidelberg being one of the obligatory destinations.[33] After the outbreak of the Thirty Years' War, the Low Countries became the focus, usually connected with a brief stop in England (for details see Gömöri, 1985), Switzerland, or Paris. The educational journeys of the Catholic nobility usually took the students from those families to universities within the Habsburg empire and were supplemented by trips to Italy (Rome, Bologna, and Padua), Flemish Brabant (Louvain), and France (Paris) (for details see Schindling, 2006).

One can precisely reconstruct the itinerary of Miklós Bethlen's three-year Cavalier's Tour (Fig. 3.8) from his rich, essayistic memoirs (Bethlen, 1955), several editions of which have been published because of their high literary and historical value. Bethlen registered at Heidelberg University in 1661, studying rhetoric,

[33]If a university was located in a town or city in which a court resided, such as Heidelberg, the activities at court and at the university could be linked quite well (Paravicini, 2005, p. 663).

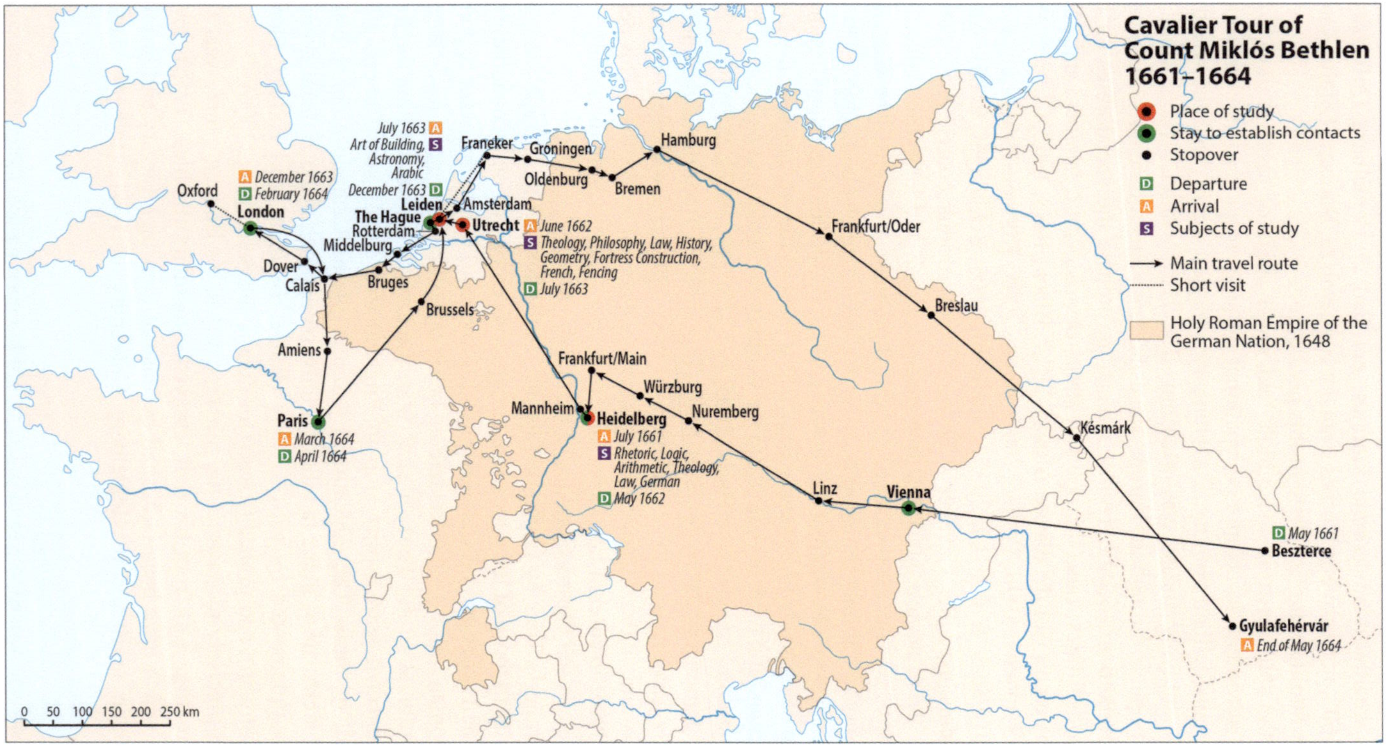

Fig. 3.8 Cavalier's Tour by Count Miklós Bethlen, May 1661 to the end of May 1664.
Source: Keserű (1989), Bethlen (1955). Design by authors. Basic map: Leibniz-Institut für Länderkunde, Leipzig (IfL). Cartography: V. Schniepp.

arithmetic, logic, theology, and law during his 11 months there. In his memoirs he mentioned receiving frequent invitations from Prince Elector Karl Ludwig to hunt and dine in Heidelberg (Bethlen, 1955; Vékás, 2001, pp. 140–141).[34] The elaborately ornamented entry of Bethlen's enrollment on the first page of the 1661 registry (Fig. 3.9) documents what Schwinges (2005, pp. 357–360) described as the ubiquitous rankings and favoritism of the nobility at the universities.[35] Differences in status were documented even for registration purposes, for the names in the registry were not listed in chronological order of the entry but rather according to the social hierarchy of the students, with the nobles placed first (Fig. 3.9). Then came the church canons, the other dignitaries and well-to-do members of the bourgeoisie, and, lastly, the *pauperes*, or students of little or no means (pp. 359–360).

After Heidelberg, Miklós Bethlen spent a year at the University of Utrecht and half a year at the University of Leiden, visiting Oxford and Paris as well. As Fig. 3.8 shows, Miklós Bethlen covered the distance from Paris back to Transylvania fairly quickly. He lingered only one or two days in the sizeable cities, less time than he had spent when going in the other direction. The reason for his haste was that he had been entrusted with a letter from the French king and wanted to deliver it to the Prince in Transylvania as soon as possible. Miklós Bethlen later became one of Transylvania's leading politicians and was charged with important diplomatic missions (Vékás, 2001, p. 140).

Figure 3.11 shows the Cavalier's Tour by Pál Teleki, the son of Mihály Teleki (1534–1690), who determined Transylvanian foreign policy as field marshal and chancellor for three decades, during which he acquired an immense fortune and gigantic estates. Pál Teleki's mother (Mihály Teleki's widow) and his professors in the Nagyenyed Academy had prepared his study and travels outside Hungary carefully and kept up a lively correspondence with him during his absence. He received a remarkable number of suggestions and basic instructions from Ferenc Pápai Páriz, a famous Heidelberg alumnus who at that time was the Teleki family's physician and a professor at the academy. The Transylvanian professors occasionally asked Teleki to buy various books, which were then to be sent to Nagyenyed. The only reason Teleki did not visit Heidelberg was that the city and its university buildings had been obliterated in 1693 during the War of Palatine Succession (the Nine Years' War, 1688–1697; for details about the destruction, see Rotzoll, 2012). The route that Pál Teleki took during his tour is traceable with the aid of his published private correspondence and his *Album amicorum*, which contains about 220 entries of his interlocutors, mostly from the professors he visited abroad and

[34]Bethlen's memoires also include humorous anecdotes about the customs of the hunt and about life at court. As a farewell present, the Prince Elector gave him two valuable pistols with ebony grips, which Bethlen appraised as the best he had ever owned (Bethlen, 1955).

[35]Unlike the academies for knights, which concentrated solely on training the nobility (Müller, 1984, p. 35), the universities accepted everyone, but the feudal distinctions between the status or rank of the families of origin were displayed at every turn, as in registries, official scholarly debates, worship services, processions, and assigned seating in lecture halls (Schwinges, 1986, pp. 346–351; 2005, pp. 357–359).

Fig. 3.9 Entry on Miklós (Nikolaus) Bethlen and his *ephorus*, Paulus Tsernatoni, first page of the Heidelberg University registry of 1661.
Source: Registry of 1661. Heidelberg University Archives. UAH, M 4 (1579–1662), fol. 184v. Reprinted with permission.

from fellow Hungarian students.[36] These sources are a trove of interesting information, but unfortunately they reveal little about what he actually studied (Keserű, 1989).

At least the students who were somewhat well-off were keen on acquiring and bringing home newly published books during their journeys, or were even commissioned to do so by their benefactors. The main private and ecclesiastical libraries of Hungary and Transylvania received most of their books through students who had studied abroad. Borne by Hungarian students returning home, the teachings of Descartes and Newton, the ideas of the European Enlightenment, the results of experimental physics and chemistry, and modern mathematics swiftly spread to Hungary (Győri, 2010, p. 254). Pál Ceglédi Szabó (1590?–1649), the Calvinist superintendent of Transdanubia (the part of Hungary west of the Danube river), had a private library of 232 books, of which 171 had been acquired abroad and 38 printed in Heidelberg, where Ceglédi Szabó had enrolled in 1613 (Zvara, 2012). This collection was not even among the largest. In Transylvania and Hungary of the late sixteenth century, there were several tens of thousands of printed books, chiefly theological works. The most often read author was Philipp Melanchthon, but the literature by ancient Greek and Roman writers and by contemporary humanists were also widely known, especially Aristotle, Erasmus, Ramus, Justus Lipsius, and Boccaccio (Köpeczi, 1990).

The scientific and cultural relations of the intellectual elites of Hungary and Transylvania were thus oriented at that time almost solely to Europe's Latin heritage. The students from that part of society mostly avoided educational tours to the Romanian princedoms, which were perceived as being economically and culturally underdeveloped,[37] dominated by the Greek Orthodox religion, and subject to a hostile, pagan Ottoman empire (Kármán, 2006). Nevertheless, certain Transylvanian cities, above all Kolozsvár (Klausenburg, today's Cluj-Napoca) and Brassó (Kronstadt, today's Braşov), were foci of some cultural exchange between the Romanian-, Hungarian-, and German-speaking populations in the sixteenth and seventeenth centuries (for details see Binder, 1975; Huttmann, 1975). Likewise, few Romanians studied at foreign universities before the nineteenth century. Most Romanians were Greek Orthodox, and they trained their future priests in their own

[36]An *Album amicorum* may contain diaries, notes on the journey, and a description of it. The handwritten mementos are occasionally accompanied by epigraphs by friends and guests, designs or, especially in earlier centuries, painted coats of arms.

[37]Brenner (1989), Gunst (1989), Stokes (1989), and many others discuss the historical origins of backwardness in some parts of eastern Europe. Legal structures, the centralization of state power, the structure of feudal society, the density and economic role of cities, the harshness and duration of serfdom, and precapitalistic property relations—all these factors bearing on economic evolution and societal change differed between western and eastern Europe.

seminaries, not at universities (Offner, 2006, p. 293).[38] The eastern and southern frontiers of Transylvania along the Carpathian mountains were thus a relatively sharp cultural divide and an effective barrier to knowledge transfer, particularly the circulation of books; the diffusion of late Humanism, the Enlightenment, and literacy; and the emergence of modern administration in the territories.

Foreign Policy Interests of the Transylvanian Princes in the Choice of Where Their Subjects Studied

After Prince Gábor Bethlen assumed power in Transylvania in 1613, many Hungarian Calvinists who later became famous and politically influential studied in Heidelberg, with the maximum annual enrollment of Hungarian and Transylvanian students ranging from 9 to 16 from 1614 to 1620 (see Teutsch, 1872, pp. 186–188). This fact owed partly to Heidelberg University's academic appeal, but politics, too, played a part (Csohány, 1994–1995). One of the reasons that Prince Gábor Bethlen sought close cultural and political relations with the Prince Elector in Heidelberg was that Heidelberg at the time was a center of the anti-Habsburg alliance of Protestant princedoms. Gábor Bethlen wanted to join it so as to stand his ground against Habsburgs and Ottomans alike. Both he and his successor, Prince György Rákóczi I (1593–1648), intensely cultivated the links to these Protestant princedoms through students who had studied in Heidelberg. Intense written correspondence between Pareus and Gábor Bethlen commenced in 1616 (Heltai, 2006, p. 73).

The long-smoldering conflict between the Protestant Union (created in 1608) and the Catholic League (created in 1609) finally ignited when, against all advice, Prince Elector Friedrich V had himself elected King of Bohemia in 1618 (for details see Pánek, 2003). This royal dignity had been offered to Gábor Bethlen and other princes as well, but they had declined because they could imagine the reaction of the Habsburgs. Friedrich's election, which the Habsburg took as an affront, triggered the Battle of White Mountain (November 8, 1620) and the subsequent rout of the Protestant army—the outbreak of the Thirty Years' War. With Heidelberg being the key political center of the Protestant Union, the city and many other settlements in the Electoral Palatinate were ravaged in the war.

With the consent of the Sublime Porte, Gábor Bethlen intervened on religious and political grounds in three field campaigns against the Habsburgs during the Thirty Years' War. His considerable military achievements and diplomatic efforts met with only moderate success, however. Contrary to all initial expectations of the Protestant powers, Habsburg imperial power was not destroyed. But Bethlen did manage to

[38]This preference changed in the nineteenth century when the University of Leipzig, for instance, "spearheaded the training of Romanian intellectuals abroad" (Heitmann, 1975, p. 124). The famous Romanian linguist Sextil Puşcariu from Kronstadt was one of the students at the Romanian Language Institute in Leipzig from 1895 to 1899.

ensure that religious freedom and the feudal constitution of Hungary stayed intact and that the Habsburg attempts at centralization failed. By joining the Haag alliance of Great Britain, Denmark, and the Low Countries through the Treaty of Westminster in 1626, he strengthened the international recognition of Transylvania's sovereignty (for details on the complicated diplomatic relations between Transylvania and the other Protestant countries during the Thirty Years' War, see Csohány, 1994–1995, and Kármán, 2013; on Bethlen's foreign policy, see Czettler, 1980–1981, and Schmidt-Rösler, 2006).

Impacts of the Thirty Years' War (1618–1648) and the War of Palatine Succession (1688–1697) on the Mobility of Hungarian Students

Heidelberg University's intellectual golden age and the institution's academic relations with Hungary and Transylvania abruptly ended during the Thirty Years' War when the town fell to Catholic troops in 1622. Many professors fled, and Heidelberg's famous Bibliotheca Palatina was taken to the Vatican as war booty (Effinger & Zimmermann, 2012; Meusburger, 2012; Neumaier, 2012a; Rotzoll, 2012). The professors remaining in Heidelberg were dismissed by the Bavarian occupation authorities in 1626, and Heidelberg University did not reopen until 1628—as a Jesuit university. During the Swedish occupation of the Palatinate (1632–1634), the Jesuit university in Heidelberg was dissolved, and the attempt was made to reestablish a Lutheran university. But Heidelberg was recaptured by the Bavarians, who reinstituted the Jesuit university, though teaching did not resume (Wolgast, 1986). With this demise of Heidelberg University, the students from Hungary and Transylvania gravitated primarily to universities in the Low Countries and England. Between 1623 and 1632 the Universities of Franeker and Leiden in the Low Countries, joined later by the newly founded Universities of Groningen, Amsterdam, Utrecht, and Harderwijk, attracted many students from Hungary and Transylvania (Bozzay, 2010, 2014; Eredics, 2008; Hotson, 2008; Murdock, 2000). Between 1621 and 1631, 52 Hungarian students attended the University of Leiden, and 66 attended the University of Franeker. Between 1631 and 1651, 190 Hungarian students were at Leiden and 200 at Franeker (Bernhard, 2015, p. 482).

In the seventeenth and eighteenth centuries more than 25% of all Hungarians and Transylvanians studying abroad were enrolled in the Low Countries, with the University of Franeker in Frisia accounting for approximately 40% of that group[39] (Bozzay, 2010, p. 215). The early predilection for the Universities of Franeker and Leiden was no coincidence. After the fall of Heidelberg, the renowned Heidelberg professor of the Old Testament, Abraham Scultetus (1566–1624), moved to the Low Countries, where he was exceptionally engaged with the University of Franeker.

[39]The University of Franeker was closed in 1811.

Scultetus helped Hungarian students a great deal and wrote them recommendations to pave their way to universities in the Low Countries. Albert Szenczi Molnár, too, was active in Leiden after fleeing from Heidelberg (Eredics, 2008). Johann Heinrich Alting (1583–1644), who was probably second only to Pareus in the number of disputations over which he had presided at Heidelberg University, taught in Groningen from 1627 to 1644 (see Nagy, 1967, p. 44). The professors who had migrated from Heidelberg were not alone in their efforts in Franeker and Leiden. Eredics (2008) found that the two towns also had several Dutch and English professors who were highly involved on behalf of Hungarian students and had major intellectual impact on them. Clearly, the place at which to study was a choice carefully made, and despite the enormous distances between locations many students did not find that moving to another university was detrimental to their educational career or that it was like leaping into the unknown (Irrgang, 2003, p. 67). Existing networks, ties between compatriots, and letters of recommendation from professors and territorial lords reduced the effect that one might expect vast spatial distances to have had on the choice of where to study. The decisive thing was "social proximity," not spatial distance: "Academic migration never dissolves existing relations; it is what really brings them into play for a person's career" (p. 67).

After the Thirty Years' War ended in 1648, Heidelberg University stayed closed for four years. It was not dissolved, but there was no longer any teaching there. Not until 1652 did the university successfully reopen, this time as a Calvinist institution again. Just 36 years later, however, the War of Palatine Succession broke out, a conflict during which the city of Heidelberg, all its university buildings, as well as numerous nearby towns and villages were utterly demolished under the scorched-earth policy of French troops (Musall & Scheuerbrandt, 1980, pp. 8–12). By 1693, not a single inhabitable building had survived in Heidelberg. The university's faculty members withdrew first to Frankfurt am Main, then in 1698 to Weinheim. The professors were able to return to Heidelberg in early 1700, but lectures did not resume for another four years. The destruction of the city, the wartime hiatuses of teaching, and the multiple forced changes of confession among the Heidelberg professors had blighted the lure of the university (Neumaier, 2012a, Rotzoll, 2012).

The registries of Heidelberg University have gaps due particularly to these wars and other disasters,[40] but other sources show that students from Hungary and Transylvania were studying at Heidelberg University in the decades after the Thirty Years' War as well. In addition to the aforementioned Miklós Bethlen, who registered at Heidelberg University in 1661, there was Ferenc Pápal Páriz (1649–1716), who studied at the Universities of Frankfurt on the Oder, Marburg, Heidelberg (where he registered in 1672), and Basle. He earned his doctorate in philosophy and theology at Heidelberg. He was even offered a chair at Heidelberg University

[40]No students were enrolled at Heidelberg University from 1632 to 1652. Teaching was suspended from 1642 to 1652 and from 1693 to 1703. The registries for the years from 1663 to 1689 are missing (lost probably in 1693). The registries of the theological faculty (Toepke, 1886, Appendix V, pp. 545–585) bear names only up to 1685, even then with gaps.

because of his prodigious achievements, but he wished to earn a doctorate in medicine in Basle, where he became a highly acclaimed assessor of medicine and was even granted the honor of delivering the address commemorating the famous professor Johannes Henricus Glaserus in 1675, with the text later being published in Latin and Hungarian many times (Vékás, 2001, p. 141). In 1677 Páriz became the personal physician to Ana Apafi-Bornemisza, the wife of Transylvania's ruling prince. From 1680 to 1716, Páriz was professor of Greek, physics, and logic at the academy in Nagyenyed and contributed much to the Cavalier's Tour by his aristocratic student Pál Teleki (Fig. 3.10). Páriz's Latin-Hungarian and Hungarian-Latin dictionaries, which impressively expanded the work by Albert Szenczi Molnár, were published in 1708 after nearly 20 years of preparation and were reprinted on several occasions. Páriz also became known through the books *Pax animae* (Peace of the Soul), *Pax corporis* (Peace of the Body), *Pax aulae* (Peace of the Court), and *Pax sepulchri* (Peace of the Grave), which were published in Kolozsvár. The volume of the greatest consequence was *Pax corporis*, written in Hungarian, in which he described a hundred different kinds of diseases and recommended various therapies. He based the other three books largely on existing publications. Páriz made a name for himself as a church historian, too. His *Rudus redivivum* (1684) described the history of the Protestant church.

Student mobility diminished for various reasons throughout Europe in the eighteenth century (for details see de Ridder-Symoens, 1996; Schwinges, 1986). One was the rise of mercantilist ideology (e.g., the state as an economic actor, state intervention in the economy, and bans on the emigration of highly skilled laborers). Another was the Enlightenment's cherished idea of education's utility (and the mercantilist corollary that the benefits of education were intended to be reaped by one's own territory, not drained through emigration). A third reason undercutting student mobility in the 1700s was the increasing territorialization and confessionalization of universities. A final factor that henceforth made study in a foreign country no longer seem as advantageous as it had been was the absolutist state's escalating need for control (de Ridder-Symoens, 1996; Hammerstein, 1995, 1996; Stagl, 1995). The university and the free movement and settlement of students were subordinated to the interests of the absolutist state of the eighteenth century (Schwinges, 1986, p. 229). For anyone seeking a position in the church or state administration, some territorial rulers at that time made it a requirement to study at the university they had created in their own lands, and they deliberately limited study elsewhere by their subjects. Other rulers even expressly forbade study at certain foreign universities (for details see de Ridder-Symoens, 1996, pp. 437–439, 443).[41]

[41]Such policies therefore heavily restricted the universality of academic degrees, which since the mid-thirteenth century had made examinations conducted at one university valid at all other universities and had established the parity of the academic degrees that they conferred (Kintzinger, 2012, p. 309).

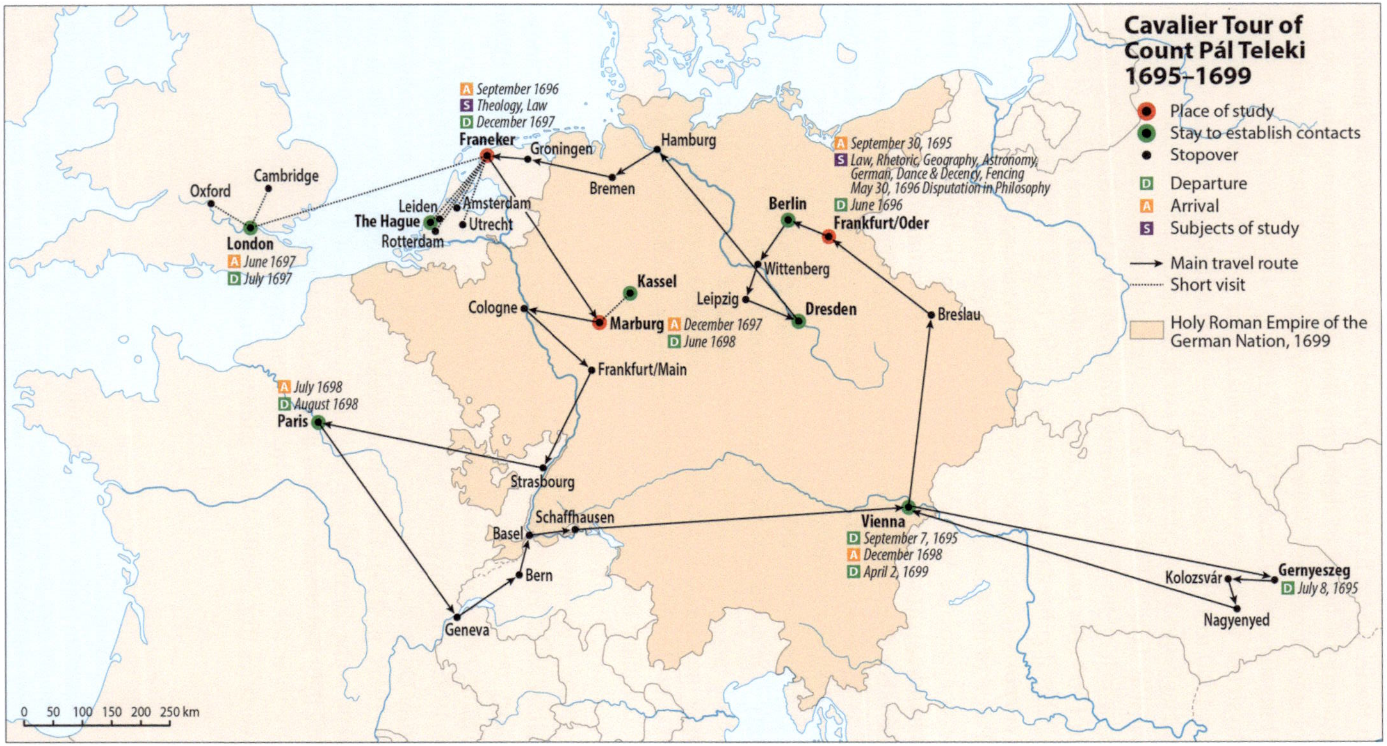

Fig. 3.10 Cavalier's Tour by Pál Teleki, July 1695 to April 1699.
Data source: Keserű (1989). Design by authors. Basic map: Leibniz-Institut für Länderkunde, Leipzig (IfL). Cartography: V. Schniepp.

The number of Hungarian students at Germany's Protestant universities dwindled in the seventeenth and eighteenth centuries also because of the re-Catholization of western and northern Hungary. Cardinal Péter Pázmány succeeded in reconverting 30 of the foremost families of the higher Hungarian nobility. He was supported by the court in Vienna, which granted privileges, honors, and high offices solely to candidates nominated by Pázmány (Bucsay, 1977, p. 166). Hungary's Catholic students thenceforth tended to opt especially for the closer, Catholic universities in Graz, Vienna, and Nagyszombat. The University of Graz had been under Jesuit influence since its founding in 1586. The University of Vienna came under Jesuits leadership in 1624, as did the university that Pázmány founded in Nagyszombat in 1635 (Andritsch, 1965, p. 2; Barta, 1937; Fata, 2004; Fata & Schindling, 2006). In the late sixteenth and first decades of the seventeenth century, most of Hungary's Catholic elite was trained at the University of Graz (Andritsch, 1965, p. 6).

As of 1725 the flow of Hungarian students to Germany abated also because of the Habsburg government's attempt to control study abroad by requiring them to possess a valid passport. As of 1748, these documents were valid for only one year (Gönczi, 2006, p. 177; Ladányi, 2001, p. 136). In the second half of the eighteenth and early nineteenth century, the Habsburg government tried to restrict study abroad by Hungarian Protestants—and the infiltration of unwanted ideas in general—by repeatedly banning attendance at specific universities, particularly Prussian ones (1756–1759, 1763–1766, 1804, 1818–1819) (Kardos, 2000, pp. 47–50; Rasche, 2006, pp. 197–204; Varga, 2001, p. 180). In 1759 Empress Maria Theresia forbade Hungary's Protestant students to collect any domestic contributions for their studies (Bozzay, 2010, p. 216).

The Decline of Heidelberg University in the Eighteenth Century

Except for the brief interludes from 1750 to 1756 and 1780 to 1783, attendance at Heidelberg University by students from the Carpathian Basin was relatively rare in the final decades of the eighteenth century (Fig. 3.7). The intellectual caliber of that university had deteriorated badly;[42] teaching was highly regimented and uninnovative; and the political authorities interfered with the recruitment of professors, favoring "native children" (Mussgnug, 2003, p. 131; Neumaier, 2012a, p. 75; Wolgast, 1986, pp. 5, 84).

[42]Friedrich Gedike, visiting the German universities on behalf of Prussian King Frederick William II, for example, wrote of Heidelberg in 1789: "Everything that I saw and heard convinced me that this university is insignificant" (cited in Wolgast, 1986, p. 84). In 1798 the rector of the university reported that "Heidelberg University exhibits the infirmities of advanced old age: dullness and inactivity" (cited in Wolgast, 1986, p. 85). In 1798 the financial plight of the university prompted a government commission to conclude that the institution had to be deemed "a terminally ill patient... whom it would be best to allow to die in peace" (cited in Mussgnug, 2003, p. 131).

The erosion of Heidelberg's importance also stemmed from clashes between Prince Elector Carl Philipp (1661–1742) and the city's inhabitants. The conflicts had prompted him to move his court to Mannheim in 1720 and leave Heidelberg bereft of meaningful political, cultural, and economic functions and largely uninteresting. When Napoleon occupied the territories west of the Rhine in 1794, Heidelberg University lost most of what it owned, for its assets had lain chiefly beyond what had just become the new frontier (Merkel, 2012, p. 47). With its economic existence jeopardized, the university fell several months behind in paying the salaries of the professors (Wolgast, 1986, pp. 83–84) and was facing closure.[43]

Halle, Göttingen, and Leipzig were Germany's most modern universities in the eighteenth century, so most of the students from Hungary and Transylvania went there. As of about 1750, the University of Göttingen was one of Europe's best universities, known for its outstanding professors. It also served many other universities as a model of reform. In the early nineteenth century, Göttingen's university library contained approximately ten times the number of volumes as that of Harvard University, the largest library in the United States at that time (Honeck & Meusburger, 2012). It is therefore not surprising that Göttingen became the leading university for Hungarian students as well in the second half of the eighteenth century (Gönczi, 2006).[44]

The Second Intellectual Heyday of Academic Relations between Heidelberg and Hungary: The Nineteenth and Early Twentieth Centuries

Reasons for the Renewed Flood of Hungarian Students to Heidelberg

In 1803, after the Electoral Palatinate had been dissolved by Napoleon and Heidelberg attached to the Margraviate (Grand Duchy as of 1806) of Baden, serious reforms ushered in Heidelberg University's reascendance in the following decades, culminating in a second phase of academic glory from the 1850s to the outbreak of World War I (Hübner, 2010; Meusburger, 2012; Meusburger & Schuch, 2010, 2012; Wolgast, 1986, 1987). In the mid-nineteenth century no German province spent a

[43]The crisis hit many universities, as did Napoleon's move to shut down German universities that he regarded as politically unreliable. Between 1786 and 1818 these two issues resulted in the closure of 20 of the 42 universities existing on the territory of the Holy Roman Empire of the German Nation (Eulenburg, 1904; Meusburger, 2012, p. 21).

[44]Baron Sándor Prónay, the curator general of the Hungarian Evangelical churches and schools, once expressed this sentiment in a letter of recommendation: "Göttingen never ceases being Hungary's teacher—the major source of our enlightenment and culture" (quoted in Gönczi, 2006, p. 176).

more sizeable share of its budget on universities and gave them more per-capita assistance than the Grand Duchy of Baden did (Pfetsch, 1974). But there were several other reasons for Heidelberg university's resurgent academic stature (for details see Engehausen, 2012; Honeck & Meusburger, 2012; Meusburger & Schuch, 2010; Wolgast, 1987).

Heidelberg University's rapidly rising scientific and intellectual appeal did not immediately open a floodgate of students from Hungary, however (for details see Fricke, Koch, Meusburger, & Preusker, 2012). In the initial decades of the nineteenth century, the Napoleonic wars, the Holy Alliance's Carlsbad Decrees of 1819 (which strengthened control over and censorship of universities and the press), the restrictions imposed during the years preceding the revolution of March 1848 (*Vormärz*),[45] and the events of the 1848–1849 revolution itself (for detail see Engehausen, 2012) continued to make study in Germany appear unworthwhile. Moreover, travel restrictions kept Hungarian students from going to Germany from 1819 until 1827 (Szögi, 2006, p. 391), so Germany's universities together registered only two to six Hungarian students per year (Szögi, 2001, p. 27, 2006, p. 391).

The measures introduced by István Széchenyi (1791–1860) and others in the 1820s and 1830s to reform and modernize Hungary's society and economy led to a rise in the number of Hungarian students in Germany after 1834. However, this trend ended when the Habsburgs and the Russian Czar put down the Hungarian revolution of 1848. Hungarian students did not start flocking to Germany until the Austro-Hungarian Compromise of 1867, which restored Hungary's political and economic independence within the Danube monarchy, facilitated numerous reforms and modernization in Hungary, and led to massive migration to Budapest. Until World War I it also sparked uncommonly dynamic development, which was also reflected by the number of Hungarian students in Germany (Table 3.1, Fig. 3.11). Szögi (2004) pointed out that Hungary sent more students to Germany between 1867 and 1919

Table 3.1 Hungarians enrolled at German universities

Period	Number registered
1694–1789	3,944
1789–1819	1,371
1819–1849	934
1849–1867	1,275
1867–1890	3,277
1890–1919	7,691
Total	18,492

Source: Szögi (2006, p. 389).

[45] To most authors the term *Vormärz* means the period between 1830 (the July revolution in France) and March 1848. Some writers, however, understand it to include only the background directly related to the Revolution of 1848—the years as of 1840.

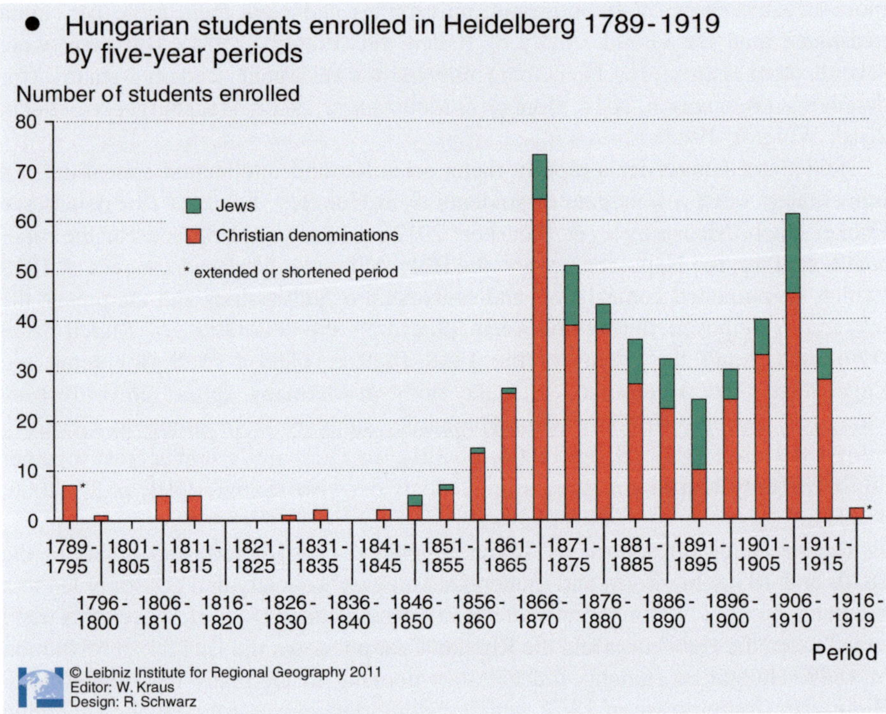

Fig. 3.11 Hungarians enrolled at Heidelberg University, 1789–1919, by five year periods. Source: Meusburger and Probáld (2012, p. 304). Reprinted with permission.

than did the Scandinavian countries (whose combined number of inhabitants was similar to Hungary's), although language affiliation and geographic distance lead one to expect the opposite.

Between 1876 and 1892 the number of Hungarians studying in Germany dipped again because of the swift expansion of the Hungarian university system, but it recovered after 1893 to swell into a second, even bigger wave that lasted until 1913 (see Szögi, 2001, pp. 27–28). However, the rise in the number of students coming to Heidelberg from Hungary (Fig. 3.11) differed from that of the same group at other German universities. In Heidelberg, the first peak of Hungarian students arriving after the Compromise of 1867 was far higher than at other German universities. But the second peak in Heidelberg—between 1906 and 1910—was much lower than in Germany at large because in that phase the metropolises Berlin and Munich were the most coveted locations for the students from Hungary.

The strong influx of Hungarian students at German universities between 1867 and 1919 had several reasons. First, the political, economic, and cultural upheaval and the profound economic and social processes of modernization occasioned in Hungary by the Comprise of 1867 coincided with a phase during which some German universities (including Heidelberg) were among the world's best (Hübner, 2010; Meusburger & Schuch, 2010, 2012; Wolgast 1986, 1987). In Hungary this reputation meant that an academic degree from a German university was exceedingly valuable.

Second, Hungarian had only recently been introduced into higher education for purposes of instruction, and many Hungarian citizens at that time were not yet adequately skilled in the language. German, by contrast, was fairly common as vernacular or as a second language of the elites and the urban population, notably among the Jews and the German minorities in Hungary. Upper-middle-class Jews in particular cultivated close ties to the German language and culture after 1867.[46] Szögi (2006) has estimated that around one third of the Hungarian nationals studying in Germany between 1789 and 1919 spoke German as a mother tongue, for several Hungarian towns, wide swathes of Transylvania, the Zips region, the Banat, and Bačka had German minorities.[47] In 1856, 56.4% of the population in Buda, Óbuda, and Pest, which were united to form Budapest in 1873, still stated they spoke German in their daily lives. Only 36.6% said they spoke Hungarian; 5%, Slovakian; and 1.7%, Serbian. In 1890, by contrast, 67.1% stated they spoke Hungarian in everyday life, whereas only 23.7% still claimed to use German anymore (Gantner, 2006, p. 461; Varga, 2003, pp. 173–174).

Third, passports were not yet required in most of Europe at that time. They were not introduced as an official document in Hungary until 1903 and were required only for visits to Romania, Serbia, the Ottoman empire, and the Russian empire (Bencsik, 2002). Fourth, the currency system, which was based on the gold standard, guaranteed the convertibility of national currencies and the stability of exchange rates. Fifth, expansion of the railroad network reached its zenith before World War I, making travel faster and cheaper than ever.

The sixth and probably decisive reason for the high numbers of Hungarian students at German universities for half a century after 1867 was that a colossal expansion of industry and services in Hungary caused a leap in the country's demand for highly educated labor. Between 1867 and 1914 the Hungarian economy grew at an annual rate of 3.5%, and national income quadrupled. With only three or four continental European countries surpassing that pace, Hungary was quickly catching up to Europe's highly developed regions (Bodrogi & Galántai, 2013). After the nobility's

---

[46]Theodor Herzl (1860–1904), one of Zionism's founders, spoke German at home. Like many other Jews, he had attended Budapest's upper-secondary Lutheran school created in 1872, which stressed the acquisition of the German language and culture and which was known for its liberal education (Gantner, 2006, pp. 461–462).

[47]The first two waves of German-speaking immigrants arrived in the Carpathian Basin in the eleventh and thirteenth centuries, mainly in some mining towns of Upper Hungary and in southern Transylvania. The third and largest period of immigration occurred between 1711 and 1780 because of a resettlement policy pursued by the Habsburgs after expulsion of the Ottomans from devastated and depopulated Hungarian territory (for details see Anderegg, 2000; Fata, 2014). In the late eighteenth century the Kingdom of Hungary had more than one million German-speaking inhabitants; at the end of World War I, almost two million.

class privileges were abolished in Hungary in 1848 and suffrage was tied to an educational or means census, people could convert schooling into political capital. This possibility enhanced the attraction of formal education as a way to rise into the influential elite, with a foreign academic degree having exceptionally high status.

The Hungarian government in those years put priority on building an effective, modern educational system. After futile earlier attempts[48] to introduce universal compulsory education, the government finally succeeded in doing so by promulgating Act XXXVIII of 1868. Only 48% of the country's 6- to 12-year-olds attended a grammar school in 1868, but by 1913 the figure had risen to 93%. Illiteracy among persons older than 7 years declined from 61% to 31% between 1870 and 1910. In the latter year 3% to 4% of the boys born in an age cohort spanning a single year attended a *Gymnasium*, completion of which entitled them to attend a university or other form of tertiary education. Between 1867 and 1917, the number of students in institutions of higher learning climbed from 4,830 to 18,033 (Pukánszky & Németh, 1996), a quadrupling that owed partly to the expansion of several universities and the founding of new ones.

From 1635 to 1871, Hungary had only one university, the Catholic university in Nagyszombat. It was transferred first to Buda in 1777 as the Hungarian state university, then to Pest in 1784. In 1871 the Technical University was added in Buda. These two institutions were among the largest in Europe and were highly regarded for their academic caliber. Other universities were founded in Kolozsvár (1872), Pozsony (1912), and Debrecen (1914).

The intensity of study in foreign countries and the choice of where to study there also depended on the Hungarian population's ethnic affiliations. Act II of 1844 abolished Latin as the official language in Hungary, replacing it with Hungarian as the language of official business and higher education. Between 1849 and 1860 the Habsburg government tried to impose German as the language of instruction but the decree issued by the reigning monarch in October 1860 reinstated the exclusive role of Hungarian. Understandably, some students from ethnic minorities whose Hungarian was not yet adequate for university academia sought to study in countries such as Austria, Germany, Serbia, or Romania, where they could receive instruction in their mother tongues. This preference partially explains why these ethnic groups were underrepresented in the Hungarian universities and other institutions of higher learning in 1900 (see Table 3.2).

There are no precise statistics on the mother tongue (nationality, ethnic affiliation) of the students from Hungary who were registered at the German universities and other institutions of higher education. However, one can infer their ethnic affiliation (mother tongue) from their surnames and provenance. Szögi (2006, p. 404) has asserted that 42.1% of the students from Hungary registered in Germany from 1789 until 1919 were

[48]Compulsory schooling was introduced in the Austrian part of the Habsburg monarchy under Empress Maria Theresia in 1774. It took many decades, however, until it could be enforced throughout the realm.

Table 3.2 Ethnic composition of the population and students at the universities and other institutions of higher learning in Hungary (in percentages), 1900

	Mother tongue						
Group	Hungarian	Romanian	Slovakian	German	Serbian	Other	Total
Population	51.4	16.6	11.9	11.9	2.6	5.6	100.0
Students	82.8	5.6	2.2	6.9	1.1	1.4	100.0

Source: Jancsó (2013, p. 37).

ethnic Hungarians; 33%, Germans; 6.1%, Slovaks; 2.1%, Romanians; and 1.3%, Serbs. He was unable to identify the nationality of 15.1% of the students.

Where the Hungarian Students in Germany Studied, 1789–1919

In the eighteenth century and the first two or three decades of the nineteenth, the universities at which the students from Hungary and Transylvania enrolled in Germany were still selected primarily according to the confessional character of those institutions. As of the mid-nineteenth century, however, the Hungarian students—except for the shrinking number of those in theology—based their choice ever less on the confessional alignment of the universities. Instead, they looked increasingly to the subject-matter specialization of the universities, the academic reputation of the professors, the cultural allure of the venue, and the scholarships and grants on offer.

The students from the upper strata of society generally wished to attend universities with a law program of high-quality. The places chosen by less-well-off students from Hungary were determined mostly by the grants and free meals that were available and by the independent organization of Hungarian students (on independent organization, see Stickler, 2006). Grants for needy students from Hungary and Transylvania were to be had from the Universities of Greifswald,[49] Tübingen, and, later, Berlin (see Alvermann, 2006, pp. 346–347; Fata, 2006, pp. 244–246). The *Evangelische Stift* in Tübingen offered a certain number of students from Hungary and Transylvania free board, student dormitories, and tuition. Between 1661 and 1830, 85 students from Hungary received an education and free meals there (Fata, 2006, p. 244). In 1829 it was decided that theology students from

[49]To help the Protestants in Hungary, a scholarship fund reserved for Hungarian students was set up by Swedish King Karl XII at the University of Greifswald in 1705, a gesture that resulted in a constant stream of Hungarian students to that place. In the nineteenth century more Hungarians than Scandinavians were studying in Greifswald (Alvermann, 2006, pp. 346–347). In the European power politics of the Swedish crown, Upper Hungary was strategically important in the seventeenth and eighteenth centuries. It was manifested, for instance, in the alliance with the Transylvanian princes György Rákóczi I and II (pp. 347–348).

Hungary should receive four scholarships of 150 Gulden each per year, a regulation that held until 1918 (Fata, 2006, p. 246). Hungarian students in Berlin enjoyed the special support of scholarships from the Hungarian state.[50] Heidelberg University had no comparable sponsorship to offer, a lack that partly accounted for the disproportionately high number of students from Hungary's upper class and the much lower number of students from financially weaker families.

During the nineteenth century the concentration of Hungarian students gradually grew in the aspiring and culturally engaging metropolises of Berlin and Munich. The recruitment policy of Ministerial Secretary Friedrich T. Althoff (1839–1908) had won Berlin the reputation of being the best and most modern German university. Many celebrated professors regarded an appointment there as the very pinnacle of their career goals, above all after 1870. For Hungarian intellectuals, Berlin became a kind of postgraduate school. The sons of the Hungarian Jewish middle class in particular headed enthusiastically for Berlin (Ganter, 2006, p. 462). In Munich, the Technical University and the University of the Arts were exceedingly popular among the Hungarians (Fig. 3.12).

Some 65.5% of the 14,548 Hungarian students registered at universities in Germany from 1789 to 1919 were enrolled at five sites: Berlin (24.8%), Munich (13.1%), Leipzig (10.5%), Jena (10%), and Halle (7.1%). Heidelberg ranked eighth (Szögi, 2001, p. 35). Between 1867 and 1890 Heidelberg University rose to sixth place thanks to its rising scientific reputation. In the nineteenth century Heidelberg University's disadvantage was that it offered no programs in engineering and technology, no business school (the nearest one was in Mannheim), and no art academy. It was one of the reasons that Heidelberg in the nineteenth century could no longer take the lead as it had in the second half of the sixteenth and early seventeenth centuries.

Analysis of the registries at Heidelberg University shows that 48.5% of all students from Hungary and Transylvania who had enrolled there between 1789 and 1919 had studied at at least one other German or Austrian university before they moved to Heidelberg; 8.0% had studied in Budapest and 0.4% in Kolozsvár; and 3.4% had enrolled at a university in Switzerland, Prague, or Paris. One can safely assume that the other 39.7% went to Heidelberg University first. The favorite universities sought out before study in Heidelberg were those in Berlin (16.4%), Vienna (13.4%), Budapest (8.0%), Leipzig (5.2%), Jena (3.6%), and Munich (1.6%). Unfortunately, the Heidelberg registries do not indicate how many of that university's enrolled Hungarian students went on to attend another university after their time in Heidelberg.

[50] After World War I Berlin gained an additional advantage: the Collegium Hungaricum, founded in 1924 for talented young Hungarian scholars (Siebe, 2006, p. 434). Germany's other university sites had no such facility.

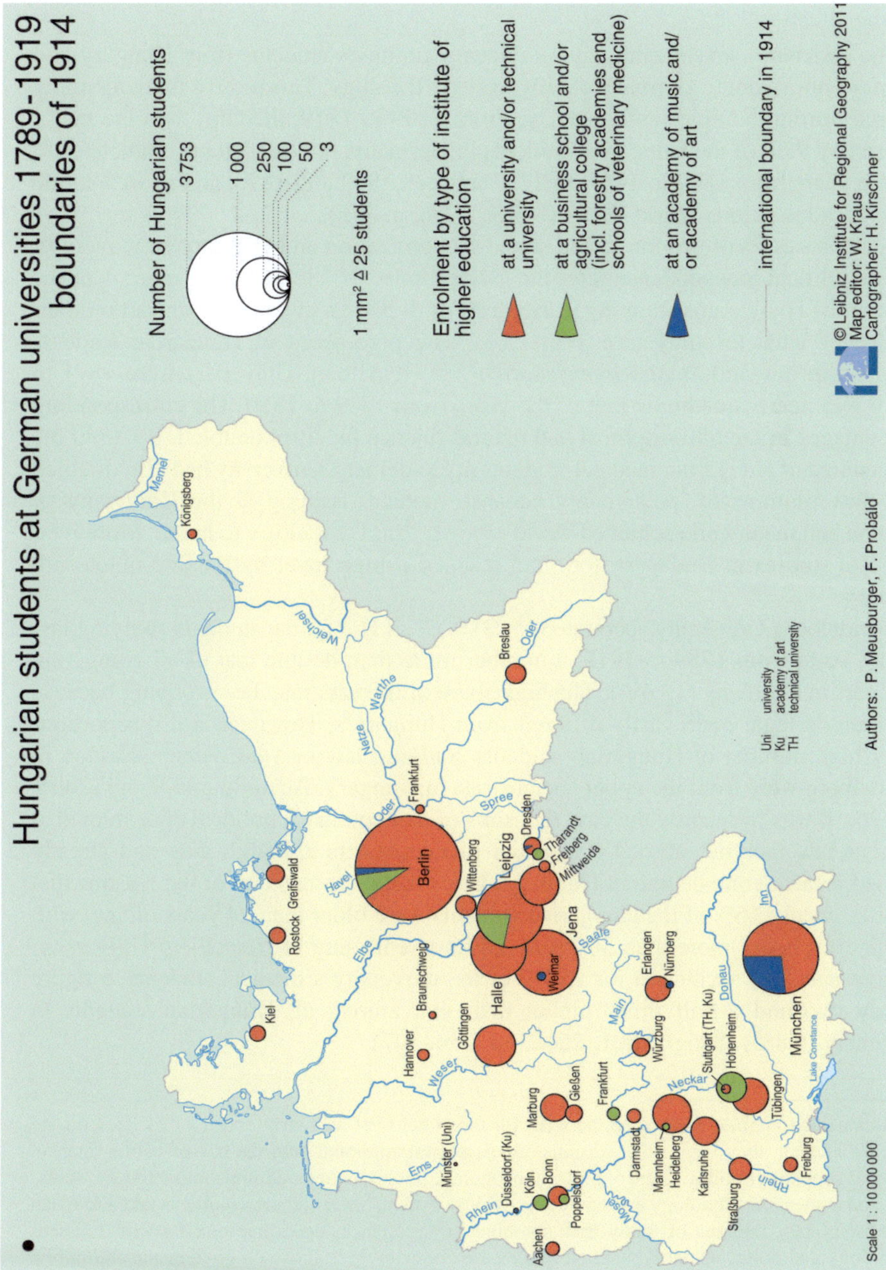

Fig. 3.12 Hungarians enrolled at German universities, 1789–1918, by type of institute of higher education. Data source: Szögi (2001, p. 34); source of map: Meusburger & Probáld, 2012, p. 305). Reprinted with permission.

Choice of Majors among Students from Hungary and Transylvania

In the sixteenth, seventeenth, and eighteenth centuries students from Hungary and Transylvania went to Germany chiefly to study theology. This motive fundamentally altered during the nineteenth century. From 1789 to 1819, theology was the major chosen by 92% of the Hungarian students in Germany, but that figure shrank to 54% by the years from 1849 to 1867, to 27% between 1867 and 1890, and to 14% in the final decades before World War I (Szögi, 2006, p. 395).

The processes of economic and social modernization and Hungary's recovery of broad political independence after the Compromise of 1867 affected more than the number of Hungarians studying in Heidelberg. It had an even larger impact on their choice of what to study (see Table 3.3). The percentage of Hungarian students enrolled in law and *Staatswissenschaften*[51] at Heidelberg University from 1871 to 1910 was nearly quadruple that of the years from 1789 to 1850. The corresponding percentages in the philosophical and natural science faculties doubled. By 1890 the percentage of Hungarian medical students at Heidelberg University had also doubled but then plummeted, presumably because medical training at the University of Vienna had meanwhile achieved world renown[52] and was closer to home. Moreover, medical studies in Budapest, too, had reached a high level by the late nineteenth century.

Heidelberg University accounted for 31% of all Hungarian students studying law in the years from 1789 to 1919, a number more than double that of all Hungarian students in Germany (13.4%). The high share of lawyers may be surprising because Germany's legal code vastly differed from Hungary's. However, a disproportionately high number of Hungarian students studying law or *Staatswissenschaften* in Heidelberg were from the upper social strata in Hungary. According to Szögi (2006, p. 396), it was frequently the case that sons of that country's political elite studied at famous law faculties abroad for one or two semesters although they had already earned an academic degree in Hungary. This practice is also one of the reasons that approximately 16% of the Hungarian students were older than 24 years of age, with 2.5% even being more than 30 years old upon enrolling at Heidelberg University. Hungarians also accounted for 15.2% of the university's medical students, a figure nearly two and a half times higher than that among all Hungarian students in Germany (6.8%) (Szögi, 2001, 2006, pp. 394–397).

---

[51]*Staatswissenschaften* used to be the term for those fields of scholarly endeavor focused on the state. In general, they were state, administrative, and international law; the law of public finance; political sociology; policy science; political economy, public finance; administrative theory; statistics; and sometimes ethnology and ethnopsychology. Today, *Staatswissenschaften* is taken to mean interdisciplinary courses of study that combine jurisprudence, economics, and social sciences (sociology and policy science). Such programs are comparable to those on governance and public policy.

[52]By 1750 the medical faculty of the University of Vienna had become one of Europe's leaders in the training of physicians.

Table 3.3 Distribution of Hungarian students across the faculties of Heidelberg University, 1789–1919

Registration year	Theology (%)	Medicine (%)	Law, *Staats wissen-schaften* (%)	Philosophy (%)	Natural sciences (%)	Total Absolute number	Total (%)
1789–1850	64.29	10.71	10.71	10.71	3.57	28	100
1851–1870	45.00	13.33	24.17	12.50	5.00	120	100
1871–1890	16.05	19.75	40.12	17.28	6.79	162	100
1891–1910	16.13	13.55	40.65	20.65	9.03	155	100
1911–1919	31.43	11.43	25.71	25.71	5.71	35	100
Total	26.80	15.20	33.80	17.40	6.80		100
Absolute number	134	76	169	87	34	500	

Note. One Hungarian student had no subject listed in the registries.
Sources of the data on individual students: Szögi (2001); Toepke & Hintzelmann (1903, 1904, 1906); Matrikelbuch der Universität Heidelberg [Heidelberg University Registry (HUR)], summer term 1871 through summer term 1872.
HUR, winter term 1872–1873 through winter term 1894–1895.
HUR, summer term 1895 through summer term 1906.
HUR, winter term 1906/1907 through summer term 1916.
HUR, wartime winter term 1916–1917 through summer term 1920.
Coding and analysis of the data on individual students: T. Schuch and P. Meusburger.

Social Origin and Choice of Subject of the Hungarian Students in Heidelberg

In this chapter the social strata of the Hungarian students attending Heidelberg University is elicited through the occupation of the father (legal guardian), which is recorded for 89.4% of the students listed in the university's registries. From 1789 through 1919, a disproportionately high percentage of the Hungarians studying in Heidelberg were from the upper social strata (see the first two rows in Table 3.4). More than a quarter (26.4%) of the students had a father who was a major landowner, politician, factory owner, entrepreneur, banker, or some other leading figure in the economy. The historically eminent aristocratic Hungarian names listed include Bánffy, Eötvös, Jeszenszky, Degenfeld-Schomburg, Szily, Teleki, Tisza, und Wesselényi. Clerical occupations (pastors or rabbis) accounted for 12.2% of the fathers, whereas 5.4% of the fathers were teachers or professors, and only 9.0% of the fathers were artisans, junior officials, or farmers.

In the first half of the nineteenth century, when studying theology was still relatively important among the Hungarian students in Heidelberg, 32% of the students came from families in which the father was a cleric or teacher. The steepest decline in Heidelberg University enrollments among students from Hungary between 1789 and 1919—from 25% to 5.6%—was among the sons of pastors. After 1850 the share of upper-class students rose rapidly, with three discernible phases in its composition. The first influx consisted of students whose fathers were major landowners, politicians, and parliamentarians. From 1851 to 1870, this group accounted for 20.8% of the Hungarian students at Heidelberg University. After 1871 the share of students consisting of the sons of Hungarian society's leading factory owners, entrepreneurs, and bankers expanded markedly. The wave from 1891 to 1910 was dominated by sons from merchant families, Jewish ones above all.

Unsurprisingly, the choice of a major or discipline within the university was largely a function of the social origin of the Hungarian students (Table 3.5). Just over 67% of the sons of major landowners and politicians studied law or *Staatswissenschaften* in Heidelberg. Among the sons of senior public officials, military officers, and self-employed academics, the share was 40.3%; among sons of clerics, only 13%; and those of teachers, 11%. The children of physicians and apothecaries also inherited high status, focusing as they did on studying medicine. By contrast, the sons from the middle and lower social strata went mostly for theology and philosophy (teaching occupations), preferring to prepare themselves for fields that promised a regular income beginning immediately after completion of a relatively short and inexpensive course of study. Theology was the leading subject of study among the sons of clerics (62.3%); of artisans, junior officials, and farmers (42.2%); of teachers and professors (40.7%); and, to a far lesser extent, of traders and merchants (10.6%). The association between upward social mobility and academic occupations, especially those of the clergy and teachers, has been verified many times in other countries, too.

Table 3.4 Social origin of Heidelberg University students from Hungary (by father's occupation)

	Year of registration						
Father's occupation	Until 1850 (%)	1851–1870 (%)	1871–1890 (%)	1891–1910 (%)	1911–1919 (%)	Absolute number	Total %
Major landowner, parliamentarian, politician	7.1	20.8	16.0	10.3	2.8	70	14.0
Entrepreneur, banker, factory owner, other leading business figure	10.7	11.7	15.4	11.6	5.6	62	12.4
Merchant (trader)	3.6	15.0	19.8	31.0	13.9	104	20.7
Senior public official, military officer, self-employed academic	0	8.3	21.0	12.3	25.0	72	14.4
Pastor, rabbi, other clerical occupation	25.0	22.5	7.4	8.4	5.6	61	12.2
Teacher, professor	7.1	5.8	3.7	3.9	16.7	27	5.4
Junior public official, artisan, farmer	0	10.8	10.5	5.2	19.4	45	9.0
Other occupation	0	0	1.2	2.6	2.8	7	1.4
No information	46.4	5.0	4.9	14.8	8.3	53	10.6
Total absolute number	28	120	162	155	36	501	
Total percentage	100	100	100	100	100		100

Note. Sources of the data on individual students: Szögi (2001); Toepke & Hintzelmann (1903, 1904, 1906); Matrikelbuch der Universität Heidelberg [Heidelberg University Registry (HUR)], summer term 1871 through summer term 1872.
HUR, winter term 1872–1873 through winter term 1894–1895.
HUR, summer term 1895 through summer term 1906.
HUR, winter term 1906/1907 through summer term 1916.
HUR, wartime winter term 1916–1917 through summer term 1920.
Coding and analysis of the data on individual students: T. Schuch and P. Meusburger.

Table 3.5 Fields of study chosen by Heidelberg University students from Hungary and Transylvania, by Social Origin, 1789–1919

	Students' major field of study						
Father's occupation	Theology (%)	Medicine (%)	Law, *Staatswissenschaften* (%)	Philosophy (%)	Natural sciences (%)	Total Absolute number	Total %
Major landowner, parliamentarian, politician	10.0	7.1	67.1	14.3	1.4	70	100
Entrepreneur, banker, factory owner, other leading business figure	30.6	17.7	30.6	12.9	8.1	62	100
Merchant (trader)	10.6	14.4	36.5	27.9	10.6	104	100
Senior public official, military officer, self-employed academic	11.1	29.2	40.3	13.9	5.6	72	100
Pastor, rabbi, other clerical occupation	62.3	3.3	13.1	18.0	3.3	61	100
Teacher, professor	40.7	18.5	11.1	18.5	11.1	27	100
Junior public official, artisan, farmer	42.2	17.8	17.8	15.6	6.7	45	100
Other occupation	28.6	0	14.3	14.3	42.9	7	100
No information	36.5	17.3	30.8	11.5	3.8	52	100
Total absolute number	134	76	169	87	34	500	
Total percentage	26.8	15.2	33.8	17.4	6.8		100

Note. The field of study could not be determined for one Hungarian student.
Sources of the data on individual students: Szögi (2001); Toepke & Hintzelmann (1903, 1904, 1906); Matrikelbuch der Universität Heidelberg [Heidelberg University Registry (HUR)], summer term 1871 through summer term 1872.
HUR, winter term 1872–1873 through winter term 1894–1895.
HUR, summer term 1895 through summer term 1906.
HUR, winter term 1906/1907 through summer term 1916.
HUR, wartime winter term 1916–1917 through summer term 1920.
Coding and analysis of the data on individual students: T. Schuch and P. Meusburger.

The fall in the share of theology students had several consequences. First, the percentage of students from Hungary's middle and lower social strata and from the country's towns and rural regions distinctly shrank, and the share of students from Budapest and other large cities swelled. As the demand for theology studies waned, as the spectrum of subjects offered in Heidelberg opened up, and as Hungary's society and economy modernized, the confessional scope also widened among the Hungarians who enrolled at Heidelberg University during the nineteenth century.

Religious Confession of the Hungarian Students Enrolled at Heidelberg University

Analysis of the religious confession of the Hungarians studying in Heidelberg sheds additional light on the fluctuating role and significance of Heidelberg University in the training of Hungarian elites and on the processes of structural transformation, liberalization, and processes of modernization in Hungarian society, particularly between 1867 and 1919. When assessing the data presented in Table 3.6, one should bear in mind that 49.3% of Hungary's population[53] in 1910 was Roman Catholic; 14.3%, Reformed (Calvinist); 12.8%, Greek Orthodox; 11%, Greek Catholic; 7.1%, Evangelical (Lutheran); 5.0%, Jewish[54]; and 0.4%, Unitarian (Központi Statisztikai Hivatal, 1920). In other words, Lutherans, Israelites, and Calvinists were the most overrepresented groups among the Hungarian students in Heidelberg, whereas Roman Catholic, Greek Catholic, and Greek Orthodox students were clearly underrepresented. Table 3.6 also reflects the change in the motives Hungarian students had for studying in Heidelberg (see immediately preceding section). During the nineteenth century, Heidelberg lost its function as a key "forge" of Hungary's Calvinist elite. In the latter half of the 1800s, Heidelberg's attraction was primarily the scientific reputation of the faculties. The data also document the social development within Hungarian society—notably the social ascendance of the Jews. Having been liberated from discrimination after the Compromise of 1867, they played a salient role in the rise of Hungary's new economic and intellectual elite, contributing impressively to the country's economic boom, the development of Hungarian culture and sciences, and German-Hungarian relations in the sciences and higher education (for details see Karády, 1997, 2004, 2012, 2013; Karády & Nagy, 2012).

From 1891 to 1910, the percentage of Jews among the Hungarian students at Heidelberg University was nearly four times higher than that from 1789 to 1850. In the five years from 1886 through 1890, the share of Heidelberg's Hungarian students who were Israelite came to 31.3% and climbed as high as 58.3% from 1891 through 1895 (see also Fig. 3.11). Between 1867 and the late 1920s, Hungarian Jews were remarkably active in German culture and science (Frank, 1999; Gantner, 2006).

[53]The immediately following statistics do not include Croatia and Slavonia.

[54]In the university's registries most Jewish students denominated themselves as Israelites—a frequently used term in the nineteenth century. Accordingly, we use this self-designation synonymously with the term Jews.

Table 3.6 Religious confession of the Hungarian students at Heidelberg University, 1789–1919 (in Percentage)

Registration Years	Lutheran (%)	Reformed (%)	Israelite (%)	Catholic (%)	Other (%)	No data (%)	Absolute number	Total %
Until 1850	28.6	60.7	7.1	3.6	0.0	0.0	28	100
1851–1870	43.3	27.5	10.0	11.7	7.5	0.0	120	100
1871–1890	40.1	13.0	22.2	21.6	1.9	1.2	162	100
1891–1910	27.1	23.9	27.7	18.1	1.3	1.9	155	100
1911–1919	41.7	22.2	16.7	5.6	8.3	5.6	36	100
Absolute number	182	116	99	80	17	7	501	
Total (%)	36.3	23.2	19.8	16.0	3.4	1.4		100

Note. Sources of the data on individual students: Szögi (2001); Toepke & Hintzelmann (1903, 1904, 1906); Matrikelbuch der Universität Heidelberg [Heidelberg University Registry (HUR)], summer term 1871 through summer term 1872.
HUR, winter term 1872–1873 through winter term 1894–1895.
HUR, summer term 1895 through summer term 1906.
HUR, winter term 1906/1907 through summer term 1916.
HUR, wartime winter term 1916–1917 through summer term 1920.
Coding and analysis of the data on individual students: T. Schuch and P. Meusburger.

An Explanation of the Soaring Number and Percentage of Jewish Students

Hungary's Israelite population expanded twelvefold between 1780 and 1910, in the latter year accounting for 5% of the country's population and up to 23% in the capital city, Budapest. This growth in the Jewish segment of the population was due essentially to the enormous immigration of Jews from Galicia, Moravia, and other regions of the Habsburg empire. The waves of Jews arriving in Hungary, notably Budapest, were triggered principally by the fact that Jews generally encountered more tolerance and better opportunities to move up the social ladder in Hungary than in many other European countries. The March 1783 Josephine decree "Systematica Gentis Judaicae Regulatio" (Systematic Regulation of the Jewish Nation), issued in the spirit of the Enlightenment, opened the way for the Israelites to study at the universities in the Habsburg empire. In the first half of the nineteenth century, the Israelites concentrated mostly on medicine, economics, and Jewish theology at the universities of Prague and Vienna; and medicine at the University of Pest as well.

Article XXIX in the law of 1840 granted Jews the right to settle in all Hungarian cities and the license to engage in trade and commerce. The Emancipation Act XVII of 1867 gave full equality to the Jews in Hungary, launching an almost seamless 50-year process of assimilation whose success was unprecedented and greatly promoted by the attitude of Hungarian society. The rapid social rise of Jews was propelled by a number of other circumstances, too. First, the urban middle class was traditionally very narrow in Hungary, so a large potential for social upward mobility existed. Second, the social and economic processes of modernization in the second half of the nineteenth century markedly escalated the demand for self-employed persons with an academic background (attorneys, physicians), scientists, and providers of services in certain fields (banking, media, cultural institutions). Third, because the Jews had an above-average level of education, they were able to meet the intense demand for highly qualified people in the economy. In that field they met little competition, for the members of the lower nobility had a manifest inclination to enter public service instead. Lastly, there were few apparent signs of discrimination against Jews in the final third of the nineteenth century. In that liberal and tolerant age the appearance of an anti-Semitic party in 1883 remained a brief and marginal political episode.[55]

Combined with various measures of the Hungarian state to strengthen Hungarian national identity, these positive conditions for Jews led to shifts in the national (cultural) identity of the Jews. In the census of 1910, more than three quarters of the Jews stated that their mother tongue was Hungarian, whereas the share of the Jews in Hungary who claimed German as their mother tongue contracted from 33.3% to 21.7% between 1880 and 1910 (Gantner, 2006, p. 461). "Acquisition of the Hungarian language and culture in Jewish middle-class families by no means meant a

[55]The prime minister, Kálmán Tisza, resolutely rejected anti-Semitism (Stokes, 1989, p. 225).

rejection of the relations to German culture. Rather, the simultaneous existence of the German and Hungarian language and culture was taken for granted" (p. 461).

As "people of the written Word," Jewry has always treasured education and intellectual occupations. As in other countries, Jews in Hungary were heavily overrepresented in universities and other institutions of higher learning. The share of Jews among students in *Gymnasien* and *Realschulen*—school types preparing 10- to 18-year-olds for higher education—rose from 9.6% (1866–1867) to 22.5% (1913–1914). They accounted for approximately 10% of the students at Hungary's universities and other institutions of higher learning in the 1860s but more than 25% by 1885, exceeding 30% during World War I (Kovács, 1922; Sebők, 2013). According to Gantner (2006), 17.7% of the students at the University of Budapest in 1867 were Jews and 29.6% by 1895. The 1891 census recorded that 32% of the student body at the University of Budapest was Jewish (p. 463). Karády (1997) stated that, on average, Jewish school students also had better grades and lower dropout rates than the average for their Hungarian peers. Furthermore, Jews chose their major subject of study and their future occupation very purposefully, flexibly adapting to demand molded by social, technological, and scientific progress (Szögi, 2004; Volkov, 1987). Jews were especially well represented among entrepreneurs (factory owners) in Hungary, in banking, and among self-employed people with an academic background. In 1910 Jews accounted for 48% of the physicians, 45% of the lawyers, and 42% of the journalists, although Jews made up only 5% of the total Hungarian population (Kovács, 1922; Sebők, 2013). Their contribution to Hungarian science and art, too, was invaluable. What they did to modernize Hungary was recognized by the country's government: "[B]etween 1900 and World War I over 200 Hungarian Jewish families were ennobled" (Stokes, 1989, p. 225).

Social Origin of the Jewish Students at Heidelberg University

Hungary's well-off Jewish factory owners, entrepreneurs, merchants, bankers, and self-employed academics, who had excellent language skills and international connections alike, naturally sought the best education for their sons and encouraged them to spend time abroad. Fully 53.5% of the fathers of the Israelite Hungarian students who had enrolled at Heidelberg University from 1789 to 1919 were businessmen (traders), 21.3% were part of the economy's leading stratum (the first two categories in Table 3.7), with 8.1% being self-employed academics or rabbis. The share of students from the middle and lower social strata was much smaller among the Jews studying in Heidelberg than among the other confessions. Compared also to the Jewish students at other German universities, the Jews studying in Heidelberg came in disproportionately large numbers from the upper classes (see Gantner, 2006, pp. 463–464).

Some 18.7% of the Lutheran students had a pastor for a father; 18.1%, a father who was a senior public servant, military officer, or self-employed academic. By contrast, the social profile of the Calvinist students, peaked at nearly opposite ends of

Table 3.7 Religious confession of Heidelberg University's registered students from Hungary, 1789–1919

Father's occupation	Lutheran (%)	Reformed (%)	Israelite (%)	Catholic (%)	Other (%)	No data (%)	Absolute number	Total %
Major landowner, parliamentarian, politician	9.3	19.8	6.1	25.0	23.5	0	70	14.0
Entrepreneur, banker, factory owner, other leading business figure	13.2	10.3	15.2	10.0	5.9	28.6	62	12.4
Merchant (trader)	15.9	6.0	53.5	12.5	23.5	14.3	104	20.8
Senior public official, military officer, self-employed academic	18.1	9.5	8.1	25.0	0	0	72	14.4
Pastor, rabbi, other clerical occupation	18.7	12.9	8.1	0	23.5	0	61	12.2
Teacher, professor	4.9	11.2	1.0	3.8	5.9	0	27	5.4
Junior public official, artisan, farmer	11.0	12.1	2.0	7.5	17.6	0	45	9.0
Other occupation	1.6	0	1.0	3.8	0	0	7	1.4
No information	7.1	18.1	5.1	12.5	0	57.1	53	10.6
Total absolute number	182	116	99	80	17	7	501	
Total percentage	100	100	100	100	100	100		100

Note. "Other confessions" includes Unitarians, Greek Orthodox, and students without a confession.
Sources of the data on individual students: Szögi (2001); Toepke & Hintzelmann (1903, 1904, 1906); Matrikelbuch der Universität Heidelberg [Heidelberg University Registry (HUR)], summer term 1871 through summer term 1872.
HUR, winter term 1872–1873 through winter term 1894–1895.
HUR, summer term 1895 through summer term 1906.
HUR, winter term 1906/1907 through summer term 1916.
HUR, wartime winter term 1916–1917 through summer term 1920.
Coding and analysis of the data on individual students: T. Schuch and P. Meusburger.

the range, with 30.1% coming from the upper strata of society and 36.2% from fathers who were clerics, teachers, artisans, junior public officials, or farmers.

Heidelberg: One of the First German Universities to Accept Jewish Students

Did the preponderance of Jews among the Hungarian students at Heidelberg University perhaps owe to the latter's reputation as a liberal, tolerant institution since the early nineteenth century (see Engehausen, 2012; Honeck & Meusburger, 2012; Meusburger, 2012; Wolgast, 1987) and the fact that its professors and students championed democratic reforms early? The university in Heidelberg was one of the first in Germany to accept Jewish students (1724), graduate Jewish students (1728), and appoint Jews as professors (1766) (Meusburger & Schuch, 2010, pp. 64, 71–74; Richarz, 1974, pp. 29–33).[56]

The first Jewish professor at Heidelberg University was Daniel Wilhelm Nebel. He completed his doctorate in Heidelberg in 1758, was appointed as an associate professor of medicine at Heidelberg University in 1766, and became a full professor in 1771. By 1860 two more Jews had become professors at Heidelberg, as had four Jews who had converted to Christianity. However, the real rise of Jewish professors at Heidelberg University did not begin until after the revolution of 1848. Between 1851 and 1870, 8.9% of Heidelberg's newly appointed professors were Jewish (for details see Meusburger & Schuch, 2010, pp. 71–74).

When interpreting these numbers, one should bear in mind Germany's 1910 census, in which Jews accounted for only 1.2% of the total population of Baden and 0.95% of the population of the Second German Reich (Titze, 1987). Thus, the proportion of Jews among Heidelberg's professors was nearly ten times that of Jews in the population as a whole (for explanations see Volkov, 1987). By comparison, the percentage of Heidelberg University's professorships that were held by Catholics in 1910 corresponded to only a quarter of the Catholic share of the German population as a whole.

The available data are inconclusive about whether the proportion of Jewish professors in the second half of the nineteenth century had a documentable effect on the share of Jews among the students from Hungary. Even German universities at which Jews accounted for only a small percentage of the professors attracted many Jewish students from Hungary after 1867. At the 15 German institutions of higher learning that recorded the religious confession of their students,[57] 23% of the students were Israelites, according to Szögi (2004). He estimated that Israelite students accounted for 21% of the Hungarian student body at German universities, 28% at the technical universities, 30% at the agricultural academies, and 50% at the

[56] A plan to appoint the philosopher Spinoza as a professor at Heidelberg University failed in 1673.

[57] These institutions accounted for approximately 37% of the Hungarians enrolled in Germany at that time.

universities for trade and commerce. The universities for trade and commerce in Leipzig, Berlin, Mannheim, and Cologne were particularly popular (Gantner, 2006, p. 464). At the Technical University of Berlin, where Hungarians usually studied chemistry and electrical engineering, Jews accounted for about 40% of the Hungarian students between 1867 and 1919. This figure roughly matched the share of Jews at the Technical University of Budapest.

Regional Origin of the Hungarians Studying in Heidelberg, 1789–1919

The places of birth or residence[58] of the 501 Hungarian students enrolled at Heidelberg University between 1789 and 1919 were distributed throughout Hungary but clearly preponderated in today's Budapest (15.2%) and in the Transylvanian cities of Nagyszeben (today Sibiu, 6.0%), Brassó (Braşov, 5.4%), Meggyes (Mediaş, 2.8%), Segesvár (Sighişoara, 2.2%), Beszterce (Bistriţa, 2.0%), Szászrégen (Reghin, 2.0%), and Kolozsvár (Cluj-Napoca, 1.8%). Kecskemét und Pozsony (today's Bratislava) each accounted for 1.6% of the Hungarian students; Miskolc and Debrecen, each 1.0% (see Fig. 3.13).

An especially high number of students were thus sent to Heidelberg by cities with a strong Protestant tradition and important administrative institutions of the Protestant church, cities with a high percentage of German-speaking or Jewish inhabitants, and cities with superior *Gymnasien*. Because of the central-place hierarchy of Hungary's settlement system and the resulting dissimilarities in the social stratification of the population, the large cities and the rural periphery differed substantially in terms of the students' social origins and chosen subjects of study. About two thirds of the sons of teachers and clerics came from towns and rural areas. Only 1.3% of the students from Budapest enrolled in theology; 58.2% in law and *Staatswissenschaften*. Towns and rural areas offered quite the opposite picture, with 31.9% of all students from those places enrolling in theology, and only 27.6% choosing law and *Staatswissenschaften*. The share of students from Transylvania who focused on theology (39.1%) was more than twice that of the students from the rest of Hungary (19.2%). The share of medical students from Transylvania (20.3%) was also much higher than in the rest of Hungary (12.0%).

The nineteenth-century shifts in social origin and choice of subject altered the spatial pattern of the students' places of origin as well. Between 1789 and 1919 the regional origin of the Hungarians and Transylvanians studying in Heidelberg moved distinctly westwards. The share of all Hungarian citizens from Transylvania enrolled at Heidelberg University declined from 54.2% (1789–1850) to 39.1% (1851–1900), ultimately sinking to 33.6% (1901–1919). Increasing vertical mobility led to a rise in

[58] Szögi (2006, p. 400) points out that the registries noted the place of residence rather than the place of birth for some of the students.

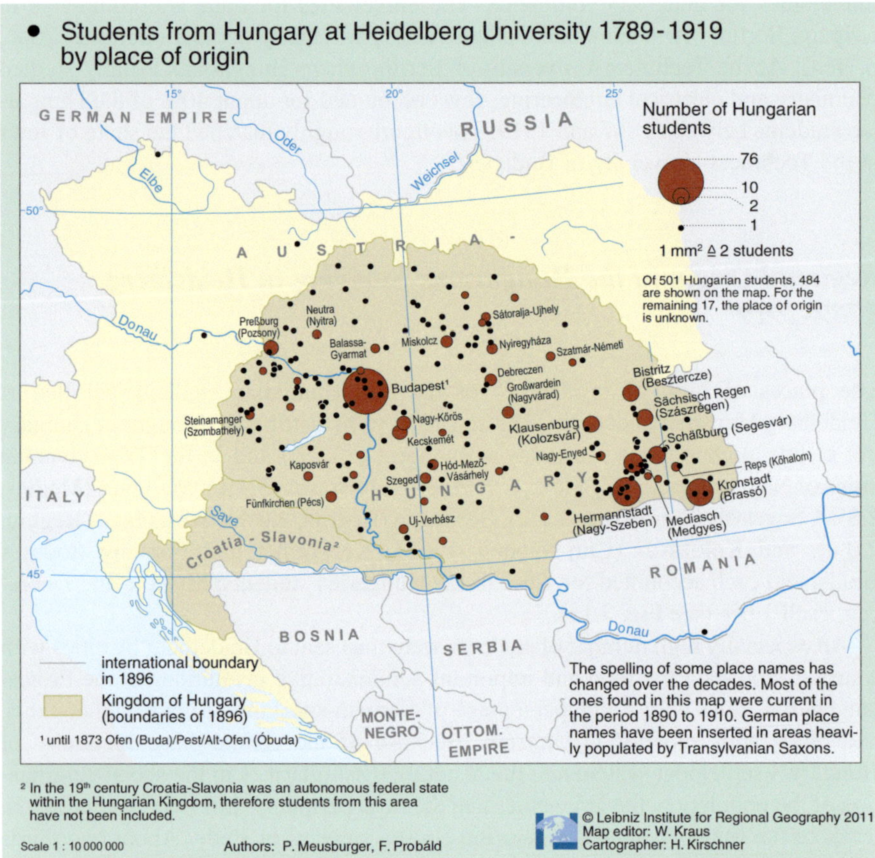

Fig. 3.13 Hungarian students at Heidelberg University, 1789–1919, by place of origin. Data source: Szögi (2001, p. 34). Source of map: Meusburger & Probáld, 2012, p. 305. Reprinted with permission.

the share of the enrolled students from small towns and rural areas—from 29.1% (1789–1850), 47.6% (1851–1900), and 54.8% (1901–1919). After 1890 many fewer children of teachers and clerics studied in Heidelberg, but more children of merchants and traders did.

Overall, the share of the Hungarian students from the urban middle class was somewhat higher in Heidelberg than the average share of all Hungarians and Transylvanians studying in Germany. A total of 40.6% of the Hungarians and Transylvanians who registered in Heidelberg came from 10 cities. According to Szögi (2001, pp. 52–53), only 33.1% of the entire number of Hungarian students in Germany were from there. Table 3.8 summarizes the key shifts in four main characteristics of the Hungarian students at Heidelberg University from the seventeenth to the nineteenth century.

Table 3.8 Representation of the social groups among Heidelberg University's students from Hungary as compared to their share of Hungary's total population

	17th century		19th century	
Characteristic	Very high	Very low	Very high	Very low
Mother tongue	Hungarian	Other	German	Romanian, Slavic
Religion	Calvinist	Other	Israelite, Protestant	Greek Orthodox
Social origin	Lower middle class	Serfs	Upper classes, land owners	Farmers, workers
Subject	Theology	Other	Law, theology	Other

Source: Design by authors.

Eminent Figures from Hungary Who Studied in Heidelberg in the Nineteenth and Early Twentieth Centuries

There are at least three reasons not only to study academic relations, student migration, and influences of universities at the structural level but also to delve into the biographies and networks of outstanding individual protagonists. First, a biographical approach can contribute knowledge about the spatial diffusion of ideas and scientific methods that a structural approach cannot yield. Analysis of biographies, letters, minutes of meetings, and personal interviews makes it possible to recognize motives that have affected decisions and to identify events (situations) and people who have promoted or impeded the learning and academic or political careers of the students in question. Academics can generally identify whom they received encouragement, suggestions, and support from while engaged in their studies or academic career and where they met resistance to their ideas or lack of interest in them (see chapter on knowledge milieus by Meusburger in this volume). The biographical method can (within certain limits[59]) trace the charisma and impact of individual professors or faculties on their students and, hence, on the spatial diffusion of ideas and methods (e.g., Eckart, Hübner, & Nawa, 2012; Hübner, 2010).

Second, a biographical approach can reconstruct networks of mutual support and emotional attachment to places, personal animosities, and gulfs between scientific schools of thought. Numerous examples illustrate how friendships, loyalties, and antipathies that form between students during their time together at a university later become decisive factors in communication processes and career decisions.

Third, only the biographical method makes it possible to grasp how the structural elements of a knowledge milieu have actually affected students and scientists. One must avoid a deterministic understanding of the weight that universities or professors have in the intellectual development and occupational careers of students. The recipient of information is always the one who decides whether or not to appropriate a locally accessible resource; whether and how to take suggestions, critique, and criticism; and how to exploit opportunities and assess risks. Much of a knowledge milieu's potential goes unrecognized or unused by many of the actors.

[59] An institute's knowledge milieu is not an independent variable exerting a particular effect. It represents a local potential that can be used or ignored (for details see the chapter by Meusburger in this volume).

German universities contributed a great deal to educating Hungarian elites in the nineteenth and twentieth centuries. According to Szögi (2006, pp. 406–407), 114 Hungarian government officials and professors of the University of Budapest studied in Germany between 1867 and 1945, most of them in Berlin, Leipzig, Göttingen, and Heidelberg. These 114 people included five Hungarian prime ministers, seventeen ministers, and two commissars of the Hungarian Soviet Republic.[60] Many alumni of German universities sat as representatives in the Hungarian parliament as well,[61] and their share of the seats mounted steadily until World War I. From 1910 to 1915, 13.8% of the members of the upper house and 10.4% of the representatives in the lower house had completed a university degree in Germany, most of them in Berlin, with the University of Leipzig and Heidelberg University following in that order (Tar, 2007, pp. 112–113). The next two sections present just a few of the Hungarian politicians and scientists who studied in Heidelberg and went on to fame.

Scholars in the Liberal Arts, Social Sciences, and Economics[62]

Hugo Meltzl von Lomnitz (1846–1908), the son of a brewery owner from Szászrégen, enrolled at Heidelberg University in November 1866, where he became a student of Karl Bartsch (1832–1888), the renowned professor of the German language and culture. Meltzl received his doctorate cum laude from Heidelberg in 1872. The University of Kolozsvár was founded in the same year, and Meltzl had the good fortune to become full professor of German there at the age of 26. He was close friends with Loránd Eötvös (see later in this chapter), with whom he had studied in Heidelberg, and translated poems by Loránd's father, József Eötvös, into German (Fassel, 2006, pp. 438, 453). Meltzl specialized in comparative literature, researched the origins of European and non-European literature as a foundation for studying comparative literary development. In 1877 he began editing the world's first comparative journal, the *Acta comparationis litterarum universarum* (Journal of Comparative Literature), which appeared from 1879 to 1888 and contained essays in 20 languages (p. 439).

Count István Tisza (1861–1918), a son of Kálmán Tisza (the prime minister of Hungary from 1875 to 1890), enrolled in law at Heidelberg University in 1879. He received a doctorate in political science when he was just 20 years old. In 1910 he became a member of the Academy of Sciences (economic sciences). As a leader of the Liberal Party, Count István Tisza was prime minister of Hungary from 1903 to 1905 and from 1913 to 1917 (Tökéczki, 2010). He staunchly endorsed the Austro-

[60]The Hungarian Soviet Republic was proclaimed on March 21, 1919, and ended 133 days later, on August 1, 1919.

[61]The Hungarian parliament (*Magyar Országgyűlés*) consisted of two chambers, the House of Magnates (*Főrendiház*), in which the members of the high nobility and the high clergy sat, and the House of Representatives (*Képviselőház*) with members elected by the votes of the electorate.

[62]Unless otherwise cited, the information in this section is based on articles in Markó (2001–2007).

Hungarian Compromise and strove to ensure the survival of the dual monarchy. He opposed Austria-Hungary's declaration of war on Serbia in July 1914 but had to yield to the overwhelming pressure in favor of it. At the end of World War I, he was made a public scapegoat and murdered in his house in October 1918.

Sándor Imre (1877–1945) studied education in Heidelberg and Leipzig (1899–1900). After 1919 he served briefly as an undersecretary in the Ministry of Culture. He was a full professor of education at the University of Szeged (1925–1934) and at the Technical University of Budapest (1934–1944). He tried to apply German social pedagogy in Hungary (Szögi, 2006, p. 407) and introduced the study of psychology at Hungarian universities. The first Institute of Psychology was founded upon his initiative at the University of Szeged in 1929.

Endre Bajcsy-Zsilinszky (1886–1944) studied law in Kolozsvár, Leipzig, and Heidelberg, where he broke off his studies in 1908 without earning a doctorate. Bajcsy-Zsilinszky cofounded the National Radical Party (Nemzeti Radikális Párt) in 1930, which he represented in the Hungarian parliament until 1935. After 1939 he represented the Hungarian Smallholders' Party until his death. During World War II, Bajcsy-Zsilinszky was among the foremost opponents of Hungary's participation in the war and the alliance with the German Reich. In parliament he successfully pressed for prosecution of General Feketehalmy-Czeydner, who had ordered mass executions in Újvidek (Novi Sad, Yugoslavia) in 1942.[63] After the German occupation of Hungary in March 1944, the Gestapo captured Bajcsy-Zsilinszky but released him in October. He went underground after the coup by Ferenc Szálasi, a Hitler loyalist and leader of the Arrow Cross Party, a National Socialist organization in Hungary. Bajcsy-Zsilinszky was arrested as the leader of the resistance movement and executed in Sopronkőhida on December 24, 1944.

Several Hungarian members of the Budapest *Vasárnapi Kör* (Sunday Circle) studied and researched in Heidelberg from 1912 until the late 1920s (for details about the Sunday Circle, see Gräfe, 2004; Karadi, 1986; Karadi & Vezer, 1985; Loewy, 1999; Wessely, 1986). This group included György Lukács, Karl Mannheim, Béla Fogarasi, and László Radványi, of whom the rest of this section highlights only Lukács and Mannheim, the two best known members. Not only did their experience in Heidelberg differ radically, they also went separate ways later.

The pivotal figure of the Sunday Circle in Heidelberg was György Lukács (1885–1971),[64] a Marxist philosopher, literary scholar, and subsequent people's commissar of the Hungarian Soviet Republic. He lived intermittently in Heidelberg from 1912 until 1917 and wrote some of his seminal works there. Lukács came from a wealthy Jewish family. His father served as a court councilor and was an important actor in finance capital as managing director of the English-Austrian Bank and

[63]The Axis Powers had occupied Yugoslavia in 1941.

[64]The following biographical details on Lukács are drawn from Bendl (1998), Benseler (1987), Borbándi (1973), Karadi (1986), Kuzias (2007), Loewy (1999), and Vorlaufer (1993).

director of the Hungarian General Credit Institute in Budapest.[65] In 1902 György Lukács began studying law and economics in Budapest, earning a doctorate in Kolozsvár in 1906. In 1907 he met Georg Simmel (1858–1918), who greatly impressed him and whose aesthetic philosophical impressionism had an effect still evident in Lukács's late theory of aesthetics. Lukács studied philosophy at the University of Berlin in the 1908–1909 academic year and completed his doctorate at the University of Budapest in 1909. When Lukács arrived in Heidelberg in 1912, he thus already had a doctorate in *Staatswissenschaften* (1906) and one in philosophy (1909). He was also already known as a writer through his essays entitled *Die Seele und die Formen* (1911; English translation, 2010: *Soul & Form*), for which Georg Simmel had been an inspiration.

He went to Heidelberg University to complete his *Habilitation* (postdoctoral work) under Wilhelm Windelband (1848–1915) and to prepare himself for a university professorship in philosophy.[66] To Lukács, the university was less a training center than an intellectual forum, a place for exchanging ideas (Gantner, 2006, p. 467). He quickly joined the circle around Max Weber (1864–1920) and took part in the "sociological discussion evenings" led by Max's brother, Alfred Weber (1868–1958) (for details see Lepsius, 2012). In 1914 Lukács married his first wife in Heidelberg, the Russian painter and anarchist Jelena Grabenko, from whom he separated in 1918. In 1914 and 1915 he delved into Hegel, Marxism, anarcho-syndicalism, and Edmund Husserl. The personal relations he forged in Heidelberg with Emil Lask (1875–1915), Stefan George (1868–1933), Friedrich Gundolf (1880–1931), Ernst Bloch (1885–1977), Karl Jaspers (1883–1969), Alfred Weber (1868–1958), Max Weber (1864–1920), Eberhard Gothein (1853–1923), Gustav Radbruch (1878–1949), Emil Lederer (1882–1939), Hans von Eckhardt (1890–1957), and other famous people not only shaped and challenged him intellectually but occasionally benefited him in other ways as well. Karl Jaspers, for instance, who was also a practicing neurologist, issued a medical certificate that exempted Lukács from military service in 1914. In 1915 Lukács did have to serve briefly as a military censor of letter correspondence, a task from which was discharged in the summer of 1916 thanks to family connections (Bendl, 1998, pp. 21–22). Lukács returned to Heidelberg in 1916 and moved to Budapest in 1917.[67]

From Budapest in 1918, he applied to Heidelberg University for the opportunity to begin the multiyear postdoctoral process of qualifying for a university chair (*Habilitation*). Alfred Weber vigorously backed Lukács's plans and personally advanced them in his interaction with various faculty members (Demm, 1990, pp. 63–64;

[65]While in Heidelberg, Lukács received annual support of 10,000 marks from his father (Loewy, 1999, p. 174).

[66]He therefore did not need to enroll at the university.

[67]Before leaving Heidelberg, Lukács packed his manuscripts, notes, diaries, and letters into a suitcase, which he deposited in a safe at the Deutsche Bank in the city, where it lay nearly 50 years until discovered after Lukács death (Bendl, 1990).

2000a). Max Weber, too, supported Lukács's wish,[68] as did the art historian Eberhard Gothein, but they could not overcome the resistance of other members of the faculty of philosophy. The application was rejected in December 1918 (for details see Bendl, 1998, p. 33; Demm, 1990; Loewy, 1999, pp. 278–279).[69]

The years in Heidelberg were highly fruitful for Lukács despite his abortive efforts to qualify for a professorship. From 1912 to 1914, he wrote a work that appeared posthumously in 1974 under the title *Heidelberg Philosophy of Art.* He himself had chosen not to publish the text, which he had produced during his pre-Marxist period. Nor did he publish a second text, "Heidelberg Aesthetics," which dates from 1916 to 1918 (Bendl, 1990).

In December 1918 Lukács joined the Communist Party of Hungary. Members of the Sunday Circle became leading cultural policy-makers under Béla Kun's Soviet Republic (on the cultural policies of the Hungarian Soviet Republic, see Borbándi, 1973; Gräfe, 2004; Karadi, 1986; Loewy, 1999). Lukács first became the commissar of trade (Gräfe, 2004, p. 894) and eventually deputy commissar of education, member of the Central Committee, chair of the Writers' Directorate, and political commissar of the Red Army's fifth division. As political commissar, Lukács ordered the execution of eight soldiers of the Hungarian Red Army in Poroszló for what in his estimation had been a failure to resist the Romanian attack on Tiszafüred (Kuzias, 2007, p. 128).[70] Lukács allegedly also had deserters of the Hungary Red Army shot near Szolnok (p. 129). In a résumé composed in Moscow on December 2, 1940, he described his situation in Hungary at that time: "Hungary's White government persecuted me on account of more than 200 murders and demanded my extradition in order to carry out the death sentence imposed on me" (Kuzias, 2007, p. 129).

After the fall of the Hungarian Soviet Republic, Lukács worked for the party illegally in Budapest and Vienna between 1919 and 1929. He was the Hungarian Communist Party delegate to the Second and Third World Congresses of the Communist International in Moscow (1920, 1921) and a staff member working for the journals *Communismus* (Communism), *Vörös Ujság* (Red News), and *Die Rote Fahne* (The Red Flag). In 1923 Lukács published *History and Class Consciousness*, which became a key work of critical western Marxism and netted him fierce criticism as an ultra-leftist. His *Theses Concerning the Political and Economic Situation in Hungary and the Tasks of the Hungarian Community Party* were published in 1928. After expulsion from Austria in 1930, Lukács emigrated to Moscow, where he worked at the Marx-Engels-Lenin Institute. He stayed in Berlin from 1931 to 1933, where he was a member of the Bund proletarisch-revolutionärer Schriftsteller (Association of Proletarian-Revolutionary Authors) and the Schutzverband

---

[68]Max Weber was particularly enthusiastic about Lukács's (1920) book *The Theory of the Novel.*

[69]In 1911 Lukács had already failed at the University of Budapest in his attempt to seek certification as a qualified candidate for a professorship (Loewy, 1999, pp. 106–107). The same endeavor came to nothing in Düsseldorf as well (p. 174). In December 1915 Lukács repeatedly asked Max Weber whether there was some easier way to qualify for a university chair than by publishing a *Habilitationsschrift*—a second major work some years after the Ph.D. thesis (p. 277).

[70]Lukács's biographer, Kadarkay (1991), reported that only six of these eight men were summarily executed. One was pardoned, and another escaped (p. 223).

deutscher Schriftsteller (Association for the Protection of German Authors) and where he worked with others on the journals *Die Linkskurve* and *Internationale Literatur* with (Benseler, 1987).

In 1933 Lukács re-emigrated to the Soviet Union. In Moscow he became a member of the Academy of Sciences of the USSR in 1934 and a staff member of the journals *International Literature*, *Literaturnyj Kritik*, *Literarische Rundschau* (as of 1936), and *Deutsche Zentral-Zeitung* (as of 1937). In the German section of the Soviet Writers' Association in Moscow, he was involved in Stalinist purges in September 1936. In 1941 he was held in the Lubjanka prison for two months by the People's Commissariat for Internal Affairs (NKWD, or secret police) and narrowly escaped the Great Terror. From 1938 to 1941 he was the chief editor of the Hungarian emigrants' journal *Uj Hang* (New Voice). In August 1945 Lukács moved back to Budapest, where he became a full professor of aesthetics and cultural philosophy. He was a member of the Hungarian parliament and the presidium of the Hungarian Academy of Sciences from 1949 to 1956. He played an inglorious role in the persecution of noncommunist intellectuals under the Stalinist regime led by Mátyás Rákosi. During the Hungarian revolution of October 1956, Lukács became the minister of education in the short-lived Communist government led by Imre Nagy, a reformist. After Soviet troops invaded Hungary on November 4, 1956, Lukács was deported to Romania but was permitted to reenter Hungary in 1957. He lived in internal emigration until 1965, ultimately rejoining the Communist Party under the Kádár regime. In his late work, translated into English as *The Ontology of Social Being*, Lukács attempted a comprehensive renewal of Marxism (Benseler, 1987).

Karl Mannheim (1893–1947), another compelling figure in the Sunday Circle, was briefly in Heidelberg during World War I.[71] He received his doctorate in Budapest in 1918 and became a professor under the Hungarian Soviet Republic. When that republic ended, he passed briefly through Vienna and Freiburg im Breisgau before arriving in Heidelberg in March 1921. He soon established himself there with the help of Emil Lederer (1882–1939), the Heidelberg professor of public policy, and was registered as a student until the end of the winter semester of 1923–1924.

Originally, he intended to write his *Habilitationsschrift* at Heidelberg University in philosophy. This hope was dashed, however, by Heinrich J. Rickert (1863–1936)—an important exponent of neo-Kantianism. Mannheim therefore wrote his *Habilitationsschrift* under Alfred Weber, entitled *Conservatism: A Contribution to the Sociology of Knowledge*. His official supervisor was Emil Lederer, but the main inspiration came from Alfred Weber, who felt greater substantive affinity to Mannheim than Lederer did. Weber defended Mannheim against both an anti-Semitic faction in the faculty and political reservations about his erstwhile

[71]He was not registered at Heidelberg University at that time, however.

part in the Hungarian Soviet Republic (Demm, 2010, 2012; Karadi, 2003).[72] After qualifying for a professorship, Mannheim taught at Heidelberg University as a *Privatdozent*[73] from 1926 to 1930. In 1930 he became a full professor of sociology and economics at the University of Frankfurt on the Main but had to leave Germany in 1933 because of his Jewish ancestry. He then took a post as professor at the London School of Economics and is known today for pioneering the sociology of knowledge (Kettler, Meja, & Stehr, 1989).

Natural Scientists

Among the outstanding natural scientists from Hungary who studied in Heidelberg in the nineteenth and early twentieth centuries, Baron Loránd Eötvös (1848–1919) is unusually prominent (Fig. 3.14). His father, József Eötvös, had become the minister of culture in the first Hungarian government in March 1848. In 1866 he became the first president of the Hungarian Academy of Sciences, which he had reorganized, and after the Compromise of 1867 he served again as the minister of culture and education (for details about the Eötvös family, see Frank, 2012; Hübner, 2010).

Loránd Eötvös first studied law in Budapest from 1865 to 1867. Károly Than[74] (1834–1908), the founder of modern chemistry in Hungary, then motivated him to study chemistry in Heidelberg. As Loránd Eötvös noted in his memoires:

> While a student of the law, I entered, in 1867, among Károly Than's students. This was a decisive step in my career. It was from his lectures that the magic light of research into natural sciences shone at me for the first time, his—then still very small and defective—laboratory was the first that opened up to me, and when I had performed the first chemical reaction, and had first sat in front of a balance, and when later my professor honoured me with allowing me an insight in his own investigations, then . . . I really learnt what—following my scientific inclination—I can be good for in this homeland. (Quoted and translated by Vámos, 2006, p. 283)

[72]Details on the faculty's internal substantive and ideological disputes concerning Karl Mannheim's process of qualifying for a university chair and on the personal relationship between Alfred Weber and Karl Mannheim are found in Demm (1999, pp. 29–44; 2000a, 2000b, 2003, 2012), Kaesler (1990), and Karadi (2003).

[73]A *Privatdozent* at that time was a university lecturer who had absolved his or her *Habilitation*, which was a prerequisite of applying for a professorship at another university. If nominated by the faculty and proposed by the Senate of Heidelberg University, a university lecturer with *Habilitation* could, after six years of service, be promoted by the government to the status of associate professor not on the state payroll. No official post was attached to this title, however.

[74]Than had worked for a time as a postdoctoral researcher under Robert Wilhelm Bunsen in Heidelberg, where he had also met the physicist Gustav Robert Kirchhoff and the mathematician Moritz B. Cantor (1829–1920). At just 26 years of age, Than acquired a position as a substitute professor at the University of Pest and was offered a full professorship in chemistry at the age of 28. He served the Royal Hungarian Society for Natural Sciences as vice president (1862–1872) and president (1872–1880) (for more details see Vámos, 2006, p. 280).

Fig. 3.14 The paramount role of Loránd Eötvös as an outstanding scientist is acknowledged in several ways. The Hungarian post office has dedicated four stamps to him (1932, 1948, 1959, and 1991), he has been nominated for the Nobel Prize three times (1911, 1914, and 1917), and Hungary's most important university was named after him in 1950.
Source: Picture collection of Postakürt Alapítvány Bélyegmúzeum, Budapest. Reprinted with permission.

Loránd Eötvös arrived in Heidelberg on October 3, 1867, and registered in chemistry on October 21. His father was well acquainted with the Heidelberg law professors Karl Adolf von Vangerow (1808–1870), Johann Kaspar Bluntschli (1808–1881), and Karl Theodor Welcker (1790–1869) as well as with Bunsen and Kirchhoff, so Loránd was well received privately, too, not least because of his manifest academic zeal (Hübner, 2010, p. 200). The younger Eötvös found the chemistry lectures by Bunsen "not exactly dazzling" but thought a great deal of him nevertheless. By contrast, Kirchhoff's personality and lectures impressed Eötvös so much that he majored in physics. Lastly, the lectures of the geologist and associate professor Gustav Alfons von Leonhard (1816–1878) sparked his interest in geophysics. Kirchhoff advised Loránd to switch to professors Franz E. Neumann (1798–1895) and Friedrich J. Richelot (1808–1875) at the University of Königsberg. The young Eötvös spent the summer semester of 1869 there, returning to Heidelberg in the fall of 1869 (Hübner, 2010, pp. 198–207). On July 8, 1870, he graduated summa cum laude from Heidelberg.

From 1870 until his death, Loránd Eötvös researched and taught at the University of Budapest. He was particularly interested in capillarity, gravitation, and geomagnetism and formulated what is now known as Eötvös's law (Eötvös, 1886), which dealt with the relation between the surface tension of fluids at different temperatures and molecular weight (see also Einstein, 1911). In 1890 Eötvös developed a dynamic method of measuring the gravitational constant, later constructed a gravimeter and the Eötvös torsion balance, which could detect the presence of crude oil in anticlines. Eötvös served as the rector of the university (1891–1892), minister of culture (1894–1895), and president of the Hungarian Academy of Sciences (1889–1905).

The physicist and cultural historian Ágost Heller (1843–1902) was already a graduate student when he registered at Heidelberg University. In 1866 he had completed training as a railroad engineer at the Joseph Polytechnic in Buda, where he worked as an assistant in the two years thereafter. He studied physics under Kirchhoff and Hermann Ludwig Ferdinand von Helmholtz in 1869. Between 1870 and 1898 he taught physics at a public secondary school in Budapest. Beginning in 1875 he worked as a librarian for the Natural Science Association. The Hungarian Academy of Sciences elected him as corresponding member in 1887 and as a full member in 1893, employing him as the head librarian as of 1894. He authored numerous textbooks, technical articles, and works of popular science. He became known especially through his two-volume *History of Physics from Aristotle to Today* (1882–1884). Heller was one of the editors and authors of the 16-volume *Pallas Nagy Lexikon*, the first Hungarian encyclopedia not based on a translation.

Kálmán Szily (1838–1924) was the son of a landed nobleman and already had a degree from the Vienna Polytechnic Institute when he studied physics, first in Berlin and then, beginning in 1865, under Kirchhoff in Heidelberg. He also distinguished himself as a scholar, cultural policy expert, and organizer. For decades he held the chair of experimental physics at the Joseph Polytechnic in Buda, which became the Technical University of Buda in 1871[75] and he its first rector. Under the president of the Hungarian Academy of Sciences, Loránd Eötvös, Szily was the academy's general secretary from 1889 to 1905. He reformed the institution's publishing and created the *Natural Science Bulletin* (Természettudományi Közlöny). He later became the head librarian of the Academy of Sciences and both the founder and president of the Hungarian Linguistic Society.

Vince Wartha (1844–1914) was born in Fiume, attended the Gymnasium in Szeged, then the Joseph Polytechnic in Buda. He thereafter studied at the Eidgenössische Technische Hochschule (Swiss Federal Institute of Technology, ETH) in Zurich, where he completed his first degree (*Diplom*) in 1864. Back in Hungary, he received an assistantship in the chemistry department of the Joseph Polytechnic. He was dissatisfied with the situation, however, and went to Heidelberg, where he graduated with a doctorate under Bunsen in 1865. After taking up residence again in Zurich, he worked as an assistant and *Privatdozent* at the Institute of Chemistry from 1865 to 1867. In 1867 he accepted the chair of mineralogy and geology at the Joseph Polytechnic. In 1871 Wartha took over the chair of chemical technology at the new Technical University, a post he held for the next 41 years. He served as dean of the Technical University of Budapest (1875–1877) and as rector (1896–1898 and 1907–1910).

Wartha's most significant innovation was the rediscovery of the long-lost secret of producing the iridescent metallic sheen for which the medieval enamels from the Italian town of Gubbio have been known for centuries. Working together with Vilmos Zsolnay, Wartha developed the technique of using eosin to refine ceramics. In Zsolnay's factory in Pécs, the process was applied to ornamental objects for the

[75]The Joseph Polytechnic, established in 1856, became the Royal Joseph University in 1862. It was among the first institutions in Europe to train engineers at the university level. In 1872 it gained full autonomy and the right to issue engineering diplomas. In 1901 it was entitled to confer the doctoral degree.

first time. The eosin enamel ceramics by Zsolnay became a world famous stylistic element of Hungarian art nouveau. Wartha also participated as a ceramics expert in founding the Museo Internazionale delle Céramiche (International Musuem of Ceramics) in Faenza, Italy. He developed a particular method of alkalimetry (the determination of the amount of alkali or base in a solution, measured by an alkalimeter or by volumetric analysis) and, together with Ignác Pfeifer, a way of determining water hardness. In 1873 he became a corresponding member of the Hungarian Academy of Sciences; in 1893, a full member of that organization; and from 1909 on, its vice president.

Alajos Schuller (1845–1920) began studying at the Joseph Polytechnic in Buda in 1863 and worked there as an assistant after graduating. He then went to Heidelberg and studied chemistry under Bunsen and physics under Kirchhoff from 1870 to 1872, serving as the latter's assistant (1871–1872). From 1872 to 1916 Schuller held the chair of experimental physics at the Technical University of Budapest. Schuller focused scientifically on calorimetric experiments, sometimes working with his friend Vince Wartha. He improved on Bunsen's calorimeter and developed an automatic mercury air pump himself. In honor of achievements, he received the status of a corresponding member of the Hungarian Academy of Sciences in 1880, and an honorary doctorate from the University of Klausenburg (Kolozsvár) in 1910.

Gyula (Julius) König (1848–1913) began studying medicine in Heidelberg in 1868, then turned to physics. After working with Helmholtz on the electrical stimulation of nerves, König completed his doctorate in 1870, writing a 24-page dissertation entitled *Zur Theorie der Modulargleichungen der elliptischen Functionen* (The Theory of Modular Equations of Elliptical Functions) under the Heidelberg mathematician Leo Königsberger. In 1874 he became a professor at the Technical University of Budapest, where he worked for the rest of his life, also serving three terms each as dean and rector. In 1889 he became a member of the Hungarian Academy of Sciences. In addition to articles on analysis and number theory, his works on algebra and Cantor's set theory were particularly important (König's theorem). His *Einleitung in die allgemeine Theorie der algebraischen Größen* (Introduction to the General Theory of Algebraic Magnitudes; König, 1903) made him a precursor of modern algebra.[76]

The high level of research and teaching at the young Technical University of Budapest was largely due to four renowned professors who had studied or had received their doctorates at almost the same time in Heidelberg: the physicists Szily and Schuller, the chemist Wartha, and the mathematician König. These four natural

[76]"In 1904, at the third International Congress of Mathematicians at Heidelberg, König gave a talk to disprove Cantor's continuum hypothesis. The announcement was a sensation and was widely reported by the press. All section meetings were cancelled so that everyone could hear his contribution. König applied a theorem proved in the thesis of Felix Bernstein; this theorem, however, was not as generally valid as Bernstein had claimed. Ernst Zermelo,, the later editor of Cantor's collected works, found the error already the next day. In 1905 there appeared short notes by Bernstein, correcting his theorem, and König withdrawing his claim." Retrieved February 2, 2017, from https://en.wikipedia.org/wiki/Gyula_K%C5%91nig.

scientists were occasionally called simply "the Heidelbergians" by their fellow academics in Hungary.

After graduating from the Joseph Polytechnic in Buda in 1866, Béla Lengyel (1844–1913) studied chemistry, first at the university in Pest and then under Bunsen in Heidelberg, where he was Bunsen's assistant (1869–1870). He qualified for a university professorship in 1870, was appointed as a *Privatdozent* in 1871, an assistant professor in 1872, and a full professor of experimental, analytical, and organic chemistry at the University of Budapest. He was dean of the philosophical faculty from 1885 to 1886 and rector of the university from 1895 to 1896. From 1863 to 1893, Lengyel was secretary of the Hungarian Natural Science Society. He became a corresponding member of the Hungarian Academy of Sciences in 1876 and a full member in 1894. He discovered carbon subsulfide (C_3S_2), among other things, and became known for his research on subjects such as the spectroscopy of gas mixtures, radioactivity, and theoretical principles underlying phenomena of light in evacuated pipes during the passage of electrical current.

Béla Issekutz (1886–1979) received a doctorate in medicine at the University of Kolozsvár and registered in 1913 as an intern in Heidelberg. He became a professor in Kolozsvár in 1918, professor and rector in Szeged in 1920, and professor and director of the pharmacological department in the medical faculty in Budapest in 1937. He introduced modern pharmacological research into Hungary and succeeded, among other things, at explaining the mechanism of action of insulin and thyroxin. He helped pioneer the use of chemotherapy against cancer.

Philipp Lenard (Fülöp Lenárd, 1862–1947) was the son of a Tirol wine dealer in Pressburg (Pozsony, today's Bratislava). After attending the Hungarian *Gymnasium* in Pressburg, he studied natural science at the universities of Budapest and Vienna for two semesters in 1880, then worked in his father's wine dealership in Pressburg. In 1883 he resumed his studies in Heidelberg under Georg Hermann Quincke (1834–1924) and Robert Bunsen. After one semester under Helmholtz in Berlin, he earned his doctorate in Heidelberg, writing his dissertation on the *Über die Schwingung fallender Tropfen* (On the oscillation of falling water drops). Lenard served briefly an assistant to Loránd Eötvös at the University of Budapest (1887). From 1887 to 1890 he was an assistant under Quincke in the physics department at the University of Heidelberg, where he continued his investigations on phosphorescence. In 1891 he went to Breslau and, as an assistant to Heinrich Hertz, moved to Bonn in 1892. There, Lenard completed his qualification for a professorship in 1893 with his postdoctoral dissertation entitled *Über die Elektricität der Wasserfälle* (On the electricity of waterfalls). He also studied cathode rays. From his own collection he supplied Conrad Röntgen with a discharge tube and a Lenard window, both of which proved decisive in the discovery of x-rays in 1895. After Röntgen became famous for this discovery, Lenard accused him of having stolen it. Lenard became an associate professor in Breslau in the winter semester of 1894–1895, a lecturer at the Technical University of Aachen in the winter semester of 1895–1896, an associate professor in Heidelberg in the winter semester of 1896–1897, and a full professor and director of the physics department at the University of Kiel in the summer semester of 1898.

Lenard received the 1905 Nobel prize for physics for his work on cathode rays and the development of electron theory. In 1907 he succeeded his teacher, Quincke, as a full professor of physics and radiology at Heidelberg University and as the director of the physics department there. In 1907 he acquired German citizenship as well. In 1913 he expanded the Heidelberg radiology department into the largest and most modern physics department in Germany. He declined a call to the University of Pressburg in 1914. In 1932 he received the Benjamin Franklin Medal, a highly endowed scientific award conferred by the American Franklin Institute in Philadelphia. Lenard was a member of the scientific academies in Heidelberg, Berlin, and Vienna and received an honorary doctorate in engineering from the Technical University of Dresden. In 1928 he became a public supporter of the racially minded, anti-Semitic convictions of the National Socialist Society for German Culture. He opposed Einstein, rejected relativity theory and quantum mechanics as too abstract and unrealistic, and espoused "German physics."

Some prominent Hungarians did not study at Heidelberg University but accepted a chair there. Their circle included two professors of ancient history—Alfred von Domaszewski (1856–1927) and Géza Alföldy (1935–2011)—anatomist Count Béla Haller von Hallerstein (1858–1914), physician Pál György (1893–1976),[77] and zoologist József Spek (1895–1964).[78]

[77]György left Heidelberg University at his own request for political reasons in 1933. In England he carried out research at the Nutrition Laboratory of Cambridge University from 1933 to 1935. In the United States in 1935, he became a visiting assistant professor of pediatrics at the medical school of the Western Reserve University in Cleveland, Ohio, and was named associate professor there in 1937. He assumed the post of associate research professor of pediatrics at the University of Pennsylvania School of Medicine in 1944 and became a full professor in 1946. From 1950 to 1957, he served as the head of pediatrics at Philadelphia General Hospital. He became internationally known through his discovery of three vitamins: riboflavin, vitamin B6, and biotin (vitamin B7, formerly known as vitamin H or coenzyme R). In 1959 he received an honorary doctorate from Heidelberg University. U. S. President Gerald Ford awarded him a posthumous National Medal of Science in the biological sciences "for his discovery of three vitamins and related research that have greatly improved human nutrition" (Retrieved March 2, 2017, from https://www.nationalmedals.org/laureates/paul-gyorgy#).

[78]Spek worked from 1932 to 1933 in the laboratory of the cellular physiologist and biophysicist Robert Chambers in New York and in Woods Hole, Massachusetts. After teaching during an interim professorship at the University of Greifswald (1943–1944) and turning down a call to the University of Halle (1946), he accepted an appointment as a full professor of zoology at the University of Rostock (1947). He was particularly interested in questions of cell structure. New detection methods that Spek developed made him a pioneer of vital staining and fluorescent marking in microscopy. As one of the leading researchers of protoplasm, he was a cofounder of the international journal *Protoplasm*.

Stagnation and a Fresh Beginning of Scientific Relations

After the National Socialists seized power in Germany in January 1933, an extraordinary intellectual period at Heidelberg University abruptly ended, the hitherto worldwide area for recruiting students and professors shrank radically, and the number of Hungarian students plummeted. But some of the origins of this demise dated back more than a decade. First, the Treaty of Trianon, signed in 1920 to formally end World War I between most of the Allies and Hungary, had stripped Hungary of two thirds of its territory and had passed those lands to neighboring countries. Students from those areas therefore no longer counted as Hungarians in the statistics on nationalities. Second, the onset of the Great Depression in 1929 made university study in Germany unattractive, a situation that only worsened under National Socialism (1933–1945) and during World War II (1939–1945).

For decades following World War II, the Communist regime blocked Hungarians from studying in the West at all. In the more than 500 years of scientific relations between Germany and Hungary, there has never been a period in which Hungarian students and scholars were as hermetically isolated from the West as they were during the Cold War. Only as of 1961 were a few Hungarian scientists of the young generation permitted to enter the Federal Republic of Germany temporarily (e.g., on a research fellowship from the Alexander von Humboldt Foundation or the German Academic Exchange Service [DAAD]). An initial, cautious rapprochement between Heidelberg University and the Eötvös Loránd University in Budapest in the early 1980s led in 1983 to an agreement on a partnership between them—Heidelberg's first such arrangement with a university in the Communist Bloc (for details, see Meusburger, 2003).

The signing of the partnership agreement opened new doors for scientists from the Eötvös Loránd University to conduct research in Heidelberg and cooperate with academics at the university there. These opportunities were taken primarily by scholars of law, medicine, and geography (for details see Meusburger, 2003). The Heidelberg visit by the Hungarian law professors Ferenc Mádl (1931–2011) and Lajos Vékás (1939–) in 1983 was to have especially intense, long-term impacts. Together with their colleagues László Sólyom (1942–) and János Németh (1933–) and the Heidelberg jurists Erik Jayme and Gert Reinhart, Mádl and Vékás launched joint seminars with German and Hungarian students of civil law, corporate law, international contract law, copyright law, and a host of other subjects that were crucial when the market economy was introduced in Hungary later. The courses proved to be a highly advantageous head start in the training of Hungary's aspiring young lawyers.

These four far-sighted Hungarian professors of law rose to the highest offices after the collapse of the Communist regime in 1989. Mádl, who came from a Hungarian-German family in Bánd (western Hungary), was the government minister (without portfolio) of science policy and the coordination of government policy. From 1990 to 1992 he also functioned as a board chairman of the state property agency, which oversaw the banking sector. He served as minister of culture from

1993 to 1994 and was elected as the president of Hungary in May 2000, although he had no party affiliation. President Mádl received an honorary doctorate from Heidelberg University in July 2001.

Lajos Vékás, one of the first contacts involved in 1983, spent 1987 in Heidelberg on a Humboldt fellowship. He became rector of the Eötvös Loránd University (1990–1993), rector of the Budapest Collegium (1992–1997), and a member of the Foundation and Academic Council of the Europe Institute Budapest. In 2014 he became vice president of the Hungarian Academy of Sciences.

László Sólyom was appointed as first president of the Hungarian Constitutional Court (1990–1998) after the democratic transformation of the political system that had begun in late 1989 and was later elected as state president (2005–2010). János Németh became the executive president of the Constitution Court (1998–2003). Kálmán Györgyi (1939–), the cooperating partner of the Heidelberg criminal law expert and criminologist Olaf Miehe (1935–), was appointed as the first attorney general of democratic Hungary after the collapse of the communist regime.

However, Hungary's cautious opening to the West before the final debacle of the communist regime was confined largely to scholars. Hungarian students had to wait several years more before they were allowed to study in Germany (partly for financial reasons, too). The 20 to 30 Hungarian nationals who studied in Heidelberg in the 1980s, most of them children of Hungarian refugees who had fled the country in 1956, were already living outside Hungary. The number of Hungarian students in Heidelberg did not recover strongly until after the opening of the East–West border in 1989 and the collapse of the communist regime in the course of 1990. It initially peaked at more than 100 students in the 2002–2003 academic year.

Conclusion

This chapter has documented that the intellectual appeal of a university and the area from which its professors and students come has historically depended on numerous scientific, political, economic, confessional, organizational, and cultural factors of influence. Heidelberg University has had heydays that made it an important intellectual and scientific center, and it has intermittently suffered catastrophes relegating it to irrelevance. Again and again, periods of internationality and outstanding scientific achievements have been succeeded by academic mediocrity; intellectual freedom, by political interference. This narrative highlights the exceptional role that the international networks and web of relations cultivated by individual famous scholars have repeatedly played over the centuries for the areas from which foreign students are recruited. It underscores how much an internationally oriented university such as Heidelberg has been marked by pan-European political power relations, foreign-policy interests, diplomatic relations, wartime events, and confessional disputes. It also emphasizes how much an intellectually powerful university can serve as a first-class innovative space capable of affecting the cultural, economic, and sometimes even political development of far-flung regions through its students and the networks of its professors.

Place, networks, and spatial relations matter. Few fields of endeavor, if any, reflect the pertinence of this statement as obviously as science does. The structural contexts, interrelations, and web of interests that frame research and teaching; the place experience of students; the ebb and flow of a university's intellectual aura and attractiveness; and the significance of universities in the development of international networks therefore pose interesting questions for research at various levels of aggregation.

Another scientific challenge is the research question of the effects that studying in a foreign country may have on the society, culture, and scientific community in the country from which the students and aspiring young academics hail. The fact that so many Hungarian students studied outside their country in different eras has clearly contributed to pronounced internationalism, superb foreign-language skills in academic occupations, and impressive cultural and scientific dynamics in Hungary during some bright periods of its history. On the other hand, totalitarian political regimes or forced isolation during hot and cold wars lead inevitably to the loss of international contacts and all the above-mentioned benefits, resulting in decay of intellectual vigor and deterioration in the quality of life as well.

References

Ábel, J. (1881). *Egyetemeink a középkorban* [Our medieval universities]. Budapest: MTA Irodalomtörténeti Bizottsága. Retrieved from http://www.mek.oszk.hu/10800/10825/

Almási, G. (2014). Touring Europe: Comparing east-central European academic peregrination in the sixteenth and seventeenth century. In G. Almási (Ed.), *Study tours and intellectual-religious relationships* (pp. 17–34). A divided Hungary in Europe: Exchanges, networks and representations, 1541–1699: Vol. 1. Cambridge, UK: Cambridge Scholar Publishing.

Altmann, J. (1997). Die mittelalterliche Universität in Óbuda (Altofen) [The medieval university in Óbuda (Altofen)]. In L. Szögi & J. Varga (Eds.), *Universitas Budensis, 1395–1995: International conference for the history of universities on the occasion of the 600th anniversary of the foundation of the University of Buda* (pp. 67–74). Budapest: Eötvös Loránd University Library and Archive.

Alvermann, D. (2006). Stipendien als strukturelle Elemente des Migrationsnetzes. Das Beispiel Greifswald [Scholarships as structural elements of the migration network: The example of Greifswald]. In M. Fata, G. Kurucz, & A. Schindling (Eds.), *Peregrinatio Hungarica. Studenten aus Ungarn an deutschen und österreichischen Hochschulen vom 16. bis zum 20. Jahrhundert* (pp. 345–384). Contubernium–Tübinger Beiträge zur Universitäts- und Wissenschaftsgeschichte: Vol. 64. Stuttgart: Steiner.

Anderegg, J.-P. (2000). Siebenbürgen: Kolonistenland am Rande Europas [Transylvania: Land of colonists at Europe's edge]. *Freiburger Geschichtsblätter, 77*, 177–190.

Andritsch, J. (1965). *Studenten und Lehrer aus Ungarn und Siebenbürgen an der Universität Graz (1586–1782)* [Students and teachers from Hungary and Transylvania at the University of Graz (1586–1782)]. Graz: Historische Landeskommission.

Asche, M. (2004). Bildungsbeziehungen zwischen Ungarn, Siebenbürgen und den deutschen Universitäten im 16. und frühen 17. Jahrhundert [Educational relations between Hungary, Transylvania, and German universities in the sixteenth and early seventeenth centuries]. In

W. Kühlmann & A. Schindling (Eds.), *Deutschland und Ungarn in ihren Bildungs- und Wissenschaftsbeziehungen während der Renaissance* (pp. 27–52). Contubernium–Tübinger Beiträge zur Universitäts- und Wissenschaftsgeschichte: Vol. 62. Stuttgart: Steiner.

Asche, M. (2010). Philipp Melanchthon als christlicher Schulhumanist und Bildungsreformer: Wittenberg und der Export des humanistischen Bildungsprogramms [Philipp Melanchthon as a Christian school humanist and educational reformer: Wittenberg and the export of the humanist educational program]. In F. Schweitzer, S. Lorenz, & E. Seidl (Eds.), *Philipp Melanchthon: Seine Bedeutung für Kirche und Theologie, Bildung und* Wissenschaft (pp. 75–94). Theologie Interdisziplinär: Vol. 8. Neukirchen-Vluyn: Neukirchener Verlagsgesellschaft

Baar-Cantoni, R. (2012). Martin Luther's disputation in Heidelberg and its widespread effects. In P. Meusburger & T. Schuch (Eds.), *Wissenschaftsatlas of Heidelberg University: Spatio-temporal relations of academic knowledge production* (pp. 56–57). Knittlingen: Bibliotheca Palatina.

Baar-Cantoni, R., & Wolgast E. (2012). Migration of professors between 1550 and 1700. In P. Meusburger & T. Schuch (Eds.), *Wissenschaftsatlas of Heidelberg University: Spatio-temporal relations of academic knowledge production* (pp. 66–69). Knittlingen: Bibliotheca Palatina.

Barta, I. (1937). Ungarn und die Wiener Universität des Jesuitenzeitalters [Hungary and the University of Vienna in the Jesuit age]. *A Gróf Klebelsberg Kunó Magyar Történetkutató Intézet évkönyve, 7,* 113–177.

Bartha, T. (Ed.). (1967). *Der Heidelberger Katechismus in Ungarn* [The Heidelberg Catechism in Hungary]. Budapest: Presseabteilung der Synodalkanzlei der reformierten Kirche in Ungarn.

Beck, A. J. (Ed.). (2016). *Melanchthon und die Reformierte Tradition* [Melanchthon and the Reformed tradition]. Refo500 Academic Studies: Vol. 6. Göttingen: Vandenhoeck & Ruprecht.

Bencsik, P. (2002). Útiokmányok, utazási lehetőségek, idegenforgalom a 20. századi Magyarországon [Travel documents, opportunities, and tourism in twentieth-century Hungary]. *Regio, 13*(2), 31–50. Retrieved from epa.oszk.hu/00000/00036/00045/pdf/02.pdf

Benda, K. (1976). La réforme en Hongrie [Reformation in Hungary]. *Bulletin de la Societé de l'histoire de Protestantisme Français, 122,* 30–53.

Bendl, J. (1990). Die wiedergefundene Heidelberger Bibliothek von Georg Lukács [The recovered library of George Lukács]. *Hungarian Studies, 6,* 173–184.

Bendl, J. (1998). Zwischen Heirat und Habilitation in Heidelberg [Between marriage and habilitation in Heidelberg]. *Jahrbuch der Internationalen Georg-Lukács-Gesellschaft 1997*, 2, 17–45.

Benseler, F. (1987). Lukács, Georg. In *Neue Deutsche Biographie: Vol. 15. Locherer–Maltza(h)n* (pp. 509–513). Berlin: Duncker & Humblot. Retrieved from http://www.deutsche-biographie.de/pnd11857518X.html

Bernhard, J.-A. (2010). Calvins Wirkung und Einfluss in Ungarn und Siebenbürgen vor 1551 [Calvin's impact and influence in Hungary and Transylvania before 1551]. In M. Fata & A. Schindling (Eds.), *Calvin und Reformiertentum in Ungarn und Siebenbürgen. Helvetisches Bekenntnis, Ethnie und Politik vom 16. Jahrhundert bis 1918* (pp. 25–62). Reformationsgeschichtliche Studien und Texte: Vol. 155. Münster: Aschendorff.

Bernhard, J.-A. (2015). *Konsolidierung des reformierten Bekenntnisses im Reich der Stephanskrone: Ein Beitrag zur Kommunikationsgeschichte zwischen Ungarn und der Schweiz in der frühen Neuzeit (1500—1700)* [Consolidation of the Reformed Confession in the lands of Saint Stephen's Crown: A contribution to the communication history between Hungary and Switzerland in the Early Modern Age (1500–1700)]. Refo500 Academic Studies: Vol. 19. Göttingen: Vandenhoeck & Ruprecht.

Bethlen, M. (1955). *Önéletírása* [Memoires]. Budapest: Szépirodalmi. Retrieved from http://www.mek.oszk.hu/06100/06152/html/bethlen0020001.html

Bierma, L. D. (2015). The theological origins of the Heidelberg Catechism. In C. Strohm & J. Stievermann (Eds.), *Profil und Wirkung des Heidelberger Katechismus: Neue Forschungsbeiträge anläßlich des 450jährigen Jubiläums. The Heidelberg Catechism: Origins, characteristics, and influences: Essays in reappraisal on the occasion of its 450th anniversary* (pp. 13–22). Schriften des Vereins für Reformationsgeschichte: Vol. 215. Gütersloh: Gütersloher Verlagshaus.

Binder, P. (1975). Kulturaustausch in Siebenbürgen auf dem Gebiet des Buchdrucks [Cultural exchange in Transylvania in the field of printing]. In P. Philippi (Ed.), *Siebenbürgen als Beispiel europäischen Kulturaustausches* (pp. 71–90). Siebenbürgisches Archiv: Vol. 12. Cologne: Böhlau.

Birkenmaier, W. (2012). World revolution, Zionism and emigration: Students from the Russian Empire. In P. Meusburger & T. Schuch (Eds.), *Wissenschaftsatlas of Heidelberg University: Spatio-temporal relations of academic knowledge production* (pp. 300–303). Knittlingen: Bibliotheca Palatina.

Black, J. (1983). France and the Grand Tour in the early eighteenth century. *Francia, 11,* 407–416.

Black, J. (1992). *The British abroad: The Grand Tour in the eighteenth century*. Stroud: Sutton.

Bodrogi, B., & Galántai, Z. (2013). *Gazdaságtörténet* [Economic history]. Budapest: Typotex.

Borbándi, G. (1973). Die Kulturpolitik der ungarischen Räterepublik [Cultural politics of the Hungarian soviet republic]. *Ungarn-Jahrbuch, 5,* 171–186.

Bozzay, R. (2010). Der Einfluss ehemaliger Studenten der Leidener Universität im 17. und 18. Jahrhundert auf Kultur und Bildung in Siebenbürgen [The influence of former students of Leiden University in the seventeenth and eighteenth centuries on culture and education in Transylvania]. In M. Fata & A. Schindling (Eds.), *Calvin und Reformiertentum in Ungarn und Siebenbürgen: Helvetisches Bekenntnis, Ethnie und Politik vom 16. Jahrhundert bis 1918* (pp. 215–238). Reformationsgeschichtliche Studien und Texte: Vol. 155. Münster: Aschendorff.

Bozzay, R. (2014). The influence of Dutch universities on the education of seventeenth-century Hungarian intellectuals. In G. Almási (Ed.), *Study tours and intellectual-religious relationships* (pp. 81–100). A divided Hungary in Europe: Exchanges, networks and representations, 1541–1699: Vol. 1. Cambridge, UK: Cambridge Scholar Publishing.

Brenner, R. (1989). Economic backwardness in eastern Europe in light of developments in the west. In D. Chirot (Ed.), *The origins of backwardness in eastern Europe: Economics and politics from the Middle Ages to the early twentieth century* (pp. 15–52). Berkeley: University of California Press.

Bucsay, M. (1977). *Der Protestantismus in Ungarn 1521–1978: Ungarns Reformationskirchen in Geschichte und Gegenwart. Teil 1. Im Zeitalter der Reformation, Gegenreformation und katholischen Reform* [Protestantism in Hungary, 1521–1978: Hungary's Reformed churches in history and presence: Part 1. In the age of Reformation, Counter-Reformation, and Catholic reform]. Studien und Texte zur Kirchengeschichte und Geschichte: Reihe 1, Vol. 3. Vienna: Böhlau.

Chaney, E. (1998). *The evolution of the Grand Tour: Anglo-Italian cultural relations since the Renaissance*. London: Cass.

Clifford, J. (1992). Travelling cultures. In L. Grossberg, C. Nelson, & P. Treichler (Eds.), *Cultural studies* (pp. 96–116). London: Routledge.

Comenius, J. A. (1668/1958). *Unum necessarium: The one thing necessary [or: The one thing Needful]* (V. H. Nelson, Trans.). Winston-Salem: Moravian Archives. Retrieved from http://moravianarchives.org/images/pdfs/Unum%20Necessarium.pdf

Comenius, J. A. (1705). Visible world, or, a nomenclature, and pictures of all the chief things in the world and mens employments therein (C. Hoole, Trans.). London: n.p. (Original work published 1658: *Orbis sensualium pictus: Hoc est, omnium princip omnium principalium in mundo*).

Courtenay, W. J. (2000). Study abroad: German students at Bologna, Paris, and Oxford in the fourteenth century. In W. J. Courtenay & J. Miethke (Eds.), *Universities and schooling in medieval society* (pp. 7–31). Education and Society in the Middle Ages and Renaissance: Vol. 10. Leiden: Brill.

Csizmadia, A. (1965). *Die Universität Pécs im Mittelalter (1367–)* [The university of Pécs in the Middle Ages (1367–)]. Studia Iuridica: Auctoritate Universitatis Pécs Publicata: Vol. 41. Budapest: Tankönyvkiado.

Csohány, J. (1994–1995). Die politischen Beziehungen von Gábor Bethlen zum reformierten Europa [Political relations of Gábor Bethlen to Protestant Europe]. *Jahrbuch für die Geschichte des Protestantismus in Österreich, 110/111,* 87–98.

Czettler, A. (1980–1981). Die Außenpolitik der Siebenbürgischen Prinzen im 16. und 17. Jahrhundert [Foreign policy of Transylvanian Princes in the sixteenth and seventeenth centuries]. *Ungarn-Jahrbuch, 11,* 91–121.

de Ridder-Symoens, H. (1996). Mobility. In H. de Ridder-Symoens (Ed.), *Universities in early modern Europe (1500–1800)* (pp. 416–448). A history of the university in Europe: Vol. 2. Cambridge, UK: Cambridge University Press.

Demm, E. (1990). *Ein Liberaler in Kaiserreich und Republik: Der politische Weg Alfred Webers bis 1920* [A liberal in the empire and the republic: The political path of Alfred Weber until 1920]. Schriften des Bundesarchivs: Vol. 38. Boppard: Boldt.

Demm, E. (1999). *Von der Weimarer Republik zur Bundesrepublik: Der politische Weg Alfred Webers von 1920 bis 1958* [From the Weimar Republic to the Federal Republic: The political path of Alfred Weber from 1920 to 1958]. Schriften des Bundesarchivs: Vol. 51. Düsseldorf: Droste.

Demm, E. (2000a). Ein „Institut der Außenseiter"? Historische Forschung und Personalpolitik am Heidelberger Staatswissenschaftlichen Institut unter Eberhard Gothein und Alfred Weber (1904–1933) [A department of "outsiders"? Historical research and personal politics at the Heidelberg Institut of Political Science under Eberhard Gothein and Alfred Weber (1904–1933)]. In E. Demm (Ed.), *Geist und Politik im 20. Jahrhundert: Gesammelte Aufsätze zu Alfred Weber* (pp. 129–183). Frankfurt am Main: Lang

Demm, E. (2000b). Karl Mannheim und Alfred Weber [Karl Mannheim and Alfred Weber]. In E. Demm (Ed.), *Geist und Politik im 20. Jahrhundert: Gesammelte Aufsätze zu Alfred Weber* (pp. 251–266). Frankfurt am Main: Lang

Demm, E. (2003). Karl Mannheim és Alfred Weber: Alfred Weber és Heinrich Rickert között [Karl Mannheim and Alfred Weber: Between Alfred Weber and Heinrich Rickert]. In L. C. Kiss, M. G. Lázár, E. Karádi, K. Mannheim, & N. Elias (Eds.), *Mannheim-tanulmányok: Írások Mannheim Károlytól és Mannheim Károlyról* (pp. 111–128). Budapest: Napvilág.

Demm, E. (2010). Alfred Weber (1868–1958): Machtkapital, Netzwerke und Lebensstil [Alfred Weber (1868–1958): Power capital, networks, and life style]. In E. Demm & J. Suchoples (Eds.), *Akademische Lebenswelten: Habitus und Sozialprofil von Gelehrten im 19. und 20. Jahrhundert* (pp. 105–135). Frankfurt am Main: Peter Lang.

Demm, E. (2012). Alfred Weber and his disciples. In P. Meusburger & T. Schuch (Eds.), *Wissenschaftsatlas of Heidelberg University: Spatio-temporal relations of academic knowledge production* (pp. 114–115). Knittlingen: Bibliotheca Palatina.

Dézsi, L. (1897). *Szenczi Molnár Albert 1574–1634* [Albert Szenczi Molnár 1574–1634]. Budapest: Magyar Történelmi Társulat. Retrieved from http://www.mek.oszk.hu/05600/05617/html/

Diener, H. (1963). Zur Geschichte der Universitätsgründungen in Alt-Ofen (1395) und Nantes (1423) [On the history of the founding of the university in Óbuda (1395) and Nantes (1423)]. *Quellen und Forschungen aus italienischen Archiven und Bibliotheken, 42–43*, 265–284.

Domonkos, L. (1997). The founding (1395) and refounding (1410) of the University of Óbuda. In L. Szögi & J. Varga (Eds.), *Universitas Budensis, 1395–1995: International conference for the history of universities on the occasion of the 600th anniversary of the foundation of the University of Buda* (pp. 19–34). Budapest: Eötvös Loránd University Library and Archive.

Dotzauer, W. (1977). Deutsches Studium und deutsche Studenten an europäischen Hochschulen (Frankreich, Italien) und die nachfolgende Tätigkeiten in Stadt, Kirche und Territorium in Deutschland [German studies and German students at European universities (France, Italy) and subsequent employment in town, church, and territory]. In E. Maschke & J. Sydow (Eds.), *Stadt und Universität im Mittelalter und in der früheren Neuzeit* (pp. 112–141). Stadt in der Geschichte: Vol. 3. Sigmaringen: Jan Thorbecke.

Driver, F. (2001). *Geography militant*: Cultures of exploration and empire. Oxford, UK: Blackwell

Driver, F. (2017). Exploration as knowledge transfer: Exhibiting hidden histories. In H. Jöns, P. Meusburger, & M. Heffernan (Eds.), *Mobilities of knowledge* (pp. 85–104). Knowledge and Space: Vol. 10. Dordrecht: Springer. doi:https://doi.org/10.1007/978-3-319-44654-7_5

Duncan, J., & Gregory, D. (1999). Introduction. In J. Duncan & D. Gregory (Eds.), *Writes of passage: Reading travel writing* (pp. 1–13). London: Routledge.

Eckart, W. U., Hübner, K., & Nawa, C. (2012). The rise of the natural sciences—Bunsen, Kirchhoff, and Helmholtz. In P. Meusburger & T. Schuch (Eds.), *Wissenschaftsatlas of Heidelberg University: Spatio-temporal relations of academic knowledge production* (pp. 96–99). Knittlingen: Bibliotheca Palatina.

Effinger, M., & Zimmermann, K. (2012). From the safes of Germany's oldest university library. In P. Meusburger & T. Schuch (Eds.), *Wissenschaftsatlas of Heidelberg University: Spatio-temporal relations of academic knowledge production* (pp. 202–203). Knittlingen: Bibliotheca Palatina.

Ehrenpreis, S. (2015). Beobachtungen zur Rolle des Heidelberger Katechismus im frühneuzeitlichen Erziehungs- und Schulwesen [Observations about the role of the Heidelberg Catechism in the early modern educational system]. In C. Strohm & J. Stievermann (Eds.), *Profil und Wirkung des Heidelberger Katechismus: Neue Forschungsbeiträge anläßlich des 450jährigen Jubiläums. The Heidelberg Catechism: Origins, characteristics, and influences: Essays in reappraisal in commemoration of its 450th anniversary* (pp. 308–319). Schriften des Vereins für Reformationsgeschichte: Vol. 215. Gütersloh: Gütersloher Verlagshaus.

Einstein, A. (1911). Bemerkungen zu dem Gesetz von Eötvös [Comments on Eötvös's law]. *Annalen der Physik*, *339*, 165–169. doi:https://doi.org/10.1002/andp.19113390109

Engehausen, F. (2012). Heidelberg University in the Vormärz and in the 1848/49 revolution. In P. Meusburger & T. Schuch (Eds.), *Wissenschaftsatlas of Heidelberg University: Spatio-temporal relations of academic knowledge production* (pp. 90–91). Knittlingen: Bibliotheca Palatina.

Eötvös, L. (1886). Über den Zusammenhang der Oberflächenspannung der Flüssigkeiten mit ihrem Molecularvolumen [On the relationship between the surface tension of liquids and molecular volume]. *Annalen der Physik, 263*, 448–459. doi:https://doi.org/10.1002/andp.18862630309

Eredics, P. (2008). *Ungarische Studenten und ihre Übersetzungen aus dem Niederländischen ins Ungarische in der Frühen Neuzeit* [Hungarian students and their translations from Dutch to Hungarian in the early modern age]. Debrecener Studien zur Literatur: Vol. 14. Frankfurt am Main: Lang.

Eulenburg, F. (1904). *Die Frequenz der deutschen Universitäten von ihrer Gründung bis zur Gegenwart* [The number of students at German universities from their founding to the present]. Abhandlungen der königlich sächsischen Gesellschaft der Wissenschaften zu Leipzig: Vol. 24, No. 2. Leipzig: Teubner.

Fassel, H. (2006). Hugo Meltzl von Lomnitz (1846–1908): Der erste Ordinarius für Germanistik in Klausenburg und seine Beziehungen zu den Universitäten in Heidelberg und Leipzig [Hugo Meltzl von Lomnitz (1846–1908): The first full professor of German studies at Klausenburg and his relations to the universities of Heidelberg and Leipzig]. In M. Fata, G. Kurucz, & A. Schindling (Eds.), *Peregrinatio Hungarica: Studenten aus Ungarn an deutschen und österreichischen Hochschulen vom 16. bis zum 20. Jahrhundert* (pp. 437–457). Contubernium–Tübinger Beiträge zur Universitäts- und Wissenschaftsgeschichte: Vol. 64. Stuttgart: Steiner.

Fata, M. (2000). *Ungarn, das Reich der Stephanskrone, im Zeitalter der Reformation und Konfessionalisierung: Multiethnizität, Land und Konfession 1500–1700* [Hungary, the lands of Saint Stephen's Crown, in the age of the Reformation and confessionalization: Multiethnicity, country, and confession, 1500–1700]. Katholisches Leben und Kirchenreform im Zeitalter der Glaubensspaltung: Vol. 60. Münster: Aschendorff.

Fata, M. (2004). Deutsche und schweizerische Einflüsse auf die Reformation in Ungarn im 16. Jahrhundert: Aspekte der frühneuzeitlich-vormodernen Identität zwischen Ethnie und Konfession [German and Swiss influences on the Reformation in Hungary in the sixteenth century: Aspects of pre- and early-modern identity between ethnicity and religious denomination]. In W. Kühlmann & A. Schindling (Eds.), *Deutschland und Ungarn in ihren Bildungs- und*

Wissenschaftsbeziehungen während der Renaissance (pp. 53–91). Contubernium–Tübinger Beiträge zur Universitäts- und Wissenschaftsgeschichte: Vol. 62. Stuttgart: Steiner.

Fata, M. (2006). Studenten aus Ungarn und Siebenbürgen an der Universität Tübingen: Eine 500 Jahre lange Beziehungs- und Wirkungsgeschichte [Students from Hungary and Transylvania at the University of Tübingen: A 500-year history of relations and impacts]. In M. Fata, G. Kurucz & A. Schindling (Eds.), *Peregrinatio Hungarica: Studenten aus Ungarn an deutschen und österreichischen Hochschulen vom 16. bis zum 20. Jahrhundert* (pp. 229–264). Contubernium–Tübinger Beiträge zur Universitäts- und Wissenschaftsgeschichte: Vol. 64. Stuttgart: Steiner.

Fata, M. (2014). *Migration im kameralistischen Staat Joseph II: Theorie und Praxis der Ansiedlungspolitik in Ungarn, Siebenbürgen, Galizien und der Bukowina von 1768 bis 1790* [Migration in the cameralistic state of Joseph II: Theory and practice of settlement policies in Hungary, Transylvania, Galicia, and Bukowina from 1768 to 1790]. Münster: Aschendorff.

Fata, M., & Millisits, M. (2010). Calvin in der ungarischen Erinnerungskultur [Calvin in the Hungarian memory culture]. In M. Fata & A. Schindling (Eds.), *Calvin und Reformiertentum in Ungarn und Siebenbürgen: Helvetisches Bekenntnis, Ethnie und Politik vom 16. Jahrhundert bis 1918* (pp. 533–564). Reformationsgeschichtliche Studien und Texte: Vol. 155. Münster: Aschendorff.

Fata, M., & Schindling, A. (2006). Peregrinatio Hungarica: Studenten aus Ungarn an deutschen und österreichischen Hochschulen vom 16. bis zum 20. Jahrhundert [Peregrinatio Hungarica: Students from Hungary at German and Austrian universities from the sixteenth to the twentieth century]. In M. Fata, G. Kurucz, & A. Schindling (Eds.), *Peregrinatio Hungarica: Studenten aus Ungarn an deutschen und österreichischen Hochschulen vom 16. bis zum 20. Jahrhundert* (pp. 3–35). Contubernium–Tübinger Beiträge zur Universitäts- und Wissenschaftsgeschichte: Vol. 64. Stuttgart: Steiner.

Fata, M., & Schindling, H. (2010). Vorbemerkung [Preliminary note]. In M. Fata & A. Schindling (Eds.), *Calvin und Reformiertentum in Ungarn und Siebenbürgen: Helvetisches Bekenntnis, Ethnie und Politik vom 16. Jahrhundert bis 1918* (pp. v–xvi). Reformationsgeschichtliche Studien und Texte: Vol. 155. Münster: Aschendorff.

Frank, G., & Treu, M. (Eds.). (2001). *Melanchthon und Europa: Bd. 1. Skandinavien und Mittelosteuropa*. Melanchthon-Schriften der Stadt Bretten: Vol. 6, Pt. 1. Stuttgart: Thorbecke.

Frank, T. (1999). Station Berlin: Ungarische Wissenschaftler und Künstler in Deutschland 1919–1933 [Berlin: Hungarian scholars and artists in Germany, 1919–1933]. *IMIS-Beiträge*, 10, 7–38.

Frank, T. (2012). Acts of creation: The Eötvös family and the rise of science education in Hungary. In M. G. Ash & J. Surman (Eds.), *The nationalization of scientific knowledge in the Habsburg Empire, 1848–1918* (pp. 118–137). Houndmills: Palgrave Macmillan.

Freller, T. (2007). Adlige auf Tour: Die Erfindung der Bildungsreise [Noblemen on tour: The invention of the educational journey]. Ostfildern: Jan Thorbecke.

Fricke, K., Koch, A., Meusburger, P., & Preusker, C. (2012). The student body 1803 to 1945. In P. Meusburger & T. Schuch (Eds.), *Wissenschaftsatlas of Heidelberg University: Spatio-temporal relations of academic knowledge production* (pp. 82–85). Knittlingen: Bibliotheca Palatina.

Gantner, B. E. (2006). Jüdische Studenten aus dem Königreich Ungarn an der Friedrich-Wilhelms-Universität in Berlin um die Wende von 19. zum 20. Jahrhundert [Jewish students from the Kingdom of Hungary at the Friedrich-Wilhelms-University in Berlin at the turn from the nineteenth to the twentieth century]. In M. Fata, G. Kurucz, & A. Schindling (Eds.), *Peregrinatio Hungarica: Studenten aus Ungarn an deutschen und österreichischen Hochschulen vom 16. bis zum 20. Jahrhundert* (pp. 459–469). Contubernium–Tübinger Beiträge zur Universitäts- und Wissenschaftsgeschichte: Vol. 64. Stuttgart: Steiner.

Garms-Cornides, E. (2005). Hofmeister auf Grand Tour [Court tutors on Grand Tour]. In R. Babel & W. Paravicini (Eds.), *Grand Tour: Adeliges Reisen und europäische Kultur vom 14. bis zum 18. Jahrhundert. Akten der internationalen Kolloquien in der Villa Vigoni 1999 und im Deutschen Historischen Institut in Paris 2000* (pp. 255–274). Beihefte der Francia: Vol. 60. Ostfildern: Jan Thorbecke.

Gebhardt, A. (2008). *Philipp Melanchthon: Praeceptor Germaniae* [Philipp Melanchthon: The teacher of Germany]. Marburg: Tectum.

Giebermann, G. (2005). Albert Molnár (1574–1634), ungarischer reformierter Theologe und Wandergelehrter, 1615–1619 Kantor und Rektor in Oppenheim [Albert Molnár (1574–1634), a Hungarian Reformed theologian and wandering scholar, 1615–1629 cantor and rector in Oppenheim]. *Oppenheimer Hefte*, *30–31*, 2–100.

Glatz, F. (Ed.). (1995). *Magyarország története térképeken elbeszélve 1: Virágkor és pusztulás. A kezdetektől 1606-ig* [The history of Hungary told in maps 1: Heyday and decay. From the beginning until 1606]. Budapest: MTA Történettudományi Intézete, História.

Gömöri, G. (1985). Hungarian students and visitors in 16th–17th century England. *Hungarian Studies, 1,* 31–50.

Gönczi, K. (2006). Juristischer Wissenstransfer von der Göttinger Universität in das Königreich Ungarn im späten 18. Jahrhundert [Knowledge transfer in the field of law from Göttingen University to the Kindom of Hungary in the late eighteenth century]. In M. Fata, G. Kurucz, & A. Schindling (Eds.), *Peregrinatio Hungarica: Studenten aus Ungarn an deutschen und österreichischen Hochschulen vom 16. bis zum 20. Jahrhundert* (pp. 175–196). Contubernium–Tübinger Beiträge zur Universitäts- und Wissenschaftsgeschichte: Vol. 64. Stuttgart: Steiner.

Gräfe, K.-H. (2004). Von der Asternrevolution zur Räterepublik: Ungarn 1918/19 [From the Aster Revolution to the Soviet Republic: Hungary 1918–1919]. *UTOPIE kreativ*, *168*, 885–900. Retrieved from http://www.rosalux.de/fileadmin/rls_uploads/pdfs/168_graefe.pdf

Gregory, D. (1999). Scripting Egypt: Orientalism and the cultures of travel. In J. Duncan & D. Gregory (Eds.), *Writes of passage: Reading travel writing* (pp. 114–150). London: Routledge.

Gregory, D. (2000). Cultures of travel and spatial formations of knowledge. *Erdkunde*, 54, 297–319. Retrieved from http://www.jstor.org/stable/25647315

Greschat, M. (2010). *Philipp Melanchthon: Theologe, Pädagoge und Humanist* [Philipp Melanchthon: Theologian, educator, and humanist]. Gütersloh: Gütersloher Verlagshaus.

Gunst, P. (1989). Agrarian systems of central and eastern Europe. In D. Chirot (Ed.), *The origins of backwardness in eastern Europe: Economics and politics from the Middle Ages to the early twentieth century* (pp. 53–91). Berkeley: University of California Press.

Győri, J. L. (2010). Zur Bedeutung des Reformierten Kollegiums Debrecen für Kultur und Politik Ungarns vom 16. bis zum 19. Jahrhundert [The importance of the Reformed College in Debrecen for the culture and politics of Hungary from the sixteenth to the nineteenth century]. In M. Fata & A. Schindling (Eds.), *Calvin und Reformiertentum in Ungarn und Siebenbürgen. Helvetisches Bekenntnis, Ethnie und Politik vom 16. Jahrhundert bis 1918* (pp. 239–259). Reformationsgeschichtliche Studien und Texte: Vol. 155. Münster: Aschendorff.

Hammerstein, N. (1995). Aufklärung und Universitäten in Europa: Divergenzen und Probleme [Enlightenment and universities in Europe: Divergencies and problems]. In N. Hammerstein (Ed.), *Universitäten und Aufklärung* (pp. 191–205). Göttingen: Wallstein.

Hammerstein, N. (1996). Epilogue: The enlightenment. In H. de Ridder-Symoens (Ed.), *Universities in early modern Europe (1500–1800)* (pp. 621–640). A history of the university in Europe: Vol. 2. Cambridge, UK: Cambridge University Press

Hebeisen, C., & Schmid, T. (2003). De Zusato, Coloniensis dioecesis: Über Herkunftsräume armer Universitätsbesucher im Alten Reich (1375 bis 1550) [De Zusato, Coloniensis dioecesis: Regional origins of poor students in the Old Empire (1375–1550)]. In R. C. Schwinges (Ed.), *Universität im Mittelalter* (pp. 28–50). Jahrbuch für Universitätsgeschichte: Vol. 6. Stuttgart: Steiner.

Heiss, G. (2005). Bildungs- und Reiseziele österreichischer Adeliger in der Frühen Neuzeit [Educational goals and destinations of Austrian noblemen in Early Modern Age]. In R. Babel & W. Paravicini (Eds.), *Grand Tour: Adeliges Reisen und europäische Kultur vom 14. bis zum 18. Jahrhundert. Akten der internationalen Kolloquien in der Villa Vigoni 1999 und im Deutschen Historischen Institut in Paris 2000* (pp. 217–235). Beihefte der Francia: Vol. 60. Ostfildern: Jan Thorbecke.

Heitmann, K. (1975). Sextil Puşcariu, Deutschland und die deutsche Wissenschaft [Sextil Puşcariu, Germany, and German science]. In P. Philippi (Ed.), *Siebenbürgen als Beispiel europäischen Kulturaustausches* (pp. 117–142). Siebenbürgisches Archiv: Vol. 12. Cologne: Böhlau.

Heller, A. (1882–1884). *Geschichte der Physik von Aristoteles bis auf die neueste Zeit* [History of physics from Aristotle to today]. 2 vols. Stuttgart: Enke.

Heltai, J. (1982). Adattár a heidelbergi egyetemen 1595–1621 között tanult magyarországi diákokról és pártfogóikról [Inventory of the Hungarian students of Heidelberg University and their patrons, 1595–1621]. *Az Országos Széchényi Könyvtár évkönyve, 1980,* 243–347.

Heltai, J. (1994). *Alvinczi Péter és a heidelbergi peregrinusok* [Péter Alvinczi and the peregrines of Heidelberg]. Budapest: Balassi Kiadó.

Heltai, J. (1999). Die Heidelberger Peregrination 1525–1621 [The Heidelberg peregrination, 1525–1621]. In A. Szabó (Ed.), *Iter Germanicum: Deutschland und die Reformierte Kirche in Ungarn im 16.–17. Jahrhundert* (pp. 169–179). Budapest: Calvin.

Heltai, J. (2006). Die Heidelberger Peregrination calvinistischer Studenten aus Ungarn und Siebenbürgen 1597–1621: Ihr Verlauf im Spiegel der Zahlen und ihre Auswirkung [The Heidelberg peregrination of Calvinist students from Hungary and Transylvania, 1597–1621: Its course as reflected by numbers and effects]. In M. Fata, G. Kurucz, & A. Schindling (Eds.), *Peregrinatio Hungarica: Studenten aus Ungarn an deutschen und österreichischen Hochschulen vom 16. bis zum 20. Jahrhundert* (pp. 65–80). Contubernium–Tübinger Beiträge zur Universitäts- und Wissenschaftsgeschichte: Vol. 64. Stuttgart: Steiner.

Henss, W. (1983). *Der Heidelberger Katechismus im konfessionspolitischen Kräftespiel seiner Frühzeit: Historisch-bibliographische Einführung* [The Heidelberg Catechism in the confessional power play of its early period: A bibliographical introduction to its history]. Zurich: Theologischer Verlag.

Höchsmann, J. (1894–1895). Zur Geschichte der Gegenreformation in Ungarn und Siebenbürgen, Teil 1 [On the history of the Counter-Reformation in Hungary and Transylvania, Part 1]. *Archiv des Vereins für Siebenbürgische Landeskunde, 26* (Neue Folge), 522–572.

Höchsmann, J. (1896–1897). Zur Geschichte der Gegenreformation in Ungarn und Siebenbürgen, Teil 2 [On the history of the Counter-Reformation in Hungary and Transylvania, part 2]. *Archiv des Vereins für Siebenbürgische Landeskunde, 27* (Neue Folge), 162–260.

Honeck, M., & Meusburger, P. (2012). American students up to 1914. In P. Meusburger & T. Schuch (Eds.), *Wissenschaftsatlas of Heidelberg University: Spatio-temporal relations of academic knowledge production* (pp. 296–299). Knittlingen: Bibliotheca Palatina.

Horn, I. (2014). Changing attitudes towards study tours among the Transylvanian elite. In G. Almási (Ed.), *Study tours and intellectual-religious relationships* (pp. 35–53). A divided Hungary in Europe: Exchanges, networks and representations, 1541–1699: Vol. 1. Cambridge, UK: Cambridge Scholar Publishing.

Hotson, H. (1995). Irenicism and dogmatics in the confessional age: Pareus and Comenius in Heidelberg, 1614. *Journal of Ecclesiastical History, 46*, 432–456. doi:https://doi.org/10.1017/S0022046900017747

Hotson, H. (2004). Irenicism in the confessional age: The Holy Roman Empire, 1563–1648. In H. Louthan & R. Zachman (Eds.), *Conciliation and confession: Struggling for unity in the age of reform, 1415–1648* (pp. 228–285). Notre Dame: University of Notre Dame Press.

Hotson, H (2008). Central Europe, 1550–1700. In D. Whitford (Ed.), *Reformation and early modern Europe: A guide to research III* (pp. 161–206). Kirksville: Truman State University Press.

Hoyler, M., & Taylor, P. J. (2012). European centres for the natural sciences, 1500 to 1900. In P. Meusburger & T. Schuch (Eds.), *Wissenschaftsatlas of Heidelberg University: Spatio-temporal relations of academic knowledge production* (pp. 76–77). Knittlingen: Bibliotheca Palatina.

Hübner, K. (2010). *Gustav Robert Kirchhoff: Das gewöhnliche Leben eines außergewöhnlichen Mannes* [Gustav Robert Kirchhoff: The ordinary life of an extraordinary man]. Schriften/Archiv und Museum der Universität Heidelberg: Vol. 16. Heidelberg: Verlag Regionalkultur.

Huttmann, A. (1975). Aspekte des Kulturaustausches in der siebenbürgischen Geschichte der Medizin und der Naturwissenschaften [Aspects of cultural exchange in the Transylvanian history of medicine and the natural sciences]. In P. Philippi (Ed.), *Siebenbürgen als Beispiel europäischen Kulturaustausches* (pp. 91–106). Siebenbürgisches Archiv: Vol. 12. Cologne: Böhlau.

Irrgang, S. (2003). Scholar vagus, goliardus, ioculator: Zur Rezeption des ‚fahrenden Scholaren' im Mittelalter [Wandering scholar, buffoon, prankster: The reception of the wandering scholar in the Middle Ages]. In R. C. Schwinges (Ed.), *Universität im Mittelalter* (pp. 51–68). Jahrbuch für Universitätsgeschichte: Vol. 6. Stuttgart: Steiner.

Jancsó, T. (2013). A magyar felsőoktatás területi szerkezetének és hallgatói létszámának alakulása 1900-tól 1945-ig [Changes in the spatial structure and number of students in Hungarian higher education between 1900 and 1945]. In Z. Bottlik (Ed.), *Önálló lépések a tudomány területén* (pp. 29–46). Budapest: Department of History, Eötvös Loránd University.

Jesse, H. (2005). *Leben und Wirken des Philipp Melanchthon: Dr. Martin Luthers theologischer Weggefährte* [Life and work of Philipp Melanchthon: Dr. Martin Luther's theological companion]. Munich: Herbert Utz.

Jöns, H., Heffernan, M., & Meusburger, P. (2017). Mobilities of knowledge: An introduction. In H. Jöns, P. Meusburger, & M. Heffernan (Eds.), *Mobilities of knowledge* (pp. 1–19). Knowledge and Space: Vol. 10. Cham: Springer. doi:https://doi.org/10.1007/978-3-319-44654-7

Juhász, T. (2010). Mirabilis est cursus verbi Dei! Die Entwicklung des Helvetischen Bekenntnisses in Ungarn und Siebenbürgen [The course of God's words is miraculous! The development of the Helvetic religious denomination in Hungary and Transylvania]. In M. Fata & A. Schindling (Eds.), *Calvin und Reformiertentum in Ungarn und Siebenbürgen: Helvetisches Bekenntnis, Ethnie und Politik vom 16. Jahrhundert bis 1918* (pp. 63–78). Reformationsgeschichtliche Studien und Texte: Vol. 155. Münster: Aschendorff.

Kadarkay, A. (1991). *Georg Lukács: Life, thought, and politics*. Cambridge, MA: Blackwell.

Kaesler, D. (1990). Mannheim, Karl. In *Neue Deutsche Biographie: Vol. 16: Maly–Melanchthon* (pp. 67–69). Berlin: Duncker & Humblot. Retrieved from http://www.deutsche-biographie.de/pnd118577190.html

Karadi, E. (1986). Der „Sonntagskreis" und die Weimarer Kultur [The "Sunday Circle" and culture in the Weimar period]. In H. Gaßner (Ed.), *Wechselwirkungen: Ungarische Avantgarde in der Weimarer Republik* (pp. 526–540). Marburg: Jonas.

Karadi, E. (2003). Karl Mannheim zwischen Kultursoziologie und Wissenssoziologie [Karl Mannheim between the sociology of culture and the sociology of knowledge]. In E. Demm & N. Chamba (Eds.), *Soziologie, Politik und Kultur. Von Alfred Weber zur Frankfurter Schule* (pp. 121–140). Frankfurt am Main: P. Lang.

Karadi, E., & Vezér, E. (Eds.). (1985). *Georg Lukács, Karl Mannheim und der Sonntagskreis* [George Lukács, Karl Mannheim, and the Sunday Circle]. Frankfurt am Main: Sendler.

Karády, V. (1997). *Felekezeti egyenlõtlenségek és iskolarendszer: Történeti-szociológiai tanulmányok* [Confessional inequalities and the school system: Studies in historical sociology]. Budapest: Replika-könyvek.

Karády, V. (2004). Der "gescheite Jude" in Ungarn vor 1919: Bildungsinvestitionen und kulturelle Assimilation [The "smart Jew" in pre-1919 Hungary: Educational investments and cultural assimilation]. In A. Kerekes, A. Millner, P. Plener, & B. Rásky (Eds.), *Leitha und Lethe. Symbolische Räume und Zeiten in der Kultur Österreich-Ungarns* (pp. 231–240). Kultur—Herrschaft—Differenz: Vol. 6. Tübingen: Francke.

Karády, V. (2012). Az értelmiség kialakulásának kezdetei a 19. századi Magyarországon. [The emergence of professional intellectuals in nineteenth century Hungary]. In V. Karády & P. T. Nagy (Eds.), *Iskolázás. értelmiség és tudomány a 19.–20. századi Magyarországon* (pp. 13–31). Budapest: Wesley János Lelkészképző Főiskola.

Karády, V. (2013). Zsidóság és modernitás [Judaism and modernity]. *Rubicon, 4,* 62–77.

Karády, V., & Nagy, P. T. (Eds.). (2012). *Iskolázás, értelmiség és tudomány a 19.–20. századi Magyarországon* [Education, intelligentsia, and science in nineteenth- and twentieth-century Hungary]. Budapest: Wesley János Lelkészképző Főiskola.

Kardos, J. (Ed.). (2000). *A magyar felsőoktatás évszázadai* [Centuries of Hungary's higher education]. Budapest: Nemzeti Tankönyvkiadó.

Kármán, G. (2006). Identitás és határok: 17. századi utazók [Identity and borders: Travelers in the seventeenth century]. *Korall, 26,* 73–95.

Kármán, G. (2013). Gábor Bethlen's diplomats at the Protestant courts of Europe. *Hungarian Historical Review, 2,* 790–823.

Keserű, B. (Ed.). (1989). *Teleki Pál külföldi tanulmányútja: Levelek, számadások, iratok 1695–1700* [The study tour of Pál Teleki: Letters, accounts, documents, 1695–1700]. *Fontes rerum scolasticarum III.* Szeged: Szegedi Oktatástörténeti Munkaközösség

Kettler, D., Meja, V., & Stehr, N. (1989). *Politisches Wissen: Studien zu Karl Mannheim* [Political knowledge: Studies about Karl Mannheim]. Frankfurt am Main: Suhrkamp.

Kintzinger, M. (2012). Universität [University]. In P. den Boer, H. Duchhardt, G. Kreis, & W. Schmale (Eds.), *Europäische Erinnerungsorte 2: Das Haus Europa* (pp. 307–312). Munich: Oldenbourg

Kohnle, A. (2001). Die Universität Heidelberg als Zentrum des reformierten Protestantismus im 16. und frühen 17. Jahrhundert [Heidelberg university as a center of Reformed Protestantism in the sixteenth and early seventeenth centuries]. In L. Szögi & M. Font (Eds.), *Die Ungarische Universitätsbildung und Europa* (pp. 141–161). Pécs: Universitas Quinqueecclesiensis.

Kohnle, A., & Wolgast, E. (2012). The Heidelberg Catechism of 1563: Origin and distribution. In P. Meusburger & T. Schuch (Eds.), *Wissenschaftsatlas of Heidelberg University: Spatio-temporal relations of academic knowledge production* (pp. 58–59). Knittlingen: Bibliotheca Palatina.

König, G. (1903). *Einleitung in die allgemeine Theorie der algebraischen Größen* [Introduction to the general theory of algebraic magnitudes]. Leipzig: Teubner.

Köpeczi, B. (Ed.). (1990). *Kurze Geschichte Siebenbürgens* [A brief history of Transylvania]. Budapest: Akadémiai Kiadó. Retrieved from http://www.mek.oszk.hu/02100/02113/html

Köpf, U. (2010a). Melanchthon und die Reform der Universität Tübingen [Melanchthon and the reform of the University of Tübingen]. In S. Lorenz, R. Rieger, E. Seidl, & K. Wiegmann (Eds.), *Vom Schüler der Burse zum "Lehrer Deutschlands": Philipp Melanchthon in Tübingen* (pp. 187–195). Tübinger Kataloge: Vol. 88. Tübingen: Universitätsstadt Tübingen, Fachbereich Kultur, Stadtmuseum

Köpf, U. (2010b). Philipp Melanchthon: Leben und Werk [Philipp Melanchthon: Life and work]. In S. Lorenz, R. Rieger, E. Seidl, & K. Wiegmann (Eds.), *Vom Schüler der Burse zum "Lehrer Deutschlands": Philipp Melanchthon in Tübingen* (pp. 13–49). Tübinger Kataloge: Vol. 88. Tübingen: Universitätsstadt Tübingen, Fachbereich Kultur, Stadtmuseum.

Köpf, U. (2010c). Der Reformer Philipp Melanchthon [Philipp Melanchthon the reformer]. In F. Schweitzer, S. Lorenz, & E. Seidl (Eds.), *Philipp Melanchthon: Seine Bedeutung für Kirche und Theologie, Bildung und Wissenschaft* (pp. 11–26). Theologie Interdisziplinär: Vol. 8. Neukirchen-Vluyn: Neukirchener Verlagsgesellschaft.

Kovács, A. (1922). *A zsidóság térfoglalása Magyarországon* [The Jewish expansion in Hungary]. Budapest: Author's edition. Retrieved from http://www.angelfire.com/zine2/judeolog/cikkek/zst1.htm

Központi Statisztikai Hivatal (KSH, Central Statistical Office of Hungary). (1920). *A Magyar Szent Korona Országainak 1910: Évi népszámlálása VI. Végeredmények összefoglalása* [Census of the states of the Holy Crown of Hungary, 1910: Vol. VI. Summary of final results]. Budapest: Athenaeum. Retrieved from https://library.hungaricana.hu/hu/view/NEDA_1910_06/?pg=0&layout=s

Kühlmann, W., Hartmann, V., & El Kholi, S. (2012). The golden age of late Humanism (ca. 1560 to 1622). In P. Meusburger & T. Schuch (Eds.), *Wissenschaftsatlas of Heidelberg University: Spatio-temporal relations of academic knowledge production* (pp. 60–61). Knittlingen: Bibliotheca Palatina.

Kühnel, H. (1964). Die adelige Cavaliers-Tour im 17. Jahrhundert [The nobleman's Grand Tour in the seventeenth century]. *Jahrbuch für Landeskunde Niederösterreichs*, 36 (Neue Folge), 364–384.

Kuzias, T. (2007). Georg Lukács und der Goethepreis: Zum folgenreichen Verhältnis von Nietzsche und Marx beim jungen Lukács [George Lukács and the Goethe award: On the portentous relationship between Nietzsche and Marx for the young Lukács]. *Beiträge zur Geschichte der Arbeiterbewegung*, 49, 111–134.

Ladányi, S. (1997). Peregrination ungarischer reformierter Studenten an westeuropäische Universitäten im 16.–17. Jahrhundert: Westeuropäische Einflüsse in der ungarischen Reformierten Kirche, mit besonderem Hinblick auf Holland [Peregrination of Hungarian Reformed students to western European universities in the sixteenth and seventeenth centuries: Western European influences in the Hungarian Reformed church with special view on Holland]. In L. Szögi & J. Varga (Eds.), *Universitas Budensis, 1395–1995: International conference for the history of universities on the occasion of the 600th anniversary of the foundation of the University of Buda* (pp. 199–209). Budapest: ELTE, Library and Archive.

Ladányi, S. (1999). Ungarische Studenten in Frankfurt an der Oder [Hungarian students in Frankfurt at the Oder]. In A. Szabó (Ed.), *Iter Germanicum: Deutschland und die Reformierte Kirche in Ungarn im 16.–17. Jahrhundert* (pp. 215–220). Budapest: Calvin.

Ladányi, S. (2001). Bedeutung und Wirkung der protestantischen Kollegs und des Auslandsstudiums im Ungarn des 16. bis 18. Jahrhunderts [Importance and impact of Protestant colleges and of studies abroad in Hungary of the sixteenth to eighteenth centuries]. In L. Szögi & M. Font (Eds.), *Die Ungarische Universitätsbildung und Europa* (pp. 133–139). Pécs: Universitas Quinqueecclesiensis.

Leibetseder, M. (2004). *Die Kavalierstour: Adlige Erziehungsreisen im 17. und 18. Jahrhundert* [The Grand Tour: Noblemen's educational travel in the seventeenth and eighteenth centuries]. Cologne: Böhlau.

Leibetseder, M. (2013). Educational journey, Grand Tour. *European History Online (EGO)*. Mainz: Leibniz Institute of European History (IEG). Retrieved from http://www.ieg-ego.eu/leibetsederm-2013-en

Leppin, V., & Wien, U. A. (Eds.). (2005). *Konfessionsbildung und Konfessionsstruktur in Siebenbürgen in der frühen Neuzeit* [Denominational education and denominational structure in Transylvania in the Early Modern Age]. Wiesbaden: Steiner.

Lepsius, R. M. (2012). Intellectual circles at the beginning of the 20th century. In P. Meusburger & T. Schuch (Eds.), *Wissenschaftsatlas of Heidelberg University: Spatio-temporal relations of academic knowledge production* (pp. 112–113). Knittlingen: Bibliotheca Palatina.

Loewy, H. (1999). *Medium und Initiation: Béla Balázs. Märchen, Ästhetik, Kino* [Medium and initiation: Béla Balázs. Fairy tales, aesthetics, cinema] (Doctoral thesis, Konstanz University, Germany). Retrieved from https://kops.unikonstanz de/bitstream/handle/123456789/3615/400_1.pdf;sequence=1

Lorenz, S. (2010). Melanchthon in Tübingen (1512–1518). In S. Lorenz, R. Rieger, E. Seidl, & K. Wiegmann (Eds.), *Vom Schüler der Burse zum „Lehrer Deutschlands": Philipp Melanchthon in Tübingen* (pp. 83–103). Tübinger Kataloge: Vol. 88. Tübingen: Universitätsstadt Tübingen, Fachbereich Kultur, Stadtmuseum.

Ludwig, U. (2009). *Philippismus und orthodoxes Luthertum an der Universität Wittenberg* [Philippism and orthodox Lutherism at the University of Wittenberg]. Reformatische Studien und Texte: Vol. 153. Münster: Aschendorff.

Lukács, G. (1911). *Die Seele und die Formen* [Soul & form]. Berlin: Egon Fleischel.

Lukács, G. (1920). *Die Theorie des Romans: Ein geschichtsphilosophischer Versuch über die Formen der großen Epik* [The theory of the novel: A historico-philosophical essay on the forms of great epic literature]. Berlin: Cassirer.

Lukács, G. (2010). *Soul & form* (J. T. Sanders & K. Terezakis, Eds.; A. Bostock, Trans.; with an introduction by J. Butler). New York: Columbia University Press. (Original work published 1911).

Maleczek, W. (1995). Deutsche Studenten an Universitäten in Italien [German students at universities in Italy]. In S. de Rachewiltz & J. Riedmann (Eds.), *Kommunikation und Mobilität im Mittelalter: Begegnungen zwischen dem Süden und der Mitte Europas (11.–14. Jahrhundert)* (pp. 77–96). Sigmaringen: Thorbecke.

Markó, L. (Ed.). (2001–2007). *Új Magyar életrajzi lexikon* [New Hungarian biographical lexicon] (Vols. 1–6). Budapest: Magyar Könyvklub-Helikon.

Marshall, R. K. (2003). Elizabeth Stuart—die Winterkönigin [Elizabeth Stuart—The winter queen]. In P. Wolf, M. Henker, E. Brockhoff, B. Steinherr, & S. Lippold (Eds.), *Der Winterkönig Friedrich V: Der letzte Kurfürst aus der Pfalz. Amberg, Heidelberg, Prag, Den Haag* (pp. 34–45). Veröffentlichungen zur Bayerischen Geschichte und Kultur: 46/03. Augsburg: Haus der Bayerischen Geschichte.

Matrikelbuch der Universität Heidelberg: Sommer-Semester 1871 bis Sommer-Semester 1872 [Heidelberg University registry, summer semester 1871 through summer semester 1872]. Retrieved from http://digi.ub.uni-heidelberg.de/diglit/matrikel1871

Matrikelbuch der Universität Heidelberg: Winter-Semester 1872/73 bis Winter-Semester 1894/95 [Heidelberg University registry, winter semester 1872–1873 through winter semester 1894–1895]. Retrieved from http://digi.ub.uni-heidelberg.de/diglit/matrikel1872

Matrikelbuch der Universität Heidelberg: Sommer-Semester 1895 bis Sommer-Semester 1906 [Heidelberg University registry: Summer semester 1895 through summer semester 1906]. Retrieved from http://digi.ub.uni-heidelberg.de/matrikel1895

Matrikelbuch der Universität Heidelberg: Winter-Semester 1906/1907 bis Sommer-Semester 1916 [Heidelberg University registry: Winter semester 1906–1907 to summer semester 1916]. Retrieved from http://digi.ub.uni-heidelberg.de/matrikel1907

Matrikelbuch der Universität Heidelberg: Kriegs-Wintersemester 1916/1917 bis Sommer-Semester 1920 [Heidelberg University registry: War-winter semester 1916–1917 to summer semester 1920]. Retrieved from http://digi.ub.uni-heidelberg.de/matrikel1916

Merkel, G. (2012). University property and income up to the end of the 17th century. In P. Meusburger & T. Schuch (Eds.), *Wissenschaftsatlas of Heidelberg University: Spatio-temporal relations of academic knowledge production* (pp. 46–47). Knittlingen: Bibliotheca Palatina.

Meusburger, P. (2003). Entwicklung der wissenschaftlichen Zusammenarbeit zwischen den Geographischen Instituten der Eötvös Loránd Universität in Budapest und der Universität Heidelberg: Einige persönlichen Anmerkungen [Thedevelopment of scientific cooperation between the Departments of Geography of Eötvös Loránd University and Heidelberg University: Some personal comments]. In J. Nemes Nagy (Ed.), *Frontiers of geography: The 20th anniversary of the partnership between Eötvös Loránd Tudományegyetem Budapest and Ruprecht Karls Universität Heidelberg* (pp. 19–31). Budapest: Department of Regional Geography, Eötvös Loránd University.

Meusburger, P. (2010). *Ungarische Studenten an der Universität Heidelberg: Einflussfaktoren und Wechselwirkungen der Bildungswanderung im Laufe der Jahrhunderte* [Hungarian students at Heidelberg University: Factors and interactions influencing educational migrations over the centuries]. Budapest: Eötvös Kiadó.

Meusburger, P. (2012). Heidelberg: Relations between a town and its university. In P. Meusburger & T. Schuch (Eds.), *Wissenschaftsatlas of Heidelberg University: Spatio-temporal relations of academic knowledge production* (pp. 18–35). Knittlingen: Bibliotheca Palatina.

Meusburger, P. (2013). Relations between knowledge and economic development: Some methodological considerations. In P. Meusburger, J. Glückler, & M. El Meskioui (Eds.), *Knowledge and the economy* (pp. 15–42). Knowledge and Space: Vol. 5. Dordrecht: Springer. doi:https://doi.org/10.1007/978-94-007-6131-5_2

Meusburger, P., & Probáld, F. (2010). Magyar diákok a Heidelbergi Egyetemen: A kulturális kölcsönhatás történeti és földrajzi tényezöi [Hungarian students at Heidelberg University: Historical and geographical factors of cultural interaction]. *Földrajzi Közlemények, 134*, 59–73.

Meusburger, P., & Probáld, F. (2012). Hungarian students at Heidelberg University up to 1919. In P. Meusburger & T. Schuch (Eds.), *Wissenschaftsatlas of Heidelberg University: Spatio-temporal relations of academic knowledge production* (pp. 304–305). Knittlingen: Bibliotheca Palatina.

Meusburger, P., & Schuch, T. (2010). From mediocrity and existential crisis to scientific excellence: Heidelberg University between 1803 and 1932. In P. Meusburger, D. N. Livingstone, & H. Jöns (Eds.), *Geographies of science* (pp. 57–93). Knowledge and Space: Vol. 3. Dordrecht: Springer. doi:https://doi.org/10.1007/978-90-481-8611-2_4

Meusburger, P., & Schuch, T. (Eds.). (2012). *Wissenschaftsatlas of Heidelberg University: Spatio-temporal relations of academic knowledge production*. Knittlingen: Bibliotheca Palatina.

Miethke, J. (1985). Die Studenten [Students]. In P. Moraw (Ed.), *Unterwegssein im Spätmittelalter* (pp. 49–70). Berlin: Duncker & Humblot.

Miethke, J. (1995). Die mittelalterliche Universität in der Gesellschaft [The medieval university in society]. In U. Weiß (Ed.), *Erfurt: Geschichte und Gegenwart* (pp. 169–188). Schriften des Vereins für die Geschichte und Altertumskunde von Erfurt: Vol. 2. Weimar: Hermann Böhlaus Nachfolger.

Miethke, J. (2004). *Studieren an mittelalterlichen Universitäten: Chancen und Risiken* [Studying at medieval universities: Opportunities and risks]. Education and Society in the Middle Ages and Renaissance: Vol. 19. Leiden: Brill.

Millisits, M. (2012). *A Heidelbergi Káté 450 éve* [The Heidelberg Catechism: 450 years]. Retrieved from http://www.reformatus.hu/mutat/7512

Módis, L. (1967). Bibliographie der ungarischen Literatur des Heidelberger Katechismus 1563–1963 [Bibliography of the Hungarian literature on the Heidelberg Catechism, 1563–1963]. In T. Bartha (Ed.), *Der Heidelberger Katechismus in Ungarn* (pp. 331–383). Budapest: Presseabteilung der Synodalkanzlei der reformierten Kirche in Ungarn.

Moraw, P. (1982). Aspekte und Dimensionen älterer deutscher Universitätsgeschichte [Aspects and dimensions of early German university history]. In P. Moraw & V. Press (Eds.), *Academia Gissensis: Beiträge zur älteren Gießener Universitätsgeschichte* (pp. 1–43). Veröffentlichungen der Historischen Kommission für Hessen: Vol. 45. Marburg: Elwert.

Moraw, P. (1983). Heidelberg: Universität, Hof und Stadt im ausgehenden Mittelalter [Heidelberg: University, court, and town in the late Middle Ages]. In B. Moeller, H. Patze, & K. Stackmann (Eds.), *Studien zum städtischen Bildungswesen des späten Mittelalters und der frühen Neuzeit* (pp. 524–552). Abhandlungen der Akademie der Wissenschaften in Göttingen: Philosophisch-historische Klasse, 3. Folge: Vol. 137. Göttingen: Vandenhoeck & Ruprecht.

Moraw, P. (1994). Einheit und Vielfalt der Universität im alten Europa [Unity and diversity of the university in old Europe]. In A. Patschovski & H. Rabe (Eds.), *Die Universität in Alteuropa* (pp. 11–27). Konstanzer Bibliothek: Vol. 22. Konstanz: Universitätsverlag.

Mühlen, K.-H. zur (1985). Die Heidelberger Disputation Martin Luthers vom 26. April 1518: Programm und Wirkung [The Heidelberg disputation of Martin Luther on April 26, 1518: Program and impact]. In W. Doerr (Ed.), *Semper Apertus: 600 Jahre Ruprecht-Karl-Universität Heidelberg 1386–1986* (Vol. 1, pp. 188–212). Berlin: Springer.

Müller, R. A. (1984). Aristokratisierung des Studiums: Bemerkungen zur Adelsfrequenz an süddeutschen Universitäten im 17. Jahrhundert [Feudalization of study at the university: Comments on the percentage of aristocrats in the student body of southern German universities]. *Geschichte und Gesellschaft: Zeitschrift für Historische Sozialwissenschaften, 10*, 31–46.

Murdock, G. (2000). *Calvinism on the frontier, 1600–1660: International Calvinism and the Reformed church in Hungary and Transylvania*. Oxford, UK: Clarendon.

Musall, H., & Scheuerbrandt, A. (1980). Siedlungszerstörungen und Festungswerke im späten 17. und frühen 18. Jahrhundert (1674–1714) [Destruction of settlements and fortifications in the late seventeenth and early eighteenth centuries (1674–1714)]. In *Kommission für Geschichtliche Landeskunde in Baden-Württemberg (Ed.), Historischer Atlas von Baden-Württemberg. Beiwort zur Karte VI-12* (pp. 1–21). Stuttgart: Kommission für Geschichtliche Landeskunde in Baden-Württemberg.

Mussgnug, D. (2003). Die Wiederemporbringung der Heidelberger Universität 1803 [The revival of Heidelberg University, 1803]. In A. Kohnle, F. Engehausen, F. Hepp, & K. L. Fuchs (Eds.), *„... so geht hervor ein' neue Zeit": Die Kurpfalz im Übergang an Baden 1803"* (pp. 131–145). Heidelberg: Regionalkultur.

Nagy, B. (1967). Geschichte und Ausgaben des Heidelberger Katechismus in Ungarn im 16. und 17. Jahrhundert [History and editions of the Heidelberg Catechism in Hungary in the sixteenth and seventeenth centuries]. In T. Bartha (Ed.), *Der Heidelberger Katechismus in Ungarn* (pp. 29–82). Budapest: Presseabteilung der Synodalkanzlei der reformierten Kirche in Ungarn.

Naylor, S., & Ryan, J. R. (Eds.). (2010). *New spaces of exploration: Geographies of discovery in the twentieth century*. London: I. B. Tauris.

Neumaier, M. (2012a). Destruction and rebuilding: The university between 1689 and 1803. In P. Meusburger & T. Schuch (Eds.), *Wissenschaftsatlas of Heidelberg University: Spatio-temporal relations of academic knowledge production* (pp. 72–75). Knittlingen: Bibliotheca Palatina.

Neumaier, M. (2012b). Regional provenance and living conditions of students and lecturers ca. 1588. In P. Meusburger & T. Schuch (Eds.), *Wissenschaftsatlas of Heidelberg University: Spatio-temporal relations of academic knowledge production* (pp. 70–71). Knittlingen: Bibliotheca Palatina.

Offner, R. (2006). Deutsche Universitäten als Ausbildungsstätten siebenbürgischer Mediziner von den Anfängen bis zum Jahr 1850 [German universities as training institutions of Transylvanian physicians from their beginning to the year1850]. In M. Fata, G. Kurucz, & A. Schindling (Eds.), *Peregrinatio Hungarica: Studenten aus Ungarn an deutschen und österreichischen Hochschulen vom 16. bis zum 20. Jahrhundert* (pp. 287–343). Contubernium–Tübinger Beiträge zur Universitäts- und Wissenschaftsgeschichte: Vol. 64. Stuttgart: Steiner.

Pánek, J. (2003). Friedrich V von der Pfalz als König von Böhmen [Friedrich V of the Palatinate as king of Bohemia]. In P. Wolf, M. Henker, E. Brockhoff, B. Steinherr, & S. Lippold (Eds.), *Der Winterkönig Friedrich V, Der letzte Kurfürst aus der Pfalz. Amberg, Heidelberg, Prag, Den Haag* (pp. 101–106). Veröffentlichungen zur Bayerischen Geschichte und Kultur: 46/03. Augsburg: Haus der Bayerischen Geschichte.

Pánek, J., & Polívka, M. (2005). Die böhmischen Adelsreisen und ihr Wandel vom Mittelalter zur Neuzeit [The Cavalier's Tours of Bohemian noblemen and their change from the Middle Ages to the Modern Age]. In R. Babel & W. Paravicini (Eds.), *Grand Tour: Adeliges Reisen und europäische Kultur vom 14. bis zum 18. Jahrhundert. Akten der internationalen Kolloquien in der Villa Vigoni 1999 und im Deutschen Historischen Institut in Paris 2000* (pp. 53–69). Beihefte der Francia: Vol. 60. Ostfildern: Jan Thorbecke.

Paravicini, W. (2005). Die Grand Tour in der europäischen Geschichte: Zusammenfassung [A summary of the Grand Tour in European history]. In R. Babel & W. Paravicini (Eds.), *Grand Tour: Adeliges Reisen und europäische Kultur vom 14. bis zum 18. Jahrhundert. Akten der internationalen Kolloquien in der Villa Vigoni 1999 und im Deutschen Historischen Institut in Paris 2000* (pp. 657–674). Beihefte der Francia: Vol. 60. Ostfildern: Jan Thorbecke.

Pareus, D. (1614). *Irenicum sive de unione et synodo evangelicorum concilianda liber votivus, paci ecclesiae et desideriis pacificorum dictatus a Davide Pareo* [Peacemaker, or an assembly for reconciling the Evangelicals, a book dedicated by David Pareus to the peace of the church and the desire of peace-loving people]. Heidelberg: Jonae Rosae librarij. Retrieved from http://reader.digitale-sammlungen.de/de/fs1/object/display/bsb10360874_00005.html

Pfetsch, F. R. (1974). *Zur Entwicklung der Wissenschaftspolitik in Deutschland 1750–1914* [The development of science policy in Germany, 1750–1914]. Berlin: Duncker & Humblot.

Pukánszky, B., & Németh, A. (1996). *Neveléstörténet [History of education]*. Budapest: Nemzeti Tankönyvkiadó.

Rasche, U. (2006). Von Fichte zu Metternich: Die Universität Jena und ihre ungarländischen Studenten um 1800. [From Fichte to Metternich: Jena University and its students from Hungary at 1800]. In M. Fata, G. Kurucz, & A. Schindling (Eds.), *Peregrinatio Hungarica: Studenten aus Ungarn an deutschen und österreichischen Hochschulen vom 16. bis zum 20. Jahrhundert* (pp. 197–226). Contubernium–Tübinger Beiträge zur Universitäts- und Wissenschaftsgeschichte: Vol. 64. Stuttgart: Steiner.

Reich, K. (2010). Johannes Stöffler: Melanchthons Tübinger Lehrer in Mathematik und Astronomie [Johannes Stöffler: Melanchthon's mathematics and astronomy teacher in Tübingen]. In S. Lorenz, R. Rieger, E. Seidl, & K. Wiegmann (Eds.), *Vom Schüler der Burse zum „Lehrer Deutschlands": Philipp Melanchthon in Tübingen* (pp. 139–151). Tübinger Kataloge: Vol. 88. Tübingen: Universitätsstadt Tübingen, Fachbereich Kultur, Stadtmuseum.

Richarz, M. (1974). *Der Eintritt der Juden in die akademischen Berufe: Jüdische Studenten und Akademiker in Deutschland 1678–1848* [The entry of Jews into the academic professions: Jewish students and academics in Germany, 1678–1848]. Tübingen: Mohr.

Rieger, R. (2010). Melanchthons Weg zur reformatorischen Theologie [Melanchthon's path to Reformed theology]. In S. Lorenz, R. Rieger, E. Seidl, & K. Wiegmann (Eds.), *Vom Schüler der Burse zum „Lehrer Deutschlands": Philipp Melanchthon in Tübingen* (pp. 153–159). Tübinger Kataloge: Vol. 88. Tübingen: Universitätsstadt Tübingen, Fachbereich Kultur, Stadtmuseum.

Röhrs, H. (1971). Die Studienzeit des Comenius in Heidelberg [Comenius as a student at Heidelberg University]. *Heidelberger Jahrbücher, 15,* 1–12.

Rotzoll, M. (2012). Plagues and wars in the early history of Heidelberg University. In P. Meusburger & T. Schuch (Eds.), *Wissenschaftsatlas of Heidelberg University: Spatio-temporal relations of academic knowledge production* (pp. 52–53). Knittlingen: Bibliotheca Palatina.

Scheible, H. (1983). Die Universität Heidelberg und Luthers Disputation [Heidelberg University and Luther's disputation]. *Zeitschrift für die Geschichte des Oberrheins, 131, 92* (Neue Folge), 309–329.

Scheible, H. (1985). Melanchthons Beziehungen zum Donau-Karpaten-Raum bis 1546 [Melanchthon's relations to the Danube-Carpathian area until 1546]. In G. Weber & R. Weber (Eds.), *Luther und Siebenbürgen: Ausstrahlungen von Reformation und Humanismus nach Südosteuropa* (pp. 36–67). Siebenbürgisches Archiv: Vol. 19. Cologne: Böhlau.

Schilling, H. (2010). Calvin und Calvinismus in europageschichtlicher Perspektive [Calvin and Calvinism from a European historical perspective]. In M. Fata & A. Schindling (Eds.), *Calvin und Reformiertentum in Ungarn und Siebenbürgen: Helvetisches Bekenntnis, Ethnie und Politik vom 16. Jahrhundert bis 1918* (pp. 1–21). Reformationsgeschichtliche Studien und Texte: Vol. 155. Münster: Aschendorff.

Schindling, A. (2006). Bildungsinstitutionen im Heiligen Römischen Reich deutscher Nation als Ziele der studentischen Migration: Wanderungen im Zeichen von Konfessionen und geistigen Strömungen [Educational institutions in the Holy Roman Empire of the German Nation as destinations of student migration: Migrations marked by denominations and spiritual trends]. In M Fata, G. Kurucz, & A. Schindling (Eds.), *Peregrinatio Hungarica: Studenten aus Ungarn an deutschen und österreichischen Hochschulen vom 16. bis zum 20. Jahrhundert* (pp. 39–54). Contubernium–Tübinger Beiträge zur Universitäts- und Wissenschaftsgeschichte: Vol. 64. Stuttgart: Steiner.

Schmidt-Rösler, A. (2006). Princeps Transilvaniae—Rex Hungariae? Gabriel Bethlens Außenpolitik zwischen Krieg und Frieden [Prince of Transylvania—King of Hungary? Gabriel Bethlen's foreign policy between war and peace]. In Duchhardt H & M Peters (Eds.), *Kalkül—Transfer—Symbol: Europäische Friedensverträge der Vormoderne* (pp. 80–98). Veröffentlichungen des Instituts für Europäische Geschichte, Mainz: Beiheft online 1. Retrieved from http://www.iegmainz.de/vieg-online-beihefte/01-2006.html

Schönebaum, H. (1926). Die ungarischen Universitäten im Mittelalter [The Hungarian universities in medieval times]. *Archiv für Kulturgeschichte, 16,* 41–59.

Schweitzer, F., Lorenz, S., & Seidl, E. (Eds.). (2010). *Philipp Melanchthon: Seine Bedeutung für Kirche und Theologie, Bildung und Wissenschaft* [Philipp Melanchton: His significance for the church and theology and for education and science]. Theologie Interdisziplinär: Vol. 8. Neukirchen-Vluyn: Neukirchener Verlagsgesellschaft.

Schwinges, R. C. (1982). Immatrikulationsfrequenz und Einzugsbereich der Universität Gießen 1650–1800: Zur Grundlegung einer Sozialgeschichte Gießener Studenten [Registration rate and catchment area of the University of Gießen, 1650–1800: Establishing the basis for a social history of Gießen's students]. In P. Moraw & V. Press (Eds.), *Academia Gissensis: Beiträge zur älteren Gießener Universitätsgeschichte* (pp. 247–295). Veröffentlichungen der Historischen Kommission für Hessen: Vol. 45. Marburg: Elwert.

Schwinges, R. C. (1984). Universitätsbesuch im Reich vom 14. zum 16. Jahrhundert: Wachstum und Konjunkturen [University attendance in the empire between the fourteenth and sixteenth centuries: Growth and fluctuation]. *Geschichte und Gesellschaft: Zeitschrift für Historische Sozialwissenschaften, 10,* 5–30.

Schwinges, R. C. (1986). *Deutsche Universitätsbesucher im 14. und 15. Jahrhundert: Studien zur Sozialgeschichte des Alten Reich*es [German university students in the fourteenth and fifteenth centuries: Studies on the social history of the old empire]. Veröffentlichungen des Instituts für Europäische Geschichte Mainz: Vol. 123, Beiträge zur Sozial- und Verfassungsgeschichte des Alten Reiches: Vol. 6. Stuttgart: Steiner

Schwinges, R. C. (1988). Migration und Austausch: Studentenwanderungen im Deutschen Reich des Mittelalters [Migration and exchange: Student migration in the medieval German empire]. In G. Jaritz & A. Müller (Eds.), *Migration in der Feudalgesellschaft* (pp. 141–155). Frankfurt am Main & New York: Campus.

Schwinges, R. C. (1994). Franken in der deutschen Universitätslandschaft des späten Mittelalters [Franconia in the German university landscape of the late Middle Ages]. In H.-A. Steger & H. Hopfinger (Eds.), *Die Universität in der Welt—Die Welt in der Universität* (pp. 1–26). Neustadt an der Aisch: Degener.

Schwinges, R. C. (1998). The medieval German university: Transformation and innovation. *Paedagogica Historica: International Journal of the History of Education, 34,* 375–388.

Schwinges, R. C. (2000). On recruitment in German universities from the fourteenth to sixteenth centuries. In W. J. Courtenay & J. Miethke (Eds.), *Universities and schooling in medieval society* (pp. 32–48). Education and Society in the Middle Ages and Renaissance: Vol. 10. Leiden: Brill.

Schwinges, R. C. (2001). Innovationsräume und Universitäten in der älteren deutschen Vormoderne [Innovative spaces and universities in early German premodernity]. In R. C. Schwinges, P. Messerli, & T. Münger (Eds.), *Innovationsräume: Woher das Neue kommt—In Vergangenheit und Gegenwart* (pp. 31–44). Zürich: vdf Hochschulverlag AG an der ETH Zürich.

Schwinges, R. C. (2005). Die Universität als sozialer Ort des Adels [The university as a social space of the nobility]. In R. Babel & W. Paravicini (Eds.), *Grand Tour: Adeliges Reisen und europäische Kultur vom 14. bis zum 18. Jahrhundert. Akten der internationalen Kolloquien in der Villa Vigoni 1999 und im Deutschen Historischen Institut in Paris 2000* (pp. 357–372). Beihefte der Francia: Vol. 60. Ostfildern: Jan Thorbecke.

Schwinges, R. C. (2006). Das Reich im gelehrten Europa: Ein Essay aus personengeschichtlicher Perspektive [The Reich in scholarly Europe: An essay from the personal history perspective]. In B. Schneidmüller & S. Weinfurter (Eds.), *Heilig–Römisch–Deutsch: Das Reich im mittelalterlichen Europa* (pp. 227–250). Dresden: Michael Sandstein.

Sebők, L. (2013). A magyarországi zsidók a számok tükrében [The Jews in Hungary in the mirror of the numbers]. *Rubiconline, 4.* Retrieved from http://www.rubicon.hu/magyar/oldalak/a_magyarorszagi_zsidok_a_szamok_tukreben/

Seebaß, G. (1983). Die Heidelberger Disputation [The Heidelberg disputation]. *Heidelberger Jahrbücher, 27,* 77–88.

Seidel, R. (2004). Der ungarische Späthumanismus und die calvinistische Pfalz [Hungarian late Humanism and the Calvinist Palatinate]. In W. Kühlmann & A. Schindling (Eds.), *Deutschland und Ungarn in ihren Bildungs- und Wissenschaftsbeziehungen während der Renaissance* (pp. 227–251). Contubernium–Tübinger Beiträge zur Universitäts- und Wissenschaftsgeschichte: Vol. 62. Stuttgart: Steiner.

Siebe, D. (2006). Studierende aus Ungarn an den Universitäten Berlin und Heidelberg zwischen 1870 und 1932/33 [Students from Hungary at the universities of Berlin and Heidelberg between 1870 and 1932–1933]. In M. Fata, G. Kurucz, & A. Schindling (Eds.), *Peregrinatio Hungarica. Studenten aus Ungarn an deutschen und österreichischen Hochschulen vom 16. bis zum 20. Jahrhundert* (pp. 409–436). Contubernium–Tübinger Beiträge zur Universitäts- und Wissenschaftsgeschichte: Vol. 64. Stuttgart: Steiner.

Sienerth, S. (2007). Leseangebot und Buchzirkulation in Siebenbürgen zwischen Humanismus und Aufklärung [Supply and circulation of books in Transylvania between Humanism and the Enlightenment]. In D. Haberland (Ed.), *Buch- und Wissenstransfer in Ostmittel- und Südosteuropa in der Frühen Neuzeit* (pp. 281–309). Munich: Oldenbourg.

Sipos, G. (2010). Zur Bedeutung des Reformierten Kollegiums Nagyenyed für die siebenbürgische Kultur vom 17. bis zum 19. Jahrhundert [The importance of the Reformed college Nagyenyed for Transylvanian culture from the seventeenth to the nineteenth century]. In M. Fata & A. Schindling (Eds.), *Calvin und Reformiertentum in Ungarn und Siebenbürgen: Helvetisches Bekenntnis, Ethnie und Politik vom 16. Jahrhundert bis 1918* (pp. 261–281). Reformationsgeschichtliche Studien und Texte: Vol. 155. Münster: Aschendorff.

Stagl, J. (1995). *A history of curiosity: The theory of travel 1550–1800*. Studies in Anthropology and History: Vol. 13. Chur: Harwood Academic Publishers.

Stempel, H.-A. (1993). Melanchthon. In T. Bautz (Ed.), *Biographisch-Bibliographisches Kirchenlexikon: Vol. 5.* Leyden, Nikolaus to Mönch, Antonius (pp. 1184–1188). Hamm: Bautz. http://www.bbkl.de.ubproxy.ub.uni-heidelberg.de/lexikon/bbkl-artikel.php?wt=1&art=./M/Mc-Me/melanchton.art

Stickler, M. (2006). Die Selbstorganisation der Studenten aus dem Königreich Ungarn an deutschen und österreichischen Hochschulen zwischen 1871 und 1918 [The self-organization of students from the Kingdom of Hungary at German and Austrian universities between 1871 and 1918]. In M. Fata, G. Kurucz, & A. Schindling (Eds.), *Peregrinatio Hungarica: Studenten aus Ungarn an deutschen und österreichischen Hochschulen vom 16. bis zum 20. Jahrhundert* (pp. 471–503). Contubernium–Tübinger Beiträge zur Universitäts- und Wissenschaftsgeschichte: Vol. 64. Stuttgart: Steiner.

Stievermann, D. (1999). Friedrich der Weise und seine Universität Wittenberg [Frederick the Wise and his University of Wittenberg]. In S. Lorenz (Ed.), *Attempto—oder wie stiftet man eine Universität: Die Universitätsgründungen der sogannten zweiten Gründungswelle im Vergleich* (pp. 175–207). Contubernium–Tübinger Beiträge zur Universitäts- und Wissenschaftsgeschichte: Vol. 50. Stuttgart: Steiner.

Stokes, G. (1989). The social origins of east European politics. In D. Chirot (Ed.), *The origins of backwardness in eastern Europe: Economics and politics from the Middle Ages to the early twentieth century* (pp. 210–251). Berkeley: University of California Press.

Strohm, C., & Becker, M. (2012). Western European influences on the Faculty of Law, 1560 to 1620. In P. Meusburger & T. Schuch (Eds.), *Wissenschaftsatlas of Heidelberg University: Spatio-temporal relations of academic knowledge production* (pp. 64–65). Knittlingen: Bibliotheca Palatina.

Strohm, C., & Hofmann, A. (2012). The university as a spiritual centre of "Reformed" Europe. In P. Meusburger & T. Schuch (Eds.), *Wissenschaftsatlas of Heidelberg University: Spatio-temporal relations of academic knowledge production* (pp. 62–63). Knittlingen: Bibliotheca Palatina.

Stupperich, R. (1990). Melanchthon, Philipp. In *Neue Deutsche Biographie: Vol. 16. Maly–Melanchthon* (pp. 741–745). Berlin: Duncker & Humblot. Retrieved from https://www.deutsche-biographie.de/gnd118580485.html#ndbcontent

Stupperich, R. (1996). *Philipp Melanchthon: Gelehrter und Politiker* [Philipp Melanchthon: Scholar and politician]. Göttingen: Muster-Schmidt.

Szabó, A. (1992). Die soziale Struktur der Universitaetsstudentenschaft im Spiegel der ungarischen Studenten zu Wittenberg [The social structure of university students as reflected by Hungarian students at Wittenberg]. In A. Buck & T. Klaniczay (Eds.), *Sozialgeschichtliche Fragestellungen in der Renaissanceforschung* (pp. 41–48). Wolfenbütteler Abhandlungen zur Renaissanceforschung: Vol. 13. Wiesbaden: Harrassowitz.

Szabó, A. (1993). Magyarok Wittenbergben 1555–1592 [Hungarians in Wittenberg, 1555–1592]. In I. Békési, J. Jankovics, L. Kosa, & J. Nyerges (Eds.), *Régi és új peregrináció—Magyarok külföldön, külföldiek Magyarországon II* (pp. 626–635). Budapest: Nemzetközi Magyar Filológiai Társaság. ScriptumKft.

Szabó, A. (1999a). Albert Szenci Molnár in Schlesien [Albert Szenci Molnár in Silesia]. In A. Szabó (Ed.), *Iter Germanicum: Deutschland und die Reformierte Kirche in Ungarn im 16.–17. Jahrhundert* (pp. 201–213). Budapest: Calvin

Szabó, A. (1999b). Ungarische Studenten in Wittenberg [Hungarian students in Wittenberg]. In A. Szabó (Ed.), *Iter Germanicum: Deutschland und die Reformierte Kirche in Ungarn im 16.–17. Jahrhundert* (pp. 155–168). Budapest: Calvin.

Szabó, A. (2003). *Szenczi Molnár Albert naplója* [Diary of Albert Szenczi Molnár]. Historia Litteraria: Vol. 13. Budapest: Universitas.

Szabó, A. (2005–2007). Akten über ungarische Studenten im Heidelberger Universitaetsarchiv [Files on Hungarian students in the Heidelberg university archive]. *Ungarn-Jahrbuch: Zeitschrift für interdisziplinäre Hungarologie, 28*, 227–236.

Szabó, A. (2006). Die Universität Wittenberg als zentraler Studienort im 16. Jahrhundert [Wittenberg University as a key place to study in sixteenth century]. In M. Fata, G. Kurucz, & A. Schindling (Eds.), *Peregrinatio Hungarica:Studenten aus Ungarn an deutschen und österreichischen Hochschulen vom 16. bis zum 20. Jahrhundert* (pp. 55–63). Contubernium–Tübinger Beiträge zur Universitäts- und Wissenschaftsgeschichte: Vol. 64. Stuttgart: Steiner.

Szabó, A. (2010). Calvinismus und Ethnie im Reich der Stephanskrone im 16. Jahrhundert [Calvinism and ethnicity in the lands of Saint Stephen's Crown in the sixteenth century]. In M Fata & A Schindling (Eds.), *Calvin und Reformiertentum in Ungarn und Siebenbürgen: Helvetisches Bekenntnis, Ethnie und Politik vom 16. Jahrhundert bis 1918* (pp. 81–89). Reformationsgeschichtliche Studien und Texte: Vol. 155. Münster: Aschendorff.

Szögi, L. (1995). *Régi magyar egyetemek emlékezete* [Memories of old Hungarian universities]. Budapest: ELTE Könyvtár és Levéltár. Retrieved from http://www.mek.oszk.hu/01800/01882/01882.htm

Szögi, L. (2001). *Magyarországi diákok németországi egyetemeken és főiskolákon: 1789–1919. Ungarländische Studenten an den deutschen Universitäten und Hochschulen: 1789–1919* [Hungarian students at German universities and colleges: 1789–1919]. Budapest: Eötvös Loránd Tudományegyetem Levéltára.

Szögi, L. (2004). Külföldi egyetemi modellek és külföldi egyetemi tanulmányok a 18. századtól a 20. század első negyedéig [University models and studies abroad from the eighteenth century to the first quarter of the nineteenth century]. In G. Palló (Ed.), *Recepció és kreativitás—Teremtő befogadás (Összefüggések. tanulságok)* (pp. 129–197). Budapest: Áron Kiadó.

Szögi, L. (2006). Studenten aus Ungarn und Siebenbürgen an den deutschen Universitäten 1789–1919 [Students from Hungary and Transylvania at German universities, 1789–1919]. In M. Fata, G. Kurucz, & A. Schindling (Eds.), *Peregrinatio Hungarica: Studenten aus Ungarn an deutschen und österreichischen Hochschulen vom 16. bis zum 20. Jahrhundert* (pp. 387–408). Contubernium–Tübinger Beiträge zur Universitäts- und Wissenschaftsgeschichte: Vol. 64. Stuttgart: Steiner.

Szögi, L. (2008). Die Universitäten in Ungarn: Gründungswellen vom späten Mittelalter bis ins 20. Jahrhundert [Universities in Hungary: Waves of foundings from late medieval times into the twentieth century]. In P. Wörster (Ed.), *Universitäten im östlichen Mitteleuropa: Zwischen Kirche, Staat und Nation. Sozialgeschichtliche und politische Entwicklungen* (pp. 235–267). Völker, Staaten und Kulturen in Ostmitteleuropa: Vol. 3. Munich: Oldenbourg.

Tar, A. S. (2007). *Magyar-német felsőoktatási kapcsolatok és kölcsönhatások a 18.–19. században* [Hungarian-German relations and interactions in higher education during the eighteenth and nineteenth centuries] (Doctoral thesis, Eötvös Loránd Unversity, Budapest, Hungary). Retrieved from http://www.doktori.btk.elte.hu/hist/tar/disszert.pdf

Taylor, P. J., Hoyler, M., & Evans, D. M. (2010). A geohistorical study of "The rise of modern science": Mapping scientific practice through urban networks, 1500–1900. In P. Meusburger, D. N. Livingstone, & H. Jöns (Eds.), *Geographies of science* (pp. 37–56). Knowledge and Space: Vol. 3. Dordrecht: Springer. doi:https://doi.org/10.1007/978-90-481-8611-2_3

Teutsch, F. (1872). Die Studierenden aus Ungarn und Siebenbürgen auf der Hochschule in Heidelberg von der Gründung derselben bis 1810 [Students from Hungary and Transylvania at Heidelberg University from its founding to 1810]. *Archiv des Vereins für siebenbürgische Landeskunde, 10* (Neue Folge), 182–192.

Thomas, M. (Ed.). (2015). *Expedition into empire: Exploratory journeys and the making of the modern world.* New York: Routledge.

Titze, H. (1987). *Das Hochschulstudium in Preußen und Deutschland 1820–1944: Datenhandbuch zur deutschen Bildungsgeschichte* [University study in Prussia and Germany, 1820–1944: Data on German educational history]. Vol. 1, Pt. 1. Göttingen: Vandenhoeck & Ruprecht.

Toepke, G. (Ed.). (1886). *Die Matrikel der Universität Heidelberg: Bd. 2. Von 1554–1662* [The registries of Heidelberg University: Vol. 2. 1554–1662]. Heidelberg: Winter. Retrieved from http://digi.ub.uni-heidelberg.de/matrikel1554

Toepke, G., & Hintzelmann, K. (Eds.). (1903). *Die Matrikel der Universität Heidelberg: Bd. 4. Von 1704–1807* [The registries of Heidelberg University: Vol. 4. 1704–1807]. Heidelberg: Winter. Retrieved from http://digi.ub.uni-heidelberg.de/matrikel1704

Toepke, G., & Hintzelmann, K. (Eds.). (1904). *Die Matrikel der Universität Heidelberg: Bd. 5. Von 1807–1845* [The registries of Heidelberg University: Vol. 5. 1807–1845]. Heidelberg: Winter. Retrieved from http://digi.ub.uni-heidelberg.de/matrikel1807

Toepke, G., & Hintzelmann, K. (Eds.). (1906). *Die Matrikel der Universität Heidelberg: Bd. 6. Von 1846–1870* [The registries of Heidelberg University: Vol. 6. 1846–1870]. Heidelberg: Winter. Retrieved from http://digi.ub.uni-heidelberg.de/matrikel1846

Tökéczki, L. (2010). Liberalismus und Reformiertentum zur Zeit des österreichisch-ungarischen Dualismus: Das Beispiel István Tiszas [Liberalism and Reformed churches during Austrian-Hungarian dualism: The example of István Tisza]. In M. Fata & A. Schindling (Eds.), *Calvin und Reformiertentum in Ungarn und Siebenbürgen: Helvetisches Bekenntnis, Ethnie und Politik vom 16. Jahrhundert bis 1918* (pp. 377–394). Reformationsgeschichtliche Studien und Texte: Vol. 155. Münster: Aschendorff.

Tonk, S. (1979). *Erdélyiek egyetemjárása a középkorban* [Peregrination of Transylvanian students in the Middle Ages]. Bukarest: Kriterion.

Tonk, S., & Szabó, M. (1993). Erdélyiek egyetemjárása a középkor és koraújkor folyamán. [Peregrination of Transylvanian students in the Middle Ages and Early Modern times]. In I. Békési, J. Jankovics, L. Kosa, & J. Nyerges (Eds.), *Régi és új peregrináció: Magyarok külföldön, külföldiek Magyarországon II* (pp. 491–500). Budapest: Nemzetközi Magyar Filológiai Társaság. Scriptum Kft.

Ugry, B. (2014). A megismerés és az információszerzés új útjain: Magyarok kora újkori európai utazásainak forrásai és jelentőségük az információtörténeti kutatások számára [On new ways of cognition and acquiring information: Sources on travels of Hungarians in the Early Modern Age and their importance for the research on information history]. *Jel-Kép*, *25*(3). doi:https://doi.org/10.20520/Jel-Kep.2014.3.6

Ulrichs, K. F. (1993). David Pareus. In T. Bautz (Ed.). *Biographisch-Bibliographisches Kirchenlexikon: Vol. 6. Moenius–Patijn* (pp. 1532–1536). Hamm: Bautz.

Vámos, E. (2006, September 6–9). Three generations of natural scientists in Hungary, 1848–1918. In M. Kokowski (Ed.), *The global and the local: The history of science and the cultural integration of Europe* (pp. 274–288). Proceedings of the 2nd ICESHS, Cracow, Poland. Retrieved from http://www.academia.edu/7596114/Michal_Kokowski_ed._The_Global_and_the_Local_The_History_of_Science_and_the_Cultural_Integration_of_Europe._Proceedings_of_the_2nd_International_Conference_of_the_European_Society_for_the_History_of_Science_Cracow_Poland_September_6–9_2006_

Varga, J. (2001). Protestantische Hochschulbildung in Siebenbürgen vom 16. Jahrhundert bis zur Mitte des 19. Jahrhunderts [Protestant university education in Transylvania from the 16th to the mid-nineteenth century]. In L. Szögi & M. Font (Eds.), *Die Ungarische Universitätsbildung und Europa* (pp. 173–182). Pécs: Universitas Quinqueecclesiensis.

Varga, P. (2003). „man sprach deutsch und fühlte nichtmagyarisch . . .": Assimilationsprozesse in der deutschen und jüdischen Bevölkerung von Pest-Buda im 19. Jahrhundert ["We spoke German and felt non-Hungarian . . .": Processes of assimilation in the German and Jewish populations of Pest-Buda in the nineteenth century]. In H. Hecker & W. Engel (Eds.), *Symbiose und Traditionsbruch: Deutsch-jüdische Wechselbeziehungen in Ostmittel- und Südosteuropa im 19. und 20. Jahrhundert* (pp. 173–182). Essen: Klartext.

Vásárhelyi, J. P. (1985). *Eszmei áramlatok és politika Szenczi Molnár Albert életművében* [Ideological trends and politics in the oeuvre of Albert Szenczi Molnár]. Humanizmus és reformáció: Vol. 12. Budapest: Akadémiai Kiadó.

Vásárhelyi, J. P. (1999). Johann Piscator und Albert Szenci Molnár [Johann Piscator und Albert Szenci Molnár]. In A Szabó (Ed.), *Iter Germanicum: Deutschland und die Reformierte Kirche in Ungarn im 16.–17. Jahrhundert* (pp. 191–200). Budapest: Calvin.

Vásárhelyi, J. P. (2006). *Szenczi Molnár Albert és a Vizsolyi Biblia új kiadásai* [Albert Szenczi Molnár and the new editions of the Vizsoly Bible]. Historia Litteraria: Vol. 21. Budapest: Universitas.

Vásárhelyi, J. P. (2014). The role of Szenci Molnár in the exchange of ideas and political knowledge among the European Calvinist principalities in the early seventeenth century. In G. Almási (Ed.), *Study tours and intellectual-religious relationships (pp. 121–144). A divided Hungary in Europe: Exchanges, networks and representations*, 1541–1699. Vol. 1. Cambridge, UK: Cambridge Scholar Publishing.

Vékás, L. (2001). Wissenschaftleraustausch und geistige Wiedervereinigung Europas [Exchange of scientists and spiritual reunification of Europe]. In Europa Institute Budapest (Ed.), *Begegnungen* (pp. 137–143). Schriftenreihe des Europa Institutes Budapest: Vol. 12. Budapest: Europa-Institut.

Verger, J. (1970). Le recrutement géographique des universités françaises au début de 15e siècle d'après les suppliques de 1403 [The recruiting catchment areas of French universities as of the early fifteenth century after the petitions of 1403]. *Mélanges d'Archéologie et d'Histoire, 82,* 855–902.

Viskolcz, N. (2010). Johann Heinrich Bisterfeld: Ein Professor als Vermittler zwischen West und Ost an der siebenbürgischen Akademie in Weißenburg 1630–1655 [Johann Heinrich Bisterfeld: A professor as intermediary between West and East at the Transylvanian academy in Weissenburg, 1630–1655]. In M Fata & A Schindling (Eds.), *Calvin und Reformiertentum in Ungarn und Siebenbürgen: Helvetisches Bekenntnis, Ethnie und Politik vom 16. Jahrhundert bis 1918* (pp. 201–214). Reformationsgeschichtliche Studien und Texte: Vol. 155. Münster: Aschendorff.

Volkov, S. (1987). *Soziale Ursachen des Erfolgs in der Wissenschaft: Juden im Kaiserreich* [Social causes of success in science: Jews in the German empire]. *Historische Zeitschrift, 245*, 315–342. Retrieved from http://www.jstor.org/stable/27625826

Vorlaufer, J. (1993). Lukács, Georg. In T. Bautz (Ed.), *Biographisch-Bibliographisches Kirchenlexikon: Vol. 5.* Leyden, Nikolaus to Mönch, Antonius (pp. 401–404). Hamm: Bautz. http://www.bbkl.de.ubproxy.ub.uni-heidelberg.de/lexikon/bbkl-artikel.php?wt=1&art=./L/Lu/lukacs.art

Wessely, A. (1986). Diskurs über die Kunst im Sonntagskreis [Discourse about arts in the Sunday Circle]. In H. Gaßner (Ed.*), Wechselwirkungen: Ungarische Avantgarde in der Weimarer Republik* (pp. 541–553). Marburg: Jonas.

Wolgast, E. (1986). *Die Universität Heidelberg 1386–1986* [Heidelberg University, 1386–1986]. Berlin: Springer.

Wolgast, E. (1987). Phönix aus der Asche? Die Reorganisation der Universität Heidelberg zu Beginn des 19. Jahrhunderts [Phoenix out of the ashes? The reorganization of Heidelberg University in the early nineteenth century]. In F Strack (Ed.), *Heidelberg im säkularen Umbruch: Traditionsbewusstsein und Kulturpolitik um 1800* (pp. 35–60). Stuttgart: Klett-Cotta.

Wolgast, E. (1996). Die Universität Heidelberg: Historische Entwicklung [Heidelberg University: Historical development]. In E Mittler (Ed.), *Heidelberg: Geschichte und Gestalt* (pp. 284–320). Heidelberg: Winter.

Wolgast, E. (1998). Späthumanismus in Heidelberg [Late Humanism in Heidelberg]. In E Wolgast (Ed.), *Reformierte Konfession und Politik im 16. Jahrhundert: Studien zur Geschichte der Kurpfalz im Reformationszeitalter* (pp. 98–114) Schriften der Philosophisch-Historischen Klasse der Heidelberger Akademie der Wissenschaften: Vol. 10. Heidelberg: Winter.

Wolgast, E. (2015). Die Universität Heidelberg in ihrer internationalen Vernetzung in den Jahren 1559–1622 [Heidelberg University in its international network, 1559–1622]. In C. Strohm & J. Stievermann (Eds.), *Profil und Wirkung des Heidelberger Katechismus: Neue Forschungsbeiträge anläßlich des 450jährigen Jubiläums. The Heidelberg Catechism: Origins, characteristics, and influences: Essays in reappraisal on the occasion of its 450th anniversary* (pp. 89–112). Schriften des Vereins für Reformationsgeschichte: Vol. 215. Gütersloh: Gütersloher Verlagshaus.

Zach, K. (2004). „Eine kleine Biblia"—Rezeption und Resonanz des reformationszeitlichen Katechismus im historischen Ungarn (1530–1640) ["A small biblia"—Reception and resonance of catechisms of the Reformation in historical Hungary (1530–1640)]. In W. Kühlmann & A. Schindling (Eds.), *Deutschland und Ungarn in ihren Bildungs- und Wissenschaftsbeziehungen während der Renaissance* (pp. 151–183). Contubernium–Tübinger Beiträge zur Universitäts- und Wissenschaftsgeschichte: Vol. 62. Stuttgart: Steiner.

Zepf, R. (2001). Fructus Uberrimi: Die Theologiestudenten von Collegium Sapientiae und Universität Heidelberg 1550–1622 [Fructus Uberrimi: Theology students of the Collegium Sapientiae and Heidelberg University, 1550–1622]. In A Kohnle & F Engehausen (Eds.), *Zwischen Wissenschaft und Politik: Studien zur deutschen Universitätsgeschichte. Festschrift für Eike Wolgast* (pp. 441–454). Stuttgart: Steiner.

Zhang, G. (2013). *Fehlgründungen von Universitäten im Spätmittelalter: Motive und Bedingungen für die Entstehung der mittelalterlichen Universität* [Abortive creation of universities in late medieval times: Motives and conditions for the rise of the medieval university]. Hamburg: Kovac.

Zvara, E. (2012). Ceglédi Szabó Pál dunántúli református püspök könyvei [The books of Pál Ceglédi Szabó, Calvinist superintendent of Transdanubia]. *Magyar Könyvszemle, 128,* 293–317.

Chapter 4
Catchment Areas and Killing Fields: Towards an Intellectual Geography of the Thirty Years' War

Howard Hotson

Part I. Introduction

Intellectual Geography

Intellectual history is traditionally text-based. Sometimes regarded as synonymous with the history of ideas, its natural starting points are the texts in which ideas are expressed and the authors who write them. Many intellectual historians now also situate authors, texts, language, and ideas in economic, social, political, and cultural contexts; and the scope of the field is now broadening further to consider the role of material objects and practices in shaping intellectual activities (Grafton, 2006; Kelley, 2002; Whatmore & Young, 2016).

Data now offers a further opportunity to enrich the field. Digital technology affords intellectual historians the capacity to extract insight from previously unmanageable quantities of highly granular data. In a manner in some ways analogous to the focus on material culture, this capacity facilitates the writing of intellectual history from the ground up. Instead of beginning from canonical texts and working downward to explore their reception, historians can potentially situate individual texts and authors within shifting landscapes of intellectual activity, animated by huge bodies of digital data previously immobilized on the printed page.

This paper explores how one source of such data—university matriculation registers—can be used to map out one kind of intellectual exchange—student migration—at a scale and pace which makes visible for the first time patterns which structured the intellectual lives of hundreds of thousands of people across vast swathes Europe. More specifically, this paper examines the densest concentration of universities in early modern Europe at the most tumultuous moment in their

H. Hotson (✉)
Faculty of History, St. Anne's College, University of Oxford, Oxford, UK
e-mail: howard.hotson@st-annes.ox.ac.uk

P. Meusburger et al. (eds.), *Geographies of the University*, Knowledge and Space 12,
https://doi.org/10.1007/978-3-319-75593-9_4

history: namely, the universities of the Holy Roman Empire in the midst of the Thirty Years' War (1618–1648). After situating the topic on a broad geographical and chronological canvas (in Part I), this paper shows how even a relatively superficial engagement with this data can reveal how the war transformed the academic geography both of the Empire itself (in Part II) and of the huge catchment area which surrounded it (in Part III). After summarizing some basic historiographical and methodological results, the paper concludes (in Part IV) by outlining a few of the prospects for a more intensive engagement with this data.

The Geography of Universities in the Early Modern Period: An Overview

The medieval and early modern history of the university in central Europe is a story of sustained growth punctuated by one enormous rupture and ended by a second (Fig. 4.1). The first major crisis to hit the German universities—the Lutheran Reformation—was profound but transient. By calling into question the utility of degrees in theology and canon law, it cut matriculations in German universities by nearly three quarters in the 1530s (Asche, 2001). The longer-term effect of the Reformation, however, was to accelerate the underlying rate of growth. From the 1540s onward, enrollment in German universities grew rapidly, as inter-confessional competition and the consolidation of political authority at the territorial level increased the demand for educated officeholders, both clerical and secular. By the

Fig. 4.1 Matriculations per decade in Germany and England.
Source: Data from Eulenburg (1904), Stone (1964). Design by author.

1550s, matriculations had regained pre-Reformation levels; by the 1560s, their rate of growth had recovered the pre-Reformation trend-line; from the 1590s they were growing even faster than before 1520; and by their peak in the 1620 they had grown eightfold in eight decades.

This sustained educational boom is particularly well documented for Germany thanks to the large number of complete matriculation registers surviving from this period, the publication of most of them in the nineteenth century, and the tabulation of their basic data contained within Franz Eulenburg's classic study of 1904.[1] But the German case was not unique. Analogous forces produced similar effects elsewhere in Europe (de Ridder-Symoens, 1996; Frijhoff, 1986a, pp. 23–63; Julia, Revel, & Chartier, 1986; for the English case see Stone, 1964). English university affairs, to provide one revealing comparison, were also dramatically disrupted by the reformations and counter-reformations of Henry VIII, Edward VI, and Mary Tudor. But growth resumed from the Elizabethan settlement onward, peaking in the 1630s, before the outbreak of civil war in 1642 caused enrollments to slump in perfect synchrony with the German figures.

Underlying these superficially similar trajectories, however, the geography of the English and German university systems could scarcely have been more different. Politically, the island nation unified early, and the ancient duopoly of Oxford and Cambridge suited the needs of the English crown. Instead of founding new universities, the English therefore populated the old ones with new colleges, which grew in number from 20 in 1500 to 35 in 1624. Within the Empire, by contrast, confessional differentiation and political consolidation took place at the level of individual princely territories. Each prince and imperial free city wanted their own institutions of higher education, adapted to serving their political, economic, and religious needs without exporting students to enrich their neighbors. The result was a proliferation of university and sub-university institutions without parallel elsewhere, instantly recognizable at the European scale. In 1500 (Fig. 4.2), universities were scattered across the European landscape relatively uniformly, from the middle of the Iberian and Italian peninsulas northward to southern Scotland and Scandinavia. By 1650 (Fig. 4.3), the dense concentration of universities and academies in the German principalities, the Swiss cantons, and the Dutch provinces had departed unmistakably from the European norm.

[1]Unless otherwise noted, matriculation data are taken from the tables provided in Franz Eulenburg (1904). Eulenburg, writing within the second German Reich, counted Königsberg amongst German universities. In the following study, *Germany* refers to the lands included in the current Bundesrepublik. The matriculations for Königsberg have therefore been excluded from the figures underlying this initial graph. A graph for the entire Holy Roman Empire would include Prague (1347), Vienna (1365), Louvain (1425), and a number of more recent foundations, but for many of these existing data in fragmentary. For an important refinement of Eulenburg's data on Germany, see Frijhoff (1986a).

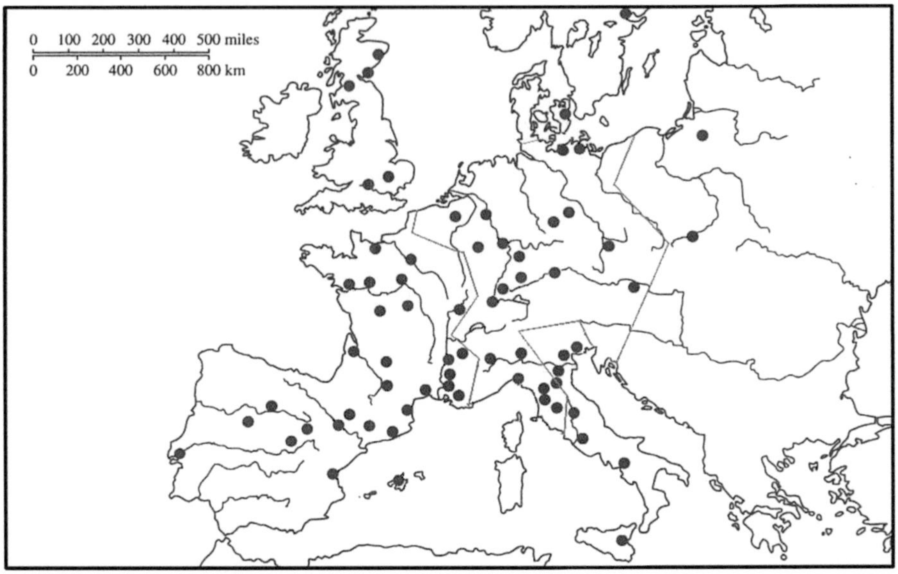

Fig. 4.2 European universities in 1500.
Source: Map derived and data taken from Frijhoff (1996, pp. 80–89). Revised map published with permission of Willem Frijhoff and the Cambridge University Press.

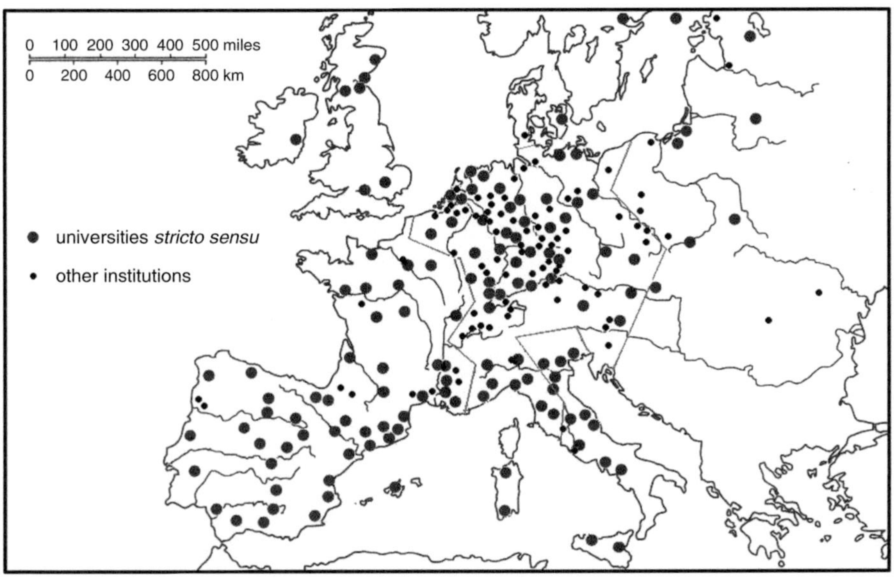

Fig. 4.3 European universities in 1650.
Source: Map derived and data taken from Frijhoff (1996, pp. 80–89). Revised map published with permission of Willem Frijhoff and the Cambridge University Press.

This unique academic geography had four significant consequences, which affected not only the Holy Roman Empire but a vast surrounding area as well. The first, immediately visible on the preceding maps, was institutional. In order to provide quasi-university education on the slender resources of a small principality or imperial free city, German educationalists were forced to recast the university itself in a constantly proliferating range of sub- and semi-university institutions. Most notable amongst these was the *gymnasium illustre*, which capped a *schola classica* or *Paedagogium* (providing pre-university education in the trivium) with a *schola publica* (providing streamlined introduction to the undergraduate arts or philosophy curriculum, often supplemented by one of more of the three higher faculties of law, medicine, and theology) (Siefert, 1996).

The second consequence was pedagogical. New kinds of educational institution required new educational methods, designed to impart the greatest quantity of useful learning in the shortest possible time. This requirement energized a long line of pedagogical reformers whose educational influences primarily derived from or were concentrated within the Rhineland corridor between Basle and the Low Countries: from the leading pre-Reformation northern humanist, Rudolph Agricola, via Melanchthon, Sturm, Ramus, Keckermann, and Alsted to the leading educational theorist of the wartime generation, Jan Amos Comenius (Hotson, 2007).

The third and fourth consequences played out on the broader canvas of Protestant Europe as a whole. On the one hand, these institutional and pedagogical innovations transformed Germany into the pedagogical laboratory of Protestant Europe. For over a century, these innovations radiated out from Germany in all directions alongside the better-studied theological impulses of the Reformation era. The *gymnasium illustre* was adapted to the needs of Prussian cities, Dutch provinces, and Scandinavian kingdoms, while the history of pedagogical practice throughout the Protestant world is, in large measure, the history of struggles to adopt, adapt, or resist the educational innovations originating from this west German tradition (Hotson, forthcoming).

While influences radiated outward, students gravitated inward, drawn by the opportunity to choose places to study at the heart of these pedagogical experiments. At its pre-war height, the German universities and academies were the center of a gigantic catchment area: their students arrived not only from German lands but from a huge arc sweeping from Hungary-Transylvania, the lands of the Czech crown, Poland-Lithuania, and the eastern Baltic to Finland, Sweden, Denmark, and Norway, and extending as far as Scotland and the Netherlands, the Swiss confederation, and even Huguenot France.[2]

[2]For Sweden: Niléhn (1983b). For Denmark-Norway: Helk (1987). For Poland: Żołądź-Strzelczyk (1996). For the Czech lands: Pešek and Šaman (1986). For Hungary: Szögi (2011).

The Thirty Years' War as a Turning Point in European University History

The second great crisis to affect the German universities, the Thirty Years' War, struck central Europe at the very apex of a great academic boom 250 years in the making. So far as we know, it also struck at the very height of its international influence as a center of academic pilgrimage and a source of innovations emulated and resisted across the Protestant world from Transylvania to New England. The academic impact of the war therefore played itself out on at two different scales. The domestic level raises a series of questions regarding the effects of the war on academic affairs within the Empire. How did the chronology of academic destruction and recovery unfold over the course of three decades? Can its effects on individual institutions be grouped together to understand its broader impact on whole regions and confessions? And how did the war affect the development of the German universities after the war? The second dimension is international. How did the disruption at the center of this huge catchment area affect the long-term development of neighboring university systems across northern and central Europe, the shifting patterns of international academic migration between them, and the intellectual influences communicated by them?

For first impressions of the magnitude of these changes, we can return to the data series surveyed already. On the domestic front, Eulenburg's graph emphatically demonstrates the status of the Thirty Years' War as a dividing line between two epochs of German university history. The left-hand side of Figure 4.1 plots two centuries of gradually accelerating growth, dented but not ultimately deflected by the Reformation. The right-hand side depicts a period of stagnation followed by decline extending throughout the eighteenth century. For a century after the war, German matriculations struggled to regain their pre-war heights, declining first as a share of the population and then in absolute terms in the latter eighteenth century. Comparison with the even more anemic English recovery suggests that this pattern was neither confined to Germany nor resulted merely from the conflict in central Europe: neither the end of the British civil wars in 1651 nor the restoration of the Stuarts in 1660 restored the English university population to its prewar level. By 1700, annual enrollments had fallen to their level in 1600, ushering in the stagnation which was to characterize Oxford and Cambridge throughout the eighteenth century and beyond. The chronology of Oxbridge college foundations tells the same tale. During the 125 years before 1624, fifteen colleges were founded in Oxford and Cambridge. In the 175 years after that data, only two.

Even more striking evidence of a sea-change in German and European university history results from tabulating the data on university foundations underlying the previous maps. As Figure 4.4 shows, the rate at which new universities were founded across Europe steadily accelerated throughout the fifteenth and sixteenth centuries. The foundation of sub-university institutions (mostly Jesuit academies and Protestant *gymnasia illustria*) likewise climbed rapidly throughout the sixteenth and early seventeenth centuries. But the foundation of universities peaked in the half-

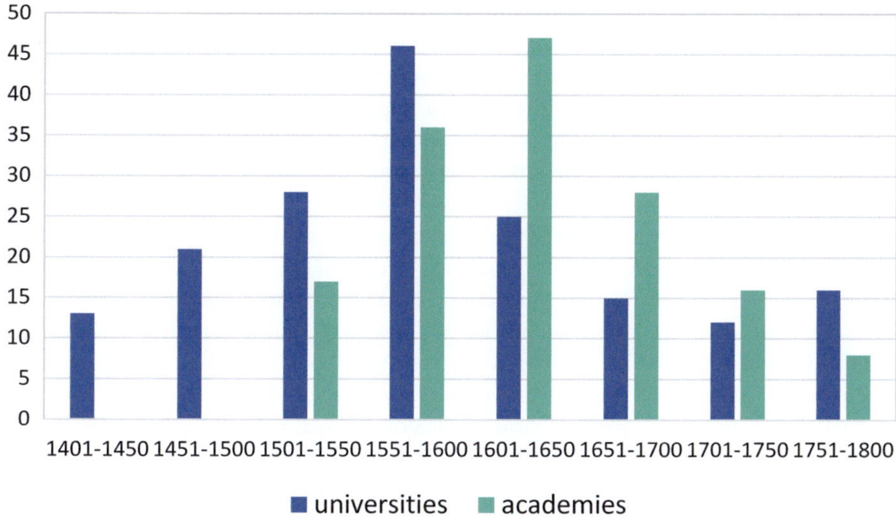

Fig. 4.4 Foundation of universities and academies in Europe, 1401–1800.
Source: Data from Frijhoff (1996, pp. 80–98). Design by author.

century before 1600, and the foundations of sub-university institutions peaked in the half-century thereafter. Both rates of foundation then fell even more sharply than they had risen, before stagnating during the eighteenth century at one third of peak levels. Whether viewed through the lens of enrollments or foundations, the confessional era, not the '*siècle des Lumières*', was the great age of higher education in early modern Europe. In helping to establish this pattern, the two great central European ruptures—the Reformation and the Thirty Years' War—were crucial parts of the broader European picture: the pre-war surge in foundations of both kinds was concentrated disproportionately in the fragmented heartland of Europe; and the wartime and post-war collapse in foundations in the Empire contributed substantially to the broader European pattern.

Methodological Presuppositions

The Thirty Years' War and the other conflicts of the mid-seventeenth century—from the Northern Wars in the Baltic to the civil wars in the British Isles—therefore help mark a major turning point in the history of the university both within and beyond the theatre in which those wars were fought. Given the magnitude of this watershed, surprisingly little effort has been expended on mapping it systematically. The overwhelming majority of work on the history of German universities in this period remains focused on individual institutions.[3] The sheer number of universities makes extrapolation from this literature difficult, and when the notorious complexity of the war itself is added to the equation it becomes obvious that grasping the scale, tempo, and texture of these changes will require a departure from the standard forms of university history. Given the diversity of experience of different confessions and regions, it is difficult to see how further case studies can form the basis of a general assessment.

Rather than diving into the archives in search of further nuance, therefore, what is most needed is to place existing historical accounts within a clear comparative framework. Such a framework can best be provided by reference to a reasonably homogeneous data set which monitors some aspect of the health of individual universities on a regular basis along a chronological axis extending before, during, and after the conflict. Since all universities need to replenish themselves on an ongoing basis, one of the most obvious indicators of institutional health is the ability to attract new students. Since universities' financial health and legal integrity also depended on keeping track of their members, matriculation records were kept meticulously throughout the German-speaking world in this period.[4]

This paper therefore explores the utility of matriculation data as an index of the impact of the war on individual universities and on broader academic landscapes. Its thesis is methodological as much as historical: it aims to show that matriculation data allow the impact of the war to be quantified in meaningful fashion, dated with some chronological precision, and therefore analyzed comparatively between institutions in a manner which can produce sound generalizations regarding the differentiated impact of the war on whole confessions and regions. At the level of individual universities, traditional narrative accounts will be aligned with graphs of matriculation data to determine whether matriculation registers, despite their imperfections, can provide sensitive barometers, widely distributed across the landscape of the Empire, of the pressure exerted by military events on individual institutions in specific times and places (Part II 1–3). The data for individual universities will then be aggregated in search of meaningful patterns distinguishing the experiences both of the three main religious confessions and of individual regions from one another (Part II 4–5). The scope of the study then expands beyond the boundaries of the Holy Roman Empire, in search of preliminary evidence that the iron age of university history within the war zone coincided with a golden age of some of the institutions immediately outside it (Part III). The ultimate objective is to sketch a rudimentary analytical framework within which the impact of the war can be assessed on the academic life, not only of the Empire as a whole, but of its gigantic catchment area in central and northern Europe.

The core data set for testing this methodology is provided by Eulenburg's tabulations of the German universities. Trier and Rinteln are lacking from his tables

---

[3]Valuable overviews are provided in Asche and Gerber (2008) and Rasche (2011). For the Thirty Years' War itself, see especially Kossert (2011), especially the contribution by Asche (2011).

[4]For a critical introduction to this genre of sources, see Asche and Häcker (2011).

because their matriculation registers have not survived. The patchy data for Marburg, Giessen, and Paderborn make their experience difficult to interpret. Surviving data for Vienna, Graz, and Salzburg help put modern Austria on the map. For the southern Low Countries and the Czech lands, the data is incomplete: the matriculation registers of Prague and Douai have not survived, Leuven's is missing before 1616, and Olomouc's is fragmentary. Stepping just outside the boundaries of the Empire, Basle (the sole university in the Swiss Confederation in this period) is tightly integrated into the German Reformed academic community and reflects its experience; but Geneva is further away, less tightly integrated, and is therefore marginal to this study. Cracow and Köngisberg provide a vivid impression of the war's impact on Catholic and Lutheran communities to the east; Copenhagen, Uppsala, Tartu, and Åbo reveal its muted influence in Scandinavia; and the data from Leiden, Franeker, and Groningen clearly tie the dark age of the German Reformed universities to the golden age of the Dutch ones. More painstaking methods will be needed before the impact of the war on Italian, French, English, or Scottish academic relations can be assessed, and the same holds for the younger Dutch ones and some of the smaller institutions to the east of the Empire.

This exploratory study has been conducted, not as an end in itself, but to provide a basis for further work. The paper therefore concludes with some methodological reflections on how this approach might be expanded in a more sustained piece of collaborative research (Part IV). Obviously, such a project vastly surpasses the capacities of this brief article. What can be offered here is proof of concept of one source and method for mapping the impact of the Thirty Years' War on the intellectual geography of the Holy Roman Empire and neighboring regions.

Part II. Universities within the Holy Roman Empire

Reformed Universities

The first German university to be affected by the war, Heidelberg, was also disrupted most drastically. In the opening decades of the seventeenth century, Heidelberg was the most international university in the Reformed world, with an intake of students rising steadily to about 200 per year, two-fifths of whom came from outside Germany. Yet, by accepting the crown of St Wenseslas from the rebellious Bohemian estates in 1619, the Elector Palatine Friedrich V provoked a massive Catholic reprisal which ruined Heidelberg for a generation. The following year, as Spanish forces occupied the far bank of the Rhine, Heidelberg's rate of matriculation was cut in half. It halved again in 1621 after the Friedrich's forces were defeated at the Battle of the White Mountain, and again in 1622, by September of which year the city, the university, and the fabled electoral library were in Catholic hands. A final handful of Reformed students matriculated during the brief half-life which followed, but by 1625 the mounting program of recatholicisation had forced all the professors either to convert or to flee. Two abortive attempts to revive the ancient university followed.

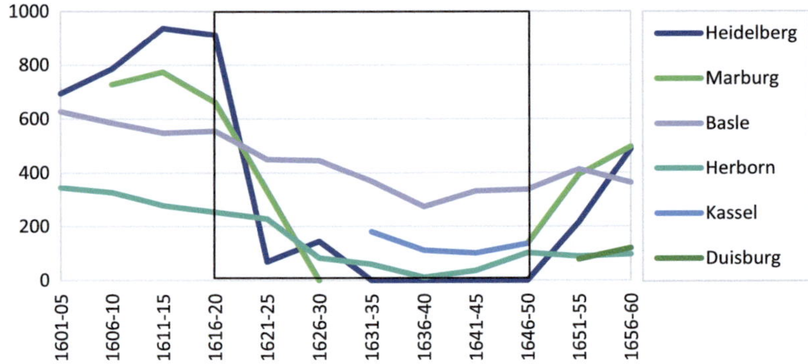

Fig. 4.5 German Reformed universities.
Source: Data from Eulenburg (1904), Falckenhainer (1893), Wachernagel (1951–1980). Design by author.

On 16 June 1629, the year after Maximilian I of Bavaria was granted hereditary status as the new Elector Palatine, Heidelberg reopened as a Catholic institution which over the next three years enrolled 145 students. But the arrival of the Swedes cut this experiment short. In October 1631, the new Jesuit professors fled the city; in December the last Catholic student matriculated; and by the end of May 1632 the city was under Swedish control. Now it was the restored Reformed regime which struggled to reassemble professors and raise revenue; but before a single student enrolled or a single lecture was held the Swedish were defeated at Nördlingen and two months later Heidelberg was re-occupied by Bavaria. In these changed circumstances, there was no thought of restoring the university: France entered the war, challenged the Bavarian electoral title and turned the Palatine lands into a battlefield once again. It was not until the Peace of Westphalia restored a fragment of Friedrich V's lands to his son that the new elector, Karl Ludwig, could refound the once great university in 1652 virtually *de novo* (Hautz, 1862–1864, Vol. 2, pp. 161–162; Press, 1985, pp. 314–370) (Fig. 4.5).

Less well known is the fact that the other Reformed university in western Germany—Marburg—was radically undermined for almost exactly the same period.[5] Ever since its Calvinization in 1605, the Hessian university had been a key source of strife between the temporarily ascendant Reformed Landgraf Moritz of Hesse-Kassel and his Lutheran kinsmen in the junior branches of Hesse. In March of 1624, the Lutheran Landgraf of Hesse-Darmstadt seized his chance. With the approbation of the emperor and the protection of Tilly, he invaded Hesse-Marburg, regained its university, and began legal proceedings to gain control of all the university's assets in Hesse-Kassel as well. Even the abdication of Landgraf Moritz

[5]For the fate of Marburg, see Hermelink & Kaehler, 1927, pp. 220–223; von Rommel, 1820–1843, Vol. 8, pp. 528–681; on the background, see Rudersdorf, 1982; Wilson, 2009, pp. 327–328, 341.

could not repair the situation, and when his son attempted to found a rival Reformed university in Kassel with less than half the means and highly unfavorable circumstances, the results were meagre. No more than six professors taught at the institution: three for theology and one each for the other three faculties. Despite letters of invitation to the chief Reformed academies in Germany, Switzerland, the Netherlands, and England, only 625 matriculations were recorded in its twenty-year history: 501 of these from within Hesse-Kassel, 94 from other German lands, and only eight from abroad (Auerbach, 1984; Falckenhainer, 1893). Not until Lutheran and Reformed universities were reunited in Marburg and Kassel was disbanded in 1652 did numbers recover to two-thirds of their average *ante bellum*.

The matriculation register of the third Reformed institution between Heidelberg and Marburg—the once thriving academy at Herborn—records not radically fluctuating fortunes but a gradual decline in numbers (beginning even before 1618) together with a catalogue of the woes which caused it (Zedler & Sommer, 1908, pp. 85, 86, 88, 92, 93, cf. 285). In 1625, it records, *pauci propter calamitatem belli tristissimi huc advenerunt*. In 1626, a fire, started behind the town hall by careless imperial soldiers, destroyed the greater part of the town (von Domarus, 1902–1903; Menk, 1981, pp. 62–89). In 1627, plague broke out among the ashes. In 1628–1629 the secularized ecclesiastical lands which financed the institution were lost to pre-emptive restitutions by the neighboring archbishoprics of Cologne and Trier. By 1630 all but three of the academy's professors had left for greener pastures (Menk, 1980). During the seven lean years from 1636 to 1642, only twenty students matriculated in Herborn. Although the academy recovered somewhat toward the end of the war, admissions levelled off at less than one third of their pre-war rate.

Just outside the effective boundaries of the Empire but firmly integrated into the German Reformed tradition, the university in Basle experienced a steady slide closely parallel to Herborn's (Wachernagel, 1951–1980). Basle's sharpest drop in matriculations came in 1633–1634, as the Swedish arrival in the south-west corner of the empire crowded 5,246 local refugees and their 1,776 cattle into the city of 11,000 inhabitants (Stritmatter, 1977, pp. 33–48, 64–68). But as the gradual downward drift of matriculations indicates, the university's problems went deeper than the ephemeral shocks of war. Basle had opened the seventeenth century as a strikingly international *Promotionsuniversität*, where not only Reformed theologians but central European students more generally took doctorates in law and medicine upon returning to the threshold of the Empire after studies in Padua, Bologna, Siena, or Montpellier (Bonjour, 1960). The demise of Heidelberg, Marburg, and Herborn cut the network of Reformed institutions which channelled theology students to Basle from the north; the rise of Leiden was to divert students of law and medicine from Italy to Holland; and Basle declined gradually to finish the century as one of the smallest and least significant universities in central Europe (Staehelin, 1957, pp. 52–57, 86–98).[6]

Catholic Universities

If the Reformed institutions were hit first and hardest, their Catholic counterparts were at first virtually unaffected. Against the background of uninterrupted imperial military success, the overall level of annual matriculations in Catholic universities remained almost perfectly steady in the years before 1630. With the arrival of the Swedes in subsequent years, however, each of the five Catholic institutions in southern Germany—Würzburg, Mainz, Ingolstadt, Dillingen, and Freiburg—saw their recruitment fall by eighty to one hundred new students per year. As in the case of the Reformed universities, it was the smaller Catholic institutions in the Rhine-Main region which suffered first and most. In the four years after the Swedes captured Würzburg and established the administration of their occupied territories in Mainz in 1631, these two universities ceased to function completely (Roberts, 1953–1958, Vol. 2, pp. 548, 558, 621) (Fig. 4.6).

The bishopric of Würzburg—long a pillar of counter-reformation in the center of Germany and a founding member of the Catholic League—surrendered after minimal resistance to the Swedish king on 15 October 1631, barely two weeks after his first great victory at Breitenfeld. 'With the enemy already approaching the city,' the Jesuit polymath Athanasius Kircher later recounted, 'the whole Jesuit college broke up within twenty-four hours' in complete confusion. Impelled by the terrible rumor that the Swedes would leave no Jesuit caught in the city alive, Kircher was swept with the tide of his brethren to Mainz, Speyer, and eventually to France. After shipping the city's rich libraries to Sweden, Gustavus combined Würzburg and the neighboring bishopric of Bamberg into a new duchy of Franconia and granted it to his leading general, Bernhard von Weimar. The new duke planned first to turn the

[6]Geneva—further removed from the Empire and less tightly integrated into the German Reformed academic world—played out a rather muted variation on Basel's theme. Before the war, its trend was manifestly downward; the first decade of the war prompted a brief recovery; but thereafter the downward trend only accelerated, and during the final decade of the war, Geneva enrolled less than 40 percent of the students enrolled during the first decade. Detailed research is needed to investigate these contradictory movements, but both the growth and the decline could be readily explained by reference to the war to the north. Between 1618 and 1628, Geneva may have benefitted from the displacement of students following the lost Heidelberg and Marburg and the demise of Herborn; but as new patterns consolidated around the Dutch universities in subsequent decades, Geneva may have found itself cut off from the academic trade routes which had previously supplied most of its non-Francophone students. Further detail will be found in Stelling-Michaud (1959–1980). I have been unable to consult the dissertation of Bernhard Troesch (1969). The bi-confessional university in Frankfurt-an-der-Oder is difficult to situate in this study. It is discussed in section II. 3 below, since its pattern of matriculations closely resembles those of its Lutheran neighbours in northern Germany.

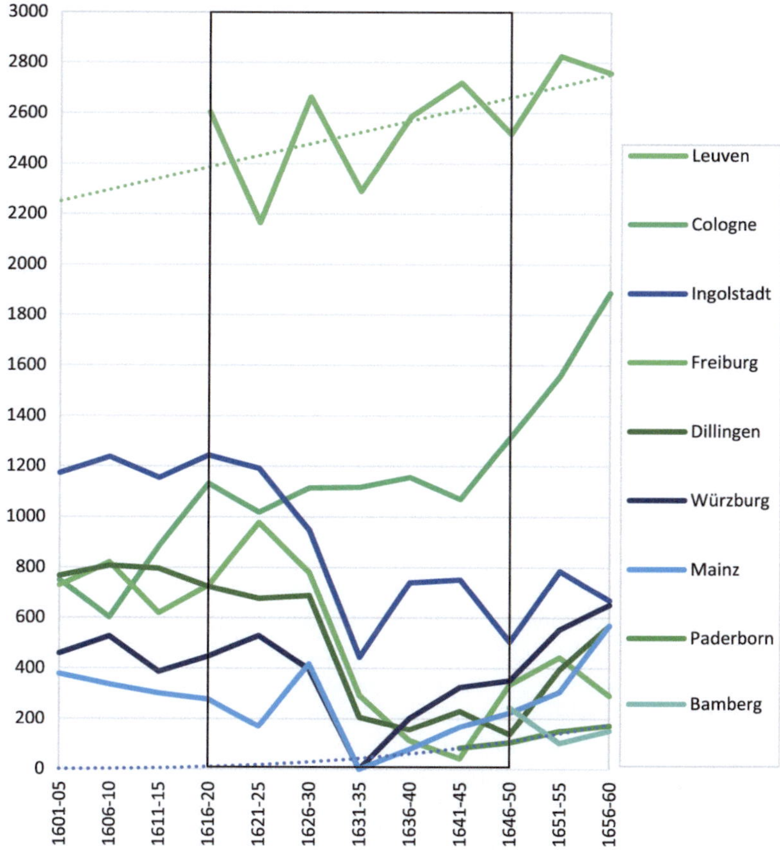

Fig. 4.6 Catholic universities in Germany and the Spanish Netherlands.
Source: Data from Eulenburg (1904), Schillings (1962–1963). Design by author.

city's university and two Jesuit colleges into an academy for educating young noblemen, to be named after the Swedish king himself. He then considered resurrecting the university and its gymnasium as a key instrument of the reintroduction of Protestantism into Franconia. But scarcely a month after the Swedish defeat at Nördlingen, Würzburg was retaken by imperial troops. The Catholic bishop, clerics, and professors returned from sanctuary elsewhere, but the reopening of the university was delayed by the need to replace some 5,000 books stolen by the Swedes from the library. Matriculations recommenced on 1 October 1636, and within a few years they had regained roughly their previous level (Schonath, 1967; von Wegele, 1882/1969, pp. 57–62, 476–479; Weber, 1979).

The story played out a few months later down-river in Mainz had a similar beginning but a less happy ending. On 18 December 1631, the elector-archbishop fled the city to join the bishop of Würzburg in exile in Cologne. Five days later, on the very day of the capitulation of the Spanish garrison, Gustavus Adolphus once again ordered the transportation of the elector's library to Sweden. In Mainz the

Swedes dug in even deeper, intending to make this state at the crossroads of Germany the capital of their new, evangelical, imperial order. When forced to leave in January 1636, they left an even more complete ruin behind them. The city lost a quarter of its houses, two fifths of its residents, and two thirds of its property; a great part of the university archives perished, and even the feeble recovery of student numbers was stalled by the French occupation of the city in 1644–1648 (Binz, 1917–1918; Brück, 1972; Frohnhäuser, 1899; Just & Mathy, 1965, pp. 21–23).

Gustavus' entry into Bavaria in 1632 was marked by a siege of the heavily fortified Ingolstadt, a bulwark of the duchy's defenses, with the duke (and now elector) Maximilian pinned down inside and his wounded general Tilly dying in the house of the city's law professor. Coming on top of the hyperinflation of the early 1620s and the disastrous harvest of 1626, Ingolstadt's intake of new students dropped to just 20 percent of its pre-war peak, but the crisis was short-lived. Unable to break through some of the strongest fortifications in Germany, the Swedes were forced to withdraw. Within two years, enrollments had bounced back to about two-thirds of their previous level, only to be dented by an outbreak of 'Hungarian fever' in 1634–1635. The systematic plundering of Bavaria in the last years of the war, coinciding with another outbreak of plague, cut matriculations to their absolute low point in 1648–1649; but very few of Ingolstadt's native sons elected to leave their impregnable home town to study in other universities during these years. Its combination of *Landesfestung* and *Universitätsstadt* made Ingolstadt a place of relative safety and stability, at least by the dismal standards of the 1630s and 1640s (Prantl, 1872, Vol. 1, pp. 376–378; Roberts, 1953–1958, Vol. 2, pp. 703–704; Schönauer, 2006, 2007, 2011).

The nearby university of Dillingen was less well defended. As the Swedes advanced, its numbers fell rapidly by 70 percent and were to remain low until the very end of the war. The city surrendered without resistance on 9 April 1632 in return for guarantees that Catholicism would be maintained. The Swedish king took the Jesuit faculty under his personal protection and relations with the Swedish governor began cordially. Lectures, services in the university chapel, and even the granting of academic degree continued without interruption throughout the war, although the latter took place without the university mace, which had been buried to protect it from the enemy. But the absence of such insignia must have been suspicious, rumors circulated that great treasures were hidden in the Jesuit college, and the Swedish governor exerted great pressure on college officials in attempting to find them.[7] The worst came—ironically but not uncharacteristically for this conflict—with the recapture of the city by imperial troops: even the university acts

[7]According to one rumour, 40 wagonloads of treasure had been removed to the Tyrol for safekeeping. According to the famous Paracelsian prophecy of the 'Lion of the North' (commonly applied to Gustavus Adolphus), at the culmination of his campaigns a stone of immeasurable value would be found somewhere between Swabia and Bavaria. The Swedish governor in Dillingen, David von Osten, believed that this treasure would be found amongst the city's Jesuits. After searching every corner of the Jesuit college without success he thrice imprisoned various clerics and officials, from the rector and episcopal chancellor on down, in attempting to press huge financial contributions from them.

complained that the Swedes inflicted nothing on the city to compare with its dreadful plundering by the Croats (Specht, 1902/1987).

When the turn of Freiburg im Breisgau arrived in 1633, it heralded a decade of pitiful enrolments. The reason was the nearby fortress of Breisach, which straddled the 'Spanish road' from Milan to Brussels, controlled access across the Rhine from France, and was therefore contested by all the main parties in the south-western theatre of the war. In 1632, the Swedes besieged both the fortress and the nearby city of Constance. In 1633, the duke of Feria, governor of Spanish Lombardy, marched 20,000 men through the Valtelline to relieve them. Freiburg—caught between the two sieges, and occupied first by the Swedes, then by imperial forces, and eventually recaptured by Sweden—saw its intake of students fall to a single pair (Beyerle, 1900; Häcker, 2011; Pagès, 1970; Parker, 1984, p. 132). Like its Lutheran neighbors in the south-western corner of the Empire, Freiburg suffered further losses with the French entry into the war: in the seven years between the French siege of Breisach in 1638 and the battle of Freiburg in 1644 (one of the war's longest and toughest), Freiburg was occupied twice more, and the university enrolled only 62 new students (Parker, 1984, pp. 152, 153, 163, 175–176; Wilson, 2009, pp. 678–684). By this time war had returned to Bavaria and dented the recovery of Ingolstadt as well. The great battle at Nördlingen in 1645, the final Swedish plundering of Bavaria in 1646–1647, and Mazarin's crushing defeat in the battle of Zusmarshausen as late as 17 May 1648 temporarily cut Ingolstadt's recruitment by 60 percent. The defenseless Dillingen was plundered by French and Swedish armies three times in the closing years of the war, and the latter did not leave entirely until 1650.

These five Catholic institutions thus shared a similar fate, yet the partial exception of Ingolstadt is noteworthy. The impression that this heavily fortified university city might mark the boundary between the main corridor of destruction and the more protected lands to the south and east is confirmed by the experience of the other universities in the region: the ancient university of Vienna, the recently-founded Jesuit university in Graz, and the Benedictines' war-time foundation in Salzburg (Fig. 4.7).

The contrast of their experience with the wildly fluctuating fortunes of their German coreligionists is immediately apparent. So too is their long-term upward trend. Yet caution is needed. Detecting the immediate, negative impact of warfare on matriculation rates is relatively simple: one must merely correlate a dramatic downturn in matriculations with a military event directly effecting the university city in question, which is often noted by contemporaries within the matriculation register itself. Demonstrating the indirect, positive impact of warfare on matriculations is a more painstaking business. Ideally, it requires knowledge both of the catchment areas of whole clusters of universities before the war and of the way in which military events disrupted those areas, undermining some institutions while benefitting others. Failing that, one can deduce the impact of the war less securely by studying changes in the geographical origins of students at a single institution. When that data is lacking, shifting matriculation rates can only suggest the likelihood that further work might reveal a more demonstrable relationship between the rise of some institutions and the decline of others.

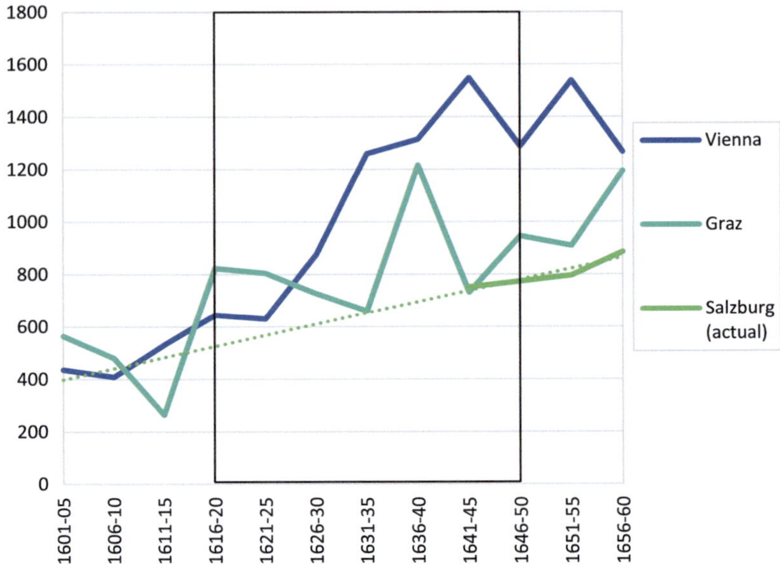

Fig. 4.7 Catholic universities in the southeast: Vienna, Graz, and Salzburg.
Source: Data from Andritsch (1977–1980), Gall and Paulhart (1974), Redlich (1933). Design by author.

Vienna's experience is a case in point. Its pattern of enrolment suggests that it benefitted from the chaos inflicted by the war to the north and west, but painstaking work will be required to confirm this hypothesis. As Swedish armies devastated Catholic universities on the Main, the Rhine, and the upper Danube in the early 1630s, Vienna's matriculations nearly doubled. After gentler growth for a further decade, Vienna was enjoying a rate of inscription two and one half times greater than before the war. Its peak of 520 matriculations in 1643 eclipsed even Leuven and Leiden to make Vienna, momentarily, the most heavily frequented university in central and north-western Europe. The synchronization of these countervailing developments—German disaster and Austrian flourishing—seems too obvious to be coincidental; yet until the Viennese register has been carefully analyzed, it will remain unclear how much of this growth results from the direct displacement of German students and how much from the reorientation of academic trade routes from the east which had previously been centered on or shared with Germany (Gall & Paulhart, 1974).

The better-studied Graz register reveals some of the trade-offs involved. Located in the duchy of Styria south of Vienna, this Jesuit university was sheltered from anti-imperial forces and also from the surge of students displaced by them. It was not until the latter 1630s that a wave of new recruits reached Graz. During this period, the number of Bavarians in Graz rose significantly; but these gains were counterbalanced by falling numbers of Swabians and Rhinelanders. Apparently, the military corridor which the Swedes had cut diagonally across southern Germany not only displaced students from within it: it also hindered mobility across it. Although the attractiveness of Graz to Catholics south and east of this corridor was increased, its accessibility to those to the north and west was diminished. Thanks to these contradictory forces, the German matriculations in Graz rose by only 13 percent during the war. Revealingly, they were then cut in half in the 1650s, as affairs to the north and west began to recover. Yet on balance Graz had a good war: a period of growth during the 1640s left it eighty percent larger at the end of the conflict than at the beginning (Andritsch, 1977, Vol. 1, p. xxix, 1980, Vol. 2, p. xxvi).[8]

To what extent this upward trend was shared by the Benedictine university founded in Salzburg in 1620, opened in 1622, and fully privileged by 1625, is impossible to say with certainty, since its matriculation register only survives from 1639 onward (Redlich, 1933). Its fortunes thereafter are not quite as serene as Figure 4.7 appears to indicate—coincidentally or otherwise, Salzburg's matriculations were cut in half the year following the conclusion of peace in Westphalia—but they had fully recovered within five years, and the overall trend is one of gentle growth. In any case, the general trajectory of this cluster is clear: annual matriculations in universities south-east of Bavaria nearly trebled from their average in the decade before the war (227 in 1611–1620) to the decade after it (646 in 1651–1660). The question is not whether they benefitted from the conflict, but whether they did so directly (from displaced German students) or indirectly (by altered patterns of international student migration), as seems more likely.

Vienna and Graz do not, however, mark the eastern limit of this effect. The matriculation register for Prague does not survive for this period, that for Olomouc is fragmentary, and the smaller universities in Poland-Lithuania and Upper Hungary await closer study.[9] But the shockwaves of the Swedish arrival in Catholic Germany were powerfully felt as far east as Cracow. In 1632, matriculations in the Jagiellonian University suddenly doubled; and they remained sky-high for another three years before suddenly falling back to their early seventeenth-century average in 1636 (Zathey & Barycz, 1950). Polish scholars have found nothing in the domestic situation to explain this dramatic rise and fall (Kaniewska, 1986, pp. 136–137, 146). Placed within a broader context, however, their origin is obvious. This spike in matriculations is by far the most dramatic of any university east of the Rhine in the seventeenth century, and probably the most dramatic anywhere in Europe in this period. It comes in the midst of the Swedish assault on Würzburg, Mainz, Ingolstadt, Dillingen, and Freiburg between October 1631 and 1633, and before the first clear evidence of rising numbers in Vienna in 1634. Cracow's huge spike therefore clearly

[8]The figures for the decades beginning in 1621, 1631, 1641, and 1651 are as follows: Bavaria: 47, 72, 66, 26; Swabia: 57, 40, 64, 48; Franconia: 16, 18, 16, 6; Rhineland: 20, 11, 12, 2; other German regions: 14, 15, 17. 9; total German matriculants: 154, 156, 175, 91.

[9]Similar effects should be sought in smaller institutions in Bratislava / Pózsony (founded 1465), Braniewo / Braunsberg (1568), Vilnius (1578/79), Zamość (1594), and Tyrnava (1635). See Lühr (1925) and Zsoldos (1990).

registers the seismic shock radiating outward from the greatest disaster in the post-Reformation history of central European Catholic higher education. The only question is whether these additional matriculants are German students fleeing the Swedes or, as seems more likely, Catholic students from Poland-Lithuania, Hungary-Transylvania, and perhaps also the lands of the Bohemian crown who might otherwise have studied further west but who gravitated in the immediate aftermath of the crisis to the safety of Cracow instead.

What then of the Catholic universities to the north and west of the corridor of destruction running diagonally from eastern Pomerania to Breisgau? Returning to Figure 4.6 suggests that they escaped the disasters visited on their neighbors, but did not benefit from them in the manner of Vienna or Cracow.

The oldest of the German Catholic universities is the crucial case. Throughout the late sixteenth century and the first years of the seventeenth, Cologne profited little from the general educational boom, but on the eve of the war its numbers grew strongly. The city's geographical position to the north-east of the main military movements subsequently shielded it from direct military entanglements, and its strenuous assertions of neutrality, backed by generous payments to passing armies, spared it the calamities visited on most other German centers. Commercially, it prospered from arms manufacture and trade, by providing credit to Catholic combatants, and through the maintenance of good relations with the Dutch Republic. Famous as a gathering point for Catholic leaders planning the recovery of assets lost to Protestant forces, it was the natural place of refuge for students and professors fleeing Würzburg and Mainz. Meanwhile, the approaches from the west for students from its traditional catchment area in the Spanish Netherlands remained open (Bergerhausen, 2010, pp. 222–223; Methuen, 1988, pp. 329, 331–332). All told, Cologne remained the least affected of all German universities of any confession throughout the conflict; but when combined with her dramatic spurts of growth immediately before and after the war, her general trend in matriculations was emphatically upward.

How much the young university of Paderborn, northeast of Cologne, shared these fortunes is unclear. Although founded just before the war, its matriculation register only survives from 1637 onward. For the next six years, it recorded an average of only twenty new students per year. Growth was detectable from 1643, after the worst had passed for the German Catholic institutions to the south but before they began to grow as well. The attempt to found a university in Osnabrück also presented a mixed picture: no sooner had it obtained papal privileges in 1629 and imperial privileges in 1632 than it closed in 1633. In 1648, the first new Catholic university since Salzburg was established in Bamberg, in the middle of what was previously the military corridor, reinforcing the now well-established upward trend and helping to mark the end of a traumatic period.

Nearly 200 kilometers due east of Cologne, Leuven's relationship with the central European conflict is not easy to perceive from matriculation rates alone. In Leuven, disruption was more evident in the pre-war period, in which the city was repeatedly occupied by foreign troops waging the war between Spain and the rebellious Dutch provinces. Between August 1569 and February 1616, not even its matriculation register survives. By the time the record resumes, however, stability had been re-established; and for the next half century, the matriculation rate fluctuated relatively little around a very gentle upward trend. Proportionately, the effect was not huge: the university was less than ten percent larger at the end of the war than at its beginning. Yet, given the large size of Leuven, this still represented a net gain of some 40 matriculants per year (Lamberts & Roegiers, 1990, p. 28; Schillings, 1962–1963, Vol. 5, pp. ix–x; Vol. 6). All told, Leuven matriculated 13,813 students between 1618 and 1648, nearly as many as the second and third-ranked Catholic universities in the Empire—Vienna (6,832) and Cologne (7,031)—put together, far more than Königsberg (8,193) and Wittenberg (10,640), more even than the fabled Leiden (13,448), and second only within the Empire to the Saxon giant: Leipzig (15,838).

The main university of the Spanish Habsburg's Netherlandish provinces, Leuven thus provides an illuminating contrast to Vienna, its counterpart in the Habsburg patrimonial lands. Vienna more than doubled in size during the Thirty Years' War because its sheltered position in the southeast benefitted not only from German students displaced by the war but also, and perhaps primarily, by accommodating students from Hungary, Poland, and the Czech lands who might otherwise have studied in Germany. Leuven's northeasterly location was situated very differently with respect to international academic trade routes. Its neighbors to the north (the predominantly Protestant Dutch Republic) and the south (a Catholic country, but one abundantly furnished with universities of its own) were not part of the German Catholic universities' natural catchment area; so the disruptions east of the Rhine did not channel fresh streams of students from northwestern Europe to Leuven during the wartime period.

Yet Leuven was not as immune from the effects of the war as its bare matriculation rate at first suggests, due to the rekindling of conflict between France and Spain. After Sweden lost its German Protestant allies in the Peace of Prague, a general cessation of hostilities favorable to the Habsburg looked likely. To prevent this, France followed its longstanding financial support for the Swedish war effort by declaring war on Spain in 1635 (Parrott, 1987). The French siege of Leuven that same year provoked the only sharp drop in Leuven's rate of matriculations during the ensuring conflict (van Nimwegen, 2014, pp. 170–171; Wilson, 2009, p. 560). The wartime peak in Leuven's enrolments in 1641 was also a consequence of warfare. Douai, the second and more southerly university of the Spanish Netherlands, found itself in the front line of war with France and suffered terribly: its estates were despoiled, its finances ruined, its damaged buildings repurposed as barracks for troops and a hospital for the wounded, and plague scattered even the permanent citizens, until the city and its environs were finally captured by the armies of Louis XIV in 1667 and permanently annexed to France in the Treaty of Aix-la-Chapelle the following year (Dehon, 1998, pp. 23–34). Unsurprisingly, as the annalist of its College of Anchin complained as early as 1640, students fled Douai in droves for the greater security of Leuven. The loss of Douai's matriculation register makes this transfer difficult to quantify without a meticulous reconstruction of Leuven's shifting catchment area in these years. Nevertheless, it seems likely that Leuven, although buffered from the effects of the war in Germany, may have benefitted, on balance, from the related conflict in Artois and Flanders.

Lutheran Universities

Even more directly than Catholic ones, Lutheran universities seemed initially to benefit from the outbreak of war. The only two universities ever founded in imperial free cities—Strasbourg and Altdorf—owed that status to adroit diplomacy in the first years of the conflict (Schindling, 1978) (Fig. 4.8). Strasbourg's strategically important location as the nearest large city to Heidelberg produced two key benefits at the outset of the war. As Catholic armies closed in on the Palatinate in 1621, skilful diplomacy won from the emperor a charter raising the city's famous academic gymnasium to full university status as a reward for Strasbourg's withdrawal from the Protestant Union (Schindling, 1977, pp. 72–77). The closing of Heidelberg the following year also provided the new university with important fresh sources of students: 50 young men from the Palatinate enrolled in Strasbourg in its first nine years—more than in any other Lutheran university, or indeed than any Reformed institution besides Leiden and Basle (Persijn, 1959, pp. 40–42 and appendix).[10] Nuremberg's location in the delicate region near the Upper Palatinate, Bohemia, and Bavaria was no less crucial, and the city's withdrawal from the Union in 1622 was soon repaid when its gymnasium in Altdorf was granted university statutes two years later (Schindling, 1978, pp. 166–167).

Even these uniquely fortunate southern Lutheran universities, however, could not escape the less benign effects of three decades of war. After eight years of sustained prosperity, Altdorf's matriculations fell briefly to zero in the summer of 1632, as Gustavus' 45,000-strong army dug in on the outskirts of the Nuremberg, Wallenstein's army attempted to starve it out, and the Swedish chancellor, Axel Oxenstierna, arrived with a third army to destroy the resources sustaining Wallenstein's horde (Parker, 1984, p. 130; Roberts, 1953–1958, Vol. 2, pp. 717–720, 726–733). Altdorf's crisis therefore arrived half a decade before the general nadir of the Lutheran universities in the latter 1620s. As the young university's fortunes vacillated over the next decade, the rector of Altdorf struggled intermittently to prevent the patricians of Nuremberg from closing the institution altogether (Ernstberger, 1966). Yet although its numbers remained modest, its overall experience was relatively benign by the terrible standards of the era.

Tübingen, the third of the smaller and more southerly Lutheran universities, retained healthy enrolments two years longer until 1634, when the young Duke Eberhard III of Württemberg, having reached majority, allied his duchy with the previously unbeatable Swedes. In the Swedish defeat at Nördlingen a few months

[10]During the same period Strasbourg replaced Heidelberg as the favourite university of students heading to or returning from Basle (Frijhoff, 1986a, p. 42).

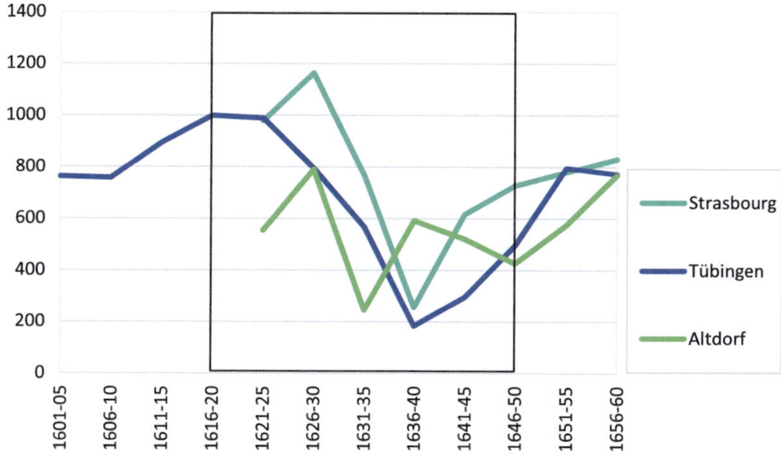

Fig. 4.8 Lutheran universities in southern Germany, 1601–1660.
Source: Data from Eulenburg (1904). Design by author.

later, the duke lost two-thirds of his army. Eberhard fled the next day with his treasury into exile in Strasbourg, and for the following four years his duchy was occupied by Habsburg forces which rigorously enforced the restitution of the rich monastic lands upon which his state and university jointly depended. In 1635, the libraries of the duke and the *Collegium illustre* were packed off to Munich and plague swept through the university, killing nine professors. By 1639, faculty numbers had been cut in half, matriculations had fallen by four-fifths, the professors' incomes had disappeared almost completely, and the university library had been decimated as well, as its most precious volumes were sold to meet the never-ending demand for contributions. French intervention in Württemberg in 1646 extended this dismal period further, and numbers in Tübingen rose only feebly before the French general, Turenne, personally returned the city to Duke Eberhard III on 27 November 1648, after which they bounced back quickly, if not quite to pre-war levels (Deckers-Hauff & Setzler, 1977, pp. 126–140; Klüpfel, 1849/1977, pp. 136–142; Schreiner, 1985; Vann, 1984, pp. 93–97; Zaschka, 1993, pp. 44–47, 155–157, 179–183). Strasbourg—more firmly ensconced within the French protectorate on the left bank of the Rhine—was to experience a later decline in 1635 and a more rapid and sustained recovery from 1641 onwards.

Of the northerly universities, the fashionable, young Helmstedt suffered first, most drastically, but also most briefly (Fig. 4.9). Caught in the middle of the Danish-Lower Saxon phase of the war, it followed a rhythm of disaster and recovery earlier than either its larger Saxon neighbors or the smaller Lutheran universities far to the south. Helmstedt's numbers faltered momentarily as early as 1622–1623, following the sudden death of three of the university's six theology professors and the very first incursion of the war into northern Germany: the skirmishes of Tilly and Christian

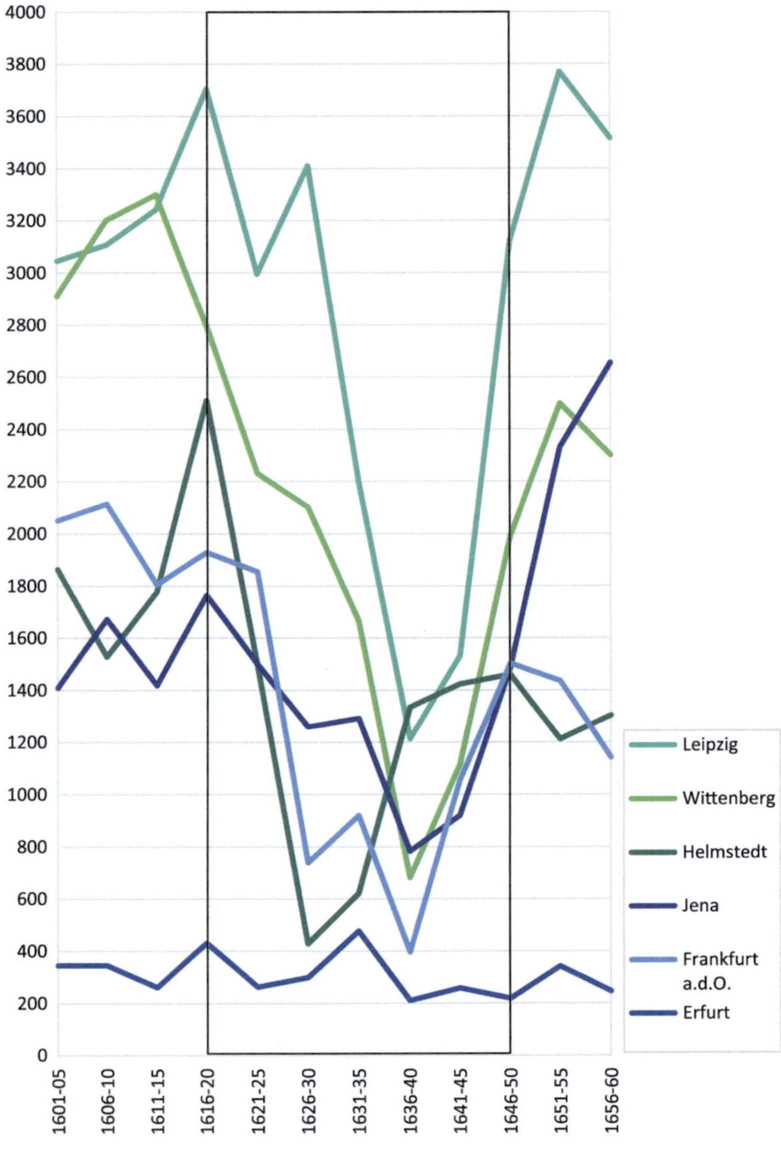

Fig. 4.9 Lutheran universities in northern Germany.
Source: Data from Eulenburg (1904). Design by author.

von Halberstadt in the Leine Valley, the Plesse, and the area around Braunschweig. Students streamed back in 1624 only to desert the university almost entirely in the following years as Christian IV of Denmark took up headquarters in nearby Wolfenbüttel, imperial forces moved northwards through the Harz, Duke Friedrich Ulrich deserted his lands for the sanctuary of the neutral free city of Braunschweig, plague broke out in Helmstedt, and the region passed into Habsburg control after the Danish defeat at Lutter. The arrival of Wallenstein's troops in 1628 and the extinction of the ruling branch of the house of Braunschweig-Wolfenbüttel in 1634 twice dented the university's recovery. But even in these trying times Helmstedt's academic reputation was bolstered through the appointment of such notable professors as the leading Lutheran irenicist, Georg Calixt, and the celebrated polyhistor, Hermann Conring. Following the recovery of dynastic stability under the passionate bibliophile, Duke August of Braunschweig-Lüneburg, in 1636, matriculations recovered strongly in 1637 and Helmstedt entered into a second period of prosperity. Despite repeated fighting over the occupied capital of Wolfenbüttel in 1638 and 1641, three-quarters as many students entered the university in the final third of the war as in the peak decade before it. Thanks to the devastation visited on nearby Saxony, Helmstedt admitted more students than any other German university in 1637, 1639, 1640, and 1643 (Alschner, 1998, pp. 65–79, 100–110; Eulenburg, 1904, p. 291; Hofmeister, 1907; Triebs, 1995, pp. 16–19).

The three large Saxon universities—Leipzig, Wittenberg, and Jena—also suffered temporarily from their proximity to the fighting in the mid–1620s, during which the armies of the Danish king as *Kreisoberst* of Lower Saxony, his allies, and his enemies moved back and forth across their territory (Parker, 1984, pp. 57–58; Richter, 2011). But their losses were relatively slight and they recovered strongly in 1627 and 1628, almost regaining their high-point of ten years earlier. Despite an outbreak of plague the year before, in 1627 the elector Johann Georg could still boast that Leipzig occupied the *first place* among evangelical universities. But just as electoral Saxony suffered worst in the years which followed, Leipzig—a major commercial crossroads in time of peace—became a favorite battleground in time of war, especially after the Edict of Restitution and the arrival of the Swedes persuaded the Saxon elector to break his earlier alliance with the Emperor in the autumn of 1631. The first and last great battles of Gustavus Adolphus's German campaign—Breitenfeld on 17 September 1631 and Lützen on 17 November 1632—took place within a few kilometres of Leipzig; a second main battle at Breitenfeld followed in 1642; and with each main conflict Leipzig was besieged. The juridical faculty found itself in a particularly unfortunate strategic position: its buildings—the *Ordinariatshaus* and the *Collegium Petrinum*—offered a semi-fortified position with a clear line of fire on Leipzig's citadel, the Pleißenburg. In 1632, the imperial general, Holke, shelled the Pleißenburg from the *Ordinariatshaus* and returning fire from the fortress wrecked havoc with neighbouring buildings. After Lützen the story was repeated with roles reversed: the Saxony artillery firing from the university premises, the imperial forces firing at them. After a fierce storm completed the devastation in 1634, the *Ordinariatshaus* was described in one bitter faculty memo as 'nicht mehr als ein Eulennest' ('no more than an owl's nest'). In 1641, even these ruins were swept away to make room for soldiers' quarters; and between these dates matriculations in Leipzig sunk to barely one quarter of their pre-war high (Friedberg, 1909, pp. 51–53; Rathmann, 1984, pp. 72–75; Rudersdorf, 2009). It was only after the Swedes recaptured and retained Leipzig in 1642 that a degree of stability allowed the university at last to begin recovering (Zirr, 2009, 2010, 2011).

Confessionally, Viadrina in Frankfurt an der Oder was something of an anomaly: a Catholic foundation, tardily Lutheranized, and then subjected to a Reformed ruler since 1613 without undergoing a 'second Reformation' (Nischan, 1994). Although it escaped the drastic fate of the Reformed institutions further south, its wartime experience was far from happy. As Figure 4.9 shows, it suffered an early collapse in the late 1620s parallel to Helmstedt's. Unlike the Guelf university, however, its recovery in the early 1630s was aborted: it had to wait for the final decade of the war to make a long and very partial recovery, and its wildly oscillating matriculation rates nevertheless indicate perhaps the most tormented history of any Germany university during the war. As a rare crossing point on the Oder, Frankfurt controlled movement east and west. Only a few kilometers down-river from the northernmost point in Silesia, it also marked the border separating the Habsburg-dominated south from the Protestant north. Add to this the awkward position of Brandenburg as a whole between the Habsburg lands to the south-east and the theatre of the Danish campaigns and Swedish landings to the north and west and the source of Frankfurt's troubles begins to become apparent. On 26 June 1626, imperial troops, moving northward in force against Denmark and her allies, took control of the city and admissions to the university dropped immediately by 80 percent (Schultze, 1964, p. 223). Scarcely had the university recovered in 1628–1629 than Gustavus Adolphus took up positions in Pomerania, Tilly responded by strengthening his garrisons at the main Oder crossing-points (p. 241), and matriculations in Frankfurt fell sharply again in 1630. The arrival of the Swedes from Stralsund down the west bank of the Oder only made matters worse. With the approach of the enemy, the imperial field-marshall, Tieffenbach, burned the city's suburbs and withdrew into freshly prepared defenses. On 13 April 1631, after intense bombardment of its defenses from the heights to one side, the city was stormed in the first main siege of the Swedish king's German campaign. The savagery of the sack was not confined to the slaughter of the imperial garrison but extended to the 'persons and property of the burgers' and to the 'seals, ornaments and maces' of the university'.[11] Again numbers bounced back, only to sink again as Wallenstein's troops retook possession of the city in 1633 (Schultze, 1964, pp. 252–253). Finally, after the failure of the Peace of Prague, Brandenburg was repeatedly fought over by Swedish and imperial armies, to the great devastation of cities and countryside alike, until the young elector Friedrich Wilhelm finally broke with the emperor and allied with Sweden on 24 July 1641, for more effective protection (Schultze, 1964, pp. 261–294; Parker, 1984, pp. 165, 168). During the years of greatest disruption (1636–1640), the university admitted only 397 students—less than one-fifth of its pre-war peak in the quinquennium 1614–1618.[12] All told, this biconfessional institution can be

[11]On the background situation, see Roberts, 1953–1958, Vol. 2, pp. 243, 261, 481; for details, see Zopf, 1927.

[12]These vicissitudes are little evident in Günter Mühlpfordt's (1983) sanguine survey (pp. 47–53).

regarded both as the most fortunate of the Reformed universities and the least fortunate of the Lutheran ones.[13]

The smallest of the three Saxon universities, Jena, by contrast, fared best. After record numbers in 1620, the war began badly, thanks to Thuringia's central strategic position and the lack of strong, consolidated political authority with which to repel plundering troops. The establishment of the Swedish military chancellery in Erfurt temporarily halted the decline; but Thuringian participation in the Peace of Prague exposed the region to years of Swedish reprisals after 1635. The leading theology professor, Johann Gerhard, fled Jena that year to escape imprisonment by the Swedes for supporting the Peace; in 1636, the Swedes thoroughly sacked his farmlands in Roßlau and in 1637 imperial forces plundered his house in Jena as well. Yet in conditions of such general misery, Jena nevertheless experienced relative good fortune. Even during its worst quinquennium, matriculations were nearly half the level of their peak, and the final armistice inaugurated a period of unprecedented growth (Richter, 2011, pp. 49–51; Steinmetz, 1958, Vol. 1, pp. 63–71).

The main exception within this cluster of universities was Erfurt, a unique institution before the war, which experienced a unique fate during it. A thriving institution in medieval times, Erfurt had suffered since the Reformation due to its anomalous situation: although theoretically subordinate to the elector-archbishop of Mainz, the city and its university were predominantly Lutheran. A chair for evangelical theology had existed there since 1566 and its Catholic equivalent had lapsed altogether between 1600 and 1627. Gustavus Adolphus therefore received a rapturous reception from the Protestant population upon entering the city on 2 October 1631: bells rang, the populace cheered, and the civic council invited him finally to reform the university entirely. The king and (after his death) his chancellor responded with lavish new endowments of secularized ecclesiastical lands, invitations to four prestigious Lutheran professors to join the theological faculty, a new set of statues, and up-dated methods of instruction which were to culminate in a botanical garden and an anatomical theatre. Matriculations (dented a few years earlier by an outbreak of plague) peaked in excess of 150 in 1633. The Swedish disaster at Nördlingen and Saxony's defection to the imperial camp after the Peace of Prague upset these grandiose plans and briefly reinstated the *status quo* of 1627; but the Swedes regained the city in 1636 and retained it until the end of the war. Although the university can hardly be said to have thrived, its figures during the war document, on balance, only the gentlest decline (see Bock, 1908, pp. 62–67; Märker, 1993, pp. 62–97; Pelizaeus, 2005; Richter, 2011, pp. 51–54; Weiß, 2008).

The wartime fortunes of these eight Lutheran institutions show a close family resemblance: all suffered a collapse in enrolments during the middle years of the war (least dramatic in the case of tiny Erfurt, most precipitous in the giant Saxon

[13] Notker Hammerstein has argued that the destruction of Heidelberg 'allowed for the development of Frankfurt on the Oder as a brilliant alternative'; but it is difficult to see how this judgement applies to the wartime period itself. See Hammerstein (1986–1987, p. 120, citing Hammerstein, 1972).

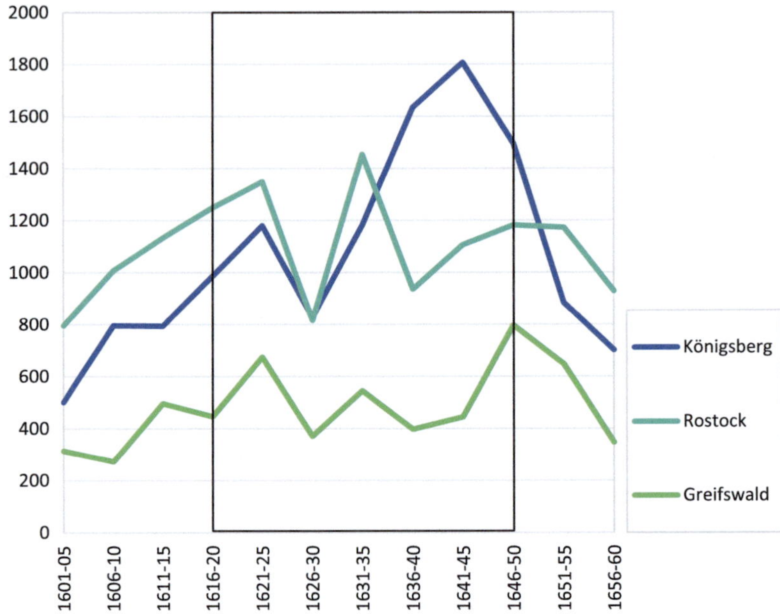

Fig. 4.10 Lutheran universities on the southern shores of the Baltic.
Source: Data from Eulenburg (1904). Design by author.

universities), and all but two (Helmstedt and Altdorf) reached their nadir during the latter 1630s and experienced little recovery until the very last years of the war. Far to the north, however, three Lutheran universities followed a different rhythm: all three grew strongly in the early years of the war, suffered their worst setbacks early, recovered quickly, and experienced positive growth over the war as a whole. All three were located on the Baltic (Fig. 4.10).

The humblest of the three—both in pre-war size and in wartime prosperity—was Greifswald. Her surge in the first years of the war—from 50 to 100 student enrolments on average—was proportionately the strongest of the three; but her fall in November 1628 was more drastic. Greifswald's occupation by imperial troops was longer even than the other much abused Pomeranian cities, and the commander during most of this period, Ludovico Francesco Perusio of Malta, was described in the university's album as an 'incarnatus diabolus'. This reputation derived, however, not from suppression the city's Lutheranism but from his uncompromising measures in defending the city from the armies of Gustavus Adolphus, who landed in Pomerania not far from Greifswald in July 1630. Despite these efforts, Greifswald was the first German university city to fall to the Swedes, almost a year later, in June 1631. The sharp recovery which followed was undermined by the return of large armies to Pomerania in 1637–1638, when, as rector Balthasar Rhau complained in the register, a deluge of rustic refugees from the miseries outside the city walls ruined both the university's rural income and the living conditions of students and professors alike. From 1639, however, the Swedes, now ruled by Queen Christina, began to invest in the university which they successfully solicited in the protracted negotiations for peace and integrated into the Swedish network of educational institutions throughout the Baltic. Greifswald ended the war with an unprecedented surge in enrolments to around 210 students. However modest her absolute numbers, she attracted 50 percent more students during the war than in the thirty-year periods before and after it (Kosegarten, 1986, pp. 237–256; Langer, 2011).[14]

Rostock's university was twice the size of Greifswald's, and its fortunes tracked those of its Pomeranian neighbor on a grander scale. Throughout the early years of the war, Rostock enrolments climbed steadily to over 200 per year, despite an outbreak of plague in 1624; but between the arrival of imperial troops in Mecklenburg early in 1627 and their capture of Rostock in October 1628, matriculations crashed by four-fifths. The importance of this Hanseatic port to Wallenstein's plans to build an imperial navy inclined him to treat the city more liberally than many (Heidorn, Heitz, Kalisch, Olechnowitz, & Seemann, 1969, Vol. 1, pp. 57–59; Stieda, 1917), and university enrolments began to recover until a new and unexpected danger emerged. On 22 January 1631, the imperial commander in Rostock, Heinrich Ludwig von Hatzfeld, was murdered in his apartments, and the culprit was quickly identified as a member of the university—a licentiate in theology named Jacob Barmeyer. Four days of inquisition under torture and strenuous assertions of loyalty from the university established, however, that the young divine had acted without accomplices; so the commander's successor satisfied the need for exemplary retribution with a most agonizing and protracted public execution and hung the theologian's four quarters over the gates of a city which had narrowly escaped a punitive sack. Further good fortune followed at the end of the year: as Rostock prepared for a long and bitter Swedish siege, the unexpected news arrived of Tilly's defeat at Breitenfeld, and on 6 October 1631 the demoralized imperial garrison capitulated without resistance rather than face the full wrath of the Swedes (Krabbe, 1863/1994, pp. 83–84, 105–125, 157–170; Roberts, 1953–1958, Vol. 2, pp. 516–517).

For the university, the worst of the war was already over. In the following six years—precisely the period in which warfare began to disrupt the more populous universities in Saxony and the south—Rostock recorded the highest entrance figures in her early history, not even surpassed in the nineteenth century; and in three of these years—1632, 1633 and 1635—she enjoyed the highest matriculation rates of any German university. Most of these students had travelled some distance to Rostock. The percentage of native Mecklenburgers in Rostock fell from almost one-half in 1630 to scarcely one-quarter in 1635 and under one-fifth in 1640. The Scandinavians who had previously constituted as much as one fifth of Rostock's students stayed away; but in their place students arrived in unprecedented numbers from lower Saxony, Brandenburg, Prussia, the Baltic countries, and other regions as

[14]On the Swedish phase, see the basic study by Ivar Seth (1956), and more recent literature including Langer (2008)

far away as Transylvania.[15] From a university which provided a landfall for Scandinavians visiting Germany, Rostock was briefly transformed into a northern refuge for Germans themselves. Even after imperial forces regained the city in 1637, letters exempting Rostock from quartering and other burdens were consistently implemented by the mild regime of Wilhelm von Calchheim (*alias* Lohausen) and numbers recovered strongly. All in all, although her intake fluctuated dramatically, Rostock too had a good war: between 1619 and 1648, Mecklenburg's university recorded about 20 percent more matriculations that in the thirty years before the war and 50 percent more than in the thirty years after it.

The clearest instance of a Lutheran university profiting from the demise of its German counterparts, however, fell far outside the boundaries of the Holy Roman Empire itself, although not outside the sphere of imperial politics. Königsberg, the capital of the duchy of Prussia, was ruled by the Elector and Margrave of Brandenburg as a fief of the Polish king. The city felt the shock of the Swedes' arrival even before their entry into the German war: at the start of the Polish-Swedish War in May 1626, Gustavus Adolphus landed in Pillau, a peninsula scarcely one day's march from the Prussian city. But the Swedish king had recently married the sister of the Prussian duke, Georg Wilhelm of Hohenzollern, whose early and unconditional promise of neutrality, backed up by the city's formidable defenses, averted a Swedish siege. By the time the Swedes king arrived in Germany four years later, Königsberg's recovery was well underway; and when disaster engulfed the Saxon universities in 1636, students from across the Lutheran world began seeking the safety which its distance from the Empire provided. In August 1638, Georg Wilhelm did the same, transferring his entire court from Berlin to the Baltic city to escape the chaos engulfing Brandenburg. For a decade toward the end of the war, this provincial *Landesuniversität* was transformed into one of the largest and most cosmopolitan university in central Europe. During 1636, 1638, 1641, 1642, 1644, and 1646, the easternmost major university in Europe also attracted more students than any other east of the Rhine (Eulenburg, 1904, pp. 84–85; Lehmann, 1929; Roberts, 1953–1958, Vol. 1, pp. 320, 322, 325–326, 338–340; von Selle, 1956, pp. 78–79).[16]

Königsberg therefore highlights a very important feature of the academic geography of northern Europe in this period: the effects of the Thirty Years' War on academic populations were not confined to the Holy Roman Empire. As already noted, the huge surge in Cracow's population during the 1630s is a dramatic evidence of this feature in the Catholic world. Königsberg provides evidence that something similar was at work in the Lutheran world as well. Part III of this paper will return to consider the Lutheran and Reformed manifestations of this phenomenon in more detail. Before doing so, however, a summary of the patterns evident within the Empire is in order.

[15]Analyzed exhaustively in Asche (2000, pp. 62–63, 170–171, 177–178, 210–211, 213–215, 535). See also the summary in Asche (1995, pp. 141–162) and Helk (1987, p. 42).

[16]While students from Prussia enrolled in roughly the same numbers as before the war, their proportion of the student body dropped to one-third as the number of Pommeranians doubled, large contingents arrived from Schleswig-Holstein, Mecklenburg, Braunschweig and especially Silesia, and smaller groups travelled from as far away as Austria, Hungary and Transylvania as well as the Baltic lands.

Regions Compared

The data set studied so far is by no means huge: sixty years of annual matriculation figures plus the tally of 12 quinquennial figures for each of 29 universities is only 2160 data points. Yet the complexity of this landscape is such that we need to aggregate these figures by region and confession to gain a clear sense of the overall patterns. In order to examine the *relative* growth or decline of different clusters of universities, the following two graphs use the pre-war size of each university as a baseline. More specifically, they express the aggregate matriculations of each cluster of universities as a percentage of their level during the five-year period 1616–1620. Matriculation levels below the 100 percent line are smaller than they were in 1616–1620, while those above the line are larger. These graphs do not, in other words, express the absolute size of these clusters relative to one another. Instead, they chart the growth and decline of each cluster relative to its size during the five-year period prior to the outbreak of large-scale war.

Figure 4.11 distinguishes the four regions whose wartime trajectories are plotted in Figure 4.12. The aim is not to divide the Empire into four equal parts but to distinguish four groups which shared similar wartime experiences. The three Catholic universities in the smallest cluster south of the Danube (the southeast quarter) share an experience strikingly different from the largest, multiconfessional cluster of twelve universities in the German lands south of the Main (the southwest). The eight

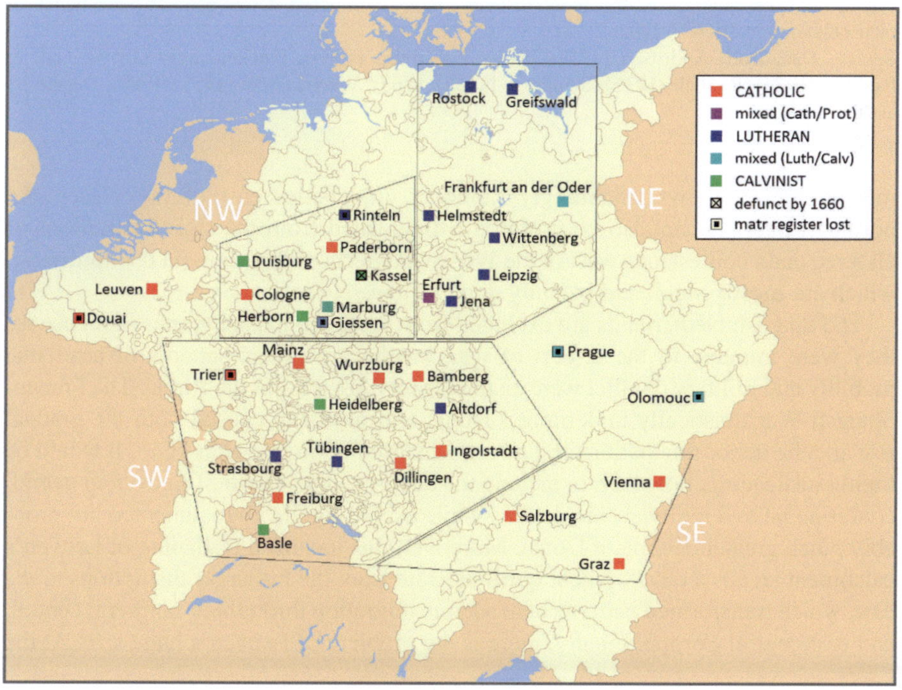

Fig. 4.11 Universities in the Holy Roman Empire, 1600–1660.
Source: Design by author. Data from Frijhoff (1996).

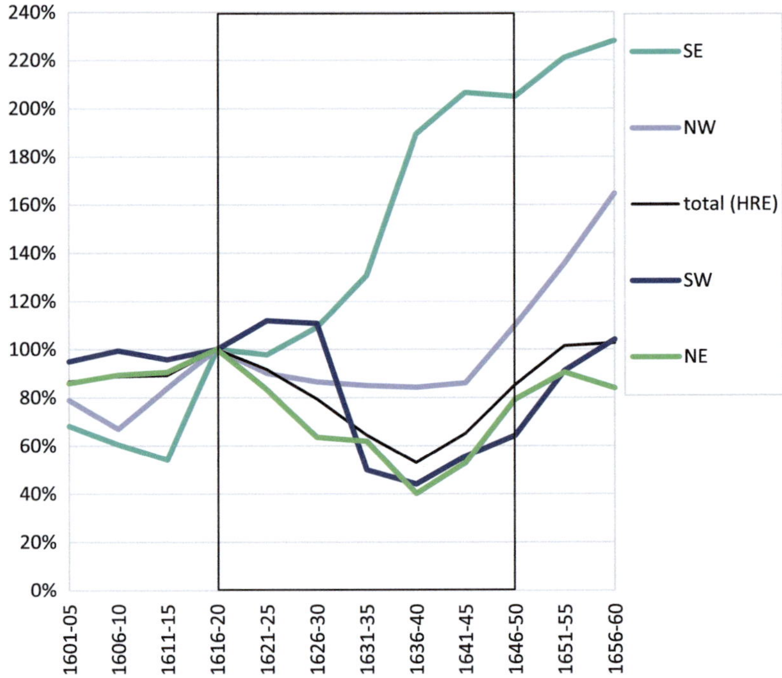

Fig. 4.12 Four quarters of the Holy Roman Empire compared. Quinquennial percentages of matriculations in 1616–1620.
Source: Data from Andritsch (1977–1980), Eulenburg (1904), Falckenhainer (1893), Gall & Paulhart (1974), Redlich (1933), Schillings (1962–1963), Wachernagel (1951–1980). Design by author.

mostly large and overwhelmingly Lutheran universities east of the Weser (the northeast quarter) can also be usefully distinguished from the eight institutions of all three main confessions west of the Weser and north of the Main (in the northwest) which are mostly small and patchily documented.

The Spanish Netherlands are excluded from these composite graphs, not because they were immune to the effects of war, but because their patterns of academic mobility and military conflict were distinct from those east of the Rhine. The Franco-Spanish War drastically undermined Douai and probably caused both the gradual rise and the major fluctuations in Leuven's matriculation rate after 1635. It would be highly misleading, however, to include the documented growth of Leuven in overall confessional and regional calculations while excluding the undocumented but probably much greater decline of Douai. Moreover, this localized reshaping of Leuven's catchment area is a development quite separate from the far larger disruptions to the east, which transformed patterns of academic migration throughout northern, central, and east-central Europe. Finally, there is the problem of Leuven's huge size. At the height of the war, Leuven matriculated more students than all the German Catholic universities put together; so its regionally-driven experience would reshape the graphs for Catholic and northwestern universities in ways not directly linked to the central European war. In short, composite graphs are more representative of the general impact of the Thirty Years' War on the Empire's universities if Leuven's data are excluded from Figures 4.12 and 4.13, and commentary on them are confined to the notes.

The most protected of these four unequal quarters, predictably, is the small southeastern cluster of Vienna, Graz, and Salzburg. During the first years of the war, their extant matriculation figures record only the tiniest contraction. After the Swedish arrival, their combined rate of matriculation skyrocketed, doubling in a decade and remaining high for at least two more. If the missing matriculation figures could be added for Salzburg between its opening in 1622 and 1639, when the surviving register begins, the resulting line would probably record even more continuous growth, beginning after 1612–1615 (when Graz's numbers temporarily dipped), receiving fresh impetus in 1622 (when Salzburg opened), and continuing in the early 1630s (immediately after the Swedish invasion to the north). In this region, therefore, the war only appears to have reinforced a sustained period of impressive growth which began well before and ended well after the conflict itself. The border between this uniquely fortunate southwestern zone and the war-torn area to the north and west was marked—geographically and militarily—by the heavily fortified Ingolstadt: as a glance back at Figure 4.6 suggests, Ingolstadt shared the fates of the other Catholic universities in the path of the Swedes but departed from the regional pattern in its partial recovery in the latter 1630s.

Diagonally opposite, the multiconfessional northwestern corner of Empire had a more varied experience and one far less confidently depicted by the surviving data. The reliable series, graphed in Figure 4.12, chart a remarkably continuous period of stability throughout the wartime period. Early in the conflict, the decline of tiny Herborn depresses the line despite the stability of the far larger Cologne. In the final years of the war, the expansion of Cologne, the partial recovery of Herborn, and the availability of matriculation figures for Paderborn begin a pronounced period of growth, to which the new university in Duisburg made a modest contribution after 1655 (Geuenich & Hantsche, 2007).[17]

Unfortunately, missing data for most of the institutions in this region undermines the reliability of these results. On the one hand, the opening of new universities in Paderborn (1616) and Rinteln (1620) would presumably have contributed modest growth in the early years of the war; but this growth cannot be graphed because Rinteln's register is missing completely and Paderborn's is lost before 1637. On the

[17]A database of the matriculation register is in preparation: https://www.uni-due.de/ub/archiv/universitaetsmatrikel.shtml.

other hand, the tumultuously interconnected histories of Marburg, Giessen, and Kassel would have disrupted this impression of stability and may well have depressed the overall trend;[18] but since Giessen's matriculation register is lost completely and Marburg's is missing from 1638 to 1652, the patchy surviving figures for Marburg and Kassel cannot contribute to a coherent general assessment and have been excluded from this graph. Since the growth of Paderborn and Rinteln and the disruption of the Hessian universities probably counteracted one another to some degree, the overall trend indicated on the graph may nevertheless be a very rough guide to the fortunes of the universities in this region. Further west, the gentle rise of Leuven was probably caused in part by the far more dramatic decline of Douai.[19] What is certain is that the well-documented Catholic universities in this region—Cologne, Paderborn, and also Leuven—can be sharply distinguished from the other German Catholic universities: while they did not benefit from the war to the degree of the three southeastern universities, they were fall less disrupted than their nearer Catholic neighbors. Helmstedt marks the eastern boundary of this comparatively protected group: as a return to Figure 4.8 suggests, its dramatic decline parallels the territories to the east, but its early partial recovery may reflect the more sheltered conditions to the west.

The relative safe havens to the far southeast and northwest contrasted sharply with the broad military corridor running diagonally from northeast to southwest. Most dramatically affected by the war was the dense population of relatively small universities of all three main confessions in the southwest.[20] Here, a far more complete dataset produces a far more coherent picture. Buoyed by positive growth in Freiburg and Würzburg and especially by the foundation of Strasbourg and Altdorf, the southern universities expanded by twelve percent in the first years of the war, despite the collapse of Heidelberg. The arrival of the Swedes then cut matriculations there by over half during the 1630s. The continuing conflict in the region allowed only a tepid recovery before the end of the conflict, after which enrolments bounced back to pre-war levels, bolstered slightly by the foundation of Bamberg.

The experience of the equally well-documented northeastern quadrant of the Empire was less dramatic but more devastating on the whole. In this overwhelmingly

[18]The Calvinist takeover of Marburg in 1605 prompted the establishment of Giessen in 1607; the Luther reconquest of Marburg in 1624 led to the closing of Giessen, followed by the establishment of Kassel in 1633; and the recovery of Marburg by the Reformed branch of the house of Hesse in 1652 provoked the closure of Kassel and the reopening of Giessen.

[19]Adding the Leuven's figures to those for Cologne, Herborn, Paderborn and Duisburg would preserve the sense of relative stability, while flattening the long-term trend, which would gradually rise from a low of 87 percent of pre-war levels in the early 1630s to a high of 126 percent in 1660.

[20]Trier is missing from this figure because its matriculation register does not survive. Basle is included because it was closely integrated into the Reformed academic system and clearly affected by the war in the southwestern corner of the Empire.

Lutheran region, the impact of war was felt in a differentiated fashion, as the main theatre shifted from place to place. Helmstedt suffered worst during the Lower Saxon phase of the war; Rostock and Greifswald during the Swedish landing; the giant Saxon universities after the failure of the Peace of Prague; and Frankfurt an der Oder was hammered twice: once in the latter 1620s and again in the latter 1630s. Overall, the modest gains experienced by Rostock and Greifswald were eclipsed by the massive losses of the larger universities further south. Due to the huge size of these northeastern universities, their aggregated matriculations closely track the trajectory of the Empire as a whole. This means that the decline of this region was far more gradual than that of the southeast. Its recovery in the final years of the war was more rapid but also only partial: unlike the other three quadrants, pre-war levels were never quite regained in this period.

The disparity of these four regions is considerable. Taking again the period 1616–1620 as the standard, the average annual rate of matriculation between 1621 and 1650 rose in the three southeastern universities by over one half (56 percent), dropped in the northwestern institutions for which we have reliable data by ten percent,[21] fell in the war-torn southwest by over one quarter (27 percent), and declined in the larger northeastern quadrant by over one third (36 percent).

Confessions Compared

An analogous graph reveals the diverse experiences of the three main confessions, although this also needs to be interpreted with care (Fig. 4.13). The collapse of Reformed matriculations (not counting the bi-confessional Frankfurt an der Oder but including Basle) was immediate and its effects widespread and long-lasting. With the loss of Heidelberg in 1622 and Marburg in 1624 and the disruption of the routes which carried students to Herborn and Basle, Reformed student numbers fell by nearly four fifths, and they remained at this very low level for a quarter century, despite the foundation of Kassel in 1633. A sharp recovery followed the reopening of Heidelberg, the restoration of Marburg, and the foundation of Duisburg in 1652; but even by 1660 the Reformed figures had only recovered two thirds of their pre-war strength. If Basle—just outside the effective borders of the Empire—is removed from the equation, the situation is even bleaker: in the quarter century between the loss of Marburg and the end of the war, matriculations in the Reformed universities within Germany fell to scarcely one tenth of their pre-war level.

A stark contrast is provided by the trajectory of Catholic enrolments within the Empire. Rock steady within the first dozen years of the conflict, they were collectively cut by one third in the years immediately following the Swedish invasion—the

[21] If Leuven is included, the loss is only six percent.

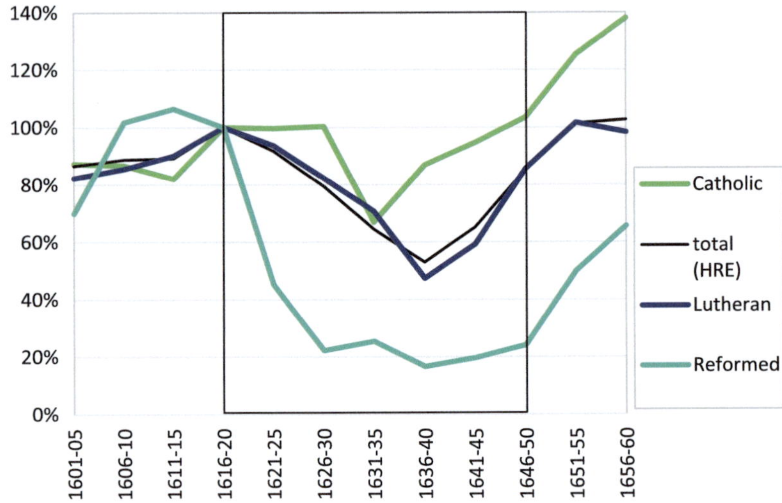

Fig. 4.13 Three main confessions within the Holy Roman Empire compared. Quinquennial percentages of matriculations in 1616–1620.
Source: Data from Andritsch (1977–1980), Eulenburg (1904), Falckenhainer (1893), Gall & Paulhart (1974), Redlich (1933), Schillings (1962–1963), Wachernagel (1951–1980). Design by author.

only period in which the Catholic universities reflected the imperial experience as a whole. Thereafter, fortunes divided sharply. Catholic institutions within the territory of modern Germany recovered very slowly until the end of the war, but returned to full strength rapidly after its conclusion. Within the Empire as a whole, however, these losses were compensated by gains to the southeast. Finishing the war marginally stronger than they began it, the Catholic figures for the Empire a whole then rose sharply during in the first post-war decade to a level 38 percent higher than their status ante bellum.[22]

The Lutheran figures again shadow the curve of the Empire as a whole, due to their huge size and an experience intermediate between the other two main confessions. After a gradually accelerating decline in enrolments from the start of the conflict, Lutheran matriculations were cut to less than half of pre-war levels by 1640; but they recovered slightly more rapidly than they declined, fully regaining their pre-wars levels shortly after the conclusion of the war.[23]

---

[22]Due to its large and relatively stable size, including Leuven flattens the curve (after 1616–1620) without transforming its overall shape: the nadir in 1631–1635 rises from 67% to 73%, and the post-war high in 1656–1660 falls from 138% to 128%. The missing data for Douai would flatten this post-war surge still further.

[23]Even if the incomplete matriculations for Giessen, Marburg, and Rinteln could be included, they would make little impact on this graph, given the large size of the other Lutheran universities. The biconfessional universities—Frankfurt an der Oder from 1613 and Marburg from 1652—are excluded.

Taking the period 1616–1620 as the standard once again, the average annual rate of matriculation between 1621 and 1650 dropped in Reformed institutions within the Empire by an average of 87 percent, in Lutheran universities by 27 percent, and in Catholic universities by just 7 percent.[24]

Part III. Universities outside the Holy Roman Empire

In all but two of the eleven years between 1632 and 1646, the universities east of the Rhine which enjoyed the largest enrolments were three of the five most northerly ones: Helmstedt, Rostock, and Königsberg. Of these, the university whose fortunes were most palpably enhanced by the war was Königsberg, which benefitted not only from its northerly position but, still more, from its location far outside the eastern boundary of the empire as well. When combined with the example of Cracow noted above, this finding opens a further line of inquiry. To what extent is the impact of the Thirty Years' War evident in universities and university systems outside the borders of the Empire?

There is obviously no room here for an attempt to provide a pan-European answer to this question, and the methods employed in this paper are in any case not adequate to such a task. There is no doubt that the influx of Protestant students to the two English universities was increased by the chaos in central Europe, at least until the outbreak of similar chaos in England after 1642; but the data needed to survey this influx does not yet exist. France may also register this effect: Orléan enjoyed a surge of foreign students during the war which subsided after it, and they came both from war-torn regions east of the Rhine and from Dutch students displaced from their traditional destinations in Germany (Brockliss, 1987, p. 18; Frijhoff, 1986b, p. 211; Julia & Revel 1989, p. 67). Further study is also needed of the extent to which the great international medical university of Padua was disrupted by the alteration of academic trade routes during the wartime period. The most obvious places to look for the extra-imperial impact of the war, however, are in two geographically and culturally adjacent regions: in Lutheran Scandinavia and the Reformed Dutch Republic.

Scandinavia

Between 1620 and 1626 Gustavus Adolphus placed the Uppsala University on entirely new foundations. An independent chancellor was appointed (the

[24]If Basle is included, the Reformed drop rises to 74 percent. If Leuven is excluded (as in Fig. 4.13), the Catholic drop increases to 8 percent. If Frankfurt an der Oder is included as a Lutheran institutions, then the Lutheran losses rise to 29 percent.

redoubtable Johan Skytte), the number of professors was more than doubled from eight to 19, the university was endowed in perpetuity with huge grants of inalienable, untaxed crown land which raised its annual revenue twelve-fold, a new charter and revised constitution were introduced, a university library founded, and new buildings erected which were to remain the principal quarters of the university for over two centuries. With the beginning of Queen Christina's personal rule in 1644, there began a second wave of activity, with a further six professorships and carefully revised statutes (Ingemarsdotter, 2011, pp. 247–250; Lindroth, 1976; Roberts, 1953–1958, Vol. 1, pp. 469–476). The renaissance in Uppsala was not, of course, such a direct result of the war as that in Königsberg, but connections between the two developments are not difficult to find. Imperial ambitions required educated officials. So did protracted involvement in the central European war. Uppsala's library was stocked almost entirely from the fruits of Sweden's campaigns in the Baltic, Germany, Poland, and Bohemia—notably with Jesuit libraries from Riga and Braunsberg and episcopal libraries from Frauenburg, Würzburg, and Mainz. The 8600 printed books and 1165 manuscripts rapidly amassed by 1640 eclipsed most university libraries on the continent, but only five of these works, it has been claimed, were Swedish. The central Europeans who filled many of the university's new chairs in this period—like the Bohemian physician Johannes Raicus who arrived in 1627 and the three Strasbourgers appointed in the 1640s—found the attractiveness of the northern university increased by the deplorable conditions at home. The same applied to a few students, but the numbers remained small: in the decade before the war only a single German inscribed his name in the matriculation register; during the war the number rose to 50 (Achelis, 1957, p. 195; Andersson, Carlsson, Sandström, Sjögren, Brenner, & Thimon, 1900, Vol. 1; Lindroth, 1976, p. 50) (Fig. 4.14).

Another direct fruit of Sweden's conquests was the foundation of a new university in the Estonian town of Tartu (Dorpat). In November 1629, Uppsala's first

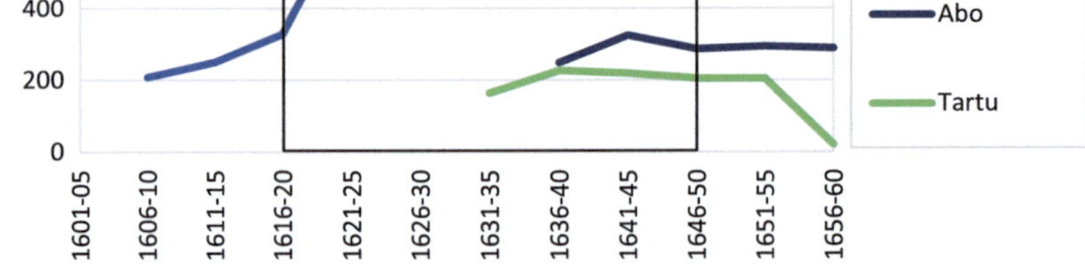

Fig. 4.14 Swedish and Danish universities.
Source: Data from Andersson, Carlsson, Sandström, Sjögren, Brenner, & Thimon (1900), Birket-Smith (1890–1912), Lagus (1889–1906), Tering (1984). Design by author.

chancellor, Johan Skytte, was appointed governor-general of the newly conquered province of Livonia together with Ingria and Käkisalmi. Realizing the need for officers for the new administration, he immediately set about founding an academic gymnasium in Tartu which the king raised to university status in 1632. Initially the institution thrived in a modest, provincial way: some 1016 students matriculated in Tartu in its first quarter-century, the greatest number (400) from Sweden but roughly 150 from Brandenburg, Pomerania, and Saxony and wandering individuals from as far afield as England, Amsterdam, Frankfurt am Main, Austria, Hungary, Transylvania, and Moscow. Numbers dipped momentarily in 1635, due to the threat of war between Poland and Sweden, but recovered immediately and remained steady for a further two decades. It was only after the end of the central European war that serious trouble set in. The Russian siege of Tartu forced the university to transfer to Tallinn in 1656, and without solid financial foundations in its new location it wasted away: only 49 new students were entered in the decade before its closure for a quarter-century in 1665 (Tering, 1984, p. 20; see also Achelis, 1957, p. 196; Roberts, 1953–1958, Vol. 1, pp. 478–479; Siilivask, 1985, pp. 22a–35b). In 1640 Queen Christina established a royal academy in the Finnish city of Turku (Swedish: Åbo) from which the modern University of Helsinki descends (Lagus, 1889–1906, Vol. 1, p. VI) (Fig. 4.14).

While matriculations plummeted to the south and soared to the north, Copenhagen's annual intake rose gradually from the first preserved lists of 1621 to peak in 1651 at 259. The Danish king's varying fortunes in the great central European war seem to have little affected this gradual rise: numbers dropped momentarily to 69 in the year of the defeat at Lutter but reached a new high of 216 the following year which even the Wallenstein's incursion into Jutland in 1627 failed significantly to reduce. Here too it was not until war began with Sweden in 1657 that the university began to feel the direct effects of the Baltic wars of the era (Birket-Smith, 1890–1912, Vol. 1; Ellehøj, Grane, & Hørby, 1979–1991, Vol. 1, pp. 248, 260). Copenhagen's experience suggests that it was the direct rather than indirect impact of the war—the arrival of enemy troops on the outskirts of the city rather than the general deterioration of social and economic conditions—which caused the drastic fluctuations of matriculation rates across central Europe in this era. The impact of the Thirty Years' War is evident in Copenhagen primarily in the changing foreign composition of the university. In the years before 1618, scarcely one German matriculated annually in Copenhagen. During the entire course of the war, only one Dutchman, one Swiss, and one Pole enrolled there; but the number of Germans rose to 160, most of them displaced between 1625 and 1630 by the Danish-Lower Saxon phase of the war and the chaos attending the arrive of the Swedes (Achelis, 1957, pp. 193–195, 196).

The Dutch Republic

Placed in relation to the catastrophic disruption of central European universities generally, the collapse of Reformed universities in particular, and the flourishing of university on the shores of the Baltic, the fabled golden age of Leiden and the other Dutch universities takes on a rather different complexion. In order to appreciate the importance of this military context, we must challenge the myth—widespread not so much in specialized literature on the history of universities as in more general treatments of Holland and her golden age—that Leiden was an international attraction of the first importance virtually from the moment of her founding by William the Silent in 1575.[25] That the university in Leiden was an important, international intellectual center from very early on is not subject to dispute: from a precocious stage the presence of great scholars such as Justus Lipsius and Joseph Scaliger made it the acknowledged capital of late humanist philology. But it does not follow that the young university established itself equally quickly as a Mecca for international student pilgrims: on the contrary, its numbers were initially rather small and its international student body numerically modest. In the quarter century between its foundation in 1575 and 1600, Leiden attracted on average 100 new students per year of which over three-quarters came from the Netherlands (north and south) and only one tenth from Germany. Such figures are substantially larger than only one-quarter of the German universities in this period, and one-fifth that of the huge Saxon institutions, Leipzig and Wittenberg. The modish university of Helmstedt, founded the year before Leiden, attracted three times as many students as Leiden in its first twenty-five years and 60 percent more in the decade immediately before the war. But as Helmstedt's numbers crashed Leiden's soared. In the quarter century between 1625 and 1650, Leiden's average annual matriculation approached 450; and of these over half were foreigners and over one quarter German— an increase, in absolute terms, of ten-fold (Colenbrander, 1925, pp. 292–294; Schneppen, 1960, p. 11).[26]

When foreign students are distinguished from domestic ones and the latter analyzed more closely according to place of origin, the relationship between Leiden's glory and central Europe's misery becomes clearer still—even when relying here on H. T. Colenbrander's relatively crude presentation of the data in terms of twenty-five-year periods.[27] Matriculations in Leiden by Dutch students peaked in the third quarter of the seventeenth century; those from Western Europe in the fourth. Matriculations from Germany, however—which constituted half or more of the foreign total throughout the seventeenth century—rose five-fold in the first

[25] A claim already firmly established in Diderot and le Rond d'Alembert (1751–1772, Vol. 9, p. 451): "Il semble que tous les hommes célèbres dans la république des Lettres s'y sont rendus pout faire fleurir, *depuis son établissement* [emphasis added] jusqu'à nos jours".

[26] Franeker—the next oldest and largest Dutch university in this period—followed a very similar trajectory, though with merely one quarter of Leiden's absolute numbers (Bots & Frijhoff, 1985).

[27] A more nuanced picture will emerge when the digitised matriculation register becomes available which Martine Zoeteman (2011) assembled for her doctoral dissertation.

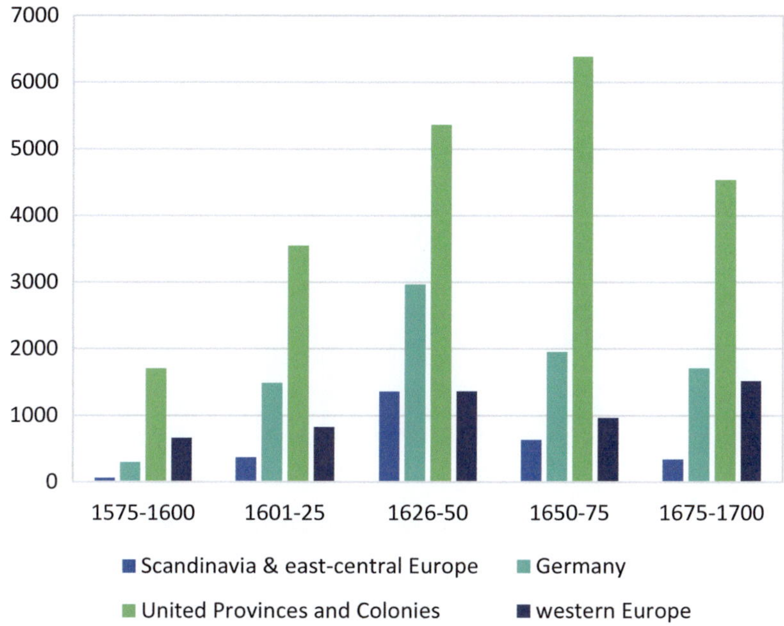

Fig. 4.15 Leiden: Major Areas of Recruitment, 1575–1700.
Source: Data from Colenbrander (1925). Design by author.

quarter of the seventeenth century, peaked at double that level in the second, and then fell by one third in the quarter-century after the war.[28] The main German Protestant cities and territories which sent students in large numbers to Leiden likewise sent most in the period 1626–1650 (Colenbrander, 1925, pp. 283–284, 292–294) (Fig. 4.15).[29]

A closer look at individual German territories establishes the point more precisely. In the decade before the sack of Heidelberg only 25 students from the Palatinate matriculated in Dutch universities (versus over three times that many

[28]Figures from Colenbrander (1925). The surge of German students into Leiden is even more impressive considering that a disproportionate share of the column for 1601–1625 represents students who matriculated in 1618–1625. The same applies, to a lesser degree, for students from Scandinavia and east-central Europe. The western European figures for 1575–1625 are inflated, on the other hand, by students born in the southern Low Countries who immigrated with their parents to the north and who were denizens of the north when they matriculated in Leiden.

[29]These figures include Emden (with 53 matriculations in this period), Westphalia (with 95), Wezel (41), Cologne (43), Hamburg (94), Holstein (94), Braunschweig (49), Brandenburg (75), Prussia (159), Danzig (87), Silesia (174), Frankfurt am Main (46), the Palatinate (90), and Basle (24). Two notable exceptions are Nuremberg, home to many Dutch commercial refugees, which peaked in the first quarter-century with 55, and Bremen, which did not begin to send its sons in large numbers to the Dutch universities until Marburg was wrested from Reformed control and Helmstedt undermined. Bremen's matriculations in Leiden did not peak until after the war was over (Prüser, 1928, pp. 245–246; Schneppen, 1960, pp. 20–22).

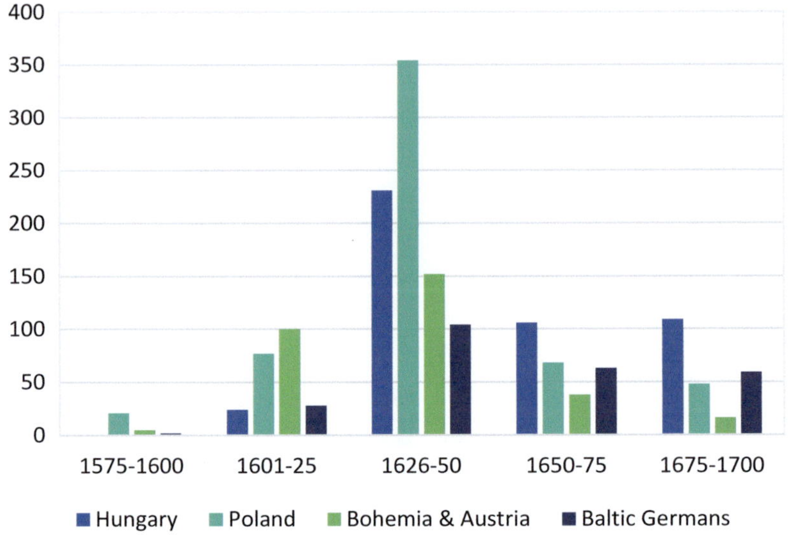

Fig. 4.16 Leiden: Matriculations from Scandinavia and east-central Europe.
Source: Data from Colenbrander (1925). Design by author.

Dutch students who enrolled in Heidelberg) (de Wal, 1886).[30] In the decade after 1622 the number nearly quadrupled to 97. During the subsequent twenty years, 124 further Palatine students followed, but in the decade following the reopening of Heidelberg in 1652 the enrollment dropped again to 21 (Persijn, 1959, pp. 11–21 and Appendix). Despite the even closer links of the counts of Nassau-Dillenburg to the house of Nassau-Orange (Lademacher, 1995; Oestreich, 1960), the case of the Wetterau counties was similar. The few Nassovian students who matriculated in Dutch universities before the war were mostly members of the House of Nassau itself; numbers surged in 1623 and remained high (by their own modest standards) throughout the war (Wolf, 1935). Hessian enrollments in the United Provinces (including many Lutheran as well as Reformed students) peaked somewhat later with the nadir of German universities generally around 1638 (Wolf, 1930). Taken together, the recruitment pattern for these three German Reformed communities appears to fluctuate with the course of the war in central Germany.

Even more dramatic was the surge of students from the northern and east-central European regions, which had previously constituted the broader catchment area of the German universities, but which had sent even fewer students to Leiden before the outbreak of the war (Colenbrander, 1925, pp. 285–286, 294). In the second quarter of the seventeenth century, the number of Polish students in Germany fell by three-

[30]De Wal's rich data must be used cautiously, since de Wal included East Frisians, Dutch emigré families, and Germans who later emigrated to the United Provinces on his list; but a conservative listing includes 77 Dutch students in the Heidelberg register in 1612–21, despite a dramatic fall in 1620 and 1621.

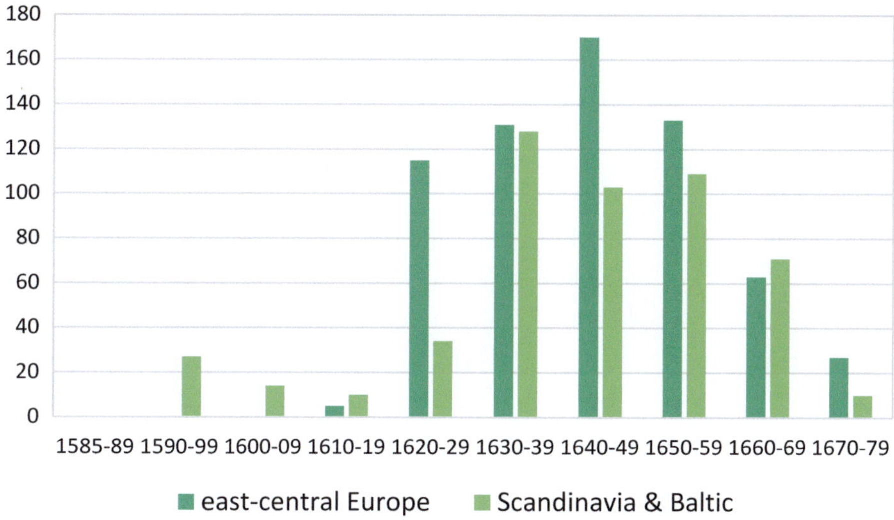

Fig. 4.17 Franeker: Matriculations from Scandinavia and East-Central Europe. Source: Data from de Ridder-Symoens (1985). Design by author.

quarters; simultaneous the number of Poles in Leiden rose five-fold, only to drop to less than its pre-war level after 1650.[31] Hungary's numbers soared ten-fold in the second quarter of the century and were then cut in half in the third. Matriculants from Austria and Bohemia both rose by one-third and then fell by 54 and 91 percent respectively. The number of Baltic Germans in Leiden grew by a multiple of four and then fell by 40 percent during the same interval (Schneppen, 1960, pp. 29–30) (Fig. 4.16).

These trends were not confined to the most populous of the Dutch universities. Franeker's experience was similar (Fockema, Sybrandus, & Meijer, 1968; de Ridder-Symoens, 1985). The number of east-central Europeans skyrocketed with the collapse of the German Reformed institutions after 1620, peaked in the 1640s, and then declined for the next thirty years. Matriculations from Scandinavia and the Baltic more than trebled in the 1620s and then nearly quadrupled again in the 1630s, as the German Lutheran universities reached their nadir, before draining away more gradually in subsequent decades (Fig. 4.17).

Comprehensive studies of student travel from Denmark and Sweden allow an even closer correlation of German losses with Dutch gains. When the first Swede matriculated in Leiden in the early 1610s, 40 of his countrymen were enrolling every year in German universities. In the middle two decades of the war, the tables turned dramatically: Swedish enrolments in German universities dropped to barely one-quarter of their pre-war level as inscriptions in Dutch universities climbed to

[31] For the German figures, see Żołądź-Strzelczyk (1996, p. 108).

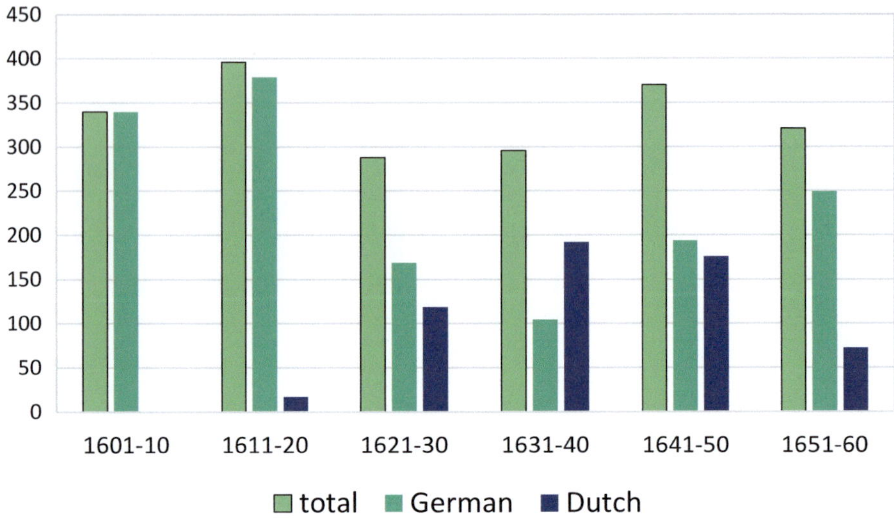

Fig. 4.18 Swedish matriculations in German and Dutch universities.
Source: Data from Niléhn (1983b). Design by author.

almost twice the German level. As German universities began to recover in the latter 1640s, the attraction of the Dutch ones fell correspondingly. Over the period 1601–60, as the majority of Swedes abroad shifted from Germany to the Netherlands and back, the total number of Swedes matriculating in these two places never varied more than 16 percent from the decennial average for this period (Niléhn, 1983a, p. 102; based on Niléhn, 1983b) (Fig. 4.18).

The Danish picture is similar. The numbers of students from Denmark-Norway in German universities mirrored the general levels of matriculations there. Attendance at Reformed institutions—representing almost one quarter of Danish matriculations in Germany in the thirty years before the outbreak of war—dropped virtually to nothing in the 1620s and only rose slightly thereafter, primarily on the strength of Basle. Attendance at Lutheran universities fell sharply in the 1630s—with the notable exception of Königsberg, which enjoyed by far its greatest Danish intake in this decade. As Danish numbers in German universities fell, those in Dutch institutions grew nearly five fold between the prewar decade and their height in the 1630s. For the final two decades of the war Danes matriculated in Dutch universities in numbers 50 to 60 percent higher than in German ones. Aside from the fact that Denmark-Norway sent more students abroad, their main difference with the Swedish case is a more enduring attachment to the Dutch institutions: Danes were among the first northerners to visit the Dutch universities in significant numbers, and their fondness for them cooled only gradually after 1648 (Helk, 1987, pp. 42–43) (Fig. 4.19).

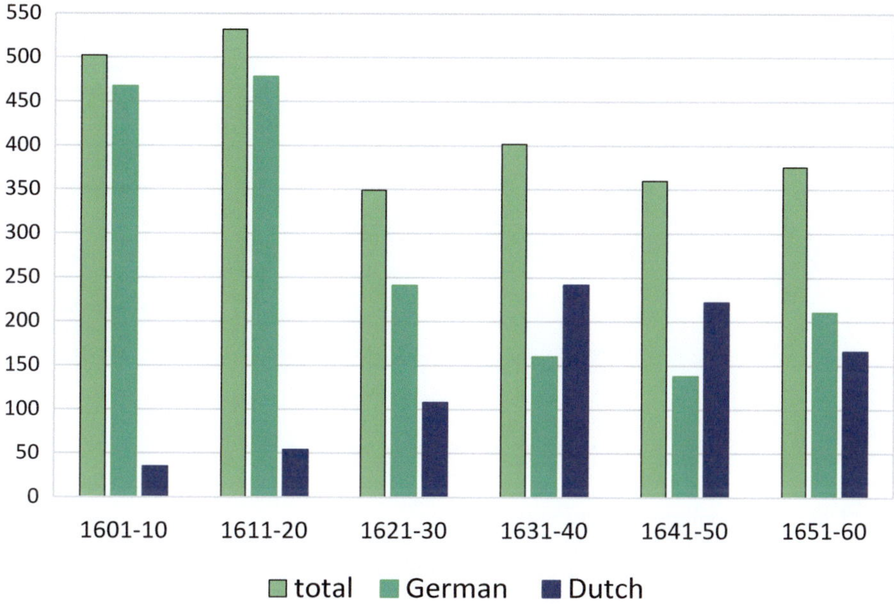

Fig. 4.19 Danish matriculations in German and Dutch universities.
Source: Data from Helk (1987). Design by author.

The salient point here is not merely that the Thirty Years' War affected patterns of student migration far beyond the borders of the Empire, nor only that the destruction of the German Reformed universities helped propel the rise of the Dutch ones. The broader point is that the Dutch and German Reformed universities appear to be integrated into a single *Universitätslandschaft* not dissimilar to the Catholic one revealed earlier. The integration of the Catholic system, discussed above, is manifested in two main ways: on the one hand, the losses suffered by the Catholic universities in southern Germany were mostly compensated by gains to those to the southest; and on the other, even the main exception to this rule—the deep trough in Catholic matriculations within the Empire as a whole manifest in the early 1630s—was mirrored by the huge spike in Cracow's recruitment between 1632 and 1636. The Catholic universities between Leuven and Cracow therefore appear to have operated as parts of a system: when military events prevented students from studying in one part of the system, they simply transferred to another part. Although the fortunes of individual institutions and whole regions fluctuated dramatically, the population of students within the system as a whole remained remarkably stable.

The Swedish and Danish-Norwegian data suggests that something similar can be said for the Reformed university system; and this is confirmed by another view of the data. Figure 4.20 plots the aggregate fortunes of the main German Reformed universities—Heidelberg, Basle, Herborn, and Marburg—as one line, and the three oldest Dutch universities—Leiden, Franeker, and Groningen—on the other

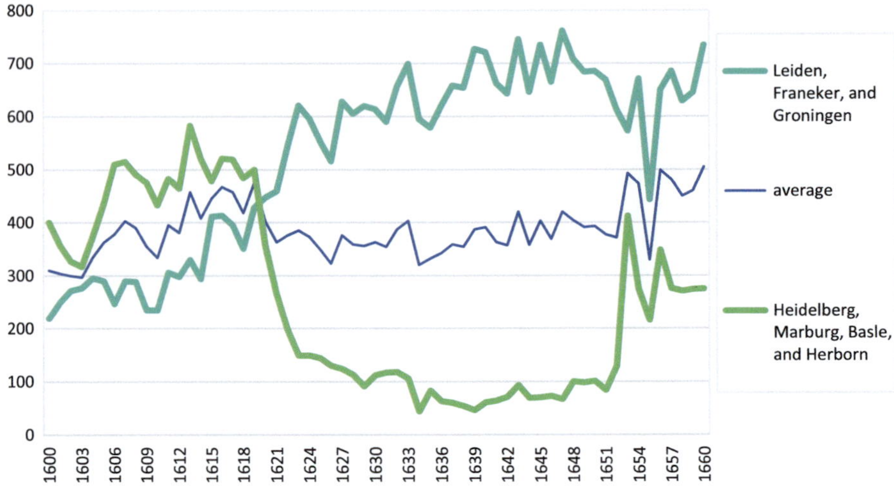

Fig. 4.20 The Reformed university system.
Source: Data from Du Rieu (1875), Eulenburg (1904), Fockema, Sybrandus, & Meijer (1968), Historish Genootschap te Groningen (1915), Wachernagel (1951–1980). Design by author.

(du Rieu, 1875). Once again, the two lines appear roughly to mirror one another. The Dutch figures leapt by 77 percent during the first five years of the war while the German Reformed numbers were cut by over two thirds. As the German Reformed line then gently settled to its low point during the second half of the war, the Dutch line gradually peaked during its final decade. Dutch numbers fall sharply at the conclusion of the conflict, without losing much of their strength in the longer term; while the German Reformed figures bounced back with the re-establishment of Heidelberg and Marburg after 1652, without regaining their pre-war level. Although these two sub-systems follow diametrically opposite courses, they appear to be interlinked: between 1621 and 1652 in particular, the total number of annual matriculations throughout this system fluctuated within a very narrow band, never diverging from the average figure for this period by more than 13 percent.[32]

[32]The main disruption to this system comes rather from Utrecht, which was raised to university status in 1636, and grew strongly after 1643, adding well over one hundred matriculations per year by the end of the war. But the new university's matriculation register is very unreliable for this period, averaging only five enrolments per year in some quinquennia and nearly 200 in others. Its arrival also heightens the need to distinguish matriculation rates from student numbers, since individual students increasingly matriculated in more than one Dutch university (as they had done previously in the German Reformed university system in its heyday). Source: Rijksuniversiteit te Utrecht (1886).

Part IV. Conclusion

Findings

Perhaps the most significant result of this preliminary survey is methodological. Within the Holy Roman Empire—where a large number of universities were distributed relatively uniformly across the landscape and where matriculation registers were systematically compiled, preserved, edited, and published—matriculation rates appear to provide a sensitive barometer of the pressure of military events on local universities and on the confessions organized around them. These rates fluctuated, to be sure, in peacetime as well as wartime; but during the Thirty Years' War the most extreme fluctuations can usually be related to specific events, many of them central to the main narrative of the war, and the same holds for the general experiences of institutions across the duration of the conflict. A relatively small and easily assembled data set—containing only 2160 data points over a sixty-year period—provides a surprisingly revealing impression of the impact of three decades of conflict across most of the Empire, aside from the lands of the Czech crown (due to the absence of adequate data) and Imperial Italy (marginal to the war and excluded from this study).

More interesting still is the comparative potential of this analysis. Grouping the trajectories of individual universities by confession sharply distinguishes the experience of the three main confessions within the Empire. Clustering institutions geographically also suggests that some regions fared much better than others. Grouping by both confession and region shows that the dramatic losses inflicted on clusters of institutions within each confession were at least partially compensated by gains to other clusters of institutions within the same confession.

The losses were naturally concentrated in specific periods in which military events directly impacted specific areas. Reformed institutions collapsed in the years after the crushing of the Bohemian Revolt. Helmstedt suffered worst during the Danish-Lower Saxon phase. Greifswald and Rostock fared worst between the Danish defeat and the arrival of the Swedes. The Catholic institutions of central and southwestern Germany were devastated by the dramatic Swedish intervention of the early 1630s. The huge Saxon universities reached their nadir in the decade after the failure of the Peace of Prague. The Franco-Spanish War disrupted Douai and Leuven after 1635.

The gains corresponding to these losses were concentrated primarily in areas spared direct military involvement, whether entirely, relatively, or during specific phases of the conflict. Finding these winners requires ranging outside the theatre of war itself and therefore also outside the boundaries of the Holy Roman Empire. The Catholic confession enjoyed the best capacity to absorb its academic refugees within the confines of the Empire, thanks to the sheltered situation of Vienna, Graz, Salzburg, and Leuven, and the relative tranquility of Cologne. The impact of the Swedish assault nevertheless registered dramatically in momentary collapse of the southwest German Catholic institutions, which coincided with the huge spike of matriculation in Cracow in 1632–1635. The Reformed confession, by contrast, possessed almost no capacity to accommodate displaced students within the Empire, so the collapse of Heidelberg and its satellites helped fuel the sharp rise of the young Dutch universities from 1622 onward. The Lutheran confession once again fell between these two extremes. Rostock, Greifswald, and Erfurt benefitted to some degree from inclusion within a Swedish protectorate, and Königsberg experienced a war-time golden age. The provision of higher education expanded in Denmark and especially in Sweden and its Baltic possessions, and some of this expansion was both the fruit of military conquest and a precondition for sustaining empire-building activities on such an ambitious scale. But the Swedish and Danish universities were not integrated fully and reciprocally into the international Lutheran educational system in a manner analogous to the Dutch universities' place within the broader Reformed system: while the destinations of Scandinavians studying abroad were powerfully redirected from Germany to the Dutch Republic, the student bodies of Copenhagen and Uppsala were not transformed by large numbers of incoming German students fleeing the war.

The academic impact of the war therefore extended over a region much larger than the war's own theatre. This is not merely because natives of the war zones fled to study in more peaceful neighboring regions: an even greater cause was probably the disruption of the international academic trade routes of the pre-war period. Before the war, the dense population of universities inside the Empire attracted students from a huge region to its north and east which possessed few full universities. The devastation of the institutions at the center of this system transformed well-established patterns of academic exchange stretching across half the continent. Leiden is the most striking beneficiary precisely because competing Reformed institutions within the Empire had been most thoroughly destroyed. Her wartime gains were not due solely to Germans and Czechs fleeing the war: they also resulted from Norwegian, Swedish, Danish, Polish, Hungarian, Transylvanian, Swiss, and indeed Dutch students opting for the safety of the Dutch Republic to the war-torn regions to the east which their countrymen had previously preferred. In all likelihood, something similar applies to the wartime gains of Vienna, Cracow, and Königsberg, and further investigations may reveal smaller-scale effects in Scotland and England, Leuven, Orléan, and Padua, to name a few likely candidates.

Prospects

The transformation of these international patterns of academic migration cannot be adequately studied, however, merely on the basis of shifting matriculation rates. When matriculations collapse at the very moment when a university city is besieged or conquered, the inference from academic effect to military cause is robust. When rates drop as military events scythe through a university's catchment area, cutting it off from an established supply of students, the inference is less direct. When the devastation of institutions in one country redirects foreign and domestic students to alternative institutions in another country, the connection between falling rates in one region and rising ones in another needs to be confirmed by more painstaking study of data harvested from in individual matriculation entries themselves.

In this regard, this preliminary study has scarcely begun to exploit the full potential of these sources. Even the most basic matriculation records typically include the matriculant's name, place of origin, and date of entry. In some cases, this is supplemented by other information, including age, social status, and subject of study. The editors of most of the German registers provided indexes, translating both surnames and place names from Latin to German, while some include additional biographical information. The vast amounts of effort already invested in creating and editing these records will only be fully exploited when this meticulous scholarship has been transformed into a homogeneous body of highly granular, instantly navigable linked data. Existing technology can facilitate most stages of this transformation. Vernacular place names can be linked to modern gazetteers in semi-automated fashion, provided with geographical coordinates, and mapped. Simple algorithms can link records of the same student matriculating in multiple universities during the course of a *peregrinatio academica*, with the process of inference and degree of certainty recorded on the system. Unlike the *Frequenzanalyse* pursued in the current paper, analyses based on individual matriculation entries can be applied to those registers surviving in only fragmentary state for this period,[33] and to new institutions founded in the region during or shortly after the war.[34]

Work on some of these tasks is already underway. Three of the largest matriculation registers relevant to this study—for Helmstedt, Rostock, and Leiden—have already been digitized and the former two published online.[35] Meanwhile, a major collaborative project, the Repertorium Academicum Germanicum, has been collecting biographical and social data on all the graduated scholars of the Holy Roman Empire between 1250 and 1550. The result, published as an online prosopoographical database and atlas, will provide comprehensive 'who's who' of late medieval scholars in the region (RAG) (Schwinges, 2015; see also chapter by Schwinges in this volume).[36] To date, nothing similar exists for the post-Reformation period.

Once coherent bodies of data have been assembled, they can be visualized and analyzed in a number of complementary ways. Analyses of the origins of students at individual universities merely requires the digitization of a single register. Analyzing competing clusters of universities in this way will show how catchment areas wax

[33] Including Copenhagen (missing before 1621), Cracow (lacking after 1641), Giessen (patchy from 1612–1649), Leuven (lacking before 1616), Marburg (lacking 1637–52), Olomouc (fragmentary in this period), Paderborn (lacking before 1637), Salzburg (lacking before 1637), and Utrecht (patchy before 1643).

[34] Including Strasbourg (1621), Altdorf (1622), Salzburg (1622), Tartu (1632), Kassel (1633), Trnava (1635), Utrecht (1636), Åbo (1640), Harderwijk (1648), Bamberg (1648), Duisburg (1654), Nijmegen (1655), and Košice (1657).

[35] Helmstedt: http://uni-helmstedt.hab.de/index.php?cPage=5&sPage=matsearch; Rostock: http://matrikel.uni-rostock.de/; and for Leiden see Zoeteman (2011).

[36] Further information and resources at http://www.rag-online.org/.

and wane in response to military events on the ground: not only sieges and occupations of the cities themselves and major battles nearby, but also the presence of armies in neighboring regions or the disruption of sister institutions further afield. Alternatively, the destinations of students from an individual city, territory, or region—whether inside or outside the Empire—could be displayed, in order to understand how these shifted in the course of the war. A third data view would combine origins and multiple destinations to reconstruct the routes followed by students who visited more than one university in the course of their *peregrinatio academica*. This mode of analysis would reveal still more clearly how the linkages between networks of institutions shifted, in some cases dramatically, as the conflict moved from one theatre to another. Such work would build on excellent existing studies of the foreign travels of students from individual countries, including Denmark-Norway, Sweden, Prussia, Hungary, and Transylvania.[37] Those matriculation registers which systematically record social status, age, or subject of study will allow even more complicated, multi-dimensional analysis. This approach could also answer the question of whether the decline in matriculations during the war represents a fall in the number of students or a decline in their mobility (Asche, 2011, pp. 162–163).[38]

This leads to a still more general, final methodological conclusion regarding the media best adapted for understanding these patterns. Print has proved an excellent means of editing matriculation registers and tabulating matriculation rates. But prose is not well adapted to analyzing the highly granular data of this kind: hence the century since 1904, during which Eulenburg's data was so poorly exploited. Simple spreadsheets, and the graphs generated by them, now help reveal general patterns with relative ease; yet the limitations of these standard tools are also readily apparent. The multitude of static graphs accompanying this article has proved a cumbersome means for analyzing even this rather small data set, and these simple means will be even less well adapted to understanding the movements of hundreds of thousands of students between thousands of places of origin and dozens of different institutions against the background of complicated physical, political, and confessional geography and constantly changing military events. The next generation of work in this field therefore needs to supplement traditional prose description and static, two-dimensional graphs and maps with a much richer variety of interactive, dynamic, animated, multi-dimensional, full color visualizations designed to allow both expert and non-expert users to explore all the dimensions of the data at a variety of different tempos and scales. Unprecedented compilations of highly granular historical data have important new stories to tell, but telling them effectively will require new means of analyzing and visualizing those data.

[37]For Denmark-Norway: Helk (1987). For Sweden: Niléhn (1983b). For Poland: Żołądź-Strzelczyk (1996). For the Czech lands: Pešek and Šaman (1986). For Hungary: Szögi (2011).

[38]The dangers of travel during wartime may have deterred students from moving from one university to another in numbers similar to the periods before and after the war, encouraging them instead to study for longer periods of time at a smaller number institutions.

References

Achelis, T. O. (1957). Deutsche Studenten auf nordischen Universitäten während des Dreißigjährigen Krieges [German students at Nordic universities during the Thirty Years' War]. *Archiv für Kulturgeschichte, 39*, 189–208. doi:https://doi.org/10.7788/akg-1957-jg11

Alschner, U. (1998). *Universitätsbesuch in Helmstedt 1576–1810: Modell einer Matrikelanalyse am Beispiel einer norddeutschen Universität* [University enrollment in Helmstedt 1576–1810: A model analysis of matriculations using the example of a north German university]. Wolfenbüttel: Braunschweigischer Geschichtsverein.

Andersson, A., Carlsson, A. B., Sandström, J., Sjögren, P., Brenner, S. O., & Thimon, G. (Eds.), (1900). *Uppsala universitets matrikel, 1595–1817: Vol. 1* [The matriculation register of Uppsala University, 1595–1817]. Uppsala: Akademiska Boktryckeriet.

Andritsch, J. (Ed.). (1977–1980). *Die Matrikel der Universität Graz.* [The matriculation register of Graz University] (Vols. 1–2). Graz: Akademische Druck und Verlagsanstalt.

Asche, M. (1995). Von einer hansischen Samthochschule zu einer mecklenburgischen Landesuniversität: Die regionale und soziale Herkunft der Studenten an der Universität Rostock in der Frühen Neuzeit [From a Hanseatic academy to a university for the duchy of Mecklenburg: The regional and social background of the students at the University of Rostock in the early modern period]. In P. Jakubowski & E. Münch (Eds.), *Universität und Stadt: Wissenschaftliche Tagung anlässlich des 575. Jubilaums der Eröffnung der Universitat Rostock* (pp. 141–162). Rostock: University of Rostock.

Asche, M. (2000). *Von der reichen hansischen Bürgeruniversität zur armen mecklenburgischen Landeshochschule: Das regionale und soziale Besucherprofil der Universitäten Rostock und Bützow in der Frühen Neuzeit (1500–1800)* [From the rich university for Hanseatic citizens to a poor provincial university for Mecklenburg: the regional and social profile of students at the universities of Rostock and Bützow in the early modern period (1500–1800)]. Stuttgart: Steiner.

Asche, M. (2001). Frequenzeinbrüche und Reformen—Die deutschen Universitäten in den 1520er bis 1560er Jahren zwischen Reformation und humanistischem Neuanfang [Matriculation crises and reforms: The German universities in the 1520s to the 1560s between the Reformation and humanistic new beginnings]. In W. Ludwig (Ed.), *Die Musen im Reformationszeitalter: Akten der Tagung der Stiftung Luthergedenkstätten in der Lutherstadt Wittenberg 14.–16 Oktober 1999* (pp. 53–96). Schriften der Stiftung Luthergedenkstätten in Sachsen-Anhalt: Vol. 1. Leipzig: Evangelische Verlags-Anstalt.

Asche, M. (2011). Der Dreißigjährige Krieg und die Universitäten im Heiligen Römischen Reich: Ein Fazit und viele offene Fragen [The Thirty Years' War and the universities of the Holy Roman Empire: Conclusions and many open questions]. In T. Kossert (Ed.), *Universitäten im Dreißigjährigen Krieg* (pp. 147–182). Militär und Gesellschaft in der Frühen Neuzeit: Vol. 15, Heft 1. Potsdam: Universitätsverlag Potsdam.

Asche, M., & Gerber, S. (2008). Neuzeitliche Universitätsgeschichte in Deutschland: Entwicklungslinien und Forschungsfelder [Modern university history in Germany: Lines of development and fields of research]. *Archiv für Kulturgeschichte, 90*, 159–201. doi:https://doi.org/10.7788/akg.2008.90.1.159

Asche, M., & Häcker, S. (2011). Matrikeln [Matriculation registers]. In U. Rasche (Ed.), *Quellen zur frühneuzeitlichen Universitätsgeschichte. Typen—Bestände—Forschungsperspektiven* (pp. 243–267). Wolfenbütteler Forschungen: Vol. 128. Wiesbaden: Harrassowitz.

Auerbach, I. (1984). Eine löbliche Hochschule, die gleich im Anfang wieder zergehen mußte: Zur Geschichte der ersten Kasseler Universität, 1633–1653 [A laudable university, which had no sooner begun than it was dissolved: On the history of the first university in Kassel, 1633–1653]. *Prisma: Zeitschrift der Gesamthochschule Kassel, 32*, 40–51.

Bergerhausen, H-W. (2010). *Köln in einem eisernen Zeitalter: 1610–1686* [Cologne in an age of iron: 1610–1686]. Geschichte der Stadt Köln: Vol. 6. Cologne: Greven.

Beyerle, K. (1900). *Konstanz im dreißigjährigen Kriege: Schicksale bis zur Aufhebung der Belagerung durch Schweden. 1623–1633* [Constance in the Thirty Years' War: Its fortunes until the raising of the Swedish siege. 1628–1633]. Heidelberg: Winter.

Binz, G. (1917–1918). Literarische Kriegsbeute aus Mainz in schwedischen Bibliotheken [Literary war booty from Mainz in Swedish libraries]. *Mainzer Zeitschrift, 12/13,* 157–165.

Birket-Smith, S. (Ed.). (1890–1912). *Kjøbenhavns Universitets Matrikel* [Matriculation register of Copenhagen University]. Vols. 1–3. Copenhagen: Gyldendal.

Bock, O. (1908). *Die Reform der Erfurter Universität während des Dreißigjährigen Krieges* [The reform of Erfurt University during the Thirty Years' War]. Hallesche Abhandlungen zur neueren Geschichte: Vol. 46. Halle: Niemeyer.

Bonjour, E. (1960). *Die Universität Basel von den Anfängen bis zur Gegenwart, 1460–1960* [Basle University from its beginnings to the present, 1460–1960]. Basle: Helbing & Lichtenhahn.

Bots, J. A. H., & Frijhoff, W. T. M. (1985). Die studentenpopulatie van Franeker academie: een kwantitatief onderzoek (1585–1811) [The student population of Franeker University: a quantitative study (1585–1811)]. In G. T. Jensma, F. R. H. Smit, & F. Westra (Eds.), *Universiteit de Franeker, 1585–1811: Bijdragen tot de geschiedenis van de Friese hogeschool* (pp. 56–72). Leeuwarden: Fryske Akademie.

Brockliss, L. W. B. (1987). *French higher education in the seventeenth and eighteenth centuries: A cultural history.* Oxford, UK: Clarendon Press.

Brück, A. P. (1972). *Mainz vom Verlust der Stadtfreiheit bis zum Ende des Dreißigjährigen Krieges (1462–1648)* [Mainz from the loss of civic freedom to the end of the Thirty Years' War (1462–1648)]. Geschichte der Stadt Mainz: Vol. 5. Düsseldorf: Rau.

Colenbrander, H. T. (1925). De Herkomst der Leidsche Studenten [The origin of Leiden's students]. In H. T. Colenbrander (Ed.), *Pallas Leidensis MCMXXV* (pp. 272–303). Leiden: Van Doesburgh.

de Ridder-Symoens, H. (1985). Buitenlandse studenten aan de Franeker universiteit 1585–1811 [Foreign students at Franeker University]. In G. T. Jensma, F. Smit, & F. Westra (Eds.), *Universiteit de Franeker, 1585–1811: Bijdragen tot de geschiedenis van de Friese hogeschool* (pp. 73–89). Leeuwarden: Fryske Akademy.

de Ridder-Symoens, H. (Ed.). (1996). *Universities in Early Modern Europe (1500–1800).* A History of the University in Europe: Vol. 2. Cambridge, UK: Cambridge University Press

de Wal, J. (1886). *Nederlanders: Studenten te Heidelberg* [Dutch students in Heidelberg]. Handelingen en Mededeelingen van de Maatschappij der Nederlandsche Letterkunde te Leiden: Vol. 1885–1886. Leiden: E. J. Brill.

Deckers-Hauff, H., & Setzler, W. (Eds.). (1977). *Die Universität Tübingen von 1477 bis 1977 in Bildern und Dokumenten* [Tübingen University from 1477 to 1977 in pictures and documents]. 500 Jahre Eberhard-Karls-Universität Tübingen: Vol. 3. Tübingen: Attempto Verlag.

Dehon, G. (1998). *L'Université de Douai dans la tourmente (1635–1765): Heurs et malheurs de la Faculté des Arts* [The University of Douai in the turmoil (1635–1765): Trials and tributations of the Faculty of Arts]. Histoire et civilisations: Vol. 609. Villeneuve d'Ascq: Presses universitaires du Septentrion.

Diderot, D, & le Rond d'Alembert, J.-B. (Eds.). (1752–1772). *Encyclopédie ou dictionnaire raisonné des sciences, des arts et des métiers* [Encyclopedia or dictionary of sciences, arts and crafts] (Vols. 1–28). Paris: Briasson, David l'ainé, Le Breton, Durand.

Domarus von, M. (1902–1903). Der große Brand der Stadt Herborn im Jahre 1626 und die Kollekten für die Abgebrannten [The great fire of Herborn in 1626 and the collections for the victims]. *Nassauische Annalen, 33,* 297–364.

Du Rieu, W. N. (Ed.). (1875). *Album Studiosorum Academiae Lugduno Batavae 1575–1875* [Matriculation register of Leiden University 1575–1875]. The Hague: Comitum apud Martinum Nijhoff.

Ellehøj, S., Grane, L., & Hørby, K., (Eds.). (1979–1991). *Københavns Universitet 1479–1979* [Copenhagen University 1479–1979], (Vol. 1–14). Copenhagen: G.E.C. Gads Forlag.

Ernstberger, A. (1966). *Die Universität Nürnberg-Altdorf während des Dreißigjährigen Krieges in ihrem Bestande bedroht* [The University of Nuremberg-Altdorf during the Thirty Years' War threatened with extinction]. Munich: Verlag der Bayerischen Akademie der Wissenschaften.

Eulenburg, F. (1904). *Die Frequenz der deutschen Universitäten von ihrer Gründung bis zur Gegenwart* [Matriculation rates of the German universities from founding to the present]. Leipzig: Teubner.

Falckenhainer, W. (Ed.). (1893). Die Annalen und die Matrikel der Universität Kassel [The annals and matriculation register of the Kassel University]. *Zeitschrift des Vereins für Hessische Geschichte und Landeskunde, n.F. 18,* 190–326.

Fockema, A., Sybrandus, J., & Meijer, T. J. (Eds.). (1968). *Album studiosorum Academiae Franekerensis (1585–1811, 1816–1844)* [Matriculation register of Franeker University (1585–1811, 1816–1844)]. Franeker: Wever.

Friedberg, E. (1909). *Die Leipziger Juristenfakultät: Ihre Doktoren und ihr Heim. 1409–1909* [The Leipzig Law Faculty: Its doctors and their home. 1409–1909]. Festschrift zur Feier des 500 jährigen Bestehens der Universität Leipzig: Vol. 2. Leipzig: S. Hirzel.

Frijhoff, W. (1986a). Grandeur des nombres et misères des réalités: la courbe de Franz Eulenburg et la débat sur le nombre d'intellectuels en Allemagne, 1576–1815 [Grand numbers and miserable realities: Franz Eulenburg's curve and the debate on the number of intellectuals in Germany, 1576–1815]. In D. Julia, J. Revel, & R. Chartier (Eds.), *Les Universités europeennes du XVIe au XVIIIe siècle: Histoire sociale des populations étudiantes* (pp. 23–63). Recherches d'histoire et de sciences sociales: Vol. 1. Paris: Édition de l'École des Hautes Études en Sciences Sociales.

Frijhoff, W. (1986b). Université et marché de l'emploi dans la République des Provinces-Unies [University and labour market in the Dutch Republic]. In D. Julia, J. Revel, & R. Chartier (Eds.), *Les Universités europeennes du XVIe au XVIIIe siècle: Histoire sociale des populations étudiantes* (pp. 205–243). Recherches d'histoire et de sciences sociales: Vol. 1. Paris: Édition de l'École des Hautes Études en Sciences Sociales.

Frijhoff, W. (1996). Patterns. In H. de Ridder-Symoens (Ed.), *Universities in Early Modern Europe (1500–1800)* (pp. 43–110). A History of the University in Europe: Vol. 2. Cambridge, UK: Cambridge University Press.

Frohnhäuser, L. (1899). Gustav Adolf und die Schweden in Mainz und am Rhein [Gustavus Adolphus and the Swedes in Mainz and on the Rhine]. *Archiv für hessische Geschichte und Altertumskunde, N.F. 2,* 1–232.

Gall, F., & Paulhart, H. (Eds.). (1974). *Die Matrikel der Universität Wien:1579/II–1658/59* [The matriculation register of the University of Vienna:1579/II–1658/59]. Publikationen des Instituts für Österreichische Geschichtsforschung. VI. Reihe: Quellen zur Geschichte der Universität Wien, 1. Abteilung. Vienna: Bohlau.

Geuenich, D., & Hantsche, I. (Eds.). (2007). *Zur Geschichte der Universität Duisburg 1655–1818: Wissenschaftliches Kolloquium, veranstaltet im Oktober 2005 anlässlich des 350. Jahrestages der Gründung der alten Duisburger Universität* [On the history of Duisburg University 1655–1818: Academic colloquium held in October 2005 on the occasion of the 350th anniversary of the founding of the old University of Duisburg]. Duisburg: Mercator-Verlag.

Grafton, A. (2006). The history of ideas: Precept and practice, 1950–2000 and beyond. *Journal of the History of Ideas, 67,* 1–32.doi:https://doi.org/10.1353/jhi.2006.0006

Häcker, S. (2011). Universität und Krieg: Die Auswirkungen des Dreißigjährigen Krieges auf die Universitäten Heidelberg, Tübingen und Freiburg [University and war. The impact of the Thirty Years' War on the universities of Heidelberg, Tübingen and Freiburg]. In T. Kossert (Ed.), *Universitäten im Dreißigjährigen Krieg* (pp. 98–123). Militär und Gesellschaft in der Frühen Neuzeit: Vol. 15, Heft 1. Potsdam: Universitätsverlag Potsdam.

Hammerstein, N. (1972). *Jus und Historie: Ein Beitrag zur Geschichte des historischen Denkens an deutschen Universitäten im späten 17. und im 18. Jahrhundert* [Law and history: A contribution to the history of historical thought in the German universities during the late 17th and 18th centuries]. Göttingen: Vandenhoeck und Ruprecht.

Hammerstein, N. (1986/1987). The University of Heidelberg in the Early Modern Period: Aspects of its history as a contribution to its Sexcentenary. *History of Universities, 6,* 105–133.
Hautz, J. F. (1862–1864). *Geschichte der Universität Heidelberg* [History of Heidelberg University]. Vols. 1–2. Mannheim: J. Schneider.
Heidorn, G., Heitz, G., Kalisch, J., Olechnowitz, K.-F., & Seemann, U. (Eds.). (1969). *Geschichte der Universität Rostock 1419–1969. Festschrift zur Fünfhundertfünfzig-Jahr-Feier der Universität.* [History of Rostock University 1419–1969. Publication celebrating the 550th anniversary of the University] (Vols. 1–2). Berlin: Deutscher Verlag der Wissenschaften.
Helk, V. (1987). *Dansk-Norske Studierejser fra reformationen til Enevaelden 1536–1660: Med en matrikel over studerende i udlandet* [Danish-Norwegian academic travel from the Reformation to Absolutism 1536–1616: With a register of students abroad]. Odense University studies in history and social sciences: Vol. 101. Odense: Odense University Publishing.
Hermelink, H., & Kaehler, S. A. (1927). *Die Philipps-Universität zu Marburg 1527–1927: 5 Kapitel aus ihrer Geshcichte (1527–1927)* [The Philipps-University at Marburg 1527–1927: Five chapters from its history (1527–1927)]. Marburg: Elwert.
Historish Genootschap te Groningen (Ed.). (1915). *Album studiosorum Academiae Groninganae* [Matriculation register of Groningen University]. Groningen: J. B. Wolters.
Hofmeister, H. (1907). Die Universität Helmstedt zur Zeit des 30jährigen Krieges [Helmstedt University during the Thirty Years' War]. *Zeitschrift des Historischen Vereins für Niedersachsen,* 1907, 241–277.
Hotson, H. (2007). *Commonplace learning: Ramism and its German ramifications, 1543–1630.* Oxford, UK: Oxford University Press.
Hotson, H. (forthcoming). *The reformation of common learning: Post-Ramist pedagogy and the reception of the New Philosophy, 1618–c. 1670.* Oxford, UK: Oxford University Press.
Ingemarsdotter, J. (2011). *Ramism, rhetoric and reform: An intellectual biography of Johan Skytte (1577–1645).* Uppsala Studies in History of Ideas: Vol. 42. Uppsala: Uppsala University.
Julia, D., & Revel, J. (1989). Les étudiants et leurs études dans la France modern [Students and their studies in modern France]. In D. Julia & J. Revel (Eds.), *Les universités européennes du XVIe au XVIIIe siècle: Histoire sociale des populations étudiantes* (pp. 25–486). Histoire sociale des populations étudiantes: Vol. 2. Paris: Editions de l'École des Hautes Études en Sciences Sociales.
Julia, D., Revel, J., & Chartier, R. (Eds.). (1986). *Les Universités europeennes du XVIe au XVIIIe siècle: Histoire sociale des populations étudiantes* [European Universities from the 16th to the 18th Century: Social History of Student Populations]. Recherches d'histoire et de sciences sociales: Vol. 1. Paris: Édition de l'École des Hautes Études en Sciences Sociales.
Just, L., & Mathy, H. (1965). *Die Universität Mainz: Grundzüge ihrer Geschichte* [Mainz University: The essentials of its history]. Mainz am Rhein: Mushake.
Kaniewska, I. (1986). La conjoncture étudiante de l'Université de Cracovie aux XVIIe et XVIIIe siécles [The student body of Cracow University in the seventeenth and eighteenth centuries]. In D. Julia, J. Revel, & R. Chartier (Eds.), *Les Universités europeennes du XVIe au XVIIIe siècle: Histoire sociale des populations étudiantes* (pp. 135–151). Recherches d'histoire et de sciences sociales: Vol. 1. Paris: Édition de l'École des Hautes Études en Sciences Sociales.
Kelley, D. (2002). *The descent of ideas: The history of intellectual history.* Aldershot: Ashgate.
Klüpfel, K. (1977). *Geschichte und Beschreibung der Universität Tübingen* [History and description of Tübingen University]. Geschichte und Beschreibung der Universität Tübingen: Vol. 2. Aalen: Scientia. (Original work published 1849)
Kosegarten, J. G. L. (1986). *Geschichte der Universität Greifswald: 2 Teile in 1 Band* [History of Greifswald University: 2 Parts in 1 Volume]. Aalen: Scientia. (Original work published 1856–1857)
Kossert, T. (Ed.). (2011), *Universitäten im Dreißigjährigen Krieg* [Universities in the Thirty Years' War]. Militär und Gesellschaft in der Frühen Neuzeit: Vol. 15, Heft 1. Potsdam: Universitätsverlag Potsdam.

Krabbe, O. C. (1994). *Aus dem kirchlichen und wissenschaftlichen Leben Rostocks: Zur Geschichte Wallensteins und des Dreißigjährigen Krieges* [From the ecclesiastical and academic life of Rostock: On the history of Wallenstein and the Thirty Years' War]. Mitteldeutsche Forschungen: Vol. 112. Weimar: Böhlau. (Original work published 1863)

Lademacher, H. (Ed.). (1995). *Oranien-Nassau, die Niederlande und das Reich* [The House of Orange-Nassau, the Netherlands and the Empire]. Niederlande-Studien: Vol. 13. Münster: Lit.

Lagus, V. G. (Ed.). (1889–1906). *Album studiosorum Academiae Aboensis MDCXL–MDCCCXXVII* [Matriculation register of Åbo University 1640–1827]. (Vol. 1–2 and Supplement). Helsingfors: Tidnings- & tryckeri-aktiebolagets tryckeri.

Lamberts, E., & Roegiers, J. (1990). *Leuven University: 1425–1985*. Leuven: Leuven University Press.

Langer, H. (2008). Die pommersche Landesuniversität Greifswald und das schwedische Reichsinteresse (1630–1720) [The Pommeranian university Greifswald and Swedish imperial interests (1630–1720)]. In P Wörster (Ed.), *Universitäten im östlichen Mitteleuropa: Zwischen Kirche, Staat und Nation. Sozialgeschichtliche und politische Entwicklungen* (pp. 85–103). Völker, Staaten und Kulturen in Ostmitteleuropa: Vol. 3. Munich: Oldenbourg Wissenschaftsverlag.

Langer, H. (2011). Die Universität Greifswald im Dreißigjährigen Krieg [Greifswald University during the Thirty Years' War]. In T. Kossert (Ed.), *Universitäten im Dreißigjährigen Krieg* (pp. 72–97). Militär und Gesellschaft in der Frühen Neuzeit: Vol. 15, Heft 1. Potsdam: Universitätsverlag Potsdam.

Lehmann, J. (1929). *Die örtliche und soziale Herkunft der Königsberger Studenten 1544–1649* [The geographical and social origin of Königsberg's students, 1544–1649] (Doctoral dissertation). Leipzig: Borna.

Lindroth, S. (1976). *A history of Uppsala University 1477–1977*. Stockholm: Alqvist & Wksell.

Lühr, G. (Ed.). (1925). *Die Matrikel des päpstlichen Seminars zu Braunsberg 1578–1798* [The matriculation register of the Papal Seminar in Braunsberg / Braniewo 1578–1798]. Quellensammlung zur Geschichte Ermland: Vol. 11. Königsberg: Meyer.

Märker, A. (1993). *Geschichte der Universität Erfurt 1392–1816* [History of Erfurt University 1392–1816]. Weimar: Hermann Böhlaus Nachfolger Weimar.

Menk, G. (1980). Das Restitutionsedikt und kalvinistische Wissenschaft. Die Berufung Johann Heinrich Alsteds, Philipp Ludwig Piscators und Johann Heinrich Bisterfelds nach Siebenbürgen [The edict of restitution and calvinist learning. The Calling of Johann Heinrich Alsted, Philipp Ludwig Piscator and Johann Heinrich Bisterfeld to Transylvania]. *Jahrbuch der Hessischen Kirchengschichtlichen Vereinigung, 31*, 29–63.

Menk, G. (1981). *Die Hohe Schule Herborn in ihrer Frühzeit (1584–1669): Ein Beitrag zum Hochschulwesen des deutschen Kalvinismus im Zeitalter der Gegenreformation* [The Herborn Academy in its early years (1584–1669): A study of German Calvinist academic affairs in the age of the Counter-Reformation]. Veröffentlichungen der Historischen Kommission für Nassau: Vol. 30. Wiesbaden: Historische Kommission für Nassau.

Methuen, E. (1988). *Die alte Universität* [The old university]. Kölner Universitätsgeschichte: Vol. 1. Cologne: Böhlau.

Mühlpfordt, G. (1983). Die Oder-Universität 1506–1811 [The Oder-University 1506–1811]. In G. Haase & J. Winkler (Eds.), *Die Oder-Universität Frankfurt: Beiträge zu ihrer Geschichte* (pp. 19–72). Weimar: Böhlau.

Niléhn, L. (1983a). Sweden and Swedish students abroad: The 17th century and its background. In G. Rystad (Ed.), *Europe and Scandinavia: Aspects of the process of integration in the 17th century* (pp. 101–122). Lund Studies in International History: Vol. 18. Solna: Esselte Studium.

Niléhn, L. (1983b). *Peregrinatio academica: det svenska samhället och de utrikes studieresorna under 1600-talet* [Peregrinatio academica: Swedish society and areas of foreign study in the 17th century]. Lund: CWK Gleerup.

Nischan, B. (1994). *Prince, people, and confession: The second Reformation in Brandenburg*. Philadelphia: University of Pennsylvania Press.

Oestreich, G. (1960). Grafschaft und Dynastie Nassau im Zeitalter der konfessionellen Kriege [The county and dynasty of Nassau in the age of confessional wars]. *Blätter für deutsche Landesgeschichte, 96,* 22–49.

Pagès, G. (1970). *The Thirty Years' War 1618–1648.* London: Adam and Charles Black.

Parker, G. (1984). *The Thirty Years' War.* London: Routledge.

Parrott, D. (1987). The causes of the Franco-Spanish War of 1635–59. In J. Black (Ed.), *The origins of war in Early Modern Europe* (pp. 72–111). Edinburgh: John Donald Publishers Ltd.

Pelizaeus, L. (2005). Die Universität Erfurt zwischen den Konfessionen 1615–1655 [Erfurt University between the confessions 1615–1655]. In P. C. Hartmann (Ed.), *Forschungen zu Kurmainz und dem Reichserzkanzler* (pp. 29–52). Mainzer Studien zur Neueren Geschichte: Vol. 17. Frankfurt a. M.: Lang.

Persijn, A. (1959). *Pfälzische Studenten und ihre Ausweichuniversitäten während des Dreißigjährigen Krieges* [Palatine students and their alternative universities during the Thirty Years' War]. (Doctoral dissertation). University of Mainz. Waldfischbach: Südwestdeutsche Verlagsdruckerei Hornberger.

Pešek, J., & Šaman, D. (1986). Les étudiants de Bohême dans les universités et les académies d'Europe central et occidentale entre 1596 et 1620 [Bohemian students at the universities and academies of central and western Europe between 1596 and 1620]. In D. Julia, J. Revel, & R. Chartier (Eds.), *Les Universités europeennes du XVIe au XVIIIe siècle: histoire sociale des populations étudiantes* (pp. 89–111). Recherches d'histoire et de sciences sociales: Vol. 1. Paris: Édition de l'École des Hautes Études en Sciences Sociales.

Prantl, C. (1872). *Geschichte der Ludwig-Maximilians-Universität Ingolstadt, Landshut, München. Zur Festschrift ihres vierhundertjährigen Bestehens im Auftrage es Akademischen Senates* [The history of the Ludwig Maximilian University Ingolstadt, Landshut, Munich. To the commemoration of its 400th anniversary on behalf of the Academic Senate]. Vols. 1–2. Munich: Christian Kaiser.

Press, V. (1985). Kurfürst Maximillian I. von Bayern, die Jesuiten und die Universität Heidelberg im Dreißigjährigen Krieg 1622–1649 [Elector Maxmillian I of Bavaria, the Jesuits, and Heidelberg University during the Thirty Years' War]. In W. Doerr (Ed.), *Mittelalter und frühe Neuzeit: 1386–1803* (pp. 314–370). Semper Apertus. Sechshundert Jahre Ruprecht-Karls-Universität Heidelberg 1386–1986: Vol. 1. Berlin: Springer.

Prüser, F. (1928). Bremen und die Universität Marburg im ersten Jahrhundert ihres Bestehens [Bremen and Marburg University during the first century of its existence]. *Bremisches Jahrbuch, 31,* 181–267.

Rasche, U. (Ed.). (2011). *Quellen zur frühneuzeitlichen Universitätsgeschichte: Typen—Bestände—Forschungsperspektiven* [Sources for the history of the early modern university: Types—collections—research perspectives]. Wolfenbütteler Forschungen: Vol. 128. Wiesbaden:Harrassowitz.

Rathmann, L. (Ed.). (1984). *Alma Mater Lipsiensis: Geschichte der Karl-Marx-Universität Leipzig* [Alma Mater Lipsiensis: History of the Karl Marx University, Leipzig]. Leipzig: Edition Leipzig.

Redlich, V. (Ed.). (1933). *Die Matrikel der Universität Salzburg: 1639–1810* [The matriculation register of Salzburg University: 1639–1810]. Salzburg: University of Salzburg.

Richter, W. (2011). Die vier mitteldeutschen Universitäten in Leipzig, Wittenberg, Jena und Erfurt im Dreißigjährigen Krieg: Eine Frequenzanalyse [The four middle-German universities in Leipzig, Wittenberg, Jena and Erfurt during the Thirty Years' War: An analysis of matriculation rates]. In T. Kossert (Ed.), *Universitäten im Dreißigjährigen Krieg* (pp. 41–56). Militär und Gesellschaft in der Frühen Neuzeit: Vol. 15, Heft 1. Potsdam: Universitätsverlag Potsdam.

Rijksuniversiteit te Utrecht, (Ed.). (1886). Album Studiosorum Academiae Rheno-Traiectinae MDCXXXVI-MDCCCLXXXVI [Matriculation register of Utrecht University 1636–1886]. Utrecht: J. L. Beijers et J. van Boekhoven.

Roberts, M. (1953–1958). *Gustavus Adolphus: A History of Sweden 1611–1632* (Vols. 1–2). London: Longman.

Rudersdorf, R. (1982). Der Weg zur Universitätsgründung in Gießen: Das geistige und politische Erbe Landgraf Ludwigs IV. von Hessen-Marburg [The way to the foundation of a university in Giessen: The intellectual and political legacy of Landgraf Ludwig IV of Hesse-Marburg]. In P. Moraw & V. Press (Eds.), *Academia Gissensis: Beiträge zur älteren Giessener Universitätsgeschichte* (pp. 45–82). Veröffentlichungen der Historischen Kommission für Hessen: Vol. 45. Marburg: Elwert.

Rudersdorf, M. (2009). Weichenstellung für die Neuzeit: Die Universität Leipzig zwischen Reformation und Dreißigjährigem Krieg 1539–1648/1660 [Setting a course for modernity: The University of Leipzig between the Reformation and the Thirty Years' War 1539––1648 / 1660]. In E. Bünz, M. Rudersdorf, & D. Döring (Eds.), *Spätes Mittelalter und frühe Neuzeit 1409–1830/31* (pp. 327–515). Geschichte der Universität Leipzig 1409–2009: Vol. 1. Leipzig: Leipzig University Press.

Schillings, A. H. (Ed.). (1962–1963). *Matricule de l'Université de Louvain* [Matriculation register of the University of Leuven]. Vols. 5–6. Collection de chroniques Belge inédites: Vol. 33. Brussels: Kiessling et Cie & P. Imbreghts.

Schindling, A. (1977). *Humanistische Hochschule und freie Reichsstadt: Gymnasium und Akademie in Straßburg 1538–1621* [Humanist higher education and imperial free city: The gymnasium and academy in Strasbourg 1538–1621]. Veröffentlichungen des Instituts für Europäische Geschichte Mainz: Vol. 77. Wiesbaden: Steiner.

Schindling, A. (1978). Straßburg und Altdorf. Zwei humanistische Hochschulgründungen von evangelischen freien Reichsstädten [Strasbourg and Altdorf. Two humanist university foundations of Protestant free imperial cities]. In P. Baumgart & N. Hammerstein (Eds.), *Beiträge zu Problemen deutscher Universitätsgründungen der frühen Neuzeit* (pp. 149–189). Wolfenbütteler Forschungen: Vol. 4. Nendeln: KTO Press.

Schneppen, H. (1960). *Niederländische Universitäten und deutsches Geistesleben: Von der Gründung der Universität Leiden bis ins späte 18. Jahrhundert* [Dutch universities and German intellectual life: From the foundation of Leiden University to the latter 18th century]. Neue münstersche Beiträge zur Geschichtsforschung: Vol. 6. Münster: Aschendorffsche Verlagsbuchhandlung.

Schonath, W. (1967). *Bibliothecae Academiae Herbipolensis Libri: Versprengte Schätze der Universitäts-Bibliothek Würzburg in Schloß Pommersfelden* [Scattered treasures of the Würzburg University Library in Pommersfelden Castle]. Quellen und Beiträge zur Geschichte der Universität Würzburg. Beiheft: Vol. 4. Neustadt an der Aisch: Degener & Co.

Schönauer, T. (2006). Die Studenten aus Ingolstadt zwischen 1392 und 1600: Eine sozialhistorisch-statistische Studie (Teil 1) [Students from Ingolstadt between 1392 and 1600: A socialhistorical-statistical study (Part 1)]. *Sammelblatt des Historischen Vereins Ingolstadt, 115*, 204–285.

Schönauer, T. (2007). *Ingolstadt in der Zeit des Dreißigjährigen Krieges: Soziale und wirtschaftliche Aspekte der Stadtgeschichte* [Ingolstadt in the era of the Thirty Years' War: Social and economic aspects of city history]. Beiträge zur Geschichte Ingolstadts: Vol. 4. Ingolstadt: Stadt Ingolstadt.

Schönauer, T. (2011). Landesfestung und Universität: Ingolstadt im Dreißigjährigen Krieg [Territorial defense and the university: Ingolstadt during in the Thirty Years' War]. In T. Kossert (Ed.), *Universitäten im Dreißigjährigen Krieg* (pp. 57–71). Militär und Gesellschaft in der Frühen Neuzeit: Vol. 15, Heft 1. Potsdam: Universitätsverlag Potsdam.

Schreiner, K. (1985). Die Katastrophe von Nördlingen: Politische, wirtschaftliche und kulturelle Folgen einer Schlacht für Land und Leute des Herzogtums Württemberg [The catastrophe of Nördlingen: Political, economic and cultural consequences of a battle for the the land and people of the duchy of Württemberg]. *Jahrbuch des Historischen Vereins für Nördlingen und das Ries, 27*, 39–90.

Schultze, J. (1964). *Von der Reformation bis zum Westfälischen Frieden (1535–1648)* [From the Reformation until the Peace of Westphalia (1535–1648)]. Die Mark Brandenburg: Vol. 4. Berlin: Duncker & Humblot.

Schwinges, R. C. (2015). Das Repertorium Academicum Germanicum (RAG): Ein digitales Forschungsvorhaben zur Geschichte der Gelehrten des alten Reiches (1250–1550) [The German Academic Repertory: A digital research project on the history of the scholars of the old Empire (1250-1550)]. In R vom Bruch, M Kintzinger, O Auge, & S Piotrowski (Eds.), Professorenkataloge 2.0. Ansätze und Perspektiven webbasierter Forschung in der gegenwärtigen Universitäts- und Wissensgeschichte (pp. 215–232). Jahrbuch für Universitätsgeschichte: Vol. 16 (2013). Stuttgart: Steiner.

Seth, I. (1956). *Die Universität Greifswald und ihre Stellung in der schwedischen Kulturpolitik 1637–1815* [The University of Greifswald and its position in Swedish cultural policy 1637–1815]. Berlin: Deutscher Verlag der Wissenschaften.

Siefert, A. (1996). Das höhere Schulwesen: Universitäten und Gymnasien [Higher education: Universities and gymnasia]. In N. Hammerstein (Ed.), *15. bis 17. Jahrhundert. Von der Renaissance und der Reformation bis zum Ende der Glaubenskämpfe* (pp. 197–374). Handbuch der deutschen Bildungsgeschichte: Vol. 1. Munich: C.H. Beck.

Siilivask, K. (Ed.). (1985). *A history of Tartu University 1632–1982.* Tallinn: Perioodika.

Specht, T. (1987). *Geschichte der ehemaligen Universität Dillingen (1549–1804) und der mit ihr verbundenen Lehr- und Erziehungsanstalten* [A history of the former University of Dillingen (1549–1804) and of the educational institutions associated with it]. Aalen: Scientia. (Original work published 1902)

Staehelin, A. (1957). *Geschichte der Universität Basel 1632–1818* [History of Basle University 1632–1818]. Vols. 1–2. Studien zur Geschichte der Wissenschaften in Basel: Vols. 4–5. Basle: Helbing & Lichtenhahn.

Steinmetz, M. (Ed.). (1958–1962). *Geschichte der Universität Jena 1548/58–1958: Festgabe zum vierhundertjährigen Universitätsjubiläum* [History of Jena University 1548/58–1958: Publication celebrating the 400 anniversary of the university]. Vols. 1–2. Jena: Gustav Fischer Jena.

Stelling-Michaud, S. (Ed.). (1959–1980). *Le livre du recteur de l'Académie de Genève: 1559–1878* [Matriculation register of the Geneva Academy: 1559–1878]. Vols. 1–6. Travaux d'humanisme et renaissance: Vol. 33. Geneva: Librairie Droz.

Stieda, W. (1917). Die Universität Rostock und Wallenstein [Rostock University and Wallenstein]. *Jahrbuch des Vereins für Mecklenburgische Geschichte und Altertumskunde, 81,* 75–88.

Stone, L. (1964). The educational revolution in England, 1560–1640. *Past and Present*, 28, 41–80. Retrieved from http://www.jstor.org/stable/649877

Stritmatter, R. (1977). *Die Stadt Basel während des Dreißigjährigen Krieges. Politik, Wirtschaft, Finanzen* [The city of Basel during the Thirty Years' War. Politics, economy, finance]. Bern: Lang.

Szögi, L. (2011). *Magyarországi diákok németországi egyetemeken és akadémiákon 1526–1700* [Hungarian students at universities and academies in Germany from 1526 to 1700]. Budapest: Eötvös Loránd Tudományegyetem Levéltára.

Tering, A. (Ed.). (1984). *Album academicum der Universität Dorpat (Tartu) 1632–1710* [The matriculation register of the University of Dorpat (Tartu) 1632–1710]. Publicationes Bibliothecae Universitatis Litterarum Tartuensis: Vol. 5. Tallinn: Valgus.

Triebs, M. (1995). *Die Medizinische Fakultat der Universitat Helmstedt (1576–1810): Eine Studie zu ihrer Geschichte unter besonderer Berücksichtigung der Promotions- und Übergangsdisputationen* [The medical faculty of Helmstedt University (1576–1810): A study of its history, with particular reference to doctoral and transitional disputations]. Repertorien zur Erforschung der frühen Neuzeit: Vol. 14. Wiesbaden: Otto Harrassowitz.

Troesch, B. (1969). *Origines géographiques et sociales des étudiants immatriculés à l'Académie de Genève (1559–1798)* [Geographical and social origins of the students matriculated at the Geneva Academy (1559–1798)]. Mémoire compl. licence sciences sociales. Geneva: Université de Genève.

van Nimwegen, O. (2014). The Dutch-Spanish War in the Low Countries 1621–1648. In O. Asbach & P. Schröder (Eds.), *The Ashgate research companion to the Thirty Years' War* (pp. 163–175). London: Ashgate Publishing.

Vann, J. A. (1984). *The making of a state: Württemberg 1593–1793*. Ithaca: Cornell University Press.
von Rommel, C. (1820–1843). *Geschichte von Hessen* [History of Hesse] (Vols. 1–8). Cassel: Perthes.
von Selle, G. (1956). *Geschichte der Albertus-Universität zu Königsberg in Preußen* (2nd ed.) [History of the Albertus University in Königsberg in Prussia]. Der Göttinger Arbeitskreis: Vol. 145. Würzburg: Holzner. (Original work published 1944)
von Wegele, F. X. (1969). *Geschichte der Universität Würzburg* [History of Würzburg University]. Vols. 1–2. Würzburg: Scientia Verlag. (Original work published 1882)
Wachernagel, H. G. (Ed.). (1951–1980). *Die Matrikel der Universität Basel* [The matriculation register of Basle University]. Vols. 1–5. Basle: Verlag der Universitätsbibliothek.
Weber, R. (1979). *Würzburg und Bamberg im Dreißigjährigen Krieg: Die Regierungszeit des Bischofs Franz von Hatzfeldt 1631–1642* [Würzburg and Bamberg during the Thirty Years' War: The rule of Bishop Franz von Hatzfeldt 1631–1642]. Würzburg: Echter-Verlag.
Weiß, U. (2008). Die Stiftungsprofessur für evangelische Theologie Augsburgischen Bekenntnisses an der älteren Erfurter Universität in ihrer Frühzeit (1566–1632) [The endowed chair in evangelical theology of the Augsburg Confession at the older Erfurt University in its early period (1566–1632)]. In J. Rogge (Ed.), *Tradieren—Vermitteln—Anwenden: Zum Umgang mit Wissensbeständen in spätmittelalterlichen und frühneuzeitlichen Städten* (pp. 189–230). Beiträge zu den Historischen Kulturwissenschaften: Vol. 6. Berlin: Akademie Verlag.
Whatmore, R., & Young, B. (Eds.). (2016). *A companion to intellectual history*. Wiley Blackwell companions to world history. Malden: Wiley-Blackwell.
Wilson, P. H. (2009). *Europe's tragedy: A history of the Thirty Years' War*. London: Allen Lane.
Wolf, K. (1930). Hessische Studenten auf niederländischen Universitäten im 17. Jahrhundert [Hessian students at Dutch universities in the 17th century]. *Hessische Chronik, 17,* 97–112.
Wolf, K. (1935). Nassauer Studenten auf den Hochschulen zu Leiden, Groningen, Utrecht, Harderwijk und Genf in der Zeit bis 1700 [Students from Nassau at the universities of Leiden, Groningen, Utrecht, Harderwijk and Geneva in the period before 1700]. *Nassauische Annalen, 55,* 139–147.
Zaschka, B. (1993). *Die Lehrstühle der Universität Tübingen im Dreißigjährigen Krieg: Zur sozialen Wirklichkeit von Professoren im vorklassischen Zeitalter* [The professorships of Tübingen University during the Thirty Years' War: On the social reality of professors in the preclassical age]. Tübingen: Attempto-Verlag.
Zathey, J., & Barycz, H. (1950). *Album studiosorum Universitatis Cracoviensis: T. 4, Continens nomina studiosorum ab anno 1607 ad annum 1642* [Matriculation register of Cracow University: Containing the names of the students from 1607 to 1642]. Cracow: Sumptibus Universitatis Jagellonicae Cracoviensis.
Zedler, G., & Sommer, H. (Eds.). (1908). *Die Matrikel der Hohen Schule und des Paedagogiums zu Herborn* [The matriculation register of the academy and gymnasium in Herborn]. Veröffentlichungen der Historischen Kommission für Nassau: Vol. 5. Wiesbaden: J. F. Bergmann.
Zirr, A. (2009). Die Schweden in Leipzig 1642–1650: Ein stadtgeschichtlicher Überblick zur Spätphase des Dreißigjährigen Krieges [The Swedes in Leipzig 1642–1650: A civic-historical survey of the late phase of the Thirty Years' War]. In M. Cottin, D. Döring, & C. Friedrich (Eds.), *Stadtgeschichte. Mitteilungen des Leipziger Geschichtsvereins e.V.: Jahrbuch 2009* (pp. 67–87). Beucha: Sax-Verlag.
Zirr, A. (2010). Stadt und Universität Leipzig im Dreißigjährigen Krieg [The city and university of Leipzig during the Thirty Years' War]. In D. Döring (Ed.), *Stadt und Universität Leipzig: Beiträge zu einer 600-jährigen wechselvollen Geschichte* (pp. 145–166). Quellen und Forschungen zur Geschichte der Stadt Leipzig: Vol. 1. Leipzig: Leipziger Universitätsverlag.
Zirr, A. (2011). Universität und fremde Besatzung: Die Alma mater Lipsiensis während der schwedischen Herrschaft über Leipzig 1642 bis 1650 [University and alien occupation: Leipzig University during the Swedish rule over Leipzig from 1642 to 1650]. In T. Kossert (Ed.),

Universitäten im Dreißigjährigen Krieg (pp. 18–40). Militär und Gesellschaft in der Frühen Neuzeit: Vol. 15, Heft 1. Potsdam: Universitätsverlag Potsdam.

Zoeteman, M. (2011). *De studentenpopulatie van de Leidse universiteit, 1575–1812* [The student population of Leiden University, 1575–1812] (Doctoral dissertation). Leiden: Leiden University Press.

Żołądź-Strzelczyk, D. (1996). *Peregrinatio academica. Studia młodzieży polskiej z Korony i Litwy na akademiach i uniwersytetach niemieckich w XVI i pierwszej połowie XVII wieku* [Studies of Polish youth from the Kingdom of Poland and Lithuania at German academies and universities in the sixteenth and first half of the 17th century]. Seria Psychologia i pedagogika: Vol. 106. Poznań: UAM.

Zopf, H. (1927). Schweden und Kaiserliche in der Mark Brandenburg anno 1631 [Sweden and Imperial forces in Mark Brandenburg in 1631]. In Landesdirektor der Provinz Brandenburg (Ed.), *Brandenburgisches Jahrbuch: 2. Band* (pp. 30–44). Berlin: Deutsche Bauzeitung.

Zsoldos, A. (Ed.). (1990). *Matricula Universitatis Tyrnaviensis 1635–1701* [Matriculation register of Trnava University 1635–1701]. Fejezetek az Eötvös Loránd Tudományegyetem történetéből: Vol. 11. Budapest: Eötvös Loránd Tudományegyetem.

Chapter 5
A Political Geography of University Foundation: The Case of the Danish Monarchy

Hanne Kirstine Adriansen and Inge Adriansen

The reasons for founding, or establishing, a university have been numerous and have varied with the raison d'être of the university. In this chapter we explore the political geography of university foundation in two ways. First, we examine how universities have been used geopolitically, then we investigate the role of university foundation in relation to nation-building. We use the geographer Klaus Dodds's (2014) conceptualization of geopolitics as a way of understanding the world and the links between power, knowledge, and geography, and we analyze the discourses related to establishing universities in certain areas from the fourteenth to the twentieth century. In early modern Europe states were not linked to well-defined territories in the same ways as they are today. Territories could have overlapping jurisdictions and mixed modes of political authority. It was an era of multiple spatialities of power in which political space could not be reduced to national territories. For example, Koenigsberger (1986) coined the term "composite state" (p. 12) to denote the monarchies in early modern Europe. Elliott (1992), referred to "composite monarchies" to describe that cultural and political construction of Europe. Gustafsson (1998) wrote of the "conglomerate state" (p. 194) to describe state formation in early modern Europe, when states consisted of several more or less well-defined territories. Territories gained new rulers through wars and alliances. In this process universities could be founded as a means of consolidating power—and universities could be perceived as a threat to rulers.

Whereas literature on education has shown mass education's bearings on efforts at nation-building by mutually competing states within the European interstate system (e.g., Ramirez & Boli, 1987; Riddle, 1996), the part that universities play

Inge Adriansen (deceased)

H. K. Adriansen (✉)
Danish School of Education, Aarhus University, Copenhagen, Denmark
e-mail: hka@edu.au.dk

P. Meusburger et al. (eds.), *Geographies of the University*, Knowledge and Space 12,
https://doi.org/10.1007/978-3-319-75593-9_5

has received less attention, although there has been an interest in the role of higher education in European regionalism (e.g., Neave, 2011; Paterson, 2001). A number of German geographers (e.g., Geipel, 1966, 1971; Mayr, 1979; Meusburger, 2012) have addressed the politics of university establishment, but English literature on university history usually passes over the spatial issues and geopolitics related to the founding of universities (e.g., de Ridder-Symoens, 1992, 1996; Pedersen, 1997; Rüegg, 2004, 2011).[1] We seek to close this gap in English scholarship by applying a geographical perspective to the history of university establishment. Because of the aforementioned complexities of state, power, and territory, we treat one state in detail rather than provide a more general overview.

Our empirical point of departure is the Danish state, also called the Danish monarchy, and its varying territorial claims from the fourteenth to the twentieth century, with the main emphasis on the nineteenth and twentieth centuries and the role of university-founding in regard to nation-building. Some readers may consider Denmark a small country for such an undertaking. The present nation-state is indeed a small example, but the composite state[2] of Denmark, which covered a much larger territory, tells a story of war and peace, territorial expansion and contraction, and fights over where knowledge creation and higher education should take place. Though focusing on Denmark, we stress that this case is not unique. Early universities in German-speaking lands, too, were founded for political reasons in the residences of the rulers (Kintzinger, 2003, 2012; Schwinges, 1998, 2008; see also chapter Schwinges in this volume).

We thus provide a geographical outlook on university history by analyzing the part that universities have played in the Danish monarchy from the founding of the University of Copenhagen in 1479 to the establishment of the University of Greenland in 1987. We first afford a territorial picture of the Danish state. It is followed by the analysis, in the first part of which we explore the geopolitics of university establishment in three selected regions. In the second part we examine the founding of four specific universities in the Danish monarchy to show how university establishment has been related to political independence and nation-building. In the final part of the analysis, we move from the Danish case to consideration of the university as part of a country's arsenal of national symbols and institutions. In the discussion and concluding remarks, we examine the role that internationalization has for universities in relation to their local or national agenda.

[1]The subdiscipline known as the geography of education was pioneered by German geographers such as Robert Geipel in the early 1960s and Alois Mayr, who focused on universities as of 1970. In 1983 the Association of German Geographers formed a working group on the geography of education (for details see Meusburger, 2015a).

[2]In this chapter we use the term *composite state*, which has become the most common one among historians in this context (e.g., J. R. Rasmussen, 1995).

The Danish State from a Territorial Standpoint

According to Weber (2004), a state is a human community that successfully claims the monopoly on the legitimate use of physical force within a given territory. So defined, territory is an important characteristic of the state. We argue that a state can be thought of more broadly as a government that has sovereignty over a geographically specific territory with a permanent population and official connections to other states. The state institutions must have exclusive rights to establish and maintain legal order within the territory. Historically, the picture is more complex. A territory could have multiple rulers (in Europe they have been emperors, kings, dukes, and counts); a state might not consist of contiguous territories; and the laws, institutions, religions, and social structures could differ from one area to the next within a state (Koenigsberger, 1986). Since the Treaties of Westphalia of 1648, which brought an end to the religious wars in Europe by dividing the area into territories ruled by representatives of the opposing religious confessions, there has been a movement toward a system of sovereign and territorially well-defined states (see chapter by Hotson in this volume). Political geographers such as Agnew (1998) have argued that state territory is not always the same as state sovereignty and have warned against a simple "Westphalian view" (p. 12), in which the world consists solely of states or territorial actors and no other types of polity. That conceptualization posits a one-to-one relationship between territory and emperor, so a territory is perceived as belonging to one ruler only. Such a framework excludes the dynamics that have shaped the present nation-state of Denmark.[3]

At present the term *Denmark* defines the small nation-state that was formed after military defeat in 1864 and the incorporation of North Schleswig in 1920. In historical literature, however, the word refers to the composite state that came into being in 1460 as a typical kind of union in the era of territorial states. This entity could also be called the Danish monarchy, the Oldenburg monarchy, or the Danish-Norwegian-Schleswig-Holstein personal union (Østergaard, 2002). These terms, however, have never come into general use either colloquially or professionally. In this chapter we refer to *the state of Denmark* and *the Danish state* to denote the composite state covering a much larger territory than the present nation-state of Denmark. That type of composite state, which was held together only by the personage of a king or queen, was the dominant type of state in Europe from the fifteenth to the eighteenth century. As long as local traditions and regional laws were respected, the composite nature of the monarchical states could persist. The apparatus of power gained its legitimacy from God, a relationship not widely questioned until the 1700s, when the legitimacy of the old monarchical states was challenged by the new bourgeois public (Adriansen, 2003).

[3]The word *natio* comes from Latin and means lineage or family. In this chapter it is used in its expanded, nineteenth-century meaning: a group of people whose identity is based on, for example, an ethnic, historical, cultural, or linguistic sense of community.

Geographically, the Danish state from 1460 to 1814 encompassed two kingdoms, Denmark and Norway; two duchies, Schleswig and Holstein (the latter was upgraded from a county to a duchy in 1474); two North Atlantic possessions, the Faroe Islands and Iceland; and four colonies: Greenland, Danish Ostindia (in India), the Gold Coast (West Africa), and the Danish West Indies (in the Caribbean). The two kingdoms constituted a personal union, as did the two duchies, and all four were bound together by their shared ruler, the Danish king, although his rank in the two duchies was only that of duke. Holstein, however, had a double status, for this duchy was also a member of the Holy Roman Empire (*Sacrum Romanum Imperium*). In Holstein the Danish king was therefore formally subject to the German-Roman emperor (known in English-language historiography as the Holy Roman emperor). To make things even more complicated, the Danish king had to promise in 1460, when he was appointed as duke of Schleswig and count of Holstein, that the two territories would be ruled as one entity (C. P. Rasmussen, 2008). The Danish state also included possessions east of Oresund in present-day Sweden. But following defeats in wars to Sweden in 1645 and 1658, the Danish state was reduced (see Fig. 5.1), losing all areas east of Oresund: Scania, Halland, Blekinge, the Baltic islands of Gotland and Saaremaa, and two areas in Norway (Härjedalen and Jämtland) (Jespersen, 2011).

The Danish state was still multinational, including many language groups or peoples. Speakers of Danish lived in the Kingdom of Denmark and the northern part of the Duchy of Schleswig, speakers of Norwegian lived in the Kingdom of Norway, and there were speakers of Frisian in western Schleswig as well as German speakers in southern Schleswig and all of Holstein. In the North Atlantic dependencies (the Faroe Islands; Iceland; and the colony, Greenland), the languages spoken were Faroese, Icelandic, and Greenlandic (Jespersen, 2011; Østergaard, 2015). In the second half of the eighteenth century, nation-building began to occur in the Danish state. Even though this process did not succeed on the periphery of the territory, it did weld different regions and ethnic groups together within the Kingdom of Denmark, which eventually emerged as one nation with major linguistic and cultural similarities (Adriansen, 2003). The gradual disintegration of the Danish composite state began in 1814. The personal union of the two kingdoms (Denmark and Norway), which had lasted for 434 years, ended in 1814 when Denmark, having fought on the losing side in the Napoleonic Wars, had to cede Norway to Sweden. In return Denmark received Swedish West Pomerania, which was soon given to Prussia in exchange for the small duchy of Lauenburg southeast of Holstein. The following year, 1815, Holstein became part of the German Confederation (*Deutscher Bund*). This separation of ways proved to be notable, for by 1848, when revolution engulfed many parts of Europe, the inhabitants of Holstein had come to feel solidarity with the German nation instead of the Danish (Adriansen & Christensen, 2015). The Danish state was further reduced after another military defeat in 1864, with Denmark having to surrender Schleswig, Holstein, and Lauenburg to Prussia and Austria.

Denmark thereby became an approximated nation-state that included only one people, the Danish—the North Atlantic dependencies notwithstanding. The duchies

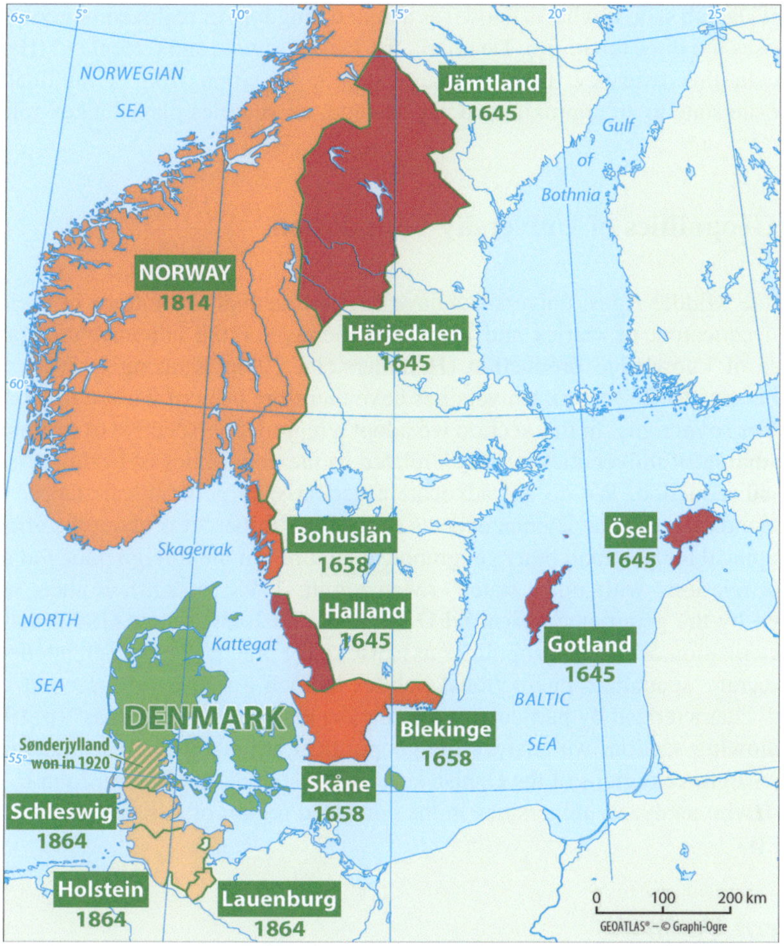

Fig. 5.1 Permanent cedings of Danish territory from the 17th to the 19th centuries. Source: Modified after Danmarks Nationalleksikon. Cartography: V. Schniepp.

became part of Prussia in 1867, but the population in North Schleswig maintained a Danish identity. In an internationally monitored referendum mandated by the Treaty of Versailles after World War I, 75% of the population in North Schleswig voted for incorporation into Denmark, and North Schleswig duly became part of Denmark in 1920 (Adriansen & Christensen, 2013).

The North Atlantic Dependencies (Iceland and the Faroe Islands) and the colony (Greenland) acquired different degrees of autonomy. Iceland became an independent state in 1918 but chose to enter into a personal union with Denmark for the first 25 years of this new arrangement. The Faroe Islands achieved home rule in 1948, with their own home-rule government. Greenland gained the same right in 1979. The present Danish state is what is known as a united kingdom or realm community, for it comprises the three territories: Denmark, the Faroe Islands, and Greenland (Harhoff, 1993). In this overview of the Danish state's transition from a multinational, composite state to an approximated nation-state, universities played a key role.

The Geopolitics of University Foundation

Since the Middle Ages, universities have been influential institutions in society as pivotal educators of clerics and loyal bureaucrats for the administration and as sources of knowledge production (Hammerstein, 1996; Kintzinger, 2003, 2012; Meusburger, 2012, 2015b). Universities have therefore been of interest to monarchs and other sovereigns. In this section we adopt a regional perspective to explore how the founding of universities has been related to the geopolitics of territorial expansion and reduction. We do not advocate a neoclassical geopolitical stance, which corresponds to what the layman expects geopolitics to be: "It is about the effects of geographical location and other geographical features on the foreign policy of a state and its relations with other states" (Mamadouh, 1998, p. 238). Rather, we are inspired by the geographers Gearóid Ó Tuathail and John Agnew (1992) and their characterization of critical geopolitics as a discursive practice "by which intellectuals of statecraft 'spatialize' international politics in such a way as to represent it as a 'world' characterized by particular types of places, peoples and dramas" (p. 192). In the following section, we analyze the geopolitical arguments for establishing universities in three regions of the Danish state. We also look to historian John E. Craig (1984), who analyzed universities in the contested region of Alsace-Lorraine.

Eastern Denmark

Scania has been the southernmost region of Sweden since 1658, but it used to be part of the region that came to make up the Danish state in the tenth century. The narrow strait, Oresund, provided a ready connection, whereas Scania was separated from the northern territories of Sweden by great, almost impenetrable forests. In the early Middle Ages Scania had greater religious and economic significance than any other part of Denmark. In 1103 the main city, Lund, became the seat of the archbishop, who was the primate not only in the Danish kingdom but also in Norway, the Faroe Islands, Iceland, and Greenland. Scania, however, was a source of discord between Denmark and Sweden and became the theater of several wars (Fig. 5.2).

The Danish king, Eric of Pomerania (1397–1439), made Copenhagen his capital. Until then, rulers had traveled around their kingdom, and there had been no permanent center of power. The idea of locating a university in Copenhagen seemed obvious because Danes wishing to pursuing academic studies had to go to foreign universities. In 1419 King Eric addressed the pope in Rome, requesting permission to establish a university in continuation of the cathedral school in Copenhagen. Pope

Fig. 5.2 Final and deliberated locations of universities in Denmark and its neighboring regions
Source: Design by authors. Cartography: V. Schniepp.

Martin V gave his permission for three faculties—law, medicine, and liberal arts. However, he declined to allow the creation of a faculty of theology, for his control would be hard to enforce in a place so far from Rome. Moreover, the permission to found a university in Copenhagen was valid for only two years, and King Eric became committed to a struggle with the Hanseatic League. In short, he was unable to carry out his educational plan (Lausten, 1991; Stybe, 1979).

Half a century later, in 1474, the Danish king, Christian I (1426–1481), made a pilgrimage to Rome to establish good relations with the pope in the hopes of gaining sway over the potentially troublesome bishops in the Danish kingdom. A generation earlier, the Danish archbishop had helped depose King Eric of Pomerania. Among the matters that King Christian discussed with the pope was renewed permission to found a university, including a faculty of theology. The charter was issued the following year (Slottved & Tamm, 2009). However, the plan was not realized until four years later, by which time Sweden, Denmark's foremost rival, had already established the first university in Scandinavia—Uppsala (1477). Upon hearing of the Danish petition in Rome, the Swedish king had decided to make a similar request, which was quickly granted by the pope and rapidly acted on in Sweden (Stybe,

1979). This example shows there can be multiple discourses for establishing a university. King Eric argued it would strengthen his new capital, King Christian sought a good relationship with the pope and permission to teach theology in Denmark in order to minimize tensions between the clergy and royal power, and the Swedish king wanted to position his country as the leading Nordic state.

It is worth noting that the papal charter of 1475 concerning the establishment of Copenhagen University was addressed to the archbishop in Lund, where a Franciscan *studium generale* (liberal arts school) had been created next to the cathedral as early as 1425. The charter stipulated that the chancellor of the university should be the archbishop or the bishop of the area where it was sited (Stybe, 1979). Nonetheless, it was not in Lund that King Christian had decided to locate his university but in Copenhagen, where he himself lived. Only when the Danish state broke with the Catholic Church after the Reformation in Denmark-Norway in 1536 was the king rather than the pope empowered to allow the founding of universities.

Following defeat in 1658, the Danish monarchy had to cede Scania to the Swedish Crown. With Scania Sweden became considerably larger, giving rise to the need for a new university (Rosén, 1968). In 1666, with many Swedish cities contending for the honor, the Swedish king allowed the establishment of a university in Lund, primarily because the cathedral chapter still administrated large estates and thus had the capacity to fund the new university, which was placed in the former cathedral school (Sanders, 2008). The founding of Lund University is often interpreted as a deliberate element of Scania's renationalization, or Swedification[4] (e.g., Asmussen, 2012; Feldbæk, 1998), a narrative that also appears on the university's home page (Lunds Universitet, 2016). There is, however, no supporting evidence. In the beginning only a third of the teaching staff consisted of Swedes, one third had been educated at the cathedral school during the era of Danish rule, and the remaining third had been Germans. It took a few decennia for Lund University to serve as a tool for strengthening the Swedish state. Analogous development took place at other European universities as well (Sanders, 2008).

The Duchies of Schleswig and Holstein

In 1460 the duchies of Schleswig and Holstein came under the Danish Crown. Schleswig was a Danish fief, whereas Holstein was a German fief with the German-Roman emperor as its feudal overlord. As of the 1630s the desire for a university arose among the prominent citizens of Flensburg, the largest city in the Duchy of Schleswig and the leading trading post in the Danish monarchy. The purpose of the university was to educate priests and civil servants for the two duchies of Schleswig and Holstein. A suitable building was identified in Flensburg, the main

[4]For this reason we have decided to mention the creation of the institution, even though Lund, as part of Scania, was already Swedish territory by that time.

part of the funding was raised, and the suggestion was endorsed by the two ruling dukes, King Christian IV of Denmark-Norway (1577–1648) and Duke Frederick III of Schleswig-Holstein-Gottorp (1597–1659). However, the plans had to be postponed when Swedish troops entered the duchies in 1643 (Battrup, 2007).

In 1658 the son of Frederick III, Duke Christian Albrecht (1641–1694), was released from his commitments to the Danish king, and this new sovereignty of his duchy was confirmed in 1660 by the Danish king (C. P. Rasmussen, 2008). In 1665 Christian Albrecht founded the University of Kiel (*Christian-Albrechts-Universität zu Kiel*). This decision probably contributed much to the visibility and staging of the duchy's new status as a sovereign state (Henningsen, 2008; Lohmeier, 1997). The creation of a university in Kiel placed the newly formed miniature state of Holstein-Gottorp on a par with the greater German states. The institution was set up in the Duchy of Holstein, even though the duke still resided in Gottorp Castle in the Duchy of Schleswig. But because Holstein was a German fief, the German-Roman emperor could be expected to offer protection against possible interference by the Danish king. In 1713 the Danish king, Frederik IV (1671–1730), took over the Gottorpian parts of Holstein, making all parts of the duchy subject to the Danish crown. Frederik was made to promise that he would strengthen the University of Kiel, which he did with a decree ensuring that all students from Schleswig and Holstein must study for two years in Kiel if they wanted employment in the duchies—even if they had completed an education at the University of Copenhagen (Tamm, 1996).

In the late eighteenth century professors at the University of Copenhagen contemplated a merger of the universities of Copenhagen and Kiel. The large new university was to be located in the city of Schleswig for three reasons. First, the physical distance from the central power in Copenhagen would increase academic freedom. Second, the distance from the temptations of the capital would enhance the students' focus on their studies. Third, the merger would strengthen the state union by building a cultural bridge between the kingdoms of Denmark and Norway and the duchies of Schleswig and Holstein. However, the suggestion was not taken seriously by the autocratic king, who was against any kind of decentralization (Albeck, 1978).

The requirement that all civil servants in the duchies had to study at least two years in Kiel came under criticism from the Danish side when the national conflict between Danish and German identity worsened from the 1830s onward (Hofmann, 1965; Tamm, 1996). The obligation was originally introduced for regional political reasons, but it acquired great national political significance when nationalism entered the stage and the university in Kiel came to play a salient role in nation-building. Modern nationalism in Schleswig and Holstein was formulated in 1815–1816 by two professors from the University of Kiel, the historian F. C. Dahlmann (1785–1860) and the lawyer N. N. Falck (1784–1850). Many students became interested in the ideology and identified closely with the German nation, turning the university in Kiel (and those elsewhere) into a hotbed of political unrest (Frandsen, 2008).

However, not all professors in Kiel were German patriots. A few of them were Danish-minded, among them Christian Paulsen (1798–1854). In 1842 he suggested that a university be founded in the Duchy of Schleswig to save civil servants from having to study for two years in Kiel and thus becoming influenced by German culture. Paulsen asserted that "the Danish spirit is in need of a scientific border fortress . . . A university will have a spiritual impact far beyond the circle of students" (Misfeldt, 1925, p. 3).[5] However, there was no immediate support for the suggestion, which was presented to the Danish king. Paulsen was right in his assessment of the spiritual impact of universities, and his concern over the ideological influence in Kiel was shared by many. In 1848 the First Schleswig War broke out, and when it came to an end in 1851 the university question was not raised again. In 1867, after the Second Schleswig War in 1864, both duchies were made part of Prussia (Adriansen & Christensen, 2013), and the University of Kiel was restructured along Prussian lines. For more than two centuries the fate of the university was closely related to the discussions and territorial disputes over the Schleswig-Holstein area.

Jutland

The following analysis is closely related to the one above, for the southern part of the Jutland peninsula is Schleswig. The proposal to establish a university in the southern part of the Danish kingdom, near the border to the duchies, was presented in 1845 by H. P. Selmer (1802–1877), a department head in the administration of the University of Copenhagen. Embedded in Denmark's national geopolitical discourse, his reasoning was that a university would have great impact on both the region where it was located and the adjacent regions. In Selmer's mind, there was no doubt that

> a major reason for the supremacy that German nationality has managed to gain in the duchies at the expense of the Danes lies in the fact that it is being fostered by the German university [in Kiel]. . . . A university in Jutland would thus provide the strongest support in the struggle against the unjustified advance of German nationality. (Albeck, 1978, p. 28)[6]

However, national disagreements after the war in 1848 prevented action in this regard.

When Denmark was planning a new university after World War I, debates arose about where it should be located. The professors at the University of Copenhagen were of the opinion that their own institution ought to be expanded and that it would prove impossible to motivate qualified researchers to settle outside the capital. The

[5]"*Den danske Aand kan trænge til en videnskabelig Grænsefæstning* [. . .] *Et Universitet kan få en aandelig Virkning langt ud over Studenternes Kreds*" (Misfeldt, 1925, p. 3). All English translations in this chapter are our own unless otherwise noted.

[6]"At en væsentlig Aarsag til den Overlegenhed som den tyske Nationalitet i Hertugdømmerne har vidst at tilvende sig paa den danskes Bekostning, netop ligger deri, at hin understøttes ved det daværende tyske Universitet [i Kiel]. . . Et Universitet i Jylland vilde saaledes give dette den kraftigste Støtte i Kampen mod den tyske Nationalitets uberettigede Fremtrængen" (Albeck, 1978, p. 28).

faculty claimed that a university on the periphery would not be viable (Albeck, 1978).

For many decades, though, a strong desire for a university in Jutland had existed, and in 1919 the Danish Ministry of Education set up a university commission. Three towns in Jutland, two of which were rather small, entered the contest to become the host of the future university, each with a different rationale for why they were appropriate candidates. Viborg had a cathedral and a regional archive and was an old center of education. Sønderborg in North Schleswig was situated close to the future Danish-German frontier and needed economic and national reinforcement. Aarhus's main advantage in the debate was that it had grown to become the second largest city in Denmark. In Viborg and Sønderborg this very fact spoke against locating the university in Aarhus. The vital interaction between teachers and between teachers and students would be facilitated much better in a smaller town, where it would also be possible to promote fruitful interaction between the university researchers and people from outside academia.

Aarhus ultimately won (Albeck, 1978). A broad range of citizens representing the business community joined forces in 1921 in The University Society, Aarhus, an organization that, together with the Aarhus city council, that eventually became the driving force in the campaign. Scientific collections were secured, scientific societies were created, and the city hospital was improved to become the leading hospital outside the capital. Private parties donated great sums, and a building plot was offered for free. Private university-level teaching in Aarhus began in 1928, and within three years the city managed to convince the parliament that it should be the second university city in the Kingdom of Denmark. The bill pertaining to Aarhus University was passed in 1931. However, it committed the central state to cover running costs only; money for the construction of university buildings had to come from other sources, including private ones. The university buildings were paid for by the Aarhus city council and private funds until the 1940s (Albeck, 1978).

In Sønderborg the abortive bid to have a university sited there led to great disappointment. One reason for the failure was a fear of national and political unrest that could be caused by an influx of German students. The border region did not acquire an institution of higher education until 1963, when an engineering college was founded in Sønderborg. It was joined by a business college in 1984, and the two institutions merged with Odense University in 1998 to become the University of Southern Denmark. The new university expanded in 2004 to include the Department of Border Region Studies.

In southern Schleswig, which remained under German rule after 1920, a college of education was changed into Flensburg university in 1994 (Ruck, 2007). Today, the universities in Sønderborg and Flensburg collaborate in numerous areas and offer joint courses. The two university campuses on either side of the border are located in peripheral regions. Likewise, the University of Southern Denmark (with it main campus in Odense) collaborates with the University of Kiel in many fields. Hence, destructive national politics has been replaced with constructive regional and cross-border politics.

University Foundation from a Geopolitical Perspective

To show how geography can contribute to university history, this section sums up the various facets of the chapter's arguments so far by supplementing existing literature with analysis of geopolitical arguments for establishing universities. Agnew (2001) showed how the national boundaries in Europe have their origins in the sixteenth century, "when political sovereignty began to shift from the personhood of the monarch to the territory of the state" (p. 7), linking authority and territory in an unprecedented manner. In the Middle Ages university students, for instance, constituted a community with their own jurisdiction and their own prisons within the university. Territorial states meant an end to overlapping jurisdictions and mixed modes of political authority. In this process of shaping nations, education became a useful tool. Although universities have been quite international from the outset, they have always been a staple of political power because universities educate clergy, bureaucrats, and other parts of the elite; because universities are houses of knowledge; and because knowledge and power are closely linked (Hammerstein, 1996; Meusburger, 2012, 2015b). Many rulers have therefore tried to consolidate their power by establishing universities and controlling their location, as the Danish state has shown on various occasions. Several chapters in Gregory, Meusburger, and Suarsana (2015) have addressed the roles that universities and scholars have in mediating knowledge and power and have examined ways in which rulers have sought to aggrandize their power through universities and academic knowledge. As the next section shows, universities have also become important for discourses on nationhood and independence, so rulers can also exercise power by *not* allowing the creation of a university.

Establishment of Universities and Nation-Building

Having analyzed geopolitical aspects of university establishment in selected regions, we now draw on specific institutional examples to illustrate the role of university-founding in relation to nation-building. As Scott (1990) put it, knowledge and nation are "two rival systems of ideas and values, beliefs and attitudes, which compete for our loyalty" (p. 1). He contended that knowledge represents reason and science, whereas nation represents instinct and custom. However, Scott also pointed out that education, particularly higher education, consists of institutions of knowledge as well as institutions that "preserve and elaborate patterns of national identity" (p. 11). Other authors, too, have noted the relationship between universities and nation-building (e.g., Cohen, 2007; Norrback & Ranki, 1996; Riddle, 1996). Although Adriansen's (2003) research on national symbolism did not include the university as a national symbol, we explore the question of whether the university can be included in the battery of national symbols and institutions. We begin by considering examples within the Danish monarchy.

The University of Oslo—Det Kongelige Frederiks Universitet

For more than 400 years (1380–1814), Norway was in personal union with Denmark. Norwegians could educate themselves only at universities outside Norway, with the University of Rostock or sometimes the University of Leiden being the preferred choice in the mid-fourteenth century (Langholm, 1996). After the Reformation in Denmark-Norway in 1536, the Danish king ruled that all Norwegians and Danes wishing to apply to foreign universities were required to study first for at least two years at the University of Copenhagen. According to the state, the university's foremost task was to educate loyal civil servants, who had to hold a diploma from the University of Copenhagen to become a priest, judge, or medical doctor in Denmark, Norway, or the North Atlantic dependencies. Because of the physical and psychological distance between northern Norway and Copenhagen, the Norwegians petitioned in 1661 to have their own university, a request that the Danish king categorically rejected. In the eighteenth century Norwegians accounted 40% of Denmark-Norway's entire population but only 15% of the student body at the University of Copenhagen. This background led to a reapplication for the creation of a Norwegian university in 1771. Reflecting the ideals of the Age of Enlightenment, the petitioners invoked the need to increase the level of knowledge and strengthen professional development (Collett, 1999). Supporters pressed for emphasis on the natural sciences and their practical application and stressed the logic of studying Norwegian history and nature in Norway (Langholm, 1996). But once again the request was rejected by the Danish king, who sought to maintain the education of civil servants in Copenhagen (Feldbæk, 1998).

The call for a Norwegian university was heard in the 1790s as well, this time on the grounds that Norway was the only European country to be denied a university. During the Napoleonic Wars Norway became isolated from Denmark because British warships prevented maritime communications between the two countries. This hiatus aided the development of independent Norwegian politics. The desire for a university, which would make Norway a nation of culture, became a central issue for the newly founded Royal Society for the Well-being of Norway, and funding was collected across the country (Collett, 1999). The Danish king feared that a Norwegian university would lead to a split between Denmark and Norway, but eventually the pressure from Norway became too intense. In 1811 the king was forced to give in. The announcement of his acquiescence sparked the first so-called national celebrations in Norway. In the cathedral of Trondheim, there was a performance of a cantata celebrating the future university with the words: "Oh Kingdom of Norway, now the foundation stone of your glory has been laid" (Bagge & Mykland, 1987, p. 295).[7]

Two years later the Royal Frederik University was inaugurated in Kristiania (present-day Oslo), and new celebrations were held to honor the university—a clear demonstration of Norwegian national self-awareness. It was the Norwegians

[7]"Lagt er da, O Norges Rige, grundvold til din herlighed" (Bagge & Mykland, 1987, p. 295).

themselves who had managed to make the university a reality as a main step toward increased independence. The following year (1814), Denmark was forced to hand over Norway to Sweden. In this new personal union, the Norwegian university became a cornerstone of Norwegian nation-building and helped secure cultural independence in the relationship with Sweden (Collett, 1999).

The University of Iceland—Háskóli Íslands

Iceland was originally a free state created by its first inhabitants, who were immigrants from Norway. In the mid-thirteenth century Iceland became a Norwegian province, meaning that Iceland was included in the personal union between Norway and Denmark in 1380. Iceland, the Faroe Islands, and Greenland remained under Danish rule in 1814, when Denmark was forced to cede Norway to Sweden. An Icelandic independence movement emerged in the 1830s, led by the scholar and expert in Icelandic saga literature, Jón Sigurðsson (1811–1879), who remained the political leader and pioneer for the Icelandic people until his death in 1879. Iceland acquired its own constitution in 1874 and a home rule agreement in 1904 (Adriansen, 2003). However, the goal for most Icelanders was a free state, and they perceived the creation of a university to be a major step in the process toward Icelandic independence from Denmark. The university was established in the Icelandic capital Reykjavik in 1911, the centenary anniversary of Sigurðsson's birth. The institution came about through a merger of three professional schools—theology (founded in 1847), medicine (1876), and law (1908)—with the addition of a faculty of arts responsible for teaching and researching the language, literature, and history of Iceland. Only seven years later, in 1918, Iceland became a free and sovereign state, which opted to enter into a personal union with Denmark. In 1944, independence was forced through by Icelandic nationalists, and the founding of the university had been part of that struggle (Haraldsson, 2003; Jóhannesson, 2013; Karlsson, 2011).

The population of Iceland in 1911 was 85,000, and the university initially had 20 teachers (11 of them full-time) and 45 male students. It was housed on the ground floor of the parliamentary building in the center of Reykjavik. In 1940 it was relocated just outside the town center with options for expansion. The university was regarded as one of the most magnificent buildings in the whole country. Its limited range of degree programs, however, led some students to study abroad, mostly in Denmark or Norway. During World War II (when Denmark and Norway were occupied by the Germans), Icelandic students stayed at home (see Hammerstein, 2004). There was thus a great need to expand the university and diversify its curriculum with degree programs such as engineering and business (Karlsson, 2011).

Through Sigurðsson a close link was forged between independence and scholarship focusing especially on Icelandic history and the sagas. At present, the University of Iceland has several programs focusing on Icelandic language, literature, history, and Norse culture and includes special courses in physical geography and geology emphasizing glaciology, volcanology, and other subjects relating to Iceland's natural landscape. Icelandic is the medium of instruction, but there are also courses taught in English, and students may submit assignments in Danish. Today, the University of Iceland is the single largest employer in Iceland. Haraldsson (2003) has argued that the strong, state university has greatly advanced the transformation of Icelandic society from traditional, dependent, fishing-and-farming structures into a modern independent state:

> The University became one of the cornerstone[s] and a symbol for the Icelanders to take the final step towards independence. That would probably have been impossible without a university, strongly emphasising the nation's cultural inheritance to strengthen the self-confidence of the people. I am certain that the Faroese people could learn a lot from our experience in that respect, and if there will be a decision to take the step all the way and seek full independence the Fróðskaparsetur [University of the Faroe Islands] should be utilized in a similar way as the University of Iceland was. (p. 21)

As the next section shows, the Faroe Islands have not yet taken that step, but the University of the Faroe Islands is an influential institution in society there.

The University of the Faroe Islands—Fróðskaparsetur Føroya

The Faroe Islands (Faroese: *Føroyar*), where the present population is approximately 50,000, have been under Norwegian and later Danish control since 1380. Unlike Greenland, however, the Faroe Islands have never been a colony; they became a Danish county in 1852. Some Faroese wanted home rule, and there has been a gradual introduction of the Faroese language in schools, churches, and legal system. Growing Faroese self-awareness was reflected in the inception and frequent use of national symbols of the Faroe Islands, not least during World War II, when the physical separation from Denmark demonstrated that Faroe Islanders could manage under other auspices. Since 1948, when the Faroe Islands were granted home rule, there has been a gradual movement toward more local autonomy. The Faroese path to greater independence is reflected in national symbols such as Faroese banknotes in 1951 and Faroese postage stamps in 1975 (Adriansen, 2003). The founding of a university can be seen as step in this process. After inauguration of The Faroese Academy of Sciences in 1952, its members promoted the creation of a university in the Faroe Islands, and in 1965 the Faroese Parliament agreed to establish an institution of higher education under the Latin name of *Academia Færoensis*. Its purpose was to carry out scientific research and teaching at the tertiary level. From the outset the goal was to promote nation-building and to prepare for nationhood (Marnersdóttir, 2003). The academy initially offered one-year courses in natural history and Faroese for schoolteachers. The limited staff focused on research on the Faroese language and folk culture. Academia Færoensis obtained official recognition as a university in 1990, and in 2008 the Faroese School of Education and the Faroese School of Nursing were incorporated into the university.

The University of the Faroe Islands currently consists of five departments: (a) Faroese Language and Literature, (b) Education, (c) Science and Technology, (d) Nursing, and (e) History and Social Sciences. There is thus great emphasis on Faroese language, literature, and history and on programs relevant to the Faroese labor market. All departments collaborate closely with universities in Denmark and other Nordic countries. Despite its small student body of approximately 600 students, the university offers bachelor's, master's, and doctoral degrees. Faroese is the medium of instruction (University of the Faroe Islands, 2016). The orientation of the university has shifted from being directed toward nation-building to being directed toward industry. Although national consciousness demands research on the Faroese language, history, and culture, financial independence requires research on economics and applied science. All of these areas of learning and research are now offered at the University of Faroe Islands (Marnersdóttir, 2003).

The University of Greenland—**Ilisimatusarfik**

Greenland (Greenlandic: *Kalaallit Nunaat*) has been inhabited by Inuit for more than 4,500 years. In the Viking Age (A.D. 793–1066), Norsemen, particularly from Norway and Iceland, settled on the southwest coast of Greenland, but they vanished in the early fifteenth century. Colonialization similar to the types practiced by other European countries began in the early eighteenth century, when Denmark and Norway asserted sovereignty over the land. It ended in 1953 when Greenland was integrated within the Danish realm. Danish citizenship was extended to the Greenlandic population, and a strategy of cultural assimilation began. However, the strategy failed against a reaffirmation of Greenlandic cultural identity among the Greenlandic elite. This resurgence led to a movement in favor of independence, which peaked in the 1970s. Greenland was granted home rule in 1979. The assimilation policies had not resulted in Greenlanders taking over administrative jobs performed mostly by skilled Danes (Loukacheva, 2007). An institution of higher education, an Inuit Institute, was suggested in 1974 by the Greenlandic County (*landsråd*), but it was not until 1981 that the home-rule government decided to set up two independent centers of higher education: the Inuit Institute, which offered two-year courses in language, literature, anthropology, sociology, and history; and an institute of theology rooted in specific Greenlandic conditions. In 1987 the two institutes merged under the name Inuit Institute. By 1989 it had been expanded to include new degree programs in social sciences and communication, and the administrative structure had been adapted to prepare for the official inauguration of the entire facility as the University of Greenland in 1989 (Marquardt, 2003).

The medium of instruction is Danish, but a few courses are taught in Greenlandic and classes by exchange lecturers are often in English. The University of Greenland is expected to provide knowledge about Greenland's past and present and to help foster Greenlandization of the society by providing academics with a Greenlandic background who can take over academic jobs from Danes. The candidates have put their distinctive mark on Greenlandic society (K. Kjærgaard & T. Kjærgaard, 2003; Marquardt, 2003). Hence, the purpose of the university is strongly linked to national interests, as in the other North Atlantic universities.

Universities as Part of the Arsenal of National Symbols and Institutions

Since the nineteenth century, a series of national European states have formed through war, subsequent territorial gains and losses, and conscious promotion of nationalism predominantly through mass education. Within a given territory, the various ethnic groups have been integrated and assimilated into the national state, or a particular ethnic group has pursued cultural separation, ending in independence and the creation of a new nation-state. These developments became apparent as of the 1840s in the Danish state, which by that point included multiple language groups: mainly Danish, German, and Icelandic, but also Frisian, Faroese, and Greenlandic. Wars and independence movements gave rise to the present Danish nation-state, the nation-state of Iceland, and two self-governing territories—the Faroe Islands and Greenland (Østergaard, 2015).

Any sovereign state has an array of national symbols and institutions that are also international signals of independence and domestic emblems of the population's support for the regime. They reinforce the citizenry's imagined sense of community and sense of belonging to the nation and the state. They can be official (e.g., the state's army, constitution, national flag, national anthem, and national day) or unofficial (e.g., monuments, events of mythology, and landscapes). The sense of national community that these different phenomena mirror suffuses the population through the educational system (Adriansen, 2003), and the symbols are used actively in the processes of state- and nation-building, in which universities, too, have figured prominently.

From the outset and until the eighteenth century, universities were fairly independent of the state in which they were located. They were once united in a large European community that used the same books and the same language, Latin (Pedersen, 1997). It may seem paradoxical that these extremely international institutions have become *national* symbols as it were, but the development of nation-states in the nineteenth century instilled the need to construct, preserve, and spread aspects of national identity, and universities were useful for that project. A common feature among the country-related examples presented above is the importance accorded to language, or rather to the mother tongue, which can be considered an unofficial national symbol. Throughout the twentieth century, there was a growing, official use of Faroese in the Faroe Islands. It became the language of administration, the church, and schools. Growing Faroese political independence is well reflected in this linguistic detachment from Danish. An analogous trend is evident in Greenland, where the use of Greenlandic as the official language has increased since the change to home rule. Evidence of this shift also surfaces in the official use of Greenlandic place names and in a general return to the use of Greenlandic personal names that had disappeared during the period of missionary influence. In Iceland the independent status of the Icelandic language has never been contested (Adriansen, 2003).

Universities have a vital role to play in the preservation of national languages. Nearly all states in Europe have a state-institutionalized national language council, a school curriculum, examinations, higher education, and a politically approved orthographic dictionary that upholds the national language. Universities, together with primary and secondary schools and museums, are likewise used for preserving unofficial national symbols such as a nation's history. Lastly, universities are often central to the study of the national landscape, a third unofficial national symbol according to Adriansen (2003). This focus is seen most clearly in the Icelandic case. Hence, although universities in themselves may not be national symbols, they can help sustain and strengthen the national symbolic world. The University of Iceland is an apt example again: "Preservation of the national cultural values and inheritance is also one of the key roles of the university" (Haraldsson, 2003, p. 20).

However, universities, too, may be national symbols, especially the first university in a small country. This standing occurred in some African countries after their independence, which granted them the possibility of controlling education and knowledge production and which was regarded as a crucial step in their development (Jensen, Adriansen, & Madsen, 2016). Israel is another interesting case, which Cohen (2007) has analyzed for the role of the Hebrew University in nation-building in prestate Israel. He also showed the university's noteworthy contribution as a cultural force in nation-building, while recognizing that universities, their employees, and their students may not always have the same intensions as each other or as the nation. The relationship between universities and nation-building was far from straightforward in nineteenth- and twentieth-century Europe. Significant academic networks and scholarly transfers between countries gave rise to a transnational academic space (Ellis, 2013; Jöns, 2008). Moreover, not all European countries were nation-states, so universities such as those in Scotland had a more composite role in relation to nation-building (Macdonald, 2009).

Discussion and Concluding Remarks

In this chapter we have explored the value of universities for nation-building and analyzed geopolitical discourses about university establishment, showing what light a geographical approach can cast on university history. There is nothing new about studying universities through the lens of geography. Since the early 1970s, geographers have studied universities in hundreds of published papers on topics ranging from location criteria and catchment areas to the mobility of students and professors (for overviews see Freytag & Jahnke, 2015; Hanson Thiem, 2009; Meusburger, 2015a). However, the history of universities and higher education in general tends to be written from a national standpoint (Ellis, 2013), with its authors beginning with the present nation-state rather than its history. For that reason Danish university history neglects what happened from the creation of the University of Copenhagen in 1479 to the creation of Aarhus University in 1928. By analyzing the composite state instead and taking a geohistorical stance, one learns about another four universities—Oslo (Norway) in 1814, Serampore[8] (India) in 1845, Kiel (Germany) in 1864, and Reykjavik (Iceland) in 1918—and about the territorial reduction that deprived the Danish state of them within a century. Hence, a geohistorical perspective can be useful, especially in analyses of universities outside the capital, where a country's first university is usually located.

Discourses of geopolitics and nationalism are not important for understanding all cases of university establishment. The University of Serampore is an example. However, discourses of geopolitics and nationalism can be important, especially for comprehending the establishment and location of universities in borderlands and the role of universities in the era of nation-building. Craig's (1984) analysis of two universities in another contested border region, Alsace-Lorraine, showed how higher education became intertwined with nationalism and struggles for cultural identity and prestige in the late nineteenth and early twentieth century. In Craig's estimation the universities in Alsace-Lorrain were exemplary in the transformation of the European university and in its adaptation to the emerging nation-states and their need for bureaucrats, teachers, political socialization, and technology. Although scholars have shown how national imperatives shaped the universities of the capitals and other old universities, Craig directed attention to universities on the periphery and their great importance for advancing the national cause. We have supplemented this work with the idea that the establishment of universities has been part of this national struggle, and we have shown how such creation has been part of geopolitical discourses before the era of the nation-states.

Despite the state's (and previously the church's) power in the founding and siting of universities, the initiative often comes first from local citizens who want to consolidate the power of a region by having a knowledge-producing institution there. The genesis of the universities in Oslo, Reykjavik, Tórshavn, and Aarhus illustrates this point. However, their university status had to be granted by the state, and in the first three of these cases the university was part of the struggle for independence from that very state. In the fourth case the struggle for a university was related to a new regional self-awareness in Jutland.

History has also shown that universities have a paradoxical function. On the one hand, they are important to the state. For instance, scholars serve as instruments of politics by providing rational arguments that can legitimate political decisions (Meusburger, 2012, 2015b). On the other hand, universities can be dangerous to the state because of their autonomy and capacity to become breeding grounds for

[8]This university was founded in 1818 and was the first in present-day India. At the time, though, Serampore was part of the Danish colonies in India and owed its status to the Danish king, Frederik VI (1768–1839). The university's establishment had little to do with Danish politics, for it had been the work of British missionaries, who were intent on training missionaries and spreading the Christian word (see Gøbel, 2014; Østergaard, 2005).

subversive ideas and conspiracies against the social order (e.g., student rejection of nationalism; see Gevers & Vos, 2004). This paradox is also intertwined with geography and center–periphery location, as in the regional examples of Jutland and the two duchies Schleswig and Holstein. Likewise, Craig (1984) has shown how these opposing forces operated in Alsace-Lorraine.

A final tension for consideration in this chapter is the international versus the local character of the university. Internationalization is currently claimed to be paramount for higher education (see chapter by Knight in this volume, for instance), but as we have shown, universities have often been created out of a need for knowledge and teaching relevant to local or national interests. The European university, since its advent in the early Middle Ages, has been very international in terms of networks, curriculum, and medium of instruction (Pedersen, 1997), but over the past two hundred years it has become increasingly oriented to a national agenda. This national turn is not exclusively a European phenomenon. In Africa, for instance, the focus of universities has shifted from the wishes of the colonial powers to a postindependence national agenda (Jensen et al., 2016). Independence means not only political independence through sovereignty but also epistemological independence, or a decolonization of knowledge. Many scholars in the Global South contend that Western knowledge has been thought of as universal and has therefore been imposed on other societies and environments (e.g., Chen, 2010; Shiva, 1993/2012). Western knowledge, however, is not universal in an epistemological sense but rather a globalized version of a local and parochial tradition originating in Europe (Madsen, Jensen, & Adriansen, 2016; Thomas-Emeagwali, 2006). Attempts to decolonize knowledge through Africanization of the curriculum can stand in contrast to internationalization efforts.

The problematic choices with which internationalization confronts universities in the Global South seem similarly difficult to resolve in the North Atlantic universities we have analyzed in this chapter. Their legitimacy is intertwined with the national agenda of using the local language as a medium of instruction, educating students for local society, and conducting research on local nature and culture (see Körber & Volquardsen, 2014). Yet these universities will lose their international reputation if they are not engaged in international networks. The difficult process of balancing the responses to these options and expectations calls for improving the knowledge of the political geography of universities and the geography of science.

References

Adriansen, I. (2003). *Nationale symboler i Det Danske Rige, 1830–2000* [National symbols in the Danish realm, 1830–2000] (Vols. 1–2). Copenhagen: Museum Tusculanum Forlag.

Adriansen, I., & Christensen, J. O. (2013). *The Second Schleswig War, 1864: Prelude, events and consequences.* Odder: Museum Sønderjylland and Tøjhusmuseet.

Adriansen, I., & Christensen, J. O. (2015). *The First Schleswig War, 1848–1851: Prelude, events and consequences.* Varde: Museum Sønderjylland and Tøjhusmuseet.

Agnew, J. A. (1998). *Geopolitics: Re-visioning world politics.* Abingdon: Routledge.

Agnew, J. A. (2001). *Reinventing geopolitics: Geographies of modern statehood*. Hettner-Lecture: Vol. 4. Heidelberg: Department of Geography, Heidelberg University.

Albeck, G. (1978). *Aarhus Universitet, 1928–1978* [Aarhus University, 1928–1978]. Acta Jutlandica: Vol. 51. Aarhus: Aarhus Universitetsforlag.

Asmussen, J. (2012). *På historisk rundrejse i Skåne, den dansk-svenske historie* [A historical roundtrip in Scania: The Danish-Swedish history]. Nærum: Globe.

Bagge, S., & Mykland, K. (1987). *Norge i dansketiden, 1380–1814* [Norway under Danish rule]. Copenhagen: Politikens Forlag.

Battrup, G. (2007). Mellem Kiel og København: Universitetsplaner i Sønderjylland/Slesvig, 1630–1930 [Between Kiel and Copenhagen: University plans in Sønderjylland/Schleswig, 1630–1930]. *PLUK—fra forskningen i Sønderjylland, 2*/3, 26–57.

Chen, K. H. (2010). *Asia as method: Toward deimperialization*. Durham: Duke University Press.

Cohen, U. (2007). University vs. society in a period of nation building: The Hebrew University in pre-state Israel. *Historical Studies in Education/Revue d'histoire de l'éducation, 19*(1), 81–110. Retrieved from http://historicalstudiesineducation.ca/index.php/edu_hse-rhe/article/download/274/691

Collett, J. P. (1999). *Historien om universitetet i Oslo* [The history of the university in Oslo]. Oslo: Universitetsforlaget.

Craig, J. E. (1984). *Scholarship and nation building: The universities of Strasbourg and Alsatian society, 1870–1939*. Chicago: University of Chicago Press.

de Ridder-Symoens, H. (Ed.). (1992). *Universities in the Middle Ages*. A history of the university in Europe: Vol. 1. Cambridge, UK: Cambridge University Press.

de Ridder-Symoens, H. (Ed.). (1996). *Universities in early modern Europe (1500–1800)*. A history of the university in Europe: Vol. 2. Cambridge, UK: Cambridge University Press.

Dodds, K. (2014). *Geopolitics: A very short introduction*. Hampshire: Oxford University Press.

Elliott, J. H. (1992). A Europe of composite monarchies. *Past and Present, 137*, 48–71. doi:https://doi.org/10.1093/past/137.1.48

Ellis, H. (2013). National and transnational spaces: Academic networks and scholarly transfer between Britain and Germany in the nineteenth century. In I. Löhr & R. Wenzlhuemer (Eds.), *The nation state and beyond* (pp. 127–148). Transcultural Research—Heidelberg Studies on Asia and Europe in a Global Context: Vol. 5. Heidelberg: Springer Verlag.

Feldbæk, O. (1998). *Nærhed og adskillelse, 1720–1814* [Proximity and separation, 1720–1814]. Danmark-Norge, 1380–1814: Vol. 4. Oslo: Universitetsforlaget.

Frandsen, S. B. (2008). *Holsten i helstaten: Hertugdømmet inden for og uden for det danske monarki i første halvdel af 1800-tallet* [Holstein in the Danish composite state: The duchy inside and outside the Danish monarchy in the first half of the 1800s]. Viborg: Museum Tusculanum Press, University of Copenhagen.

Freytag, T., & Jahnke, H. (2015). Perspektiven für eine konzeptionelle Orientierung der Bildungsgeographie [Perspectives for a conceptual orientation of the geography of education]. *Geographica Helvetica, 70*(1), 75–88. doi:https://doi.org/10.5194/gh-70-75-2015

Geipel, R. (1966). Angewandte Geographie auf dem Feld der Bildungsplanung [Applied geography in the field of educational planning]. In F. Monheim & A. Beuermann (Eds.), *Tagungsbericht des Deutschen Geographentages 1965 in Bochum* (pp. 448–457). Wiesbaden: Steiner.

Geipel, R. (1971). Die Universität als Gegenstand sozialgeographischer Forschung [The university as an object of research in social geography]. *Mitteilungen der Geographischen Gesellschaft in München, 56*, 17–31.

Gevers, L., & Vos, L. (2004). Student movements. In W. Rüegg (Ed.), *Universities in the nineteenth and early twentieth centuries (1800–1945)* (pp. 269–362). A history of the university in Europe: Vol. 3. Cambridge, UK: Cambridge University Press.

Gregory, D., Meusburger, P., & Suarsana, L. (2015). Power, knowledge, and space: A geographical introduction. In P. Meusburger, D. Gregory, & L. Suarsana (Eds.), *Geographies of knowledge and power* (pp. 1–18). Knowledge and Space: Vol. 7. Dordrecht: Springer.

Gustafsson, H. (1998). The conglomerate state: A perspective on state formation in early modern Europe. *Scandinavian Journal of History, 23*, 189–213. doi:https://doi.org/10.1080/03468759850115954

Gøbel, E. (2014). Radiating from Tranquebar: Education in Serampore, 1800–1845. In E. Fihl & Ā. R. Vēṅkaṭācalapati (Eds.), *Beyond Tranquebar: Grappling across cultural borders in South India* (pp. 556–578). New Delhi: Orient BlackSwan.

Hammerstein, N. (1996). Relations with authority. In H. de Ridder-Symoens (Ed.), *Universities in early modern Europe (1500–1800)* (pp. 113–153). A history of the university in Europe: Vol. 2. Cambridge, UK: Cambridge University Press

Hammerstein, N. (2004). Epilogue: Universities and war in the twentieth century. In W. Rüegg (Ed.), *Universities in the nineteenth and early twentieth centuries (1800–1945)* (pp. 637–672). A history of the university in Europe: Vol. 3. Cambridge, UK: Cambridge University Press.

Hanson Thiem, C. (2009). Thinking through education: The geographies of contemporary educational restructuring. *Progress in Human Geography, 33*, 154–173. doi:https://doi.org/10.1177/0309132508093475

Haraldsson, G. G. (2003). Háskóli Íslands [The University of Iceland]. In Fróðskaparsetur Føroya (Ed.), *Universitet í útnorðri* (pp. 18–26). Tórshavn: Fróðskaparsetur Føroya.

Harhoff, F. (1993). Rigsfællesskabet [The community of the Danish realm]. Århus: Klim.

Henningsen, L. N. (2008). Die Herzöge von Gottorf [The dukes of Gottorp]. In C. P. Rasmussen (Ed.), *Die Fürsten des Landes: Herzöge und Grafen von Schleswig, Holstein und Lauenburg* (pp. 142–185). Neumünster: Wachholtz.

Hofmann, E. (1965). *Allgemeine Entwicklung der Universität* [General development of the university]. Geschichte der Christian-Albrechts-Universität Kiel, 1665–1965: Vol. 1. Neumünster: Wacholtz.

Jensen, S., Adriansen, H. K., & Madsen, L. M. (2016). Do 'African' universities exist? Setting the scene. In H. K. Adriansen, L. M. Madsen, & S. Jensen (Eds.), *Higher education and capacity building in Africa: The geography and power of knowledge under changing conditions* (pp. 12–37). Routledge Studies in African Development: Vol. 6. Abingdon: Routledge.

Jespersen, K. J. V. (2011). *A history of Denmark* (2nd ed.). Palgrave Essential Histories Series: Vol. 18. Basingstoke: Palgrave.

Jóhannesson, I. Á. (2013). Excellence, innovation and academic freedom in university policy in Iceland. *Icelandic Review of Politics and Administration, 9*, 79–99. doi:https://doi.org/10.13177/irpa.a.2013.9.1.4

Jöns, H. (2008). Academic travel from Cambridge University and the formation of centres of knowledge, 1885–1954. *Journal of Historical Geography, 34*, 338–362. doi:https://doi.org/10.1016/j.jhg.2007.11.006

Karlsson, G. (Ed.). (2011). *Aldarsaga Háskólá Íslands 1911–2011* [The history of the University of Iceland, 1911–2011]. Reykjavík: Háskólaútgáfan.

Kintzinger, M. (2003). *Wissen wird Macht: Bildung im Mittelalter* [Knowledge becomes power: Education in the Middle Ages]. Darmstadt: Jan Thorbecke Verlag.

Kintzinger, M. (2012). Universität [University]. In P. de Boer, H. Duchhardt, G. Kreis, & W. Schmale (Eds.), *Das Haus Europa* (pp. 307–312). Europäische Erinnerungsorte: Vol. 2. Munich: Oldenbourg

Kjærgaard, K., & Kjærgaard, T. (2003). *Ny Herrnhut i Nuuk 1733–2003: Missionsstation, rævefarm, embedsbolig, museum, universitet*. [New Herrnhut in Nuuk, 1733–2003: Missionary building, fox farm, official residence, museum, university.] Nuuk: Forlaget Atuagkat/Ilisimatusarfik.

Koenigsberger, H. G. (1986). *Politicians and virtuosi: Essays on early modern history*. London: The Hambledon Press.

Körber, L-A., & Volquardsen, E. (2014). *The postcolonial North Atlantic: Iceland, Greenland and the Faroe Islands*. Berlin: Nordeuropa Institut der Humboldt Universität.

Langholm, S. (1996). The university and the nation in early 19th century Norway. In M. Norrback & K. Ranki (Eds.), *University and nation: The university and the making of the nation in northern Europe in the 19th and 20th centuries* (pp. 139–152). Studia Historica: Vol. 53. Helsinki: Conference on the History of Universities 1994.

Lausten, M. S. (1991). Københavns Universitet i middelalderen 1479–ca. 1530 [The University of Copenhagen in the Middle Ages, 1470 to about 1530]. In S. Ellehøj, L. Grane, & K. Hørby (Eds.), *Universitetet 1479–1979* (Vol. 1, pp. 1–77). Copenhagen: Gads Forlag.

Lohmeier, D. (1997). *Kleiner Staat ganz groß: Schleswig-Holstein-Gottorf* [Small state quite large: Schleswig-Holstein-Gottorp]. Heide: Westholsteinische Anstalt Boyens.

Loukacheva, N. (2007). *The Arctic promise: Legal and political autonomy of Greenland and Nunavut.* Toronto: University of Toronto Press.

Lunds Universitet. (2016). *Universitetets historia* [The university's history]. Retrieved from http://www.lu.se/lunds-universitet-350-ar/om-jubileet/universitetets-historia

Macdonald, C. M. M. (2009). 'To form citizens': Scottish students, governance and politics, 1884–1948. *History of Education, 38*, 383–402. doi:https://doi.org/10.1080/00467600902855439

Madsen, L. M., Jensen, S. & Adriansen, H. K. (2016). Dilemmas and paradoxes of capacity building in African higher education: Concluding remarks. In H. K. Adriansen, L. M. Madsen, & S. Jensen (Eds.), *Higher education and capacity building in Africa: The geography and power of knowledge under changing conditions* (pp. 239–250). Abingdon: Routledge.

Mamadouh, V. D. (1998). Geopolitics in the nineties: One flag, many meanings. *GeoJournal, 46*, 237–253. doi:https://doi.org/10.1023/A:1006950931650

Marnersdóttir, M. (2003). Fróðskaparsetur Føroya [The University of the Faroe Islands]. In Fróðskaparsetur Føroya (Ed.), *Universitet í útnorðri* (pp. 46–59). Tórshavn: Fróðskaparsetur Føroya.

Marquardt, O. (2003). Ilisimatusarfik [The University of Greenland]. In Fróðskaparsetur Føroya (Ed.), *Universitet í útnorðri* (pp. 38–45). Tórshavn: Fróðskaparsetur Føroya.

Mayr, A. (1979). *Universität und Stadt. Ein stadt-, wirtschafts- und sozialgeographischer Vergleich alter und neuer Hochschulstandorte in der Bundesrepublik Deutschland* [University and city: Old and new university locations in the Federal Republic of Germany from the perspective of urban, economic, and social geography]. Münstersche Geographische Arbeiten: Vol. 1. Paderborn: Schöningh.

Meusburger, P. (2012). Heidelberg: Relations between a town and its university. In P. Meusburger & T. Schuch (Eds.), *Wissenschaftsatlas of Heidelberg University: Spatio-temporal relations of academic knowledge production* (pp. 18–35). Knittlingen: Bibliotheca Palatina.

Meusburger, P. (2015a). Education, Geography of. In J. D. Wright (Ed.), *International encyclopedia of the social & behavioral sciences*: *Vol. 7.* (pp. 165–171). 2nd ed. Oxford, UK: Elsevier

Meusburger, P. (2015b). Relations between knowledge and power: An overview of research questions and concepts. In P. Meusburger, D. Gregory, & L. Suarsana (Eds.), *Geographies of knowledge and power* (pp. 19–74). Knowledge and Space: Vol. 7. Dordrecht: Springer Netherlands. doi:https://doi.org/10.1007/978-94-017-9960-7_2

Misfeldt, H. H. (1925). Det jydske universitet [The University of Jutland]. In *Det Nye Universitet og Genforeningsarbejdet* (pp. 3–27). Kolding: n.p.

Neave, G. (2011). Patterns. In W. Rüegg (Ed.), *Universities since 1945* (pp. 31–69). A history of the university in Europe: Vol. 4. Cambridge, UK: Cambridge University Press.

Norrback, M., & Ranki, K. (Eds.). (1996). *University and nation: The university and the making of the nation in northern Europe in the nineteenth and twentieth centuries.* Studia Historica: Vol. 53. Helsinki: Conference on the History of Universities.

Ó Tuathail, G., & Agnew, J. (1992). Geopolitics and discourse: Practical geopolitical reasoning in American foreign policy. *Political Geography, 11*, 190–204. doi:https://doi.org/10.1016/0962-6298(92)90048-X

Østergaard, U. (2002). Den danske stat—territorialt betragtet [The Danish state—A territorial view]. In T. B. Jørgensen & K. K. Klausen (Eds.), *Territorial dynamik—streger på landkort, billeder i vore hoveder* (pp. 45–60). Aarhus: Aarhus Universitetsforlag.

Østergaard, U. (2005). College for the instruction of Asian and other youth in Eastern literature and European science in Serampore. In G. Oommen & H. R. Iversen (Eds.), *It began in Copenhagen: Junctions in 300 years of Indian-Danish relations in Christian mission* (pp. 204–220). Delhi: Indian Society for Promoting Christian Knowledge.

Østergaard, U. (2015). Nation-building and nationalism in the Oldenburg Empire. In S. Berger & A. Miller (Eds.), *Nationalizing empires* (pp. 461–510). Budapest: Central European University Press.

Paterson, L. (2001). Higher education and European regionalism. *Pedagogy, Culture, & Society, 9*, 133–160. doi:https://doi.org/10.1080/14681360100200117

Pedersen, O. (1997). *The first universities: Studium generale and the origins of university education in Europe*. Cambridge, UK: Cambridge University Press.

Ramirez, F. O., & Boli, J. (1987). The political construction of mass schooling: European origins and worldwide institutionalization. *Sociology of Education, 60*, 2–17. doi:https://doi.org/10.2307/2112615

Rasmussen, C. P. (2008). Die dänischen Könige als Herzöge von Schleswig und Holstein [The Danish kings as dukes of Schleswig and Holstein]. In C. P. Rasmussen (Ed.), *Die Fürsten des Landes. Herzöge und Grafen von Schleswig, Holstein und Lauenburg* (pp. 72–109). Neumünster: Wachholtz.

Rasmussen, J. R. (1995). The Danish monarchy as a composite state, 1770–1830. In N. A. Sørensen (Ed.), *European identities: Cultural diversity and integration in Europe since 1700* (pp. 23–36). Odense: Syddansk Universitetsforlag.

Riddle, P. (1996). The university and political authority: Historical trends and contemporary possibilities. *Research in Sociology of Education and Socialisation, 11*, 43–62.

Rosén, J. (1968). *Lunds universitets historia: Vol. 1.* 1668–1709 [History of Lund University: Vol. 1. 1668–1709]. Lund: Gleerup.

Ruck, M. (2007). 60 år med Flensborg Universitet (1946–2006): Dybde, differentiering, ekspansion [Sixty years with Flensburg University (1946–2006): Depth, differentiation, expansion]. *PLUK—fra forskningen i Sønderjylland, 2/3*, 16–25.

Rüegg, W. (Ed.). (2004). *Universities in the nineteenth and early twentieth centuries (1800–1945).* A history of the university in Europe: Vol. 3. Cambridge, UK: Cambridge University Press.3.

Rüegg, W. (Ed.). (2011). *Universities since 1945.* A history of the university in Europe: Vol. 4. Cambridge, UK: Cambridge University Press.

Sanders, H. (2008). *Efter Roskildefreden 1658. Skånelandskapen och Sverige i krig och fred* [After the Peace of Roskilde, 1658: Scanian provinces and Sweden in war and peace]. Munkedal: Makadam.

Schwinges, R. C. (1998). The medieval German university: Transformation and innovation. *Paedagogica Historica: International Journal of the History of Education, 34*, 375–388. doi: https://doi.org/10.1080/0030923980340204

Schwinges, R. C. (Ed.). (2008). *Universität im öffentlichen Raum* [University in public space]. Veröffentlichungen der Gesellschaft für Universitäts- und Wissenschaftsgeschichte: Vol. 10. Basle: Schwabe.

Scott, P. (1990). *Knowledge and nation.* Edinburgh: University Press.

Shiva, V. (2012). *Monocultures of the mind: Perspectives on biodiversity and biotechnology.* Penang: Third World Network. (Original work published 1993)

Slottved, E., & Tamm, D. (2009). *The University of Copenhagen: A Danish centre of learning since 1479* (T. McTurk, Trans.). University of Copenhagen. Retrieved May 27, 2017, from the University of Copenhagen website http://www.e-pages.dk/ku/196/

Stybe, S. E. (1979). *Copenhagen University: 500 years of science and scholarship.* Copenhagen: Royal Danish Ministry of Foreign Affairs.

Tamm, D. (1996). Professors and politics—An outline. In M. Norrback & K. Ranki (Eds.), *University and nation: The university and the making of the nation in northern Europe in the 19th and 20th centuries* (pp. 21–29). Studia Historica: Vol. 53. Helsinki: Conference on the History of Universities, 1994.

Thomas-Emeagwali, G. (2006). *Africa and the academy: Challenging hegemonic discourses on Africa*. Trenton: Africa World Press.
University of the Faroe Islands, The. (2016). *The University of the Faroe Islands*. Retrieved from https://setur.fo/top-menu/in-english/about-theuniversity-of-the-faroe-islands/
Weber, M. (2004). *The vocation lectures* (R. Livingstone, Trans.; ed., with an introduction by D. Owen & T. B. Strong). Illinois: Hackett Books.

Chapter 6
"A Small Town of Character": Locating a New Scottish University, 1963–1965

Michael Heffernan and Heike Jöns

The ancient Scottish universities of St. Andrews (founded in 1410), Glasgow (1451), Aberdeen (1495), and Edinburgh (1583) are among Europe's most celebrated institutions of higher learning, renowned for their liberal commitment to religious and social inclusion. Established centuries before the union between Scotland and England in the early eighteenth century, the Scottish universities occupy a unique and distinctive position in British higher education. Although the three oldest were religious foundations, Edinburgh was established by the city's council, which remained the governing authority until 1858. Two years later, King's College Aberdeen combined with Marischal College, a separate institution established in the same city in 1593, to create a powerful quartet of Victorian universities that educated in the mid-nineteenth century proportionately twice as many Scottish students as the similar number of English universities educated from the population south of the border (Anderson, 2006, pp. 12–13; Whyte, 2015, pp. 4–5, 32).

Scotland's relative preeminence in British higher education began to wane in the late nineteenth and early twentieth century when six new "redbrick" universities were established in the larger industrial cities of Manchester, Birmingham, Liverpool, Leeds, Sheffield, and Bristol, mainly inspired by the civic higher educational ideals developed in London and Edinburgh rather than the ecclesiastical traditions of Oxford and Cambridge. Further new university colleges, emerging in smaller cities and towns such as Nottingham, Reading, and Southampton and also initially offering degrees validated by the University of London, reinforced the changing geography of British higher education (Beloff, 1969).

M. Heffernan (✉)
School of Geography, University of Nottingham, Nottingham, UK
e-mail: mike.heffernan@nottingham.ac.uk

H. Jöns
Department of Geography, Loughborough University, Loughborough, UK
e-mail: h.jons@lboro.ac.uk

P. Meusburger et al. (eds.), *Geographies of the University*, Knowledge and Space 12,
https://doi.org/10.1007/978-3-319-75593-9_6

Plans to expand the network of British universities in response to a rising population, a growing demand for graduates in science and technology, and a political desire to increase participation rates among geographically and socially disadvantaged communities were extensively debated during the period of the Labour government after 1945 (Anderson, 2006, pp. 131–133). Alan Barlow's (1946) report on "scientific man-power" included one of several proposals for a new, publicly funded university (Shattock, 2012, p. 44). By the late 1950s, after almost a decade of Conservative rule, public finances had sufficiently recovered from World War II for these plans to bear fruit. Between 1961 and 1968, 24 new universities were chartered in the United Kingdom, including ten preexisting, local-authority-controlled Colleges of Science and Technology that were expanded and redesigned as independent universities. Based on a generous, but means-tested, system of student grants, the proportion of 19- to 20-year-olds attending university rose sharply from circa 5% in 1960 to 14% ten years later (Robertson, 2010, p. 19).

It was during this period of unprecedented expansion that the principal characteristics of British university life were established based on a presumption of high student mobility and the separation of home and campus (Anderson, 2006, pp. 139–141; Committee on Higher Education [hereafter CHE], 1963, p. 162). The most eye-catching developments were the seven new "plateglass" universities established *ab initio* during the 1960s on green-field sites adjacent to Brighton, Norwich, York, Canterbury, Colchester, Warwick, and Lancaster (Tight, 1987). These proposals were agreed on separately by the British government between 1958 and 1961 without parliamentary scrutiny and based solely on the recommendations of the University Grants Committee (UGC), a government quango established in the aftermath of World War I and chaired from 1953 to 1963 by Keith Murray (1903–1993), a Scottish agricultural economist (Beloff, 1969; Shattock, 1994).[1]

The UGC's Subcommittee on New University Colleges, created in 1959 to stimulate discussion on the nature and location of new institutions of higher education, received dozens of proposals for new universities from towns and cities across the country. This list was eventually whittled down to the seven projects recommended to government by UGC members and the small circle of Treasury civil servants, university vice-chancellors, Oxbridge academics, and industrialists with whom Murray confided at regular dinners in London's Athenaeum Club. The "back-room" nature of these decisions, and their bias toward the more attractive county towns and cities of England, a country still perceived to be inadequately served by its existing universities, caused understandable resentment in Scotland, Wales, and Northern Ireland (Shattock, 2012, pp. 16, 43–54).

In July 1960, the government agreed to establish "a committee of persons outside the Government service, under an independent chairman of high status . . . to study

[1]The National Archives of the UK, University Grants Committee [hereafter TNA, UGC] 7/238. Fifth university for Scotland: Extract from Hansard, June 20, 1961, p. 935.

the fundamental long-term problems arising in the field of full-time higher education."[2] The Committee on Higher Education (CHE), set up in February 1961 under the chairmanship of the LSE economist Lionel Robbins (1898–1984), was intended to provide firmer evidence on which to guide future decisions (Carswell, 1985; King & Nash, 2001; Layard, King, & Moser, 1969; Robbins, 1966; Shattock, 2014). The Robbins committee consisted of twelve members, including two Scottish individuals: Robbins; four university representatives from different disciplines (Oxford humanist Helen L. Gardner; Cambridge-trained educationalist H. Lionel Elvin—director of the Institute of Education at London University; Scottish-born Edinburgh psychologist James Drever—educated in Edinburgh and Cambridge; and chemist Patrick Linstead—Rector of his former alma mater Imperial College London); two heads of independent schools (Kitty Anderson and A. Chenevix-Trench); an industrialist (Edward Herbert, who died in April 1963); and representatives of local authorities (Harold C. Shearman), the Vice-Chancellors Committee (Philip Morris), technical research (Scottish-born David Anderson), and technical education (R. B. Southall).[3] The committee meetings were also attended by P. S. Ross of the Treasury as secretary, Keith Murray (UGC), and Antony Part (Ministry of Education). Claus Moser and Richard Layard (both LSE) were responsible for the statistical data collection and analysis (Stewart, 1989, p. 333).

Far from initiating the expansion of British higher education, as is often claimed, the lengthy deliberations of the Robbins committee between 1961 and 1963 temporarily halted the ad hoc process of expansion that had been gathering momentum under Murray's chairmanship of the UGC, a tenure that ended shortly before Robbins published his recommendations (for personal recollections, see Robbins, 1971; Annan, 1982; Moser, 1987). The Robbins report, published on October 23, 1963, contained a mass of statistical data on every aspect of British higher education, ranging from projected student numbers and likely employment demand for graduates in different sectors of the economy to the most effective means of university governance based on international comparisons (e.g., Callender, 2014; Scott, 2014; Shattock, 2014).

The report recommended the immediate further expansion of the existing system and the creation of up to six new universities to be located in or near to large cities and with an emphasis on science and technology (CHE, 1963, pp. 163, 272). To counter the earlier bias toward England, Robbins endorsed the expansion and reorganization of Scotland's existing universities, specifically the redesignation of Glasgow's Royal College of Science and Technology as Strathclyde University (1964), and indirectly also the transformation of Edinburgh's Heriot-Watt College into Heriot-Watt University (1966), as well as the conversion of Queen's College

---

[2]TNA, UGC 7/237. Fifth university for Scotland: Committee meeting, November 2, 1960, p. 1.

[3]TNA, Department of Education and Science [hereafter ED] 116/11. Committee on Higher Education (Robbins committee): Agenda and minutes: Composition of the committee. University expansion in the 1960s was shaped by very few women. All individuals cited in this text with their first-name initials were men.

Dundee, previously part of St. Andrews, into an independent university (1967; see CHE, 1963, pp. 132–133; Paterson, 2003). The Robbins report specified that at least one of the six new universities should be established in Scotland (CHE, 1963, p. 284).

Drawing on previously unused documents in the UK National Archives, we consider the debates about the location of the new Scottish university recommended by Robbins and analyze the decision to award this institution to the county town of Stirling. As a geographical inquiry into the factors that influenced the location of a major British university, this chapter may rectify the paradoxical absence of research on the basic geographies of the working environments in which most professional geographers earn their living, mindful of the enormous economic, social, and cultural advantages that university towns and cities have enjoyed since the 1960s compared to otherwise similar urban centers that lack institutions of higher education (Cochrane & Williams, 2013; Goddard & Vallance, 2013; Lawton Smith, 2007; Parkinson et al., 2006; Tight, 1987, 1996).

Early Initiatives

Following the 1946 Barlow report, several local authorities approached the government with the aim of securing a new university. To promote Scotland's case, Joseph Westwood, Secretary of State for Scotland in the Labour government, organized a meeting of representatives from five Scottish boroughs—Dumfries, Inverness, Oban, Perth, and Stirling—on May 2, 1947, to discuss a proposal he had previously circulated for a fifth Scottish university. Although these exchanges ended with a cautious, financially motivated decision to focus on the expansion of existing universities, Westwood "seemed to think that we would need to fight England for the university and was quite pleased that he had, in a sense, stolen a march on England with the early propaganda."[4]

In the event, nothing of any significance happened for more than a decade until 1960 when, in the midst of the UGC debates about new universities for England, the Scottish case was revived. On April 28, Alan Thompson, Labour MP for Dunfermline, tabled a parliamentary question to John Scott Maclay, Secretary of State for Scotland in Harold Macmillan's second Conservative government, demanding to know why none of the new universities mentioned in the media were located north of the border. In early August, a group of local MPs, town councilors, and local government officials made a more concerted attempt to promote the case for a new

[4]TNA, UGC 7/237. Fifth university for Scotland: Ernest Fyfe, Provost of Dumfries, to Colonel J. G. Crabbe, Dumfries; The Reverend J. A. Fisher, Castle Douglas; and Hugh S. Gladstone, Penpont, May 6, 1947; see also Mr. Parker, Scottish Education Department, to H. A. de Montmorency, UGC secretary, April 28, 1947.

University of East Stirlingshire somewhere in the vicinity of Falkirk–Grangemouth–Larbert, midway between Edinburgh and Glasgow.[5]

Some of the ensuing correspondence, including Maclay's responses outlining his interpretation of the UGC criteria for assessing the suitability of proposed locations, was leaked to the press, triggering a wave of counterproposals from other Scottish towns.[6] On September 15, a *Glasgow Herald* article on "How to start a new university?" described how Falkirk, Inverness, Dumfries, and Ayr might develop proposals as persuasive as those successfully advanced by Brighton, York, and Norwich, adding—in a follow-up article on September 30—that the Highlands "will not lack well-wishers if they undertake the quest."[7] Over the next two months, the UGC received requests for further information on submitting proposals from Dumfries (October 6), Falkirk (October 18), Stirling (October 19), Inverness (November 1), and Perth (November 8), and the secretary of town council in Ayr contacted Thomas Moore, the town's veteran Conservative MP, on November 1, asking him to lobby the government on the town's behalf.[8]

On September 21, Malcolm MacPherson, Labour MP for Stirling and Falkirk, wrote to Maclay expressing his concern about the "undignified rivalry among several localities for the one prize."[9] In MacPherson's view, Scotland needed a more united front. One way to achieve this, he argued, was to ask the existing universities to sponsor the new institution, having decided on its location, just as the universities of Oxford, Manchester, and Birmingham had done in the case of Keele University in England—a suggestion forwarded by Maclay to the UGC. Despite a reserved response from UGC secretary Cecil Syers, who noted that new proposals would have to await the "Government's decision on the general expansion question," Murray—a native of Edinburgh and an alumnus of the city's university—agreed to meet representatives from the six alternative Scottish locations at a hastily arranged gathering in St. Andrew's House in Edinburgh between Christmas and New Year 1960.[10]

While insisting that there was no need for a new Scottish university given the capacity to expand existing institutions, Murray acknowledged that the Robbins

[5]TNA, UGC 7/237. Fifth university for Scotland: Alexander Duncan, Secretary of the Proposed University for East Stirlingshire Campaign Committee, to Malcolm MacPherson, Labour MP for Stirling and Falkirk, August 3, 1960; Duncan to Maclay, August 8, 1960; H. H. Donnelly, Scottish Education Department, to Cecil Syers, UGC secretary, August 29, 1960.

[6]TNA, UGC 7/237. Fifth university for Scotland: Niall MacPherson, Parliamentary Under-Secretary of State for Scotland, to Arthur Woodburn, Labour MP for Clackmannan and East Stirlingshire, August 20, 1960; Maclay to M. MacPherson, September 8, 1960.

[7]TNA, UGC 7/237. Fifth university for Scotland: Making it five: The Glasgow Herald Leader Article, September 30, 1960.

[8]TNA, UGC 7/237. Fifth university for Scotland.

[9]TNA, UGC 7/237. Fifth university for Scotland: M. MacPherson to Maclay, September 21, 1960.

[10]TNA, UGC 7/237. Fifth university for Scotland: Syers to F. M. M. Grey, Scottish Education Department, October 4, 1960, p. 1; Duncan to Murray, November 17, 1960; J. E. Fraser, Office of the Secretary of State for Scotland, to Murray, December 19, 1960.

committee might revise this assessment. He therefore encouraged representatives from each town to submit full applications to the UGC within six months, based on the successful applications from Norwich and York. Each town was instructed to demonstrate sufficient economic vitality, including the ability to generate the necessary financial support for a new university, estimated to be roughly 3% of annual capital expenditure during the development phase; the excellence of existing facilities, including transport links with other centers of learning across the UK; the housing supply for academic staff and students, based on the assumption that a new university would need to accommodate about 60% of its circa 3,000 students; the capacity of local schools to accommodate an influx of new pupils; and the community's wider cultural vitality and support for the proposal. Submissions would also need to identify a green-field site of at least 200 acres.[11]

No full submissions were made within the next six months, but Murray received the following response to a question he asked his Scottish colleagues about the Airthrey estate that Stirling representatives had mentioned in the meeting:

> . . . the Department of Health have decided that about 190 acres of this site should be offered to the County Council as the previous owners of this site . . . if there was any serious intention to use the Airthrie [*sic*] site for a university, the best plan would probably be to remove the existing maternity hospital altogether so that the whole of the estate would be available for university purposes . . . the happiest solution would presumably be for the Town Council of Stirling and the County Council to agree to keep the land available for use for a university, if the local people want to pursue the idea of a university at Stirling.[12]

This reassurance about an available site of about 200 acres or more might have been the moment in which Murray chose to support the idea of a University of Stirling. The matter certainly remained on the policy agenda through several questions about a new Scottish university in parliament (June and December 1961, March 1962) as well as respective memoranda sent to the UGC by the Scottish Union of Students (August 1961); the campaign committees for East Stirlingshire (December 1962, March 1963) and Stirling (August 1963); and the National Committee for a New University in Scotland of the Educational Institute of Scotland (February 1963). The latter sent its memorandum to both the UGC and the Robbins committee and was able to arrange a meeting with the chief secretary of the Treasury in April 1963.[13] As John Rankin, a Labour MP from Glasgow, had expressed the wider mood that calling "attention to the need for another university in Scotland . . . deplores the fact that when seven universities are provided for England none is considered necessary for Scotland,"[14] it suggests itself that this asymmetry became a central concern for the Robbins committee. This body's meetings were in fact

[11]TNA, UGC 7/238. Fifth university for Scotland: Meetings with the chairman of the UGC about a new Scottish university, January 26, 1961.

[12]TNA, UGC 7/238. Fifth university for Scotland: William Murrie, Permanent Under-Secretary of State in the Scottish Office, to Murray, January 4, 1961.

[13]TNA, UGC 7/238. Fifth university for Scotland; TNA, UGC 7/239. Fifth university for Scotland.

[14]TNA, UGC 7/238. Fifth university for Scotland: Extract from Hansard, June 30, 1961, p. 935.

attended by UGC chairman Keith Murray, who, according to Shattock (2012), exerted "a powerful influence on the Committee" (p. 39).

The task of the Robbins committee was to advise the government on what principles its long-term development of higher education should be based (CHE, 1963, p. iii). Over a period of two and a half years, the committee held 111 meetings, consulted more than 400 written submissions, conducted over 120 interviews, and visited several institutions at home and abroad, thus providing the basis for a new phase of evidence-based planning in higher education (CHE, 1963, p. 1). As an outcome of these deliberations, the Robbins report raised the target number of students for the early 1970s from the previous UGC figure of 170,000 to 218,000 and recommended institutional expansion through four strategies:

- the foundation of six further new universities;
- the upgrading of ten Colleges of Advanced Technology (CAT) and some ten Regional Colleges, Central Institutions, and Colleges of Education to universities;
- the development of five Special Institutions for Scientific and Technological Education and Research (S.I.S.T.E.R.);
- the establishment of at least one of the six new universities in Scotland (CHE, 1963, pp. 281, 284).

In a press statement, the government endorsed most of the Robbins report's recommendations, including the formulation of a ten-year program for university expansion and the foundation of a new university in Scotland. Yet by February 1965, when the government suddenly announced that no more new universities would be needed for about ten years (Mountford, 1966, p. 43), only part of the post-Robbins expansion program had been pursued. The ten CATs in England and Wales were about to be upgraded to universities; the Royal College of Science and Technology in Glasgow had been transformed into the University of Strathclyde, thus turning the search for the fifth Scottish university into the sixth; and the new Scottish university was to be founded in Stirling.[15]

A New University for Scotland

The post-Robbins expansion program was implemented promptly by the UGC's new chairman, John Wolfenden (1906–1985), an Oxford graduate in philosophy (Queen's College, 1928), who subsequently held a Henry P. Davison scholarship at Princeton University (1928–1929) and a philosophy fellowship at Magdalen College Oxford (1929–1934) before serving on invitation as a school headmaster at Uppingham School (1934–1944) and Shrewsbury School (1944–1950). Wolfenden was appointed Vice-Chancellor of the University of Reading in 1950

[15]The decisions on chartering Heriot-Watt University, the University of Dundee, and the new University of Ulster in Coleraine, Northern Ireland, were still pending.

and remained in this position until chairing the UGC from October 1963 to 1968. He is probably best known for recommending the decriminalization of homosexuality in Britain as the chair of the government committee that now bears his name and that reported in 1957 (Weeks, 2004).

Only two weeks after the publication of the Robbins report, on November 8, 1963, Wolfenden wrote to all towns and cities that had previously inquired about the possibility of establishing a new university, encouraging them to submit a full proposal by the end of the year. A separate letter was addressed to Imperial College London, Manchester College of Science and Technology, and the Royal College of Science and Technology in Glasgow regarding the possible development as S.I.S.T.E.R. institutions, asking for a ten-year development program with estimated costs. The following letter was sent to six towns and cities in Scotland: Ayr, Dumfries, Falkirk (East Stirlingshire), Inverness, Perth, and Stirling:

> As you will have seen from the Government's statement about the Robbins Report which was published on 24th October (Cmnd. 2165), the University Grants Committee have been asked for an early report on the specific recommendation in the Robbins Report that a new university should be located in Scotland. The Committee have given some preliminary consideration to this matter and, in order that they may have the fullest possible information before them when they come to advise the Government, they have agreed that the promoting bodies in each place concerned should be given an opportunity of supplementing, if they so wish, the representations already made. I am accordingly writing to the appropriate people in all the places in Scotland which have been suggested as possible sites for new universities and inviting them to submit any additional information, which they may wish the Committee to have, by the end of the year.[16]

Although the UGC had discussed in a meeting on November 7, 1963, "how they should handle the Government's request for an early report on the proposal in the Robbins report for the foundation of a new university in Scotland" and what "action they might take with regard to the areas which have requested interviews," the letters addressed to interested locations in England and Wales and to potential S.I.S.T.E.R. institutions were sent despite the Robbins committee's suggestions about five further new universities and five S.I.S.T.E.R. institutions not having been formally taken forward by the government via the UGC.[17]

Subsequently, Wolfenden's efforts focused entirely on founding a new university in Scotland. On January 27, 1964, the government formally accepted the UGC's endorsement of the Robbins committee's recommendation that at least one new university was required in Scotland to absorb the rising number of qualified students, asking the UGC "to advise on the choice of a location from those that have been suggested."[18] For making a decision in this semipublic round of bid evaluation—

[16]TNA, UGC 1/172. Subcommittee on New University Colleges: Report of the Committee on Higher Education: Letters regarding the recommendation of new universities and of special institutions for scientific and technological education and research, November 21, 1963, Annex b.

[17]TNA, UGC 7/239. Fifth university for Scotland: Report of the Committee of Higher Education, New University in Scotland, November 7, 1963, p. 4.

[18]TNA, UGC 7/239. Fifth university for Scotland: J. P. Carswell, Treasury, to E. R. Copleston, UGC secretary, January 24, 1964, Note.

Cumbernauld New Town had joined the competition in December 1963—Wolfenden chaired a UGC subcommittee consisting of four other UGC members—the historians Asa Briggs (Pro Vice-Chancellor of Sussex University, educated in Cambridge and previously based in the Universities of Oxford, 1945–1955, and Leeds, 1955–1961) and Lucy Sutherland (Principal of Lady Margaret Hall, Oxford, and Pro Vice-Chancellor of Oxford University); the physicist Francis Arthur Vick (Director of the Atomic Energy Research Establishment Harwell, first Vice-Principal of the University College of North Staffordshire at Keele, 1950–1959); and Iain M. Stewart (Chairman of Thermotank Ltd. in Airdrie, Lankarkshire). Briggs and Vick had also been members of Murray's earlier UGC Subcommittee on New University Colleges (1959–1961).

Apart from Stewart, Wolfenden's UGC subcommittee lacked Scottish expertise, which prompted the suggestion by Ronald Edwards, chairman of The Electricity Council, to co-opt additional members with Scottish background or experience such as Lord Polwarth, a Scottish representative peer.[19] To take Scottish views into account, the UGC subcommittee had invited colleagues from the Scottish Office's Education Department (Mr. Graham and J. A. M. Mitchell) and Development Department (James McGuinness and Robert Grieve) to give their views on a suitable new university location during the UGC subcommittee's first meeting on February 28, 1964. However, when the press reported that Wolfenden's subcommittee was to visit seven potential university locations in Scotland in April and May 1964, Gordon Wilson, secretary of the Scottish National Party, complained that the visitors included six "English people" and only "one Scots gentleman":

> When one considers that the Scottish Universities were founded long before the two countries were joined together and that their organisation, their tradition and their whole method of operation is based on a completely different system from the English system, this would appear to be nothing less than a deliberate attempt on the part of the Government to destroy yet another aspect of Scotland's national life in the interests of uniformity throughout the United Kingdom.[20]

In the context of increasing national sentiments, Wilson felt that a committee of Scottish people with one English representative, or at least a majority of Scots knowledgeable of "the real feelings of the people of Scotland,"[21] would be much more appropriate for the fundamental task at hand, so he asked Wolfenden to resign from the subcommittee before its return to Scotland on May 17. In his response, Wolfenden referred to the UGC's need to carry out the duty of advising the government and assured Wilson that the subcommittee received plenty of advice "on particularly Scottish problems."[22]

---

[19]TNA, UGC 7/239. Fifth university for Scotland: Edwards to Wolfenden, February 28, 1964.

[20]TNA, UGC 7/239. Fifth university for Scotland: Wilson to Wolfenden, May 1, 1964, p. 2.

[21]TNA, UGC 7/239. Fifth university for Scotland: Wilson to Wolfenden, May 1, 1964, p. 3.

[22]TNA, UGC 7/239. Fifth university for Scotland: Wolfenden to Wilson, May 6, 1964.

The Competing Locations

The seven Scottish locations competing for a new university after publication of the Robbins report were—in alphabetical order—Ayr, Cumbernauld, Dumfries, Falkirk, Inverness, Perth, and Stirling (Fig. 6.1). Dumfries, Stirling, Inverness, and Perth were local authorities that had jointly approached the UGC about the possibility of a new university for the first time in 1947 and separately again in October and November 1960. Falkirk had started the second wave of lobbying for a new university in Scotland in August 1960, whereas Ayr expressed its interest first in November 1960, and the new town of Cumbernauld in December 1963. In addition, Dunkeld and Kinross (both Perthshire), the county town of Duns (Berwickshire), and Scotland's first new town East Kilbride (South Lanarkshire) had been suggested as suitable sites for a new university in Scotland but not been pursued further.[23]

Stirling

The small historic county town of Stirling, located on the northern fringe of Scotland's central population belt, within the 26-mile commuter radius of Glasgow and 35 miles from Edinburgh, had about 28,000 inhabitants. It offered to use the Airthrey estate, a site of 303 acres, including the late-eighteenth century Airthrey Castle, for the new university. This site was situated about one mile from the Stirling town center in a picturesque landscape with a central loch (which can be walked around in one hour) and bordering hills. Stirling promoted itself as a pleasant community in the heart of Scotland within easy reach of other centers of learning, hosting an annual festival, and providing a sufficient number of schools, staff housing, cultural amenities, as well as financial support from local authorities and the community at large. Although no specific total of such contributions was mentioned, £25,000 per year were promised by Lanark County Council. The new university was seen as incorporating the broad fields of the arts, social sciences, and pure sciences, with industrial facilities located nearby for applied sciences.[24]

Stirling had a strong advocate in Thomas Erskine Wright (1902–1986), a former fellow of Queen's College Oxford (1928–1948) based at St. Andrews (1948–1962). Wright was not only a native of Stirling but was also a member of the UGC until December 1963. Wright wrote on several occasions to both Murray

[23]TNA, UGC 7/237. Fifth university for Scotland: The Reverend Bruce Robertson, Dunkeld, to Murray, December 28, 1960; R. R. Kydd, Kinross, to Murray, December 29, 1960; TNA, UGC 7/239. Fifth university for Scotland: Report of the Committee of Higher Education, New University in Scotland, November 7, 1963, p. 3.

[24]TNA, UGC 7/241. Fifth university for Scotland: Submissions: Royal Burgh of Stirling, Submissions to the UGC for the establishment of a university in and adjacent to the Royal Burgh of Stirling, n.d.; Royal Burgh of Stirling, Proposed university at Airthrey estate, Supplementary memorandum by the sponsoring committee, n.d.

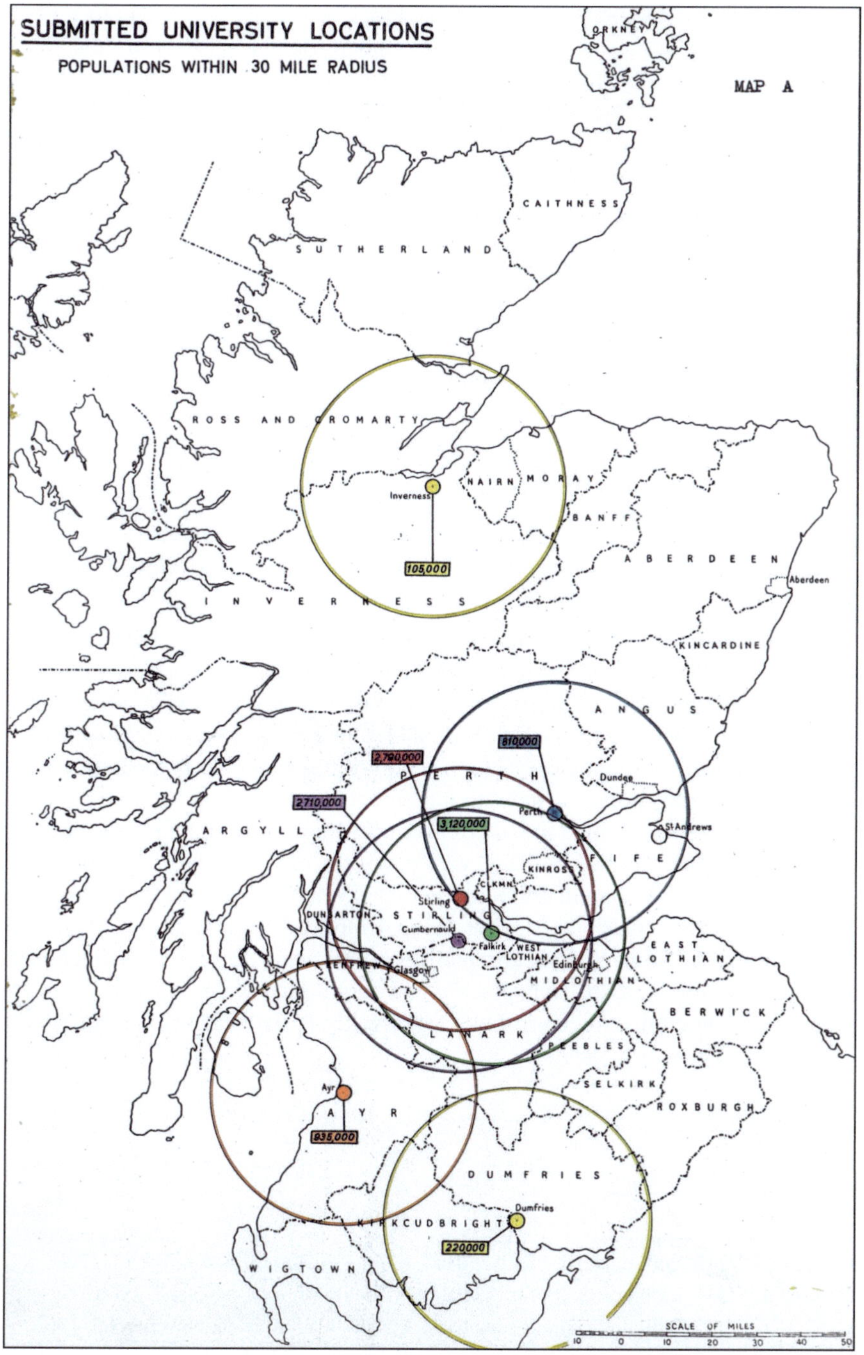

Fig. 6.1 Map on the locations and population catchment areas of places applying for a new university in Scotland (April 1964). Source: The National Archives, Kew, University Grants Committee, UGC 7/244. Fifth university for Scotland: Action following UGC visit: Proposed university locations, Map A. Reprinted with permission.

(August 8, early September—letter "Retained by Chairman"[25]—, November 12 and 22, 1960) and Wolfenden (October 26, November 3 and 11, 1963) in their role as UGC chairmen. In Wright's first letter, written only a few days after Falkirk supporters had revived their expression of interest in a new university in correspondence with the Scottish Office and local MPs on August 3, he suggested that if another Scottish university were ever to be set up,

> . . . the best location would seem to be Stirling (centre of communications, residential area, history & general amenities): in saying this, I am not, I think, influenced by partiality for my native town—in fact, it was Arbuckle who first urged its claims upon me![26]

Pursuing the idea of William Arbuckle, Secretary of the Scottish Education Department, further over the coming four years, Wright used his subsequent letters to both UGC chairmen for strengthening the case of Stirling, mainly by discrediting Falkirk's campaign for a university in East Stirlingshire. This intention is most evident in the surviving correspondence with Wolfenden:

> . . . there is some resentment that Falkirk has been rather blatantly trying to 'beat the gun'. There is also some suspicion of the people who are campaigning for 'East Stirlingshire'. I was told that the Trust Fund was constituted for a new university in Scotland (no location specified), but that it is in fact run by the same people as are campaigning for East Stirlingshire.[27]

In an attached note, Wright referred to his informal discussions with representatives of Stirling County Council, who faced the dilemma that two towns within their boundaries—Stirling and Falkirk—were competing for a new university. He explained that he advised them to declare their support for a University of Stirlingshire and to remain neutral in regard to the two locations in order to raise the chances that the UGC would choose a site within their county. However, he subtly added in parentheses that "they seemed in fact to favour Stirling" and concluded the note with a stunning request: "as the meeting was a private one, my name should not be mentioned in their subsequent deliberations."[28] Wolfenden seemed overly receptive to his former fellow Oxonian's blunt tactical moves in his response to all three letters, writing of "your zealous compatriots"; promising "a tour of Scotland with discussions on the spot in each place"; and looking forward "to receiving your further advice on all this on Thursday," presumably at the next UGC meeting.[29]

---

[25]TNA, UGC 7/237. Fifth university for Scotland: Murray to Wright, September 13, 1960.

[26]TNA, UGC 7/237. Fifth university for Scotland: Wright to Murray, August 8, 1960.

[27]TNA, UGC 7/239. Fifth university for Scotland: Wright to Wolfenden, November 3, 1963.

[28]TNA, UGC 7/239. Fifth university for Scotland: Wright to Wolfenden, November 3, 1963, Note on an informal meeting with the Convener & Clerk of Stirling County Council, November 2, 1963.

[29]TNA, UGC 7/239. Fifth university for Scotland: Wolfenden to Wright, November 5, 1963.

Inverness

Inverness, the small cultural and administrative center of the Scottish Highlands, had circa 28,700 inhabitants, medieval origins, and a representative nineteenth-century castle and cathedral. It was frequently portrayed as a remote tourist destination in the economically deprived and depopulating, but environmentally very attractive, Highlands along the northeast coast of Scotland. Accordingly, the promotion committee's submission to the UGC stressed the importance that a new university would have for economic development, industrial expansion, and sociocultural enrichment in the Highlands. Accommodation existed for 1,100 students in town, and for a further 400 students in the unoccupied Cameron Barracks. Local financial contributions of 10% were in place for a university of 1,500 students, amounting to circa £225,000 a year for ten years.[30]

Inverness was put forward as an ideal place for studying biology, botany, zoology, geography, geology, forestry, medicine, and Celtic studies, which represented a more specialized range of studies than that envisioned elsewhere. Moreover, the inclusion of subjects such as geography addressed the fact that they had been excluded from some of the more experimental curricula in the new English universities for lack of backing from learned societies and powerful lobbyists (Johnston, 2004). In addition to the 70 acres of the Cameron Barracks, a site of 80 acres was immediately available within a ten-minute walk, and there was a possibility of obtaining more land in the future. Given that the UGC subcommittee was after a site of at least 200 acres, this limitation might have been a competitive disadvantage.[31]

Inverness received the most support in writing from a diverse group of interested parties, starting in 1960 with the Scottish Union of Students informing the Secretary of State for Scotland that, in their view, the new university should be a residential one and located in Inverness because locations in the central belt would be too close to the existing institutions that were serving this area well.[32] Just after the publication of the Robbins report, Margaret Cohen, who held an M.A. in English Literature from Edinburgh and a Ph.D. from Cambridge, urged the UGC to consider Inverness as a location for the new university because it would "strengthen the whole cultural life of the north" and enable Highlanders to study in their home region.[33] A month later, the medieval historian Lionel Butler (University of St. Andrews) reminded Wolfenden of their meeting in his Oxford college All Souls in 1956 or 1957 with

[30]TNA, UGC 7/241. Fifth university for Scotland: Submissions: Proposed erection of university in Inverness, December 27, 1963.

[31]TNA, UGC 7/241. Fifth university for Scotland: Submissions: Proposed erection of university in Inverness, December 27, 1963.

[32]TNA, UGC 7/237. Fifth university for Scotland: D. John Parker, Vice-President of the Scottish Union of Students, to Maclay, November 30, 1960.

[33]TNA, UGC 7/239. Fifth university for Scotland: Cohen to Copleston, October 25, 1963.

the aim of arranging a meeting with him and his colleague R. J. Adam, a native from the Scottish Highlands keen to press the case for Inverness as a private person.[34]

Wolfenden politely declined the invitation to be hosted at All Souls but invited written comments, in which the historian Adam included a three-page memorandum that he also sent to the Inverness promotion committee. In this document, he elaborated on five main arguments: the suitability of the town; the advantages of the surrounding region; the establishment of faculties for the humanities, the sciences, and environmental studies; the possibility of adding a new scheme of first-degree studies; and the relevance of a university to the Highland situation.[35] Other backers added to these arguments the new university's importance for preserving a distinct Scottish heritage.[36]

Further support arrived from the Presbytery of Inverness, the Presbytery of Chanonry and Dingwall, and the local Conservative MP, Neil McLean. In addition, Ex-Provost Robert Wotherspoon, chairman of the local sponsoring committee, met with Wolfenden on December 13, 1963, to strengthen the case of a new university in Inverness.[37] Most lobbyists argued that increased student mobility would attract students to the Highlands and that teaching staff would reside in the place of the university rather than spending time on commuting.[38] Yet, not all lobbying was supportive. Wolfenden was sent articles from *The Inverness Courier*, in which the author expressed the highly critical, but clearly narrow-minded, view of those ignoring the significant economic and sociocultural benefits generated by incoming student mobility:

> . . . we . . . see absolutely no benefit to either the Highlands or Highlanders coming from a university at Inverness. On the contrary, we know that for young people to have all their education in the one area, and particularly in the one town, is the most stultifying thing that could happen to them, and parochial narrow-mindedness is the main disastrous effect, as can be seen in the case of certain of the already established Scottish universities.[39]

Ayr, Dumfries, Perth, Falkirk, and Cumbernauld

The other five locations competing for the new Scottish university submitted memoranda on their case to the UGC by the deadline of December 1963 but received far less written endorsement than Stirling and Inverness. Additional supporting letters were received from the Presbytery of Ayr for Ayr, and from Prime Minister Alec

---

[34]TNA, UGC 7/239. Fifth university for Scotland: Butler to Wolfenden, November 24, 1963.

[35]TNA, UGC 7/239. Fifth university for Scotland: Wolfenden to Butler, November 26, 1963; Adam to Wolfenden, December 7, 1963, including memorandum.

[36]TNA, UGC 7/239. Fifth university for Scotland: The Reverend John A. Muirden, Rosskeen, to Wolfenden, May 12, 1964.

[37]TNA, UGC 7/239. Fifth university for Scotland.

[38]TNA, UGC 7/239. Fifth university for Scotland: The Reverend A. Gordon McGillivray, Inverness, to Wolfenden, December 7, 1963.

[39]TNA, UGC 7/239. Fifth university for Scotland: The Inverness Courier, Friday, May 8, 1964.

Douglas-Home, acting in his capacity as the Conservative MP for Kinross and West Perthshire, for the site offered by Lord Mansfield near Perth. Representatives of these destinations were also treated differently by Wolfenden, who met with Wright as a supporter of Stirling in the context of the UGC meetings in November and December 1963, and arranged to see the chairman of the local sponsoring committee of Inverness also in December 1963, but declined similar requests from Dumfries and Perth.[40]

Ayr, a county town of circa 45,000 inhabitants, originated as a vibrant medieval port town and was located 32 miles south of Glasgow on the west coast of Scotland. The sponsoring committee of a new university emphasized Ayr's excellent social, cultural, recreational, and educational facilities; the presence of local and regional industries; local community support; and famous Ayrshire men such as the Scottish poet Robert Burns. Four sites were available for purchase, the largest one comprising 204 acres adjacent to the foreshore, and local authority contributions to the finances were estimated at 2% to 3% per year.[41] By the end of November 1963, the town clerk of Ayr was concerned about press reports quoting Tam Dalyell, the Labour MP for West Lothian, as saying that the new university "is almost certain to be sited just outside Falkirk." The clerk therefore asked the UGC secretary whether preparing a full submission, as invited by Wolfenden earlier in the month, was still sensible—an incident that may have reinforced Wolfenden's emerging reservations against Falkirk.[42]

Dumfries—known as "The Queen of the South"—was the second claimant for a new university located south of the central belt of Scotland, at the edge of the Southern Uplands, circa 80 miles south of Glasgow and 35 miles north of the English city Carlisle. The small historic county town of 27,000 inhabitants represented itself as a growth point in the southwest of Scotland, situated in unspoiled countryside with many historic landmarks and proximity to the north of England. Two sites owned by the Secretary of State for Scotland encompassed an impressive 700 and 1,900 acres, but the minimum amount of listed contributions from the local authority was relatively low—£8,180 per year for ten years—even though more was promised. The regional center underlined its thriving cultural bodies; rich opportunities for tourism and leisure; links with university education and scientific and technical facilities for special studies; modern industrial facilities, some of which related to agriculture; excellent communication links; and sufficient private and public housing for students and staff. The submission maintained that the large prospective sites and the small community would be well suited for developing a truly residential university. This assessment was considered important, for according to the Robbins report, half of English students but only 13% of Scottish

[40]TNA, UGC 7/239. Fifth university for Scotland.

[41]TNA, UGC 7/241. Fifth university for Scotland: Submissions: Proposed establishment of new university, Submissions by Ayr Town Council to the UGC, December 30, 1963.

[42]TNA, UGC 7/239. Fifth university for Scotland: Robert C. Brown, town clerk of Ayr, to Copleston, November 25, 1963.

students resided in halls. The submission argued that higher education facilities would contribute in significant ways to the recent government policy of creating growth and development in the southwest of Scotland and would thereby address key concerns of regional development and planning.[43]

The city of Perth, situated along the River Tay and the railway line from London to Inverness, about 22 miles west of Dundee, represented itself as a historic county town situated within a scenic agricultural area within easy reach of Scotland's population centers and as a gateway to the Scottish Highlands. Serving as Scotland's capital and one of the richest merchant towns in the Middle Ages up until the fifteenth century, Perth was the place in which John Knox precipitated the Reformation in Scotland by encouraging iconoclasm in 1559. The large burgh of circa 41,000 inhabitants offered a readily available and purchasable site of 200 acres located about 1 mile northeast of the town center. It was argued that Perth provided rich cultural, unrivaled recreational, and excellent educational facilities and that no problems were anticipated in regard to available housing for students and staff. Local authorities promised £11,400 per year for ten years, and a public appeal for funds was considered if the campaign to found a university in Perth were successful. In many ways comparable to the proposals submitted by the historic market towns of Ayr, Dumfries, and Stirling, the document presented by the promotion committee of Perth did not specify what subjects the members envisioned the university should emphasize.[44]

Falkirk differed from the other contenders because it was the only industrial town with a heritage of heavy industry that had originated in the late eighteenth and flourished in the nineteenth century. Situated like Stirling in the Forth Valley, the larger town of circa 40,000 inhabitants was located further southeast and thus closer to both Edinburgh (23 miles) and Glasgow (26 miles). New industrial strengths were foregrounded in the adjacent Grangemouth area, with a focus on dyestuff, pharmaceutical, and petrochemical industries; and in Falkirk, with an accent on light casting-industries as well as coach- and caravan-building. The proposed University of East Stirlingshire was supposed to focus on science and technology in an area of industrial growth. The statement that a faculty for the social sciences but none for the arts was envisioned for the new university might not have been well received by the humanists on the UGC subcommittee, and it was also a disadvantage in the light of Glasgow's newly upgraded University of Strathclyde.

The submission of the largest sponsoring committee of all claimant towns, headed by James Drever from the University of Edinburgh, who had been a member of the Robbins committee, offered the Callendar estate at Falkirk as a site of up to 800 acres for the new university. It was stated that 105 acres of the Callendar estate

[43]TNA, UGC 7/241. Fifth university for Scotland: Submissions: Dumfries area university committee, Proposal to site a new Scottish university in the Dumfries area, December 1963.

[44]TNA, UGC 7/241. Fifth university for Scotland: Submissions: Robert Ritchie, Lord Provost of the city and Royal Burgh of Perth and chairman of the university promotion committee, to Wolfenden, December 27, 1963.

were owned by the town council, and were readily available, and that other parts could be bought from supportive private owners. The financial support from ten local authorities amounted to £100,000 a year for ten years, and the professionally produced memorandum not only discussed the usual advantages and amenities but also contained a number of supporting statements evidencing local, regional, and national enthusiasm for the project.[45]

The county council of Stirling remained neutral in regard to the two competing sites in Stirlingshire, as recommended by UGC member Wright, but this stance changed shortly before the UGC subcommittee's field visits in the spring of 1964, when the county clerk of Stirling offered to assist if Stirling were chosen over Falkirk, for the Airthrey estate was located outside the burgh in the landward area owned by the county council.[46] The direct rivalry between the two locations was very evident because East Stirlingshire's original submission warned that locating "a university in some quiet, exclusive residential backwater confers privilege only on the few,"[47] whereas Stirling's supplementary memorandum, sent to the UGC shortly before the field visits, evoked the weather for taking a dig at Falkirk—"The district is free from the fog which often in winter covers the industrial area to the south."[48]

Cumbernauld New Town, located 13 miles to the northeast of Glasgow city center and thus halfway to both Stirling and Falkirk, was also very different because its center had been under construction since 1963, for a town with a targeted population size of circa 70,000 inhabitants. Of the seven contenders, it was the only place that did not send a full submission to the UGC by the December 1963 deadline but only a preliminary report of eight pages and the new town's official development brochure. This lack of a full submission prompted the Cumbernauld Development Corporation to explain to the UGC in March 1964 that it was in touch with the responsible Education Committee of Dunbarton County Council about a full submission, implying that complicated planning processes underpinned the new town development.[49] The Cumbernauld Development Corporation suggested that a site of 300 acres with potential for expansion could be fitted into the new town's master plan, stressing both ample time for planning the town with an integrated

[45]TNA, UGC 7/241. Fifth university for Scotland: Submissions: Promotion committee for the proposed University of East Stirlingshire, Submission to the UGC, December 1963.

[46]TNA, UGC 7/239. Fifth university for Scotland: James D. Kennedy, county clerk of Stirling, to Wolfenden, December 2, 1963; TNA, UGC 7/241. Fifth university for Scotland: Submissions: Kennedy to Wolfenden, March 12, 1964.

[47]TNA, UGC 7/241. Fifth university for Scotland: Submissions: Promotion committee for the proposed University of East Stirlingshire, Submission to the UGC, December 1963, p. 9.

[48]TNA, UGC 7/241. Fifth university for Scotland: Submissions: Royal Burgh of Stirling, Proposed university at Airthrey estate, Supplementary memorandum by the sponsoring committee, n.d., p. 2.

[49]TNA, UGC 7/239. Fifth university for Scotland: G. R. B. MacGill, General manager of the Cumbernauld Development Corporation, to Copleston, March 6, 1964.

university and the unique chance of the new town and new university growing up together.[50] Yet Cumbernauld's chances seemed slim given that the new town had still to be built, whereas student places were needed imminently.

The Decision

Throughout the decision-making process, Wolfenden gauged opinions on the potential locations for the new university during lunch and dinner meetings and through correspondence with individuals such as Wright, a tireless advocate for Stirling, and Douglas Douglas-Hamilton, the Duke of Hamilton and Brandon and the Chancellor of St. Andrews from 1948 to 1973. Considering the distribution of population with an emphasis on the west of Scotland and the prospect of two universities in Glasgow and four in the east, Douglas-Hamilton argued that the new university should not be established further east than Stirling. He supported Inverness's claim with a view of countering depopulation in the Highlands but also suggested his family's property, the High Parks at Hamilton in South Lanarkshire, as a possible university site located in the southern agglomeration of Glasgow. In response, Wolfenden specified the key challenge of the decision-making as the contrasting opinions of those who felt that the central belt already offered enough university places and those who argued that saving a remote depopulating area should not be the main reason for locating a new university there.[51]

At the first meeting of the UGC Subcommittee on New Universities, held on February 28, 1964, the members were informed by Wolfenden that one of the factors to be taken into consideration when deciding on the new university was regional planning. For that reason McGuinness and Grieve from the Scottish Development Department had been invited to give their views on the matter. The Robbins report had made an argument for choosing large cities or places in their vicinity as locations for new universities, whereas the previous UGC Subcommittee on New University Colleges, which had approved six of the seven new English universities, regarded local support, student lodgings, a site of not less than 200 acres, and the attraction to well-qualified staff as the most important criteria because staff required pleasant surroundings and good facilities for themselves and their families.

Most of these UGC and Robbins criteria gave weight to the claims of Cumbernauld, Falkirk, and Stirling because they were located in Scotland's central belt (Fig. 6.1), whereas considerations of regional development and planning were treated in a highly ambiguous way. McGuiness from the Scottish Development

[50]TNA, UGC 7/241. Fifth university for Scotland: Submissions: Cumbernauld Development Corporation, A new university in the new town of Cumbernauld, Preliminary report, December 12, 1963.

[51]TNA, UGC 7/239. Fifth university for Scotland: The Duke of Hamilton and Brandon, Lord Steward of the Household and Chancellor of the University of St Andrews, to Wolfenden, February 7, 1964; Wolfenden to The Duke of Hamilton and Brandon, February 10, 1964.

Department underscored that academic success was more important than subordinate regional considerations, whereas his colleague Grieve first supported the case of Inverness in many ways (e.g., influx of new money, students counterbalancing seasonality through tourism) but then called it an isolated area. Furthermore, a clear statement in the minutes on the local and regional economic benefits of a new university in remote areas was immediately extenuated as follows:

> A university placed elsewhere than in the central belt would have a considerable effect in helping to regenerate the outer areas, e.g., in Dumfries or Inverness. Nevertheless, the Development authorities would not support a university in a particular area merely in order to provide an economic boost.[52]

It was further brought to attention that Inverness lacked local labor and resources for a university catering for circa 3,000 students, and that Dumfries would be better placed to offer both but was short on staff and student accommodation. Another argument for a location in the central belt was the possibility for Cumbernauld, Falkirk, and Stirling to recruit students from the local population and a wider commuter area. In response to the chairman's question about building a university into the new town of Cumbernauld, McGuinness explained that the terrain was not well suited or very attractive and that inconvenience was expected for the first five years. The narrative of the minutes then centered squarely on the choice of Falkirk and Stirling, culminating in a final paragraph on the perceived advantages of Stirling:

> Stirling was situated at the "superior" end of the central belt and was within comparatively easy reach of Edinburgh and Glasgow. It was an attractive place and came nearest to achieving the best of both worlds.[53]

This first UGC subcommittee meeting concluded with agreement on two visits over two days to the seven aspirant university locations in the spring of 1964. On April 28 and 29, the UGC subcommittee, consisting of Wolfenden, Briggs, Stewart, Sutherland, and Vick (not on April 29), and two UGC colleagues visited Dumfries and Ayr on the first day and Cumbernauld and Falkirk on the second. The committee members spent between two-and-a-half and four hours in each place and continued their inspections of suitable sites—according to Wolfenden "the pleasant end of the task"[54]—two weeks later in beautiful weather with visits to Stirling and Perth (May 12) and Inverness (May 13). All visits attracted wide coverage in the local and national press, especially in Inverness and Dumfries, the latter of which had tried to obtain a university in the seventeenth, nineteenth, and mid-twentieth centuries and had been unsuccessful on all occasions.[55] Dumfries's representatives specifically

[52]TNA, UGC 7/239. Fifth university for Scotland: UGC Subcommittee on New Universities, Minutes of a meeting held on February 28, 1964, p. 3.

[53]TNA, UGC 7/239. Fifth university for Scotland: UGC Subcommittee on New Universities, Minutes of a meeting held on February 28, 1964, pp. 3, 7.

[54]TNA, UGC 7/239. Fifth university for Scotland: The Glasgow Herald, Wednesday, May 13, 1964.

[55]TNA, UGC 7/237. Fifth university for Scotland: The case for a new Scottish University, November 2, 1960, p. 6.

accentuated the town's advantages over Inverness as a rural area, maintaining that the latter was too remote, whereas Ayr was confident that its connection to Robert Burns would help generate increased financial support if the town were chosen for the new university.[56]

Cumbernauld highlighted its unique chance of building up the university together with the new town, thereby benefitting from cheap land and good housing. Falkirk tried to enthrall Wolfenden and his UGC colleagues through a helicopter flight over Callendar House estate, resulting in the evaluation that "Falkirk has by far the most active and widely based Promotion Committee and are extremely keen contenders for the University."[57] Yet this intensity also generated very critical comments in the UGC subcommittee's report on the field visits, which called the flight somewhat unnecessary, identified "fundamental dissensions" among the more than 25 attending members of Falkirk's promotion committee, and characterized the discussions disrespectfully as "widely discursive."[58] At a dinner with six colleagues from the Scottish Development Department and the Scottish Education Department, the UGC subcommittee members stated not only that they found Dumfries too remote, and the local authority in Ayr too unreliable, but also that Falkirk did not possess any advantages that neighboring Cumbernauld would not have.[59]

The impressions noted after the second site visits in mid-May seemed to be more positive and much fairer, especially in regard to Stirling, where the promotion committee stressed that the facilities could be adapted to any type of university desired. The only hesitant remarks in an overly lengthy UGC report section on Stirling claimed that the readily available site was "very undulating and wooded and the siting of the university within the estate might not be easy," whereas Perth was acknowledged to be a very attractive place offering a site that was "largely treeless with gentle slopes where it is not completely flat," yet its closeness to Dundee and St. Andrews was seen critically.[60] After a three-and-a-half hour drive to Inverness the next morning, the UGC party was "somewhat bemused by scenery" before being treated to twelve different speeches, during which the novelist Eric Linklater compared Britain in terms of its distribution of population and higher education to "a

[56]TNA, UGC 7/244. Fifth university for Scotland: Action following UGC visit: New university in Scotland, UGC visits, April 27–28, 1964, pp. 2–3.

[57]TNA, UGC 7/239. Fifth university for Scotland: Sixth Scottish University, n.d., p. 1.

[58]TNA, UGC 7/244. Fifth university for Scotland: Action following UGC visit: New university in Scotland, UGC visits, April 27–28, 1964, pp. 6–7.

[59]TNA, UGC 7/244. Fifth university for Scotland: Action following UGC visit: Note of discussion with Scottish Development Department and Scottish Education Department Officials, Edinburgh, April 28, 1964.

[60]TNA, UGC 7/244. Fifth university for Scotland: Action following UGC visit: New university in Scotland, UGC visits, May 12–13, 1964, pp. 2–3.

Fig. 6.2 UGC subcommittee visit to the Airthrey Castle estate in Stirling (from left to right: William MacFarlane Gray, former Provost of Stirling; Michael Kelly, Provost of Stirling; John Wolfenden, UGC chairman; G. W. Norman, town clerk of Stirling; and Lucy Sutherland, UGC subcommittee member). Source: The National Archives, Kew, University Grants Committee, UGC 7/239. Fifth university for Scotland: The Glasgow Herald, Wednesday, May 13, 1964. Reprinted with permission.

badly loaded ship," arguing that a University of Inverness "would help to redress the balance."[61]

The Glasgow Herald eagerly compared the value of the sites in Stirling, Perth, and Inverness, capturing some of the subcommittee members' great enthusiasm about Airthrey Castle estate at Stirling during the field visit (Fig. 6.2). Apparently, such moments of inspection seemed to be less enjoyable in Inverness, for it was reported in the press—and to some extent confirmed by the UGC notes blaming the speeches—that

> Sir John and his six-strong party were in such a hustle after an hour-long meeting that they did not have time to step out of their cars and examine the four sites pointed out to them. They drove slowly past possible sites at Holme Mains, Beechwood, round the Cameron Barracks, and peered out from a vantage point window at the barracks towards the spacious Longman. Then they rushed a Press Conference of less than five minutes in time to catch a south-bound plane from Dalcross.[62]

[61]TNA, UGC 7/244. Fifth university for Scotland: Action following UGC visit: New university in Scotland, UGC visits, May 12–13, 1964, pp. 3–4.

[62]TNA, UGC 7/239. Fifth university for Scotland: The Press and Journal [Inverness], Thursday, May 14, 1964.

Fig. 6.3 UGC subcommittee visit to the Cameron Barracks in Inverness (Brigadier Maitland Makgill-Crichton chatting with Wolfenden surrounded by other subcommittee members and the local sponsoring committee). Source: The National Archives, Kew, University Grants Committee, UGC 7/239. Fifth university for Scotland: The Press and Journal (Inverness), Thursday, May 14, 1964. Reprinted with permission.

Wolfenden's brief discussion with a brigadier of the Cameron Highlanders apparently added confusion about the Cameron Barracks' full availability for the new university,[63] so it might be possible that Lucy Sutherland's rather grim facial expression during this exchange preempted the UGC subcommittee's final decision that was considerably shaped by her views, which she subsequently communicated to her colleagues in writing (Fig. 6.3).

After the field visits, additional information requested by the UGC subcommittee members was submitted by the town clerk of Ayr on where students from the district were studying (68% at the University of Glasgow), by the town clerk of Dumfries on the availability of lodgings within a half-an-hour travel radius, and by the Cumbernauld Development Corporation on the availability of student accommodation.[64] Wolfenden received further letters, from the Earl of Elgin and Kincardine in support of Falkirk and from Norman Petch, a well-known Scottish scientist who was Cochrane Professor of Metallurgy at the University of Newcastle upon Tyne, in

[63]TNA, UGC 7/244. Fifth university for Scotland: Action following UGC visit: New university in Scotland, UGC visits, May 12–13, 1964, p. 5.

[64]TNA, UGC 7/239. Fifth university for Scotland; TNA, UGC 7/244. Fifth university for Scotland: Action following UGC visit.

favor of Stirling. Wolfenden responded to these approaches very differently by reassuring Petch that the subcommittee would keep his considerations in mind and explaining to Elgin that the decision would be made "in the full knowledge that inevitably there will be six times as many people disappointed as satisfied."[65]

The subsequent second meeting of the UGC subcommittee, during which the locational decision was made, took place on June 5, 1964. Lucy Sutherland, an influential Oxford historian, soon to be elected principal of St. Hilda's College, had outlined her evaluations in writing because she could not participate in the subcommittee's post-visit meeting. She ruled out both Dumfries, which was "likely to become a second St. Andrews in its draw to English students from over the border," and Inverness, which exhibited "too much Highland regionalism," as too inaccessible, thus reiterating press statements that raised concerns about recruiting and retaining staff in remote regions.[66]

In Ayr, Sutherland observed a lack of local support beyond the request of gaining material advantage. She also argued against Falkirk and Cumbernauld because of their closeness to Edinburgh and Glasgow, respectively. She reasoned that Falkirk's site lacked adequate building space—presumably because of some protected woodland and the relatively small share that was owned and ready to be built on—and that the university's "chief advantages lie on the technological side, while the student demand is likely to be for the Arts."[67] Cumbernauld would be an interesting yet "eccentric choice" due to the early stages of what she called an experiment with an unknown social future, the main reason why she favored "a more orthodox choice."[68]

Eventually, Sutherland chose Stirling over Perth—between which she did not see "a great deal of difference"—because the latter was too close to Dundee and Stirling would be "more generally accessible." However, she asked whether the university could be sited within the historic setting of this "small town of character" rather than miles from the town center.[69] Seen together with subsequent discussions of the subcommittee, this reasoning confirms our argument that Stirling was chosen over the rival Scottish locations partly because Sutherland and her colleagues, many of whom had been based at Oxbridge at some point during their education and career, wished to reproduce the creative setting of the ancient universities.

The assumption that "In Stirling [the area] has a town which is the historian's delight, while dons might turn the Bridge of Allan, for better or worse, into another

[65]TNA, UGC 7/240. Fifth university for Scotland: Wolfenden to the Earl of Elgin and Kincardine, Grand Master Mason of Scotland, June 5, 1964.

[66]TNA, UGC 7/242. Fifth university for Scotland: Visits to prospective sites: Sutherland to Wolfenden, May 27, 1964, p. 1.

[67]TNA, UGC 7/242. Fifth university for Scotland: Visits to prospective sites: Sutherland to Wolfenden, May 27, 1964, p. 1.

[68]TNA, UGC 7/242. Fifth university for Scotland: Visits to prospective sites: Sutherland to Wolfenden, May 27, 1964, p. 2.

[69]TNA, UGC 7/242. Fifth university for Scotland: Visits to prospective sites: Sutherland to Wolfenden, May 27, 1964, p. 2.

North Oxford"[70] had already been made in the autumn of 1960 by two journalists of *The Glasgow Herald*, Robert D. Kernohan and James Holburn, in their memorandum on a new Scottish university. Their suggestion will not only have appealed to the Oxford humanists on the UGC subcommittee but also seemed to have been visionary in regard to how Bridge of Allan has actually developed ever since. Yet although the town of Stirling afforded the much longed-for "dreaming spires" of Oxford, the reality of the 1960s modernist campus turned out to be architecturally quite different in style than the historic town center and thus became an experiment in its own right—albeit a very successful one (Neave, 1976).

For their decision, the members of the subcommittee could draw on Sutherland's letter and different memoranda about the visits; one-page summaries on the contender's location, population, site, and local industries; an ordnance survey map of the proposed site's or sites' locations in relation to the respective town (scale 1:25,000); and two quite sophisticated thematic maps that enabled them to compare the locations of proposed universities in regard to (a) the size of their populations within a 30-mile radius (Fig. 6.1) and (b) potential student populations in areas within one hour of travel by train or bus (Fig. 6.4). Transparent overheads allowed the presenter to superimpose on these maps a second layer conveying the corresponding information for each of the four existing university cities. From the perspective of regional policy, both maps suggest that Inverness, Dumfries, and Ayr would have been the most suitable locations for the new university, if the government had indeed wanted to stimulate regional development and economic growth in Scotland's more rural and remote areas, to support equivalent living conditions, and to encourage geographically disadvantaged strata of the population to access higher education.

Yet in their final, unanimous decision, the members of the subcommittee largely followed Sutherland's reasoning when recommending to the government—via the UGC—that the new university in Scotland should be located at Stirling—a recommendation formally accepted and publicly announced by Quintin Hogg, the Conservative Secretary of State for Science and Education, in the House of Commons on July 17, 1964.[71] Stirling's location, site, and sponsorship seemed to guarantee effective growth of the new academic institution, to be attractive to students and staff from elsewhere, and to act on the Robbins report's suggestion to place new universities within the vicinity of large population centers (90% of Scotland's population concentrated on the industrial belt between Glasgow and Edinburgh at the time). Against any consideration of regional development and planning, the subcommittee argued that their geographical location outside of Scotland's central population belt spoke against Ayr, Dumfries, Inverness, and Perth as sites for the new university. Among the three central locations, Stirling and nearby Bridge of

[70]TNA, UGC 7/237. Fifth university for Scotland: The case for a new Scottish university, November 2, 1960, p. 5.

[71]TNA, UGC 7/245. University of Stirling: Wolfenden to John Spencer Muirhead, chairman of the New University Sponsoring Committee at Stirling, July 18, 1964.

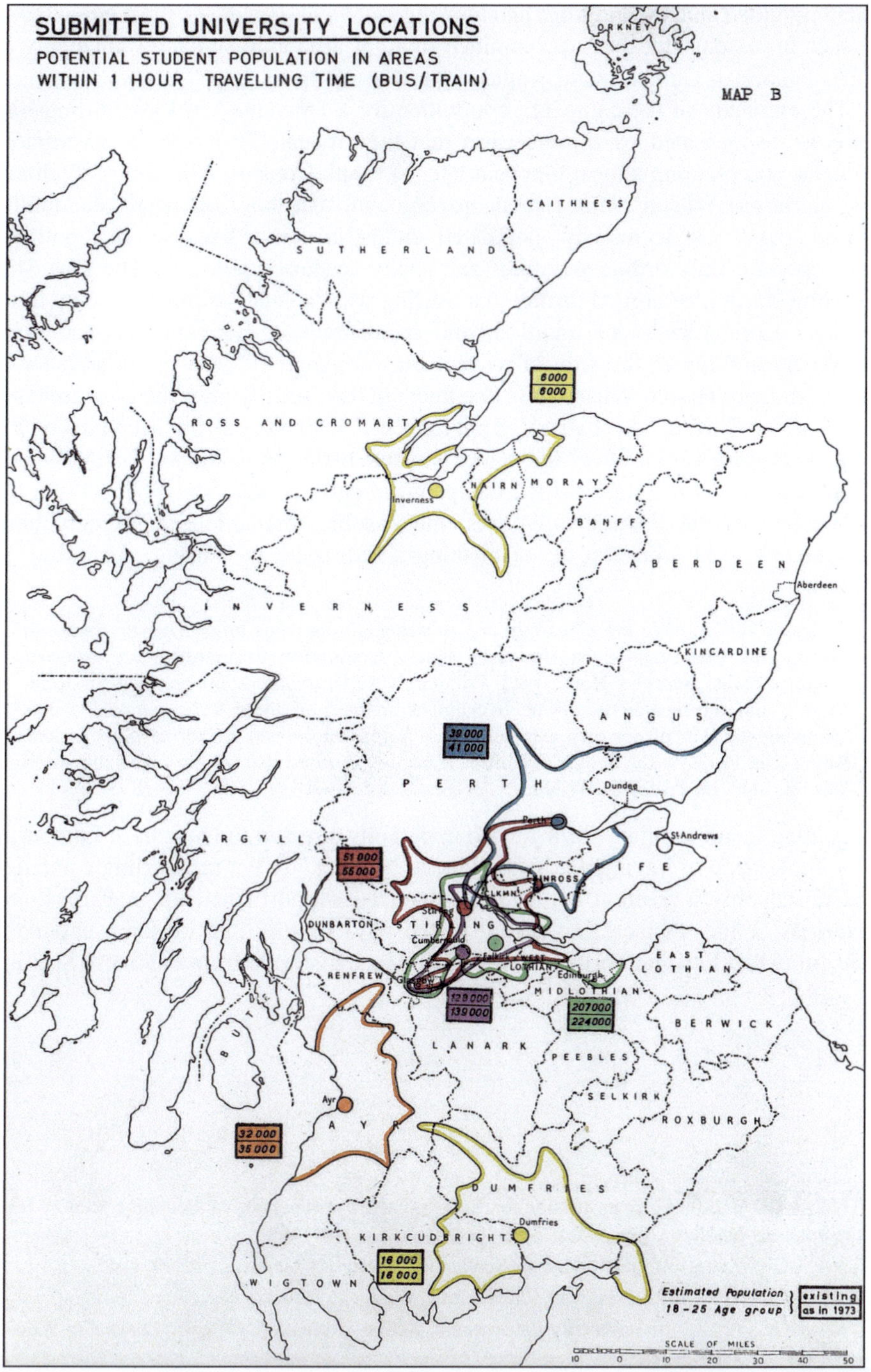

Fig. 6.4 Map on the locations and student catchment areas of places applying for a new university in Scotland (April 1964). Source: The National Archives, Kew, University Grants Committee, UGC 7/244. Fifth university for Scotland: Action following UGC visit: Proposed university locations, Map B. Reprinted with permission.

Allan provided suitable lodgings and housing and were seen as the most attractive—to such an extent that the subcommittee thought academic staff at a university in either Falkirk or Cumbernauld would want to reside in Stirling.[72]

The members of the campaign committee for a University in East Stirlingshire were so disappointed by this decision that they repeatedly wrote to government officials, complaining almost one year later to Frank Cousins, Minister of Technology in Harold Wilson's first Labour government, that they had been "shamefully passed over" as a heavily populated industrial area because of "political manoeuvring, both ordinary politics and shady academic politics." The East Stirlingshire committee argued further that Stirling was "a small exclusive county town in the historical claimant tradition, and incidentally a claimant that had done absolutely nothing in the way of respectable campaign."[73] In a previous letter to Prime Minister Harold Wilson, it is documented that, shortly after the government's decision for Stirling, the Falkirk supporters had learned that the secretary of the Falkirk campaign committee, Alexander Duncan, had been instructed to visit Wright at St. Andrews at the start of their campaign in early November 1960.[74] From the perspective of the Falkirk supporters, the ensuing interaction might only have generated Wright's interest in establishing a university in Stirling. According to Duncan,

> Professor Wright gave me a hearing, and subsequently used his influence with Sir Keith Murray, the Chairman of the University Grants Committee, to arrange for a series of meetings at St. Andrews House [*sic*], Edinburgh, in December 1960 between Sir Keith Murray and representatives of the 6 claimant areas, 5 of these having staked claims following the start of our own campaign. . . . A year ago, when the formal Submissions were being made by the Scottish claimant areas, it transpired that the Stirling Submission was 'ghosted' by Professor Wright.[75]

Adding insult to injury, Murray had apparently stressed to Duncan in Edinburgh "the importance of having pledged financial backing."[76] Yet the Stirling campaign had raised only a fraction of the local financial support lined up in Falkirk and Inverness, a fact that created a row in the days of the final decision about the possibility that local-authority backing should be transferred from Falkirk to Stirling,

---

[72]TNA, UGC 7/240. Fifth university for Scotland: New University in Scotland, Report from Subcommittee on New Universities, June 18, 1964.

[73]TNA, UGC 7/240. Fifth university for Scotland: Duncan to Cousins, April 10, 1965.

[74]TNA, UGC 7/237. Fifth university for Scotland: Wright to Duncan, November 10, 1960.

[75]TNA, UGC 7/244. Fifth university for Scotland: Action following UGC visit: Duncan to Wilson, November 13, 1964, p. 1.

[76]TNA, UGC 7/244. Fifth university for Scotland: Action following UGC visit: Duncan to Wilson, November 13, 1964, p. 1.

if the latter were chosen as the site for the new university.[77] With an undertone of gratification, Wolfenden then announced the government's choice to the six defeated campaigns with a daring appeal: "I hope that you will now feel able to lend your full support to the new Scottish university at Stirling and to encourage those who had expressed their support for your own proposal to rally round the new foundation."[78]

This correspondence, surviving in its full lucidness, therefore confirms the impression that a group of former and present Oxonians did not take the risk of choosing a novel type of university site—or of following the Redbrick tradition expressed in the Robbins committee's preference for populated industrial areas. Instead, they aimed to reproduce an exclusive historic and environmentally attractive university setting similar to that so familiar to the UGC members from their alma mater. The involvement of Oxford alumni in the decision-making process had undeniably been impressive—in 1929 Murray, Wolfenden, Sutherland, and Wright had been all at Oxford University in one capacity or other, and Sutherland was still working there in 1964. Coincidentally, the Stirling campaign's leader, the renowned solicitor John Spencer Muirhead, who had graduated from Oriel College with a double first B.A. degree in 1912, had been made honorary fellow there in 1962—in the same Oxford college in which Murray had worked from 1929 to 1932.

Conclusions

> *On the other hand it might be held that the location* [Cumbernauld New Town] *would be relatively unattractive both to students and staff, apart perhaps from social scientists.*[79]

This chapter has begun to contribute to a new research agenda on the histories, politics, and geographies of British university expansion in the 1960s. By reconstructing the debates and decisions about locating a new university in Scotland, we have shown that, after the recommendation made by the Robbins report in October 1963 to found at least one new university in Scotland, government policy on university expansion continued to be shaped by what Shattock (2012) called "a common 'Oxbridge' culture built up particularly through the War years which bound together senior university figures, the UGC, Treasury officials and (some) politicians" (p. 15). This key finding underlines the abiding significance that closely knit personal networks across different economic sectors and that informal exchanges

[77]TNA, UGC 7/244. Fifth university for Scotland: Action following UGC visit: Duncan to Wolfenden, June 6, 1964.

[78]TNA, UGC 7/244. Fifth university for Scotland: Action following UGC visit: Wolfenden to Drever, Provost of the East Stirlingshire Sponsoring Committee, July 18, 1964.

[79]TNA, UGC 7/244. Fifth university for Scotland: Action following UGC visit: UGC, Subcommittee on New Universities, New university in Scotland, Points for consideration, June 5, 1964, p. 3.

and gentle(wo)men's agreements had for evidence-based policy-making in the post-Robbins era.[80]

Our analysis has revealed that devious lobbying and counterlobbying practices had discredited the campaign for a University of East Stirlingshire in Falkirk since November 1960, and repeatedly from October 1963 onward, with the sole aim of promoting the Stirling campaign. This situation made the efforts of Ayr, Cumbernauld, Falkirk, Dumfries, Inverness, and Perth seem doomed right from the start. The key person able to influence the thinking of both Murray and Wolfenden as the successive chairmen of both the UGC and the UGC subcommittee responsible for choosing the locations of new universities was fellow UGC member Wright, who had been born in Stirling and lived there. He had also been based at Oxford at the same time as other key decision-makers, especially Murray, Sutherland, and Wolfenden, whose undergraduate college Wright entered as a fellow in 1928, the year Wolfenden graduated.

In conjunction with other archival documents, the correspondence with Wright suggests that Stirling had already been chosen in the UGC chairman's mind when Wolfenden wrote to the seven claimant locations in early November 1963 to request supporting statements by the end of the year, and also when the government officially asked the UGC in January 1964 to advise on the location of the new Scottish university from among the places that had previously expressed an interest in such an institution. In this context, Anderson's (2006) characterization of the University of Stirling as one of "eight new universities on greenfield sites" (p. 136) created by the UGC under Keith Murray, whose chairmanship ended in October 1963, contains—perhaps intuitively—more truth than the subsequent work of Wolfenden's UGC subcommittee and their field visits implies.

This argument gains further traction when one considers that Shattock (2012) characterized Murray as "powerful" and Wolfenden as "a much less committed figure" (p. 39). Scottish-born Murray might indeed have pondered Stirling as the most suitable location for a new Scottish university since his meetings with local sponsoring committees in December 1960. Having formed an "unofficial inner group" (Shattock, 2014, p. 114) with Robbins, Morris, and Linstead, Murray might also have actively pursued this agenda during the Robbins committee meetings, for example, by assuring that the new Scottish university could be located *near to* rather than only *in* large population centers, for it was he whom Wolfenden

[80]Similarly dense intersectoral networks were reconstructed in Britain during the 1950s and 1960s by Craggs and Neate (2017) in regard to former colonial administrators becoming New Town general managers.

recruited as the chairman of Stirling's Academic Planning Board after consultation with the Scottish Education Department and the local sponsoring committee.[81]

Like the earlier debates about the locations of the new English universities, approved before Robbins began his deliberations, discussions about Scotland's new university were essentially ad hoc and informal and thus fitted the bigger picture of how Shattock (2012, p. 3) characterized British higher education policy in the second half of the twentieth century. Partly as a result, discussions about the location of the new Scottish university recommended by Robbins were as unsystematic as the debates about the new English universities. Although the Robbins report was based on a mass of statistical research on the necessary scale and character of Britain's higher education system, minimal attention went to the long-term economic, social, and cultural implications of a new university in the various locations considered. Even less thought was given to the potential of different towns and cities to influence the teaching and research environments of a new institution of higher education.

The Robbins report also did not address the wider national geography of the proposed expansion in any significant way, an omission that can be linked to Johnston's (2004) point that geography had few advocates in British policy circles during the 1950s and 1960s for lack of engagement and collective action by the learned societies and leading academic geographers. The absence of geographers in the decision-making might have made it easier for the UGC subcommittee to invert prevailing aims of regional development and planning. British regional policy, originating from the 1930s and shifting its focus from full employment to economic growth in the 1960s, was at that time attending to three problem regions: rural, depressed industrial, and congested areas (McCrone, 1969). Under such circumstances, the strong contender of Falkirk, in need of a university to support industrial transformation in the Falkirk–Grangemouth–Larbert area, was sidelined as much as the remoter rural locations that required economic and cultural development, especially Ayr, Dumfries, and Inverness as the capital of the Scottish Highlands.

In the tradition of the English "Baedeker towns" that had been awarded universities under Murray's UGC chairmanship and were developing their new institutions during the 1960s—all run by Oxbridge-educated Vice-Chancellors—the choice came down to a site featuring a picturesque landscape near a historically appealing county town in Scotland's central population belt.[82] We therefore argue that this choice of location reproduced the setting of the ancient universities rather than considering the long-term economic, social, and cultural implications of different university locations for Britain's urban network and experimenting with new types of places for learning, teaching, and research—as the new universities in England did with their innovative curricula (Briggs, 1991). From 1965 to 1975, such geo-

[81]TNA, UGC 7/245. University of Stirling: Wolfenden to MacFarlane Gray, former Provost of Stirling and member of the New University Sponsoring Committee at Stirling, August 25, 1964.

[82]TNA, UGC 7/241. Fifth university for Scotland: Submissions: Royal Burgh of Stirling, Proposed university at Airthrey estate, Supplementary memorandum by the sponsoring committee, n.d., p. 1.

graphical experimentation occurred in West Germany, where the government followed the advice of geographers and regional planners to adopt a rather rationalist planning perspective when locating new universities in old industrial and rural areas, as with the universities of Bochum and Constance, respectively (Mayr, 1979). The success of this strategy subsequently led to the recommendation "to choose smaller towns as locations of universities in order to intensify the regional educational and economic activities and potential in underdeveloped areas" (Mayr, 1979, p. 324).

Our analysis has demonstrated that the new Scottish university only partly modernized an ancient university system. The all-important locational decision to select Stirling over other possible towns reaffirmed a traditional, Oxonian view of an appropriate setting for a university, an outcome only partially offset by the architectural choice of using modern brutalist architecture, albeit in a modest way. From this perspective, the Wolfenden era seemed to continue the Murray era that favored small historic towns for new arts-based universities. This perpetuation contrasted with the idea of universities developing the natural, technical, and social sciences in larger industrial cities, a view reminiscent of the Redbrick era and epitomized by the Robbins report. In terms of location and subject orientation, the rivalry between the Stirling and Falkirk campaigns, headed by an honorary fellow of Oriel College Oxford and a member of the Robbins committee, respectively, arguably exemplified the competition between an arts-based Oxford model and a science-based LSE view on where new universities should be located and how they should be designed.

These opposing views reflected wider, multilayered tensions in British society, such as those between Oxbridge and the civic universities (Anderson, 2006, pp. 136–137). A second strata of tensions evident in the divergent perspectives discussed in this chapter existed between a more arts-based Oxford University with close connections to politics, as illustrated by the dominance of Oxford-educated UGC members, and a more science-oriented Cambridge University, whose alumni figured more prominently on the Robbins committee. This difference was reinforced throughout the twentieth century by the practices underlying these institutions' conferment of honorary degrees (Heffernan & Jöns, 2007, p. 414). A third frictional interface was that between old-school humanists and increasingly powerful modern scientists, as expressed in the Robbins report's support for science and technology and a "belief in the 'white heat of technology' in the electoral programme of Harold Wilson" (Anderson, 2006, p. 150). Yet Murray, as the new chairman of Stirling's Academic Planning Board, strategically bridged these varying cultures by asking Robbins to become the first chancellor of the new University of Stirling, an appointment the LSE economist accepted after some consideration (Howson, 2011, p. 987).

Lastly, we argue that the chance to establish a second new university in Scotland—as tentatively suggested by the Robbins committee (CHE, 1963, p. 155)—and thus to use a new university for stimulating economic growth in deprived regions was sadly missed by the British government. This option might have been prevented by looming financial austerity that prompted the government to put a hold on further expansion plans in April 1964, just before the Scottish field visits took place. After adoption of the view that the existing universities would be able to accommodate the target numbers of students, it took another ten months, and a change of government, before Anthony Crosland, the new Labour Secretary of State for Education and

Science, announced on February 24, 1965, that no additional new universities would be needed for about ten years (Mountford, 1966, p. 43). This decision ended the ambitions of over 40 promotion committees that were still arguing for new universities in England and Wales at the time, and whose stories remain to be told.

Acknowledgement We are grateful to Peter Meusburger, Robert Anderson, Holger Nehring, and David Antal for insightful comments on an earlier version of our chapter.

References

Anderson, R. D. (2006). *British universities: Past and present*. London: Hambledon Continuum.

Annan, N. (1982). British higher education, 1960–80: A personal retrospect. *Minerva, 20*, 1–24. Retrieved from http://www.jstor.org/stable/41820484

Barlow, A. (1946). *Scientific man-power: Report of a Committee appointed by the Lord President of the Council*. London: His Majesty's Stationerye Office. Retrieved from http://www.educationengland.org.uk/documents/barlow1946/barlow1946.html

Beloff, M. (1969). *The plateglass universities*. London: Secker & Warburg.

Briggs, A. (1991). A founding father reflects. *Higher Education Quarterly, 45*, 311–332. doi: https://doi.org/10.1111/j.1468-2273.1991.tb01575.x

Callender, C. (2014). Students numbers and funding: Does Robbins add up? *Higher Education Quarterly, 68*, 164–186. doi:https://doi.org/10.1111/hequ.12042

Carswell, J. (1985). *Government and the universities in Britain: Programme and performance, 1960–1980*. Cambridge, UK: Cambridge University Press.

Cochrane, A., & Williams, R. (2013). Putting higher education in its place: The socio-political geographies of English universities. *Policy and Politics, 41*, 43–58. doi:https://doi.org/10.1332/030557312X645775

Committee on Higher Education. (CHE). (1963). *Higher education: Report of the committee appointed by the Prime Minister under the chairmanship of Lord Robbins, 1961–63* (Robbins Report Cmnd. 2154). London: Her Majesty's Stationery Office.

Craggs, R., & Neate, H. (2017). Post-colonial careering and urban policy mobility: Between Britain and Nigeria, 1945–1990. *Transactions of the Institute of British Geographers, 42*, 44–57. doi: https://doi.org/10.1111/tran.12147

Goddard, J., & Vallance, P. (2013). *The university and the city*. Regions and Cities: Vol. 63. London: Routledge.

Heffernan, M., & Jöns, H. (2007). Degrees of influence: The politics of honorary degrees in the universities of Oxford and Cambridge, 1900–2000. *Minerva, 45*, 389–416. doi:https://doi.org/10.1007/s11024-007-9065-8

Howson, S. (2011). *Lionel Robbins*. Cambridge, UK: Cambridge University Press.

Johnston, R. (2004). Institutions and disciplinary fortunes: Two moments in the history of UK geography in the 1960s I: Geography in the "plateglass universities". *Progress in Human Geography, 28*, 57–77. doi:https://doi.org/10.1191/0309132504ph470oa

King, D., & Nash, V. (2001). Continuity of ideas and the politics of higher education expansion in Britain from Robbins to Dearing. *Twentieth Century British History, 12*, 185–207. doi:https://doi.org/10.1093/tcbh/12.2.185

Lawton Smith, H. (2007). Universities, innovation, and territorial development: A review of the evidence. *Environment and Planning C: Government and Policy, 25*, 98–114. doi:https://doi.org/10.1068/c0561

Layard, P. R. G., King, J., & Moser, C. A. (1969). *The impact of Robbins*. Harmondsworth: Penguin Books.

Mayr, A. (1979). *Universität und Stadt: Ein stadt-, wirtschafts- und sozialgeographischer Vergleich alter und neuer Hochschulstandorte in der Bundesrepublik Deutschland* [University and city: Old

and new campuses in the Federal Republic of Germany from the perspective of urban, economic, and social geography]. Münstersche Geographische Arbeiten: Vol. 1. Paderborn: Schoeningh.
McCrone, G. (1969). *Regional policy in Britain.* University of Glasgow Social and Economic Studies: Vol. 15. London: Allen & Unwin.
Moser, C. (1987). The Robbins report: Twenty-five years after. *Contemporary Record, 1*, 33–35. doi:https://doi.org/10.1080/13619468808580930
Mountford, J. F. (1966). *British universities*. Oxford Paperbacks University Series: Vol. 14. Oxford, UK: Oxford University Press.
Neave, G. (1976). The University of Stirling and the central region of Scotland: The adaptation of university to regional needs. *Paedagogica Europaea, 11*, 15–39.
Parkinson, M., Champion, T., Evans, R., Simmie, J., Turok, I., Crookston, M.,. .. Wood, P. (2006). *State of the English cities: A research study: Vol. 1.* London: Office of the Deputy Prime Minister. Retrieved from https://www.liverpool.ac.uk/media/livacuk/publicpolicypractice/State,of,Cities,V1.pdf
Paterson, L. (2003). *Scottish education in the twentieth century*. Edinburgh: Edinburgh University Press.
Robbins, L. (1966). *The university in the modern world and other papers on higher education.* London: Macmillan.
Robbins, L. (1971). *Autobiography of an economist*. London: Macmillan.
Robertson, S. L. (2010). *Globalising UK higher education.* London: Centre for Learning and Life Chances in Knowledge Economies and Societies. Retrieved from https://susanleerobertson.files.wordpress.com/2012/07/2010-robertson-globalising-higher-ed.pdf
Scott, P. (2014). Robbins, the binary policy and mass higher education. *Higher Education Quarterly, 68*, 147–163. doi:https://doi.org/10.1111/hequ.12040
Shattock, M. (1994). *The UGC and the management of British universities*. Buckingham: Society for Research into Higher Education & Open University Press.
Shattock, M. (2012). *Making policy in British higher education, 1945–2011*. Maidenhead: Open University Press.
Shattock, M. (2014). Remembering Robbins: Context and process. *Higher Education Quarterly, 68*, 110–124. doi:https://doi.org/10.1111/hequ.12044
Stewart, W. A. C. (1989). *Higher education in postwar Britain.* Basingstoke: Macmillan.
Tight, M. (1987). The location of higher education. *Higher Education Quarterly, 41*, 162–183. doi:https://doi.org/10.1111/j.1468-2273.1987.tb01774.x
Tight, M. (1996). The re-location of higher education. *Higher Education Quarterly, 50*, 119–137. doi:https://doi.org/10.1111/j.1468-2273.1996.tb01695.x
Weeks, J. (2004). Wolfenden, John Frederick, Baron Wolfenden (1906–1985). In *Oxford Dictionary of National Biography*. Oxford, UK: Oxford University Press. Retrieved from http://www.oxforddnb.com/view/article/31852
Whyte, W. (2015). *Redbrick: A social and architectural history of Britain's civic universities.* Oxford, UK: Oxford University Press.

Part II
The University, Knowledge, and Governance

Chapter 7
Knowledge Environments at Universities: Some Theoretical and Methodological Considerations

Peter Meusburger

The Spatiality of Learning, Research, and Academic Careers

Until the early 1980s most research on scientific creativity had focused on the personal attributes of scholars. Few authors found it necessary for their theoretical considerations to include the social, cultural, and scientific environment—the seat of socialization, learning, control, and encouragement by supervisors, research activities, and creative processes. Spatial differences in scientific practices and research cultures; spatial inequalities of career opportunities, financial resources, academic freedom, and other factors influencing the scientific achievements of scholars; and the attractiveness and reputation of universities received little attention. Science was assumed to be placeless. At best, it was admitted that talented individuals could not develop their creativity in repressive societies.

In the 1980s and 1990s researchers studying creativity increasingly accepted that talent, wealth of ideas, motivation, endurance, and other personal traits are not the only factors influencing how well a person's scholarly pursuits and academic career develops. What people call creative is never the result of individual action alone. The talented student and the creative scientist do not work in a social, cultural, and economic vacuum. A stimulating knowledge environment and a talented individual must come together and interact before a creative process can occur (Amabile, 1979, 1983a, 1983b, 1996; Amabile, Conti, Coon, Lazenby, & Herron, 1996; Amabile, Goldfarb, & Brackfield, 1990; Amabile, & Gryskiewicz, 1989; Hennessey & Amabile, 1988; Meusburger, 2009; Sternberg & Lubart, 1991, 1999). No one can study scientific achievements and academic careers effectively by isolating scholars and their works from the political, social, cultural, and historical milieus in which

This chapter greatly expands on Meusburger (2015, 2016).

P. Meusburger (✉)
Department of Geography, Heidelberg University, Heidelberg, Germany

P. Meusburger et al. (eds.), *Geographies of the University*, Knowledge and Space 12,
https://doi.org/10.1007/978-3-319-75593-9_7

their academic socialization and research has taken place and in which their achievements have been evaluated, legitimated, or rejected by evaluators and gate-keepers of their research field.

A second line of arguments underlining the importance of places, settings, environments, and spatial relations for research and scientific careers has been put forward by the geography of science (Livingstone, 1995, 2000, 2002, 2003; Withers, 2001, 2002, 2010; Withers, Higgitt, & Finnegan, 2008) and science studies in other disciplines (Collins, 1983, 2008; Collins & Evans, 2002; Collins & Pinch, 1998; Gieryn, 1983, 2000, 2002; Knorr-Cetina, 1999; Matthiesen, 2006, 2007). "Scientific knowledge is a geographical phenomenon. It is acquired in specific sites; it circulates from location to location; it transforms the world" (Livingstone, 2010, p. 18). The geography of science is interested in a variety of questions: Why has a specific research field first developed at university A and not at universities B or C? Why have some universities produced and attracted a host of outstanding scholars, whereas others have not? Which factors, expectations, or institutional logics inhibit or promote the scientific creativity and academic career of scholars? Why do scientific results travel in the selective way they do? Why does it sometimes take decades until outstanding scientific findings are accepted by other scholars? Why have important texts been read and interpreted so differently in specific places and environments? What can and cannot be said in particular venues? What effects do political power, cultural politics, and the geopolitics of science have on the generation and spatial diffusion of science? Why have outstanding centers of research lost their importance in the course of history? How do international networks of scientists evolve and which long-term consequences do they have? What are the preconditions and primary sources of paradigmatic change in various disciplines and places of science? Which local (regional) barriers and facilitators of change can be identified?

A third strand of arguments paving the way for research on the effects that environments have on action and performance originated in organizational research, business studies, and institutional theory. Scholars in these fields became interested in the psychological climate of organizations, in institutional logics, in rules and informal norms that structure behavior, and in organizational learning (Battilana, 2006; Berthoin Antal, Meusburger, & Suarsana, 2014; Bitektine & Miller, 2015; Cummings, 1965; Denison, 1996; Glückler, Lazega, & Hammer, 2017; Lawrence, Suddaby, & Leca, 2011; Mintzberg, 1979; Suddaby, 2010; Suddaby & Greenwood, 2005). They found that creative persons are keenly sensitive and responsive to formal and informal sets of mutual expectations between people, to quality conventions and assessment tools prevailing at their place of work. Institutional theorists have recognized the constraining effect that institutional norms, conventions, and meaning systems have on the emergence of new theories, new research paradigms, and research methods (Bitektine & Miller, 2015, p. 117). Some authors in organizational studies have discussed the social construction of fields (Fligstein, 2013), inhibitory constraints in organizational behavior (Lind & van den Bos, 2013), adaptive leadership theory (DeRue, 2011), the influence of local communities on organizations (Marquis & Battilana, 2009), and implications of the physical work environment on organizations (Zhong & House, 2012).

A key contribution came, fourth, from environmental psychology (Barker, 1968; Bechtel, 1997; Graumann, 1978, 2002a, 2002b, 2002c; Graumann & Kruse, 1990, 2003; Kruse, 1986). Environmental psychologists no longer saw the behavior of the individual agent as the main focus of research in psychology. They were convinced that a behavior could be predicted more from knowledge about the spatiomaterial and social situation in which it occurs than from the knowledge and traits of the actor (Kruse, 1986, p. 135). Further insights have come from psychological studies on complex problem-solving, the decisive role that the environment has in it, the limitations of the corresponding psychological experiments, the tensions between field and laboratory research, and the contextual dependency of strategies (Dörner & Funke, 2017, pp. 4–5; Funke, 2012; Runco, 1994; Runco & Okuda, 1988).

> [T]here is no strategy that is so universal that it can be used in many different problem situations. . . . Complex phenomena require complex approaches to understand them. The complex nature of complex systems imposes limitations on psychological experiments: The more complex the environments, the more difficult is it to keep conditions under experimental control. (Dörner & Funke, 2017, pp. 4–5)

Fifth, since the 1980s human geography has developed new concepts of space and place that have facilitated interfaces with the social and behavioral sciences and have helped avoid spatial determinism and ecological fallacies (Harvey, 1969, 2005; Klüter, 1986, 1999, 2003; Lefebvre, 1974/1991; Lippuner, 2005; Lippuner & Lossau, 2004; Massey, 1985, 1999a, 1999b, 2005; Meusburger, 2008; Meusburger & Werlen, 2017; Schmid, 2005; Soja, 1980, 1985; Weichhart, 1996, 1999, 2003; Werlen, 1983, 1987, 1993, 1995, 1996, 1997a, 1997b, 2010a, 2010b).

A place can be defined as a "discursively constructed setting" (Feld & Basso, 1996, p. 5) having a symbolic and emotional meaning; providing an identity; and communicating a complex history of events, cultural memories, and emotional attachments (Canter, 1977; Furnham, 1985; Giddens, 1984; Manzo, 2005; Rowles, 2008a, 2008b; Scannel & Gifford, 2010; Werlen, 1983, 1987, 1997a). People are rooted in and attached to places, and places are "metaphorically tied to identities" (Feld & Basso, 1996, p. 11). A place of work often signifies social status or a position in a hierarchy.

Place names of famous universities serve as a kind of shorthand or metonomy[1] for the inscrutable complexity of research processes taking place at these locations. The achievements and academic reputation earned by generations of individual scholars are ascribed to or projected onto their university. Through these projections, university locations become symbolically charged. Because nobody can unravel the complexity of interaction, learning processes, and institutional decision-making, the appraisal of a scientist's potential is closely tied to the places at which that

[1]A metonomy is a figure of speech in which a word, phrase, thing, or concept is substituted for another with which it is closely associated. When people refer to complex and nontransparent situations, they often use place names. *Wall Street* stands for the worldwide center of financial services; *Berlin*, for the government of Germany; and *Berkeley*, *Stanford*, or *Cambridge*, for outstanding levels of science.

person earned his or her doctorate, did postdoctoral work, or received an offer of a professorship. A scholar who has been trained at a distinguished research department has credibility from the outset because the scientific reputation of a place or institution reflects on the scholars working there. The scientific reputation built by scientific institutions stems from their ability to appoint the best candidates available. The identity of scholars is defined not only by what they do but also by *where* they do it. The scientific career of most scholars is distributed across more than one university. However, the move from one university to the next is rarely random; it is rather a "contingent series of events" (Welskopp, 2002, p. 79)[2] that follow a definite logic and stem from particular mechanisms.

A place is like a screen on which possibilities, expectations, benefits, and hopes—or fear and anxieties—are projected. People trust the academic reputation of excellent universities or departments even if unfamiliar with any of the scholars working there. The symbolic meaning, reputation, and attractiveness of a university lie not only in its present merits and achievements but also in those of previous scholars no longer belonging to the institution's knowledge environment. In many university towns these collective memories and narratives are reinforced by regular self-reenactments (e.g., anniversary celebrations and graduation ceremonies) or an iconography of the urban space. Plaques on buildings, museums, and memorials constantly remind people of past achievements by famous scientists or of important local events in the history of science. An awareness of the city's iconography and recurring contact with places important in the history of science can instill in scholars and students an emotional bond[3] and can support personal identification with their university. A university is a junction or intersection where the career paths of academics of various disciplines, generations, and provenances meet, where scholars communicate and (it is hoped) interact with each other for a certain period.[4] Universities and other research institutions are not simply locations but social spaces, epistemic venues, nodes of scientific discourse, enablers or impediments of actions, and carriers of reputation.

In summary, recent theoretical and methodological trends in a number of scientific disciplines underline the importance of environments for action, learning processes, and problem-solving. This chapter focuses on the role of knowledge environments at universities and their impact on research processes and academic careers. I focus on five questions: What does the term *knowledge environment* mean? Which components constitute a local knowledge environment? In what way can a local knowledge environment affect goals, decisions, learning, research processes, and the careers of academics? Which theoretical concepts can contribute to the understanding of the interaction between knowledge environment and individual

---

[2]All translations are my own unless otherwise stated.

[3]"Any kind of learning is connected with some tradition," and names of scholars, places, and scientific practices "already suffice to form a collective bond" (Fleck, 1935/1979, p. 52).

[4]Mager (2012, pp. 251, 253) gives an example of the representation of career paths in a time–space model.

scholar? How can the consequences of a knowledge environment be verified? The main goals of this chapter are to shed light on possible research agendas about knowledge environments and to discuss some of the methodological and theoretical issues connected with knowledge environments.[5]

What Is a Knowledge Environment and Which Caveats Should Be Considered?

Research Needs Specific Environments

Scholars depend on research infrastructure, financial resources, face-to-face contacts with distinguished members of various disciplines, critique and support of peers, and experienced key persons. There is no doubt that cognition, aspirations, motivations, and emotions of scholars can be influenced by the ambience in which they act. New scientific results have to be accepted and legitimated by other scholars. Local knowledge environments can enthuse, support, or frustrate scholars. In each discipline, departments, faculties, and universities differ in resource endowment, the availability of sophisticated and expensive research infrastructure,[6] the career opportunities they open up to academics, and the extent to which their scholars are integrated into important national and international scientific networks.

The history of science is rich in examples of creative scholars and innovative research projects having met with rejection, incomprehension, neglect, and hostility at particular universities. However, there are also many examples of scholars of slowly changing or disrupting the knowledge environment of their place of work by altering the rules, conventions, and style of leadership, by upgrading the resources of their institution, and by hiring new academic staff. A knowledge environment is never stable; it is constantly in motion.

Sociology of knowledge, environmental psychology, geography of science, creativity studies, institutional theory, research on organizational behavior, and other fields have studied a great number of particular parameters, conditions, and structures affecting learning and research. Scholars studying knowledge environments build on this kind of research, but they are more interested in the connectivity, interdependencies, interrelations, and interactions of relevant variables *at a certain place or environment.* These academics focus especially on the scientific outcomes of different environments.

In summary, a university's knowledge environment is definable as the result of systemic interdependencies and causal interactions of personal, financial, material and nonmaterial resources relevant to the generation, diffusion, and application of

[5]See Meusburger and Schuch (2010, 2012) for empirical studies.

[6]Research infrastructure is more than technical equipment. It also encompasses privileged access to documents, unpublished census data, and private archives.

scientific knowledge at a specific place or environment. A knowledge environment is a multiscale phenomenon. From the perspective of an individual scholar, it is a subject-centered space of cognition, interaction, and learning. From a systemic perspective, analysis focuses on mutual interdependencies of processes and parameters that influence the production of knowledge. Parameters effective at the local, regional, national, and global level intermingle with each other at a specific place in a particular way and evolve into an inimitable local knowledge environment. Saying that the knowledge environment of a university is a *case sui generis* does not preclude the generalizability of specific processes and interrelations characterizing knowledge environments. As Mayntz (2002) put it,

> systemic interdependences are about mutual dependence and influence in the relations between different simultaneous processes or between different institutions. The discrete processes or institutions are parametrically linked with each other as it were; that is, they can reciprocally change important basic conditions for each other. (p. 33)

Because creativity is domain-specific, agents working in different domains (e.g., science, humanities, arts, industry, and financial services) need different knowledge environments. Some scientific disciplines depend more on research infrastructure and financial resources than others do. Theoretical physicists, for instance, are less place-dependent in their research than experimental physicists, who may find only three locations in Europe where they can conduct their costly experiments.

The intensity to which scholars need or depend on their knowledge environment also varies over time. In some phases of their research projects, scholars seek intense interaction with their environment, have many face-to-face contacts with experts of other fields, engage as many researchers as possible in multifaceted negotiations, depend on support and critique from peers, "borrow metaphors and notations from epistemic sources as distant as possible to maximize innovative variance" (Fiedler, 2004, p. 126), and seek both controversy with epistemic opponents and inspiration through serendipitous interaction with scholars from other research fields. In other phases scholars try to avoid the noise of their environment, retreating to concentrate on their manuscripts without interference. These processes of opening and narrowing the areas of contact are similar to what Kelly (1955/1991) called the "Creativity Cycle": "The Creativity Cycle is one which starts with loosened construction and terminates with tightened and validated construction" (Vol. 2, p. 7). As Fiedler (2004) explains, a creativity cycle

> encompasses two stages, loosening and tightening. By analogy to evolution theory, loosening corresponds to the production of random variation . . . , whereas tightening corresponds to a subsequent selection stage that serves to separate the strong and weak products of the preceding loosening stage. (p. 124)

Research on Knowledge Environments of Universities Should Address Various Caveats

First, a knowledge environment should not be thought of as an independent variable that directly influences all actors through direct cause and effect (if A, then B). It is rather a locally available potential or local offer of resources, opportunities, incentives, challenges, stimulations, and support networks. What makes a location attractive is its potential for high-level interactions, its possible or imagined advantages, not just the realized ones. This local potential can be used, ignored, or rejected by scholars working at the relevant place. A knowledge environment can operate as it should only if the actors involved appropriate and use the local resources, take advantage of the opportunities, understand and adopt the available knowledge, use the available research infrastructure proficiently, and interact with each other appropriately. Only interested, competent, and talented individuals will be able to take full advantage of a given knowledge environment.[7] A knowledge environment does not come out of the blue; it is constantly molded and remolded by scholars and other people responsible for a research institution.

Second, the outcomes of human interactions and experiences in life are always indeterminate. No one can predict the results of learning, appropriation, and interaction, the ways in which the relationships between individual scholars of different age groups and disciplines will develop, or whether and how often the local potential for integrating diverse viewpoints and knowledge bases will be activated. It is relatively easy to analyze some of the preconditions for scientific creativity (e.g., financial, personal, and material resources) by multivariate methods,[8] but a knowledge environment's significance and effect in a given period are explicable only ex post—after events have taken place, after the scientific careers and research results associated with a place, territory, or environment have become evident.

Third, students of knowledge environments must cope with the contradiction that knowledge environments are both volatile and remarkably robust. Many important elements of a knowledge environment are fleeting—eminent scholars and talented students come and go, research questions and methods change, the creativity of scholars varies over time, financial resources (especially external funding) fluctuate, and internal conflicts or imprudent appointments of professors may detract from a department's scientific achievements and reputation. Conversely, the scientific quality and reputation of an institution's knowledge environment in many cases show a remarkable robustness over time. Excluding wars (see Meusburger, 2012a, pp. 19–20, 22; Neumaier, 2012; Rotzoll, 2012, and the chapter by Hotson and Meusburger & Próbald in this volume) and other political disasters (see Mußgnug, 2012), many countries have a remarkably stable hierarchy of universities as self-

[7]This experience is one of the main reasons why recruitment processes of scholars are crucial for the quality of a knowledge environment.

[8]Contextual studies of knowledge environments do not replace multivariate analyses but rather contribute additional benefits and new insights.

organizing systems. The hierarchy of universities admits of some upward and downward mobility, but it is rare that a university soars from the lowest to the top rank of a nation's university hierarchy or plummets from the top to the bottom. This persistence, or "dynamic robustness,"[9] is usually due to universities' differential ability to regulate themselves and adapt to new challenges, the so-called Matthew effect ("the more you have, the more you get"), and other factors contributing to a university's scientific attractiveness.

Fourth, it is necessary to take the time dimension into account. Short-term evaluations of a knowledge environment may be quite biased. With hindsight, most people judge persons, situations, processes, and environments differently than immediately after a certain event has occurred. A senior professor might evaluate the knowledge environment of his or her institution differently than a doctoral student does. A knowledge environment is characterized by multiple realities (Graumann, 1983).

Of What Components Does a Local Knowledge Environment Consist?

Capacities, Characteristics, and Behavior of Persons Significantly Involved in the Activities of a Scientific Institution

The most important element of a knowledge environment at a given place are its human beings (e.g., scholars, students, technical, and administrative staff) with their cognitive capacities, competencies, skills, experiences, scientific reputation, personal networks, emotions, and manner of interacting with other people. It depends mainly on these stakeholders how the other elements of a knowledge environment (including resources, infrastructure, organizational culture, and scientific reputation) develop and interrelate with each other.

The key quality of a knowledge environment is its power to attract internationally renowned scholars and highly talented students. To become and remain attractive and internationally competitive, a university's knowledge environment must continually be replenished by the best available candidates bringing in new ideas, new research questions, innovative scientific methods, new constructive criticism on existing research, new bodies of knowledge, professional experience with other settings and research cultures, high quality standards in teaching and research, international personal networks, high social competencies, and appropriate professional value systems. The input of newly appointed scholars may influence the dynamics of the university's capacity of self-regulation and the range of international networks and communication spaces. Scholars socialized in other knowledge environments may question or subtly change some of the prevailing institutional logics at

[9] I borrow this term from Robert Panitz (Department of Geography, Heidelberg University).

their new place of work. Scholars criticizing mainstream theoretical discussions in their discipline will have an interest in counterhegemonic practices (see Clegg, 2010, p. 8), and some of them will even be able to affect governance structures in important ways (Dacin, Goodstein, & Scott, 2002, pp. 51–52).

There are several key questions for the success of this process of continuous intellectual renewal: According to which quality standards are academics appointed? How transparent are the recruitment processes? What personal, financial, and material resources is a university able to offer to its newly appointed scholars? Which levels of scholars is a university able to attract? Some universities will be more successful than others in the competition to hire the best candidates.

Intellectual renewal comes about not only through the appointment of new scientists but also through what is known as circular mobility (for details see Jöns, 2003, 2007, 2008, 2009, 2017), such as brief periods of study abroad, the invitation of visiting scholars, scientific cooperation with other institutions, and other ways of exchanging and testing new knowledge (e.g., participation in conferences). Mobility is equivalent to gathering experience in different knowledge environments, tackling new challenges, and familiarizing oneself with issues and methods one has not encountered at previous places of work. Mobility can trigger new ways of learning and thinking, enrich a scholar's capacities, and bring about new cooperation in the world of science.

Financial and Material Resources

Whether distinguished scholars and talented students are attracted to a given university or not partly depends on its basic material wherewithal—its financial strength and the quality of its research infrastructure (libraries, laboratories, computing centers, archives, and other rare or expensive research facilities). Many innovative research questions can be studied empirically only if the latest research equipment and analytical methods are on hand or if a scholar gains access to hitherto unused data sources or archives. The more expensive such research infrastructure is, the fewer universities there are that can acquire it.

To attract the famous chemist Robert Bunsen (1811–1899) to Heidelberg University in 1852, the government of the Grand Duchy of Baden funded a state-of-the-art chemistry laboratory building, in those days the most modern in the world. In the second half of nineteenth century, the Grand Duchy of Baden spent about 4% of its state budget on its two universities and a technical college, and its per capita expenditures on universities exceeded those of any other German state (Pfetsch, 1974). Robert Bunsen was a prime reason why other exceptional scientists, including Hermann Helmholtz (1821–1894) and Robert Kirchhoff (1824–1887) joined Heidelberg University and created a research center of worldwide reputation (for details see Eckart, Hübner, & Nawa, 2012; Hübner, 2010; Meusburger, 2012a, b,

Meusburger & Schuch, 2010). Because the government provided Bunsen with eight assistants[10] (most of his colleagues had only two), and because he generously allowed them to pursue their own research interests, his department pioneered a unique variety of new research fields that made Heidelberg the power center of chemistry for some decades.

This example confirms that financial and material resources should not be isolated from the scientific achievements and reputation of a university's scientists. In most countries research money does not flow automatically. It must be acquired by scholars in keen national and international competition or negotiated during an appointment process. The fact that only a small share of distinguished scientists will be successful in these hard competitions for research grants and expensive research equipment feeds the ascending spiral of attractiveness and the effect of "the more you have, the more you get." This is one of the main reasons why the university system has a relatively stable hierarchic structure—at both the national and global levels.

Empirical data from almost every country documents that most research funding goes to only a few universities. In Germany, for instance, the 40 universities (9.4% of all the country's universities) that were most successful at securing external funds from the German Research Foundation (DFG) received 86.6% of all DFG funding from 2011 through 2013 (Deutsche Forschungsgemeinschaft, 2015, p. 60). Of the approximately 420 German universities (average from 2011 through 2013), 230 still did not have the right to award doctorates in 2013, which of course has a negative effect on research performance (Deutsche Forschungsgemeinschaft, 2015, p. 33). Expensive new research infrastructure is usually first introduced at leading research universities, so these repeatedly gain advantage, enhancing the Matthew effect.

Organizational Structures, Institutional Rules and Logics, and Scientific Cultures

Although above-average financial and material endowment is crucial to a university's attractiveness, it by no means guarantees excellent scholarship. Organizational structures, institutional logics, scientific or epistemic[11] cultures, and other immaterial factors influencing communication, interaction, practices, and scientific ethics at a scientific institution or among the members of a research group can have great impact on the success of research processes and on the academic careers of junior scientists.

Evaluation of scientific cultures and the organizational behavior of departments or universities raises some decisive questions: What kind of positive and negative role models exist in a department or faculty and how influential are they? What kind

[10]Today they would be called assistant professors or lecturers.

[11]For details see Knorr-Cetina (1999).

of motivation, intellectual encouragement, and emotional and financial support do doctoral students and junior scientists receive in their department? How do research groups cope with dynamic complexity and uncertainty? What are the quality standards for research, teaching, and administration?[12] What kind of publication culture prevails in a department or faculty? What kind of assessment tools are used to measure the quality and impact of research?[13] Is research evaluation institutionalized by a standard protocolization of assessments (see overview by Wouters & Costas, 2012) or is there respect for diversity of assessment tools? How do key decisions on structural changes, resource distribution, and new research fields come about within a department, faculty, or university, and what criteria apply? How are conflicts of interests solved? To what extent do established scholars exploit the research of their doctoral students and junior scientists? How great is the scientific autonomy of doctoral students and junior scientists? To what extent are they free to follow their own research interests and to choose their theoretical concepts and methodologies?[14] The influence of assessment practices on publication strategies can hardly be overestimated. In some departments, unfortunately, publication is tied less to scientific principles than to the number of points it will bring the authors for their next assessment or promotion.

Studies in business administration, organizational and institutional theory, and complex problem-solving yield interesting findings about the impact of organizational climate and institutional expectations on performance in industrial firms and the service industry. However, the decisive question is the extent to which these findings can be transferred to basic research of top universities—especially in the early phases of a creative process. Universities should not be confounded with industrial companies. They cannot be organized, coordinated, and evaluated like a company manufacturing sausages (for details see Beck, 2005, pp. 99–100; Weichhart, 2012a, pp. 14–15). The problems with which universities must cope are much more ill defined[15] than those confronting industrial firms. The issues that

[12]Weichhart's (2012b) sarcastic essay lists several shortcomings frequently observable in assessment practices focusing on formal criteria rather than the innovativeness, originality, and quality of research.

[13]In digital societies the number of assessment practices and citation and impact measures has increased dramatically despite the highly problematic nature of most of these indices from a methodological and statistical point of view (for details see Franzen, 2011, 2015; Wouters & Costas, 2012).

[14]The last three questions of this paragraph will be valued differently in specific research fields. What is everyday practice in department (or discipline) A will be frowned upon in department (or discipline) B. The key questions are which of the mentioned organizational cultures, assessment practices, and types of scientific relationships expand the latitude that junior scientists have in order to develop their creativity and pursue a successful academic career and which of those tend to maximize the benefits of senior scientists instead.

[15]"Well-defined problems have a clear set of means for reaching a precisely described goal state. . . . Ill-defined problems have no clear problem definition, their goal state is not defined clearly, and the means of moving towards the (diffusely described) goal state are not clear" (Dörner & Funke, 2017, p. 1).

universities grapple with incorporate all the typical attributes of complex systems described in detail by Dörner and Funke (2017):

> According to Funke (2012), the typical attributes of complex systems are (a) complexity of the problem situation which is usually represented by the sheer number of involved variables; (b) connectivity and mutual dependencies between involved variables; (c) dynamics of the situation, which reflects the role of time and developments within a system; (d) intransparency (in part or full) about the involved variables and their current values; and (e) polytely (Greek term for "many goals"), representing goal conflicts on different levels of analysis. (p. 2)

The university is a complex, dynamic system characterized by intransparency, a high degree of self-governance of its single elements (e.g., departments and research groups), and a vast multiplicity of goals.[16] Unlike top managers of industrial firms or public administration, who set forth clearly defined objectives to all its actors (at least for specified period), a rector, dean, or the administration of a university cannot prescribe the research topics its scholars are to study, which methods they are to use, or which cooperations and networks they have to join. Basically, each scholar can follow his or her own objectives or, in commercial terms, can pursue his or her own product and marketing as desired. Scholars often try to solve on their own the complex problems that they have found or created and that were not known before (for details about problem-solving, see Dörner, 1996; Dörner & Funke, 2017; Funke, 2012).[17]

Croissant (chapter in this volume) draws attention to a famous dictum of Albert Einstein: *If we knew what we were doing, it wouldn't be research.* This statement applies especially to high-level basic research in the first phase of the innovation S-curve—where pioneering scientists study important topics only very few people, if any, have tackled. This kind of pioneering basic research must be distinguished from applied research, which has explicit goals, and from basic research in late phases of the S-curve on topics that have been on the table for many years and are already being tackled by large clusters of scientists all over the world.[18]

Goal-oriented, applied research can be managed and organized to a greater extent than pioneering, high-level basic research, which involves much more uncertainty (e.g., uncontrollable conditions and unexpected events). The problematic term

[16]University actors pursue many different, sometimes quite contradictory goals. The problem-solver must therefore set priorities and make compromises.

[17]The problems to be solved are "self-created" because they would not exist if it were not for the scientific curiosity of researchers.

[18]Distinguished research universities have all types of basic and applied research. Universities ranking lower in the national hierarchy predominantly conduct applied research or basic research in a late phase of the S-curve.

"organized creativity"[19] should be used only, if at all, for research conducted in the late phases of the S-curve. Pioneering, high-level basic research in the first phase of the S-curve is difficult to manage or coordinate because most upcoming research questions, methodological problems, and outcomes of research processes are not yet foreseeable. Applied researchers at least know their main research questions. In high-level basic research, it is possible to organize financial resources, research infrastructure, and safety regulations for some of the work processes but not the generation of new ideas, the outcome of interactions, or the behavior and scientific ethics of supervisors.[20]

The key issue for creating an optimal research culture is to find a balance between two positions that initially seem to contradict each other. On the one hand, certain institutional rules, guidelines, best practices, and evaluation mechanisms are indispensable for enforcing and sustaining high-quality research and teaching and for avoiding misbehavior and unethical practices. On the other hand, creative scholars need a great deal of academic freedom and personal autonomy to achieve outstanding scientific achievements. Without this freedom and autonomy, without being allowed phases of search and orientation that may fail (Weichhart, 2012a, p. 15), they will not have the courage to abandon outmoded thought structures; to forsake beloved scientific routines; or to criticize departmentally esteemed, but superficial, assessment tools[21] that may even dominate the core of their discipline. Creative scholars stand out not just by having new, original, and valuable research results but also by showing the courage to break through institutional constraints and

[19]Unfortunately, some authors in business studies, economics, and urban and economic geography do not use the term *creative* as it has been defined in the field of creativity studies (Boden, 1994, 2004). Florida (2002, 2005) has introduced a rather colloquial meaning of the term, but it has nothing to with the scientific definition of creativity as developed in psychology. Creativity research distinguishes between at least two types of creativity: psychological and historical. Psychological creativity is ubiquitous. A three-year-old child who discovers something new and valuable for him- or herself is creative. Every profession imaginable is at least occasionally creative in the psychological sense. Terms such as *creative industry* or *creative class* (Florida, 2002, 2005) are therefore contradictions in themselves. Whether a person, process, or product is *historically* creative can be judged only after the results have become apparent.

[20]Supervisors who are convinced that they are entitled to exploit the research results of their doctoral students can scarcely be organized in a way that changes their behavior. Established scientists who have neglected important research in neighboring disciplines or in languages than other English are hard to convince that they should change their gridlocked behavior.

[21]Wouters and Costas (2012) discuss the limitations of established and emerging impact metrics and the interplay of technologies of narcissism (self-assessment), information-filtering, and technologies of control that shape science in the digital age (see also Franzen, 2011, 2015, 2017; Meyer & Gupta, 1994). In digital societies the number of companies offering citation and impact measures (F1000, PaperCritic, Google Scholar, Google Citations, Arnetminer, JISC Open Citations, Mendeley, CiteUlike, Zotero, Scopus, Readermeter, ScienceCard, and many others) has increased dramatically, although most of these indices are extremely problematic from a methodological and statistical standpoint. Such organizations are booming because they oblige both the control mania of institutions and scholars' greed for reputation and status (for details see Franzen, 2011, 2015; Wouters & Costas, 2012).

expectations of their social environment. They do so not from ignorance of the rules or leading paradigms but rather from their personal experience that those rules and paradigms are no longer adequate (helpful) in their field and now only stifle creativity or obstruct efforts to advance scientific knowledge and insight. Being creative thus also means having courage to resist the mainstream in their discipline or least being able to endure for a while without the applause of colleagues.

External Factors of Influence

Universities have never been able to withdraw from the influence of national and local politics. Private universities may be less affected by the science policy of a state than public universities are, but no university is able to remain unaffected by politics. In some countries the decision-making community tends to distribute funding for public universities according to criteria shaped by industrial policy rather than science. Politicians often lack the willingness to acknowledge the difference in quality between public universities and to take account of it when allocating financial resources. Many times, the bargaining process in governments seems to be guided by the motto, "If your region receives money, then my region has to get some, too." This approach is defended with the argument that the allocation of research funding is intended to offset disequilibria in regional development. Such regional policy has had little success and has hurt the good universities in international competition for scientific standing.

Equally important are the attitudes of the local population (and their politicians) toward science in general and toward *their* university in particular, local principles of urban development planning, potential expansion of an existing campus, a university's relations to the region's economy, the support of a university by philanthropists, and many other factors.[22]

Paradoxically, the knowledge society of the twenty-first century is characterized by the simultaneous and unprecedented increase in both scientific knowledge and ignorance (not-knowing). The gaps between experts and nonexperts are widening dramatically. Leading scientific institutions must cope with more ignorance and lack of understanding from some local and national politicians than ever before. In the twenty-first century presidents of important nations deny climate change, and vice presidents deny evolution and adhere to creationism. In Heidelberg some politicians of the town council (e.g., members of the leftist party *Bunte Linke*) maintain that scientists no longer need face-to-face contacts because they communicate primarily via the Internet. Scientific institutions would, therefore, no longer need proximity to each other; they could be localized anywhere. The attitude of local politics and urban

[22]This volume's chapters by den Heijer and Curvelo Magdaniel; Etzkowitz; Goddard; Heffernan and Jöns; and Zillich focus on some of these external factors, so it is unnecessary to repeat their arguments here.

planning toward the needs of scientific institutions will become an increasingly important locational factor.

Another important location factor for universities are the activities of philanthropists in their region. According to Glückler and Ries (2012), Heidelberg University has long enjoyed privileged status in this respect because most of the local philanthropists identify with the oldest university in Germany even if they are not Heidelberg University alumni.

> Almost half of all the grants [of philanthropists and local foundations] went straight into academia and research (48%) and another 41% favoured charity and society. The area of culture and arts received no more than 11%, respectively. This distribution of philanthropy contradicts with general trends in the non-profit sector. At the national and European level philanthropic investments largely flow into charitable social welfare projects. Only 13% of all foundations in Germany declare science and academia as their primary target group (Bundesverband Deutscher Stiftungen, 2010). And according to a recent survey on European foundations in seven EU member states, only 3% of all grants benefit the realm of academia (European Foundation Centre, 2008). (Glückler & Ries, 2012, p. 522)

These statements are additional evidence that place matters and that the identification of high-level local decision-makers with *their* university is a valuable asset.

How Can a Knowledge Environment Affect Learning and Research Processes at a Given Place?

Universities Offer Unequal Learning and Research Opportunities

In principle, cognition and thought processes of individuals can take place everywhere. Why should a situation, setting, place, venue, spatial configuration, or environment affect cognition, motivation, learning, and research processes? The answer is relatively simple. Different places and work environments open different opportunities to learn, different possibilities for face-to-face contact, different resources for research and access to different networks, and different incentives and obstacles enabling or impeding specific learning processes and actions.

Universities vary in academic reputation, scientific achievements, the resources they offer, domain-specific authority, innovativeness, and the attractiveness of their location. Different departments (faculties, universities) consist of different research cultures. They train students in different fields and methodologies, exert different amounts of control, allow different degrees of autonomy, and present different truths. Some departments belong to the innovators or early adopters of a new research field, theory, or methodology; others, to the late adopters. Some of these actors do not even realize that they have overlooked important developments in their field of research.

At the top of the hierarchy are research universities and nonuniversity research institutions that enjoy an excellent international scientific reputation, have extensive research funds, and can attract the best talent worldwide. At the base are universities that have little or no research funding, are unable to engage scientifically renowned scholars, are not entitled to submit applications to important research foundations, have no right to award doctorates, primarily teach, and draw a large percentage of their students from a relatively small catchment area. The higher the scientific reputation of a university, the more it acts as a magnet attracting the best candidates available and the more impressive are its global networks and recruitment areas. Highly successful scholars can choose between many job opportunities[23] and vote with their feet.

Do outstanding candidates accept a university's call because the university has a sterling scientific reputation and provides an attractive research infrastructure? Or does the university have a high reputation and an attractive research infrastructure because it has had outstanding success for decades in appointing outstanding scholars? These questions indicate that two intertwined mechanisms are responsible for the quality and development of a knowledge environment. If a faculty or university is capable of recognizing[24] and attracting outstanding scholars over a lengthy period, then the probability is very high that these scholars will create the resources, incentives, and research cultures necessary for an attractive knowledge environment and an upward spiral of attractiveness. If many search and appointment procedures fail to produce the desired positive results, then the knowledge environment of that institution will deteriorate. Mediocre scholars will later recruit mediocre colleagues. If this downward spiral continues for long and falls below a certain level, the university will no longer receive applications from promising applicants.

Environment Related Mechanisms and Processes that Influence Learning and Research?

"Introduction to a field of knowledge is a kind of initiation that is performed by others" (Fleck, 1935/1979, p. 96). A teacher–student relationship is characterized by asymmetries of power, experience, and knowledge, but it is also subject to cultural influences.[25] Many doctoral students depend on financial resources and research

[23]This regularity does not apply to small disciplines (e.g., Assyriology, Medieval Latin) that have a notably restricted job market. Another consideration is the market cycles of disciplines that influence the number of appointments at universities.

[24]Daily experience and the history of science in universities prove that nowhere near all recruitment committees, faculties, and senates of universities are capable of detecting which of the many candidates are the most talented, competent, and promising. Committees and university presidents relying on superficial and questionable impact factors (e.g., the H-factor) have made terrible mistakes that have brought disaster upon the departments involved (see Weichhart, 2012a).

[25]It is quite interesting that in German-speaking countries the person who oversees or guides a graduate student's work on a doctoral dissertation is known as a doctor father (*Doktorvater*) or doctor mother (*Doktormutter*), whereas the common term is *supervisor* in Anglo-Saxon countries.

infrastructure to which senior scientists give them access. Doctoral students and junior scientists are subject to a certain measure of disciplinary control and are influenced by what the prominent members of their department or discipline view as established knowledge, appropriate methods acceptable experiments, and desirable forms of publication.

Which elements and processes of the environment are able to affect learning processes? Supervisors, reviewers, role models, critical audiences, institutional regulations, informal expectations of their social environment, and a multitude of other factors shape the goals, research interests, and fields of interaction of young scholars. "Actors perceive the meaning of institutions and infuse their actions with meaning based upon these perceptions" (Dacin et al. 2002, p. 47). Very few prospective scientists can ignore a dominant paradigm[26] or thought style and the expectations communicated in their social environment, at least not in the first years of their academic career. Therefore, students and young scholars operate in an environment of "preorganized knowledge" (Knoblauch, 1995, p. 15), they are being socialized in epistemic and scientific communities and specific "thought styles" (as defined by Fleck, 1935/1979; Trenn & Merton, 1979).

> A thought style functions by constraining, inhibiting, and determining the way of thinking. Under the influence of a thought style, one cannot think in any other way. It also excludes alternative modes of perception. Accordingly, no proper communication can arise between different thought styles. A thought style functions at such a fundamental level that the individual seems generally unaware of it. It exerts a compulsive force upon his thinking, so that he normally remains unconscious both of the thought style as such and of its constraining character. Yet such a style can be revealed in practice by an examination of how it is applied. The existence of stable thought collectives suggests the presence of a rather permanent thought style. (Trenn & Merton, 1979, p. 159)

According to Kuhn (1962), the process by which scientific disciplines change and develop is not characterized by a cumulation of knowledge but rather by fundamental crises and discontinuities, each of which eventually causes one paradigm to be replaced by another. There is broad agreement today, however, that several paradigms can coexist and compete with each other in a given period (for details see Schurz, 1998; Weichhart, 2012b, pp. 54–55).

Particular research groups and subdisciplines are repeatedly marked by the emergence of dogmatic convictions indisputable and seemingly unassailable among their adherents but sharply criticized by proponents of other paradigms. Paradigms thus not only reflect developmental processes of sciences, they also have great societal significance (Weichart, 2012b, p. 56). They promote cohesion and loyalty within research groups and can simultaneously exclude representatives of other paradigms.

[26] A paradigm is understood to mean a research-guiding perspective or viewpoint on which consensus builds for a given period and group of scholars. Paradigms offer scientific conceptions of the world that facilitate the formulation of and solution to problems in specific ways that the experts accept without question (Weichhart, 2012b, p. 54).

Established scientists who are convinced that their theoretical and methodological approaches are superior to those of others tend to exert a measure of control within their research group or intellectual community. Within a thought collective (e.g., among the members of a research group) doctoral students and junior scientists repeatedly confirm that their group is on the right track and will be successful with its research. "A special feeling of dependence therefore dominates all communication of thought within a collective" (Fleck, 1935/1979, p. 106).

> Between two members of the same thought collective on the same mental level, there is always a certain solidarity of thought in the service of a super-individual idea which causes both intellectual interdependence and a shared mood between the two individuals. (p. 51)

Intellectual solidarity and mutual confirmation reduce uncertainty about the value of one's research, but they also reduce the willingness of self-criticism and the ability to learn.

The Significance of Social Spaces

Livingstone (1995, 2000, 2002, 2003, 2005, 2010) has called attention to the critical significance of *social spaces*—especially *spaces of speech* and *spaces of reading*—for the generation and distribution of new research results. "[S]cience is made and remade by how it is talked about" (Livingstone, 2010, p. 18).

> What can and cannot be said in particular venues, how things are said, and the way they are heard are all implicated in the production of knowledge spaces. In different arenas there are protocols for speech management; there are subjects that are trendy and subjects that are taboo. In public spaces and in camera, in formal gatherings and in private salons, in conferences and consultations, in courtrooms and churches, in clinics and clubs—in all these venues different things are speakable (and unspeakable) about scientific claims. In every case the setting sets limits on what can be spoken; the social space conditions what is heard. . . . [T]he control of speech space is intimately connected with the maintenance of identity. Spaces of speech, of course, are also spaces of silence. There are always voices that are absent, or are not allowed to speak, or are denied access. In colonial societies, as Scott (1985) powerfully reminds his readers, the oppressed can rarely let their voices be heard. No doubt for different reasons, but with not dissimilar effects, those people marginalized in scientific debates find their voices unwelcome in science's privileged sites. (p. 16)

A textbook example often cited in science studies (e.g., Livingstone, 2003, 2010) concerns the reception of Darwin's main work, *On the Origin of Species* (1859). The book was interpreted differently in Belfast than in Edinburgh. In Charleston it was seen as an argument for racism; in Auckland, as one against it. "Darwinian language was embraced to a much greater degree in Canada than it was in Russia" (Livingstone, 2010, p. 10). "In Boston, the students of Louis Agassiz could talk about the new Darwinian theory only in secrecy for fear of their teacher's ire" (p. 17). Meusburger (2012b) has continued the narrative:

> Immediately after the book was published, the Heidelberg paleontologist Heinrich Georg Bronn (1800–1862), who was also the prorector of Heidelberg University at the time, translated *On the Origin of Species* into German, a version that appeared in 1860. Bronn enjoyed a good personal relationship with Darwin. Their bond, though, did not keep Bronn from reading Darwin's work very selectively and incorporating elements of his own perspectives into the translation. (p. 16)

Darwin's breakthrough in German-speaking countries was accomplished by Haeckel (1866, especially 1868). In a lecture entitled "Darwin's Theory of Evolution" and delivered in Stettin on September 19, 1863, at the meeting of the Society of German Naturalists and Physicians (founded in 1822), Haeckel made it clear that science was undergoing a profound change of paradigm.

The institutional, cultural, and social context in which a young person studies or a young scientist does research and teaches can be elemental in determining the research topics they will find of interest, the methodological skills they will acquire, the scientific networks they are able to join, the kind of critique and support they will encounter, and the interesting, unsolved problems they will notice. The knowledge environment of a university can have a bearing on whether and how soon new scientific concepts, practices, or technical innovations are accepted and acted upon and whether it is possible to discuss contested ideas, conduct expensive experiments, become part of important international networks, hear promptly of crucial developments, meet with agreement or criticism upon airing new ideas, or get away with straying from rigid regulations of a discipline's gatekeepers (Meusburger, 2012b, pp. 12–14).

Support from key figures in their discipline will affect the initiative and resilience of these young academics and will shape the subsequent perception they have of their roles as lecturers or researchers (Meusburger, 2012b, p. 14). Prominent scholars indisputably play a major part in shaping the development and careers of talented students. The history of science is replete with evidence of how a head start translates into a crucial advantage in a scholar's later career and creates important path dependencies. Just as certain time periods offer unique windows of opportunities, certain places and knowledge environments engender superior learning opportunities and role models, specific value systems, specific challenges, and outstanding prospects for professional careers.

Daily interaction with positive (or negative) role models and personal relationships have lost none of their significance in the age of the internet. On the contrary, the overwhelming flood of information and questionable publications have only increased the importance of face-to-face contacts with leading thinkers from the relevant disciplines (Meusburger, 2012b, p. 14).

Theoretical Concepts Contributing to the Understanding of How Knowledge Environments Function

"Knowledge Travels in a Selective Way"

The functioning of a knowledge environment depends, first, on the quality of the information exchange at the given place. This statement may seem trivial, but a local cluster of illustrious and globally networked scientists does not automatically equate with creative environments, knowledge exchange, scientific cooperation, or mutual inspiration across disciplines. Although proximity facilitates unplanned, spontaneous, face-to-face contact, it does not guarantee interaction and exchange of knowledge. Scientific insights and acquisition of scientific knowledge rely on dialogue, scholarly debate, and other forms of communication and interaction. What counts is whether and to what extent the scientists at a university interact and engage in cross-disciplinary discussion (Christmann, 2016; Keller, Knoblauch, & Reichertz, 2013; Knoblauch, 2016) and how they manage their social relations in that environment. A knowledge environment does not emerge until the actors communicate and interact with each other; come to grips with the structural conditions, institutional regulations, and organizational cultures in which they find themselves; and prove themselves willing and able to activate the locally available potential.

The key words are not *cluster* or *proximity* but *communicative interaction* and *ways to transcend the silo principle*. The silo mentality is a condition or attitude marked by an absence of open and cross-disciplinary communication and a disinterest or unwillingness to learn from other disciplines and thought communities.[27] It is combined with so-called boundary-work[28] and the persistence of staunch commitments to retaining and preserving established scientific concepts and methods of one's own research field. The difficulties posed by knowledge exchange between scholars of different disciplines, thought communities, or languages are vastly underestimated. The exchange of information between sender and recipient must overcome a number of filters. Even if a piece of information is understood by the recipient and might be valuable for his or her research, it may be rejected for various—also unscientific—reasons (see communication model in Meusburger, 2017, pp. 38–43).

International knowledge exchange, scientific cooperation, and scientific travel are influenced by even more factors than individual communication, including state

[27]In Frey and Iselin (2017) 71 authors discuss ideas and concepts used in (traditional) economics that need to be corrected or call for modification or elimination. Examples are the idea of *homo oeconomicus*, the way in which the term *rationality* is used, the assumed effect of prices on behavior, and the disregard of intrinsic motivation and job satisfaction. Most of this critique—especially of rationality and behavior—was formulated in social and behavioral science decades ago. From the perspective of science studies, the interesting question is why and how the silo mentality of economists was able to repel this critique from other disciplines for so many decades.

[28]Boundary-work is "a combination of rhetorical and social organizational devices to exclude some people and their knowledge claims from science" (Gieryn, 1983, p. 786).

fellowships, selective international research cooperation, old colonial relations, the quality of diplomatic relations, political goals, and shared denominations (Ellis, 2017; Heffernan, 1994; Heffernan & Jöns, 2013; Jöns, 2009, 2017; see also chapters by Hotson and Meusburger & Probáld in this volume).

Concepts Focusing on Structure, Agency, Interaction, and Interrelations

This chapter is not the place to discuss and value all the theoretical concepts that can help answer the question of how knowledge environments function. A brief overview will have to suffice. The first group of approaches centers on structure and agency and on relations between environment and behavior, social relations, interactions, and networks of the scholars working at a given place. Focusing on these facets, researchers study the "dialectical interplay between . . . *actions* (practices and structures), *meanings*, and *actors*" (Zilber, 2002, p. 235; emphasis in the original; see also Dacin et al., 2002, p. 48). The best known of these concepts are structuration theory (Giddens, 1979, 1984; Werlen, 1997a, 1997b, 2017), actor-network theory (Latour, 1987, 1999), communicative constructivism (Berger & Luckmann, 1967; Christmann, 2016; Keller et al., 2013), environmental psychology (Clitheroe, Stokols, & Zmuidzinas, 1998; Graumann, 1978; Kruse, Graumann, & Lantermann, 1990), concepts of appropriation (Graumann, 1996, 2002a, 2002b, 2002c; Graumann & Kruse, 2003), environmental phenomenology (Seamon, 1982, 1987), interaction analysis (Argyle, 1957, 1969, 1991; Scharpf, 2002), network analysis (for an overview see Glückler et al., 2017), and organizational and institutional theory (Lawrence et al., 2011; Mintzberg, 1979; Suddaby, 2010; Suddaby, Elsbach, Greenwood, Meyer, & Zilber, 2010). However, each of these theories or concepts leads to only a partial understanding of how knowledge environments function; each has its specific strengths and shortcomings. Many of these studies remain abstract and are place-blind, that is, they ignore not only the historicity of places but also what Raffestin (1989) calls the *geographicity* of knowledge production. Moreover, they do little to address the question of why norms, conventions, mutual expectations, and institutional logics may vary in the spatial dimension and, especially, why their correlations with other variables differ according to types of environments.

This is evident in institutional theory, which provides "new understandings of the manner in which institutions are created, transformed, and extinguished and the way in which institutional processes interact to affect institutional change" (Dacin et al., 2002, p. 45).

> The concept of institutional work insists on the need to consider the permanent recursive and dialectical interaction between agency and institutions. This invites researchers to not only account for the institutional embeddedness of actors but also for their capacity to reflect on this embeddedness, relate to their own self, and develop conscious intentionality. Agency is neither just an effect of the actors' institutional embeddedness nor isolated from this

> embeddedness. It is an ongoing activity whereby actors reflect on and strategically operate within the institutional context where they are embedded. (Lawrence et al., 2011, p. 55)

A major shortcoming of institutional theory is the fact that its macroperspectives largely exclude issues of power and dominance (see Hudson, Okhuysen, & Creed, 2015; Willmott, 2015).

> In most institutional theory, examination of the relatedness of knowledge and power is deemed to be largely irrelevant; the self-understanding of institutional theory as politically neutral is self-evident; and the silence of institutional theory on contentious issues, such as the institutionalization of domination, oppression, and exploitation in modern work organizations, passes unnoticed. (Willmott, 2015, p. 106)

Theoretical Concepts Focusing on Life Worlds and Holistic Interpretations

Because the numerous components of a knowledge environment are strongly interwoven and interconnected, a knowledge environment is more than the sum of its single elements. Studying venues, environments, and social spaces of academic knowledge generation opens new horizons for explaining interrelations and path dependencies of academic careers and academic achievements.

Given the complexity and historicity entailed, the students of knowledge environments are less concerned with the discovery of causalities in the sense of universal laws than with the causal reconstruction of the network of interactions (see Mayntz, 2002, p. 22). The main interest lies in describing and interpreting a system's functionality and in identifying the "normative epistemological and discursive conditions" (Welskopp, 2002, p. 78) under which scientific achievement was scored. What is sought in the empirical analysis of social macrophenomena—such as knowledge environments—is

> not abstraction and maximal simplification but rather specification and adequate complexity of the explanation. . . . Causal reconstruction is not about seeking statistical relations between variables but rather about explaining the macrophenomenon by identifying the processes and interdependencies involved in its having arisen. (Mayntz, 2002, p. 13)

Phenomenological approaches hold that cognitive processes should be thought of holistically and that one cannot describe place experience, knowledge environments, or life worlds by precisely dissecting them into constituent parts and then quantitatively measuring and modeling the significance of the discrete variables.

> Gestalt psychology views perception and other mental processes as holistic rather than atomistic in nature. That is, complex perceptual and cognitive patterns are grasped in their entirety as integral wholes (gestalts), rather than being pieced together from simpler sensations or "atoms of experience." A musical melody, for example, is not simply a summation of the individual notes that it contains. The melody is given in experience as a whole, and the character of each individual note is determined by its place within that whole. . . . (Schroeder, 2007, p. 295).

> The starting point for Fuller's (1990) phenomenological account of gestalt psychology is the life world of human experience. 'Life world' (*Lebenswelt*) is a term coined by Husserl and Schütz to denote the original, meaningful, everyday world in which people actually live. The task of phenomenological description and interpretation is to explicate hidden structures of life world meaning that hitherto have been lived only in an implicit, unreflective way. (Schroeder, 2007, p. 299)

Seamon (1987) characterizes environmental phenomenology as "striving for a holistic view of place, in which the various parts of environmental experience are understood as reciprocally related to each other and to the larger geographical whole to which they belong" (p. 20). He describes the phenomenological essence of place as "a psycho-social environmental whole larger than the sum of its parts" (p. 20; see also Keen, 1975; McCall, 1983; Schroeder, 2007, pp. 295–296; Seamon, 1982, 1987).

How Can the Quality and Influences of a Knowledge Environments be Verified?

Research about knowledge environments uses at least two different perspectives. The first one is the life world of individual scholars: How do individuals subjectively cognize, experience, and evaluate the knowledge environment of their research group, department, or university?

According to Welskopp (2002) "All structural interrelations must . . . be tracked through the consciousness of the participants as it were, although multiple discerning rounds of translation and decoding will be necessary" (p. 76). Depending on the objectives, cognitive abilities, motivation, scientific productiveness, and social relations of an individual, that person will find his or her own widely diverse positive role models or deterrent examples, selectively follow up on various ideas, and draw different lessons from personal experience but will also be able to deal flexibly with obstacles and personal liabilities.

Most scholars have been influenced by more than one knowledge environment during their academic career. Even so, scientific careers are not a "random sequence of events" but rather a "contingent series of events" (Welskopp, 2002, p. 79) that follow a definite logic and rest on particular mechanisms. In individual cases, certain events are therefore linked with each other, and a rationale can be found for the sequence of events (e.g., the steps in a scientist's career).

The career and mobility of a scholar represents a first important hint at his or her scientific reputation (for details see Pietsch, 2017; Weick, 1995). If two or more faculties have confirmed that the scientific achievements, personal competence, and other merits of a scholar warrant that person's appointment as professor, then the uncertainty about that persons scientific potential is much lower than if the confirmation were only issued by the same faculty where the candidate had studied and graduated. From experience it can be assumed that an academic who has worked successfully at more than one university has a greater wealth of experience, more extensive personal networks, a broader methodological spectrum, and a better academic reputation in general than one who has worked in only one institutional setting. The analysis of a scholar's professional career is also a precondition to understand his or her networks and scientific interests. There exist many studies about the professional careers of scholars (Meusburger & Schuch, 2010, 2012; Weick, 1995) but very few about the question of how they have personally experienced and evaluated the knowledge environments at various stages of their career (Demm, 2010; Hübner, 2010).

This approach involves the attempt to reconstruct the influences and watersheds that shaped the academic aspirations, research interests, methodological expertise, and career trajectory in various phases of the scholar's career. Most (senior) academics are able to recall with relative accuracy who decisively influenced and inspired them as students; what personal strengths and weaknesses their professors had; what intellectual standard the doctoral seminars had; the times, places, and people of the critical junctures in their professional success or failure; the time and circumstances relating to their development of a new scientific idea; the way in which their immediate social environment and their academic discipline responded to new ideas; and the reasons that particular research projects could not be pursued. The research taking this approach is based on interviews, memoires, and correspondence; the scholar's own account of his or her professional career; biographies and autobiographies; minutes of meetings; and reports by persons from the social context. This approach has a long tradition in the history of science but has rarely been used to describe knowledge environments.

The second perspective is the view from the outside. Researchers adopting it scrutinize a knowledge environment's particular elements, structures, and interrelations; analyze the long-term historical, political, and economic influences that have contributed to a department's or university's reputation and knowledge environment. They try to discern how a specific knowledge environment has come about over time and who has been and is responsible for specific positive or negative developments.

This approach can be further differentiated into two procedural methods. First, the quality of a knowledge environment of an academic institution can be studied for the scientific accomplishments, professional success, or academic reputation attained by its scholars or graduates in a given period. There are numerous questions to pursue in this regard. For instance, what pioneering publications have originated in a particular institute? What new ideas, methods, and theoretical concepts have come from a given department or faculty? How many of a department's graduates have been appointed to professorships at other universities or prominent positions outside academia? How many of a university's scholars have received outstanding scientific awards (e.g., Nobel Prize, Fields Medal,[29] or honorary doctorates)? How often have its members been engaged as reviewers by major institutions of research funding or in important evaluation procedures? This first procedure allows inferences about the

[29]International Medal for Outstanding Discoveries in Mathematics.

quality of the knowledge environment without opening the black box of causes and relations and without discussing the personal experiences of the given scholars. The method communicates the external perspective and is useful especially at the highest level of aggregation or with comparative studies that cannot delve into the biographies of numerous individual scholars. Figure 7.1 shows, for example, that former doctoral candidates and assistants of Heidelberg chemist Robert Bunsen came to occupy numerous chairs and other prominent positions in the chemistry departments throughout Europe (Eckart et al., 2012, p. 98). In addition, a number of prominent U.S. chemists had studied under Bunsen (Honeck & Meusburger, 2012).

If so many successful scientists hail from one department, one may legitimately conclude that a very stimulating and unusually creative knowledge environment had existed there during a certain period. Such recruitment of scientists to other universities not only promotes knowledge transfer and creates important networks but generally intensifies the sway that the home institutes have over their discipline. Figure 7.1 also suggests what an international power center Heidelberg must have been in the field of chemistry during the second half of the nineteenth century. Such internationally important nodes of scientific networks often dominate the review process in appointment procedures and have a strong influence on scientific journals.

A second approach turns attention to interactions, social relations, networks of scholars, and processes of appropriation. It raises a different set of questions. For example, how do (or did) scholars interact with each other? For what reasons have some scholars been included in and other excluded from important local networks or regular *jours fixes*? How much are the life worlds of various disciplines at a university or in a knowledge city communicatively interlinked? Which persons or departments are the ones bridging between disciplines or serving as the key nodes in networks? How do scholars get along with the basic local conditions and power structures under which they try to reach their objectives? Processes of appropriation should not be taken for granted.

Many universities offer local resources, support networks, and opportunities that some of their academics fail to use to their own advantage. Research equipment whose potential is not fully exploited, important data bases that go unused, outstanding publications that remain unread, and potential contacts that are not activated cannot have an effect on learning and research processes. The earlier mentioned silo mentality of some academics often prevents interaction and appropriation. Therefore, the second approach focuses on personal friendships and networks of scholars (Meusburger & Schuch, 2010), on scientific circles (Lepsius, 2012), joint publications, and various other forms of more or less regular interaction.

Conclusion

Research on knowledge environments will replace neither traditional science studies nor studies about the biographies and careers of individual scholars. However, there is a need for concepts focusing on social macrophenomena, holistic explanations,

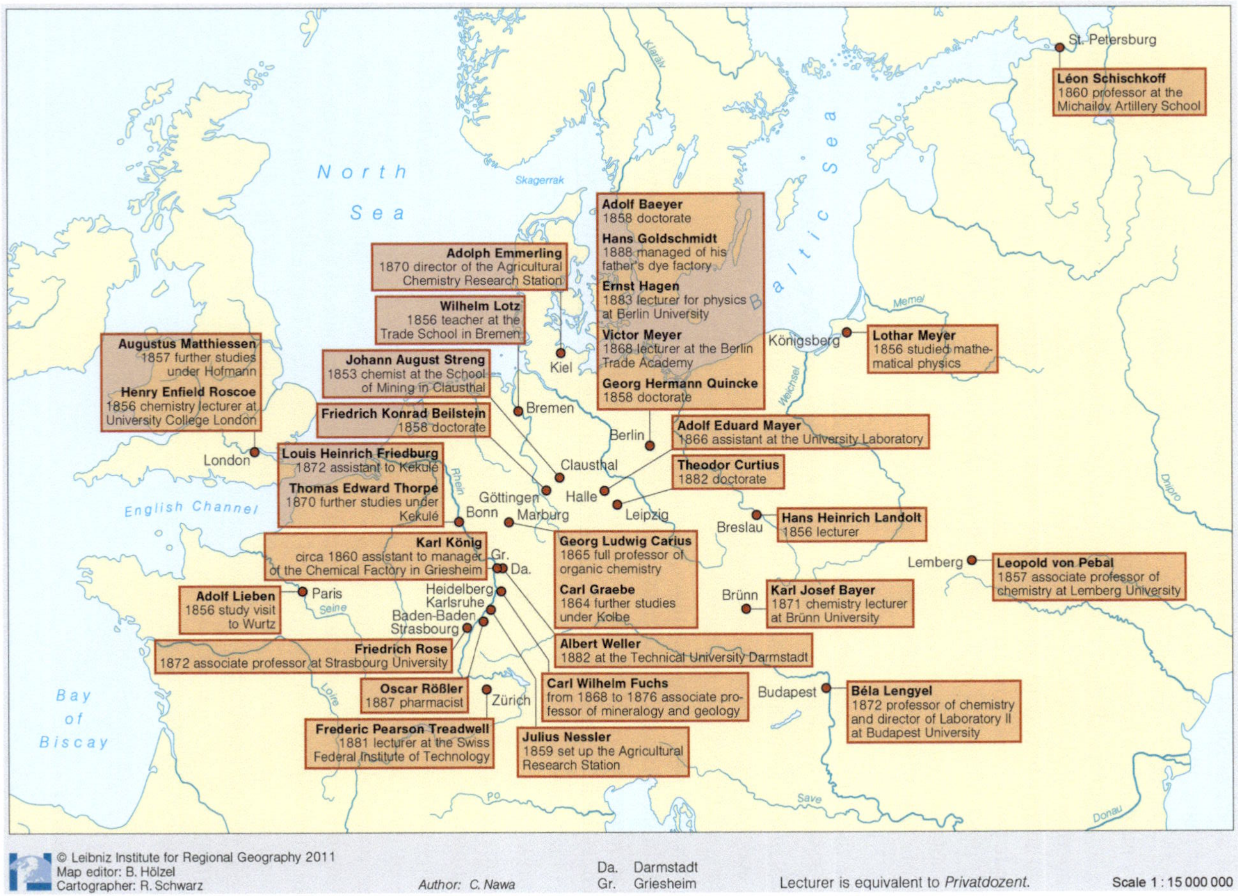

Fig. 7.1 Chairs and other prominent research positions of former doctoral candidates and assistants of Heidelberg chemist Robert Bunsen (1811–1899). Source: Eckart et al. (2012, p. 98). Reprinted with permission from Heidelberg University and IfL Leipzig.

and environmental clues in the sense of Gigerenzer and his coauthors (Gigerenzer & Gaissmaier, 2011, 2015; Gigerenzer & Selten, 2001). Authors studying knowledge environments should be able to deal with the tension between a knowledge environment's conditionality, which is in principle historical and local, and "efforts to formulate generalizing concepts and generic explanations" (Esser, 1999, p. 128; 2002).

With the knowledge-environments approach, it is possible to analyze and reconstruct the conditions under which outstanding scientific achievements were achieved or prevented in an institute or at a university over an extended period. One begins by analyzing the recruitment of professors (Pietsch, 2017; Weick, 1995) and continues by asking what human, financial, and infrastructural resources are available and what paradigms and research cultures are dominant. It is also necessary to inquire how hostile the assessment is toward scientific project proposals and research results at certain universities (or entire states) and to what extent standardized assessment tools have curtailed the originality and creativity of academics.[30]

> Project proposals and manuscript submissions alike have the best chances of being accepted if they conform as closely as possible to whatever the mainstream paradigm is, not contradicting it in any way and taking care not to suggest anything incommensurate with the reviewers' horizon of expectation defined by that paradigm. (Weichhart, 2012b, p. 20)

As with historical theories, knowledge environments can be credited with "a retrospective forecasting ability" (Welskopp, 2002, p. 83). In other words, it is possible to formulate a post hoc explanation of how relational networks between human and institutional agents have affected observable scientific practice (pp. 76–77). The reconstruction and description of knowledge environments after the events have taken place, however, is tied to "epistemological and discursive conditions of the present" (p. 78) and can therefore be revised the moment new information surfaces or scientific standards and criteria of judgment change.

Acknowledgments I thank my colleagues Bernhard Eitel (Heidelberg University), Joachim Funke (Heidelberg University), Peter Weichhart (University of Vienna), and Benno Werlen (University of Jena) for critical comments and valuable suggestions.

References

Amabile, T. M. (1979). Effects of external evaluation on artistic creativity. *Journal of Personality and Social Psychology, 37*, 221–233. doi:https://doi.org/10.1037//0022-3514.37.2.221

Amabile, T. M. (1983a). *The social psychology of creativity*. New York: Springer.

[30]"At times of the citation index, creativity, innovation, [and] genuine breakthroughs that risky hypotheses require necessarily get suppressed in favor of mainstream science" (Mocikat, 2009, p. 102; see also Liessmann, 2006, p. 131; Weichhart, 2012a, p. 21). Achieving high scores for the next assessment seems to have higher priority than the ambition to do innovative, but risky, research.

Amabile, T. M. (1983b). The social psychology of creativity: A componential conceptualization. *Journal of Personality and Social Psychology, 45*, 357–376. doi:https://doi.org/10.1037/0022-3514.45.2.357
Amabile, T. M. (1996). *Creativity in context*. New York: Westview Press.
Amabile, T. M. Conti, R. Coon, H. Lazenby, J., & Herron, M. (1996). Assessing the work environment for creativity. *Academy of Management Journal, 39*, 1154–1184. doi:https://doi.org/10.2307/256995
Amabile, T. M., Goldfarb, P., & Brackfield, S. C. (1990). Social influences on creativity: Evaluation, co-action, and surveillance. *Creativity Research Journal, 3*, 6–21. doi:https://doi.org/10.1080/10400419009534330
Amabile, T. M., & Gryskiewicz, N. D. (1989). The creative environment scales: The work environment inventory. *Creative Research Journal, 2*, 231–254. doi:https://doi.org/10.1080/10400418909534321
Argyle, M. (1957). *The scientific study of social behavior*. London: Methuen.
Argyle, M. (1969). *Social interaction*. London: Methuen.
Argyle, M. (1991). *Cooperation: The basis of sociability*. London: Routledge.
Barker, R. G. (1968). *Ecological psychology: Concepts and methods for studying the environment of human behavior*. Stanford: Stanford University Press.
Battilana, J. (2006). Agency and institutions: The enabling role of individual's social position. *Organization, 13*, 653–676. doi:https://doi.org/10.1177/1350508406067008
Bechtel, R. B. (1997). Environment and behavior: An introduction. Thousand Oaks: Sage.
Beck, U. (2005). *Was zur Wahl steht* [What the choice is]. Frankfurt am Main: Suhrkamp.
Berger, P. L., & Luckmann, T. (1967). *The social construction of reality: A treatise in the sociology of knowledge*. New York: Doubleday.
Berthoin Antal, A., Meusburger, P., & Suarsana, L. (Eds.). (2014). *Learning organizations: Extending the field*. Knowledge and Space: Vol. 6. Dordrecht, The Netherlands: Springer. doi:https://doi.org/10.1007/978-94-007-7220-5
Bitektine, A., & Miller, D. (2015). Methods, theories, data, and the social dynamics of organizational research. *Journal of Management Inquiry, 24*, 115–130. doi:https://doi.org/10.1177/1056492614546897
Boden, M. A. (1994). What is creativity? In M. A. Boden (Ed.), *Dimensions of creativity* (pp. 75–117). Cambridge, MA: MIT Press.
Boden, M. A. (Ed.). (2004). *The creative mind: Myths and mechanisms* (rev. and expanded 2nd ed.). London: Routledge.
Bundesverband Deutscher Stiftungen. (2010). Stiftungen in Zahlen: Errichtungen und Bestand rechtsfähiger Stiftungen des bürgerlichen Rechts in Deutschland im Jahr 2009 [Foundations in numbers: Creation and stock of foundations under civil law in Germany in 2009]. Berlin: Bundesverband Deutscher Stiftungen. Retrieved from https://www.stiftungen.org/uploads/tx_leonhardtfebecm/downloads/Jahresstatistik.pdf
Canter, D. (1977). *The psychology of places*. London: Architectural Press.
Christmann, G. B. (2016). Das theoretische Konzept der kommunikativen Raum(re)konstruktion [The theoretical concept of the (re)construction of communicative space]. In G. B. Christmann (Ed.), *Zur kommunikativen Konstruktion von Räumen: Theoretische Konzepte und empirische Analysen* (pp. 89–117). Wiesbaden: Springer VS.
Clegg, S. (2010). The state, power, and agency: Missing in action in institutional theory? *Journal of Management Inquiry, 19*, 4–13. doi:https://doi.org/10.1177/1056492609347562
Clitheroe, H. C., Stokols, D., & Zmuidzinas, M. (1998). Conceptualizing the context of environment and behavior. *Journal of Environmental Psychology, 18*, 103–112. doi:https://doi.org/10.1006/jevp.1998.0091
Collins, H. M. (1983). The sociology of scientific knowledge: Studies of contemporary science. *Annual Review of Sociology, 9*, 265–285. Retrieved from https://www.jstor.org/stable/2946066
Collins, H. M. (2008). Actors' and analysts' categories in the social analysis of science. In P. Meusburger, M. Welcker, & E. Wunder (Eds.), *Clashes of knowledge: Orthodoxies and*

heterodoxies in science and religion (pp. 101–110). Knowledge and Space: Vol. 1. Dordrecht, The Netherlands: Springer. doi:https://doi.org/10.1007/978-1-4020-5555-3_4
Collins, H. M., & Evans, R. (2002). The third wave of science studies: Studies of expertise and experience. *Social Studies of Science, 32*, 235–296. doi:https://doi.org/10.1177/0306312702032002003
Collins, H. M., & Pinch, T. (1998). *The Golem: What everyone should know about science* (New ed.). Cambridge, UK: Cambridge University Press.
Cummings, L. L. (1965). Organizational climates for creativity. *Academy of Management Journal, 8*, 220–227. doi:https://doi.org/10.2307/254790
Dacin, M. T., Goodstein, J., & Scott, W. R. (2002). Institutional theory and institutional change: Introduction to the special research forum. *Academy of Management Journal, 45*, 45–56. doi: https://doi.org/10.5465/AMJ.2002.6283388
Darwin, C. R. (1859). *On the origin of species by means of natural selection, or the preservation of favoured races in the struggle for life*. London: John Murray.
Demm, E. (2010). Alfred Weber (1868–1958): Machtkapital, Netzwerke und Lebensstil [Alfred Weber (1868–1958): Power capital, networks, and lifestyle]. In E. Demm & J. Suchoples (Eds.), *Akademische Lebenswelten: Habitus und Sozialprofil von Gelehrten im 19. und 20. Jahrhundert* (pp. 105–135). Frankfurt am Main: Peter Lang.
Denison, D. R. (1996). What is the difference between organizational culture and organizational climate? A native's point of view on a decade of paradigm wars. *Academy of Management Review, 21*, 619–654. doi:https://doi.org/10.2307/258997
DeRue, D. S. (2011). Adaptive leadership theory: Leading and following as a complex adaptive process. *Research in Organizational Behavior, 31*, 125–150. doi:https://doi.org/10.1016/j.riob.2011.09.007
Deutsche Forschungsgemeinschaft. (2015). *Förderatlas 2015: Kennzahlen zur öffentlich finanzierten Forschung in Deutschland* [Atlas of public funding, 2015: Indicators about publicly funded research in Germany]. Weinheim: Wiley-VCH Verlag.
Dörner, D. (1996). *The logic of failure: Recognizing and avoiding error in complex situations*. New York: Basic Books.
Dörner, D., & Funke, J. (2017). Complex problem solving: What it is and what it is not. *Frontiers in Psychology, 8*(1153), 1–11. doi:https://doi.org/10.3389/fpsyg.2017.01153
Eckart, W. U., Hübner, K., & Nawa, C. (2012). The rise of the natural sciences—Bunsen, Kirchhoff, and Helmholtz. In P. Meusburger & T. Schuch (Eds.), *Wissenschaftsatlas of Heidelberg University: Spatio-temporal relations of academic knowledge production* (pp. 96–99). Knittlingen: Bibliotheca Palatina.
Ellis, H. (2017). Collaboration and knowledge exchange between scholars in Britain and the Empire, 1830–1914. In H. Jöns, P. Meusburger, & M. Heffernan (Eds.), *Mobilities of knowledge* (pp. 141–155). Knowledge and Space: Vol. 10. Cham, Switzerland: Springer. doi:https://doi.org/10.1007/978-3-319-44654-7_8
Esser, H. (1999). *Situationslogik und Handeln* [Situational logic and action]. Soziologie: Spezielle Grundlagen: Vol. 1. Frankfurt am Main: Campus.
Esser, H. (2002). Was könnte man (heute) unter einer '*Theorie mittlerer Reichweite*' verstehen? [What could one take a "middle-range theory" to mean today?]. In R. Mayntz (Ed.), *Akteure—Mechanismen—Modelle: Zur Theoriefähigkeit makro-sozialer Analysen* (pp. 128–150). Schriften des Max-Planck-Instituts für Gesellschaftsforschung Köln: Vol. 42. Frankfurt am Main & New York: Campus.
European Foundation Centre. (2008). Foundations in the European Union: Facts and figures: Report on work by EFC Research task force. Brussels: European Foundation Centre.
Feld, S., & Basso, K. H. (1996). Introduction. In S. Feld & K. H. Basso (Eds.), *Senses of place* (pp. 3–11). Santa Fe: School of American Research Press.
Fiedler, K. (2004). Tools, toys, truisms, and theories: Some thoughts on the creative cycle of theory formation. *Personality and Social Psychology Review, 8*, 123–131. doi:https://doi.org/10.1207/s15327957pspr0802_5

Fleck, L. (1979). *Genesis and development of a scientific fact* (F. Bradley & T. J. Trenn, Trans.; T. J. Trenn & R. K. Merton, Eds.; with a Foreword by T. S. Kuhn). Chicago: University of Chicago Press. (Original work published 1935)

Fligstein, N. (2013). Understanding stability and change in fields. *Research in Organizational Behavior, 33*, 39–51. doi:https://doi.org/10.1016/j.riob.2013.10.005

Florida, R. (2002). *The rise of the creative class: And how it's transforming work, leisure, community and everyday life*. New York: Basic Books.

Florida, R. L. (2005). *Cities and the creative class*. New York: Routledge.

Franzen, M. (2011). *Breaking News: Wissenschaftliche Zeitschriften im Kampf um Aufmerksamkeit* [Breaking news: Scientific journals in the battle for attention]. Wissenschafts- und Technikforschung: Vol. 5. Baden-Baden: Nomos.

Franzen, M. (2015). Der Impact Faktor war gestern: Altmetrics und die Zukunft der Wissenschaft [The impact factor was yesterday: Altmetrics and the future of science]. *Soziale Welt, 66*, 225–242. doi:https://doi.org/10.5771/0038-6073-2015-2-225

Franzen, M. (2017). Digitale Resonanz: Neue Bewertungskulturen fordern die Wissenschaft heraus [Digital resonance: New evaluation cultures challenge science]. *WZB Mitteilungen, 155*, 30–33. Retrieved from https://www.wzb.eu/sites/default/files/publikationen/wzb_mitteilungen/30-33franzenwm155web.pdf

Frey, B., & Iselin, D. (Eds.). (2017). *Economic ideas you should forget*. Cham, Switzerland: Springer.

Fuller, A. R. (1990) *Insight into value: An exploration of the premises of a phenomenological psychology*. Albany: State University of New York Press.

Funke, J. (2012). Complex problem solving. In N. M. Seel (Ed.), *Encyclopedia of the sciences of learning*: Vol. 2 (pp. 682–685). Heidelberg: Springer.

Furnham, A. (Ed.). (1985). *Social behavior in context*. London: Allyn and Bacon.

Giddens, A. (1979). *Central problems in social theory: Action, structure and contradiction in social analysis*. London: MacMillan.

Giddens, A. (1984). *The constitution of society: Outline of the theory of structuration*. Cambridge, UK: Polity Press.

Gieryn, T. F. (1983). Boundary-work and the demarcation of science from non-science: Strains and interests in professional ideologies of scientists. *American Sociological Review, 48*, 781–795. doi:https://doi.org/10.2307/2095325

Gieryn, T. F. (2000). A space for place in sociology. *Annual Review of Sociology, 26*, 463–496. doi:https://doi.org/10.1146/annurev.soc.26.1.463

Gieryn, T. F. (2002). Give place a chance: Reply to Gans. *City & Community, 1*, 341–343. doi: https://doi.org/10.1111/1535-6841.00034

Gigerenzer, G., & Gaissmaier, W. (2011). Heuristic decision making. *Annual Review of Psychology, 62*, 451–482. doi:https://doi.org/10.1146/annurev-psych-120709-145346

Gigerenzer, G., & Gaissmaier, W. (2015). Decision making: Nonrational theories. In J. D. Wright (Ed.), *International encyclopedia of the social & behavioral sciences: Vol. 5* (2nd ed., pp. 911–916). Amsterdam: Elsevier. doi:https://doi.org/10.1016/B978-0-08-097086-8.26017-0W

Gigerenzer, G., & Selten, R. (Eds.). (2001). *Bounded rationality: The adaptive toolbox*. Cambridge, MA: MIT Press.

Glückler, J., Lazega, E., & Hammer, I. (Eds.). (2017). *Knowledge and networks*. Knowledge and Space: Vol. 11. Cham, Switzerland: Springer. doi:https://doi.org/10.1007/978-3-319-45023-0

Glückler, J., & Ries, M (2012). Why being there is not enough: Organized proximity in place-based philanthropy. *Service Industries Journal, 32*, 515–529. doi:https://doi.org/10.1080/02642069.2011.596534

Graumann, C. F.(Ed.). (1978). *Ökologische Perspektiven in der Psychologie* [Ecological perspectives in psychology]. Bern: Huber.

Graumann, C. F. (1983). On multiple realities. *International Social Science Journal, 35*, 309–321.

Graumann, C. F. (1996). Aneignung [Appropriation]. In L. Kruse, C.-F. Graumann, & E.-D. Lantermann (Eds.), *Ökologische Psychologie*: Ein Handbuch in Schlüsselbegriffen (pp. 124–130). Weinheim: Psychologie Verlags Union.

Graumann, C. F. (2002a). Auf dem Wege zu einer humanökologischen Psychologie: Zu Florian Kaisers Replik [On the way to human-ecological psychology: An answer to Florian Kaiser's reply]. *Umweltpsychologie, 6,* 82–83.

Graumann, C. F. (2002b). The phenomenological approach to people–environment studies. In R. B. Bechtel & A. Churchman (Eds.), *Handbook of environmental psychology* (pp. 95–113). New York: John Wiley & Sons.

Graumann, C. F. (2002c). Zwischen den Disziplinen: Dilemma und Chancen der Umweltpsychologie [Between the disciplines: Dilemma and opportunities of environmental geography]. *Umweltpsychologie, 6,* 154–161. Retrieved from http://www.umps.de/php/suche.php?heftid=11

Graumann, C. F., & Kruse, L. (1990). The environment: Societal construction and psychological problems. In H. T. Himmelweit & G. Gaskell (Eds.), *Societal psychology* (pp. 221–229). Newbury Park: Sage.

Graumann, C. F., & Kruse, L. (2003). Räumliche Umwelt: Die Perspektive der humanökologisch orientierten Umweltpsychologie [Spatial environment: The perspective of an environmental psychology oriented to human ecology]. In P. Meusburger & T. Schwan (Eds.), *Humanökologie: Ansätze zur Überwindung der Natur-Kultur-Dichotomie* (pp. 239–256). Erdkundliches Wissen: Vol. 135. Stuttgart: Steiner.

Haeckel, E. (1866). *Generelle Morphologie der Organismen* [General morphology of organisms]. 2 Vols. Berlin: Reimer.

Haeckel, E. (1868). *Natürliche Schöpfungsgeschichte: Gemeinverständliche wissenschaftliche Vorträge über die Entwickelungslehre im Allgemeinen und diejenige von Darwin, Goethe und Lamarck im Besonderen* [Natural history of creation in general and that of Darwin, Goethe, and Lamarck in particular]. Berlin: Reimer.

Harvey, D. (1969). *Explanations in geography.* London: Arnold.

Harvey, D. (2005). Space as a key word. In D. Harvey (Ed.), *Spaces of neoliberalization: Towards a theory of uneven geographical development* (pp. 93–115). Hettner-Lecture: Vol. 8. Stuttgart: Steiner.

Heffernan, M. (1994). A state scholarship: The political geography of French international science during the nineteenth century. *Transactions of the Institute of British Geographers, New Series, 19,* 21–45. Retrieved from https://www.jstor.org/stable/622444

Heffernan, M., & Jöns, H. (2013). Research travel and disciplinary identities in the University of Cambridge, 1885–1955. *The British Journal for the History of Science, 46*, 255–286. doi: https://doi.org/10.1017/S000708741200074X

Hennessey, B. A., & Amabile, T. M. (1988). The conditions of creativity. In R. J. Sternberg (Ed.), *The nature of creativity: Contemporary psychological perspectives* (pp. 11–38). Cambridge, UK: Cambridge University Press.

Honeck, M., & Meusburger, P. (2012). American students up to 1914. In P. Meusburger & T. Schuch (Eds.), *Wissenschaftsatlas of Heidelberg University: Spatio-temporal relations of academic knowledge production* (pp. 296–299). Knittlingen: Bibliotheca Palatina.

Hübner, K. (2010). *Gustav Robert Kirchhoff: Das gewöhnliche Leben eines außergewöhnlichen Mannes* [Gustav Robert Kirchhoff: The ordinary life of an extraordinary man]. Schriften/Archiv und Museum der Universität Heidelberg: Vol. 16. Heidelberg: Verlag Regionalkultur.

Hudson, B. A., Okhuysen, G. A., & Creed, W. E. D. (2015). Power and institutions: Stones in the road and some yellow bricks. *Journal of Management Inquiry, 24*, 233–238. doi:https://doi.org/10.1177/1056492614565240

Jöns, H. (2003). *Grenzüberschreitende Mobilität und Kooperation in den Wissenschaften Deutschlandaufenthalte US-amerikanischer Humboldt-Forschungspreisträger aus einer erweiterten Akteursnetzwerkperspektive* [Transnational mobility and cooperation in the sciences: U.S. Humboldt Research Award winners in Germany from an expanded actor-network

perspective]. Heidelberger Geographische Arbeiten: Vol. 116. Heidelberg: Selbstverlag des Geographischen Instituts.
Jöns, H. (2007). Transnational mobility and the spaces of knowledge production: A comparison of global patterns, motivations and collaborations in different academic fields. *Social Geography, 2,* 97–114. doi:https://doi.org/10.5194/sg-2-97-2007
Jöns, H. (2008). Academic travel from Cambridge University and the formation of centres of knowledge, 1885–1954. *Journal of Historical Geography, 34*, 338–362. doi:https://doi.org/10.1016/j.jhg.2007.11.006
Jöns, H. (2009). 'Brain circulation' and transnational knowledge networks: Studying long-term effects of academic mobility to Germany, 1954–2000. *Global Networks, 9*, 315–338. doi:https://doi.org/10.1111/j.1471-0374.2009.00256.x
Jöns, H. (2017). The University of Cambridge, academic expertise, and the British Empire, 1885–1962. In H. Jöns, P. Meusburger, & M. Heffernan (Eds.), *Mobilities of knowledge* (pp. 185–210). Knowledge and Space: Vol. 10. Cham, Switzerland: Springer. doi:https://doi.org/10.1007/978-3-319-44654-7_10
Keen, E. (1975). *A primer in phenomenological psychology*. New York: Holt, Rinehart & Winston.
Keller, R., Knoblauch, H., & Reichertz, J. (2013). Der Kommunikative Konstruktivismus als Weiterführung des Sozialkonstruktivismus: Eine Einführung in den Band [Communicative constructivism as a continuation of social constructionism: An introduction to the volume]. In R. Keller, H. Knoblauch, & J. Reichertz (Eds.), *Kommunikativer Konstruktivismus: Theoretische und empirische Arbeiten zu einem neuen wissenssoziologischen Ansatz* (pp. 9–21). Wiesbaden: Springer VS.
Kelly, G. A. (1991). *The psychology of personal constructs*. London: Routledge. (Original work published 1955)
Klüter, H. (1986). *Raum als Element sozialer Kommunikation* [Space as an element of social communication]. Gießener Geographische Schriften: Vol. 60. Gießen: Selbstverlag des Geographischen Instituts.
Klüter, H. (1999). Raum und Organisation [Space and organization]. In P. Meusburger (Ed.), *Handlungszentrierte Sozialgeographie: Benno Werlens Entwurf in kritischer Diskussion* (pp. 187–212). Erdkundliches Wissen: Vol. 130. Stuttgart: Steiner.
Klüter, H. (2003). Raum als Umgebung [Space as environment]. In P. Meusburger & T. Schwan (Eds.), *Humanökologie: Ansätze zur Überwindung der Natur-Kultur-Dichotomie* (pp. 217–238). Erdkundliches Wissen: Vol. 135. Stuttgart: Steiner.
Knoblauch, H. (1995). *Kommunikationskultur: Die kommunikative Konstruktion kultureller Kontexte* [Communication culture: The communicational construction of cultural contexts]. New York: de Gruyter.
Knoblauch, H. (2016). Über die kommunikative Konstruktion der Wirklichkeit [On the communicative construction of reality]. In G. B. Christmann (Ed.), *Zur kommunikativen Konstruktion von Räumen: Theoretische Konzepte und empirische Analysen* (pp. 29–53). Wiesbaden: Springer VS.
Knorr-Cetina, K. (1999). *Epistemic cultures: How the sciences make knowledge*. Cambridge, MA: Harvard University Press.
Kruse, L. (1986). Drehbücher für Verhaltensschauplätze—oder Scripts für Settings [Scripts for behavioral scenes—or scripts for settings]. In G. Kaminski (Ed.), *Ordnung und Variabilität im Alltagsgeschehen* (pp. 135–153). Göttingen: Hogrefe.
Kruse, L., Graumann, C. F., & Lantermann, E-D. (Eds.). (1990). *Ökologische Psychologie: Ein Handbuch in Schlüsselbegriffen* [Ecological psychology: A handbook in key words]. Munich: Psychologie Verlags-Union.
Kuhn, T. S. (1962). *The structure of scientific revolutions*. Chicago: University of Chicago Press.
Latour, B. (1987). *Science in action: How to follow scientists and engineers through society*. Cambridge, MA: Harvard University Press.
Latour, B. (1999). *Pandora's hope: Essays on the reality of science studies*. Cambridge, MA: Harvard University Press.

Lawrence, T. B., Suddaby, R., & Leca B. (2011). Institutional work: Refocusing institutional studies of organization. *Journal of Management Inquiry, 20,* 52–58. doi:https://doi.org/10.1177/1056492610387222

Lefebvre, H. (1991). *The production of space* (D. Nicholson-Smith, Trans.). Oxford, UK: Basil Blackwell. (Original work published 1974)

Lepsius, R. M. (2012). Intellectual circles at the beginning of the 20th century. In P. Meusburger & T. Schuch (Eds.), *Wissenschaftsatlas of Heidelberg University: Spatio-temporal relations of academic knowledge production* (pp. 112–113). Knittlingen: Bibliotheca Palatina.

Lind, E. A., & van den Bos, K. (2013). Freeing organizational behavior from inhibitory constraints. *Research in Organizational Behavior, 33*, 79–95. doi:https://doi.org/10.1016/j.riob.2013.10.001

Liessmann, K. P. (2006). Theorie der Unbildung: Die Irrtümer der Wissensgesellschaft [Theory on the lack of education: The errors of the knowledge society]. Vienna: Zsolnay.

Lippuner, R. (2005). *Raum—Systeme—Praktiken: Zum Verhältnis von Alltag, Wissenschaft und Geographie* [Space, systems, practices: On the relation between everyday life, science, and geography]. Sozialgeographische Bibliothek: Vol. 2. Stuttgart: Steiner.

Lippuner, R., & Lossau, J. (2004). In der Raumfalle: Eine Kritik des spatial turn in den Sozialwissenschaften [Caught in the space trap: A critique of the spatial turn in the social sciences]. In G. Mein & M. Riegler-Ladich (Eds.), *Soziale Räume und kulturelle Praktiken: Über den strategischen Gebrauch von Medien* (pp. 47–64). Bielefeld: transcript.

Livingstone, D. N. (1995). The spaces of knowledge: Contributions towards a historical geography of science. *Environment and Planning D: Society and Space, 13*, 5–34. doi:https://doi.org/10.1068/d130005

Livingstone, D. N. (2000). Making space for science. *Erdkunde, 54*, 285–296. doi:https://doi.org/10.3112/erdkunde.2000.04.01

Livingstone, D. N. (2002). Knowledge, space and the geographies of science. In D. N. Livingston (Ed.), *Science, space and hermeneutics* (pp. 7–40). Hettner-Lecture: Vol. 5. Heidelberg: Department of Geography, Heidelberg University.

Livingstone, D. N. (2003). *Putting science in its place: Geographies of scientific knowledge.* Chicago: University of Chicago Press.

Livingstone, D. N. (2005). Text, talk and testimony: Geographical reflections on scientific habits—An afterword. *British Journal for the History of Science, 38*, 93–100. doi:https://doi.org/10.1017/S000708740400648X

Livingstone, D. N. (2010). Landscapes of knowledge. In P. Meusburger, D. Livingstone, & H. Jöns (Eds.), *Geographies of science* (pp. 3–22). Knowledge and Space: Vol. 3. Dordrecht, The Netherlands: Springer. doi:https://doi.org/10.1007/978-90-481-8611-2_1

Mager, C. (2012). Heidelberg Nobel Prize winners. In P. Meusburger & T. Schuch (Eds.), *Wissenschaftsatlas of Heidelberg University: Spatio-temporal relations of academic knowledge production* (pp. 250–253). Knittlingen: Bibliotheca Palatina.

Manzo, L. C. (2005). For better or worse: Exploring multiple dimensions of place meaning. *Journal of Environmental Psychology, 25,* 67–86. doi:https://doi.org/10.1016/j.jenvp.2005.01.002

Marquis, C., & Battilana, J. (2009). Acting globally but thinking locally? The enduring influence of local communities on organizations. *Research in Organizational Behavior, 29,* 283–302. doi:https://doi.org/10.1016/j.riob.2009.06.001

Massey, D. B. (1985). New directions in space. In D. Gregory & J. Urry (Eds.), *Social relations and spatial structures* (pp. 9–19). New York: St. Martin's Press.

Massey, D. B. (1999a). Imagining globalisation: Power-geometries of time-space. In D. B. Massey (Ed.), *Power-geometries and the politics of space-time* (pp. 9–23). Hettner-Lecture: Vol. 2. Heidelberg: Department of Geography, Heidelberg University.

Massey, D. B. (1999b). Philosophy and politics of spatiality: Some considerations. In D. B. Massey, *Power-geometries and the politics of space-time* (pp. 27–42). Hettner-Lecture: Vol. 2. Heidelberg: Department of Geography, Heidelberg University.

Massey, D. B. (2005). *For space*. London: Sage.

Matthiesen, U. (2006). Raum und Wissen: Wissensmilieus und Knowledge Scapes als Inkubatoren für zukunftsfähige stadtregionale Entwicklungsdynamiken? [Space and Knowledge: Knowledge milieus and knowledge scapes as incubators for sustainable urban development dynamics]. In D. Tänzler, H. Knoblauch, & H.-G. Soeffner (Eds.), *Zur Kritik der Wissensgesellschaft* (pp. 155–188). Erfahrung—Wissen—Imagination: Vol. 12. Constance: UVK.

Matthiesen, U. (2007). Wissensmilieus und KnowledgeScapes [Knowledge milieus and knowledgescapes]. In R. Schützeichel (Ed.), *Handbuch Wissenssoziologie und Wissensforschung* (pp. 679–693). Erfahrung—Wissen—Imagination: Vol. 15. Constance: UVK.

Mayntz, R. (2002). Zur Theoriefähigkeit makro-sozialer Analysen [The theory-generating ability of macrosocial analyses]. In R. Mayntz (Ed.), *Akteure—Mechanismen—Modelle: Zur Theoriefähigkeit makro-sozialer Analysen* (pp. 7–43). Schriften des Max-Planck-Instituts für Gesellschaftsforschung Köln: Vol. 42. Frankfurt am Main & New York: Campus.

McCall, R. J. (1983). *Phenomenological psychology: An introduction*. Madison: The University of Wisconsin Press.

Meusburger, P. (2008). The nexus of knowledge and space. In P. Meusburger, M. Welker, & E. Wunder (Eds.), *Clashes of knowledge: Orthodoxies and heterodoxies in science and religion* (pp. 35–90). Knowledge and Space: Vol. 1. Dordrecht, The Netherlands: Springer. doi:https://doi.org/10.1007/978-1-4020-5555-3_2

Meusburger, P. (2009). Milieus of creativity: The role of places, environments, and spatial contexts. In P. Meusburger, J. Funke, & E. Wunder (Eds.), *Milieus of creativity: An interdisciplinary approach to spatiality of creativity* (pp. 97–153). Knowledge and Space: Vol. 2. Dordrecht: Springer. doi:https://doi.org/10.1007/978-1-4020-9877-2_7

Meusburger, P. (2012a). Heidelberg: Relations between a town and its university. In P. Meusburger & T. Schuch (Eds.), *Wissenschaftsatlas of Heidelberg University: Spatio-temporal relations of academic knowledge production* (pp. 18–35). Knittlingen: Bibliotheca Palatina.

Meusburger, P. (2012b). Wissenschaftsatlas of Heidelberg University—An introduction. In P. Meusburger & T. Schuch (Eds.), *Wissenschaftsatlas of Heidelberg University: Spatio-temporal relations of academic knowledge production* (pp. 12–17). Knittlingen: Bibliotheca Palatina.

Meusburger, P. (2015). Knowledge environments in universities. *Hungarian Geographical Bulletin, 64*, 265–279. doi:https://doi.org/10.15201/hungeobull.64.4.1

Meusburger, P. (2016). Zur Bedeutung und Wirkung von Wissensmilieus [The importance and effectiveness of knowledge milieus]. In A. Froese, D. Simon, & J. Böttcher (Eds.), *Sozialwissenschaften und Gesellschaft: Neue Verortungen von Wissenstransfer* (pp. 263–306). Bielefeld: transcript.

Meusburger, P. (2017). Spatial mobility of knowledge: Communicating different categories of knowledge. In H. Jöns, P. Meusburger, & M. Heffernan (Eds.), *Mobilities of knowledge* (pp. 23–50). Knowledge and Space: Vol. 10. Cham, Switzerland: Springer. doi:https://doi.org/10.1007/978-3-319-44654-7_2

Meusburger, P., & Schuch, T. (2010). Karrieren, soziale Herkunft und räumliche Mobilität der Heidelberger Professoren 1803–1932 [Careers, social origin, and spatial mobility of Heidelberg University's professors, 1803–1932]. In E. Demm & J. Suchoples (Eds.), *Akademische Lebenswelten. Habitus und Sozialprofil von Gelehrten im 19. und 20. Jahrhundert* (pp. 217–249). Frankfurt am Main: Peter Lang.

Meusburger, P., & Schuch, T. (Eds.). (2012). *Wissenschaftsatlas of Heidelberg University: Spatio-temporal relations of academic knowledge production*. Knittlingen: Bibliotheca Palatina.

Meusburger, P., & Werlen, B. (2017). Knowledge, action, and space: An introduction. In P. Meusburger, B. Werlen, & L. Suarsana (Eds.), *Knowledge and action* (pp. 1–30). Knowledge and Space: Vol. 9. Cham, Switzerland Springer. doi:https://doi.org/10.1007/978-3-319-44588-5_1

Meyer, M. W., & Gupta, V. (1994). The performance paradox. *Research in Organizational Behavior, 16*, 309–369.

Mintzberg, H. (1979). *The structuring of organizations: A synthesis of the research.* Englewood Cliffs: Prentice Hall.

Mocikat, R. (2009). Die Diktatur der Zitatenindizes: Folgen für die Wissenskultur. Reaktion auf V. Winiwarter und H.-J. Luhmann. 2009. Die Vermessung der Wissenschaft. GAIA 18/1:1 [The tyranny of citation indices: Implications for the culture of knowledge. Reaction to V. Winiwarter and H.-J. Luhmann. 2009. The survey of science. *GAIA, 18*/1:1]. *GAIA, 18,* 100–103. Retrieved from http://www.adawis.de/admin/upload/navigation/data/GAIA.pdf

Mußgnug, D. (2012). Dismissal and persecution of lecturers under National Socialism. In P. Meusburger & T. Schuch (Eds.), *Wissenschaftsatlas of Heidelberg University: Spatio-temporal relations of academic knowledge production* (pp. 136–137). Knittlingen: Bibliotheca Palatina.

Neumaier, M. (2012). Destruction and rebuilding: The university between 1689 and 1803. In P. Meusburger & T. Schuch (Eds.), *Wissenschaftsatlas of Heidelberg University: Spatio-temporal relations of academic knowledge production* (pp. 72–75). Knittlingen: Bibliotheca Palatina.

Pfetsch, F. R. (1974). *Zur Entwicklung der Wissenschaftspolitik in Deutschland 1750–1914* [On the development of science policies in Germany, 1750–1914]. Berlin: Duncker & Humblot.

Pietsch, T. (2017). Geographies of selection: Academic appointments in the British academic world, 1850–1939. In H. Jöns, P. Meusburger, & M. Heffernan (Eds.), *Mobilities of knowledge* (pp. 157–183). Knowledge and Space: Vol. 10. Cham, Switzerland: Springer. doi:https://doi.org/10.1007/978-3-319-44654-7_9

Raffestin, C. (1989). Théories du reél et de geographicité [Theories of the real and of geographicity]. *Espaces Temps, 40,* 26–31. Retrieved from http://www.persee.fr/doc/espat_0339-3267_1989_num_40_1_3454

Rotzoll, M. (2012). Plagues and wars in the early history of Heidelberg University. In P. Meusburger & T. Schuch (Eds.), *Wissenschaftsatlas of Heidelberg University: Spatio-temporal relations of academic knowledge production* (pp. 52–53). Knittlingen: Bibliotheca Palatina.

Rowles, G. D. (2008a). The meaning of place. In E. B. Crepeau, E. S. Cohn, & B. A. Boyt Schell (Eds.), *Willard and Spackman's occupational therapy* (11th ed., pp. 80–89). Philadelphia: Wolters Kluwer/Lippincott Williams & Wilkins.

Rowles, G. D. (2008b). Place in occupational science: A life course perspective on the role of environmental context in the quest for meaning. *Journal of Occupational Science, 15,* 127–135. doi:https://doi.org/10.1080/14427591.2008.9686622

Runco, M. A. (Ed.). (1994). *Problem finding, problem solving, and creativity.* Norwood: Ablex.

Runco, M. A., & Okuda, S. M. (1988). Problem discovery, divergent thinking, and the creative process. *Journal of Youth and Adolescence, 17,* 211–220. doi:https://doi.org/10.1007/BF01538162

Scannel, L., & Gifford, R. (2010). Defining place attachment: A tripartite organizing framework. *Journal of Environmental Psychology, 30,* 1–10. doi:https://doi.org/10.1016/j.jenvp.2009.09.006

Scharpf, F. W. (2002). Kontingente Generalisierung in der Politikforschung [Contingent generalization in political science]. In R. Mayntz (Ed.), *Akteure—Mechanismen—Modelle: Zur Theoriefähigkeit makro-sozialer Analysen* (pp. 213–235). Schriften des Max-Planck-Instituts für Gesellschaftsforschung Köln: Vol. 42. Frankfurt am Main & New York: Campus.

Schmid, C. (2005). *Stadt, Raum und Gesellschaft: Henri Lefebvre und die Theorie der Produktion des Raumes* [City, space, and society: Henri Lefebvre and the theory of the production of space]. Sozialgeographische Bibliothek: Vol. 1. Stuttgart: Steiner.

Schroeder, H. W. (2007). Place experience, gestalt, and the human–nature relationship. *Journal of Environmental Psychology, 27,* 293–309. doi:https://doi.org/10.1016/j.jenvp.2007.07.001

Schurz, G. (1998). Koexistenzweisen rivalisierender Paradigmen: Eine begriffsklärende und problemtypologisierende Studie [Coexistence of rival paradigms: A study on terms and types of problems]. In G. Schurz & P. Weingartner (Eds.), *Koexistenz rivalisierender Paradigmen: Eine post-kuhnsche Bestandsaufnahme zur Struktur gegenwärtiger Wissenschaft* (pp. 1–51). Opladen: Westdeutscher Verlag.

Scott, J. C. (1985). *Weapons of the weak: Everyday forms of peasant resistance*. New Haven: Yale University Press.

Seamon, D. (1982). The phenomenological contribution to environmental psychology. *Journal of Environmental Psychology, 2*, 119–140. doi:https://doi.org/10.1016/S0272-4944(82)80044-3

Seamon, D. (1987). Phenomenology and environment-behavior research. In E. H. Zube & G. T. Moore (Eds.), *Advances in environment, behavior, and design: Vol. 1* (pp. 3–27). New York: Plenum Press.

Soja, E. W. (1980). The socio-spatial dialectic. *Annals of the Association of American Geographers, 70*, 207–225. doi:https://doi.org/10.1111/j.1467-8306.1980.tb01308.x

Soja, E. W. (1985). The spatiality of social life: Towards a transformative retheorisation. In D. Gregory & J. Urry (Eds.), *Social relations and spatial structures* (pp. 90–127). London: Macmillan.

Sternberg, R. J., & Lubart, T. I. (1991). An investment theory of creativity and its development. *Human Development, 34*, 1–31. doi:https://doi.org/10.1159/000277029

Sternberg, R. J., & Lubart T. I. (1999). The concept of creativity: Prospects and paradigms. In R. J. Sternberg (Ed.), *Handbook of creativity* (pp. 3–15). Cambridge, UK: Cambridge University Press.

Suddaby, R. (2010). Challenges for institutional theory. *Journal of Management Inquiry, 19*, 14–20. doi:https://doi.org/10.1177/1056492609347564

Suddaby, R., Elsbach, K. D., Greenwood, R., Meyer, J. W., & Zilber T. B. (2010). Organizations and their institutional environments: Bringing meaning, values, and culture back in: Introduction to the special research forum. *Academy of Management Journal, 53*, 1234–1240. doi: https://doi.org/10.5465/AMJ.2010.57317486

Suddaby, R., & Greenwood, R. (2005). Rhetorical strategies of legitimacy. *Administrative Science Quarterly, 50*, 35–67. doi:https://doi.org/10.2189/asqu.2005.50.1.35

Trenn, T. J., & Merton, R. K. (1979). Commentary and annotation. In L. Fleck, *Genesis and development of a scientific fact* (pp. 149–165). Chicago: University of Chicago Press. (Original work published 1935)

Weichhart, P. (1996). Die Region: Chimäre, Artefakt oder Strukturprinzip sozialer Systeme? [The region: Chimera, artifact, or structural principle of social systems]. In G. Brunn (Ed.), *Region und Regionsbildung in Europa: Konzeptionen der Forschung und empirische Befunde* (pp. 25–43). Schriftenreihe des Instituts für Europäische Regionalforschung: Vol. 1. Baden-Baden: Nomos.

Weichhart, P. (1999). Die Räume zwischen den Welten und die Welt der Räume: Zur Konzeption eines Schlüsselbegriffs der Geographie [The spaces between the worlds and the world of spaces: On the inception of a key concept of geography]. In P. Meusburger (Ed.), *Handlungszentrierte Sozialgeographie: Benno Werlens Entwurf in kritischer Diskussion* (pp. 67–94). Erdkundliches Wissen: Vol. 130. Stuttgart: Steiner.

Weichhart, P. (2003). Gesellschaftlicher Metabolismus und Action Settings: Die Verknüpfung von Sach- und Sozialstrukturen im alltagsweltlichen Handeln [Social metabolism and action settings: The link between technical and social structures in everyday action]. In P. Meusburger & T. Schwan (Eds.), *Humanökologie: Ansätze zur Überwindung der Natur-Kultur-Dichotomie* (pp. 15–44). Erdkundliches Wissen: Vol. 135. Stuttgart: Steiner.

Weichhart, P. (2012a). "Slow Science" versus Exzellenzstalinismus: Vom Nutzen wissenschaftlicher Reflexion abseits der Evaluierungsbuchhaltung ["Slow science" versus excellence Stalinism: About the usefulness of scientific reflections beyond evaluation bookkeeping]. In M. M. Seebacher (Ed.), *Raumkonstruktionen in der Geographie: Eine paradigmenspezifische Darstellung gesellschaftlicher und fachspezifischer Konstruktions-, Rekonstruktions- und Dekonstruktionsprozesse von „Räumlichkeit“* (pp. 7–38). Abhandlungen zur Geographie und

Regionalforschung: Vol. 14. Vienna: Department of Geography and Regional Research, University of Vienna.
Weichhart, P. (2012b). Wie „funktioniert“ ein Paradigma? [How does a paradigm “function”?]. In H. Fassmann & T. Glade (Eds.), *Geographie für eine Welt im Wandel* (pp. 53–64). Göttingen: Vienna University Press.
Weick, C. M. (1995). *Räumliche Mobilität und Karriere: Eine individualstatistische Analyse der baden-württembergischen Universitätsprofessoren unter besonderer Berücksichtigung demographischer Strukturen* [Spatial mobility and career: A statistical analysis of university professors in Baden-Württemberg with special focus on demographic structures]. Heidelberger Geographische Arbeiten: Vol. 101. Heidelberg: Department of Geography, Heidelberg University.
Welskopp, T. (2002). Die Theoriefähigkeit der Geschichtswissenschaft [The theory-generating ability of historical science]. In R. Mayntz (Ed.), *Akteure—Mechanismen—Modelle: Zur Theoriefähigkeit makro-sozialer Analysen* (pp. 61–90). Schriften des Max-Planck-Instituts für Gesellschaftsforschung Köln: Vol. 42. Frankfurt am Main: Campus.
Werlen, B. (1983). Zur empirischen Forschung in der Sozialgeographie [Empirical research in social geography]. *Veröffentlichungen der Schweizerischen Geographischen Kommission, 7,* 11–22.
Werlen, B. (1987). *Gesellschaft, Handlung und Raum: Grundlagen handlungstheoretischer Sozialgeographie* [Society, action, and space: Principles of social geography based on action theory] (3rd ed.). Erdkundliches Wissen: Vol. 89. Stuttgart: Steiner.
Werlen, B. (1993). *Society, action and space: An alternative human geography* (G. Walls, Trans., with a preface by A. Giddens). London: Routledge.
Werlen, B. (1995). *Sozialgeographie alltäglicher Regionalisierungen: Bd. 1. Zur Ontologie von Gesellschaft und Raum* [Social geography of everyday regionalizations: Vol. 1. On the ontology of society and space]. Erdkundliches Wissen: Vol. 116. Stuttgart: Steiner.
Werlen, B. (1996). Die Geographie globalisierter Lebenswelten [The geography of globalized life worlds]. *Österreichische Zeitschrift für Soziologie, 21*(2), 97–128.
Werlen, B. (1997a). Social geography. In L. Embree, J. Kokelmans, & R. Zaner (Eds.), *The encyclopedia of phenomenology* (pp. 645–650). Contributions to Phenomenology: Vol. 18. Dordrecht: Kluwer.
Werlen, B. (1997b). *Sozialgeographie alltäglicher Regionalisierungen: Bd. 2.* Globalisierung, Region und Regionalisierung [Social geography of everyday regionalizations: Vol. 2. Globalization, region, and regionalization]. Erdkundliches Wissen: Vol. 119. Stuttgart: Steiner.
Werlen, B. (2010a). *Gesellschaftliche Räumlichkeit: Bd. 1.* Orte der Geographie [Societal spatiality: Vol. 1. Places of geography]. Stuttgart: Steiner.
Werlen, B. (2010b). *Gesellschaftliche Räumlichkeit: Bd. 2.* Konstruktion geographischer Wirklichkeiten [Societal spatiality: Vol. 2. Construction of geographical realities]. Stuttgart: Steiner.
Werlen, B. (2017). Action, knowledge, and social relations of space. In P. Meusburger, B. Werlen, & L. Suarsana (Eds.), *Knowledge and action* (pp. 1–30). Knowledge and Space: Vol. 9. Cham, Switzerland: Springer. doi:https://doi.org/10.1007/978-3-319-44588-5_2
Willmott, H. (2015). Why institutional theory cannot be critical. *Journal of Management Inquiry, 24,* 105–111. doi:10.1177/1056492614545306
Withers, C. W. J. (2001). *Geography, science and national identity: Scotland since 1520.* Cambridge, UK: Cambridge University Press.
Withers, C. W. J. (2002). The geography of scientific knowledge. In N. A. Rupke (Ed.), *Göttingen and the development of the natural sciences* (pp. 9–18). Göttingen: Wallstein.
Withers, C. W. J. (2010). Geographies of science and public understanding? Exploring the reception of the British Association for the Advancement of Science in Britain and in Ireland, c.1845–1939. In P. Meusburger, D. Livingstone, & H. Jöns (Eds.), *Geographies of science* (pp. 185–197). Knowledge and Space: Vol. 3. Dordrecht, The Netherlands: Springer. doi:https://doi.org/10.1007/978-90-481-8611-2_10

Withers, C. W. J., Higgitt, R., & Finnegan, D. A. (2008). Historical geographies of provincial science: Themes in the setting and reception of the British Association for the Advancement of Science in Britain and Ireland, 1831–c.1939. *British Journal for the History of Science, 41,* 385–415.

Wouters, P., & Costas, R. (2012). *Users, narcissism and control: Tracking the impact of scholarly publications in the 21st century.* Utrecht: SURFfoundation. Retrieved from http://research-acumen.eu/wp-content/uploads/Users-narcissism-and-control.pdf

Zhong, C.-B., & House, J. (2012). Hawthorne revisited: Organizational implications of the physical work environment. *Research in Organizational Behavior, 32,* 3–22. doi:10.1016/j.riob.2012.10.004

Zilber, T. B. (2002) Institutionalization as an interplay between actions, meanings, and actors: The case of a rape crisis center in Israel. *Academy of Management Journal, 45,* 234–254. doi:https://doi.org/10.2307/3069294

Chapter 8
Innovation Governance: From the "Endless Frontier" to the Triple Helix

Henry Etzkowitz

U.S. Government Role in Innovation

During World War II the U.S. Office of Scientific Research and Development (OSRD) spent significant sums at universities to support advanced weapons development, often on projects proposed by academics. Although this direct intervention authority was dismantled upon the close of hostilities, a revised format for government role in innovation was set in motion in response to the perception of an *innovation gap* that arose upon the Soviet Union's lofting of Sputnik in 1957. This shock was reinforced by Japan's postwar economic success in developing technological industries that threatened U.S. leadership in electronics by the 1970s. In succeeding decades, the governance of innovation emerged as a triple helix of university-industry interactions behind the facade of a "hands off" ideology derived from the *Endless Frontier* report, an iconic early postwar document that was widely read to legitimate only a limited government role focused on basic research, despite a series of support chapters delineating various practical research objectives, including housing, defense, and health.

The concept of *national innovation systems* was derived from the leading role of the Japanese governmental Ministry of Trade and Industry in selecting and directing limited resources to particular technical areas while closing down others and supporting innovation at the firm level (Freeman, 1987). In the United States, an alternative innovation model, often characterized by bottom-up and lateral initiatives, was developed under conditions in which government was typically constrained from playing an overt and direct role (Mazzucatto, 2013). Macro- and microeconomic policies have been balanced by mesolevel triple-helix initiatives in which government plays a significant cooperative role that is relatively hidden in U.S. circumstances where government is ideologically suspect and often operates

---

H. Etzkowitz (✉)
International Triple Helix Institute (ITHI), Palo Alto, CA, USA

P. Meusburger et al. (eds.), *Geographies of the University*, Knowledge and Space 12,
https://doi.org/10.1007/978-3-319-75593-9_8

sub rosa within a research rather than an innovation framework. Thus, innovation studies are much more highly developed in Europe, while indirect measures to induce innovation are a U.S. forte. Ironically, the pressure to downplay, even denigrate, government's role has the effect of encouraging U.S. intervention to the more risky, longer-term implications for disruptive innovation of basic research, while in Europe a general acceptance of government's role allows programs that intervene close to the market, thereby emphasizing incremental innovation.

By acting indirectly and involving universities in the formulating and execution of innovation policy, the United States arguably developed a more creative innovation model, exemplified by the rise of Route 128 and Silicon Valley. In the following, I discuss the changing role of government in innovation and the emergence of a *triple-helix* innovation system under laissezfaire conditions in the United States in comparison to statist regimes. I argue that university and government are indispensable elements in national innovation systems and demonstrate how and in which ways government is crucial to initiate and mediate systematic innovation and economy (Edquist, 2003).

Sources of the Assisted Linear Model

Despite barriers, a de facto innovation policy is created through pressures on government to act in crises. The World War II Office of Scientific Research and Development (OSRD), originated at the initiative of academic scientists, was active across the spectrum of research areas of potential military use. Under wartime conditions R&D, testing, manufacturing, and customer demand were integrated into a *seamless web*, ignoring traditional boundaries. Government negotiated R&D contracts with universities, accepting their argument that it *should* contribute to the infrastructural costs of the project. In the postwar period, practices of academic initiative and subsidization of the university were generalized into a metaphorical "contract" in which government assumed responsibility for support of science with public funds. The linear model of an automatic transmission belt from science to society was thus invented (Godin, 2006).

The role of government in innovation is, of course, longstanding; both to carry out traditional state functions, such as defense and enumeration of the population, as well as additional tasks, upon petition by the citizenry, such as agricultural and industrial advance and cure of diseases. Government has employed various means to achieve this objective: offering prizes for results, for example, for a method to calculate longitude to improve navigation of ships and reduce the risk of shipwreck in eighteenth-century Britain (Sobel, 1998); establishing laboratories to achieve specific objectives, such as improvement in weapons, sanitation, and farming practices in nineteenth-century United States (Rossiter, 1975); purchasing equipment such as the Hollerith card sorter to speed analysis of census data in the early twentieth century and transistors to reduce the size and weight of battlefield communications equipment during the early post-World War II period; and in the course of the second half of the twentieth century until the present, granting funds for research that government research officials and their peers in the scientific community expect will be of practical as well as theoretical import (Stokes, 1997).

Nevertheless, despite demonstrated success, such government actions have met with skepticism in a society characterized by a laissezfaire ideological orientation. Under such stringent sociopolitical conditions, strong justifications are required to carve out a role for government, typically by arguing that market forces are ill-suited to address an issue that is societally important. Emergencies such as those created by wartime also provide an impetus for conservatives who would usually oppose governmental action to strongly support and even take the lead in their initiation. Such was the case at the advent of the World War II, when academics who had been skeptical of government support of research during the Depression took the lead in creating government agencies to develop new science-based weaponry (Baxter, 1946).

The world view of many academic scientists was also transformed by their World War II research experience. For example, physicists who had put aside their basic research interests to work as engineers on weapons development projects soon found that they had ideas for basic research that they would pursue after the war. This rediscovery of the interconnection between the practical and theoretical, and the experience of working with virtually unlimited resources at their disposal, transformed academic scientists' antigovernment attitudes that had led them to refuse support in the depths of the 1930s Depression. Support was feared to bring control with it, but with the introduction of an organizational model that put scientists in charge of a significant portion of the decision-making process through peer review, such fears tended to dissipate and even morph into a converse euphoria, exemplified by the endless frontier metaphor.

With the return of peace, universities and companies appeared to return to their previous boundaries, with an important difference: the heritage of their wartime experience of cooperation and collaboration. Prewar opposition to government funding at the universities was reversed as universities sought government funds to support research (Leslie, 1993). In addition to ad hoc appropriations for individual projects, a more systematic approach was sought. A rationale was needed to continue government funding of science after the war. Although a quid pro quo of public benefit was part of the "contract," few research results were actually translated into useful innovations, even given an extended timeframe. Two evaluation studies, carried out in the late 1960s, produced somewhat contradictory findings but the overall assessment was that a more structured approach would produce greater outcomes. This conclusion encouraged advocates of government playing an enhanced role in innovation (National Science Board, 1982). A series of initiatives were set in motion, balancing individual initiative in science and entrepreneurship with collective principles of regulation for public benefit, producing an immanent framework for innovation governance.

The Origins of the Endless Frontier

In 1944, Vannevar Bush, the head of the wartime Office for Scientific Research and Development (OSRD) had persuaded President Roosevelt to write a letter commissioning the report that was issued immediately after the war as *Science: The Endless Frontier*. However, he calibrated the report in accordance with his conservative view of government rather than his life experience as a scientific entrepreneur and reverse-linear theorist. The report contained an implicit concept of science as a self-regulating mechanism, operating according to a strict forward-linear progression: Put in the money at one end and the results will flow out at the other. World War II R&D successes such as radar, the proximity fuse, and the atom bomb encouraged the belief that similar results could be achieved in civilian areas of the economy and other countries picked up the model. On August 28, 1944, Vannevar Bush wrote a letter to the Secretaries of War and Navy and suggested that U.S. forces should gather any available "German technical information of an industrial nature" (Gimbel, 1990, p. 5). In autumn 1945 the so-called document program started. Its task was to photograph in 20,000 German industrial companies "secret patent applications, documents in original manuscript form, documents covering processes, formulas, and techniques not generally known in the United States" (p. 62).[1]

Practical results from basic research were the premise for the funding flow. Although the timeframe was generous, it was expected that scientific findings with practical applications would easily move to firms without an intermediary support structure (Mirskaya & Rabkin, 2004). This hands-off version of tight, "whatever-it-takes" wartime cooperation, while in accord with laissezfaire ideological predilections, was divorced from the realities of technology transfer practice. A reverse-linear model was also instituted: When a firm could not address a technical problem internally, it was encouraged to seek research externally, through acquisition of patent licenses from universities, purchase of startups, or collaboration with other firms, a strategy that has come to be called "open innovation" (Chesbrough, 2003). The resulting technologies would then be returned to the firm and taken to market. The former approach relies more on the university and the latter on the firm.

Nevertheless, despite the different emphases, both classic linear models relied on government to supply research funding. Business consortia, whether in the United States or Europe, often include university partners, with access to government funding that in Europe may more often be received directly through European Union (EU) Framework Programs. Government, however, has found it necessary to revise its role and play a more active part *downstream*, by crafting innovation policies and programs to insure that research results, however generated, are actually put into practice. However, there are massive differences between European countries not only in the coordination of the relations between government and

[1]According to Gimbel (1990) the Office of Technical Services had the task of distributing the some 500,000 documents (5 million pages) microfilmed in Germany about scientific scientific results, technologies, technical processes, and patents.

universities but also regarding the volume and scale of resource flow. Today about 85% of the aggregate budget of German universities are publicly provided subsidies from the state, whereas in the United Kingdom only 30% of the resources come from the state (OECD, 2014).

What is the optimum role of government in innovation? A Swedish university liaison director recently asked, "Why a triple helix; why not a 'double helix' of university–industry?" The answer is that it is only possible to develop university–industry relations up to a point, without considering the role of government. In the late 1970s, the U.S. Secretary of Health Education and Welfare rescinded the authority that had been developed by precedent in the National Institutes of Health to transfer intellectual property rights to universities on a case-by-case and university-by-university basis. Stable conditions for disposition of intellectual property arising from federally funded research were reestablished by law in 1980 and technology transfer assumed the format of a business-like activity between university and industry. Government's role in establishing a legitimate framework for technology transfer was the basis of this relationship (Berneman & Denis, 1998).

While the United States has established a series of programs and a regulatory environment to facilitate technology transfer in order to reap the benefits of munificent research funding that followed from World War II, the role of the state in innovation is also most clearly apparent in countries such as Mexico, where state-sponsored industry sector associations and university consultative councils coordinate these spheres. The Singapore government organized the transition to high-tech manufacturing and then to knowledge-based economic development. Other countries, such as Sweden, with a high rate of R&D spending and relatively low rates of economic return, have undertaken parallel steps, restructuring a series of basic research and technology programs into an innovation agency, VINNOVA.

Beyond the Endless Frontier

Despite numerous *eloges* for the linear model, especially in its forward format, a trend toward increasing reliance on science-based innovation can be identified in different countries. In Sweden, it is represented by a new research-funding sector focused on *strategic* science, in Singapore by the founding of a science-based institute sector, and in the United States by interagency research initiatives, such as those in nanotechnology, supported by the Department of Energy and the National Science Foundation. It is expected that many of the results of these initiatives will be introduced into the economy, not by firms tied to existing industries, but by new firms seeded by government funding as the basis of future industrial sectors.

The *innovation state* attempts to regenerate the sources of productivity, through such investments in science and technology, and by changing the rules of the game, through legal and administrative adjustments, to encourage the creation and growth of new firms. The state increasingly undertakes these tasks, not as a sole actor but through new forms of cooperative relations with industry and university. The innovation state is the successor to the capitalist, Keynesian, and welfare states, with their respective foci of assisting existing industry, promoting general economic advance, and securing the basic conditions of a good life for all the population.

The innovation state builds upon these various bases, incorporating elements of each of these models into a broader framework to support their realization under changed conditions of global competition. Neither socialism nor capitalism as an isolated national model is feasible but it is difficult to give up policies to realize these goals. In the laissez faire model of separate institutional spheres, moving beyond Keynesian macroeconomic policies arising from the 1930s depression, such as central bank adjustments of interest rates or money supply, was also a difficult transition. Similarly, in statist societies, the relaxation of the total state, based upon central planning, to a more modest role of incentivizing innovation, without going all the way to inaction, was also a difficult transition (e.g., Meusburger & Jöns, 2001).

Thus, one path to the innovation state is from a top-down model of bureaucratic control, with the state devolving its authority to various degrees. The other is from a standpoint of modest participation by central government in which case the pathway is to increased activity. The two different starting points intersect at some midpoint, where government, industry, and university assume relatively equal status as interdependent institutional spheres. It is especially at the intersection of the spheres that enhancements of innovation systems may be constructed that ultimately strengthen the sources of product innovation.

The Triple Helix

In industrial society the university was a support structure supplying trained people and knowledge; in a knowledge-based society the university moves from a secondary status to become an equal player with government and industry as the source of growth poles for economic development based on new knowledge that it generates or existing knowledge whose transfer it facilitates (Miao, Benneworth, & Phelps, 2015). Also, in these triple helixes one often sees each of the institutional spheres "taking the role of the other": Universities participate in forming firms, playing a classic industrial role; government acts as a venture capitalist, providing the funds for these entrepreneurial ventures; industry raises the levels of its training and research activities in university-like units.

The factor that differentiates different forms of the triple helix is the relative presence or absence of civil society. The ability to freely associate, brainstorm new approaches to innovation and form new organizations to realize them is the basis of a vibrant triple helix. Rather than a fourth helix, civil society is a platform on which enhanced triple-helix innovation regimes may be built. The optimal triple-helix model (Figure 8.1) may be viewed as a classic Venn diagram, with intersection of three interrelated and overlapping spheres, each taking the role of the other and working closely together, with free association in civil society encouraging lateral

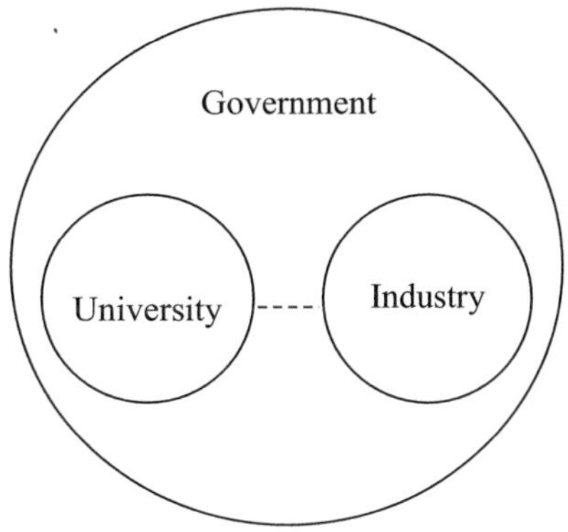

Fig. 8.1 The optimal triple-helix model.
Source: Design by author.

and bottom-up interactions. Under these conditions civic entrepreneurship flourishes along with business entrepreneurship, with each enhancing the other.

The *second academic revolution* is the translation of research into use, the development of an entrepreneurial format in which it becomes an explicit mission of the university to create new firms and networks with industrial and government stakeholders. This process is nonlinear and progressive, with the first revolution, the inclusion of research as an academic mission usually preceding the second one. Nevertheless, both revolutions sometimes happen simultaneously. Indeed, the second revolution can even precede the first when a teaching university engages in entrepreneurship utilizing existing knowledge. From an entrepreneurial base such a university may raise its level of resources and undertake research. I think it is possible to also see a bi-evolution of university missions from a traditional focus on individuals, in seminar rooms, classrooms, and lecture halls. However, university is not only training individuals, it is also training organizations in its incubation facilities and experiential education programs.

We may thus experience an expansion of the physical university rather than its contraction even as lecture courses are delivered electronically. Instead of students using the classroom for 90 minutes three times a week, they need availability of tools, materials, and locale on a 24/7 basis for collaborative project work. Some of these needs can be addressed outside the university. For example, Radicand is a Stanford spinoff, a consortium of mechanical engineering and design PhDs who have rented space in a former industrial building for their consulting organization. One of their objectives is to provide space for larger student projects from the university that cannot be handled in the normal space available to the Mechanical Engineering 310, the university's graduate interdisciplinary design and prototyping course.

A bi-evolution of the university is taking place in the entrepreneurial university's tripartite missions of education, research, and economic and social development: This is seen in research, the movement from the traditional dyadic relationship of professor and students in humanities to the research group. The organization of professors, postdoctoral students, and technicians in teams with lateral relationships and bottom-up initiatives, as well as in the traditional hierarchical relationship, has become commonplace. In the new economic development mission of the university, moving from an initial focus on capturing value from individual pieces of intellectual property through technology transfer, the university has come to play a broader role in innovation by becoming involved in efforts together with other stakeholders to enhance the future development of the region.

A key feature of a triple helix, building upon the work of classic sociologist Georg Simmel on the formal properties of different numbers in social interaction, is that triadic interactions allow for mediation and nurturance of novelty (Wolff, 1950). A two-sided relationship is subject to the inherent love/hate properties of a dyad. On the one hand, agreement to a proposal made by a prestigious person or organization may be made without full consideration. On the other, there is a tendency to fall into conflict over goals and objectives. A three-sided relationship moderates these tendencies by introducing possibilities for mediation, coalition building, and indirect bonding. A triple helix raises Simmel's analysis of the potential of the triad from the individual level of personal and familial relationships to the institutional and organizational levels of university–industry–government interaction. A triple helix provides a format for innovation in a knowledge-based society that takes shape in different formats.

A triple-helix statist model (Figure 8.2), where government encompasses industry and academia, with direction top-down can accomplish great projects. However, the downside is that a triple helix coordinated entirely by the state may neglect potential contributions from other spheres. Under these circumstances, government may take initiatives without consulting others; it is not the most productive form of triple-helix relationship since ideas are coming only from one source, the central government. Conversely, if the state is absent from the innovation picture, then coordination, regulation, and funding necessary to encourage improvements may be insufficient. There is no single answer to finding an appropriate balance between intervention and nonintervention. However, the previous history of the role of the state in society will set some bounds and also determine whether it is most useful for the state to intervene directly or indirectly, acting through other institutional spheres.

In statist societies direct intervention is expected while, under laissezfaire conditions, only indirect approaches may be possible. The laissezfaire triple helix with institutional spheres far apart, interacting across strong borders, which is said to be the way that the United States operates, is largely an ideological model. I always say to Europeans, especially those looking to the States for an innovation model, who are told by their American colleagues that government should desist from taking a role in university-industry interactions: "Don't look at what your U.S. colleagues say. Look at what the U.S. does and you will find a very strong role for government in innovation." However, the relationship is likely to be indirect and thus obscured, such as with the Bayh-Dole Act of 1980, incentivizing universities to move closer to industry without actually providing direct support for doing so. But the universities can charge administrative expense for technology transfer and patenting to the overhead rate that they receive for research grants. Without direct appropriations, the relationship is supported indirectly. It is hidden and indirect in agencies, such as

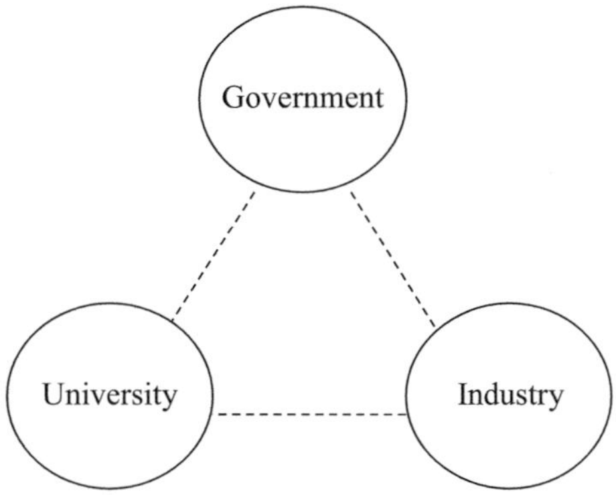

Fig. 8.2 The statist triple-helix model.
Source: Design by author.

the Defense Advanced Research Projects Agency (DARPA), which has academics seconded for short periods of time as program officers. They bring in new ideas, such as the Internet, and then move back to academia and a new set of people come in to take a temporary role in government.

What are the implications of triple helix for policy and practice? The triple helix is an analytical and normative concept derived from the changing role of government and university in relation to industry in different societies. Interaction among university–industry–government, as relatively independent, yet interdependent, institutional spheres is the key to improving the conditions for innovation in a knowledge-based society (Etzkowitz, 2003). This approach was derived from a response of the political infrastructure of New England after recruiting the academic and industrial leadership of the region to formulate and implement a response to the region's loss of industry in the early twentieth century. Triple-helix interactions have since been identified or put into practice in a variety of settings in industrialized and industrializing countries and regions as an entrepreneurial strategy to fill gaps in clusters or even create them ab initio. From Amsterdam to Niteroi, Brazil; Linkoping, Sweden, to Linyi, China, a variety of efforts, involving the three classic actors and substitutes, as necessary, have been identified (Etzkowitz & Zhou, 2017).

Crossnational comparisons are instructive to explain why the same organizational mechanism can produce different outcomes in different contexts. For example, a Mexican researcher at a conference asked: "Why has the incubator been successful in Brazil but not in Mexico?" One explanation is that an incubator movement arose in Brazil, with industry associations and local and national government supporting university initiatives, rather than being an isolated top-down initiative. The Mexican government has a program to provide funds to universities to start incubators. However, it is a relatively limited project with a very narrow base of support, rather than a movement that has spread throughout the entire society. In Brazil, the incubator initiative was part of the revival of civil society in the postmilitary era, with various institutional spheres involved, and national government, only one among several sources of support.

The existence of an organization infrastructure to receive a new element also explains why transfer may *take place* or be rejected. An initial attempt to introduce

CONNECT, a local level networking format from San Diego to Sweden, made by members of the local biotechnology association in Skane region did not succeed, lacking sufficient support from the region and the university. A later effort undertaken by the prestigious Academy of Engineering in Stockholm attracted support from regional officials and universities across Sweden and several CONNECT networks, linking entrepreneurs, business advice providers, patent lawyers, accountants, and "angels," were successfully established (Walshok, 1995). The cultural carryover of a top-down tradition of initiative was decisive.

Different state capacities affect both the trajectory and visibility of a triple helix, whether it is organized openly and transparently or is routed through hidden channels. In *high-state* societies, where triple-helix relationships have traditionally been directed top-down, bottom-up initiatives appear in conjunction with the emergence of regions and the growth of civil society. In *low-state* societies with a laissezfaire tradition, the emergence of the triple helix is associated with a strengthening of the role of the state, acting together with university and industry. Activating regional levels of government to become innovation actors, and creating such levels when they are lacking, becomes a key issue in creating such a system.

Devolution of Responsibility for Innovation

There has been a significant devolution of powers in recent years in countries, such as Great Britain, France, and Sweden, lacking a strong regional level of governance (Greyson, 2002). Formerly central government operated through regional levels that mandated common policies. Increasingly, it is seen that it is necessary to have policies specific to the competencies and capacities of different areas. Given the lack of an activist regional tradition, an initial step from the center may be to incentivize regional actors to come together and develop new initiatives. Sweden's Innovation Agency, VINNOVA, has taken this approach in a funding competition based on triple-helix actors developing joint proposals for science-based innovation.

On the other hand, there is a need to knit together different local initiatives that might otherwise be at odds with each other in the complex of innovation initiatives that have been established by Swedish government agencies and foundations. Government has provided the universities with *holding companies* to transfer technology and help start new firms, but it is only a modestly funded initiative. Technology Bridge foundations were established in several regions with a significantly higher level of funding, for much the same purpose. The triple-helix model provides a rationale to cooperate and aggregate resources to a common end and reduce friction among what otherwise might be a set of small competitive projects. VINNOVA's Vinnvext program incentivized regional innovation actors to come together in triple-helix coalitions and create joint initiatives as a condition of further funding (Etzkowitz & Klofsten, 2005).

The Changing Role of Government

Top-down models have been highly successful in organizing large military and space projects in both socialist and capitalist regimes. In countries with a planning system, government kept the entire innovation process under its control. Thus, in the former Soviet Union and Eastern Europe, a system of research institutes focused on industry problems. However, the results could only be implemented if they were centrally approved, although there were always informal exceptions to the rule. Nevertheless, bureaucratic controls were an impediment to the introduction of inventions. Although research and production were formally linked by intermediary organizations, industry's focus was on quantity production, not qualitative innovation and local technology transfer.

Transition from Statism to Laissez Faire

In the postsocialist era, top-down coordination was removed and each element in the former system was left to fend for itself, with sharply reduced funds from the state. The abrupt reconfiguration from a statist to laissezfaire regime left a question mark where the state had formerly played a leading role (Figure 8.3). Science and technology policy had formerly been the centerpiece of regimes legitimated by a thesis of a *scientific-technological revolution*. Given the discrediting of government it was difficult to justify more than a minimalist state, confined to basic security and welfare activities. Science and technology policy was barely a legitimate activity, no longer a priority in postsocialist countries. Nevertheless, after more than a decade of hands off, it is said that "government officials have come to their senses and realize that Government should stimulate . . . reforms" in Russian science and the academic community (Mirskaya & Rabkin, 2004, p. 13).

Fig. 8.3 The laissezfaire triple-helix model. Source: Design by author.

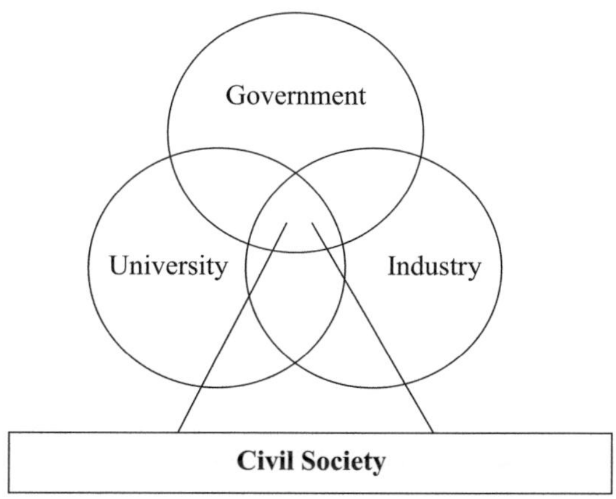

Transition from Statism to Civil Society

The possibility of individuals and groups to freely organize, debate, and take initiatives, is the basis for a triple helix including bottom-up as well as top-down initiatives. This can be seen most clearly in countries that are just emerging from military dictatorships. Bottom-up initiatives became possible in Brazil with the recreation of civil society that took place when the military gave up control in the early 1980s. University science and technology researchers introduced the concept of the incubator from the United States. In succeeding years, various levels of government as well as industry and civil associations took up the incubator concept and spread it throughout Brazilian society, applying it to a variety of problems from raising the level of low tech industry to creating jobs for the poor (Etzkowitz, Mello, & Almeida, 2005).

Transition to an Innovation State

The Finnish case is a focused version of the linear model of R&D pump priming, with funding opportunities focused on a relatively few areas of IT and biotechnology identified as having future economic potential. Much less research intensive in the early 1990s than Sweden, Finland has moved ahead by using monies from privatization of public enterprises to sharply raise the level of R&D spending (Benner, 2003). In relatively few years, the Helsinki region has come close to Stockholm in its concentration of biomedical research. The city of Tampere by the early 2000s was home to 3,000 information technology researchers in contrast to few dozen in the early 1990s. Nokia rose and fell but the skilled people left behind then became the initiators of new firms. To insure that the Nokia success was not an isolated instance, Finland made innovation a direct responsibility of the prime minister's office.

Transition to an Interventionist State

The United States is often misperceived as a laissezfaire society where innovation is left up to industry. Thus, there is reluctance to recognize that a plethora of specific policies and programs accumulated over the past half century constitutes a U.S. innovation policy. Government is playing a greater role in promoting innovation, often utilizing the university to reach its objectives. Given the resistance to government action at the federal level, when intervention is decided upon, it is typically occurs indirectly, utilizing universities as an interface between government and industry. In response to ideological constraints, the trajectory of immanent industrial policy formation creates networks and initiatives that cut across the institutional spheres.

Behind the laissezfaire presumption of the linear model that academic research results would seamlessly pass to industry through graduated students taking employment and industrial researchers following the journal literature, a more focused organizational approach to technology transfer, utilizing the patent system, had grown from its origins at the Massachusetts Institute of Technology in the early twentieth century. According to a George Washington University technology transfer official, "the national innovation strategy is to put federally-funded R&D on a conveyer belt that gets the R&D commercialized either by tech transfer to established companies or by wrapping the R&D into a university start-up" (Stanco, 2004). Previous governmental foci included regulation of exchange in the market to maintain competition, manipulation of macroeconomic aggregates through monetary policies, and redistribution of the results of productivity to redress inequalities (Hirst, 1994).

Innovative Governance

The basic precepts of *innovation governance* are set forth in a series of propositions about the transformation of traditional state functions to promote innovation:

1. Establishment of legitimate control of violence within a territory, promoting stability and reducing uncertainty as the basis for public authority, is extended from the public to the private sphere.
2. Corollary: Government guarantees are given to private capital so that, with this insurance, it may take greater risks in investing in new ventures.
3. Levying of taxes to support the protection of the nation and promotion of the general welfare is extended by using the tax system in a targeted fashion to provide special incentives and benefits.
4. Corollary: R&D tax credits and reduced capital gains taxes are made available to promote innovation.
5. Establishment of rules to structure economic life including procedures to charter firms and foundations, regulate the conduct of markets and currency systems.
6. Corollary: New agencies are established to promote innovation, including hybrid public/private entities.
7. Use of legal system to establish special rights such as patents as temporary monopolies to promote innovation.
8. Corollary: Universities are granted control of intellectual property rights from government-funded research, incentivizing them to become involved in technology transfer and innovation (United States, 1980; Denmark, 2000; Germany, 2002). Universities in Sweden are subsidized through the holding company initiative to encourage them to implement the *third mission*.
9. Provision of basic research funding to establish a linear model of innovation.
10. Corollary: Provision of public venture capital to create an assisted linear model of innovation.

U.S. Innovation Policy

Despite ideological structures against industrial policy and disbelief in the efficacy of attempting to "pick winners," the United States has arguably the world's strongest innovation policy, comprising bottom-up pressures from aspiring research universities, less research-intensive states, federal agencies under pressure to show practical results from research funding, and increasing international competition. No single impetus is decisive; rather it is the interaction of the various forces and initiatives that has generated an active innovation model wherein, if one level is forced to be inactive, say the federal government in supporting stem cell research, the state level picks up the slack, as in California's proposal to support research on this topic with a 3 billion dollar bond issue.

During the postwar period in the United States, high overhead payments became a method of funding the major research universities directly from the federal government, without explicitly acknowledging an elitist federal higher education policy. These universities thus supported were clustered in relatively few parts of the country, on the east and west coasts, with a few in the Midwest. This disparity was not a major issue as long as academic institutions were primarily seen in their traditional role as educational and research institutions. As a few universities with concentrations of research became foci of economic development, other less research-intensive regions wished to emulate their success.

Pressure has increased on the federal government to increase research spending and to distribute it more broadly, eschewing peer-review mechanisms instituted in the early postwar period to focus federally funded research on a relatively small group of schools. Nevertheless, regions with low levels of federal R&D spending are unwilling to depend upon modest set-asides, instituted to reduce pressures for equalization, or slowly building up their capabilities with local funds. Modest set-asides in the National Science Foundation (NSF) budget targeted at less research-intensive states in the Established Program to Stimulate Competitive Research (EPSCOR) program were insufficient to slake the thirst for R&D funds.

A de facto innovation program was instituted that works the same way as appropriations for roads or bridges or any local improvement that a Senator or Congressperson wants to obtain for their district. A legislator attaches a provision for a research center for a local university to a funding bill for another purpose, the so-called "earmark." When earmarks were eventually discouraged at the federal level, the states took up some of the slack. New York's Governor Cuomo recently announced that a 50 million dollar state program to encourage spinoffs from universities around the state would be doubled to 100 million. When one state takes the lead in a field where others feel they have strength, there is pressure to match the leader. Thus, when California developed a 4 billion dollar stem cell initiative in 2004, New Jersey and Massachusetts developed their versions in the hundreds of millions.

Competitive pressures drive the innovation system. Universities that have been outside of the federal research funding arena but want to increase their research strength have become active in seeking targeted federal and state funds. Typically as this emerging group of research universities enhances their capabilities, through such targeted measures, they then begin to compete successfully for peer-reviewed funds through the normal research funding channels. It is this increase in competition from universities across the country that has given the older research universities the feeling, indeed it is a reality, that competition for research funds has increased even as federal research budgets have risen significantly, especially in health and security.

Bottom-up Activates Top-Down

As new industrial areas arose from an academic research base in molecular biology and computer science in a few locations, other parts of the country became aware of the significance of universities as engines of economic development and wished to follow this model: first in North Carolina in the 1950s. By the end of the 1980s, virtually every state had some kind of science and technology agency focused on creating economic growth from research, typically by expanding research at local universities. In addition to state governments funding, research focused on economic development rose to a total of 3 billion by the turn of the new century, although it has since declined somewhat due to budgetary pressures (Berglund & Coburn, 1995).[2]

Strategies typically follow from the industrial and scientific condition of the state. Less research-intensive states attempt to build up the research capabilities of local universities in fields related to local resources. A longer-term goal is to create new firms from this research. Oklahoma and Georgia offer research funds to professors who have developed successful research groups to entice them to relocate, thereby improving the competitive chances of the state's universities in the federal research funding system. On the other hand, research-intensive industrial states, such as Michigan and New York, fund their universities to develop research capacities related to existing industries.

The Emergence of a "Bottom-Up Planning System"

Although states can be explicit, the federal government can only set very general outlines in civilian innovation policy for fear that it will be accused of attempting to pick winners. Government is ideologically perceived as naturally and inevitably incompetent, despite manifest success in military, health, and agricultural

[2]Direct state investment in science and technology policy remains a relatively small sum in comparison to other sources of R&D spending. Nevertheless, it made a significant, if modest contribution to the increase in academic R&D from $7 billion in 1980 to $17 billion in 1993 (in 1987 dollars). During the same period industry funding of academic R&D expanded by 265% from $334 million in 1980 to $1.2 billion in 1993 (Berglund & Coburn, 1995).

innovation. Nevertheless, it is advisable to watch what the United States does rather than what it says with respect to government's role in innovation. Even the most conservative politicians are activists when it comes to creating new knowledge-based industry in their locality.

Since the federal government is precluded from playing a direct role in civilian technological innovation, it often seeds other institutions with ideas and develops them collaboratively. For example, ATP program officers regularly made visits to companies and held national and regional conferences to encourage firms to work together with universities and government labs. Brainstorming sessions at these meetings typically included representatives of large and small companies, academics, and government technology experts. The objective of the discussion is to reduce the general category of a critical technology to a particular point, at which the people who are closest to the technology agree that a blockage exists. White papers were encouraged around these strategic points and the funding competition is thus made much more specific.

The result was a bottom-up planning process, an immanent triple helix arising across strong boundaries, with both top-down and bottom-up features. Nevertheless, the official mandate given by the first Bush administration was for a government–industry model in response to the European Union's emerging Framework programs. Although universities participated and even played a covert leadership role in an underground fashion, the overt prominence of government made the program vulnerable to attack, including charges of "corporate welfare" and eventually resulted in its defunding (Etzkowitz, 2003). Other initiatives, such as the Industry/University Cooperative Research Centers and Small Business Innovation Research (SBIR) programs, with universities in a prominent role and industry and government relatively subordinate had a better survival rate and SBIR has even received increasing resources.

There are positive implications of innovation policy makers having to deal with ideological resistance to government interventions. Since it is only considered to be legitimate for government to intervene in the event of clear *market failure*, such as when it is needed to support basic research, policy measures are forced upstream toward the research frontier. This tends to lead to a focus on startups and in creating new industries rather than providing input into existing industries. Policy measures are typically instituted as an extension of basic research funding programs that take the form of grants rather than loans. Thus, a higher degree of business risk can be taken. Again this is conducive to supporting the early stages of bringing advanced technology to the market. On the other hand, it is more difficult for the federal government to take steps to support later stages of firm growth and development, with the important exception of military and security related technologies.

Indirect Industrial Policy

Increased international competition has called attention to the role of government in innovation. During the economic downturn of the 1970s there were proposals for government to become directly involved in aiding existing industries and building up new ones, but these were quickly defeated. Instead, government went through the universities to reach industry. The patent system was reorganized to give intellectual property rights from federally funded research to the universities, with the condition that they had to take steps to put them to use. After 1980, technology transfer mechanisms, which had only been utilized by a relatively few universities, were diffused throughout the research university system.

Since the late 1970s the federal government has played an indirect role in encouraging academic-industry cooperation by changing the legal framework for federally funded research at universities. Laws such as the 1980 Bayh-Dole Act were passed to tie government-supported basic research at universities more closely to industry by creating a series of incentives and requirements to encourage universities to transfer technology deriving from federally funded research on campus. The new framework requires and offers incentives to encourage academic institutions to commercialize their research.

This policy has led to the emergence of a technology transfer profession, primarily based at universities. Their task is to negotiate agreements to move research across boundaries from one sector to the other. The university licenses intellectual property to a company, sometimes with little continuing of academic involvement apart from consultation. Increasingly, a more intensive effort takes place locally through the founding of a firm, with continuing academic participation in research and product development, at least in the early stages of firm growth. The 1980 Stevenson-Wydler Act extended the effort to transfer technology from research organizations to industry by mandating technology transfer efforts by the laboratories and research agencies of the federal government. The objective was to take intellectual property or capacities that exist within government laboratories and infuse them into companies as a part of a strategy to improve their international commercial competitiveness.

Government→Industry Initiatives

In addition to regulatory changes, basic research funding models were adapted to close the gap between research and innovation, the so-called "valley of death." To solve this problem, programs have been created that provide support for the early stages of firm formation. This *public venture capital* does not take minority ownership but fulfills the seed capital function in all but name (Etzkowitz, Gulbrandsen, & Levitt, 2000). For example, the NSF program officers who founded the SBIR program created a neutral language for direct government intervention in the economy. They delineated a three-phase model of the entrepreneurship process, with a transition from public to private funding in the third phase.

The strong focus on scientific and technical criteria provided a resemblance to (previously justified) basic research. Finally, *small business* can be viewed as a strong ideology by itself that very few people oppose. An advocate of SBIR and similar initiatives said:

> We definitely see the programs as a de facto industrial policy, but we cannot use that term, so we usually call it R&D policy and things like that instead; but it [SBIR] is a federal program that has created a whole lot of new industrial activity.

Nevertheless, SBIR offically operates as a grant-based federal R&D procurement program for small business. The growth of partnerships between large and small firms, between large firms and startup firms and between university and government laboratories, is also encouraged. The Advanced Technology Program (ATP) was founded in the late 1980s, partly in response to concerns that the EU Framework programs would encourage large U.S. corporations to move significant portions of their research to Europe to take advantage of subsidies. Although, ATP consortia programs were expected to provide a counter attractant, the focus on large corporations was strongly attached as corporate welfare, from both sides of the political spectrum. Nevertheless, many of these projects were actually initiated by startups or universities who found firms to apply on their behalf. After deep cuts in funding, the ATP shifted its strategy to partnerships of universities and startups, conforming to the politically acceptable U.S. model of focusing on small firms and university-industry interactions.

Several other federal programs, such as the Manufacturing Extension Centers of the National Institute of Standards and Technology (NIST), provide support that has allowed state-initiated programs to expand their efforts. The focus of these new collaborative arrangements is on translating research capacities into economic development. Building upon the 1980 Bayh-Dole Act's incentivizing universities to transfer technology concomitant with their receipt of federal research funds; the 1984 Stevenson-Wydler Act mandated government agencies with R&D capabilities to follow suit. The Department of Agriculture had long been involved in technology transfer in its field and NASA, seeking to increase support for spaceflight by demonstrating uses for its technological capabilities closer to home, eagerly took up the task.

A cooperative R&D agreement mechanism (CRADA) has been created to encourage and subsidize collaborative technology development projects between federal laboratories and industry. Consortia of companies, large and small, and universities are being supported to develop new technologies in response to international competition. These or similar formats have been utilized in a variety of general purpose innovation programs, such as the U.S. NSF Engineering Research Centers, NSF Industry-University Research Centers Program, NIST Advanced Technology Program, and Defense Advanced Research Project Agency (DARPA), at varying levels of funding, as well as in special purpose projects, such as the SEMATECH rescue effort mounted to save the U.S. semiconductor industry from Japanese competition in the 1980s.

Industry→ Government Initiatives

Government-industry R&D cooperation is emerging as an overlay upon a hidden industrial policy of encouraging academic-industry ties. Incentivized by relaxation of antitrust laws U.S. industry initiatives are largely confined to joint projects among firms. However, in certain circumstances, where the challenge is great and industry feels it cannot succeed by itself, and the danger of loss is too great, industry calls upon government assistance. Two instances exemplify industry's relation to government: The formation of SEMATECH and Joint Venture Silicon Valley, at the national and regional levels. During the early 1990s decline of Silicon Valley, an industry-initiated consortia invited participation of local governments and universities in what was initially a series of brainstorming meetings to generate ideas for renewal of the region. A venture-capital approach of selecting a few ideas for further development was followed and an organization was formed to support the initiatives, led by a local politician.

When the semiconductor industry was at risk due to intense Japanese competition from the 1970s, industry leaders developed a strategy to jointly develop a new generation of production technology and sought significant government support from the Department of Defense. The Reagan administration overcame its ideological objections and approved the project. Thus, an industry-initiated university-industry government collaboration was created at the national level. University research centers, such as the Center for Integrated Systems at Stanford, brought together academic and industrial researchers who moved the larger project forward. SEMATECH contributed to the revival of the U.S. industry and later dropped government funding to admit foreign firms and reorganize as an international consortium for precompetitive research to support the industry as a whole.

These instances exemplify an evolving knowledge and innovation infrastructure that is increasingly constructed from elements of the triple helix. University research centers explicitly adopt industrial models of research management to provide a support framework for academic research groups, balancing these coordination and control elements with research autonomy, including the right of graduate students to have a considerable say in the formulation of research projects through negotiation with their mentor. Similarly, startup firms are a hybrid creature, embodying academic, industrial, and government elements rather than a pure business model, even though they are legally constituted as firms.

Conclusion: Endless Frontier and Triple Helix

The intersection of dual linearities induces nonlinearity and interactive innovation through networked rather than isolated entities. When the forward-linear model of university research groups and its support structure of technology-transfer offices and incubator facilities meets the reverse-linear model with its clusters of learning firms, each infuses the other. It is increasingly recognized that government plays an important, if sometimes hidden, role in innovation in market economies and that government programs have an important role to play, not only from the national level—top-down—but also from the local level—bottom-up, in a movement from a hands-off linear to an assisted linear model of innovation.

Beyond the development of new products, innovation is the creation of new configurations among the institutional spheres. University-industry-government interactions are increasingly the basis of economic and social development strategy in both advanced industrial and developing societies. Heretofore, innovation was expected to largely take place within industry with other institutional spheres playing only a limiting contributing role, government, for example, acting only when clear market failures could be identified. Innovation is being transformed from a relatively simple set of linear and reverse linear processes within industry, extending from research to the market and vice versa, to a nonlinear process in the transition to a knowledge-based society

The insertion of Bacon's (1620/2011) practical vision of a series of institutional mechanisms realizing the promise of science was the way forward. An assisted linear model, consisting of a series of organizational mechanisms, such as technology-transfer offices and programs to explore the practical implications of research, inserted between the producers and users of research emerged through a series of government initiatives at the national and regional levels. By combining top-down and bottom-up approaches, arising from opposite starting points in laissezfaire and statist societies, the promise of the endless frontier is realized.

References

Bacon, F. (2011). *The great instauration*. Calgary: Theophania. (Original work published 1620)

Baxter, J. P. (1946). *Scientists against time*. Boston: Little, Brown.

Benner, M. (2003). The Scandinavian challenge: The future of advanced welfare states in the knowledge economy. *Acta Sociologica, 46*, 132–149. doi:https://doi.org/10.1177/0001699303046002004

Berglund, D. R., & Coburn, C. M. (1995). *Partnerships: A compendium of state and federal cooperative technology programs*. Columbus: Battelle Press.

Berneman, L. P, & Denis, K. A. (1998). Evolution of academic-industry transfer in the U.S.A. *Industry and Higher Education, 12*, 202–205. doi:https://doi.org/10.1177/095042229801200403

Chesbrough, H. W. (2003). The era of open innovation. *MIT Sloan Management Review, 44*(3), 35–41. Retrieved from http://sloanreview.mit.edu/article/the-era-of-open-innovation/

Edquist, O. (2003). Layered science and science policies. *Minerva, 41*, 207–221. doi:https://doi.org/10.1023/A:1025509629159

Etzkowitz, H. (2003). Innovation in innovation: A triple helix of university-industry-government. *Social Science Information, 42*, 293–337. doi:https://doi.org/10.1177/05390184030423002

Etzkowitz, H., Gulbrandsen, M., & Levitt, J. (2000). *Public venture capital*. New York: Harcourt Brace.

Etzkowitz, H., & Klofsten, M. (2005). The innovating region: Toward a theory of knowledge-based regional development. *R&D Management, 25,* 243–255. doi:https://doi.org/10.1111/j.1467-9310.2005.00387.x

Etzkowitz, H., Mello, J., & Almeida, M. (2005). Towards "meta-innovation" in Brazil: The evolution of the incubator and the emergence of a triple helix. *Research Policy, 34*, 411–424. doi:https://doi.org/10.1016/j.respol.2005.01.011

Etzkowitz, H., & Zhou, C. (2017). *Triple Helix* (2nd ed.). London: Routledge.

Freeman, C. (1987). *Technology policy and economic performance: Lessons from Japan*. London: Pinter.

Gimbel, J. (1990). *Science, technology, and reparations. Exploitation and plunder in postwar Germany*. Stanford: Stanford University Press.

Godin, B. (2006). The linear model of innovation: The historical construction of an analytical framework. *Science, Technology, & Human Values, 31*, 639–667. doi:https://doi.org/10.1177/0162243906291865

Greyson, L. (2002). Give us back our regions. *Innovation Policy Review, 4*(2), 1.

Hirst, P. (1994). *Associative democracy: New forms of economic and social governance*. Cambridge, UK: Polity Press.

Leslie, S. W. (1993). *The Cold War and American science: The military-industrial-academic complex at MIT and Stanford*. New York: Columbia University Press.

Mazzucato, M. (2013). *The entrepreneurial state: Debunking public vs. private sector myths*. London: Anthem Press.

Meusburger, P., & Jöns, H. (Eds.) (2001). *Transformations in Hungary: Essays in economy and society*. Heidelberg: Physica-Verlag.

Miao, J. T., Benneworth, P., & Phelps, N. A. (Eds.). (2015). *Making 21st century knowledge complexes*. Regions and Cities: Vol. 86. London: Routledge.

Mirskaya, E. Z., & Rabkin, Y. M. (2004). Russian academic scientists in the first post-Soviet decade: Empirical study. *Science and Public Policy, 31*, 2–14. doi:https://doi.org/10.3152/147154304781780145

National Science Board. (1982). *U.S. university-industry research relations*. Washington DC: National Science Foundation

OECD. (2014). *Education at a glance 2014: OECD Indicators*. Paris: OECD. doi:https://doi.org/10.1787/eag-2014-en

Rossiter, M. W. (1975). *The emergence of agricultural research*. New Haven: Yale University Press.

Sobel, D. (1998). *Longitude: The true story of a lone genius who solved the greatest scientific problem of his time*. London: Fourth Estate.

Stanco, T. (2004). Start-up discussions. George Washington University techno-l, 180604.

Stokes, D. E. (1997). *Pasteur's quadrant: Basic science and technological innovation*. Washington: Brookings Institution Press.

Walshok, M. (1995). *Knowledge without boundaries: What America's research universities can do for the economy, the workplace, and the community*. San Francisco: Jossey Bass.

Wolff, K. (Ed. & Trans.) (1950). *The sociology of Georg Simmel*. New York: Free Press.

Chapter 9
Quality Cultures in Higher Education Institutions—Development of the Quality Culture Inventory

Christine Sattler and Karlheinz Sonntag

Theoretical Background

The implementation of quality tools and procedures of quality management represents one of the major challenges of today's globally operating universities. Higher education institutions face an increasingly competitive environment, leading to elevated demands for quality in teaching and research as well as in service and administration. Quality assurance and quality development have therefore been central issues of policy discussions in higher education for many years now, especially since the beginning of the Bologna Process in 1999. The main objective of the Bologna Process is to create a European higher education area by improving mobility, instituting comparable university degrees and credit point systems, and developing comparable criteria and methods for quality assurance (Bologna Declaration, 1999). Framed in that context, the design and implementation of measures to ensure quality constitute key aspects of the Bologna Process.

Extensive debates on quality assurance have served as a starting point for introducing the concept of quality culture, which expands on classical approaches of quality assurance by drawing on organizational psychology, adding that field's perspective to the structural-formal side of quality management. It is no longer only a question of assessing quality by means of hard facts, such as the number of publications or the amount of third-party funding, but also of discerning the extent to which quality is actually subscribed to and lived by members of a higher education institution. Whereas academics disagree on the comparability of quality criteria (e.g., bibliometric indicators), quality culture could well become a concept with which they can all identify, regardless of their discipline.

C. Sattler (✉) · K. Sonntag
Institute of Psychology, Heidelberg University, Heidelberg, Germany
e-mail: christine.sattler@psychologie.uni-heidelberg.de;
karlheinz.sonntag@psychologie.uni-heidelberg.de

P. Meusburger et al. (eds.), *Geographies of the University*, Knowledge and Space 12,
https://doi.org/10.1007/978-3-319-75593-9_9

The quality culture approach is closely related to the well-known concept of organizational culture. According to Schein (2010), organizational culture comprises three distinct levels:

- Artifacts: tangible elements of culture (e.g., furniture, dress code), which are visible to nonmembers of an organization
- Espoused values of an organization (e.g., customer orientation)
- Shared basic assumptions: unconscious beliefs that guide the behavior of organizational members and that are difficult to decipher

Schein's conceptualization of organizational culture provides valuable information about different levels that need consideration when cultural aspects of an organization are being operationalized. To assess quality culture, it is essential not only to take account of visible quality artifacts within an organization (e.g., quality assessment tools) but also to analyze its quality values and shared basic assumptions (e.g., commitment) pertaining to quality. The quality culture approach thereby goes far beyond classic ranking procedures, which are limited primarily to the assessment of artifacts that distinguish quality.

The first comprehensive definition of quality culture relating to the construct of organizational culture was given by the European University Association (2006):

> Quality culture refers to an organizational culture that intends to enhance quality permanently and is characterized by two distinct elements: on the one hand, a cultural/psychological element of shared values, beliefs, expectations and commitment towards quality and, on the other hand, a structural/managerial element with defined processes that enhance quality and aim at coordinating individual efforts. (p. 10)

In this definition quality culture consists of two distinct levels. First, it is objectively tangible in terms of the tools and procedures (artifacts) of quality management. Second, quality culture encompasses organizational-psychological aspects (e.g., espoused values, expectations, and commitment to quality, that is, shared basic assumptions), which are rather difficult to capture.

Despite the increasing number of qualitative research papers on quality culture (e.g., European University Association, 2005a, 2005b; Loukkola & Zhang, 2010), empirical approaches operationalizing this phenomenon have not been developed sufficiently. This chapter summarizes selected results from the project entitled "*heiQUALITY Cultures*," which aimed to create an empirical instrument for the organizational diagnosis of quality culture within the context of higher education (Sonntag, Stegmaier, & Schaper, 2016).

The *heiQUALITY Cultures* Project

The heiQUALITY Cultures Project ("Development and Testing of an Instrument for the Description and Assessment of Quality Cultures at Higher Education Institutions") was carried out between April 2012 and May 2015. The foremost objectives of the research project were to:

1. develop a comprehensive definition and assessment model that considers structural-formal and organizational-psychological aspects of quality culture;
2. develop a Quality Culture Inventory (QCI) that enables higher education institutions to analyze their current state of quality culture autonomously and empirically;
3. analyze strengths and developmental potential of current quality cultures within the higher education context; and
4. derive target-oriented recommendations for quality development and improvement.

The following section offers a detailed overview of the methodology used in the heiQUALITY Cultures Project in order to achieve these objectives.

Methods

The heiQUALITY Cultures Project represents the first empirical approach operationalizing quality culture within the higher education context. Milestones of the project are presented in Figure 9.1.

In the first step a comprehensive literature review was conducted to identify previous qualitative and empirical studies focusing on quality culture and its operationalization. The literature review included one interdisciplinary and two disciplinary databases—with a focus on organizational psychology and other branches of that field—including publications up to December 2012. Strikingly, only 3 out of 786 publications focused on the operationalization of quality culture directly (Ali & Musah, 2012; Trivellas & Dargenidou, 2009; Zeitz, Johannesson, & Ritchie, 1997). These studies applied very heterogeneous methodological approaches, underscoring the relevance of our research objective of promoting additional empirical research in this field (for a detailed review of additional literature, see Sattler et al., 2016).

The literature review served as a sound basis for developing a previous assessment model of quality culture, which was subsequently challenged and discussed in 41 international expert interviews. In order to qualify for an expert interview, prospective partners had to meet at least one of the following criteria:

- Practical experience working for an accredited quality assurance agency (e.g., evalag[1])
- Practical experience working for an independent organization of higher education (e.g., the European University Association, the European Students Union)

[1]Evaluationsagentur Baden-Württemberg.

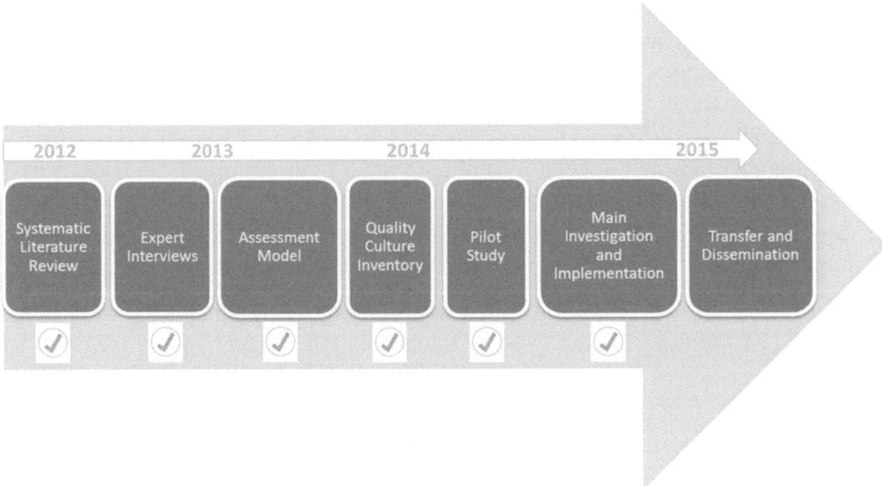

Fig. 9.1 Milestones of the heiQUALITY Cultures Project. Source: Adapted from Sattler, Sonntag, and Götzen (2016, p. 46).

Table 9.1 Sample topics and corresponding questions of the semistructured interview guideline

Sample topics	Sample questions
Professional background	Could you please elaborate on current core areas of your work? In which regard do you deal with the topic of quality in your occupation?
Quality culture: Theoretical considerations	What do you associate with the term *quality culture*? What constitutes a quality culture at higher education institutions in your opinion?
Quality culture in practice	From your point of view, how can a quality culture be furthered sustainably? What hindrances or resistance can you think of when trying to further quality culture?

Source: Design by authors.

- Practical experience concerning quality assurance, quality management, or both within higher education institutions (e.g., quality managers)
- Research publications addressing quality culture within the context of higher education

Most of the interviews were face-to-face ($n = 35$). For practical reasons (e.g., travel time) the remaining expert interviews ($n = 6$) were conducted by telephone. Women accounted for 37% of the sample ($n = 15$); international interview partners, for 17% ($n = 7$). Almost all the interviews ($n = 40$) were audiotaped and professionally transcribed for further systematic analyses. The interview length averaged 60 minutes, resulting in approximately 41 hours of material.

All interviews followed a semistructured guideline based on an approach that had been used in the "learning culture" project by Sonntag, Stegmaier, Schaper, and

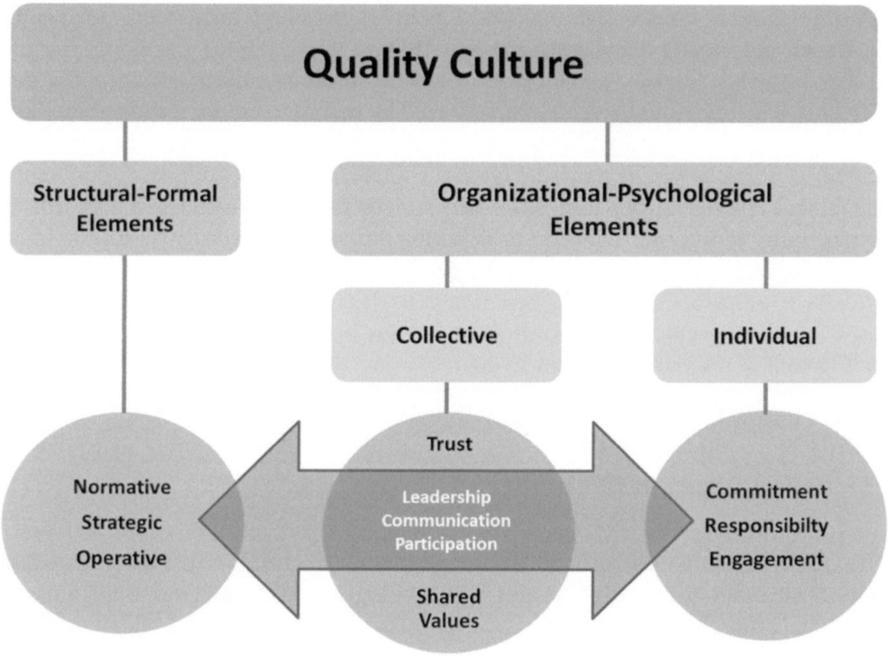

Fig. 9.2 Final model for assessing quality culture. Source: Adapted from Sattler et al. (2016, p. 49).

Friebe (2004) and Sonntag, Schaper, and Friebe (2005). Table 9.1 shows sample topics and corresponding questions of the interview guideline.

When asked to elaborate on relevant dimensions of quality culture, 39 experts (95.1%) referred to quality-oriented leadership and communication. More than 70% of the experts emphasized the importance of commitment, participation, and the development of quality objectives. Quality values, mutual trust, individual responsibility, recognition, and information ranked among the ten most frequently mentioned elements of quality culture, with an agreement rate of more than 65% (for details on additional results of the expert interviews, see Sattler et al., 2016).

Experts' suggestions led to minor revisions in the initial model of quality culture, resulting in the final model for assessing quality culture (Fig. 9.2). According to this model, quality culture can be described on a structural-formal and an organizational-psychological level. The structural-formal level comprises normative, strategic, and operative elements, which represent heterogeneous aspects of quality assurance and quality management. In the model by Bleicher (2011), normative aspects of quality management are expressed by an organization's quality goals, its mission statements, or both. Responsibilities for quality assurance are defined at the strategic level (e.g., a quality assurance unit). Specific quality tools and measures (e.g., student evaluation, controlling) are located at the operative level. It is assumed that all these structural-formal aspects are important in order to adopt approaches to quality assurance and quality management successfully.

The organizational-psychological level is made up of collective and individual elements. The individual level is characterized by commitment to, responsibility for, and engagement in quality. At the collective level it is hypothesized that trust and shared values function as a mutual basis for quality-oriented leadership, communication, and participation. The latter three elements are illustrated as an arrow, representing a dynamic connection between the structural-formal and individual levels. For example, participation in the development of quality assurance measures is likely to enhance individual commitment to these measures. The final model for assessing quality culture served as an empirical basis for the development of the QCI, which is presented in detail in the following section.

Operationalization of Quality Culture

The QCI consists of two questionnaires, one for the structural-formal level of quality culture; the other, for the organizational-psychological level. Both questionnaires are based on comprehensive literature reviews focusing on previous questionnaires that operationalized the proposed dimensions of quality culture.

Structural-Formal Questionnaire

The structural-formal questionnaire focuses on the operationalization of structural-formal aspects of quality culture. The questionnaire is used as a guideline for structured interviews with subject-matter experts on quality assurance within higher education institutions. Seventy-three items were constructed on the basis of a sound literature review focusing on normative, strategic, and operative aspects of quality assurance. The chief objective of the structural-formal questionnaire is to describe the status quo of quality assurance within higher education institutions. It covers six core areas: general information, institutional structures, teaching and learning,

Table 9.2 Design of the structural-formal questionnaire on quality culture assessment

Core areas	Sample dimensions	No. of items
General information	Significance of QA	13
Institutional structures	QA responsibility	17
Teaching and learning	Quality goals	14
Research	QA concept	9
Young scientists	QA instruments	9
Administration and service	Evaluation	11
Total		**73**

Source: Design by authors. *Note*. QA = Quality Assurance.

research, young scientists, and administration and service (see Table 9.2). The structural-formal survey was piloted and positively evaluated by four quality assurance experts within the higher education context.

Organizational-Psychological Questionnaire

Unlike the structural-formal questionnaire, the organizational-psychological questionnaire is addressed to *all* members of higher education institutions (HEI members): the university leaders, academic staff (professors, academic assistants), and nonacademic staff (administrators, secretaries, and service personnel). The questionnaire contains a set of 53 items about various aspects of quality culture. The individual dimensions (commitment, engagement, and responsibility) are represented by 4 items each. So are participation, shared values, and trust, which represent collective elements of quality culture. Leadership is assessed by 12 items; communication, by 9. We also developed 8 items assessing "global aspects" of quality culture, which require the respondent to evaluate aspects of quality culture that pertain to the entire institution of higher education. Using a 6-point Likert scale ranging from 1 (*does not apply at all*) to 6 (*fully applies*), respondents indicated their level of agreement with the statements about quality culture. Table 9.3 presents sample items of the organizational-psychological questionnaire.

To answer relevant scientific and practical questions related to the construct of quality culture, the survey additionally includes several demographic characteristics (e.g., age, gender) along with potentially moderating and dependent variables (e.g., conscientiousness, satisfaction with quality culture). The online questionnaire contains 97 items (duration: approximately 15 minutes).

Data Collection and Sample Characteristics

The QCI was piloted and conducted at three institutions of higher education in Germany, each representing a different type: (a) universities, (b) universities of applied sciences, and (c) cooperative or dual universities.[2] These higher education institutions differ considerably, with their educational tasks allowing for differentiated analyses of institution-specific quality cultures.

Participants were contacted via email distribution lists. In the first step the QCI was administered to participants in a pilot sample (N = 93 HEI members) and

[2]Dual universities offer students the opportunity to combine their academic studies with professional work (dual studies).

Table 9.3 Design and sample questions of the organizational-psychological questionnaire on quality culture assessment

Dimension	Sample item	Source	Item	α[a]
	Individual level			
Commitment	"I am particularly intent on supporting the quality development of [name of HEI[b]]."	Adapted from Jackson (2004) (affective subscale)	4	.603
Engagement	"I am willing to make additional effort to meet the quality demands of my work."	Adapted from Jackson (2004) (behavioral subscale)	4	.696
Responsibility	"I feel that I am jointly responsible for the quality development of [name of HEI]."	Adapted from Jackson (2004) (cognitive subscale)	4	.358
	Collective level			
Leadership	"It is important to me to appreciate good working results adequately."	Adapted from Heinitz & Rowold (2007)	12	.935
Communication	"Ideas concerning quality improvement are openly discussed in our department."	Adapted from Brodbeck, Anderson, & West (2000)	9	.871
Participation	"I keep myself up to date concerning new developments at [name of HEI]."	Adapted from Staufenbiel and Hartz (2000)	4	.716
Shared values	"Quality values of [name of HEI] are actually put into practice."	Own development	4	.772
Trust	"I have full confidence in my employee's skills."	Adapted from Zeitz et al. (1997)	4	.734
Global aspects	"[Name of HEI] is characterized by high quality awareness."	Own development	8	.889
Total			53	

Source: Design by authors. [a]Cronbach's Alpha. [b]Higher education institution.

Table 9.4 Sample sizes of the pilot study and main investigation on quality culture assessment

	Pilot study		Main investigation	
Group	*N*	%	*N*	%
University leaders	2	2.2	3	0.4
Professors	23	24.7	138	17.5
Academic assistants	46	49.5	289	36.6
Administrators	6	6.5	162	20.5
Secretaries	12	12.9	84	10.6
Service personnel	4	4.3	113	14.3
Total	93	100	789	100

Source: Design by authors.

slightly modified after evaluation. The main investigation was completed by 789 HEI members (see Table 9.4).

As expected, academic staff represented the largest participating target group (17.5% professors, 36.6% academic assistants), followed by administrators (20.5%), service personnel (14.3%), and secretaries (10.6%). The sample was characterized by a significantly higher share of female participants (59.4%; χ^2 (1, N = 727) =

25.817, $p < .001$) than male. In total, 37.1% of the respondents held a temporary employment contract. The duration of employment varied from less than one year (5.8%) to 1–4 years (26.8%), 5–9 years (27.8%), 10–19 years (23.1%), and more than 20 years (16.4%), representing heterogeneous knowledge and experience with the quality culture of the participating institutions.

Selected Results: Evaluation Options

Organizational-Psychological Questionnaire

Data from the organizational-psychological survey offer the opportunity to create differentiated profiles of individual (e.g., commitment, engagement) and collective (e.g., leadership, communication) elements of quality culture. These profiles of quality culture allow for institution-specific analyses of strengths and developmental potentials. Figure 9.3 illustrates the profiles of quality culture in different departments within a higher education institution.

In this example profiles of quality culture differed significantly from one department to the next. Whereas department B reported the highest level of participation, it scored lowest in quality-oriented leader behavior and leader expectations. These results indicate that heterogeneous characteristics of quality culture may exist even within one institution. Creating awareness about the existence of such differences

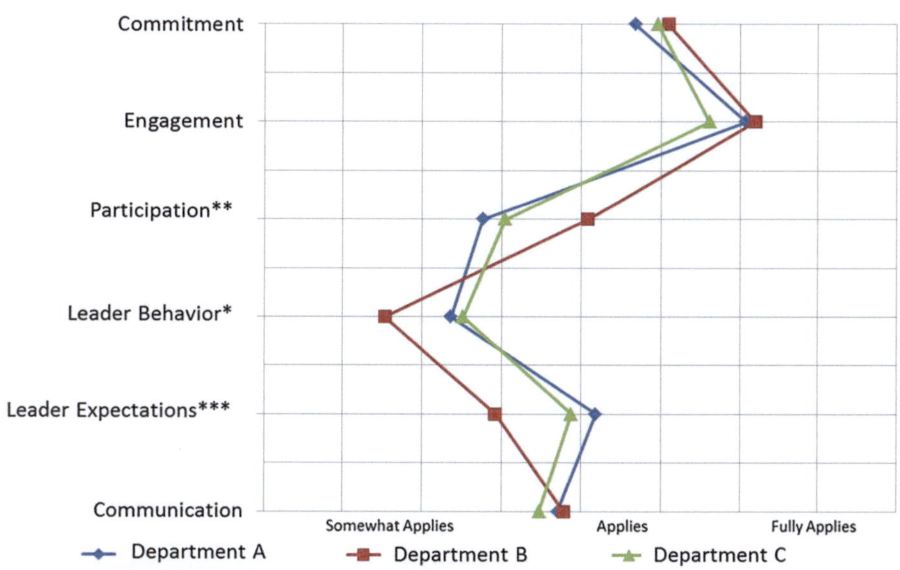

Fig. 9.3 Example of quality culture profiles (*$p \leq .05$; **$p \leq .01$; ***$p \leq .001$). Source: Design by authors.

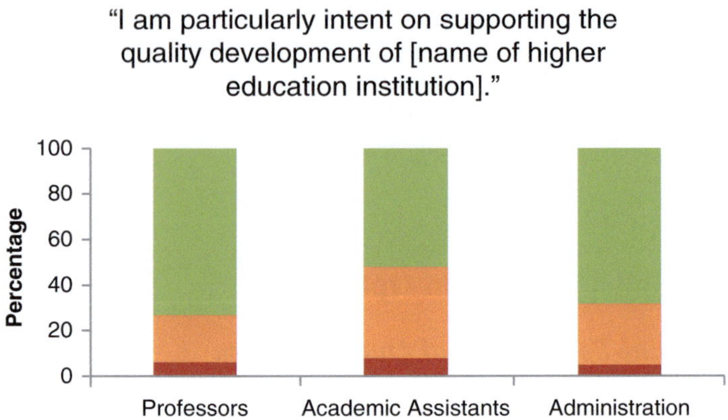

Fig. 9.4 Quality Commitment: Group Comparison. Level of high (green), middle (yellow), and low (red) agreement, whereby χ^2 (4) = 26.591, $p < .001$. Source: Design by authors.

may lead to the development of department-specific interventions addressing heterogeneous developmental needs and potential. In our example the quality culture of department B is likely to benefit from leadership-oriented interventions, whereas interventions focusing on participation may be more beneficial within departments A and C.

Another interesting option of evaluation is to find out the extent to which our groups differed in their scores on individual items (e.g., commitment). Figure 9.4 shows a sample analysis of an item operationalizing quality commitment.

The comparison between professors, academic assistants, and administrators shows significant differences between their levels of agreement. Whereas professors and administrators indicated a similarly high level of agreement, academic assistants were more likely to show an intermediate level of agreement. This result may be due to a relatively high share of temporary employment contracts across the members of this staff group (66.5%).

Whereas the level of commitment to quality tended to be very high among all HEI members, the rates of agreement with statements assessing the overall quality of an institution's culture ("global aspects") turned out to be considerably lower (see Fig. 9.5). Again, academic assistants showed significantly lower levels of agreement than did the other two groups.

Data of the organizational-psychological questionnaire may be used for various further analyses, too. Demographic characteristics allow for the creation of differentiated profiles of quality culture for gender, age, or duration of affiliation, for instance. The questionnaire also provides valuable information about the level of satisfaction with quality culture. These results serve as an empirical foundation on which to base target-oriented recommendations for improving quality.

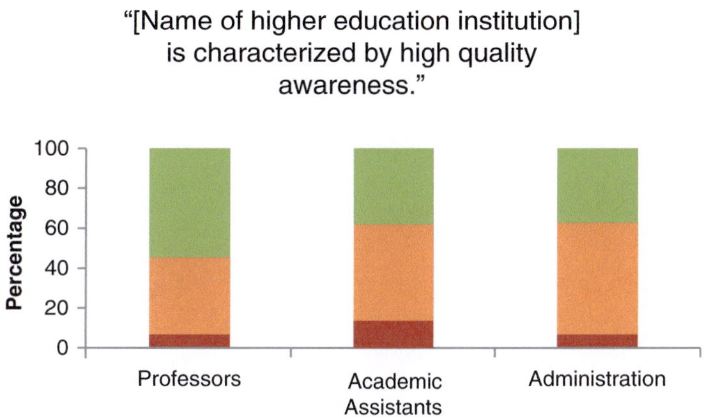

Fig. 9.5 Global aspects of quality culture. Level of high (green), middle (yellow), and low (red) agreement, whereby χ^2 (4) = 12.706, $p < .05$. Source: Design by authors.

Structural-Formal Questionnaire

The structural-formal questionnaire of the quality culture inventory allows for a systematic assessment of quality-oriented structures within an institution of higher education. The questionnaire differentiates between core areas of these institutions: the institutional level, teaching and learning, research, young scientists, and administration and service. It contains items about the definition of competencies and quality goals, for instance. The questionnaire also assesses the existence of quality assurance concepts and quality control loops. As with the organizational-psychological questionnaire, results of the structural-formal questionnaire make it possible to create structural-formal profiles for specific institutions (see Fig. 9.6).

In this example competencies in assuring quality are defined at almost all levels of the higher education institution. At the same time, information about quality assurance and for the definition of quality goals is available at almost all levels of the higher educational institution. Potential for structural-formal development can be identified for applying concepts of quality assurance and using quality-control loops. A regular evaluation of quality goals might represent a useful strategy for improvement at the structural-formal level.

Data of the structural-formal questionnaire may also be used for numerous other analyses. For example, the questionnaire gives a systematic overview of quality assurance measures that are applied within a higher education institution (see Table 9.5).

The questionnaire also allows for a differentiated insight into options that different status groups have to contribute to measures for developing quality. Furthermore, it is possible to assess quality-oriented communication structures. The results of the structural-formal questionnaire offer a sound foundation for analyses of strengths, weaknesses, opportunities, and threats (SWOT) followed by an optimization of structural-formal aspects of quality culture.

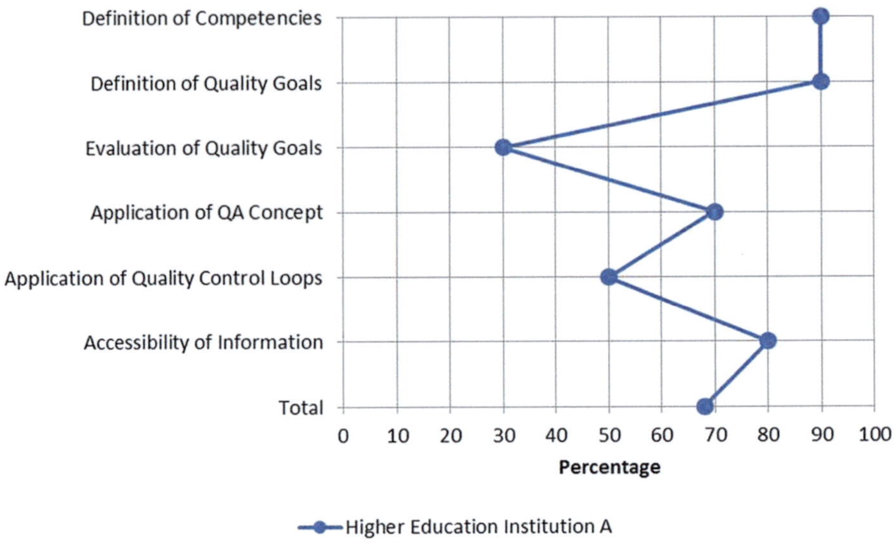

Fig. 9.6 Example of structural-formal profiles. Source: Design by authors. *Note*. QA = Quality Assurance.

Discussion and Future Prospects

The QCI represents a sound, economic tool with which to describe the current state of quality culture within institutions of higher education. The results lay an empirical foundation for discussions about strengths, weaknesses, and potential measures for improving quality. So far, three institutions of higher education have taken part in the QCI, bringing about intense exploration of quality culture. Within a relatively short time, the instrument has generated valuable data on this subject, contributing to a focused discourse about enhancing quality at the participating institutions. Using the QCI requires the openness of all participants and the willingness to debate controversially, but fairly, about the institution's quality culture. It is therefore critically important to promote acceptance of the QCI by addressing both the institution's leaders and staff before administering the questionnaire. The leaders need to be convinced of the great gains possible through the QCI, and it is essential that staff members know their answers will be taken seriously and can make a difference. Fortunately, the motivation to meet both of these requisites can grow from the distinctly practical benefits that the QCI offers. It facilitates the analysis of the status quo of quality culture and quality-oriented leadership as well as the analysis of strengths and weaknesses of quality culture profiles. It can effectively guide the formulation of recommendations for quality improvement and can thereby shape quality assurance and quality development.

None of these advantages will come of their own, however. Continued effort is needed to reap them. The number of higher education institutions included in future

Table 9.5 Quality assurance measures: Which quality assurance measures are applied in the following core areas of higher education? (Multiple responses possible, √ = yes)

Measures	Teaching & Learning	Research	Young Scientists	Administration/ Service	Total
Procedures & process descriptions	√		√		2
Compliance management					
Controlling					
Monitoring	√	√	√		3
Performance review					
Target agreement		√	√		2
Performance-related resource allocation	√	√	√		3
Standardized appointment of professors		√	√		2
SWOT[a] analysis	√	√	√		3
Benchmarking		√	√		2
Evaluation system	√				1
Self-evaluation					
Peer-evaluation	√	√	√		3
Student evaluation	√				1
Satisfaction survey	√			√	2
Improvement management	√			√	2
Key performance indicators	√	√	√		3
Meta-evaluation	√				1
Other instruments					
Total	11	8	9	2	30

Source: Design by authors. [a]Strengths, weaknesses, opportunities, and threats.

studies of the kind presented in this chapter must increase if the QCI is to provide a valid benchmarking option. Moreover, the QCI needs to be professionally translated into English in order to broaden the range of international institutions of higher education and for-profit organizations that assessments of quality culture can reach. Lastly, longitudinal investigations are desirable in order to identify antecedents and consequences of quality culture. With this clear agenda for further conceptual and empirical work on the QCI, the prospects for this instrument's future—and that of the people and institutions it may serve—look dynamic indeed.

Acknowledgments The heiQUALITY Cultures Project was funded by the Innovation and Quality Fund of the Ministry of Sciences, Research, and Arts, Baden Württemberg, Germany (Innovations- und Qualitätsfonds (IQF) des Ministeriums für Wissenschaft, Forschung und Kunst, Baden-Württemberg, Deutschland). Term: 04/2012–05/2015.

References

Ali, H. M., & Musah, M. B. (2012). Investigation of Malaysian higher education quality culture and workforce performance. *Quality Assurance in Education, 20,* 289–309. doi:https://doi.org/10.1108/09684881211240330

Bleicher, K. (2011). *Das Konzept Integriertes Management. Visionen—Missionen—Programme* [Integrated management: Visions—Missions—Programs]. Campus: Frankfurt.

Bologna Declaration. (1999). The Bologna Declaration of 19 June 1999. Joint Declaration of the European Ministers of Education. Retrieved from http://www.eua.be/eua/jsp/en/upload/OFFDOC_BP_bologna_declaration.1068714825768.pdf

Brodbeck, F. C., Anderson, N., & West, M. (2000). *TKI: Teamklima-Inventar* [Team climate inventory]. Göttingen: Hogrefe.

European University Association (2005a). *Developing an internal quality culture in European universities: Report on the quality culture project, 2002–2003.* Brussels: European University Association. Retrieved from http://www.eua.be/eua/jsp/en/upload/QC1_full.1111487662479.pdf

European University Association (2005b). *Developing an internal quality culture in European universities: Report on the quality culture project, Round II—2004.* Brussels: European University Association. Retrieved from http://www.eua.be/eua/jsp/en/upload/QCII%20Report%2030.03.05.1115967574238.pdf

European University Association. (2006). *Quality culture in European universities: A bottom-up approach: Report on the three rounds of the quality culture project 2002–2006.* Brussels: European University Association. Retrieved from http://www.eua.be/eua/jsp/en/upload/Quality_Culture_2002_2003.1150459570109.pdf

Heinitz, K., & Rowold, J. (2007). Gütekriterien einer deutschen Adaptation des Transformational Leadership Inventory (TLI) von Podsakoff [Psychometric properties of a German adaption of the Transformational Leadership Inventory (TLI) by Podsakoff]. *Zeitschrift für Arbeits-und Organisationspsychologie A&O*, 51, 1–15. doi:https://doi.org/10.1026/0932-4089.51.1.1

Jackson, P. R. (2004). Employee commitment to quality: Its conceptualisation and measurement. *International Journal of Quality and Reliability Management, 21*, 714–730. doi:https://doi.org/10.1108/02656710410549073

Loukkola, T, & Zhang, T. (2010). *Examining quality culture: Part 1—Quality assurance processes in higher education institutions.* Brussels: European University Association. Retrieved from http://aic.lv/bolona/2010/Reports/Examining%20Quality%20Culture%20Part%201.pdf

Sattler, C., Sonntag, K., & Götzen, K. (2016). The Quality Culture Inventory (QCI)—An instrument assessing quality-related aspects of work. In B. Deml, P. Stock, R. Bruder, & C. M. Schlick (Eds.), *Advances in ergonomic design of systems, products, and processes* (pp. 43–56). Heidelberg: Springer.

Schein, E. H. (2010). *Organizational culture and leadership* (4th ed.). San Francisco: Jossey-Bass.

Sonntag, K., Schaper, N., & Friebe, J. (2005). Erfassung und Bewertung von Merkmalen unternehmensbezogener Lernkulturen [Assessment of organizational learning cultures]. In Arbeitsgemeinschaft Betriebliche Weiterbildungsforschung e.V./Projekt Qualifikations-Entwicklungsmanagement (Eds.), *Kompetenzmessung im Unternehmen: Lernkultur- und Kompetenzanalysen im betrieblichen Umfeld* (pp. 19–340). Münster: Waxmann.

Sonntag, K., Stegmaier, R., & Schaper, N. (2016). Organisationsdiagnose: Strukturale und kulturelle Merkmale [Diagnosis of organizations: Structural and cultural features]. In K. Sonntag (Ed.), *Personalentwicklung in Organisationen: Psychologische Grundlagen, Methoden und Strategien* (pp. 255–294). Göttingen: Hogrefe.

Sonntag, K., Stegmaier, R., Schaper, N., & Friebe, J. (2004). Dem Lernen im Unternehmen auf der Spur: Operationalisierung von Lernkultur [Learning in organizations: Operationalizing learning culture]. *Unterrichtswissenschaft, 32,* 104–127. Retrieved from http://www.pedocs.de/frontdoor.php?source_opus=5809

Staufenbiel, T., & Hartz, C. (2000). Organizational Citizenship Behavior: Entwicklung und erste Validierung eines Messinstruments [Organizational citizenship behavior: Development and validation of a measurement instrument]. *Diagnostica, 46*, 73–83. doi:https://doi.org/10.1026//0012-1924.46.2.73

Trivellas, P., & Dargenidou, D. (2009). Organisational culture, job satisfaction and higher education service quality: The case of Technological Educational Institute of Larissa. *The TQM Journal, 21*, 382–399. doi:https://doi.org/10.1108/17542730910965083

Zeitz, G., Johannesson, R., & Ritchie, J. E. (1997). An employee survey measuring total quality management practices and culture: Development and validation. *Group and Organization Management, 22*, 414–444. doi:https://doi.org/10.1177/1059601197224002

Chapter 10
Agnotology: Ignorance and Absence, or Towards a Sociology of Things that Aren't There

Jennifer L. Croissant

If we knew what we were doing, it wouldn't be research.
Albert Einstein

Studying ignorance invites bad puns and awkward moments of self-reflection. Proctor (1995) claims that we "know very little about ignorance" (p. 1), and the case studies in the important volume *Agnotology: The Making and Unmaking of Ignorance* (Proctor & Schiebinger, 2008) are meant to encourage thinking about the "structural production of ignorance" (p. 3). Like new work "making ignorance an ethnographic object" (Mair, Kelly, & High, 2012, p. 1), this chapter is meant to be a continuation of that inquiry, another contribution to the conversation on ignorance. It is meant, however, to expand the problems of ignorance, particularly those which are matters of absent knowledge, to be a more specific set of cases in the consideration of absences more generally. Or conversely, considering other things that aren't there sheds light on some finer distinctions that might be made within the emerging framework of agnotology, particularly the distinction between absent knowledges as forms of non-knowledge in relation to other agnoses, such as alternative, controversial, illusive, rejected, or otherwise erroneous knowledges (see Machlup, 1980, pp. 144–152, for these categories of what he terms "negative knowledge") which are not matters of absence per se. This chapter is organized into two parts: The first considers agnotology and other studies of ignorance from their various disciplinary origins, continuing with a discussion of privatives and other forms of absence. The end result is a set of clarifications that are meant to enhance the study of ignorance and absences through examining their points of contact and divergence.

This chapter is a revised and updated version of its original publication (Croissant, 2014). With permission from Taylor and Francis. See http://www.tandfonline.com/doi/full/10.1080/02691728.2013.862880

J. L. Croissant (✉)
Department of Gender and Women's Studies, University of Arizona, Tucson, AZ, USA
e-mail: jlc@email.arizona.edu

P. Meusburger et al. (eds.), *Geographies of the University*, Knowledge and Space 12,
https://doi.org/10.1007/978-3-319-75593-9_10

Agnotology: Properties of Ignorance

In the emerging field of studies of ignorance, tracing its relation to synonyms is traditionally the first step in the project (Bernstein, 2009a, 2009b; Gross, 2010; Proctor, 2008, p. 2; Rescher, 2009), and thus I feel free to continue that fine practice and point out some of the interesting associations that the term ignorance has, as a connotative and denotative referent. Terms include benightedness, bewilderment, blindness, callowness, crudeness, darkness, denseness, disregard, dumbness, empty-headedness, fog, half-knowledge, illiteracy, incapacity, incomprehension, innocence, inscience, insensitivity, lack of education, mental incapacity, naiveté, nescience, oblivion, obtuseness, philistinism, rawness, sciolism, shallowness, simplicity, stupidity, unawareness, unconsciousness, uncouthness, unenlightenment, unfamiliarity, unscholarliness, and vagueness. Technical terms that have emerged include *nichtwissen*, *negative knowledge*, and *non-knowledge*, and, of course, *agnotology*. Known unknowns are an understood feature of standard scientific practice in universities—expected knowledge not yet verified through hypothesis-testing or discovery processes.[1] The presence of terms that have contrasting connotations within the list of synonyms (innocent and uncouth, for example), as well as the vast interdisciplinarity of the scholarship on the topic lead to a great deal of confusion and complexity in studying ignorance. Fields include library and information science, psychology, cognitive science, philosophy, sociology, and history, plus popular commentary and disciplinary approaches in the humanities.

Agnotology is a term rechristened from *agnatology* by historians Proctor and Schiebinger (2008), who so handily encapsulated some new and important cases of problems of ignorance in a potential program of study. The categories and examples reported by Proctor and Schiebinger include the identification of ignorance as a native state, a lost realm, selective choice, passive construct, strategic ploy or active construct, and as moral caution or mode of resistance. Throughout this chapter the terms *non-knowledge*, *ignorance*, and *agnosis* will be used. Like Gross (2010), Smithson (1989), and others (Kerwin, 1993; Ravetz, 1993), I am seduced into yet another typological work on ignorance but wish, like Gross, to avoid a linear typology and to integrate multidisciplinary tools and a comparison with theories of absence to improve clarity in thinking about ignorance.

Smithson (2008) argues that a typology of ignorance must make consistent distinctions and be consequential (pp. 211–212), and thus I propose here five important attributes that might be applied to case studies of ignorance or non-knowledge. My goal is not some fully formed generalizable theory of ignorance,

[1]As institutions of knowledge production, universities have an undertheorized role in ignorance production. Disciplinarity is taken to task, and interdisciplinarity or trans- or multidisciplinarity are seen as solutions. Similarly, methodologies such as participatory action research are meant to bring the emic knowledge of people's day-to-day lives into closer contact with emic, analytical, professional, or disciplinary knowledge. The connections between the disciplines, non-knowledge, and the institution as organizations are, however, not sufficiently theorized.

but rather a framework which allows cross-case comparisons and systematic sociological analysis across different domains, even as the challenges of postmaterialist (Latour, 2007) and postrepresentational theories (Barad, 2007) and anthropological specificity (Bille, Hastrup, & Sørensen, 2012; High, Kelly, & Mair, 2012) warn against such a project. (1) To speak of the ontology of ignorance refers to questions about the presence or absence of knowledge, both in relative terms, such as when one person knows something someone else is ignorant about, but especially in regards to the question of uncertainty. As I discuss in greater depth below, there are forms of uncertainty that are the results of as yet unrefined models, and others that are foundational to stochastic or probabilistic phenomena. (2) Chronicity refers to issues of time and the prospective and retrospective elements of knowledge and non-knowledge identification. (3) Granularity refers to the *texture* of the (non) knowledge—concrete facts of specific purview having a fine granularity, while broad statements of knowledge or domains of inquiry have a coarse granularity. (4) Scale, related to but independent of granularity, has more to do with the assignment of the level of analysis and causality in the assessment of ignorance. This can range from individual information or cognitive processes; institutions and organizations such as universities or workplaces; to the workings of cultural formations and ideologies which shape broad matters of interest, inquiry, and explanation. The final dimension discussed has to do with (5) intentionality, whether from direct intent such as fraud or hoaxes, and various forms of censorship, nondisclosure, or knowledge nontransmission, to inadvertent ignorances unconsciously produced through various effects. I review each of these attributes below.

Ontology and Epistemology

Oreskes and Conway (2008) discuss the ignorance surrounding the issue of global warming. As outlined in their work, there has been enough scientific consensus and a multiplicity of evidence since the mid-1990s to be able to state empirically that anthropogenic global warming is occurring. However, conservative think tanks have stressed the *uncertainty* of the models and the evidence. This represents a conflation of the meaning of uncertainty as a probabilistic outcome, and uncertainty meaning unknown or unreliable. Smithson (2008) also argues that uncertainty and ignorance are often conflated in their respective literatures (p. 214; Tapp, 2000), while Michaels (2008) similarly situates problems of agnotology in the realm of problems of probabilistic outcomes, rather than problems of the existence of knowledge in contestations over toxic chemicals and public health (see also Japp, 2000). This points us in the direction of a distinction between knowledges that are uncertain as unknown-at-this time, to be solved with more research or better modeling, versus knowledges that are fundamentally based on stochastic processes by which probability, and thus uncertainty, are endemic to the system.

Rescher (2009) frames this important distinction as between ignorances which are epistemological versus those which are ontological (pp. 100–101). That is, he helps us to distinguish between those agnoses that are what he calls “cognitively inaccessible” because we do not yet know, where ignorance is a matter of epistemology, versus things that are unknowable as a matter of property, that is, “developmentally open” via causal indeterminacy or contingencies of “choice, chance, or chaos” (p. 101) as matters of ontology. Gross (2010) capitalizes on the concept of the uncertain in his analysis of surprises: “Ignorance and surprise belong together” (p. 1). The management of ignorance and its resulting surprises is a major feature of knowledge-based societies (Beck, 2007/2009). Gross (2010) clarifies Merton’s (1987) specified ignorance in relation to the terms nescience, non-knowledge, and negative knowledge. In Gross’s (2010) typology, non-knowledge is related to Merton’s (1987) specified ignorance, the ignorance of a discipline which knows where the unknowns are. Where is that damned Higgs boson?[2] Thus, a hypothesis is a kind of specified ignorance, based on the presumed existence of a known unknown.

So, as outlined by many scholars, certain kinds of ignorance are the after-effects of knowledge processes, including the identification of known-unknowns and future work, or Merton’s (1987) specified ignorance. For example, Jacklin, Robinson, and Torrance’s (2006) discovery of a lack of data about children in public care here qualifies as a form of specified ignorance. In the spirit of Foucault, forms of ignorance are the necessary dual effects of knowledge productions. Relational ignorances can be matters of nontransmittal, also perhaps the result of dual effects, but more specifically knowledge that is available in one realm of social action but absent in another, whether by intention or not.

The question of the ontology of ignorance thus requires clarification as to a specific agnosis’s relationality and epistemological features: Someone somewhere knows something, someone elsewhere does not. Someone knows there is something to be known. That which is to be known may be based on probability or stochastic processes which have a residual uncertainty. These knowings and non-knowings are not patternless, but neither are they completely specified or structured.

Chronicity

The epistemological relationality of ignorance is closely paralleled by but not identical to the issues of time in assessing knowledge and agnosis. If the epistemology of ignorance is in part locative in terms of spaces (metaphoric social spaces, literal geographies), then time needs to be figured carefully in discussions of ignorance, as there are forms of agnosis which figure as the not yet known, and

[2]When this text was first presented as a conference paper in 2009, the Higgs-boson particle was an important missing part of establishing the Standard Model in physics. It has since been discovered (2012) and subsequently verified, hence, the problem of chronicity and timekeeping in ignorance studies. See https://home.cern/topics/higgs-boson.

others as the forgotten or obliterated. In addition, social power is operative in similar but not identical ways in producing epistemological ignorances co-constituted with the production and use of ignorance over time.

To theorists such as Giddens, or Simmel, non-knowledge can only be determined in retrospect (Gross, 2007). Nescience is sometimes considered the vocabulary for evaluating ignorance identified in retrospect (Knorr-Cetina, 1999). However, in the same way as historians of science struggle with anachronism in attributing knowledge and non-knowledge to prior regimes of thought and connecting past to present, the problem of time in prospection and retrospection produces specific challenges of inference and attribution. Consider Simon (2002) and his account of cold fusion as an "undead science." Cold fusion, in mainstream accounts and across much of the scientific (particularly physics) community, is dead and discredited. It is, however, still researched (if under a different name), producing rare and difficult-to-explain effects in reputable scientific labs. The possible futures of cold fusion as (in)credible knowledge produce difficulties in reading its past or stating in the present what is known or unknown. This indeterminacy is a matter of epistemology over time (rather than ontology based on probabilistic uncertainty). It is also of a different kind of indeterminacy than the puzzling out of "who knew what when," in cases of negligence, or the case of tobacco companies' obfuscations about harm (Oreskes & Conway, 2010).[3] Nonetheless, establishing chronology and location are essential projects for agnotology.

Granularity

Galison (2008) discusses the classified universe of restricted documents and the processes of classification of secret materials. His insight is the obverse of the usual "knowledge transmission" or replication questions of science studies (e.g., Collins, 1981, 1992), instead about the mechanisms in the prevention of knowledge transmission, as well as an inquiry into what I will call the *granularity* of knowledge. Concrete, factual statements have a higher (or fine) granularity. Galison (2008, p. 52) refers to specific statements as *punctiform* which can be subject to specific forms of censorship, even though in their formulation much can be deduced that is of more theoretical or of lower (or coarse) granularity, and vice versa. The highest granularity of knowledge might be considered like Bertrand Russell's and Ludwig Wittgenstein's ideas for "atomic propositions" (p. 50), some smallest units of intelligible meaning. While specific statements are the most clearly identifiable, and presumably the most easily managed through censorship practices, Galison

[3]Ignorance in time or space is what makes mystery writing work. Someone, if no one else but the author, knows "who done it," while some combination of characters and the reader may or may not be in on the secret, creating suspense. Similarly, most but not all stage magic works through withholding information—how did they do that?—in conjunction with distraction and dissembling.

argues that the excision of punctiform or high-granularity knowledge quickly expands to broad scope and impossible censorship of knowledge domains, because facts are not independent of theory in any straightforward way, thus a tendency to decreasing granularity in classification. The concepts of granularity and the concept of entropy as error/ignorance are the two places at which this framework intersects with information theory and computing. Absent bits of information in communication streams or storage media are discrete elements of knowledge, referred to in terms of granularity as a measure of their size. While censorship is intentional, disciplinary and departmental boundaries produce nontransmissions of knowledge via structural means, which can have high or low granularity.

Scale

In matters of scale, questions of agnotology need to consider both origins or causal processes, and consequences, or the reach of ignorance. If granularity refers to the size of the knowledge to be transmitted, scale refers to the components and systems in which that knowledge, whatever its granularity, might circulate. Is it an individual's knowledge, or lack thereof, that is in question, and in what relation does it stand to various other assemblages which might constitute knowledge or agnoses? The agnoses of disciplines in universities, especially between disciplines, the blind spots of research paradigms, and cultural formations are examples of scales of ignorance.

There are several research traditions focused on ignorance and error at various levels of analysis. For example, Rescher (2009) is focused on ignorance in relation to error as matters largely of cognition based in an analysis of logic from philosophy. For Rescher, ignorance is a matter of individual reasoning, and there is ample work in this area. Similarly, Watts (2011) provides a critique of common sense which is a popularized discussion of similar issues: the conflation of correlation and causation, the cultural and contextual specificity of common sense, and the problems of confirmatory bias and the inadequacy of folk sociology. Watts is concerned with the inability of social science to be predictive or to produce laws in the way that physics does (but see Cartwright, 1983), but his contribution to this discussion of ignorance is his review of the systematic errors in inference produced by common cognitive processes, such as retrospective inference and confirmatory biases.

Organizational theory represents a middle range for the exploration of ignorance, connecting individual cognition to organizational forms and processes. Ten Bos Rene (2007) observes that there are two frameworks for the question of stupidity and organization: "The older one claims that organization in fact needs a certain dose of stupidity and the newer one takes it that stupidity should be banned from organization" (p. 140). Pollitt (2000) observes that re-organization, personnel change, archival practices and changes in storage media, and organization fads produce an organizational amnesia, a set of forgettings that lead to wheels being reinvented, an inability to learn from past lessons, inefficiencies, and ineffectiveness within organizations.

Vaughan (1999) and, of course, Perrow (1984) (and back to Merton (1987) and Weber (1922/1964)) argue that all organizational forms have pathologies. Vaughan's (1999) perceptive analysis of mistake, misconduct, and disaster points to the communication components of ignorance, that is, that organizations produce ignorance, and thus the possibility of mistakes, through compartmentalization and structural secrecy. But there are other structural components, too. For example, centralization has its trade-offs; routine-following can produce error through oversubscription or misapplication of rules. In addition, Vaughan notes that "all judgments are made under conditions of imperfect knowledge, thus routine non-conformity is a normal by-product of techno-scientific work" (p. 279).

The framework for understanding organizations and ignorance can be extended through the examination of organizational cultures, such as described by Eden (2004) or Vaughan (1996), where intraorganizational processes lead to blind spots, prioritizations of data, and intra- and interorganizational competition for resources and prestige leads to ignorances of omission or distortion. For example, Eden (2004) examines the lack of knowledge of fire effects from nuclear weapons as a matter of professionals and organizations focusing on what they do well and excluding that which eludes them, leading to substantial misrepresentations of the world in which they work. In Eden's analysis, the agnosis about fire effects after nuclear detonation led to a mass overproduction of strategic nuclear weapons. In the case of post-Katrina environmental contamination, testing protocols are sedimented into disciplinary regimes and organizational practices, producing ignorance about ecological and sociohistorical contexts and thus the distribution of risks across the landscape (Frickel & Vincent, 2007).

And, of course, there are ignorances of a broader scale, wrapped up in economic, political, cultural, and ideological processes. As Hess (2009) articulates, "social change agents face . . . an often lopsided field of scientific research" (p. 306). Social movements often confront an area of "undone science" which would be useful to them but remains underfunded. His study of civil society research, such as environmental nongovernmental organizations (NGOs) providing research reports in support of movement goals, suggests an alternative to traditional routes for research agenda-setting in science, which is dominated by for-profit and governmental funding organizations. Research in universities is expensive, and access to knowledge production is thereby limited.

Oodshorn (2003) and Daniels (2006) both provide case studies which examine the ways in which configurations of masculinity have led to a lack of technological development of male birth control options for the former, and a lack of research on male reproductive health, particularly its environmental constituents, for the latter. Proctor's (1995) work is concerned with the politics and economics of cancer research shaping what is known and unknown, as the project of "curing cancer" has far more prestige and resources in relation to the project of "preventing cancer."

Finally, there are the frameworks which articulate deep epistemological rifts in knowledge, such as Kosofsky Sedgwick's (1990) analysis of the ways in which the homo/heterosexual binary produces non-knowledges that shape understandings of sexuality and subjectivity. And while Foucault (1994) rarely mentions ignorance or non-knowledges explicitly, his archaeologies and genealogies are rife with examples of things/bodies/identities elided by epistemological formations. Similarly, Butler's (1993) conception of the abject, while not specifically about ignorance per se, suggests the production of zones of unintelligibility where might be found that which escapes, exceeds, or is cast out of normative modes of being/knowledge.

Scholars from postcolonial studies (de Sousa Santos, 2016) and critical race theory (Mills, 2008) provide examples of epistemological ignorances on a broad scale, in the impossibility of knowing the "Other", particularly under conditions of subjugation. Marx's dialectics of the master-slave relationship inform feminist theory (de Beauvoir, 1949/2009), colonial relations (Fanon, 1952/2008), and black feminist thought (Collins, 2000) in that the subordinated can and must as a matter of survival have knowledge of the master who cannot know the other and cannot be interested in knowledge of the subjected, for that would require recognizing the humanity of the other. Other postcolonial theorists (Said, 2003) similarly articulate the production of ignorance in disciplinary and popular representations of "Others" under colonial relations. One of the major attributes of ignorance requiring articulation is thus the matter of scale, examining processes by which knowledge and agnoses are constituted across assemblages based on size and complexity and overdetermined by power relations.

Intentionality

Beck (2007/2009, p. 126) develops a typology which focuses on the intentionality of the knower: willful ignorance in relation to a conscious inability to know (we know we don't know). The next types are the unconscious non-knowing that "does not reflect on its own limits" and finally the unknown unknown, which provides the "element of surprise." Beck uses as an example the willful ignorance of denying the effects of global climate change to discuss "side-effects" as things that might be unknown, but when known and not acted upon can intensify the effects the system producing the (side-)effects.

There are numerous examples of intentionally produced agnoses. Tuana's (2008) analysis of the erasure of knowledges about the female orgasm and the structure and function of the clitoris, or Schiebinger's (2004) study of the nontransmission of knowledge of the abortifacient properties of bird of paradise plants from colonial contexts to the metropoles of the "long eighteenth century" are examples where race, gender, and culture produce absences of knowledge through nontransmission. Mayor (2008) describes the suppression and neglect of native American and related indigenous groups' paleontological knowledge of fossils as a result of the dismissal by the colonizers of native knowledges as mere myths and legends of barbaric others. Moore and Tumin (1949) posited the functionality of ignorance, for example in preserving privileged positions such as between experts and consumers or competitors. Their framework is ambivalent about the relationship between function and intention—most of their examples, such as producing anxiety about performance through withholding feedback to spur greater productivity in competitive arenas (pp. 793–794) suggest the production and maintenance of ignorance can be an organizational or interpersonal strategy.

Dismissals and suppressions of knowledges are not identical with the production of falsehoods as non-knowledges, such as fraud, hoaxes, or propaganda (Bernstein, 2009a, 2009b), which are intentional distortions of knowns, although they may be similarly motivated by a multiplicity of factors linked together by considerations of social power.[4] Thus, ignorance has its uses, as the utility of non-knowledge is produced in relation to the intention of its locutors. For example, the sites of the Salem witch trials in seventeenth-century New England were effaced, as are many locations of violent crime, primarily in shame (Foote, 1990).

Social conventions, particularly around privacy and politeness (Smithson, 2008), produce intentional nondisclosures, whether they are of the "too-much-information" variety, or things we really don't want to know about or disclose to our conversational partners and mere acquaintances. Surprise parties require withholding information, although revealing the ending of a story does not necessarily mean ruining it (Leavitt & Christenfeld, 2011). Frickel and Vincent (2007) discuss strategic not-wanting-to-know with regards to Hurricane Katrina, as real estate values and environmental justice outcomes will be shaped by the potential (non)identification of toxic accumulations in the soil.

Ignorance is useful. Bernstein (2009a, 2009b) reminds us that non-knowledge and nonsense are frequently found in the realm of literature and philosophy, such as in Bataille (2001). Knorr-Cetina (1999) identifies *nichtwissen* as knowledge where the limits of knowledge are important to future action and planning, as opposed to negative knowledge which is a deliberate choice not to engage knowledge in a particular direction (as it is presumed to be unimportant) (Gross, 2007, p. 749). The productive nature of non-knowledges is identified in surrealism, for creativity and spiritual enrichment, and for innovation. Smithson (2008) reminds us that some form of ignorance is necessary for creativity and problem-solving. The production of ignorance is part of the work of ideologies and propaganda, and to conspiracy theorists, a necessity for the ever-oppressive state. So perhaps a refinement of the Enlightenment dictum that *knowledge is power*, already turned on its head by Foucault's (1994) power/knowledge formulation, must be refined by considering power/agnosis in its various manifestations.

With the properties of chronicity, scale, granularity, ontology, and intentionality as ways of describing ignorances, we are in a better position to consider a broader range of comparisons across case studies, and to include the dynamism and relationality that undergirds many, but not all, forms of ignorance. Further refinement to our consideration of ignorance has to do with its points of connection to interdisciplinary scholarship on absences.

---

[4]The problem of *fake news* and propaganda in recent U.S. political discourse can benefit from the framework presented here, but a comprehensive discussion is beyond the scope of this chapter.

Fig. 10.1 "Cold in the Abstract": Lay versus professional understandings of temperature.
Source: *New York Times*, December 19, 1886.
Copyright: Public domain.

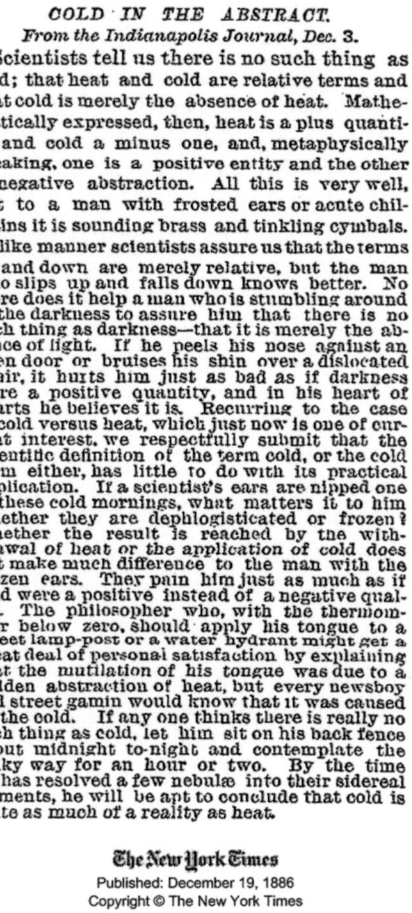

COLD IN THE ABSTRACT.
From the Indianapolis Journal, Dec. 3.

Scientists tell us there is no such thing as cold; that heat and cold are relative terms and that cold is merely the absence of heat. Mathematically expressed, then, heat is a plus quantity and cold a minus one, and, metaphysically speaking, one is a positive entity and the other a negative abstraction. All this is very well, but to a man with frosted ears or acute chilblains it is sounding brass and tinkling cymbals. In like manner scientists assure us that the terms up and down are merely relative, but the man who slips up and falls down knows better. No more does it help a man who is stumbling around in the darkness to assure him that there is no such thing as darkness—that it is merely the absence of light. If he peels his nose against an open door or bruises his shin over a dislocated chair, it hurts him just as bad as if darkness were a positive quantity, and in his heart of hearts he believes it is. Recurring to the case of cold versus heat, which just now is one of current interest, we respectfully submit that the scientific definition of the term cold, or the cold term either, has little to do with its practical application. If a scientist's ears are nipped one of these cold mornings, what matters it to him whether they are dephlogisticated or frozen? Whether the result is reached by the withdrawal of heat or the application of cold does not make much difference to the man with the frozen ears. They pain him just as much as if cold were a positive instead of a negative quality. The philosopher who, with the thermometer below zero, should apply his tongue to a street lamp-post or a water hydrant might get a great deal of personal satisfaction by explaining that the mutilation of his tongue was due to a sudden abstraction of heat, but every newsboy and street gamin would know that it was caused by the cold. If any one thinks there is really no such thing as cold, let him sit on his back fence about midnight to-night and contemplate the milky way for an hour or two. By the time he has resolved a few nebulæ into their sidereal elements, he will be apt to conclude that cold is quite as much of a reality as heat.

The New York Times
Published: December 19, 1886

Absences, More Generally

Figure 10.1 is an editorial clipped from the December 19, 1886, *New York Times*, reprinted from the *Indianapolis Journal* of December 3 of that year. It encapsulates for us two important issues in study of absences in general and ignorance in particular, the issue of privatives and the issue of symmetry. Coldness, and darkness or silences, are understood scientifically as privatives—as negative abstractions rather than positive entities. As noted in the figure, these abstractions, while nice in theory, are irrelevant to the person with their tongue stuck to the flagpole or stumbling around in the dark.[5]

[5]Fantasist Terry Pratchett writes of darkness, and silence, as positive quantities with specific properties: "Old Tom" the bell tolls silences on the hour. See Pratchett (1990) for other literalizations like this.

Embedded within this clipping are issues concerning the public understanding of science, particularly invoking a problem of symmetry (discussed below). Considering forms of ignorance that are matters of absence calls us to think about the problems of absences more generally. As for ignorance, the possibility of cross-cultural and cross-disciplinary studies of absence is rather daunting. For example, Portuguese speakers have the term *saudade*, which is a feeling of nostalgic longing for that which is lost. Derrida (1993/1994) coined the term *hauntology* to describe a nostalgia for an imagined past and the ways in which Marxism will continue to haunt the West. Baudrillard (1981/1994) describes the *simulacrum* as the copy for which there is no original. Both hauntology and simulacrum are both taken as pejorative, in some way inauthentic even as there is no *real* referent which is doing the haunting or is the original for the simulation. The negative connotation of inauthenticity is challenged by Deleuze and Guattari (1980/1987) in positing a *double becoming* of both the referent and the simulation, both as productions of culture machines. These multiple valences of positive and negative valuation—is it real? Does it matter?—add to the complexity of studying absences.

The first point of contact between agnotology and studies of absence I will explore is the question of the privative, a specific form of absence in relation to a known presence. The privative, as a specific kind of absence, encapsulates a number of problems for the social study of ignorance and absences particularly vividly. I will explore various modalities of absence that have been studied across disciplines, focusing on silence and invisibility as key sensory systems from which things might go unperceived. Then finally this section will interrogate the problems of stupidity and symmetry, each being a special problem at the intersection of agnotology and absence.

Privatives and Absences

The Buddhist term, *avidyā,* is exactly a privative, specifically an *alpha privative* formed by the addition of an *a-* prefix. This term reflects the inherent limitations in human knowledge, and is not a lack of knowledge or scholarship, *per se*, but a lack of the knowledge of *being*, and as such the foundation for human misery. But our concern here, besides a cross-cultural interest in a consequence of ignorance, is the formulation of the negation. As a privative, *avidyā* is specifically formulated as an absence of an available (if difficult to achieve) knowledge. Of course, agnosis and agnotology are thus alpha privatives, as is absence.

Why should we care if people misunderstand privatives such as cold? Consider the general "misunderstanding" of how thermostats work. If a house is chilly, the thermostat need only be set to the desired temperature plus maybe one degree. Many, many people, however, add 10 degrees with the idea that the house will heat faster. Most people treat their thermostat as the setting to a pump, which at a higher setting will pump more heat into the house. This is not how they work. This leads to measurable waste in energy as the furnace will overshoot the desired end temperature, running for longer than necessary. Or consider variations on the joke that circulates about trying to puzzle out how a thermos knows whether to keep its contents hot or cold. That misunderstanding can only come from misunderstanding the properties of the vacuum in the walls of the thermos, and the properties of temperature in materials. This misunderstanding is a form of ignorance: individually enacted within a cultural context, but it is not an absence of knowledge *per se* as it does encapsulate a folk model of physics. It is ignorance *about* a privative, not a privative in itself.

Ignorance may be the privative of knowledge, or, more specifically, certain kinds of ignorance are the privative of knowledge, while others are not, and this distinction is important. Privative forms of ignorance should then be amenable to the same kinds of analysis as other forms of absences, and vice versa, while ignorances that are not the result of an absence but of various forms of misinformation or error require slightly different analytic considerations.

Theories and Modalities of Absences: Silences and Invisibilities

The analysis of things absent to two of our senses, sight and hearing, produce much of the scholarship on absences. There are various forms of silences and silencing practices. There are, as Bourdieu (1972/1977) notes, the things that go without saying, the *doxa* which limit thought, action, and social mobility. There are things unspeakable and forbidden, explicitly through censorship and social convention, and there are the things discursively unthinkable and thus unspeakable inhabiting abject spaces. Open secrets (Kosofsk Sedgwick, 1990) are those things that everyone knows but must not be said, in contrast with what Bourdieu (1972/1977) calls the "complicitous silence" (p. 188), the silence that sustains ideologies. Speech withheld, whether refusing interpellation or in other forms of resistance, has revolutionary potential, and yet speech withheld has been identified as "reactionary" in certain contexts (Habermas, 2002, p. 67). This points to a relational quality to absences, as absences *from* one domain in relation to another—a question of the ontology and epistemology of an absence as well that absence's consequences.

Besides being silent, things can be invisible. We know that there are invisible colleges, that the poor are rendered invisible in public life and media, that transgendered people are invisible, and that many other kinds of bodies and or facts about them are erased. Secrets, considered either as unspoken words or nondisclosed representations, are situated at an awkward nexus of individual rights to privacy, at least for those able to protect personal information as property, and the need for transparency (e.g., visibility) as a part of the social contract of contemporary public life in liberal democracies (Harris, 2009).

Casper and Moore (2009) articulate the ways in which bodies can be invisible, as matters of representation, or figure as missing in more literal ways. Calling for an ocular ethic to complement Rose's (2008) somatic ethic, Casper and Moore (2009) ask us to consider why some bodies are valued for their invisibility, and others valued for their hypervisibility. Similarly, Gordon (2008) recognizes "[v]isibility is a complex system of permission and prohibitions, of presence and absence, punctuated alternately by apparitions and hysterical blindness" (p. 193). Like Rapp (2000) and the concept of stratified reproduction, Casper and Moore (2009) suggest that complex webs of valuation, most evidently by sex/gender, sexuality, race, and class but also by perceived ability, attractiveness, and other ascribed and achieved characteristics, are tied to processes of valuation of bodies in relation to their potentials for representation and their capacity for sustaining existence.

As a postcolonial scholar, de Sousa Santos (2016) argues that several logics undergird the production of nonexistence: "Non-existence is produced whenever a certain entity is disqualified and rendered invisible, unintelligible, or irreversibly discardable. What unites the different logics of production of non-existence is that they are all manifestations of the same rational monoculture" (p. 172). Here tying nonexistence to ignorance, de Sousa Santos (2008) argues that the Western monoculture of knowledge with the elevation of science as the sole arbiter of knowledge, produces "non-existence . . . in the form of ignorance or lack of culture" (p. 238). The other logics include the monoculture of *linear time*, of *classification* which naturalizes differences and hierarchies, of privileging the global and erasing the local, and the logic of *productivity* which privileges growth and market logics. De Sousa Santos argues these are "forms of non-existence produced by hegemonic epistemology and rationality" (p. 239) to be confronted by a sociology of absences.

Looking briefly at silence and invisibility, I note that the properties of scale, granularity, chronicity, ontology, and intentionality that provide a framework for comparing and contrasting studies of ignorance can similarly organize and inform studies of absences. These studies are not identical, however, and each produces problems related to symmetry, formulated as epistemological (Collins, 1981), methodological (Bloor, 1976), or the generalized symmetry of Latour (1992).

Symmetry and Stupidity

Bernstein (2009a, 2009b) argues for a symmetric approach to the categories of knowledge and non-knowledge, despite the argument by Ten Bos Rene (2007), who considers stupidity "an independent quality with a logic all its own" (p. 147; see also van Boxsel, 1999/2003). For example, "Terms associated with knowledge at all levels can usually be matched with approximate counterparts in the domain of nonknowledge" (Bernstein, 2009a, p. 27). However, Tuana (2008) warns that "while the movements and productions of ignorance often parallel and track particular knowledge practices, we cannot assume that their logic is similar to the knowledges that they shadow" (p. 110). Bernstein's (2009a) goal is a classificatory one in support of libraries and their need to order knowledges: Where would knowledge about non-knowledges be classified?

Ignorance may be useful, or may have a socially or psychologically adaptive mechanism, and much the same can be said for stupidity: "All our organizations work by virtue of stupidity. Our world revolves around fantasies and around fools who believe in them. Stupidity is useful" (van Boxsel, 1999/2003, p. 43). However, both ignorance and stupidity are often taken as problematic, with stupidity in particular framed as non-knowledge that is self-defeating (Welles, 1986). "Stupidity is the talent of acting unwittingly against your own best interests, with death as the ultimate consequence" (van Boxsel, 1999/2003, p. 31).

Garcia (1997) also points to a potentially moral dimension to ignorance—ignorance in decision processes can be an indication of choosing stupidity by avoiding responsibility. Burt (2005) describes how public attribution of stupidity, particularly in public political discourse, has the effect of "an advanced Orwellian double-speak in which the stupid masquerades as the smart, the zealot as the skeptic" (p. 30). Public attributions of stupidity, then, are political. Take, for example, a recent argument that young people are too stupid to vote, or other vitriolic exchanges of attributions of stupidity in the public sphere (Thomas, 2012; see also Hardy & Clark, 2005; Moore, 2002). The role of social power in attributions of stupidity, and as one of the objectives of those attributions, points to a challenge to symmetry in the study of ignorances and absences.

The conventional model of analyzing lay or public understandings of science, identified as the deficit model, can be criticized for treating lay nonuse of canonical science as matters of absence, a deficit, or as matters of distortion (Wynne, 1995). Consider again the newspaper clipping from Figure 1. How might the difference between public understandings of cold and thermodynamic understandings of cold be treated symmetrically, especially when in this context they have approximately the same behavioral outcome: Do not stick your tongue to freezing cold metal posts.

As Christensen (2008) notes, symmetry as a journalistic norm for reporting *both sides* of a controversy can produce ignorance, as equal weight ends up given to knowledge statements either intentionally misleading or otherwise marginal or discreditable. "Knorr-Cetina and I [Michael J. Smithson] have accurately identified the main problem here, namely[,] that anyone referring to ignorance cannot avoid making claims to know something about who is ignorant of what" (Smithson, 2008, p. 210). High et al. (2012) are not concerned with knowledge gaps as recognized or adjudicated by social science analysis. Instead, they focus on ignorances that are culturally recognized by participants. As anthropologists, they claim there is little to be done to draw universal conclusions about ignorance or its relations to "comparable phenomena such as stupidity, error, and confusion" (p. 17).

In the sociology of scientific knowledge, Bloor's (1976) argument for symmetry in the sociological explanation of both true and false beliefs helped shepherd in transformations of social studies of science. Symmetry is similarly demanded of actor-network theory, although the methodological principle is that the distinction between the technical and the social (or political) is an outcome of actors' articulations and not an a priori attribution. However, through the ostensible collapse of the social (Latour, 2007) as an explanans, there still remains the problem of sorting out whether or not the network and assemblages are the explanans or the explanandum, assuming explanation is in fact the goal of agnotology and related studies of absences.

Conclusions: Studying Things that Aren't There

> Absence is therefore not just a theoretical concept implied as the default logical antonym to presence; it is also a corporeal, emotional, and sensuous phenomenon articulated in discretely concrete, political, and cultural registers. (Bille, et al., 2012, p. 12)

The projects of agnotology and absence (should that be absentology?) require a great deal of taxonomic work, and this chapter is meant to add to the conversation, not as a matter of lexical policing, but as a necessary step in theory-building and developing the capacity for cross-case comparisons in studies of ignorance, as well as to articulate a possible framework for studies of other kinds of absences. Unpacking the distinctions within the framework of agnotology provides insight into the multiple forms of ignorance, especially those which are indeed forms of absent knowledge. This provides a way of interrogating things that are absences more generally, illustrating some particular challenges for the social studies of science and knowledge. For example, identifying an agnosis, especially, but not solely, privative agnoses, requires a suspension of traditional epistemological symmetry. Like the case of the scientific understanding of cold as the absence of heat rather than a substance in itself, the identification of things as privatives, or identifying ignorances as either absences or misunderstandings, requires claiming positionality as to knowing the properties of the primary referent or elemental "truth" identifying the *gnosis* to which the *a-* might be attached.

Full exploration of the challenges to symmetry will need to be taken up elsewhere, but it is clear that strict epistemological symmetry generating sociological explanations for both "true" and "false" beliefs cannot hold in studying either ignorance or absences. But neither can a generalized symmetry which eschews social causes for explanation at all: Each perhaps is to be replaced by a more modest methodological toolkit which maintains integrity across comparable levels of analysis in various case studies. We do not want to return to the pre-Bloor (1976) days of explaining "false beliefs" with sociological explanations and apparently "true beliefs" with "just-so stories," reducing the power of science studies to muckraking journalism.

Methodology

Like physicists who study black holes by their effects, sociologists and other theorists have an emerging repertoire for studying absences. A black hole is not visible, although it is not absent, and its effects on light and nearby masses are measurable. Vacuums are an absence of matter in space, and while not an object of study in themselves, as an absence in which things might be made present, vacuums highlight properties of those things: the properties of light in a vacuum, for example. Gordon (2008) adopts the term hauntology to describe the ways that various kinds of absences linger and trouble discourses as present absences. Slavery in the Americas, or the disappeared in Argentina are both absences (missing persons) and knowledge about those made absent. These haunt rationality and consciousness, both subjective and public: Hauntings are analyzed through their a-effects. Structural holes (Burt, 1995) are absences in a network or between networks. They are measurable as network phenomena: nodes or linkages that might be expected given all the mathematically possible connections in a network but are not present. These absent network features need to be explained, as do the eventual apperception and capitalization of these absences by some participants in the network, and the lack of perception of the possibilities of structural holes by other participants.

In their theory of knowledge and culture, Deleuze and Guattari (1980/1987) suggest the metaphor of the rhizome as a new model of knowledge and subjectivity, as a poststructuralist orientation that does not reproduce dichotomies of knowledge/power. They specifically oppose the rhizome, think bamboo and its structure and proliferation, to the tree, as a model of knowledge. The rhizomic principles of (1 and 2) connection and heterogeneity and (3) multiplicity means rhizomes are epistemically flat, like Latour's (2007) assemblages and networks. We might think of rhizomes through a fourth principle, what Deleuze and Guattari frame as the asignifying rupture, a way of tracing knowledges as de- and re-territorializations that are "drawings in" of features. With this drawing in, knowledges are more than simple additive collections. Deleuze and Guattari also suggest what they call aparallel evolution as their fifth and sixth principles, proposing cartography and decalcomania (a form of tracing). That is, there is no regularity in the reproduction of rhizomic extensions, and rather than representational maps knowledge should be conceived of as nonrepresentational tracings. (See, for example, maps of the London underground.) This is articulated in Barad (2007) as a post-representationalist theory of performance and functionality in knowledge production.

However, these are theories of knowledge, and ignorance and absence are themselves absent from Deleuze and Guattari's (1987/1980) method. I argue that the tracings of rhizomes nonetheless produce spaces between the lines of the rhizome or the tracings of routes as empty and as potential agnoses. In addition, they argue that rhizomes are nonhierarchical and antigenealogical, that there is no deep structure to knowledges and that "the rhizome connects any point to any other point" (p. 21) and the rhizome is not "overcoded." However, rhizomes do have dimension, length, and mass. They are not, themselves, structureless, nor are the spaces between: They have granularity. This suggests that a fractal or holographic metaphor may be more apt than that of the rhizome: that the structure of the macro is reproduced through diminishing scale (or vice versa), or that the whole is present, even if at degraded resolutions, in the parts and fragments. Similarly, like the network theory underpinning structural holes, Deleuze and Guattari make the assumption that all network relations—or rhizomes—are equally possible, which may be mathematically true. Thus, what explains the lack of rhizomes or network positions or knowledge that might have otherwise been expected in a fully articulated network?

Along with these suggestive models and metaphors, there are caveats for an emerging methodology of the sociology of absences. For example, Collins (2007) warns against the use of counterfactuals as a methodology for historiography or historical sociology. This "thought experiment" is the projection of the presence or absence of a person, place, thing, or specific event changing "the course of history." Collins finds the use of counterfactuals to interrogate historical events as misrepresenting historical causality and the scale and scope of historical forces. To shift the frame slightly to apply to a methodology for absences, looking for absences as causal features of social life must be done carefully to avoid anachronistic fallacies and related logical errors produced by counterfactual thinking.

Two other hesitations warrant consideration at this time for our emerging methodology for the analysis of ignorance: The first is the aphorism that "absence of evidence is not evidence of absence" Like black swans and other absences, the not-yet-ness of evidence challenges the easy attribution of knowledge and non-knowledge. Originally appearing in print to justify long-term investment in searches for extraterrestrial intelligences, the aphorism is considered a logical fallacy and is frequently deployed to shift the burden of proof.[6] For example, for the stereotypical conspiracy theorist, the absence of evidence for conspiratorial activities is taken, at face value, as evidence *of* the conspiracy. The second problem not easily resolved is the matter of imputation: Based on the presumed inaccessibility of others' minds, our agnoses of each other's motives, the imputation of intentionality, while an important dimension of the politics of agnotology and absence, is a fraught project. What would a "symmetric" analysis of these attributions of knowledge and motive to others in the absence of evidence look like?

What do we know about ignorance and absence?

Ignorances can be distinguished by kind and by degrees, requiring attention to the factors identified above as granularity, chronicity, scale, intentionality, and ontology. Some forms of ignorance are absences, some are errors, and those that are privatives can benefit from some of the theoretical and methodological resources from fields concerned with the identification and study of absences. Conversely, applying ideas such as granularity, intentionality, ontology, chronicity, and scale can inform inquiries into the production and structuration of absences. Not every one of these properties will be of the most analytic or political interest for making cross-case comparisons, but these properties do provide some traction in theorizing agnoses and absences.

While physicists worry about their "theory of everything" and the integration of general relativity and quantum physics remains elusive (itself a form a disciplinary specified ignorance which is epistemological and of low granularity), what I propose

---

[6]The "absence of evidence" quote is attributed to astrophysicist Martin Rees and quoted in Sagan (1995, p. 213).

is the articulation of the sociology of nothing,[7] or a project similar to that of agnotology which will articulate methodological parameters necessary for studying things that aren't there. What do we know about the things that aren't there? That they can take many modalities based on their absence from our senses and discursive practices; that they are constituted in systems of stratification and valuation which render these absences (il)legible; that they are institutionalized; and that we can study them by their effects, ever attentive to the complexity of inferences about absences.

By moving back and forth between ignorance and absence, many of the conceptual tools for studying agnotology may help to shape a framework for connecting the diverse studies of absence and its causes, and the studies of absence illuminating studies of ignorance, particularly those forms of ignorance which are absences, especially privative agnoses. Moving back and forth between the two, we will need to remind ourselves that ignorance and absence are produced, and productive, situated in time and reflecting the regimes of knowledge and legibility that constitute an episteme.

Postscript

Croissant and Smith-Doerr (2008) review the state of research on university–industry research relations (UIRRs), noting that location plays an important role in the establishment and effectiveness of these collaborations. Location can figure literally, where research parks and other forms of spatial proximity can have positive impacts on likelihood of UIRR establishment and success. In addition, location can figure metaphorically (p. 697), as the social location of institutions, in terms of prestige, for example, greatly affect their perceived desirability as research partners. A literal absence, geographically speaking, is the establishment of research parks which remain, despite optimistic models (Etzkowitz & Leydesdorff, 2000), large tracts of open space and underutilized capacity.[8] An absence of knowledge haunts the optimistic assessments of places like Silicon Valley or Research Triangle Park, and that lack of knowledge about the factors which actually influence geographic concentration of UIRRs and if they can be manipulated through policy challenges the reproduction of these sites. The multiple versions of space, empty and full, of location and relationships, both present and missing, and of knowledge, both present and absent, provide a lens, if not always geographical, then at least spatial in

[7]This sociology of nothing is different from, but not incompatible with, Ritzer's (2003) "globalization of nothing," which is a critique of global commodity fetishism and its intentional stripping of meaning from products for mass consumption. The stripping away of local meanings is a kind of ignorance production, constituting an agnosis that allows commodities to circulate without controversy (see also de Sousa Santos, 2008).

[8]On the absence of successful UIRR establishment, see, for example, the decades-old empty lot that serves as the annex to the University of Arizona "Tech Park" (University of Arizona, 2015).

allowing us to think about absent projects and absent knowledge. Geography and landscape become one kind of scale for thinking about knowledge-distribution (non) processes, and for conceiving of them as a metaphor for the the ecological system of disciplines and interdisciplinarity of the academy and beyond. Boundaries and bridges, as metaphors for exclusionary and inclusionary processes, become ways of understanding the relational features which produce or inhibit knowledge transmittal. The literal and metaphoric use of geography provides another dimension and potential methodological resources for conceptualizing and assessing agnoses and absences in various aspects of contemporary knowledge systems.

References

Barad, K. (2007). *Meeting the universe halfway: Quantum physics and the entanglement of matter and meaning*. Durham: Duke University Press.

Bataille, G. (2001). *The unfinished system of nonknowledge* (M. Kendall & S. Kendall, Trans.). Minneapolis: University of Minnesota Press.

Baudrillard, J. (1994). *Simulacra and simulation* (S. F. Glaser, Trans). Ann Arbor: University of Michigan Press. (Original work published 1981)

Beck, U. (2009). *World at risk* (C. Cronin, Trans.). Cambridge, UK: Polity Press. (Original work published 2007)

Bernstein, J. (2009a). Nonknowledge: The bibliographical organization of ignorance, stupidity, error, and unreason: Part one. *Knowledge Organization, 36,* 17–29. Retrieved from http://academicworks.cuny.edu/cgi/viewcontent.cgi?article=1017&context=kb_pubs

Bernstein, J. (2009b). Nonknowledge: The bibliographical organization of ignorance, stupidity, error, and unreason: Part two. *Knowledge Organization, 36,* 249–260. Retrieved from http://academicworks.cuny.edu/cgi/viewcontent.cgi?article=1041&context=kb_pubs

Bille, M., Hastrup, F., & Sørensen, T. F. (Eds.). (2012). *An anthropology of absence: Materializations of transcendence and loss*. New York: Springer.

Bourdieu, P. (1977). *Outline of a theory of practice* (R. Nice, Trans.). Cambridge, UK: Cambridge University Press. (Original work published in 1972)

Bloor, D. (1976). *Knowledge and social imagery*. Chicago: University of Chicago Press.

Burt, R. S. (1995). *Structural holes: The social structure of competition.* Cambridge, MA: Harvard University Press.

Burt, R. S. (2005). Stupid shit: (In)security in the age of twilightenment. *ArtUS, 11,* 29–37.

Butler, J. (1993). *Bodies that matter: On the discursive limits of 'sex'*. New York: Routledge.

Cartwright, N. (1983). *How the laws of physics lie*. Oxford, UK: Oxford University Press.

Casper, M. J., & Moore, L. J. (2009). *Missing bodies: The politics of visibility*. New York: New York University Press.

Christensen, J. (2008). Smoking out objectivity: Journalistic gears in the agnotology machine. In R. N. Proctor & L. Schiebinger (Eds.), *Agnotology: The making and unmaking of ignorance* (pp. 266–282). Stanford: Stanford University Press.

Collins, H. M. (1981). Stages in the empirical programme of relativism. *Social Studies of Science, 11,* 3–10.

Collins, H. M. (1992). *Changing order: Replication and induction in scientific practice*. Chicago: University of Chicago Press.

Collins, P. H. (2000). *Black feminist thought: Knowledge, consciousness, and the politics of empowerment*. New York: Routledge.

Collins, R. (2007). Turning points, bottlenecks, and the fallacies of counterfactual history. *Sociological Forum, 22,* 247–269. doi:https://doi.org/10.1111/j.1573-7861.2007.00030.x

Croissant, J. L. (2014). Agnotology: Ignorance and absence or towards a sociology of things that aren't there. *Social Epistemology, 28*, 4–25. doi:https://doi.org/10.1080/02691728.2013.862880

Croissant, J. L., & Smith-Doerr, L. (2008). Organizational contexts of science: Boundaries and relationships between university and industry. In E. J. Hackett, O. Amsterdamska, M. Lynch, & J. Wajcman (Eds.), *The handbook of science and technology studies* (3rd ed., pp. 691–719). Cambridge, MA: MIT Press.

Daniels, C. R. (2006). *Exposing men: The science and politics of male reproduction*. New York: Oxford University Press.

de Beauvoir, S. (2009). *The second sex* (C. Borde & S. Malovany-Chevailler, Trans.). London: Cape Publishing. (Original work published 1949)

Deleuze, G., & Guattari, F. (1987). *A thousand plateaus: Capitalism and schizophrenia* (B. Massumi, Trans.). Minneapolis: University of Minnesota Press. (Original work published 1980)

Derrida, J. (1994). *Specters of Marx: The state of the debt, the work of mourning & the New International* (P. Kamuf, Trans.). New York: Routledge. (Original work published 1993)

de Sousa Santos, B. (2008). The World Social Forum and the global left. *Politics & Society, 36*, 247–270. doi:https://doi.org/10.1177/0032329208316571

de Sousa Santos, B. (2016). *Epistemologies of the South: Justice Against Epistemicide*. New York: Routledge.

Eden, L. (2004). *Whole world on fire: Organizations, knowledge, and nuclear weapons devastation*. Ithaca: Cornell University Press.

Etzkowitz, H., & Leydesdorff, L. (2000). The dynamics of innovation: From national systems and 'Mode 2' to a Triple Helix of university–industry–government relations. *Research Policy, 29*, 109–123. doi:https://doi.org/10.1016/S0048-7333(99)00055-4

Fanon, F. (2008). *Black skin, white masks* (R. Philcox, Trans.). New York: Grove Press. (Original work published 1952)

Foote, K. (1990). To remember and forget: Archives, memory, and culture. *The American Archivist, 53,* 378–392. doi:https://doi.org/10.17723/aarc.53.3.d87u013444j3g6r2

Foucault, M. (1994). *The order of things: An archaeology of the human sciences*. New York: Vintage. (Original work published 1966)

Frickel, S., & Vincent, M. B. (2007). Hurricane Katrina, contamination, and the unintended organization of ignorance. *Technology in Society, 29*, 181–188. doi:https://doi.org/10.1016/j.techsoc.2007.01.007

Galison, P. (2008). Removing knowledge: The logic of modern censorship. In R. N. Proctor & L. Schiebinger (Eds.), *Agnotology: The making and unmaking of ignorance* (pp. 37–54). Stanford: Stanford University Press.

Garcia, G. N. (1997). *Why we choose to be stupid* (Unpublished doctoral dissertation). Montana State University, Bozeman.

Gordon, A. F. (2008). *Ghostly matters: Haunting and the sociological imagination*. Minneapolis: University of Minnesota Press.

Gross, M. (2007). The unknown in process: Dynamic connections of ignorance, non-knowledge, and related concepts. *Current Sociology, 55*, 742–759. doi:https://doi.org/10.1177/0011392107079928

Gross, M. (2010). *Ignorance and surprise: Science, society, and ecological design*. Cambridge, MA: MIT Press.

Habermas, J. (2002). Transcendence from within, transcendence in this world. In E. Mendieta (Ed.), *Religion and rationality: Essays on reason, God, and modernity* (pp. 67–94). Cambridge, MA: MIT Press.

Hardy, D. T., & Clark, J. (2005). *Michael Moore is a big fat stupid white man*. New York: William Morrow.

Harris, V. (2009). Against the grain: Psychologies and politics of secrecy. *Archival Science, 9,* 133–142. doi:https://doi.org/10.1007/s10502-009-9096-z

Hess, D. J. (2009). The potentials and limitations of civil society research: Getting undone science done. *Sociological Inquiry, 79*, 306–327. doi:https://doi.org/10.1111/j.1475-682X.2009.00292.x

High, C., Kelly, A. H., & Mair, J. (Eds.). (2012). *The anthropology of ignorance: An ethnographic approach*. New York: Palgrave Macmillan.

Jacklin, A., Robinson, C., & Torrance, H. (2006). When lack of data is data: Do we really know who our looked-after children are? *European Journal of Special Needs Education, 21,* 1–20. doi:https://doi.org/10.1080/08856250500491773

Japp, K. P. (2000). Distinguishing non-knowledge. *The Canadian Journal of Sociology, 25*, 225–238. doi:https://doi.org/10.2307/3341824

Kerwin, A. (1993). None too solid: Medical ignorance. *Science Communication, 15,* 166–185. doi:10.1177/107554709301500204

Knorr-Cetina, K. (1999). *Epistemic cultures: How the sciences make knowledge*. Cambridge, MA: Harvard University Press.

Kosofsky Sedgwick, E. (1990). *Epistemology of the closet*. Berkeley: University of California Press.

Latour, B. (1992). One turn after the social turn: Easing science studies into the non-modern world. In E. McMulin (Ed.), *The social dimensions of science* (pp. 272–292). Studies in Science and the Humanities from the Reilly Center for Science, Technology, and Values: Vol. 3. Notre Dame: Notre Dame University Press.

Latour, B. (2007). *Reassembling the social: An introduction to actor-network theory*. Oxford, UK: Oxford University Press.

Leavitt, J. D., & Christenfeld, N. J. S. (2011). Story spoilers don't spoil stories. *Psychological Science, 22*, 1152–1154. doi:https://doi.org/10.1177/0956797611417007

Machlup, F. (1980). *Knowledge: Its creation, distribution, and economic significance*. Knowledge and Knowledge Production: Vol. 1. Princeton: Princeton University Press.

Mair, J., Kelly, A. H., & High, C. (2012). Introduction: Making ignorance an ethnographic object. In C. High, A. H. Kelly, & J. Mair (Eds.), *The anthropology of ignorance: An ethnographic approach* (pp. 1–32). New York: Palgrave Macmillan.

Mayor, A. (2008). Suppression of indigenous fossil knowledge: From Claverack, New York 1705 to Agate Springs, Nebraska, 2005. In R. N. Proctor & L. Schiebinger (Eds.), *Agnotology: The making and unmaking of ignorance* (pp. 163–182). Stanford: Stanford University Press.

Merton, R. K. (1987). Three fragments from a sociologist's notebooks: Establishing the phenomenon, specified ignorance, and strategic research materials. *Annual Review of Sociology, 13,* 1–29. doi:https://doi.org/10.1146/annurev.so.13.080187.000245

Michaels, D. (2008). Manufactured uncertainty: Contested science and the protection of the public's health and environment. In R. N. Proctor & L. Schiebinger (Eds.), *Agnotology: The making and unmaking of ignorance* (pp. 90–107). Stanford: Stanford University Press.

Mills, C. W. (2008). White ignorance. In R. N. Proctor & L. Schiebinger (Eds.), *Agnotology: The making and unmaking of ignorance* (pp. 230–249). Stanford: Stanford University Press.

Moore, M. (2002). *Stupid white men*. New York: Harper.

Moore, W. E., & Tumin, M. M. (1949). Some social functions of ignorance. *American Sociological Review, 14,* 787–795. Retrieved from http://www.jstor.org/stable/2086681

Oudshoorn, N. (2003). *The male pill: A biography of a technology in the making*. Durham: Duke University Press.

Oreskes, N., & Conway, E. M. (2008). Challenging knowledge: How climate science became a victim of the Cold War. In R. N. Proctor & L. Schiebinger (Eds.), *Agnotology: The making and unmaking of ignorance* (pp. 55–89). Stanford: Stanford University Press.

Oreskes, N., & Conway, E. M. (2010). *Merchants of doubt: How a handful of scientists obscured the truth on issues from tobacco smoke to global warming*. New York: Bloomsbury.

Perrow, C. (1984). *Normal accidents: Living with high-risk technologies*. New York: Basic Books.

Pollitt, C. (2000). Institutional amnesia: A paradox of the 'Information Age'? *Prometheus: Critical Studies in Innovation, 18,* 5–16. doi:https://doi.org/10.1080/08109020050000627

Pratchett, T. (1990). *Eric*. London: Victor Gollancz/Corgi Press.

Proctor, R. N. (1995). *Cancer wars: How politics shapes what we know and don't know about cancer*. New York: Basic Books.

Proctor, R. N. (2008). Agnotology: A missing term to describe the cultural production of ignorance (and its study). In R. N. Proctor & L. Schiebinger (Eds.), *Agnotology: The making and unmaking of ignorance* (pp. 1–35). Stanford: Stanford University Press.

Proctor, R. N., & Schiebinger, L. (Eds.). (2008). *Agnotology: The making and unmaking of ignorance*. Stanford: Stanford University Press.

Rapp, R. (2000). *Testing women, testing the fetus: The social impact of amnicentesis in America*. New York: Routledge.

Ravetz, J. R. (1993). The sin of science: Ignorance of ignorance. *Science Communication, 15,* 157–165. doi:https://doi.org/10.1177/107554709301500203

Rescher, N. (2009). *Ignorance: On the wider implications of deficient knowledge*. Pittsburgh: University of Pittsburgh Press.

Ritzer, G. (2003). The globalization of nothing. *SAIS Review, 23,*189–200. doi:https://doi.org/10.1353/sais.2003.0053

Rose, N. (2008). The value of life: Somatic ethics & the spirit of biocapital. *Daedelus, 137*(1), 36–48. doi:https://doi.org/10.1162/daed.2008.137.1.36

Sagan, C. (1995). *The demon-haunted world: Science as a candle in the dark.* New York: Ballantine.

Said, E. (2003). *Orientalism* (25th anniversary ed.). New York: Vintage.

Schiebinger, L. (2004). *Plants and empire: Colonial bioprospecting in the Atlantic world*. Cambridge, MA: Harvard University Press.

Simon, B. (2002). *Undead science: Science studies and the afterlife of cold fusion*. New Brunswick: Rutgers University Press.

Smithson, M. (1989). *Ignorance and uncertainty: Emerging paradigms*. New York: Springer.

Smithson, M. J. (2008). Social theories of ignorance. In R. N. Proctor & L. Schiebinger (Eds.), *Agnotology: The making and unmaking of ignorance* (pp. 209–229). Stanford: Stanford University Press.

Tapp, K. P. (2000). Distinguishing non-knowledge. *The Canadian Journal of Sociology, 25*, 225–238. doi:https://doi.org/10.2307/3341824

Ten Bos Rene (2007). The Vitality of Stupidity. *Social Epistemology, 21*(2), 139–150. doi:https://doi.org/10.1080/02691720701393442

Thomas, G. (2012, May 12). Jonah Goldberg: Young People are 'so frickin' stupid'. [Video]. *The Daily Caller*. Retrieved September 27, 2017, from http://dailycaller.com/2012/05/12/jonah-goldberg-young-people-are-so-frickin-stupid-video/.

Tuana, N. (2008). Coming to understand: Orgasm and the epistemology of ignorance. In R. N. Proctor & L. Schiebinger (Eds.), *Agnotology: The making and unmaking of ignorance* (pp. 108–148). Stanford: Stanford University Press.

University of Arizona (2015). *Tech Parks Arizona: Welcome to the UA Tech Park at the Bridges*. Retrieved September 27, 2017, from https://techparks.arizona.edu/welcome-visitors/welcome-the-ua-tech-park-the-bridges

van Boxsel, M. (2003). *The encyclopedia of stupidity* (A. Pomerans & E. Pomerans, Trans.). London: Reaktion. (Original work published 1999)

Vaughan, D. (1996). *The challenger launch decision: Risky technology, culture, and deviance at NASA*. Chicago: University of Chicago Press.

Vaughan, D. (1999). The dark side of organizations: Mistake, misconduct, and disaster. *Annual Review of Sociology, 25*, 271–305. doi:https://doi.org/10.1146/annurev.soc.25.1.271

Watts, D. J. (2011). *Everything is obvious: Once you know the answer: How common sense fails us*. New York: Crown Business.

Weber, M. (1964). *The theory of social and economic organization* (A. M. Henderson & T. Parsons, Trans.). Glencoe, Ill: The Free Press. (Original work published 1922)

Welles, J. F. (1986). *Understanding stupidity: An analysis of the premaladaptive beliefs and behavior of institutions and organizations*. Orient, NY: Mount Pleasant.

Wynne, B. (1995). Public understanding of science. In S. Jasanoff, G. E. Markle, J. C. Petersen, & T. Pinch (Eds.), *Handbook of science and technology studies* (pp. 361–391). Thousand Oaks: Sage.

Part III
The University and the City

Chapter 11
The Civic University and the City

John Goddard

In the context of the ongoing globalization of the economy and society—a process in which higher education is an active player—questions are being asked in many circles about the contribution that universities can make to the public good, not least in the places where they are located. More specifically, not only what is a particular university "good at" in terms of the quality of its research and teaching (as reflected in national and international ranking tables) but also what is it "good for" in terms of its active contribution to the wider society globally and locally.

The local dimension is particularly relevant when universities are directly or indirectly funded from the public purse and where governments are accountable to their electorates via territorially based governance systems. Politicians might be heard to ask: "I have a university *in* my constituency or local authority area but how does it actively contribute to the development *of* my area." A typical response is that although the university is not formally bound to a particular area it can be a key link for that area to the wider world, connecting the global and the local.

This response chimes with a growing recognition of the link between globalization and localization. As the leader of the UNESCO Global Universities Network for Innovation points out "Although communication is now global, location, proximity, and uniqueness still matters" (Grau, 2014, p. 2). He quotes the distinguished urbanist Manuel Castells, who notes that:

> The network society diffuses selectively, working on the pre-existing sites, organizations and institutions which still make most of the material environment of people's lives. The social structure is global but most of human experience is local, both in territorial and cultural terms. (Grau, 2014, p. 2)

J. Goddard (✉)
Centre for Urban and Regional Development Studies (CURDS), Claremont Bridge, Newcastle University, Newcastle upon Tyne, UK
e-mail: john.goddard@ncl.ac.uk

P. Meusburger et al. (eds.), *Geographies of the University*, Knowledge and Space 12,
https://doi.org/10.1007/978-3-319-75593-9_11

As key institutions in society, all universities have a unique location and cannot avoid a relationship with the myriad of other institutions and communities also inhabiting that place, particularly others also involved in the production and distribution of knowledge and public bodies, such as local authorities, responsible for the place in the round and the wellbeing of its citizens.

This chapter explores the changing nature of links between the university and the city in both theory and practice, outlining a model of the civic university engaged with global challenges that have an urban dimension, and illustrates the arguments with reference to experience of universities working with cities in England.

Universities as Urban Anchor Institutions

In promoting dialogue between universities and city authorities the notion of the university as an *anchor* institution can be helpful. Anchor institutions might be characterized as not just in the place but of the place. The U.K. think tank The Work Foundation defines anchor institutions as:

> large locally embedded institutions, typically non-governmental public sector, cultural or other civic institutions that are of significant importance to the economy and the wider community life of the cities in which they are based. They generate positive externalities and relationships that can support or "anchor" wider economic activity in the locality. Anchor institutions do not have a democratic mandate and their primary missions do not involve regeneration or local economic development. Nonetheless their scale, local rootedness and community links are such that they can play a key role in local development and economic growth representing the "sticky capital" around which economic growth strategies can be built. (The Work Foundation, 2010, p. 3)

In the case of universities, their main location, in comparison with private firms, is fixed within the current home location. Notwithstanding possible expansion to other nearby or faraway campuses, it is where they have considerable sunk investment in buildings and research infrastructure and strong identification with place through the name of the institution. On past experience universities have generally been immune to institutional failure or sudden contractions in size. They can therefore act as a source of stability in local economies, buffering against the worst effects of periodic downturns. They are particularly important as anchor institutions in weaker economies (Goddard, Coombes, Kempton, & Vallance, 2014).

What does *anchoring* imply for universities? Being anchored in a particular location does raise normative questions for the university about the requirement for academic practice to be of relevance to the place in which academics live and work as citizens. The former director of the London School of Economics, Craig Calhoun, in a famous paper entitled "The University and the Public Good" made an important point when he wrote:

> We treat our opportunities to do research not as a public trust but as a reward for success in past studies. Rewards for research are deeply tied up with the production of academic hierarchy and the relative standing of institutions. (Calhoun, 2006, p. 19)

But, significantly, Calhoun goes on to say: "Public support for universities is based on the effort to educate citizens in general, to share knowledge, to distribute it as widely as possible in accord with publically articulated purposes" (Calhoun, 2006, p. 19).

More recently in his treatise on *The Public Value of the Social Sciences,* John Brewer (2013) has unpacked the word *public*:

> Use of the adjective "public" not only implies fundamental questions about accountability but also poses additional queries about to whom should we as social scientists feel accountable . . . Public social science has both a research and teaching agenda and involves a commitment to promote the public good through civic engagement. (Brewer, 2013, p. 6)

Although neither of these authors is specifically writing about territorial issues or indeed all disciplines within the university, they are relevant to a narrative about the civic university and its relation to the wider society locally as well as globally. In relation to the local, much academic writing on territorial development recognizes that the city cannot only be viewed as an economic engine or physical place—which it is—but also as a node in a network of local and global, social, cultural, and political interactions. Put more simply the development of the city is about businesses that generate jobs, the people who live there, and the institutions of urban governance connecting these domains. The civic university is therefore engaged with the city in the round.

The University and the Development of the City in the Round

How are universities actively contributing to place-making, to innovation, economic and social development? Thomas Bender (1988) in his seminal book on the university and the city referred to campuses as "semi-cloistered spaces in the midst of the city to meet the work and leisure needs of students and academic communities" (p. 290). In terms of place-making the expansion of universities has led to demand for more space. In some cases, university sites have been dispersed all over a city, reducing their impact. Science parks developed to accommodate businesses linked to universities have often been established on the urban periphery. However, there has been recent and growing pressure to open out university campuses to the city. Even science parks have been experiencing an urban turn toward sites that are more mixed in function and integrated into the fabric of the city. In this trend universities have become involved in local regeneration projects and the development of initiatives such as cultural quarters, science zones, and media hubs.

In terms of the contribution of universities to business innovation, the way innovation takes place is changing from a linear model to a coproduction model highlighting the important role of users, service, and open and social innovation. According to the European Commission open innovation can be defined as:

> a new paradigm based on a Quadruple Helix Model where government, industry, academia and civil participants work together to co-create the future and drive structural changes far

> beyond the scope of what any one organization or person could do alone. This model encompasses also user-oriented innovation models to take full advantage of ideas' cross-fertilisation leading to experimentation and prototyping in real world settings. (European Commission, 2015)

This model refers to a wider range of knowledge inputs, additional entrepreneurs, and different selection mechanisms and ways of allocating capital and people to projects. A range of partners, including local authorities, public service organizations (health providers, schools, etc.), charities and social enterprises, and universities can be involved. This new reality for innovation gives even greater salience to the role of personal contacts between a wide range of actors and agents, underscoring the advantages of urban agglomeration. Students can be a key part of this mix. They can act as knowledge transfer agents through work placements linked to their courses. If these students are subsequently employed in the organization, this will establish the social relations with their teachers on which subsequent links can be built.

Turning to social development, universities cannot avoid the inequalities present in most large cities, where they are located, not least because of its likely impact on attracting students and staff from elsewhere. They are also expected to recruit more students from disadvantaged backgrounds and this can be done by work with schools within the city. Cities are also under fiscal stress and expected to deliver more services in a joined-up way to the local population. Social innovation can be seen as one focus for university collaboration with the city.

The influential European Commission's Board of European Policy Advisors (BEPA) has defined social innovation as:

> innovations that are social in both their ends and their means. Specifically, we define social innovations as new ideas (products, services and models) that simultaneously meet social needs (more effectively than alternatives) and create new social relationships or collaborations. They are innovations that are not only good for society but also enhance society's capacity to act. The process of social interactions between individuals undertaken to reach certain outcomes is participative, involves a number of actors and stakeholders who have a vested interest in solving a social problem. (BEPA, 2010, pp. 9–10)

This can be boiled down into three perspectives: first, a social demand perspective in terms of the needs of vulnerable groups traditionally not met by the market and where there is a strong role for social entrepreneurs; second, a societal challenge perspective through which societal problems are addressed through new coalitions and where the boundaries between the economic and social blur; and third, a systematic change perspective where social innovation is reshaping society itself.

Social innovation implies extending the dominant model for university external collaboration from the so-called *triple helix* of university (see chapter by Etzkowitz in this volume), business, and government to a *quadruple helix* that embraces civil society. More specifically, to quote two reports for the European Commission:

> The Quadruple Helix, with its emphasis on broad cooperation in innovation, represents a shift toward systemic, open and user-centric innovation policy. An era of linear, top-down, expert driven development, production and services is giving way to different forms and

> levels of coproduction with consumers, customers and citizens. (Arnkil, Järvensivu, Koski, & Piirainen 2010, p. 6)

> The shift toward social innovation also implies that the dynamics of ICT [information communication technology]-innovation has changed. Innovation has shifted downstream and is becoming increasingly distributed; new stakeholder groups are joining the party, and combinatorial innovation is becoming an important source for rapid growth and commercial success. Continuous learning, exploration, co-creation, experimentation, collaborative demand articulation, and user contexts are becoming critical sources of knowledge for all actors in R&D [research and development] & Innovation. (Information Society Technology Advisory Group, 2011, p. 5)

According to Arnkil et al. (2010) the quadruple helix model can have four variants, depending on whether the focus is on citizens, firms, the public service sector, or simply the better commercialization of university research by testing products and services with users; first, a triple helix model with users added on; second, a firm-centered *living lab* model; third, a public-sector-centered living lab model; and finally, a citizen-centered model.

Although the role of digital technologies is central to the quadruple helix, this does not necessarily mean that geography no longer matters. Indeed, the city as a living lab for testing new ways of organizing the delivery of services in a sustainable and inclusive way, for example, to an ageing population, is influencing public policy all over Europe.

Societal Challenges and the Civic University

Part of the growing expectation of universities is that they will contribute to the major challenges facing society. Such an approach characterizes the European Union's Horizon 2020 program designed to contribute to the Europe 2020 Lisbon Treaty agenda of "smart sustainable and inclusive growth" (European Union, 2007). Many of the themes within the program, such as health, demographic change, and well-being; smart, green, and integrated transport; and inclusive, innovative, and secure societies, have an explicit or implicit territorial dimension.

Horizon 2020 also has a cross-cutting theme of *Science with and for Society*, which recognizes that "betting on technology acceptance by way of good marketing is no longer a valid option . . . Early and continuous iterative engagement with society in research and innovation is key to innovation adequacy and acceptability" (Science With and For Society Advisory Group, 2014, p. 7).

With these points in mind the European Commission has endorsed the concept of Responsible Research and Innovation (RRI):

> RRI is a process where all societal actors (researchers, citizens, policy makers, business) work together during the whole R&I process in order to align R&I outcomes to the values, needs and expectations of European society. . . . There is a need for a new narrative drawing on a broad-based innovation strategy encompassing both technological and non-technological innovation at all levels of European society, and with a stronger focus

> on the citizen and responsible and sustainable business—a quadruple helix and place-based approach to science, research and innovation. (Science with and for Society Advisory Group, 2014, p. 8)

These principles have been embodied in the Rome Declaration adopted by the European Council in December 2014, which calls upon public and private research and innovation performing organizations, including universities, to implement institutional change that fosters RRI by:

- Reviewing their own procedures and practices in order to identify possible RRI barriers and opportunities at organization level;
- Creating experimental spaces to engage civil society actors in the research process as sources of knowledge and partners in innovation;
- Developing and implementing strategies and guidelines for the acknowledgment and promotion of RRI;
- Adapting curricula and developing training to foster awareness, know-how, expertise, and competence of RRI;
- Including RRI criteria in the evaluation and assessment of research staff.

Although not specifically referring to the civic university, traditional universities seeking to pay regard to RRI and perform a civic role may need to implement significant changes in the way they work and collaborate with the city. There may well be tensions in this change process.

Tensioned Themes

Developing a quadruple helix and RRI approach to science, research, and innovation within the city is not without both challenges and opportunities. This is inevitable. To once again refer to Thomas Bender (1988):

> I propose that we understand the university as semi-cloistered heterogeneity in the midst of uncloistered heterogeneity (that is to say the city . . .). Because of this difference, relations between the two are necessarily tense, and they cannot be assimilated into one another. To do so, either practically or conceptually, is to empty each of its distinctive cultural meaning and falsify the sociology of each. (p. 290)

In terms of physical development there may be tensions between the optimal strategy for the expansion of the university estate in terms of location and function and with projects that have an urban development or regeneration focus targeted at the needs of the city. This includes issues around student housing.

Universities as institutions partly protected by public funding can be sources of "slack" in metropolitan innovation systems. By virtue of harboring noncommercial activities that cannot be supported by the local private sector, universities can potentially add to the adaptive capacity of the metropolitan economy, particularly small and medium-sized enterprises (Vallance, 2016). But this potential is tensioned against the immediate opportunities of working with the best companies regardless of location and the (low) level of absorptive capacity of local businesses.

These specific tensions are underpinned by those between the external civic role of the university and the internal processes within the university, which are heavily influenced by the higher education policy environment within which it operates. Public universities are principally influenced by national (or federal) governments. A city may have several higher education institutions within its boundary but no powers to develop a city- or region-wide higher education system to meet a range of local needs. It could be said that this is because the work of a university is not bounded by any specific territory. It operates within a national higher education system that does not have an explicit concern with territorial development issues. Because higher education is now a global business, a key driver for many universities is position in national and international ranking tables. These are heavily weighted in favor of recognition for research, with its very straightforward metrics of citations, and pay little regard to contributions to civil society where the metrics are much more complex.

Although city interests might expect a corporate response from "the university," this does not recognize that the *traditional* university is a loosely coupled organization composed of discipline-based units driven by higher education metrics and with only limited horizontal or vertical coordination. In such universities responding to external needs may be easier at the level of the academic unit than the entire university. This raises questions around business models of the university.

Business Models of the University

One well-established model is that of the entrepreneurial university model outlined by the American sociologist Robert Burton Clark (1998). This was designed to help the traditional university become a more corporate and outward facing institution, hence its subtitle "organizational pathways to institutional transformation." His model consists of a strengthened steering core (or what would now be called an executive board), an enhanced developmental periphery (composed of intermediate organizations like science parks and centers for continuing professional development), a diversified funding base (reducing dependence on state funding), and a stimulated and more entrepreneurial academic heartland. It is this model that underpins the triple helix framework extolled of universities, business, and the state and now adopted by governments across the world (Etzkowitz & Leydesdorff, 2000).

However, the shortcomings of this model as it has been adopted in policy and practice are increasingly being recognized, not least for its focus on research in science and technology and links to business. It neglects teaching except in the field of student entrepreneurship, the role of humanities and social sciences, place-based communities, and civil society more generally. An alternative model for the civic university is proposed here and this is best introduced by defining first a non-civic university (Fig. 11.1).

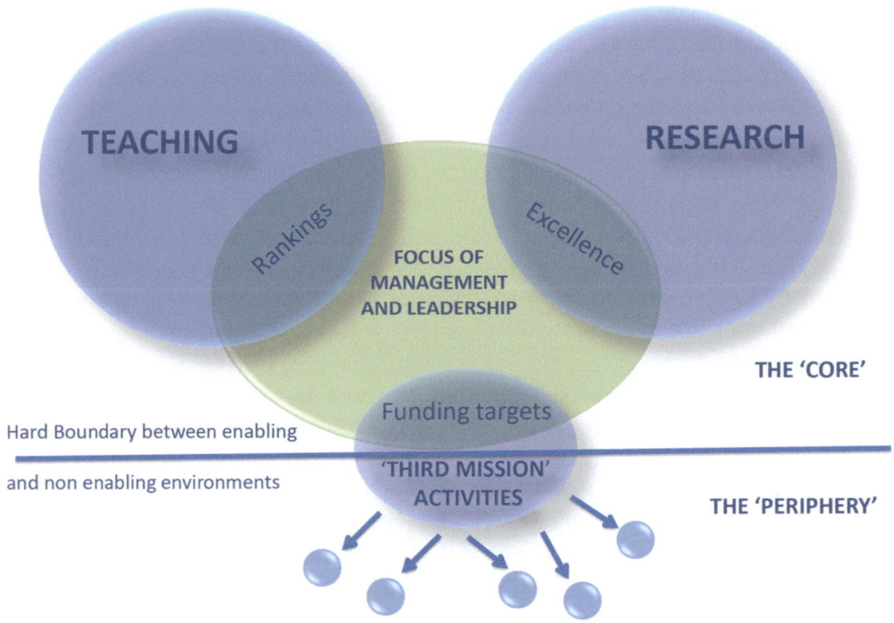

Fig. 11.1 The non-civic university. Source: Design by author.

Such a university maintains a strict separation of its teaching and research, with research performance judged by academic publications in peer-reviewed journals and teaching judged by student-satisfaction scores. Third mission activities are only seen as *core* when there are hard funding targets attached. Activities outside the core areas of focus are not enabled through incentives and others kinds of support, so are often seen as "below the radar" of management. The outcome of this is that the results of this work is not absorbed back into the teaching or research taking place in the university and impacts are not tracked or measured.

In contrast, Goddard, Hazelkorn, Kempton, and Vallance (2016) have developed an alternative model of the civic university that integrates teaching, research, and engagement with the outside world such that each enhances the other (Fig. 11.2). In the civic university, research has socioeconomic impact designed in from the start and teaching has a strong community involvement with the long-term objective of widening participation in higher education. Most importantly there is a soft, flexible boundary between the institution and society.

To turn this into a practical way in which institutional leaders and managers can appraise their own organizations seven dimensions of the civic university can be suggested. These are:

1. It is *actively engaged* with the wider world as well as the local community of the place in which it is located.
2. It takes a *holistic approach* to engagement, seeing it as institution-wide activity and not confined to specific individuals or teams.

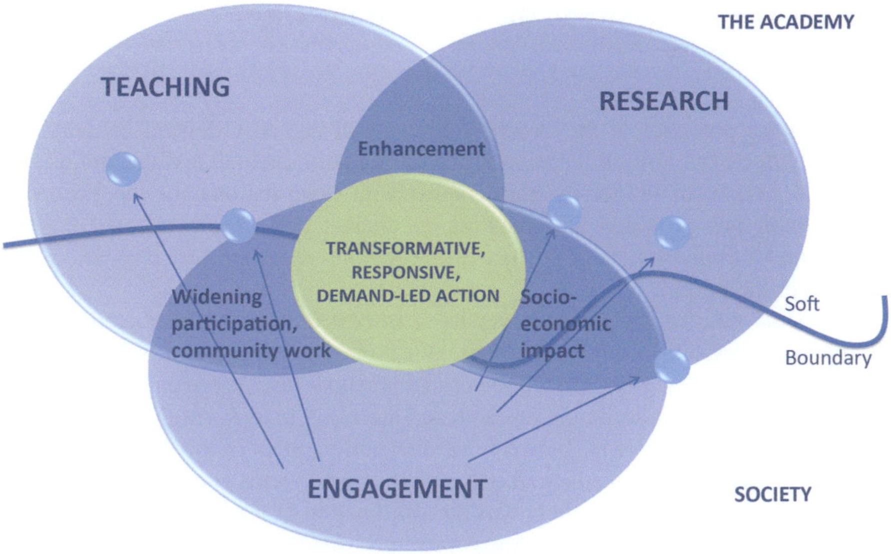

Fig. 11.2 The civic university. Source: Design by author.

3. It has a strong *sense of place*—it recognizes the extent to which is location helps to form its unique identity as an institution.
4. It has a *sense of purpose*—understanding not just what it is good at, but what it is good for.
5. It is *willing to invest* in order to have impact beyond the academy.
6. It is *transparent and accountable* to its stakeholders and the wider public.
7. It uses *innovative methodologies* such as social media and team building in its engagement activities with the world at large.

Those traditional universities that are seeking to enhance their civic role are likely to be on a journey of institutional transformation and may position themselves at different points along a spectrum against each of these dimensions, from embryonic to fully embedded in the customs and practices of the institution. In an international comparative study on the leadership and management of aspiring civic universities this framework is used as a means of developing a shared understanding among the participating institutions of the challenges they may be confronting on this journey and how these might be overcome.

Linking the University to the City and the City to the University

Realizing the potential of the civic university will not only depend on what the university does, but also on the capacity of its city partners in the public and private sector. A review of university partnerships with their regions for the European Commission has provided a framework to characterize the "connected region" (European Commission, 2011). Most of the regions reviewed had city-based universities at their core.

As in the case of the civic university it is best to start by characterizing the disconnected region. In terms of higher education universities were seen as *in* the region but not *of* the region. Their policies and practices discourage engagement with a focus on rewards for academic research and teaching. In terms of the public sector there was a lack of coherence between national and regional or local policies, a lack of political leadership, and a lack of a shared voice and vision at city region level. In the case of the private sector there was no coordination or representative voice with which universities could engage; firms were motivated by narrow self-interest and short-term goals and had low demand or absorptive capacity for innovation. Lastly, in terms of the mechanisms for connecting higher education into the development of the city and region, there were no *boundary spanning* people; relations with universities focused on supply side, transactional links; ineffective or non-existent partnerships; no shared understandings about the challenges, and last but not least entrepreneurs being locked out of regional planning.

By way of contrast in the connected city, the university is generating intellectual and human capital assets for the city region. The public sector is developing coherent policies that link territorial development to innovation and higher education and the private sector is investing in people and ideas that will create growth.

The U.K. Experience: Universities and Sustainable, Healthy, and Creative Cities

In the United Kingdom nineteenth-century institutions that were the predecessors of the so-called *redbrick* universities evolved to meet the needs of a rapidly evolving industrial society. This included not only support for key industrial sectors such as mechanical engineering but also hospitals contributing to a healthy workforce (and which later became the foundation for university medical schools). These institutions depended to a large degree on local public support. During the twentieth century these local links weakened with increasing central government support and influence over local government, the nationalization of higher education, and the concentration of banking and corporate headquarters in London. As a consequence, many of these civic institutions turned their backs on their host cities. However, in the twenty-first century some of these universities are seeking to reinvent themselves as civic institutions in the context of a globalization of both the economy and higher education, an urban renaissance, and of devolution to city regions (Goddard, 2009).

More specifically, how are the universities in four English cities—Newcastle, Manchester, Sheffield, and Bristol—meeting contemporary urban challenges of environmental sustainability, health, and cultural development? To set these responses to urban challenges in context, it is possible to compare the promise and the practice of one aspect of civic engagement—research—by drawing on the evidence of a coauthored online survey of a 1-in-3 random sample of academics in all disciplines at the six universities in these cities (Newcastle, Northumbria, Sheffield, Sheffield Hallam, Bristol, and the University of West of England) regarding the intended impact of their research. These data related to both the older redbrick universities and the former polytechnics given university status in 1992. The coauthored survey had 700 responses, a response rate of 30% (Goddard & Vallance, 2013).

Respondents were asked to distinguish between the direct and indirect impacts of their research in terms of the intended primary and secondary beneficiaries. Not surprisingly the principal focus of most academics was on knowledge creation, followed by the transfer of this through education. Impact on the economy and society across a wide range of areas from public policy through to cultural enrichment was a secondary concern (Fig. 11.3). It makes sense that the primary intended beneficiaries of most academics' research were peers in their own discipline, followed by their own students (Fig. 11.4). Notwithstanding the triple helix rhetoric only 10% of academics intended their research to have direct impact on private businesses. And only 20% saw their work as directly contributing to technological development.

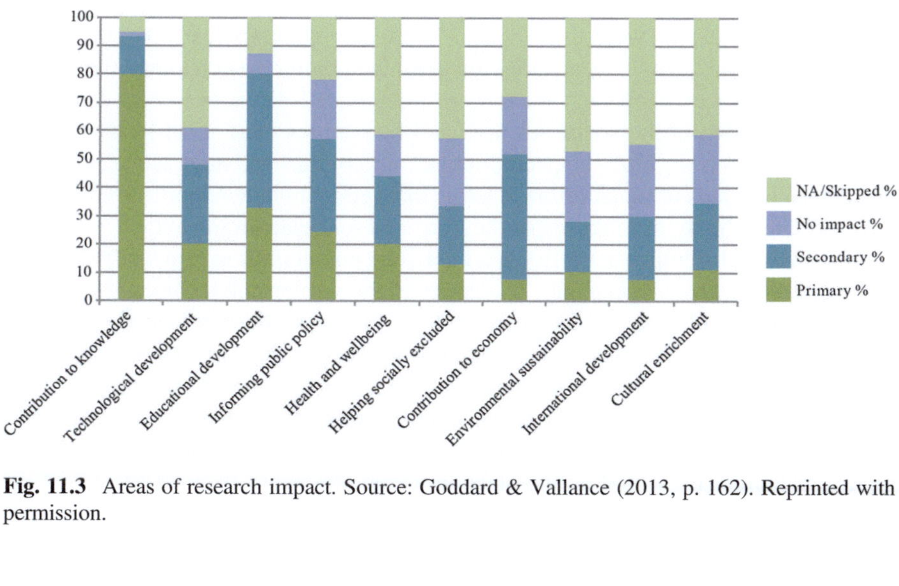

Fig. 11.3 Areas of research impact. Source: Goddard & Vallance (2013, p. 162). Reprinted with permission.

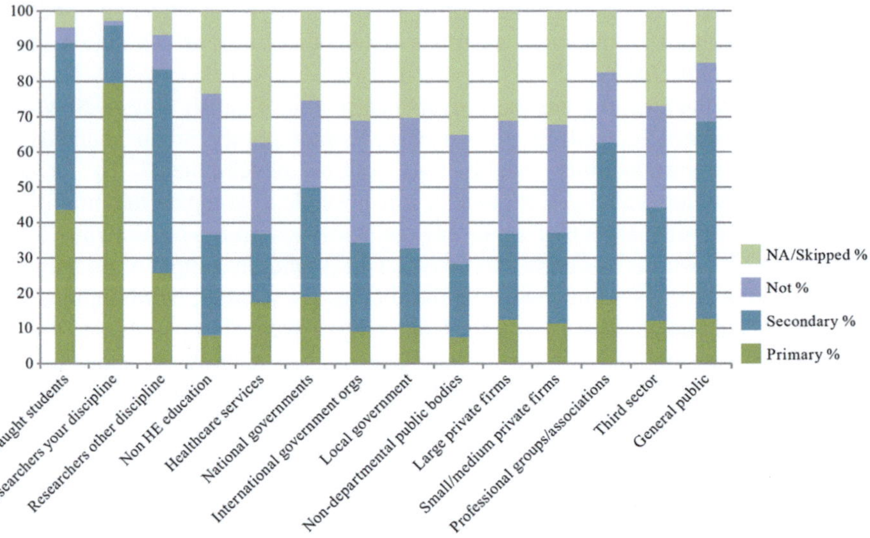

Fig. 11.4 Beneficiaries of research. Source: Goddard & Vallance (2013, p. 163). Reprinted with permission.

However, academics anticipated their research to have a secondary impact on a wide range of beneficiaries in civil society, most notably professional associations, the third sector, and the general public. This lends support to the quadruple helix model. Moreover, when those who said the intended impact of their research was on one of our urban challenge themes were separated out, it became apparent that those academics were more likely to be seeking an impact on other disciplines and civil society across the board.

But to what extent were these intended impacts geographically targeted? Not surprisingly the majority of academics did not intend their research to have an impact on particular places. However, there were pronounced differences between disciplines. Academics in the social sciences and humanities were most likely to want their research to have a place-specific impact. In contrast the hard sciences, which have been the focus of much effort in terms local economic development initiatives, were even less likely to look to specific locations for research impact.

There were also important differences between universities in terms of geographical focus. Again not surprisingly, academics in the former polytechnics in the three cities covered in the survey were more likely to want their research to have a geographically specific impact. Interestingly this orientation across both types of university was greatest in the northern cities, which have a lower level of prosperity than Bristol, which is arguably an extension of southeast England's *golden triangle*. This lends weight to the view that some academics are influenced in their priorities by the challenges presented by the place in which they work (Fig. 11.5).

Reviewing the documentary evidence, it is clear that universities in the four U.K. cities considered are working hard to minimize the environmental footprint

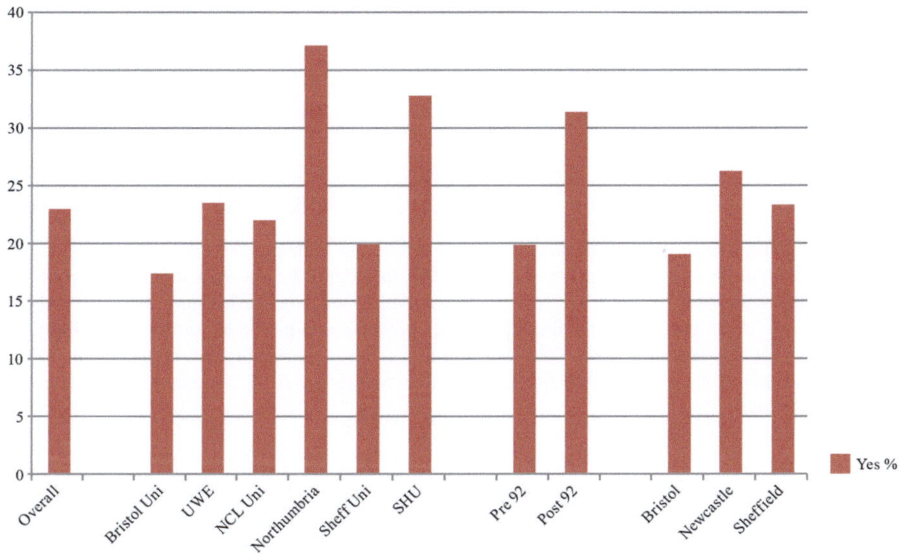

Fig. 11.5 The geography of research impact. UWE = University of West England; NCL Uni = Newcastle University; Sheff Uni= Sheffield University, SHU = Sheffield Hallum University. Pre 92 = universities existing prior to 1992; Post 92 = institutions granted university status in or after 1992. Source: Goddard & Vallance (2013, author's data).

of their estates. More significantly they are involved in economic development and regeneration initiatives involving the public sector that have a strong environmental dimension—among others the Science Central site in Newcastle and the Manchester Low Carbon Economic Area and the Manchester Corridor. Academics through their national and international roles are influencing the debate about sustainable cities and the regulatory environment within which energy production, distribution, and consumption has operated. In this process they are contributing to what can be referred to as multilevel governance and anchoring their agenda setting roles in their home university and city. Academics from different disciplines are engaging with the city as an urban laboratory. The city is simultaneously the object of study, the setting for field research, and the site for collaboration, experimentation, and intervention. For example, one senior academic reported:

> The notion of treating our city and its region as a seedbed for sustainability initiatives is a potent one . . . the vision is of academics out in the community, working with local groups and businesses on practical initiatives to solve problems and promote sustainable development and growth. This necessitates that we proceed in a very open manner, seeking to overcome barriers to thought, action and engagement; barriers between researchers and citizens, between the urban and the rural, between the social and natural sciences, between teaching research and enterprise. (Goddard & Vallance, 2013, p. 148)

Turning to the health challenge facing cities there is a mutual dependence of public health services and university medical faculties. They are in separate governance domains but joined together by many types of organizational and personal linkages of a financial and informal character. There is a well-established work-based learning model for medical students and the hospital and local population acts a living laboratory for clinical academics. Although acute medicine and public health are in different universities, the latter is now a key function of local government. This is leading to three-way partnerships.

In terms of public health, work-based learning is a key mechanism linking one university to the city. According to one interviewee in Sheffield Hallam University

> We're continually revising our curriculum, in partnership with our stakeholders—the strategic health authorities, the acute trusts, the PCTs—in order to be one step ahead in terms of anticipating the need. . . . We are very much wedded to work-based learning delivery, and particularly when you're talking about part-time, postgraduate [students] our unique selling point is that you learn by using your day-job, and so the assignments are actually around projects that will take your organization forward as well as yourself. (Goddard & Vallance, 2013, p. 116)

Finally, in relation to the contribution of universities to the creative city, the diversity of the cultural sector in cities is mirrored by the diversity of creative and artistic disciplines taught, researched, and practiced in universities—visual arts, music, drama, creative writing. The academic units in the universities and the constituent communities of students and staff have a strong identity with and connection to urban cultural life. These are fields where the hierarchy of research ratings between "old" and "new" universities does not apply—practice led research and teaching used in art, design and media fits particularly well with the mission of new universities. The campus provides cultural venues—university museums, theatres, art galleries, media labs and also the shared use of off campus sites where practice, teaching and research are linked. According to one interviewee in Northumbria University:

> I think what we are attempting to do is to try and crack that nut that a lot of fine art departments have to crack, which is how do you work in a professional practice environment that's recognized by students and postgraduates, but also works to the needs of a research culture. . . . What kinds of resources do you need? . . . Really the model you want to put forward is a sort of relationship of art and the city; so very metropolitan, very urban. It's not on campus, its right in the middle of town. (Goddard & Vallance, 2013, p. 135)

In the digital media area and according to one interviewee in the University of the West of England complementary temporalities can be seen:

> We as academics are really planning for five to ten years ahead, people in business are usually planning for the next quarter or the next six months or the next year. There are different temporalities, and one of the things that we can do is try to use our expertise to catch some of the things that they don't really have time to reflect on, or have the analytical purchase on, and play it back to them, and help them enrich their own process. (Goddard & Vallance, 2013, p 144)

And according to one interviewee in the Watershed's Pervasive Media Studio:

> I think one of the benefits of working with academics is that they provide a kind of stability in the way we work. . . . There is a space in the middle where they can collaborate which is the work that might come out in 2 to 3 years. And then there's the horizon work, which the academy is in a much better place to look at, because it hasn't got the commercial

> constraints. . . . But the studio acts as a kind of gearing mechanism to try and help those timescales, agendas, cash flows, find each other and work together. . . . There is a 5 year collaboration agreement between the three [organizations] at a corporate level, which we are calling a creative technologies collaboration. It's for research, innovation and teaching in what we are broadly calling creative technologies; so that cross—over space between what you would normally call creative content and what you would normally call digital computing. It is a mixed up space that none of us quite understand. . . . So it is an active collaborative space, which adds value to what the universities can do in their own faculties, on their premises, and on their own. (Quoted in Goddard & Vallance, 2013, p. 145)

All of these examples refer to ongoing collaboration and short- to medium-term horizons in regard to city development. Nevertheless, some of the challenges are longer duration, raising the question of the contribution a university can make to long-term thinking and planning about the future of the city where it is located using the methodologies of science foresight.

Anchoring Universities in Cities through Urban Foresight: The Civic University in Action

Foresight projects examine either an important public policy issue where science might be part of the solution, or a scientific topic where potential applications and technologies are yet to be realized. The projects involve critical thinking concerning long-term developments, debate and effort to create wider participatory democracy, and shaping the future, especially by influencing public policy. City-based foresight activity can be one means by which the global knowledge base and influence of the academy are unlocked for the benefit of the city where they are based. This can be achieved by universities fostering networks across and between the public, private, voluntary, and community sectors; identifying gaps in intelligence within cities, sector by sector; facilitating the exchange of intelligence and data between different agencies; developing long-term scenario options; synthesizing and mapping the varying strands of intelligence and data that exist within each city region; and mediating between government and citizens by actively engaging with service providers and others.

As part of a national foresight project on the Future of U.K. Cities to 2065, Newcastle University—which badges itself as a "World Class Civic University"—has sought to mobilize the academic expertise of the two universities in the city to work with partners in the public, private, and voluntary sectors on the long-term future of the city region (Tewdwr-Jones, Goddard, & Cowie, 2015). This approach involved applying national foresight methodology locally by establishing a Lead Expert Group and a wider Stakeholder Group undertaking the following activities:

- Baseline evidence—the current picture
- Newcastle City region research and literature database
- Stakeholder workshops
- Delphi survey of key actors

- Newcastle City Futures Exhibition—an Urban Room
- Scenario building

More specifically this process involved over 100 experts and stakeholders from diverse disciplines and organizations in the North East of England, covering public, private, community, and voluntary sectors; over 100 pieces of evidence contained in official reports and academic papers, as well as a wide range of ongoing reviews and studies; the opinions of approximately 2500 members of the public expressed at a specially convened *futures facing city* pop-up exhibition and events series, generating over 100 comment cards and ideas alongside feedback in 24 public forum events. This work fed into scenarios around the following dimensions:

- Technological (e.g., digital)
- Economic (e.g., globalization)
- Environmental (e.g., climate change)
- Political (e.g., devolution)
- Social (e.g., ageing population)
- Values (e.g., individualism)

Three scenarios emerged: Continuation of present socioeconomic trends (business as usual); London implodes: rebalancing the national economy; and Newcastle finds its niche: test bed city. The last of these found most support. More specifically the city has developed as a demonstrator platform for a range of scientific and technological future-facing public–private projects and programs that are socially inclusive. Social and cultural developments and consumer services support this platform role. A City Futures Development Group involving the universities and the City Council with participation from the other parts of the public and private sector has been established to sustain the activity. The Group is committed to improving services, quality of life, and economic growth by utilizing existing academic and industry excellence; creating opportunities for research and product development by facilitating access to infrastructure and residents; and overall ensuring that Newcastle is seen as a test bed for innovation providing further chances for research, investment, and business growth.

Going forward the Group will seek to use a city futures perspective to get around all of the "here and now" challenges of collaboration; appoint a dedicated city futures partnership manager jointly between the universities and local authorities with access to senior offices in each organization; create a value-added knowledge base by linking primary research in the universities with policy and practice research produced by the public and private sectors; launch a professional development program for key individuals expected by institutional leaders to play a boundary-spanning role between higher education and the city region, covering the *know what* and *know how* of futures work; develop an *action learning* program for those individuals around selected mid-term projects; and link up with other cities and universities nationally and internationally to create a community of practice around city futures.

Conclusion

Across the world universities are increasingly being expected to be active contributors to city development—in place-making, in business innovation, and in economic and social development in the round. With society increasingly facing complex challenges (for example ageing and climate change) that have both local and global dimensions, the role of universities in addressing these problems must come to the fore. To meet these demands universities will need to work in new ways. Frameworks and methodologies, such as the quadruple helix, social innovation, living laboratories, and city futures, are just some emerging tools for the new forms of multidisciplinary and transpartner working that can help.

Developing a quadruple helix approach to science, research, and innovation within the city will not be easy. There will be tensions between the external civic role of the university and its internal processes, with the latter being heavily influenced by the higher education policy environment in which it operates, one which in many countries is quiet detached from other policy areas, not least those relating to city and regional development. Addressing societal challenges requires an institutional response from a wide range of disciplines and clear institutional leadership. This raises questions around business models of the university. A new set of models may therefore be needed, of which the civic university is one.

The civic university should be characterized by its ability to integrate its teaching, research, and engagement with the outside world in such a way that each enhances the others without diminishing their quality. Civic research will have socioeconomic impact designed in from the start and teaching will have a strong community involvement with the long-term objective of widening the participation in higher education of disadvantaged groups and producing civic-minded graduates. Most importantly, taken together this will require a soft, flexible boundary between the institution and society.

Nevertheless, realizing the potential of the civic university will not only depend on what the university does, but also on the capacity of its city partners. Where there is weak leadership, ineffective partnerships, and lack of a shared vision the university may need to take a leadership role and over the long term help other public and private institutions in the city and beyond to build their capacity to absorb knowledge generated within the academy, to coproduce knowledge, and to articulate knowledge demands. Or to put another way, to both anchor the university in the city and the city in the university.

References

Arnkil, R., Järvensivu, A., Koski, P., & Piirainen, T. (2010). *Exploring quadruple helix: Outlining user-oriented innovation models.* Finland: Work Research Centre, Institute for Social Research, University of Tampere. Retrieved from https://tampub.uta.fi/handle/10024/65758

Bender, T. (1988). Afterword. In T. Bender (Ed.), *The university and the city: From medieval origins to the present*. Oxford, UK: Oxford University Press.

BEPA (Bureau of European Policy Advisors for the European Commission). (2010). *Empowering people, driving change: Social innovation in the European Union*. Luxembourg: Publications Office of the European Union. Retrieved from http://www.ess-europe.eu/fr/publication/empowering-people-driving-change-social-innovation-european-union

Brewer, J. D. (2013). *The public value of the social sciences: An interpretative essay.* London: Bloomsbury.

Calhoun, C. (2006). The university and the public good. *Thesis Eleven*, *84*(1), 7–43. doi:https://doi.org/10.1177/0725513606060516

Clark, R. B. (1998). *Creating entrepreneurial universities: Organizational pathways of transformation.* Kidlington: Pergamon/Elsevier.

Etzkowitz, H., & Leydesdorff, L. (2000). The dynamics of innovation: From national systems and mode 2 to a triple helix of university-industry-government relations. *Research Policy, 29*, 109–23. doi:https://doi.org/10.1016/S0048-7333(99)00055-4

European Commission. (2011). *Connecting universities to regional growth: A practical guide*. Brussels: European Commission. Retrieved from http://ec.europa.eu/regional_policy/sources/docgener/presenta/universities2011/universities2011_en.pdf

European Commission. (2015). *Digital single market: Open innovation 2.0*. Retrieved from https://ec.europa.eu/digital-agenda/en/growth-jobs/open-innovation

European Union. (2007). Treaty of Lisbon: Amending the Treaty on European Union and the Treaty Establishing the European Community. *Official Journal of the European Union, 2007/C 306/1*. Retrieved from http://eur-lex.europa.eu/legal-content/EN/TXT/HTML/?uri=CELEX:12007L/TXT&from=EN

Goddard, J. (2009). *Re-inventing the civic university*. London: National Endowment for Science, Technology and the Arts. Retrieved from https://www.nesta.org.uk/sites/default/files/reinventing_the_civic_university.pdf

Goddard, J., & Vallance, P. (2013). *The university and the city.* Abingdon: Routledge.

Goddard, J., Coombes, M., Kempton, L., & Vallance, P. (2014). Universities as anchor institutions in cities in a turbulent funding environment: Vulnerable institutions and vulnerable places in England. *Cambridge Journal of Regions, Economy and Society, 7*, 307–325. doi:https://doi.org/10.1093/cjres/rsu004

Goddard, J., Hazelkorn E., Kempton, L., & Vallance, P. (2016). (Eds.). *The civic university: The policy and leadership challenges*. London: Elgar.

Grau, F.-X. (2014). *The glocal university.* Barcelona: Global University Network for Innovation. Retrieved from http://www.guninetwork.org/files/glocal-university_fx_grau_2014.pdf

Information Society Technology Advisory Group. (2011). *Orientation of EU ICT R&D and innovation beyond 2013: Report from the Information Society Technology Advisory Group (ISTAG).* Brussels: European Commission. Retrieved from http://cordis.europa.eu/fp7/ict/istag/documents/istag-key-recommendations-beyond-2013-printed.pdf

Science With and For Society Advisory Group, (2014). *Strategic opinion for research and innovation in the Horizon 2020, 2016–17 Work Programme*. Science With and For Society Advisory Group. Brussels: Economic Commission. Retrieved from http://www.google.co.uk/url?sa=t&rct=j&q=&esrc=s&source=web&cd=1&cad=rja&uact=8&ved=0ahUKEwj_7JX6x4jTAhVISZoKHZ8FAn8QFggdMAA&url=http%3A%2F%2Fwww.eesc.europa.eu%2Fresources%2Fdocs%2Fstrategicopinion-for-research-and-innovation_11-july-2014.doc&usg=AFQjCNG8ctafMtJvmG4Gjh5otexSN-SC2w

Tewdwr-Jones, M., Goddard, J., & Cowie, P. (2015). *Newcastle city futures: Anchoring universities in urban regions through cities foresight.* Newcastle: Newcastle Institute for Social Renewal.
Vallance, P. (2016). Universities, public research, and evolutionary economic geography. *Economic Geography, 92*, 355–377. doi:https://doi.org/10.1080/00130095.2016.1146076
The Work Foundation. (2010). *Anchoring growth: The role of 'anchor institutions' in the regeneration of UK cities*. London: The Work Foundation Alliance Limited.

Chapter 12
City and University—An Architect's Notes on an Intriguing Spatial Relationship

Helmut Bott

The university as a European institution has grown in the complex relationship of local, national, and even pan-European interests and power structures (see Bott, 2015, p. 12). From the beginning, universities have had a distinctive internal orientation with their rituals and regulations, privileged by a kind of academic autonomy. At the same time, however, they have been an important element of local society, culture, and economy, having a strong relationship to the place they are located. Lastly, yet importantly, universities have always been nodes in a network of science. Going beyond the local bonds, their orientation is to the international scientific community. Universities have thus developed within multidimensional relationships, of which the polarity between territorial exclusivity outside urban society and integration into urban structures is just one. This chapter deals with the change in the architectural concepts of university buildings and in the spatial relationship between university, town, and landscape over the centuries up to the present.

The Early European University Within the Power Structure of Town, Court (Government), and Pope

Higher learning revived during the High Middle Ages, when private schools of cities were founded and complemented exclusive institutions such as cathedral and monastery schools of the pan-European Roman Catholic Church, along with court schools and palace schools. The papal administration declared some of these private schools to be legal and granted them special rights, such as tax exemptions,

H. Bott (✉)
Institute of Urban Planning and Design, University of Stuttgart, Stuttgart, Germany
e-mail: helmut.bott@si.uni-stuttgart.de

P. Meusburger et al. (eds.), *Geographies of the University*, Knowledge and Space 12,
https://doi.org/10.1007/978-3-319-75593-9_12

municipal services, or special rights of judicial practice. However, conflicts between legalized universities and local governments were not resolved forever by these acts. They flared up from time to time.

European universities were founded as of the eleventh century, first in Bologna and Paris, somewhat later in Oxford and Cambridge, spreading throughout the Roman Catholic European empires and countries (see Bott & Teodorovici, 2015, p. 24; Verger, 1992). Because universities provided a pool of educated graduates from which feudal courts and ecclesiastical institutions could draw their qualified staff, the founding of universities became a matter of vital importance and crucial interest throughout western Europe. At the same time, it posed a serious conflict between Church and kingdom.

Universities were founded, funded, legalized, and protected by feudal clerical or secular powers (Fig. 12.1), who appointed rectors and controlled these institutions of higher education. University life was initially similar to monastic life but also had elements of self-government as practiced in medieval municipalities, craft guilds, and associations of those towns in which professors, scholars, and students lived (Bott & Teodorovici, 2015, p. 25; Nardi, 1992).

Internal academic controversies about dogmas and theoretical principles were commonplace and could escalate into violent conflicts, even to the final exodus of a group of scholars from famous universities. In fortunate cases these disputes led to new foundings that also eventually became eminent academic institutions (e.g., Bologna—Padua, Oxford—Cambridge). From the outset, universities had both this strong inward orientation to their academic world and necessarily to the relation with urban life, but they were simultaneously part of the international world of the Latin–Christian sciences. Latin was the lingua franca, and all texts had to be written in that language. Diplomas of universities recognized by the pope had to be accepted everywhere. Academic mobility, such as that of students of Canon and Roman law at the renowned law school of Bologna, was nothing extraordinary. Traveling scholars were common in the medieval world (Bott & Teodorovici, 2015, p. 25; de Ridder-Symoens, 1992; see chapter by Meusburger & Probáld in this volume).

Until the early modern era universities were not research universities, which constantly seek new scientific knowledge. On the contrary, the main task of their scholars was to interpret the Bible and to read, comment, and explain to students well-known works by celebrated theologians and texts by ancient philosophers (Bott & Teodorovici, 2015, p. 27).

Urban Integration and Architecture of Early Universities

At first, lessons were held in small public houses. Prominent and successful schools grew house by house, plot by plot. Whenever spatial and financial conditions improved, school buildings were erected. They were arranged similarly to urban monasteries: mostly four wings around a central courtyard with entrance gate, chapel or prayer rooms, classrooms, library, dormitories for professors and students, dining

Fig. 12.1 Inauguration ceremony in the Cathedral of Basle, Switzerland, 1460. Bishop Johann von Venningen appoints provost Georg von Andlau (front left) as the first rector of the University of Basle. However, he gives the papal founding bull to the mayor of Basle. This anonymous drawing shows the institutional involvement of power at different levels: local (mayor), regional (bishop as sovereign of the canton), and international (pope). Very specific indeed was the fact that this papal bull was signed by an antipope whose reign was brief and recognized by only a few states, including the Swiss Confederation.
Source and copyright: Basle University Library AN II 3 (n.d.). Reprinted with permission.

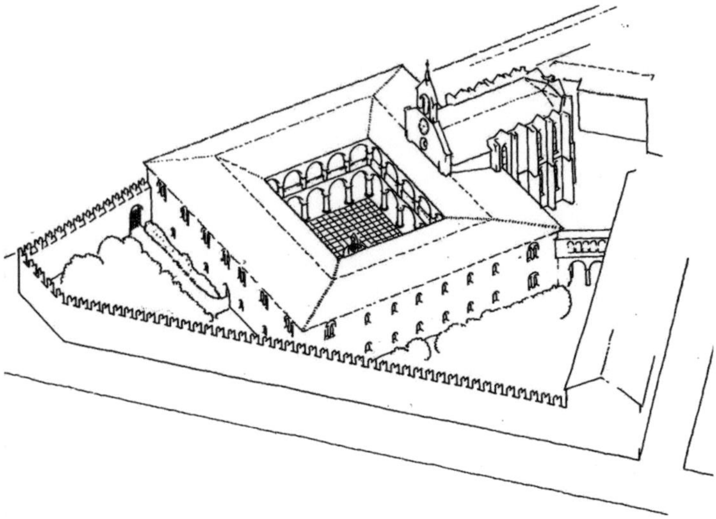

Fig. 12.2 Spanish College, Bologna, Italy.
Source: Rückbrod (1977, image 18). Copyright by K. Rückbrod. Reprinted with permission.

hall, kitchen, and ancillary rooms. Furniture and equipment were poor (Bott, 2015, p. 26; see also Schwinges, 1992).

The first collegium following a new concept was the Spanish College in Bologna (1365–1367), which became the future model for European collegiate buildings (see Capitani, 1987; Teodorovici, 2015a). The four wings of the school complex facing the inner courtyard were clearly arranged in a functional manner: western wing with classes, assembly room, and dining hall; east wing with chapel, offices, and rectorate (Fig. 12.2). The central cloister courtyard demonstrates the introverted community living and working like monks. In many cases students had to live in rented rooms in citizens' houses or in hostels outside the collegiums. For them, university life and urban life were tightly interwoven.

In the late Middle Ages and the early modern period, university colleges could permeate whole quarters, as in Paris (Latin Quarter), or the whole townscape in smaller settlements (such as Oxford and Cambridge). They could even constitute the city crown (*Stadtkrone*), as in the Portuguese city of Coimbra (Erl & Teodorovici, 2015; Rodrigues, de Almeida, & de Alburquerque, 1990).

Medieval Paris

In the Middle Ages, Paris became one of Europe's leading centers of higher education and academic discourse (Teodorovici, 2015b; Tuilier, 1994). As at many early universities, teaching started there in small private schools, extending room by room, house by house, taking place in the houses of private citizens and in

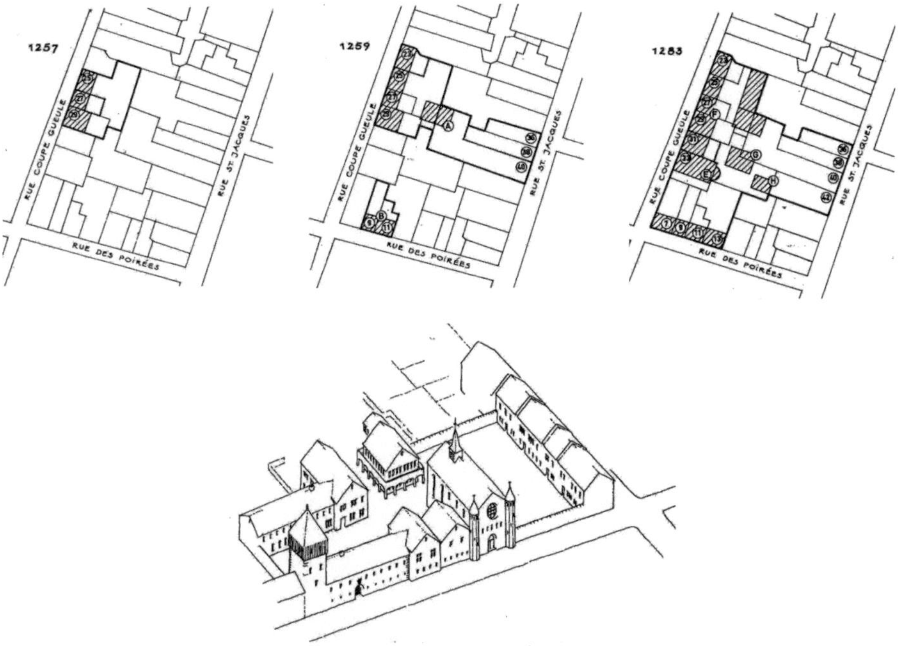

Fig. 12.3 Growth of the Sorbonne, Paris, France.
Source: Rückbrod (1977, images 15–17). Copyright by K. Rückbrod. Reprinted with permission.

former convents. A network of schools and reputed teachers grew steadily, attracting more and more scholars and students. When the University of Paris finally received approval by Pope Gregory IX in 1231 and was provided with funds and privileges, the southern bank of the Seine river experienced the rise of a rich scene of Latin schools and colleges populated by a host of scholars. The area still goes by the name it acquired at that time—the Quartier Latin (Fig. 12.4).

In 1257 Robert de Sorbonne (1201–1274), a theologian and one of the French king's intimate advisors, founded a college to host the faculty of theology (Fig. 12.3). It was named after him, the Sorbonne. It came to be the university's most illustrious department, with the name becoming synonymous with the University of Paris itself. Step by step, the building complex of this institution expanded, forming an inner courtyard. Because of its enormous success, small two-story medieval buildings gradually gave way to larger and higher buildings. A Baroque church replaced the old chapel, and a block of surrounding university buildings shaped an interior courtyard as enclosure. The church, however, was designed as a hybrid building with two main façades, one facing the urban public space, today Place de la Sorbonne; the second, perpendicular to the first, facing the courtyard. To me, it is an expression of the polyvalent character of universities in Europe, with their intense internal academic life and rituals belonging to the international world of science yet simultaneously integrated into public spatial urban patterns.

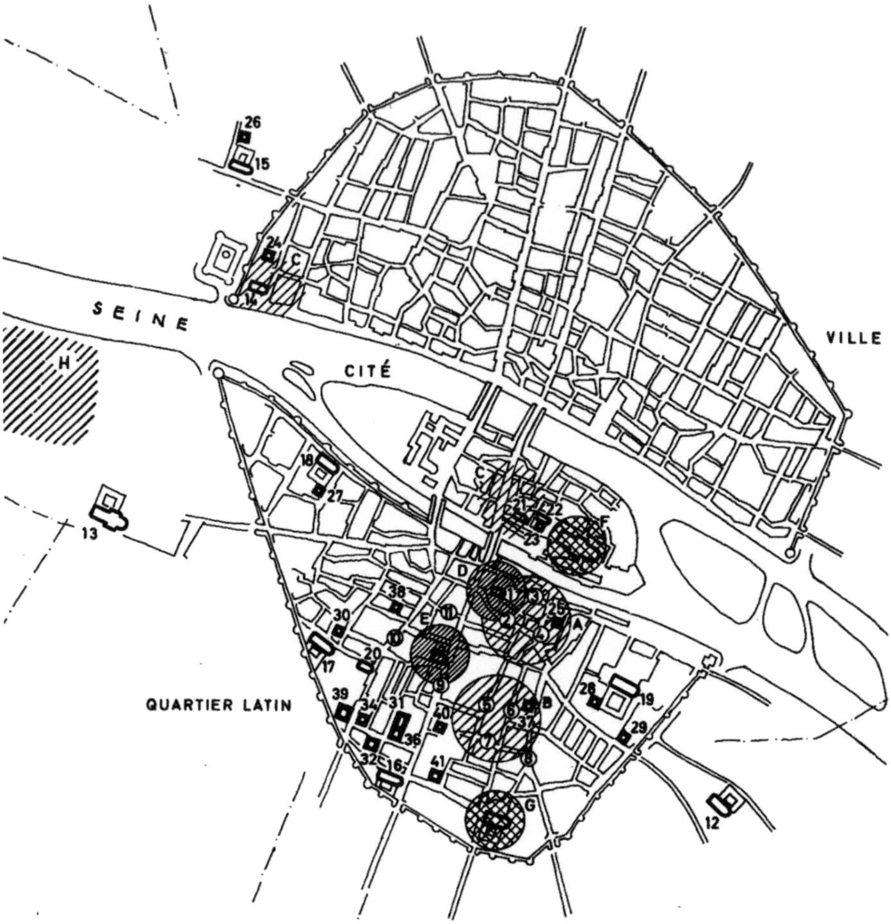

Fig. 12.4 Quartier Latin, Paris, 1200–1300.
Source: Rückbrod (1977, image 8). Copyright by K. Rückbrod. Reprinted with permission.

Medieval Oxford and Cambridge

As in Paris, teaching at the early English universities Oxford and Cambridge started in rented rooms for about a decade for lack of college buildings (Teodorovici, 2015c). The first Oxford College, Merton College (1264), was an ensemble of heterogeneous structures, positioned around an interior patio. The *New College* (1379) became the first English university complex, erected in a homogeneous style shortly after the Spanish College in Bologna (Fig. 12.5). The *New College*

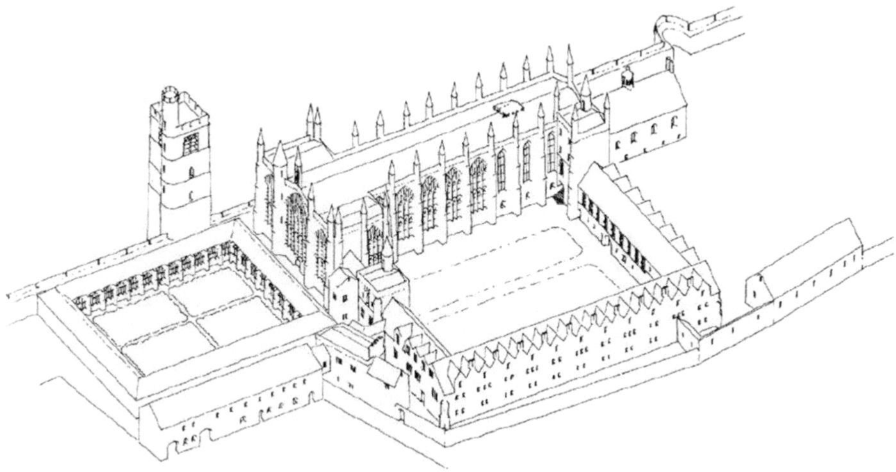

Fig. 12.5 New College, Oxford, England.
Source: Rückbrod (1977, image 20). Copyright by K. Rückbrod. Reprinted with permission.

had massive dimensions and a decisive impact on the further development of English college buildings. The rigid, precise gothic façades outside and inside the wide, green courtyards became a typical Anglo-Saxon university style—later complemented by classical architecture.

Compared to Bologna and Paris, places such as Oxford and Cambridge were small towns. The university's influence on their urban life and townscape was therefore much more powerful and enduring. The college buildings permeated the small town's corpus, establishing a very specific relationship between private houses and university institutions. Whereas university facilities in Paris and Bologna were positioned in the city center, far from the surrounding landscape, most colleges in the two English cities, particularly in Cambridge, had a direct spatial connection to the open landscape. They were not separated from it by a town wall (Fig. 12.6). The flood plain of the lovely Cam river became an integral spatial element of daily university life and of the scholar's rituals and customs. A series of colleges, and later the impressive library by Christopher Wren (1632–1722), formed a wonderful town façade facing the Cam green belt. This close interweaving of landscape and university facilities eventually created an early spatial model that ultimately became such a characteristic feature of American campus universities (Fig. 12.7).

Medieval Vienna

It is instructive to contrast this spatial setting with the medieval university quarter in a much larger central European university town such as of Vienna. The university buildings and dormitories were spread throughout that space (Fig. 12.8). Whereas

Fig. 12.6 Map of Cambridge, England, dated 1574. Source: Braun and Hogenberg (1575). Drawing of Cambridge attributed to William Smith, engraving after Richard Lyne, 1574. Via Wikimedia Commons. Copyright: Public domain. Retrieved from https://upload.wikimedia.org/wikipedia/commons/f/fc/Cambridge_1575_colour.jpg.

Fig. 12.7 Aerial view of Cambridge city center from the west. The picture shows the direct connection between the Cam river green belt and the famous colleges.
Source: © CMG Lee (2013), via Wikimedia Commons. Used under Attribution-ShareAlike 3.0 Unported (CC BY-SA 3.0). Retrieved from http://commons.wikimedia.org/wiki/file:cmglee_Cambridge_aerial.jpg.

the basic studies (*artes liberales*) and lecture rooms were located in the central buildings, the main courses for medicine and law were held in smaller buildings some blocks apart. Furthermore, university facilities such as the students' hospitals, a detention cell, and the library were dispersed in the quarter. One can easily imagine that scholars and normal citizens unavoidably mingled with each other in their everyday lives in this area of old Vienna. After all, university members and neighborhood craftsmen, traders, and servants alike had to use public streets to reach the venues of their daily work (Fig. 12.9).

Universities in the Early Modern Era

The Renaissance saw the rise of science and the evolution of the mechanical arts (*artes mechanicae*) from crafts into applied science. Medicine slowly transformed from theoretical narratives into an experimental science based mainly on the dissection of corpses, negating papal interdicts. The transition from text work to experimentation, from theory to empiricism, and from theorizing to practicing required new, specialized university facilities and, hence, new architectural elements. Last but not least, the invention of letterpress printing in the mid-fifteenth century multiplied

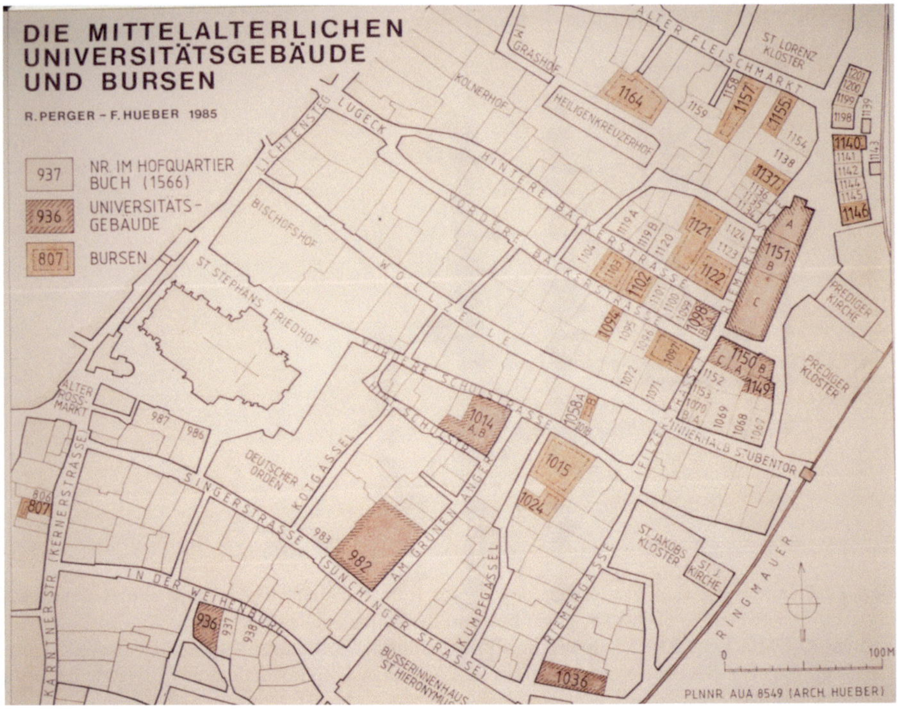

Fig. 12.8 Medieval university quarter of Vienna, Austria, prior to the Jesuits' transformation under Emperor Ferdinand II after 1623. "Original plan No. AUA 8549, by Richard Perger Friedmund Hueber, Vienna University archive. The plan shows the city quarter at the Stuben Gate, with house numbers in the area as listed in the 1566 court registry. The colleges, student dormitories, and accommodations for poor students were located there.
Source: Vienna University Archive, No. 109.3, http://geschichte.univie.ac.at. Reprinted with permission.

the number of scientific books in circulation, expanding library inventories tremendously. New major libraries became a new building typology for universities.

The reformation also brought about a drastic change of the international university system. Protestant dukes and kings no longer accepted the papal oversight of the universities. This resistance dissolved the unique integrated European system of higher education.

The Turn Toward Research Universities

After the Renaissance numerous universities lagged behind extramural institutions, for circles of scholars at the courts of high sovereigns (e.g., Florence, Milan, Rome, and Urbino), and later at Royal academies, provided scientists with superior research

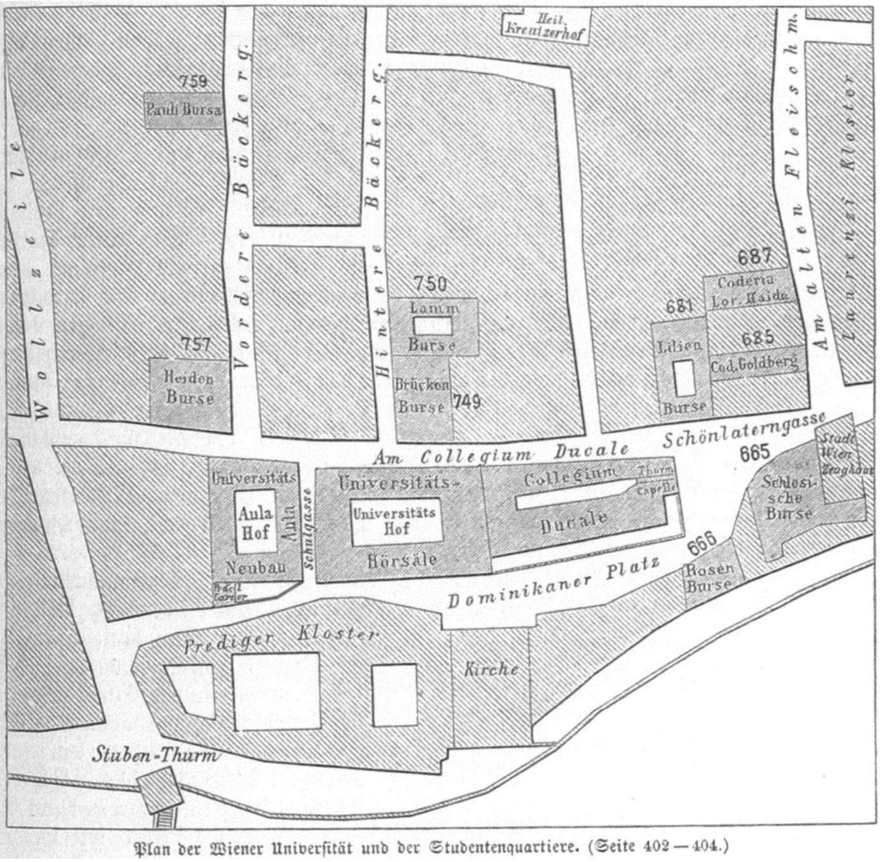

Fig. 12.9 Old University quarter in Vienna before the urban reconstruction by the Jesuits. Source: Bermann (1880, p. 425), via Wikimedia Commons. Copyright: Public domain.

facilities such as telescopes, more funding, and freedom from teaching.[1] Universities could hardly offer similar conditions, although they struggled to develop scientific research, improving the technical equipment as well as introducing experimental methods by creating chairs of natural science and new humanities. Establishing research universities was a long process. In Heidelberg, for example, lectures on physics had been based on Aristotele since 1387 as part of the fundamental *art*

[1]Such centers included the Collège de France (1530), the Académie française (1653), the Académie royale des sciences (1666), the Royal Society London (1661), and, in Schweinfurt, an institution later named the Academia Leopoldina (1652). Many famous scientists such as Galileo Galilei (1564–1642) and Johannes Kepler (1571–1630) tried to work for sovereigns and leave the university.

studies. In 1556 a chair of physics was created, but it was not until 1752 that a chair of experimental and mathematical physics was established there.[2]

In 1609 Johannes Hartmann (1568–1631) received the first professorship for *chymiatry* (chemistry) at the University of Marburg. The new post was in the faculty of medicine. He founded Germany's first chemical university laboratory where students could carry out experiments. The Protestant University of Leiden in The Netherlands built the first astronomical observatory around 1623. The Bavarian university of Ingolstadt, led by the Jesuits, erected the first observatory at a German university in 1637. Thanks to the successes of great researchers in the natural sciences and humanities since the Renaissance, many universities started to extend the canon of subjects to history, geography, natural law, and experimental natural sciences during the Age of the Enlightenment. Lectures started to be given not only in Latin but in national languages, too, improving the accessibility and practicability of research results.[3] The purpose of university education became more practically oriented and was intended to serve the needs of state and society. As a result, cameralism was introduced, the optimization of the absolutist state's public administration—a predecessor of economics.

Over the course of the seventeenth and eighteenth centuries, great scholars and researchers improved scientific theories, a broad spectrum of the humanities evolved, and applied sciences drew the close attention of famous scholars. Schools for higher technical education were founded, such as the Bergakademie in Freiberg (1765), a higher school for education in mining and metallurgy in Saxony, where the famous Alexander von Humboldt was educated.

The Integration of New Research Facilities: From Convent to Palace

The *teatro anatomico* at Padova (1594) was the first lecture hall where students seated in steeply raked rows could follow medical professors' demonstrations of their pathology work. This design can be considered the beginning of specialized architecture for research and teaching, an application of an archetypal representation and spectacle like that in classical theaters.

Another specific university facility was the botanical garden, which had its predecessors in herbal gardens. All botanical systematization began in such gardens, in which medicinal plants were organized in two-dimensional arrangements as a *tableau* according to similarity and their effects on diseases—one of the roots of systematic biology in the faculty of medicine. By this time anatomical theaters, botanical gardens, laboratories, and astronomical observatories were specialized

[2]On the history of the faculty of physics and astronomy, see www.physik.uni-heidelberg.de

[3]However, this shift also expedited the dissolution of the entire European scientific world into national science traditions.

Fig. 12.10 Archiginnasio. Bologna, Italy (1562). Painting by Contardo Tomaselli e Onofrio Zanotti, La facciata del palazzo dell'Archiginnasio, 1849 C.
Source: Bibliotheca Communale dell'Archiginnasio (n.d.). Copyright: Public domain. Retrieved from http://www.archiginnasio.it/facciata.htm.

institutions for teaching and research, which clearly distinguished the equipment of early modern universities from medieval collegiate buildings. But these facilities were seldom seen from the outside. Although architectural style changed to Renaissance and later Baroque, the courtyard archetype for colleges remained dominant during the early modern era up to the eighteenth century. New Protestant universities founded in the sixteenth century, such as Marburg (the first one ever) and Leiden (the famous and oldest university in The Netherlands) used former convent buildings. Most of the details of the well-preserved university of Würzburg (1582–1591) clearly date from the Renaissance. Because of the courtyard type and the traditional steep rafter roof, however, it still looks like a medieval university.

Archiginnasio Bologna

A highly instructive example of the spatial relationship between the city and the university during the Renaissance is the Archiginnasio in Bologna, Italy (1562), erected by Antonio Morandi (1508–1568) directly in the city center close to the main church and near the central square, Piazza Maggiore (Fig. 12.10). All ground-floor rooms open completely onto the public urban space. Remarkably, the ground floor houses primarily commercial units, such as shops and trattorias. Only the main

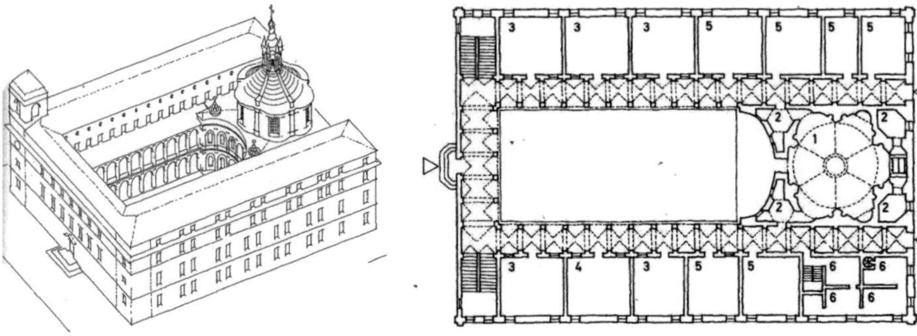

Fig. 12.11 Isometry and floor plan of La Sapienza, Rome.
Source: Rückbrod (1977, images 25, 24). Copyright by K. Rückbrod. Reprinted with permission.

entrance leading to the inner courtyard points to the university. A chapel and some administrative rooms open onto this courtyard. Via two separate staircases one reaches either the rooms of the *Legisti* (jurists) on the right, students of canon and civil law in the main course, or the rooms of the *Artisti*, the students of the basic general studies. The spatial structure thus expresses the institution's hybrid character: (a) the famous university's successful integration into the center of the rich city of Bologna, closely related to commercial urban life; (b) the inwardness of academic life with its regulations and rituals as reflected by the courtyard and the collegium's internal organization.

La Sapienza

The university La Sapienza in Rome, Italy, designed by Giacomo della Porta (1532–1602) in 1585 and completed by Francesco Borromini (1599–1667) a few decades later, shows an interesting evolutionary intermediate step (Fig. 12.11). The architecture is late Renaissance, whereas Borromini's church is Baroque. Della Porta's design follows the traditional medieval patio-type of university building but shows a kind of opening to one side, for the courtyard has only a narrow gallery toward the Corso del Rinascimento. This gallery gives a spatial direction to the west—and vice versa a processional space leading to the church, weakening the strict interior orientation of the courtyard, although it remains an enclosure. The Sapienza building thereby foreshadows the triple-wing concept.

Altdorf University

Altdorf University, founded in 1571 by the Protestant free imperial city of Nuremberg in Bavaria, Germany, is unpretentious architecture with Renaissance elements (Fig. 12.12). Given the buildings and roof proportions, however, the overall impression was still medieval. Yet Altdorf was already a triple-wing system, no longer a

Fig. 12.12 Altdorf with university center (C), (at the southern edge of the town), the botanical garden outside the town wall (B). Engraving by Merian, 1648, *Topographia Franconiae*. (This publication is one of the 16 volumes of the *Topographia Germaniae* edited by Mathias Merian). Source: Merian (1648), via Wikipedia. Copyright: Public domain. Retrieved from https://de.wikipedia.org/wiki/Altdorf_bei_N%C3%BCrnberg#/media/File:De_Merian_Frankoniae_027.jpg.

Fig. 12.13 *The Diligent Student*, Johann Georg Puschner, 1725. Fourteen 14 copper engravings by Johann Georg Puschner (1680–1749). Source: Via Wikipedia. Copyright: Public domain. Retrieved from https://de.wikipedia.org/wiki/Johann_Georg_Puschner#/media/File:Der_Fleissige_Student.jpg.

patio enclosed by four wings. The front courtyard was separated from the public space by a wall and a gatehouse but was much more open to the city than the introverted collegiate buildings of the Middle Ages had been. A later etching, *The Diligent Student* (1725), depicts a notable relation between town and university court, a view onto the University of Altdorf (Fig. 12.13), with university members streaming out of the open gate and mixing in public town life.

This Lutheran university installed a botanical garden in 1626, an anatomical theater in 1650 (Fig. 12.14), an observatory in 1657, and a superbly equipped *laboratorium chymicum* in 1682 (Fig. 12.15). As a modern research university at that time, it provided all the facilities necessary for empirical science.

Fig. 12.14 Johann Georg Puschner. *Theatrum Anatomicum* in the Collegio of Altdorf, side view during a lecture.
Source: Braunschweig, Herzog-Anton-Ulrich-Museum, Inv.-Nr. JGPuschner AB 3.1. Copyright: Public domain.

Uppsala University

Another, even more stunning example of modern equipment can be found at the University of Uppsala. This oldest Scandinavian university (1477) commissioned a new building in 1620, funded by King Gustav Adolph (1594–1632), hence its name, *Gustavianum* (Figs. 12.16 and 12.17). It is an modest linear edifice, but in 1662 it was spectacularly "crowned" by the famous physician, Professor Olof Rudbeck (1630–1702). He had a cupola enclosing an anatomic theater constructed on the rooftop center, proudly creating a landmark of modern science,[4] visible far and wide.

Vienna University

An interesting roof superstructure is found at the Vienna University. Between 1623 and 1655, the medieval university and all student houses were demolished in order to build a Jesuit college on this site, an act of Counter-Reformation by Emperor Ferdinand II of Habsburg (1578–1637), who gave the Jesuits the supervision over

[4]http://www.uu.se/en/about-uu/history.

Fig. 12.15 Laboratorium Chymicum [Chemical laboratory, seventeenth century]. University of Altdorf (Bavaria, Germany). Engraving by Johann Georg Puschner.
Source and copyright: University Library, Göttingen. Reprinted with permission.

the faculties of theology and philosophy. This new university ensemble was built in the Baroque style and comprised a church, lecture halls, a new library, an observatory, a botanical garden inside the patio, a refectory, a Latin school, dormitories, a hall for theater performances and festivals, and even stables—a complete building program. Set atop the new main building, the observatory was indeed an outstanding construction.

No fewer than 17 privately owned houses of citizens had to be bought and destroyed, along with most of the old mediaeval university buildings such dormitories, the library, and the detention cell. All in all, it was a colossal project of urban reconstruction inside the city walls of the Catholic imperial city of Vienna while the Thirty Years' War was raging in the central and northern parts of the Holy Roman Empire of the German Nation.[5] Under the rule of Maria Theresa (1717–1780), the university supervision was withdrawn from the Jesuits and transferred to state authorities. New buildings were added, including an auditorium and an anatomical theater (Fig. 12.18). The whole complex illustrates the concept of a university integrated into a major city during the early modern period and the Age of Enlight-

[5]http://geschichte.univie.ac.at/artikel/das-akademische-kolleg.

Fig. 12.16 Gustavianum, built in Uppsala in the 1620s and once the main representative building of Uppsala University, now the *Museum Gustavianum*. The anatomical theater inside the cupola was built by Olof Rudbeck the Elder.
Source: Photo by the author.

enment as well as the state's growing focus on science in the age of enlightened absolutism (Fig. 12.19).

Coimbra University (Portugal)

A truly fascinating university ensemble with Renaissance and Baroque buildings is the old University of Coimbra (founded in 1290). It was reconstructed and expanded with wonderful buildings between the seventeenth and eighteenth century. Set on a hilltop, the whole university crowns the city like an Acropolis (Fig. 12.20) and is organized around several courtyards positioned on platforms at different levels connected by stairs, ramps, and arched gates (Fig. 12.21). The Joanina Library (1728), one of the most fascinating libraries of the early modern age, is positioned at the edge of the main courtyard. Users enter the library from there, coming into the main rooms on the third floor, with archives and other facilities being are located down the hill.

Fig. 12.17 Anatomical Theater Gustavianum, 1662. Source: Photo by the author.

Trinity College in Cambridge, England

Libraries became a particularly interesting construction task requiring wide-span structures, heavy load, and ample light. Very impressive indeed is the library of the Trinity College in Cambridge, designed by Sir Christopher Wren and completed in 1695 (Fig. 12.22). It is a rationally designed building with a strictly serial façade and a spacious, bright interior, an early example of classicism. The totally geometric design shows the structural rationalism of the intellectual scientist Wren, who became an architect as an autodidact

Triple-wing university buildings as new paradigm

Although courtyard designs were still common for university buildings, albeit on a larger scale than medieval patios, the eighteenth century saw the final breakthrough of the triple-wing building type as paradigm of university buildings (Fig. 12.24). Universities had developed from an introverted convent to a palace type, opening a forecourt toward the public space. A clear example of this new type is the

Fig. 12.18 The new auditorium of Vienna University, 1700.
Source: Image from the Vienna University Archive, No. 106.I.3287. Reprinted with permission.

Seminarium Carolinum (built 1751–1752) at Heidelberg University (Hoffmann, 1985) (Fig. 12.23). At the same time, overall university life had changed from a monastic lifestyle to a much more luxurious leadership education of young noblemen (Figs. 12.25 and 12.26).

An Ideal Plan for a University

In 1750 the architect and illustrator Giovanni Battista Piranesi (1720–1778) published a book with engravings, including a design for an ideal university. It is a geometric and axially symmetric figure generating extremely complex spatial sequences with substantial numbers of rooms, halls and stairs, eight bridges, and many small patios. The entire system's center is a circle island with seminar rooms and dormitories, separated by a channel from a surrounding second circle containing halls and central university facilities. The outer circle merges in an encompassing square flanked by two halls, a theater, and an arena, like a sport hall. A longitudinal

Fig. 12.19 The new auditorium of Vienna University with an observatory on the roof, 1700. Source: Image from the Vienna University Archive, No. 106.I.3373. Reprinted with permission.

axis crosses these circles from the main entrance with the rector's seat to the church on the opposite side of the complex (Fig. 12.27).

Piranesi thus combined religion, science, art, and aspects of chivalric fitness. All these matters were part of a whole educational system, and riding, fencing, vaulting, and dancing were components of aristocratic physical training. This exuberant design opens a wide field of interpretation. It shows a complex total geometric system. It is a late example of 250-year tradition of geometric ideal urban design figures extending from Filaretes Sforzinda (1464) to the design for Karlsruhe (1715) (Bott, 2013).

Fig. 12.20 Coimbra, Portugal, with university and library (top right).
Source: © Kondephy (2013), via Wikipedia. Used under Creative Commons Attribution-ShareAlike 3.0 Unported (CC BY-SA 3.0). Retrieved from https://commons.wikimedia.org/wiki/File:Coimbra_from_River.jpg.

Fig. 12.21 University of Coimbra, central courtyard with Renaissance and Baroque buildings.
Source: © Andreas Trepte (2007), via Wikipedia. Used under Creative Commons Attribution-ShareAlike 2.5 Generic (CC BY-SA 2.5). Retrieved from https://de.wikipedia.org/wiki/Coimbra#/media/File:University-of-Coimbra.jpg.

In many cultures, including European antiquity, the circle represents heaven or the universe, whereas the square symbolizes earth. Niccolò Tartaglia (1499–1557) used the circle to incorporate the realm of science. A frontispiece of his treatise *La Nova Scienza* (1537) showed two circles. The first one, depicting the entrance gate controlled by Euclid, encompasses the *artes liberales* and applied sciences. From there, passing the gate controlled by Plato and Socrates, the procession continues to the second circle, encompassing philosophy. That is, to understand Creation one must learn geometry and mathematics, then study the general fundamentals of science (*artes liberales*) to apply this knowledge and deepen it through professional

Fig. 12.22 Cambridge Trinity College, 1676–1695, Christopher Wren (1632–1722). Source: Photo by the author.

studies. The person who is successful and even goes on to study the ancient philosophers may become wise and reach the realm of philosophy. Piranesi seems to follow this tradition, but his procession through the first circle of study leads to the church cupola circle at the end of the central axis. His university still stands under the supervision of the Roman Church, using the Bible to explain the Creation, whereas north of Rome the Enlightenment had dawned. At that time even the Catholic Habsburg emperors revoked the Jesuits' supervision of the University of Vienna.

Early European University Export

Latin America

Spain and Portugal's establishment of expansive colonial empires in southern and central America, blessed and regulated by the papal bull *Inter caetera divinae* by Pope Alexander VI. in 1493, brought with it, among other things, the spread of Catholic, especially Jesuit, educational institutions in that part of the New World. In 1538 the University of Santo Domingo was founded, privileged by Pope Paul III.

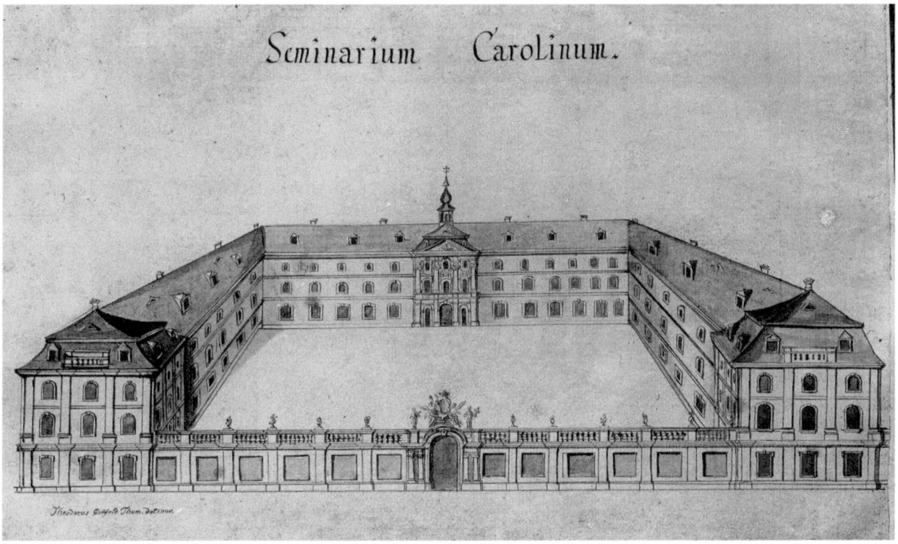

Fig. 12.23 Seminarium Carolinum (originally college for the Jesuits), Heidelberg University, 1751. Source: Thesaurus Palatinus. Landesarchiv Baden-Württemberg, Abt. Generallandesarchiv Karlsruhe 498-1 Nr. 2826 Bild 1. Copyright: Public domain.

Fig. 12.24 The University of Fulda, Germany (1734), a triple-wing type with a forecourt, 1887. Source: Laverrenz (1887, p. 124), via Wikipedia. Copyright: Public domain. Retrieved from https://de.wikipedia.org/wiki/Datei:Adolphiana_Universit%C3%A4t_Fulda_1887.jpg.

Fig. 12.25 University lectures in the late the fourteenth century. Source: Grandes Chronique de France. Castres, bibliothèque municipal Cours de philosophie à Paris Grandes chroniques de France. Copyright: Public domain. Retrieved from https://commons.wikimedia.org/wiki/File:Philo_mediev.jpg.

Fig. 12.26 The scuffling student. Johann Georg Puschner, 1725. Source: Via Wikipedia. Copyright: Public domain. Retrieved from https://de.wikipedia.org/wiki/Johann_Georg_Puschner#/media/File:Der_Rauffende_Student.

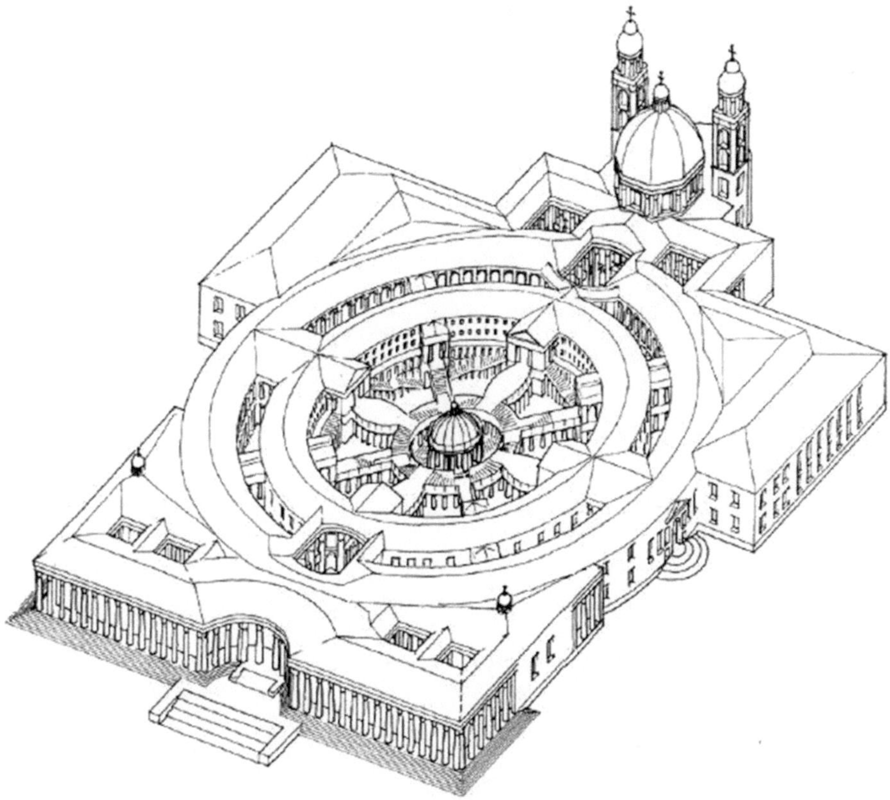

Fig. 12.27 Ideal plan of a university by Giovanni Battista Piranesi, 1750. Reconstruction by K. Rückbrod.
Source: Rückbrod (1977, image 33). Copyright by K. Rückbrod. Reprinted with permission.

The University of Lima followed in 1551, the oldest university on the South American mainland.

In 1553 the representative of the Spanish Crown, Viceroy of New Spain Luis de Velasco (1539–1617), opened the first University of Mexico, on which the Spanish king conferred the same privileges as Salamanca, its mother university. The first building was erected in the historic center of Mexico City, near by the main square, where the principal cathedral and the representative offices of the colonial empire were located. Close to this oldest building a complex of monasteries, colonial palaces, and colleges was built from 1595 onward (Colegio Máximo de San Pedro y San Pablo, cuatro seminarios: San Pedro y San Pablo, San Bernardo, San Miguel y San Gregori). Unfortunately, most of them were destroyed in the twentieth century.

Approval by the pope and institutional organization were not all that was taken over from Europe when Latin American universities were founded. Even the introverted collegiate type of building surrounding a patio stems from the European tradition that the Iberian invaders brought to the Latin American colonies.

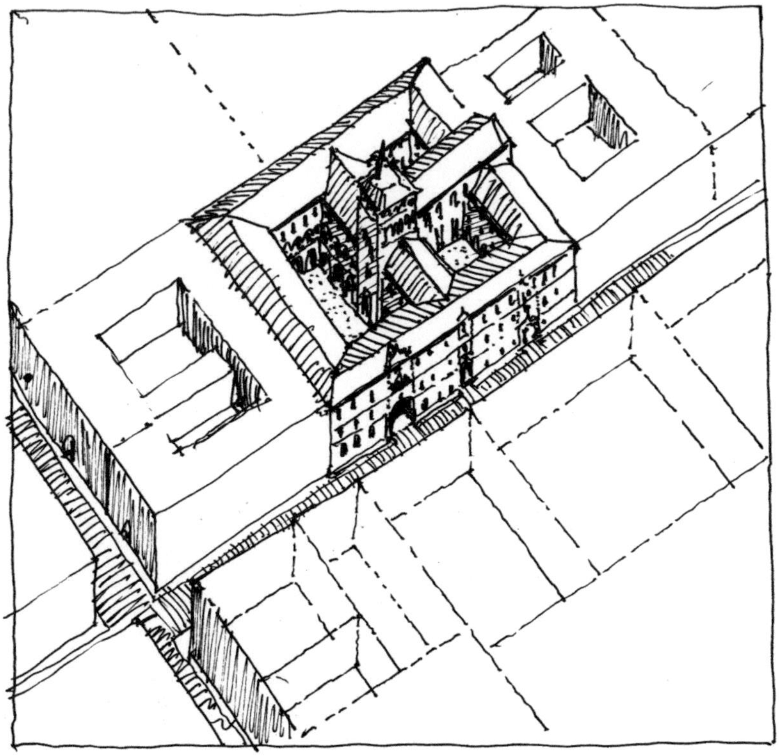

Fig. 12.28 Principles of South American university types, interpretative sketch by Helmut Bott. Source: Design by author.

Compared to central and western European universities, the southern American universities were huge buildings that filled entire urban blocks (Fig. 12.28). But they were of southern European origin, for they had many courtyards of different sizes that seemed to be carved out of the solid, homogeneous mass of buildings.

Universities in colonial towns used the medieval European architectural concepts for closed courtyard buildings of colleges, sometimes for series of different courtyards. However, Spanish colonial town planning followed Renaissance designs with geometrical rectangular street patterns and rigid regulations of eave heights and street alignment. In many cases college buildings have thus been totally integrated into the urban fabric of the block system.

North America

About one century later, the first Protestant university was founded on the east coast, in the Boston region of New England. Harvard, in Cambridge, Massachusetts,

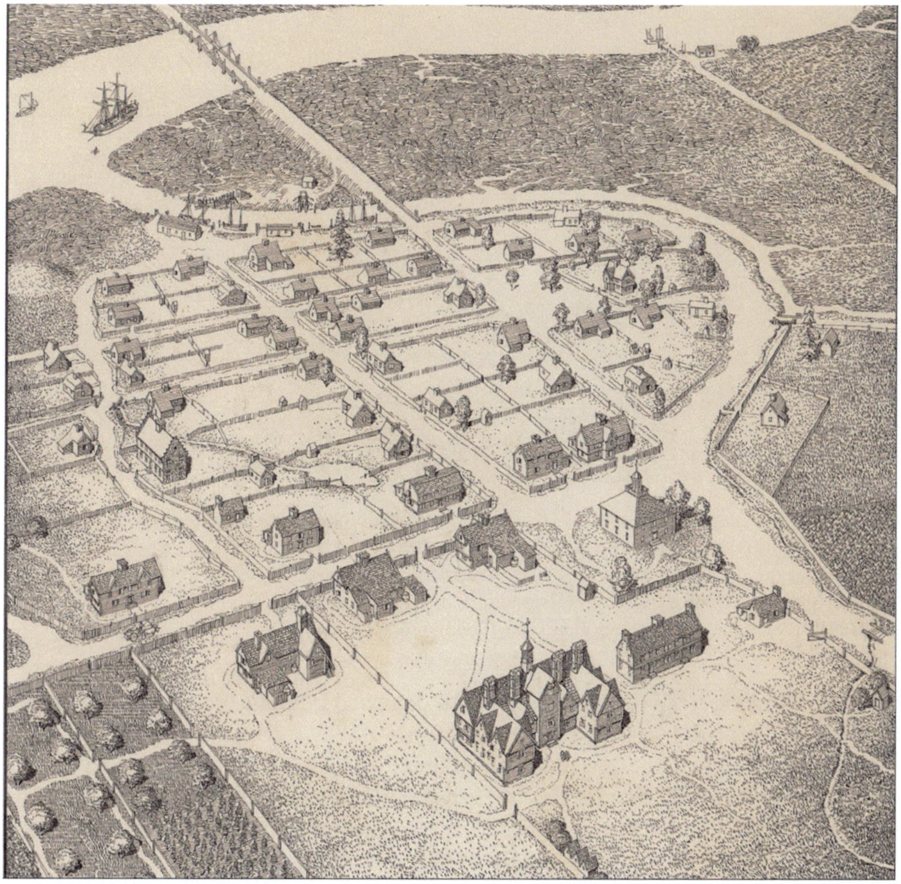

Fig. 12.29 Collegium Harvardianum Cantabrigiae in Nova Anglia [A Conjectural View of Harvard College in Cambridge, New England, 1668]. Map (1935) by Harold R. Shurtleff. HUV 2038, olvwork374306. Harvard University Archives. Reprinted with permission.

started in 1636 as a school of theology. From the beginning, private donations and private initiative were an important element of its organization.

The first college buildings of Harvard University were neither directly connected to a town nor integrated into a dense, urban structure of a city center, nor did they follow the traditional European courtyard type. Harvard's nucleus was a group of detached Georgian timber buildings surrounding a former meadow—called *the yard*. The ensemble looked less like a European university than it did a rural village or a manor with different annexes, auxiliary buildings, and the houses of farm workers (Fig. 12.29). A new university type had been born. In the following centuries up to the early 1900s, brick façades partly covered with ivy shaped the characteristic appearance of today's most famous university in the world. This edifice has become an image of university life acknowledged around the globe, with innumerable

Fig. 12.30 Harvard University. Eliza Susan Quincy's drawing of the September 1836 procession of Harvard alumni leaving the First Parish Meeting House and walking to the Pavilion. Eliza Susan Quincy was the daughter of Josiah Quincy, President of Harvard University (1829–1845). Source: Quincy (1840/1977), via Wikipedia. Copyright: Public domain {{PD-Art}}. Retrieved from https://en.wikipedia.org/wiki/Harvard_College_v._Amory.

movies depicting a university campus with detached buildings in open parks where students sit or walk by.

Even later, when the university positioned outside of the town lost its rural character and became encircled by the expansion of Cambridge, mostly by detached buildings with suburban appearance, Harvard preserved its personality as a park university and proudly continued calling its center the yard (Fig. 12.30).

Toward a Modern Research University

Universities in the Nineteenth Century

Political, social, and economic changes in the nineteenth century triggered a radical reform of the university system (see Bott, 2015, p. 100; Rüegg, 2004). The original system of only four faculties (arts, theology, medicine, and law) was further divided so that new faculties and subjects were created. In revolutionary France the specialized civil colleges were founded from which the *Grandes écoles* emerged. One of them, the *École Politechnique*, became the model of the new type of technical colleges.

The radical, liberal reform of the university, which the Prussian diplomat and scholar Wilhelm von Humboldt (1767–1835) launched in 1810 by founding the University of Berlin, fostered the development of humanities and natural sciences in the course of the nineteenth century in Germany and beyond. Through the interconnection and freedom of research and teaching and through self-administration, the Berlin University became one of the prototypes of modern research universities.

The natural sciences, for instance, had evolved from the natural scientific *quadrivium* of the liberal arts as well as from medicine. But it was not until the nineteenth century that independent faculties of natural sciences were established. The humanities developed from the *trivium*, whereas the social sciences grew out of the law faculty, which already included cameralism. This prodigious diversification led to new, more complex structures in the vast majority of universities, to a remarkable proliferation of professorships, and to a large building program with specializations in the lecture hall and laboratory equipment.

The Rise of Technical Universities

State schools for public institutions and commercial enterprises, such as academies for medical officers and engineers in the military service, mining, and metallurgy, had existed since the eighteenth century. Those institutes were mostly founded in residential cities, for the trainees were needed in the army and bureaucracy of the state apparatus. In the nineteenth century, the public sector expanded enormously, especially in central Europe, where, quite unlike England, the state promoted the industrial revolution. The need for qualified civil servants and engineers soared in the technical sectors of construction and military, induced by the rapid development of technology and its use in factory production.

The number and quality of schools for higher technical education therefore had to increase. Polytechnic schools often originated from predecessors, as was the case with the first German technical university in Karlsruhe, founded by the engineer Johann Gottfried Tulla (1770–1828) and the architect Friedrich Weinbrenner (1766–1826). Tulla had become acquainted with the *École Polytechnique* in Paris. Studies there began in the university tradition with basic scientific and theoretical training (e.g., mathematics, geometry, technical mechanics), with engineering as the specific professional orientation coming later. This curriculum became the principle of education in engineering throughout much of Europe—until the Bologna reform inverted it. In 1865 the Karlsruhe Polytechnic School received a new organizing statute with a full university governance structure, but it was not until 1885 that the polytechnic could be named *Technische Hochschule* (Institute of Technology). Almost all states of the German Confederation followed suit with their polytechnic schools, resulting in a highly dense and regionally spread network of technical colleges. This change eventually proved advantageous for promoting further development of technology and industry in Germany.

Whereas the Swiss Federal Institute of Technology (ETH) in Zurich was founded (1854) as a university with technology, humanities, and social science, the dispute between the classical universities and the new polytechnic schools concerning the academic status and subjects, continued in Germany throughout the second half of the nineteenth century. The technical colleges reinforced natural sciences and the mathematical-theoretical base of their teaching and research, finally claiming the right to award doctorates. In 1899 this right was granted upon intervention of Emperor William II.

State, City, and University in the Nineteenth Century

As the number of chairs, staff, students, and research facilities grew astoundingly in the 1800s, so did the dimension of university buildings, especially compared to that in the centuries before. Laboratory buildings and technical facilities became more and more important. In some instances special architecture departments for university planning were founded in the second half of the century. In central Europe the new university buildings were set up at central locations within the city, for they were seen as important institutions of urban and national culture. They marked and sometimes even shaped whole urban districts.

Humboldt University, Berlin

Humboldt University was founded in 1810 in an empty palace of a deceased Prussian prince (Gandert, 1985). It was first named the *Friedrich-Wilhelms-Universität* after the Prussian king who founded it, and after World War II it was renamed as Humboldt University by the government of the Germany Democratic Republic. Neither the urban location nor the architecture of the triple-wing building contradicted contemporary ideas of universities. The forecourt opens toward the monumental public space of the grand boulevard *Unter den Linden*. The university's central position within the urban fabric, close to the Prussian Royal Palace and midst other buildings of highest importance, proves the high reputation the universities had gained by that time. The urban ensemble composed of the university, the opera house (*Staatsoper*), the royal library, St. Hedwig's cathedral, and the Academy of Sciences, formed an impressive cultural forum (Fig. 12.31). The university courtyard and opera square shaped a major spatial accent on the main urban axis leading from the Brandenburg Gate to the royal palace. The building was extended to the rear into the former palace garden. More and more university functions were allocated to the neighborhood, creating a unique urban university district that still exists (Figs. 12.32 and 12.33).

Ludwig Maximilian University, Munich

The situation in Munich is quite comparable with that in Berlin. In 1840 Friedrich von Gärtner (1791–1847) designed the new building for the Ludwig Maximilian University at the main axis *Ludwigstrasse* (Hederer, 1942, 1964). This monumental

Fig. 12.31 Ground plan of Berlin and surroundings. Drawn by Böhm, Lieut. a. D.; engraved by Carl Jättning. Berlin 1848. Source: Via Wikipedia. Copyright: Public domain. Retrieved from https://commons.wikimedia.org/wiki/File:Boehm_Grundriss_von_Berlin_mit_n%C3%A4chster_Umgebung_1849.jpg.

Fig. 12.32 Friedrich Wilhelm University, with equestrian statue of Frederick the Great (Carse, ca. 1850). Picture by A. Carse (1770–1843), steel engraving by A. H. Payne.

boulevard leads from the historical core of Munich and the Royal Bavarian palace to the northern gate, a triumphal arch. At the last northern segment of the grand boulevard, the street opens toward the university square, formed by the university on the western edge and other educational buildings on the eastern edge (Fig. 12.34).

The building is a triple-wing-palace type, but its open space in front is no longer a fenced forecourt any more but rather a public urban square. The access to the imposing entrance hall is directly from the square, without steps or ramps. The public space flows freely into the entrance lounge. Much as in Berlin, the university is an important spatial element of a magnificent, central urban space, situated right in the heart of the political center of a German royal residence. Soon the main building was expanded and, together with new buildings in the neighborhood, formed an urban university quarter (Fig. 12.35).

University of Vienna

In 1884, a new main university edifice, designed by Heinrich Ferstel (1828–1883), was erected on Vienna's *Ringstraße*, where the most important buildings of the Habsburg imperial dynasty and of the aristocratic-bourgeois urban culture were placed: the new imperial residence with the opposite pole of the parliament, the opera house, the *Burgtheater* (imperial theater), museums, the town hall, and last but

Fig. 12.33 Berlin, aerial view Opernplatz, Unter den Linden. January 1, 1935.
Source: Bundesarchiv Bild 146-2007-0017. Used under Creative Commons Attribution-ShareAlike 3.0 Germany (CC BY-SA 3.0 DE). Reprinted with permission.

Fig. 12.34 Munich 1841; View from University Square south along Ludwigstrasse toward the city center. Sketch by H. Adam 1841.
Source: Hederer (1942, p. 61). Public domain.

Fig. 12.35 Aerial view of Ludwigstrasse, in the foreground the *Siegestor* (triumphal arch) and university square. Source: Hederer (1942, p. 92). Copyright: Public domain.

not least, the new university (Fig. 12.36). They were arranged as monumental ensembles demonstrating the ideas of the late nineteenth century's urban planning for an imperial residence on a gigantic scale. Compared to the Vienna university of the seventeenth and eighteenth centuries, the dimensions of the new university building and the extension of the urban space surrounding it reflect an amazing leap of scale. The academic institution was torn from its traditional busy quarter with its narrow streets and small squares and placed into a majestic capital's ceremonial space. It seems to be a commitment of state and local authorities to the appreciation

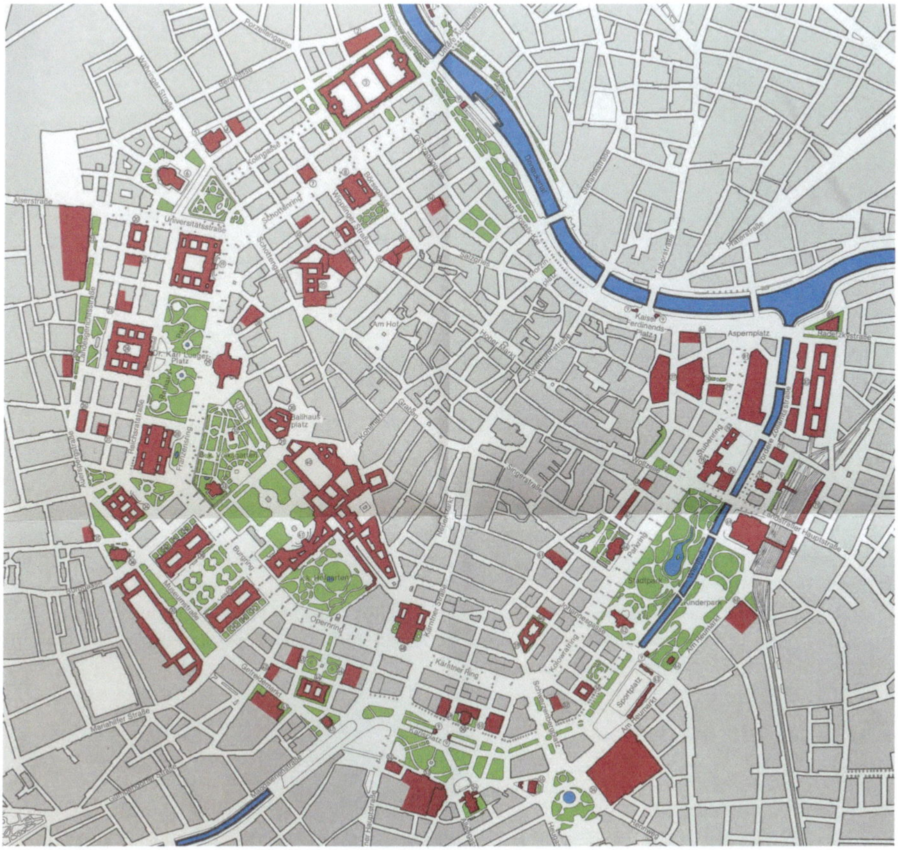

Fig. 12.36 Site plan of the *Ringstraße* (ring boulevard) in Vienna, 1910. All public buildings are marked in red. The university building is in the northwest; public parks are in green. Source: Mollik, Reining, & Wurzer (1980, map 81). Reprinted with permission.

of science and suggests a late demonstration of power by the Habsburg dynasty in its decline.

The university itself is set on a pedestal. A central staircase and side ramps lead into the capacious entrance hall with its pretentious architecture. It is elevated and highlighted in a noble manner and sumptuously decorated. The vast building forms a hermetic block structured by a great central courtyard and four lateral smaller courtyards. Its dimensions compete with the town hall and exceed even those of the national parliament (Fig. 12.37).

Fig. 12.37 Vienna University and part of the *Rathaus Park* in about 1900; in the background the *Votivkirche.* Author unknown.
Source: Original image: Photochrom print (color photo lithograph) Reproduction number: LC-DIG-ppmsc-09214 from Library of Congress, Prints and Photographs Division, Photochrom Prints Collection. Copyright: Public domain.

The Campus University

With the founding of the University of Virginia in 1819, Thomas Jefferson (1743–1826) conceptualized a new paradigm contrary to the European university tradition (O'Neal, 1969/1980). The masterplan of the first American university without a theology faculty was conceived by him to be on the green field as an entire entity independent of an urban settlement structure. The architecture of the university is completely neoclassical, related to antiquity and humanism, according to the Anglo-American interpretation with brick façades and white columns. Its main building is the library with its dome-shaped construction modeled on the Pantheon in Rome, though about half its size. Jefferson wanted the architectural language to correspond to the classical-humanist educational ideas. A large, green area forms the central space. It lies on a soft hill ridge and is structured by four slope terraces that are axially oriented to the library on the hilltop (Fig. 12.38).

Other relevant structures, such as faculty buildings and classrooms, are shaped like mansions flanking the central green space. They are connected by a one-story colonnaded pavilion on either side of the central space, which is a reference to an antique forum (Fig. 12.39). However, the space in between is neither a busy square nor a courtyard, but a lawn. Jefferson referred to his university as an *academic village*.

Fig. 12.38 University of Virginia, Charlottesville. Engraving by B. Tanner (1827). Source: O'Neal (1969/1980, p. 10). Reprinted with permission.

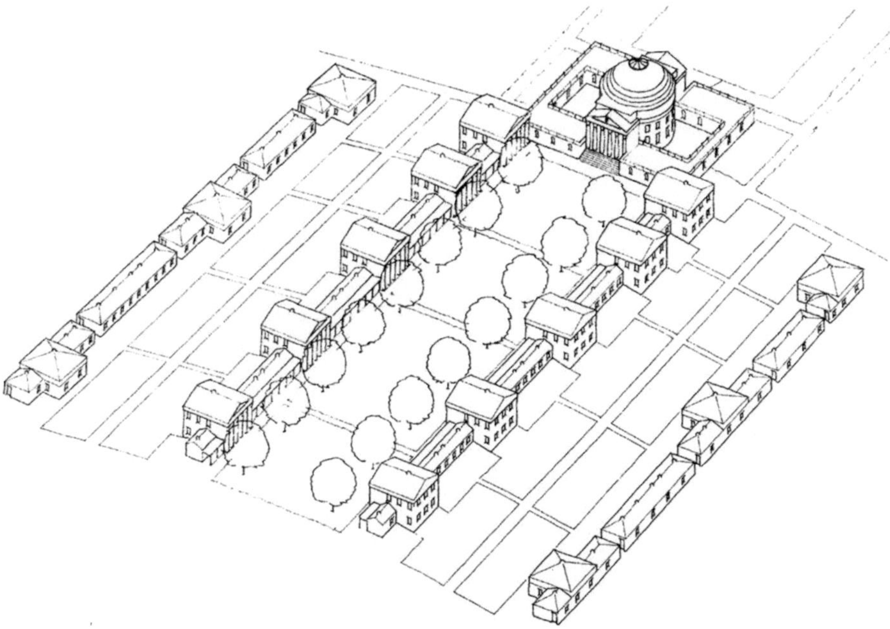

Fig. 12.39 University of Virginia, Charlottesville.
Source: Rückbrodt (1977, image 42). Copyright by K. Rückbrodt. Reprinted with permission.

Lawn and *village*, or *yard* at Harvard, and *campus* in general—all these notions have rural connotations and evoke scenic, picturesque associations. As Dober (1992) observed:

> The fabled 19th century hill-top colleges that populated the once rural mid-America are fine examples of simple architecture made prominent by a commanding site. Thomas Jefferson's University of Virginia lawn, bricked-in gardens, and buildings (considered by many as an epitome of campus planning and design) is informed by the gentle sloping of Charlottesville's terrain. (p. 31; also cited in Paulus, 2010, p. 506)

The Charlottesville concept has additional references. The young French nobleman and officer Pierre Charles L'Enfant (1754–1825), who joined the American army and became a friend of George Washington, was commissioned by the latter to plan the new capital city of the United States. He applied all elements and principles he knew from Europe, especially from France—a geometric layout of streets, diagonals, and a long axis, as in a European residential town. The most important ceremonial space, however, is the long green axis leading to the capitol (Fig. 12.40). L'Enfant designed an immense ceremonial open green space west of Capitol Hill, *The Mall*. It became the inversion of the relation palace—private park behind a palace. Jefferson applied a similar principle to his Charlottesville university design—albeit on a smaller scale.

In London the *Mall* leads from Trafalgar Square via the Admiralty Arch as a ceremonial space to Buckingham palace, the palace of the United Kingdom's royal family since the early eighteenth century. Initially, the *Mall* had been an open green space outside the city, used as playground to play *pall–mall*, an early modern croquet-like lawn game played by noblemen. The central green space of the Charlottesville campus, leading axially to the main building, which is emphasized by a cupola on a hill, can be seen as a miniature of *The Mall*, an Anglo-Saxon and Anglo-American archetype of a green ceremonial space.

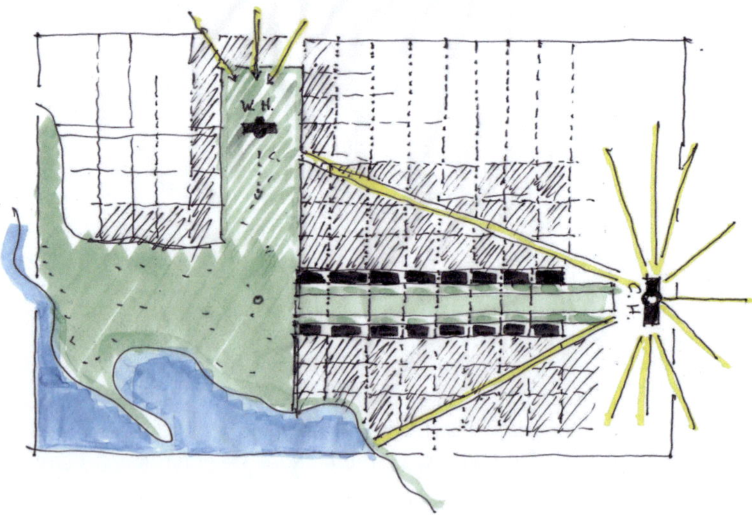

Fig. 12.40 The Mall, an Anglo-American archetype. Interpretation sketch of L'Enfant's design, by H. Bott.

Central and south European universities, however, have mostly grown over centuries inside cities, interwoven with the urban texture, always restricted by shortage of building plots. Thus, they are ensembles of buildings in different style from different ages. By contrast, campus universities have been planned outside cities on larger plots interconnected with the landscape, all according to an overall masterplan.

Another feature of the American campus universities is the stronger interior orientation of the buildings' front façades to the internal green center. True, there are also many North American universities integrated into urban structures. Columbia University in New York, for example, fits exactly into the measuring system of the New York block grid. However, it has the same principle of a strong internal orientation to the central green area, culminating as it usually does in the strict axial symmetry of the main neoclassical building standing in the central axis.

The term of *academic village*, beside its rural connotations, implies a social model of a small community with its own internal life embedding all villagers into the social network far away from urban anonymity. A young student, once integrated into the social network by rituals and a rich village culture of sport, art, and all manner of clubs and friendships during a formative period of life, remains a member of the village community as long as he lives, especially if the village is renowned. The village concept thus paves the way to the creation of a wealthy alumni network, the basis for successful fundraising.

Worldwide Expansion of the University System

By the end of World War I, the United States had gained worldwide importance as an industrial and military power through technical inventions as well as by science and research. The American university system became an equally significant global model, paralleling the European universities and later surpassing them.

Universities were founded in nearly all regions of the world in the late nineteenth and early twentieth century, mostly under colonial influence, initially from Europe, and later by the United States, too. For example, in the late phase of the Qing Dynasty in China, the first modern universities were founded on European-American principles in 1877. The prestigious Beida University, today's Peking University, was set up during a short reform phase of the last dynasty. Tongji University in Shanghai was established in 1904 by German physicians as a medical school and was financially and organizationally supported by the government of the German empire. Tsinghua University in Beijing was launched in 1911 with support from Americans. At some Chinese universities one therefore finds elements of the campus idea combined with traditional architectural concepts dating from the Ming and Qing dynasties.

The British colonial authorities established the University of Cape Town in 1829 and the first Australian university in Sydney 1850. In India they also created the

University of Calcutta in 1857, the University of Bombay in 1858, and three others in the late 1800s.

In Japan during the era of Meiji reform, Tokyo University was founded in 1877 on the basis of on older traditional educational institutions. There was no military intervention or even direct political influence from Europe or the United States in Japan. However, the structures and principles of the university systems from western regions with a longer university tradition were adopted.

The situation was different in Latin America, where the colonial administration established 21 universities from the sixteenth century to the eighteenth century. After the successful wars of independence between 1808 and 1825, the new states at last founded independent, national universities.

A very specific case of the colonial impetus behind the establishment of tertiary educational institutions outside Europe and the United States is found in Indonesia. Its first institute of higher academic education, the Technische Hogeschool (TH), was founded in Bandung by the Dutch authorities in 1920 and later became the Institute of Technology Bandung (ITB).

Universities in the Twentieth Century

Modernity and Traditionalism in the First Half of the Twentieth Century

The Weimar Republic: A short intermediate period in Germany

During the Weimar Republic (1919–1933) only one university, that of Hamburg, was newly established (1919). The University of Cologne, which had existed once before, was refounded in the same year. However, the number of students nationwide soared from 60,000 in 1914 to 120,000 in 1919. This explosion in the student population was only partly due to the war-related postponement of enrollments. In 1933 the student population reached 133,000, compelling the universities to accept considerably more students than in the past. The young democratic state faced serious problems (inflation, economic crises) funding the universities, almost all of them public. Despite these difficulties though, the quality of Germany's universities remained high. A special institution in this academic landscape was the *Bauhaus*, a small art and industrial design school that went on to have profound influence on university architecture and design alike.

Architectural transition phase: Hamburg, Cologne, and the Bauhaus

A comparison of the University of Hamburg and the University of Cologne, both created during the interwar period, delineates the transition phase from traditional European university design to principles of the Modern Movement in architecture.

Fig. 12.41 University of Hamburg in the early 1920s.
Source and copyright: University of Hamburg. Arbeitsstelle für Universitätsgeschichte. Reprinted with permission.

The two institutions still have their main entrance on important streets, forging a strong relation to the urban public space.

The University of Hamburg, designed by Hermann Distel (1875–1945) and Ernst Ludwig Grubitz (1876–1936) and built in 1919, still complies with the traditional building typology of a symmetric main façade on a major street and a portico in front of the entrance hall. It has large inner courtyards, pitched roofs, and a classical façade layout (Fig. 12.41), retaining the characteristics of a nineteenth-century building.

The University of Cologne, planned in the late 1920s and opened in 1934, incorporates some principles of the Modern Movement (Kantner, 1969). The Bauhaus and Congrès internationaux d'architecture modern (International Congresses of Modern Architecture), or CIAM,[6] had already proclaimed new principles of architecture and urban design. The idea was that space and buildings should no longer be bound to and organized in traditional block figures but rather were to be dissolved into detached, solitaire buildings with flowing space between them. Going somewhat in this direction but still looking back to traditional patterns, the Cologne building, designed by Adolf Abel (1882–1968), is oriented with the main entrance facing an

[6]CIAM was an international group of young modern-movement architects who established principles of town planning starting in 1928. Among other things, they proclaimed the separation of functions.

Fig. 12.42 University of Cologne, main building. Architect Adolf Abel.
Source: Photo Kreyenkamp in the 1930's. Rheinisches Bildarchiv. Reprinted with permission.

important road passing the area, but its back opens toward a park belt (Fig. 12.42). It still forms an entire shaped configuration of linked wings, with courtyard-like interspaces. However, the wings are in the process of becoming detached from the entire figure and moving into the flowing space (Fig. 12.43).

Since 1925 the Bauhaus Design School at Dessau has embodied this concept of the free-standing ensemble, rejecting traditional European principles of streets and blocks, of front façade and backyard (Fig. 12.44). True, it is only on a small scale, but impressive nonetheless and has had massive impact on the further development of architecture and urban design. Its short, but conflictual, history (1919–1932) reflected the swift rise of the Modern Movement in Germany under the Weimar Republic, during which it emanated radical new design ideas for the industrial age internationally. But it was shut down as early as 1932 by Nazi intervention. Its last director, Ludwig Mies van der Rohe (1886–1969), later emigrated to the United States, where he was able to apply Bauhaus ideas to his design of the new campus for the Illinois Institute of Technology in Chicago.

Fig. 12.43 University of Cologne, Main Building, 1929–1934. Architect Adolf Abel. Aerial view. Source: Photograph by Aero-Lux Frankfurt am Main, 1952. Rheinisches Bildarchiv. Reprinted with permission.

Fig. 12.44 Bauhaus Dessau, 2009, Architect: Walter Gropius (1925/1926), south-eastern view. Source and copyright: Stiftung Bauhaus Dessau, Photo: Yvonne Tenschert (2009). Reprinted with permission.

Traditionalism and gigantomania

However, modernism of the first half of the twentieth century was rather an appetizer than the main course, for it seemed to be looking back. Even in Germany, the center of modernism in the 1920s, modernism was not the dominant style as measured by the number of erected buildings. And it was abruptly interrupted after 1933.

Fig. 12.45 Edmund Zintl Institute, built in 1942 at the Technical University Darmstadt. Source and copyright: Hessisches Staatsarchiv Darmstadt, Signatur R 4 Nr. 5020UF. Reprinted with permission.

Fascist states preferred a kind of neoclassic gigantomania combined with traditionalism. Academic humanistic education was just the opposite of Nazi educational ideals, which required soldiers. Few extensions of existing universities were erected, such as the Zintl Institute built at the Technical University of Darmstadt (Fig. 12.45).

The New Sapienza in Rome

Unlike the Modern Movement in Germany, its counterpart in Italy—Futurism and Rationalism—was partly affiliated with the fascist party. One of the great architects of this period, Guiseppe Terragni (1904–1943), was a splendid modernist familiar with the German Modern Movement but was a convinced fascist party member all the same.

The new campus for Sapienza University at Rome became one of the paradigms of fascist architecture (Bodenschatz, 2012). Marcello Piacentini (1881–1960) received the commission to design the New Sapienza in 1932. He mixed a traditional, neoclassic layout of the urban design and giant proportions with a modernist, purist design of façades, construction, and details (Fig. 12.46). He carried out only some of the buildings himself, enough to demonstrate the typical Italian attitude of the Mussolini period. Establishing a rigid urban design guide, he quite successfully

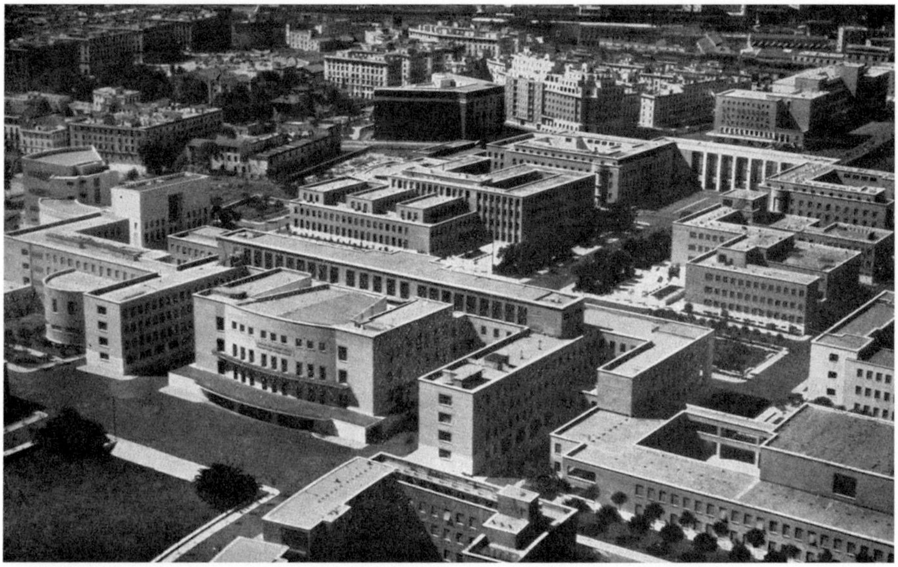

Fig. 12.46 The new campus of the Sapienza University in Rome (1938).
Source: Unknown photographer. Copyright: Public domain. Retrieved from http://wikipedia.org/wiki/Marcello_Piacentini#/media/File:Cittauniversitaria.

directed some colleagues to create an entire ensemble. The spatial concept was not remote from historic urban fabrics in residential towns, whereas the architecture tended to embrace purism. Proportions, however, sometimes morphed into gigantomania, albeit more tolerably than in Germany. Piacentini did not work only in Italy. He also received planning commissions in other countries such as Portugal and Brazil, proving that his ideas were well-known internationally. They met the conservative imagination of that period.

Lomonosov State University, Moscow

During the short Leninist period, the young Soviet Union was an experimental field of art and architecture. In Stalin's time, however, the Soviet Union favored architectural concepts similar to those in the fascist countries. The new Lomonosov State University was an important element of the 1935 masterplan for the urban reconstruction and expansion of Moscow, intended to help make the city a modern capital of the first socialist country (Brumfield, 1993; Huber, 2007; Summerfield, 1998). The idea for the new university was to make it the nucleus of the main southwest development. It was to be the highest and most important of seven new planned high-rise landmarks. The monumental building complex rose west of the Moskva River between 1949 and 1953. Boris Iofan (1891–1976), who delivered the first draft, fell into political disgrace in 1948 and the design work was handed over to Lew Rudnev (1885–1956).

The widely ramified building system unfolds on a rectangular cross, according to strict hierarchical neoclassical axial planning. It is a giant palace with one main and two lateral forecourts, with dimensions larger than any absolutist palace ever, flanked by, and integrated into a vast geometric park area. The complex system is dominated by a central 36-story triply terraced tower culminating in a spire that holds the Soviet star aloft. This building, designed in what is called *confectioner's style*, is a dictator's showing off, claiming to be in the center of future world communism. Using historical details in a rude manner and transforming them into a giant verticalism, the jagged silhouette is designed to recall historical Russian city shapes. The construction system was a steel skeleton frame filled in with brick, covered with natural stone slabs, and adorned with monumental stone sculptures.

Inside it was equipped with modern Soviet technology of heating and vertical transportation. The high-rise building houses three faculties (geography, geology, and mathematics), their museums, as well as the university museum, the university library, 23 lecture halls, 125 group workrooms, and 700 lab workplaces. The side wings provide living space for 6,000 students and doctoral candidates. The main buildings of the side wings have apartments for 200 professors. The auditorium has 1,500 seats. Lastly, the building complex includes other museums, shopping, and leisure facilities, including an indoor pool (Huber, 2007; Summerfield, 1998).[7]

As previously seen in Rome, universities were accustomed to demonstrating state and party power, visualizing ideological principles, exhibiting architectural positions of a dictatorial power by using giant proportions and putting them into spatial limelight (Fig. 12.47). This was for the time being the end of a central European process to separate universities more and more from their urban neighborhood, emphasizing their importance and make them to crucial state affairs. In the late nineteenth century universities started to be elevated, put on pedestals, lifted above the normal urban level. This kind of conservative neoclassical architecture was used for university planning on a giant scale in many countries of the world (Columbia University in New York, for example), however, European fascist and Stalinist designs topped all.

The Illinois Institute of Technology—An icon of the Modern Movement in architecture

Ludwig Mies van der Rohe (1886–1969), a leading figure of the European Modern Movement in architecture and urban design during the 1920s, came in 1938 to the United States, where he was appointed professor at The Illinois Institute of Technology (IIT) in Chicago. In 1940 he was commissioned to plan the new IIT campus.

[7]This high-rise university type became a model for the former Eastern bloc states during the Stalinist period. A similar building stands in Warsaw.

Fig. 12.47 Lomonosov State University, Moscow.
Source: Max Pixel. Copyright: Public domain. Retrieved from http://maxpixel.freegreatpicture.com/Lomonosov-Architecture-Stalin-Moscow-University-1378927.

He brought in his ideas and experience of modernism about how to use industrial methods and modern materials for design and construction. His famous project combined the American university concept with the ideas of European modernism. It became an icon of modernity.

The solitaire, detached buildings form both a free-flowing space and a central interior green open space—the campus university concept. But the cubic buildings themselves had a totally new expression, for they are mostly steel and glass constructions, precise like machines, designed on an underlying technical grid of measures. The campus institutes were built in the following years up to the 1950s. Among them, the Crown Hall, a transparent steel-and-glass construction with its flat roof hanging from mammoth exposed steel frames, became a masterpieces of Mies van der Rohe and the Modern Movement itself (Blaser, 2001). The Crown Hall is still used by the architecture department, in which Mies was the most influential teacher for 20 years. Today, the buildings designed by him still look modern. It seems they are expressing the timeless prosaic modernity of rationality—technology's promise to solve the future problems of the human race forever, dissolving history in rationalism and functionalism. A wrong promise, as it soon turned out. And a joke of history was that one of the master's best scholars at IIT, Helmut Jahn, later became a leading architect of Postmodernism, designing

Fig. 12.48 Mies van der Rohe, Project for the Illinois Institute of Technology campus in Chicago. Final scheme, 1940.
Source: Johnson (1947, p. 135). Copyright: Public domain.

decorative skyscrapers in a kind of art deco style far from Mies's "less is more." History is never dissolved.

The urban design of IIT seems to create a central lawn and a nearly symmetrical figure based on the line of East 33rd Boulevard, which crosses the lawn. All areal outlines are totally integrated into the street lines of the neighboring quarters (Fig. 12.48). At the same time, the design aims to shape a central interior green space following the American tradition as exemplified by Columbia University in order to integrate the campus into its surrounding urban fabric. This idea, however, was rather difficult to achieve because of the streets and railway lines passing by. But looking at the realized building design, one finds no implementation of this concept, for the building's access and orientation counteract rather than enhance the urban fabric. The Crown Hall's main entrance lies at the averted side of the lawn, affording a pleasant view of it but actually showing its back to it (Fig. 12.49). The building is accessible only via small stairs from the main lawn. Nearly all building entrances on the campus are separated from the attached lawns by streets or bushes, a design

Fig. 12.49 Crown Hall at the Illinois Institute of Technology, Chicago, Illinois. Designed by Ludwig Mies van der Rohe; completed 1956.
Source: © Joe Ravi (2011), via Wikimedia Commons. Used under Creative Commons Attribution-ShareAlike 3.0 Unported (CC BY-SA 3.0). Retrieved from https://de.wikipedia.org/wiki/Datei:Crown_Hall_1.jpg.

leaving it bereft of the charming character it could have and making it into a prosaic working sector for white-collar engineers. Mies van der Rohe would surely have considered this statement to be a compliment.

Development of University Planning and Architecture in the Second Half of the Twentieth Century

After World War II, the process of decolonization led to the emergence of many new states in Africa, the Middle East, and Southeast Asia. Practically all new countries founded national universities, which were regarded as a prerequisite of economic and cultural development. Since the 1960s, a network of universities therefore spans the globe, though the knots of this net vary in distance from each other and in quality in different parts of the world.

The Sputnik crisis in 1957 triggered a fundamental discussion about the educational system and research in the western countries. As a result, the United States and most European countries multiplied research budgets and improved schooling. West Germany even declared its educational system to be a catastrophe and began inquiring into it. In the following years and decades, the university system greatly expanded. Growth of the spatial dimensions of universities and the number of enrolled students far surpassed that in the first half of the century. Throughout the

Fig. 12.50 Central library of the National Autonomous University of Mexico on the Mexico City campus (built by Juan O'Gorman).
Source: Régis Lachaume (2006), via Wikipedia. Copyright: Public domain
Retrieved from https://de.wikipedia.org/wiki/Nationale_Autonome_Universität_von_Mexiko.

world, university planning in many cases became an experimental field of the avant-garde movement in architecture and urban design.

The new campus completed in 1954 for the Universidad Nacional Autónoma de México (UNAM) in Mexico City, for instance, is a striking example of an attempt to combine architecture of the International Modern Movement with traditional Mexican elements of art, which, in turn, were a superimposition of pre-Columbian and Hispanic culture. UNAM's library, designed by Juan O'Gorman (1905–1982), is a paradigm of this approach. It uses modern construction materials and exhibits the cubic style of the international Modern Movement, but the façades are covered with reliefs in a kind of Aztec style and Mexican wall paintings (Fig. 12.50). The content of the graphics, however, deals with history and science.

A very bold project for a new university was realized by Oscar Niemeyer (1907–2012), who designed most of the official buildings for Brazilia, the planned city that became Brazil's capital in 1960. He designed a curved linear building 700 meters long (nearly half a mile), constructed with prefabricated concrete elements as a radical serial composition, adapting the design concept by Lucio Costa (1902–1998) for the whole town (Fig. 12.51).

Fig. 12.51 Central Institute of Sciences—University of Brasilia, Brazil.
Source: Nossedotti (2011), via Wikipedia. Copyright: Gemeinfrei. Retrieved from https://de.wikipedia.org/wiki/Datei:ICC_UnB.jpg.

In Europe famous architects like Alvar Aalto (1898–1976, masterplan and many buildings for the new campus of Helsinki University of Technology since 1955), Giancaro di Carlo (1919–2005, masterplan and many institutes and dormitories for the new campus of the University of Urbino since 1965), and Ralph Erskine (1914–2005, buildings for the Frescati campus Stockholm since 1974) designed masterpieces of contemporary architecture for universities. The long list of such architecture grows each year, but this chapter's discussion of the concepts of new campus development for the modern mass university focuses on the West German example illustrating extremely rapid expansion of the university system.

Urban and architectural design of the mass university, Germany's example

Only a few universities were founded immediately after World War II. The French occupation administration created one Mainz in 1946 and another in Saarbrücken in 1948, both initially using abandoned German barracks. In the U.S. sector of West Berlin, the Free University of Berlin was founded, for Humboldt University was under Soviet administration.

Many university towns tried to extend the existing university areal into the city, following the European tradition. Because building plots were exceptionally rare at their historical sites, most of which were situated in old centers or at the edge of the inner city, the American paradigm of campus universities outshined urban alternatives. Only the outskirts provided enough space for new universities. The discourse about spatial visions and planning sites for rapid, large-scale expansions or new foundings of universities thus soon led to the ideal of the campus university. The United States had become the dominant cultural power, and German scholarship holders, returning from North America, reported enthusiastically about U.S. university life. An important influence on the discourse came from the Central Archive for University Planning (*Zentralarchiv für Hochschulbau*), established in 1963 at the University of Stuttgart and headed by Horst Linde. This institute analyzed American universities and declared the IIT campus to be exemplary. Up to that time, the notion of campus was not common in Germany, for universities had always been a part of urban culture and more or less integrated into the urban fabric and daily life.

Only a few years after the *Sputnikschock* in Germany, several new universities were founded: Bochum (1962), Regensburg (1962), Constance (1966), Bielefeld (1969), Kaiserslautern (1970), Bremen (1971), Kassel (1971), Bayreuth (1972), and Oldenburg (1973). Nearly all are campus universities laid out on large coherent areas on the outskirts according to a masterplan. Most of them were constructed very quickly, many buildings in the same style with the same material—mostly exposed concrete and a high percentage of prefabricated components. Compared to the American archetype of campus, they lack the rural charm of field (campus), yard, or lawn.

The labels used for the spatial elements are proving to be a conceptual confusion. Instead of a green center, for example, there is the idea of a forum, which means precisely the opposite—a central urban space. Hence, Bochum, Constance, and Regensburg have each a stone covered central open space, which looks much more like contemporary pedestrian areas or the satellite shopping centers of the 1960s and early 1970s. Constance reflects these contemporary ideas perfectly (Fig. 12.52). All details inside and outside the buildings are designed affectionately as a total artwork. Architecture, interior design, applied art, and landscaping work together, modeling an enormous sculpture integrated into the gentle moraine landscape of Lake Constance. However, it is hardly a campus in its original meaning but rather an artful "urban" space with closely related landscape. Bochum and Bielefeld have the charm of *learning factories* designed in the late style of classic modernism. Krefeld even has a covered passage like a shopping arcade, called a *communication axis*.[8]

This first generation of universities after World War II. followed neither the old campus model nor the IIT campus design by Mies van der Rohe. Urban design paradigms of the Modern Movement had been moving from solitary ensembles to huge spatial figures since the late 1960s. It seems that the familiar conceptual notions

[8]Nevertheless, the Ruhr-University Bochum is listed as a landmark illustrating a typical comprehensive university (*Volluniversität*) idea in that German era.

Fig. 12.52 The University of Konstanz, Germany, aerial view.
Source: Universität Konstanz (2012), via Wikipedia. Used under Creative Commons License Attribution-Share Alike 4.0 International (CC BY-SA 4.0).
Retrieved from https://de.wikipedia.org/wiki/Universität_Konstanz.

rooted in the urban European heritage of universities could not abandoned all at once. That legacy continued to glow under modern cubes and surfaces of the machine age, obscured by wrong notions.

In the late 1960s and early 1970s, however, some areas of university expansion and new foundings were planned as campuses, far removed from the historical sites on the city periphery (Würzburg–Hubland Campus; Darmstadt–Lichtwiese, and the University of Bayreuth), with the centers henceforth becoming real green areas (Fig. 12.53). Bayreuth is an evident example of a green center surrounded by faculties and a wide range of facilities. But even at the University of Bayreuth, the front façades of the buildings lack a clear internal orientation to the center. The main access is from a road surrounding the central green space. The campus design is obviously influenced by functional concepts for contemporary district centers of that time—a ring road for cars and supply surrounding the campus, pedestrian areas inside of it. In a nutshell, the American concept had become the beacon of the era but was not really adapted to its spatial concepts. Everybody still uses the label *campus* unthinkingly, even for the historical sites: the urban campus. what an odd contradiction.

The idea of monofunctional university areas in Europe has another root, too. It was wholly in accordance with the ideals of functionalism as set out in the Athens Charter of 1933. Each land use should find its very special place and space, according to its specific requirements (functions). Architecture and urban structures

Fig. 12.53 Campus Lichtwiese, Technical University of Darmstadt, Germany, developed since 1967, status of 2015.
Source and copyright: Nikolaus Heiss. Retrieved from www.tu-darmstadt.de/universitaet/orientierung/anreise/index.de.jsp.
Reprinted with permission.

could thereby be optimized for those special needs: purely residential areas, purely commercial and industrial areas, recreation areas, infrastructure facilities, all separated from each other and connected by the transportation system. Large campus facilities were consistent with these planning concepts. The small-scale European mix of city and university facilities seemed to contradict the contemporary idea of functional organization and optimization of processes in the modern industrial society.

Reurbanization of Universities?

Old European universities such as Bologna, Paris, Cambridge, Oxford, and Montpellier expanded gradually on a smaller scale in different urban areas. Today's Bologna still has all its facilities in the inner city areas. Many of its new university institutions are housed in restored landmarks, such as monasteries or formerly abandoned cinemas. University life there is reminiscent of past European times, with people walking in the well-restored city center; sitting in bars to discuss and prepare for the next lecture or seminar; taking an *aperitivo* after the last lecture;

watching others, especially nice young students, famous writers, hip musicians, or even well-known politicians and beggars passing by, as in novels about earlier European urban life—or at least the way it is imagined to have been.

Other universities set in old city centers built campuses on the outskirts but kept their historic buildings within the city and added new buildings as far as possible. This situation pertains in almost all of the old European university cities, including Heidelberg, Tübingen, Marburg, Prague, Padua, Pavia, Leiden, and Amsterdam, to name only a few. Natural sciences, engineering, and university clinics have mostly moved to outside because of the special needs that their laboratories and machine halls entail, whereas the rector's office, administration, humanities, and social sciences have remained at the historical locations.

Conversion of former military, industrial, and infrastructural areas

The economic changes brought about by globalization have made it necessary to abandon many large industrial areas and infrastructural facilities of the nineteenth century and early twentieth century. Projects to convert and reuse those tracts and buildings spawned new concepts featuring an urban mixture of functions and novel spatial patterns, sometimes integrating university facilities. An early example was the University of Kassel, founded in 1971 in a remote campus area. By the late 1970s, long before the effects of globalization, the architecture faculty was accommodated in what had once been an administration building of a former heavy-industry firm neighboring on the city center. Over the following decade, a new high-density university district with a traditional block structure was built on the adjacent abandoned factory terrain. In Heidelberg, a recent example of this architecture is the *Bahnstadt*, a new district situated on former railyard terrain connected to the main train station. The plan for the space is based on a mixed use concept that will integrate university functions. Even more spectacular is the new *Hafencity* university under construction on the waterfront of a vacated port area in Hamburg. After the new opera house, it has become the second important landmark of the new quarter.

Universities are thus no longer seen as bulky institutions for which it is difficult to find space. Rather, they are used as an initial investment, a driving force behind the urban development of large fallow land and problematic districts. Examples are the University of Milano-Bicocca, built on the site of the former Pirelli factory (Fig. 12.54), and the new University of Torino, both planned by the architectural firm Gregotti Associati. They incorporate abandoned areas and reused, partly historical industrial facilities. The architectural concept follows traditional European urban patterns such as street, square, and block. Some elements of Gregotti Associati's projects recall Piaccentini's university town, though on a smaller spatial scale, making university planning an integrated part of urban renewal, the reuse of industrial heritage, and the upgrading of run-down districts.

Fig. 12.54 University of Milan-Bicocca, Italy. View of the Trivulzana Sqaure, 2013. Source and copyright by Antonella Sgobba. Reprinted with permission from A. Sgobba.

Campus universities captured by city development

Many of the universities built on the outskirts of cities in the 1960s or early 1970s have long since been incorporated into urban or suburban structures. Those campus areas are now regarded as obsolete. Despite their high-rises, their population density is low, and their monofunctionality makes them empty, uninviting districts on evenings and weekends. Both characteristics have been harshly criticized. Moreover, modern requirements for fire protection, escape routes, and energy-saving make it extremely costly to maintain and modernize buildings constructed in the 1960s and early 1970s.

Adopting a concept for densifying the once isolated, monofunctional Hönggerberg campus and for reclassifying it for mixed use, the Swiss Federal Institute of Technology (ETH) in Zurich launched a phase of urbanization and urban integration of large, peripheral university facilities in 2004. Since then, nearly all sprawling campus areas on the outskirts have undergone critical analysis, reconsideration, renewal, and remodeling, which in many instances has also improved and expanded the historical university areas.

Even in the United States, the motherland of the originally introverted and remote campus concept, there are trends toward transferring research and teaching facilities to the urban context.

> Especially in the United States, universities are now moving away from separating teachers' places of life from those of students and toward selectively interweaving their range of research activities and educational options with the city. In 2010 Google bought a 2,700,000 square-meter former warehouse and administration building in Chelsea, one of New York's hottest neighborhoods, for $1.8 billion. The main reasons were the opportunities offered by the city, but also the preferences of the employees, who no longer want to live and work in suburbia. Universities also see the advantage of an urban location for the recruitment of leading scientists and paying students. This was not always so. Forty years ago, most of the cities in North America were characterized by emigration, disintegration, crime, and poverty, and universities are still the most important institutional anchors of centers in cities such as Cleveland, Baltimore, and St. Louis. New York, on the other hand, has been growing steadily since the early 1990s, and "NYC" has now become an academic trademark. In 2012, according to the New York State Department of Education, more than half a million students were enrolled in the 102 colleges and universities located in the city, a good 6% of the population—and 11% more than in 2007. (Schindler, 2013, p. 25)

Universities in the Age of Globalization and Digital Information-Processing

International relations and cooperation have steadily increased in recent decades, not only through student and academic exchange, exchange networks, and cooperative bilateral and multilateral research networks but also through the founding of international universities (see chapter by Knight in this volume). Students and scholars are required to be mobile and to enhance international relations, an expectation intended to expand their habitat, contacts, and reference points across many cultures and, analogously, to economic global chains.

Some routine functions of teaching and communication may indisputably be transferred to the Internet, depending on the field of study involved. However, a wide range of practice at working with material (basic engineering and science courses, design, art and architecture, music, and performance) and of internationally renowned research will still need personal contact, face-to-face critique, and discourse in seminars and symposia.

In my opinion, universities have to solve a fundamental predicament that they themselves have created in science and research—the increasing specialization and isolation of research and, even more, the issues of applying research and technical inventions to nature and society. Worldwide academic life requires specialization and focus on ever more detailed topics, as in medicine, and analysis of ever smaller particles of matter and energy, as in physics. Exploring and explaining the complexity of interdependencies of various interventions in nature, the economy, and society is a challenge for researchers, especially because it might provoke conflicts with powerful political and economic interests. Meanwhile, the Anthropocene has begun, as a growing number of scientists claim, with most technical interventions creating so many side effects on nature, humanity, and culture that research must intensify efforts to develop transdisciplinary research methods as complex as the mounting problems of the real world.

The right place and space of that future kind of complex, multidisciplinary research may be the best universities, if they intend it. It is there that highly qualified professors with superior equipment are internationally and transdisciplinarily networked and open minded to learn from each other. These complex methods of research also need personal contact and creative situations. Excellent universities situated in inspiring towns and sites may provide that creative space and those networks. I am looking forward to university design in the future.

Summary

The architecture typology of university buildings developed in Europe from medieval urban courtyard collegiums to triple-wing palace buildings up to the eighteenth century, became monumental blocks with several courtyards in the late nineteenth century, and was always integrated into urban patterns connected to public streets and squares. In most cases they formed an internal semipublic space and had a main façade overlooking the public space. Beginning in in North America in the seventeenth century, however, there arose a new typology featuring ensembles of detached buildings set in pleasant countryside locations outside of towns, no longer centered on enclosures and small courtyards but rather in yards or lawns. Thomas Jefferson even called the wonderful Charlottesville university, founded and designed by himself, an academic village.

The relation to landscape and countryside became a characteristic spatial feature of Anglo–American universities. Whereas urban settings were the framework within which everyday life at continental European universities took place, spatial elements of landscape and rural connotations became a specific feature of life at Anglo–American universities, affecting concepts all over the world. Spatial university concepts thereby acquired rural associations. University lifestyle became related to landscape, park, and nature. Such imagery and notions affected nearly all cultures by the end of World War II, when the United States at last achieved supremacy in the western hemisphere. The campus concept (academic village) with its intensely inward orientation may explain the much stronger lifelong bonds between former students and their alma mater. The word *experience* is *Er-fahr-ung* in German, derived from *fahren* ‘to move, to ride’. In past ages many students used to travel from one university to another and never developed ties as strong as those in England or the United States, where alumni networks are an important part of the university organization and, of course, fundraising.

As it appears now, ideas about university life and its spatial manifestations are already shifting again to a kind of reurbanization. The campus today is envisioned more as urban life with its dense array of opportunities and options than as landscape and rural sceneries. The word is used by a wide variety of organizations dealing with information, knowledge, and technology and has lost its original meanings and associated spatial visions.

References

Basle University Library AN II 3 (n.d.). *Rektoratsmatrikel: Basle University. Vol. 1 (1460–1567)*. doi:https://doi.org/10.5076/e-codices-ubb-AN-II-0003

Bermann, M. (1880). *Alt- und Neu-Wien: Geschichte der Kaiserstadt und ihrer Umgebungen seit dem Entstehen bis auf den heutigen Tag und in allen Beziehungen zur gesammten Monarchie* [The old and the new Vienna: A history of the imperial city and its environs from the beginning to the present and in all relations to the entire monarchy]. Vienna: Hartleben. Retrieved from https://commons.wikimedia.org/wiki/Category:Alt-_und_Neu-Wien._Geschichte_der_Kaiserstadt_und_ihrer_Umgebungen_(1880)_by_BERMANN#/media/File:BERMANN(1880)_p0455_Plan_der_Wiener_Universtät_und_Studentenquartiere.jpg

Blaser, W. (2001). *Mies van der Rohe—Crown Hall: Illinois Institute of Technology, Chicago. The Department of Architecture/Architektur Fakultät*. Basel: Birkhäuser.

Bodenschatz, H. (Ed.). (2012): *Städtebau für Mussolini. Auf der Suche nach der neuen Stadt im faschistischen Italien* [Urban design for Mussolini: In search of the new city in fascist Italy]. Berlin: DOM.

Bott, H. (2013). *Vom Leib zum großen Plan: Über den Entstehungszusammenhang räumlicher Ordnungs- und Gestaltungsprinzipien* [From body to town-planning: On the context of the origins of spatial principles of structural organization and design]. Materialien des Internationalen Zentrums für Kultur- und Technikforschung [IZKT] der Universität Stuttgart, Vol. 12. Stuttgart: IZKT.

Bott, H. (Ed.). (2015). *UniverCity—Stadt und Universität* [UniverCity—City and university]. Catalogue of the exhibition at the Faculty of Architecture and Urban Planning of the University of Stuttgart, February 2013. Stuttgart: Städtebau-Institut, University of Stuttgart.

Bott, H., & Teodorovici, D. (2015). Die Entstehung der europäischen Universität [The emergence of the European university]. In H. Bott (Ed.), *UniverCity—Stadt und Universität* (pp. 24–27). Catalogue of the exhibition at the Faculty of Architecture and Urban Planning of the University of Stuttgart, February 2013. Stuttgart: Städtebau-Institut, University of Stuttgart.

Braun, G., & Hogenberg, F. (Eds). (1575). *Civitates orbis terrarum: Vol II* [Cities of the world: Vol. II]. Cologne: Braun.

Brumfield, W. C. (1993). *A history of Russian architecture*. Cambridge, UK: Cambridge University Press.

Capitani, O. (1987). *L'Università a Bologna: personaggi, momenti e luoghi dalle origini al XVI secolo* [The University of Bologna, personalities, dates, and places from its creation to the fourteenth century]. Bologna: Cassa di Risparmio.

Carse, A. H. (ca 1850). *Berlin und seine Kunstschätze* [Berlin's art treasures]. Leipzig and Dresden: Payne.

de Ridder-Symoens, H. (Ed.). (1992). *Universities in the Middle Ages*. A history of the university in Europe: Vol. 1. Cambridge, UK: Cambridge University Press.

Dober, R. P. (1992). *Campus design*. New York: Wiley.

Erl, T., & Teodorovici, D. (2015). Universidade de Coimbra, Portugal [University of Coimbra, Portugal]. In H. Bott (Ed.), *UniverCity—Stadt und Universität* (pp. 44–45). Catalogue of the exhibition at the Faculty of Architecture and Urban Planning of the University of Stuttgart, February 2013. Stuttgart: Städtebau-Institut, University of Stuttgart.

Gandert, K.-D. (1985). *Vom Prinzenpalais zur Humboldt-Universität: Die historische Entwicklung des Universitätsgebäudes in Berlin mit seinen Gartenanlagen und Denkmälern* [From the prince's palace to Humboldt University in Berlin: The historical development of the university building with its gardens and monuments]. Berlin: Henschel.

Hederer, O. (1942). *Die Ludwigstraße in München* [The Ludwigstrasse in Munich]. Munich: Franz Eher.

Hederer, O. (1964). *Leo von Klenze: Persönlichkeit und Werk* [Leo von Klenze: The man and work]. Munich: Callwey.

Hoffmann, W. (1985). Das ehemalige Seminar Carolinum [The former Seminarium Carolinum]. In P. A. Riedl (Ed.), *Die Gebäude der Universität Heidelberg: Textband* (pp. 159–177). Semper apertus: Sechshundert Jahre Ruprecht-Karls-Universität Heidelberg, 1386–1986: Vol. 5. Heidelberg: Springer.

Huber, W. (2007). *Moskau—Metropole im Wandel: Ein architektonischer Stadtführer* [Moscow—A metropolis in flux: An architectural guide]. Cologne: Böhlau.

Johnson, P. C. (1947). *Mies van der Rohe*: *Exhibition catalogue*. New York: The Museum of Modern Art.

Katner, W. (Ed.). (1969). *Die Universität zu Köln 1919–1969* [The University of Cologne, 1919–1969]. Basel: Länderdienst.

Laverrenz, C. (1887). *Die Medaillen und Gedächtniszeichen der deutschen Hochschulen. Ein Beitrag zur Geschichte der Universitäten Deutschlands. Teil II* [The medals and commemorations of German universities: On the history of Germany's universities, Part II]. Berlin: J. L. V. Laverrenz.

Mollik, K., Reining, H., & Wurzer, R. (1980). *Planung und Verwirklichung der Wiener Ringstraßenzone* [Planning and realization of Vienna's Ringstrasse zone], Wiesbaden: Steiner.

Nardi, P. (1992). Relations with authority. In H. de Ridder-Symoens (Ed.), *Universities in the Middle Ages* (pp. 77–107). A history of the university in Europe: Vol. I. Cambridge, UK: Cambridge University Press.

O'Neal, W. B. (1980). *Pictorial history of the University of Virginia*. Charlottesville: University Press of Virginia. (Original work published 1969)

Paulus, S. (2010). *Vorbild USA? Amerikanisierung von Universität und Wissenschaft in Westdeutschland 1945–1976* [The United States of America as a model? The Americanization of the university and science in West Germany, 1945–1976]. Munich: Oldenbourg.

Quincy, J. (1977). *The history of Harvard University*. 2 vols. New York: Arno. (Original work published 1840)

Rodrigues, M. A., de Almeida, M. L., & de Alburquerque, L. (1990). *A Universidade de Coimbra e eus seus reitores: Para uma história da instituição* [The University of Coimbra and its rectors: On the history of the University of Coimbra]. Coimbra: Arquivo de Universidade de Coimbra

Rückbrod, K. (1977). *Universität und Kollegium, Baugeschichte und Bautypen* [University and Collegium, architectural history, and types of buildings]. Darmstadt: Wissenschaftliche Buchgesellschaft.

Rüegg, W. (Ed.). (2004). *Universities in the nineteenth and early twentieth centuries (1880–1945)*. A History of the University in Europe: Vol. 3. Cambridge, UK: Cambridge University Press.

Schindler, S. (2013). Aus dem Campus in die Stadt [From campus to city]. *Werk, Bauen + Wohnen, 100*(9), 21–26. Retrieved from http://www.e-periodica.ch/digbib/view?pid=wbw-004:2013:100#6

Schwinges, R. C. (1992). Admission. In H. de Ridder-Symoens (Ed.), *Universities in the Middle Ages* (pp. 171–194). A history of the university in Europe: Vol. I. Cambridge, UK: Cambridge University Press.

Summerfield, C. (1998). Lomonosov University. In C. Summerfield & M. E. Devine (Eds.), *International dictionary of university histories* (pp. 280–284). Chicago: Fitzroy Dearborn.

Teodorovici, D. (2015a). Universitá di Bologna, Italien [University of Bologna, Italy]. In H. Bott (Ed.), *UniverCity—Stadt und Universität* (pp. 30–31). Catalogue of the exhibition at the Faculty of Architecture and Urban Planning of the University of Stuttgart, February 2013. Stuttgart: Städtebau-Institut, University of Stuttgart.

Teodorovici, D. (2015b). Université Paris–Sorbonne, Frankreich [University of Paris–Sorbonne]. In H. Bott (Ed.), *UniverCity—Stadt und Universität* (pp. 33–35). Catalogue of the exhibition at the Faculty of Architecture and Urban Planning of the University of Stuttgart, February 2013. Stuttgart: Städtebau-Institut, University of Stuttgart.

Teodorovici, D. (2015c). University of Oxford, Großbritannien [Oxford University, Great Britain]. In H. Bott (Ed.), *UniverCity—Stadt und Universität* (pp. 36–37). Catalogue of the exhibition at the Faculty of Architecture and Urban Planning of the University of Stuttgart, February 2013. Stuttgart: Städtebau-Institut, University of Stuttgart.

Tuilier, A. (1994). *Histoire de l'Université de Paris et de la Sorbonne* [History of the University of Paris and of the Sorbonne]. Paris: Nouvelle Librairie de France.

Verger, J. (1992). Patterns. In H. de Ridder-Symoens (Ed.), *Universities in the Middle Ages* (pp. 35–74). A history of the university in Europe: Vol. I. Cambridge, UK: Cambridge University Press.

Chapter 13
Campus–City Relations: Past, Present, and Future

Alexandra C. den Heijer and Flavia T. J. Curvelo Magdaniel

The relation between the campus and the city is important for university strategies and urban ambitions. City–university partnerships are being encouraged because they are regarded as mutually beneficial to the stimulation of innovation, which is a common goal of municipalities and universities in the knowledge-based economy (Curvelo Magdaniel, 2016; den Heijer, 2011). These partnerships take place as joint initiatives such as network platforms, learning programs, entrepreneurial activities, and projects to improve their cooperation in tackling societal challenges.[1] To stimulate innovation for socioeconomic development, attracting and retaining talented students and highly skilled workers is arguably the most important joint task of universities and cities (den Heijer, 2011; van den Berg, Pol, van Winden, & Woets, 2005). Creating smart, healthy, inspiring, and appealing environments is therefore crucial for both organizations in the global competition for talent. Quality-of-life factors—such as affordable and desirable housing, diversity of people and functions, convenient commuting, efficient transportation, and cultural and green amenities—contribute to a city–university capacity to draw and keep talent (Drucker & Goldstein, 2007; Fernández-Maldonado & Romein, 2008; Florida, 2002; O'Mara, 1999; van den Berg et al., 2005). In this context the ways in which campuses and cities relate to each other become important because they collectively shape the particular dynamics related to innovation, society, and the economy by bringing in and retaining talent and by creating and applying knowledge for socioeconomic improvement of cities and regions.

[1]Examples of city–university partnerships are the EUniverCities Network launched in Europe in 2012 and the MetroLab Network started in the United States in 2015.

A. C. den Heijer (✉) · F. T. J. Curvelo Magdaniel
Faculty of Architecture and the Built Environment, Delft University of Technology, Delft, The Netherlands
e-mail: a.c.denheijer@tudelft.nl; f.t.j.curvelomagdaniel@tudelft.nl

P. Meusburger et al. (eds.), *Geographies of the University*, Knowledge and Space 12,
https://doi.org/10.1007/978-3-319-75593-9_13

The locations of universities play an important role in the competitive profile of cities and regions in the knowledge economy (Baltzopoulos & Broström, 2013; Florida, 2014). However, the simple presence of universities and their human capital is not enough to stimulate innovation and create wealth in cities. There are challenges for cities in exploiting and managing the provision of human capital as economic assets. Accordingly, managing the interaction between universities, industry, and governments is considered the essence of remaining competitive in the knowledge economy (Laursen, Reichstein, & Salter, 2011). This task involves managing the relationships among stakeholders within each of these organizational spheres, which are place-based. Cities and regions have the ability to optimize the cooperation between these spheres through different activities and at different levels (e.g., from strategic to operational). City–university partnerships can be considered instructive examples of strategic approaches.

At the operational level, investing in the development and management of physical infrastructure that supports the creation, diffusion, and application of knowledge can be seen as a way to strengthen these relationships (van Winden, 2008). In global policies, for instance, the organizational spheres and the infrastructure that support their activities are regarded as national science systems (OECD, 1996). The physical infrastructure—including the built environment—is thus an essential part of these systems. Florida (2010) outlines it in a general way as an enabler of innovation. He conceives of technology, education, and transportation as large-scale systems infrastructures that are needed to support the current demands driven by innovation, velocity, and flexibility. Similarly, he regards the physical infrastructure as a common supportive ground for these systems.

The perception of innovation as a process driven by the exchange of ideas has influenced the physical and functional ways in which campuses relate to cities. In this perception social dynamics are inherent in the early phases of knowledge creation, where ideas are developed and shared as tacit knowledge (Simmie, 2005). In regional studies there is the assumption that having firms and people with complementary intellectual backgrounds in close geographical proximity is also vital to knowledge creation (Audretsch & Feldman, 1996, 2004; Beaudry & Schiffauerova, 2009; Porter, 2008; van Oort & Lambooy, 2014). The more one facilitates social interaction, the greater the potential becomes for collaboration or the cultivation of ideas. These concepts have spread in urban studies because cities are seen as natural sources of diversity (of people and functions) and of positive environments for innovation (Florida, 2008; Glaeser, Kallal, Scheinkman, & Shleifer, 1992; Jacobs, 1961). Proponents of innovation districts as a new urban agenda also embrace the city as an optimal place for innovation and criticize the science park model (Katz & Wagner, 2014). These considerations are influencing the way in which campus location and campus functional mix are perceived in the knowledge economy. That perception is especially important because there are many different types of physical and functional campuses (den Heijer, 2011), each of which may have different capacities to stimulate innovation (Curvelo Magdaniel, 2016).

Raising awareness of the dynamic and diverse physical and functional relations between the campus and the city can help stakeholders in universities and cities improve the decisions they make, specifically the strategic decisions that efficiently exploit the physical and functional resources the campus and city share and that effectively support the mutual goals of universities and cities. We aim to deepen the understanding of dynamic campus–city relations by asking two questions: What are the past, present, and future trends in the physical settings and functional mix of campuses? How can universities and cities act upon these trends?

To elaborate on the physical and functional relation between campus and city, this chapter combines components of two dissertations (den Heijer, 2011; Curvelo Magdaniel, 2016) and builds on the findings reported in a journal article (den Heijer & Curvelo Magdaniel, 2012) and in research involving 39 case studies worldwide (Curvelo Magdaniel, 2016, 2017). The first section provides our theoretical framework. In the second section we describe the methodology used to answer our questions. The third section conveys our results through descriptions of past, present, and future trends in physical and functional relations between campus and city. In the fourth section we discuss the results and their implications for practice and theory. The final section offers this chapter's main conclusions in response to our central questions.

Conceptual Framework

To operationalize campus–city relations, it is necessary to define their physical and functional characteristics. To do so, we use components of earlier research (Curvelo Magdaniel, 2016; den Heijer, 2011; den Heijer & de Vries, 2007) and combine the resulting physical and functional typologies in a conceptual framework that we then use to position the past, present, and future campus–city relations of 39 cases as explored and assessed by Curvelo Magdaniel (2016).

Defining Physical Campus–City Relations

The term *campus* is often associated with a greenfield site (outside the city) or an area that is isolated from the urban setting (sometimes even gated). In practice this description does not necessarily apply. At universities the word campus is often used to designate where university activities take place. It also increasingly refers to a virtual campus or a downtown café (den Heijer, 2011). In this chapter we define the university campus as the sum of locations with predominantly university or university-related functions (see den Heijer, 2011). In other words, a collection of inner-city university buildings can be called an inner-city campus even though the borders are not altogether distinct. The typology in Fig. 13.1 illustrates this definition, identifying three different spatial configurations: (a) the greenfield campus,

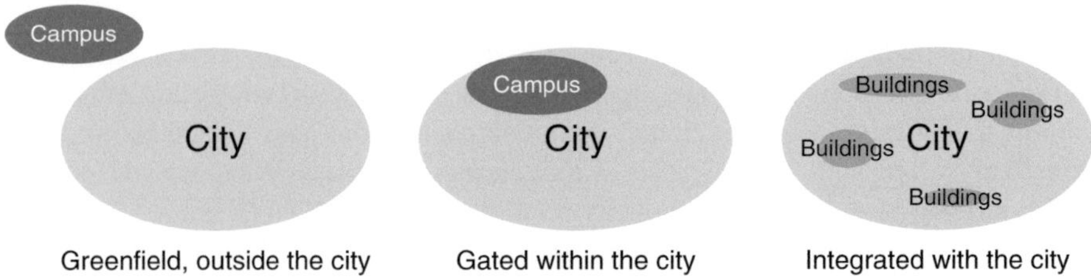

Fig. 13.1 Three different spatial configurations of physical campus–city relations. Source and copyright: Den Heijer (2011, p. 53). Adapted and reprinted with permission.

outside the city, (b) the campus (gated) within the city and (c) the campus integrated into the city.

Whereas the term campus used to refer only to *university* land and buildings, more and more types of institutions have come to use it to refer to their territory. The terms *corporate campus* and *high-tech campus* (Hoeger & Christiaanse, 2007, p. 188), for example, refer to a specific location. In this study, however, the university campus can refer to more than one location or spatial configuration. In fact, recent research shows that many of the 14 Dutch research universities combine two or even three models to accommodate their rapid growth (TU Delft, 2016).

In analyzing location characteristics, Curvelo Magdaniel (2016) described the position of technology campuses in relation to their host cities (or regions). Her study showed that technology campuses entail a variety of built environments designed to accommodate technology-driven research activities of multiple organizations (e.g., science parks, campuses of universities of technology. and corporate R&D parks). Topology helped identify a set of five relationships that the campuses and the cities can have with each other (see Table 13.1).

Linked to specific changes in their temporal and social contexts, most of these relationships are dynamic. Most campuses studied by Curvelo Magdaniel (2016) are considered "touched by the city" (p. 441) because they are at the edge thereof. These locations could have come about in different ways. For instance, some campuses were built outside the city, whose expansion due to urbanization ultimately reached their peripheries. Perhaps these campuses also induced urban developments in their vicinities. Conversely, some campuses may have been built in inner-city locations where their full urban integration was impeded by particular urban configurations (e.g., a waterfront or natural features, which happen to be both geographic and administrative boundaries). Depending on each development and campus–city configuration, some campuses categorized as Touches may eventually evolve into Contains or Overlaps. Furthermore, Curvelo Magdaniel (2016) found that nine of the 39 campuses she studied have at least two relationships with the city. These locations can be considered campuses in transition,[2] for this duality has resulted

[2]Campuses in transition are those perceived as having two physical campus–city relationships simultaneously because of constant spatial transformations and individual campus–city features in terms of relative size, infrastructure systems, and/or natural elements.

Table 13.1 Typology of five physical relations between the city and technology campuses (N = 39)

Relationship	Description	Cases
Equals 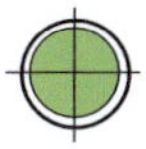	*City is the same as the campus.* It includes those areas that were newly built as towns or cities. They were built and planned from scratch to accommodate clusters of technology. They are located only in Asia.	4
Disjoints 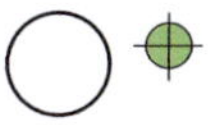	*City shares nothing with the campus.* It includes those areas located outside the city limits but not distinguished as independent cities.	8
Touches 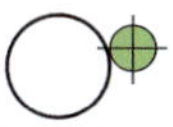	*City touches the campus.* It includes those areas bordering on the city. In most cases they and the city are tangent. Touches and the city are usually tangent, but in some cases they are separated by a river, highway, or some other feature).	17
Contains 	*City contains the campus.* It includes those areas that are inside the urban fabric, but they are perceived of as a distinct campus with borders (e.g., roads, fences, waterfronts, or natural features).	12
Overlaps 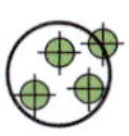	*City and campuses have multiple points in common.* It includes those areas integrated into the urban fabric, and in many cases the boundaries between the sites and the rest of the city are not clearly defined or perceived.	6

Source: Adapted from Curvelo Magdaniel (2016, p. 114). Copyright by Curvelo Magdaniel. Adapted and reprinted with permission.

from certain campus characteristics in relation to changes in urban features (e.g., a growing campus with a large surface area and dispersed arrangement in a relatively small city).

Generally, the university campus in this research refers physically to the location (s) of the university or the ensemble of buildings that are either used or owned (or both) by the university and have a role in achieving the institutional goals. Three different types of locations are identified and used in the conceptual framework of this chapter: greenfield (the city disjoints[3] or touches the campus), gated in the city (city contains the campus), and integrated into the city (city overlaps the campus).

[3]For our chapter's conceptual framework, two of the five types of campus–city relationships identified by Curvelo Magdaniel (2016)—Equals and Disjoints—are merged into one (Disjoint) because their main difference lies in the scale of the development, which is not relevant for the descriptive purpose of this study.

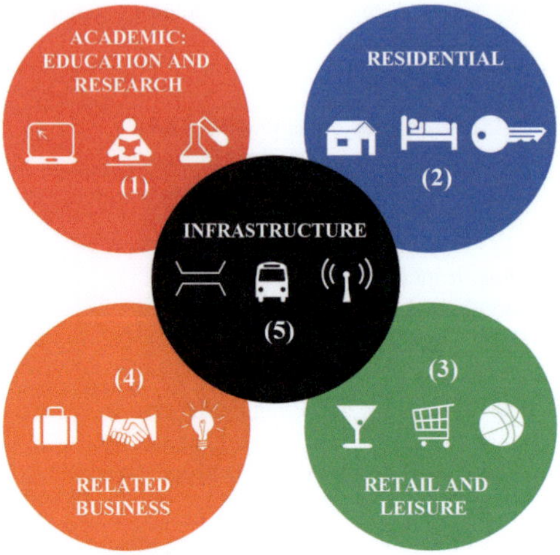

1. Academic function for education & research

2. Residential function: housing for students and staff, hotels

3. Retail and leisure function: sports, cultural and catering facilities

4. Related business function: space for partners linked to the academic goals and supporting processes

5. Infrastructure function: from parking to accessibility

Fig. 13.2 Space types on campus—the required functional mix for the future university. Source: Den Heijer (2011, p. 181). Adapted and reprinted with permission.

Defining Functional Campus–City Relations

The functional campus–city relation describes how dependent the university is on the city's functions: How many functions do the campus and city share? Or is the campus an autonomous, self-contained city that is functionally independent of the city?

Developments in university strategies show that the university is becoming increasingly dependent on the presence of nonacademic types of space in their vicinity (Chapman, 2006; den Heijer, 2011; TU Delft, 2016). Examples are housing and apartments for foreign students or hotel capacity for visiting professors for promoting goals of internationalization. Trendy coffee bars and sports facilities are important for creating a lively campus and a place to meet on campus. To assure knowledge transfer, which most universities mention as their third strategic goal, it is crucial to welcome businesses that combine learning and working, incubators for entrepreneurs, and breeding grounds for young artists (Wissema, 2009; Worthington, 2009). Lastly, quality infrastructure and adequate parking space should guarantee the university's accessibility to students, staff members, and many visitors. The functional campus models are based on these five required aspects of university processes and goals (see Fig. 13.2), which have been specified by Dutch campus managers in workshops (den Heijer & de Vries, 2007) and confirmed in recent research (TU Delft, 2016).

The types of space and associated functions identified in Fig. 13.2—education and research, residence, retail and leisure facilities, related businesses, and infrastructure—are elaborated in Table 13.2, which explains the extent to which the city's facilities complement those of the campus in each functional category. The functional mix required by the university need not be supplied on campus. Depending on

Table 13.2 Required university functional mix specified by campus managers and supplied and managed by a university, municipality, or third party

Functions	Who manages/own/uses?			Similar city functions
	University	Municipality	3rd party	Alternative available in city? Examples
ACADEMIC • EDUCATION AND RESEARCH				
Classrooms and studio spaces (small groups)	X			
Lecture halls (large groups)	X			Movies, theaters
Office space academic staff	X			
Office space support staff	X			
Laboratories	X		X	R&D facilities of large companies
Study places for individual use/small groups	X			Inner city coffee bars
Library	X			Community library
Special places for ceremonies (graduation)	X	X		City halls, churches
Special conference facilities	X		X	Conference center
Special educational facilities (dance, media, arts)	X	X		Theaters, studios, museums
Academic hospital			X	Other hospital
Medical school			X	
RESIDENTIAL				
Student housing: national			X	Social housing in city
Student housing: international – short stay	X		X	Hotels or apartments
Alumni housing: young potentials, creative class			X	Housing supply in city
Faculty housing			X	Housing for expats
Housing for support staff			X	Housing supply in city
Hotel facilities			X	Hotels in city
Short stay apartments for visiting professors	X		X	
RETAIL AND LEISURE				
Sports facilities	X		X	Sport facilities in city
Book stores			X	Book stores in city
Coffee bars	X		X	Espresso bars in city
Student associations and societies/fraternities			X	
Restaurants (lunch)	X		X	Restaurants (lunch)
Restaurants (dinner)	X		X	Restaurants (dinner)
Bars	X		X	Bars
Theaters			X	Theaters
Jazz clubs			X	Jazz clubs
Cultural center, museum	X		X	Cultural center
Dry cleaning, day care center, supermarkets			X	Existing city facilities
RELATED BUSINESS				
Incubators (academic spin-off)	X		X	Office supply in city
R&D facilities of large companies			X	Business campuses
Related services (service spin-off)			X	Office parks in city
Business that combine learning and working			X	
Artists, creative professions			X	(Vacant) industrial buildings
INFRASTRUCTURE				
Parking space	X	X		Existing parking facilities
Transport on campus (trolleys)	X			
Accessibility (by car)	X	X		Car transport network city
Accessibility (by public transport)	X	X		Public transport network city
Public space (bicycles, pedestrians)	X	X		Bicycle paths in city

Source: Den Heijer (2011, p. 184), based on research results from den Heijer and de Vries (2007). Adapted and reprinted with permission.

the location of the campus(es) in the city, the urban functional mix can help meet the needs of the university.

The functional specifications in Table 13.2 also show that supplying and managing the required university functions is, in practice, not a management task of the university alone. When asked who manages, owns, or uses this function, the respondents from many universities indicated that it was the municipality or a third party instead of the university. In 2006 the university's collaboration with the municipality and third parties was quite common as assessed by den Heijer and de Vries (2007) in workshops with both campus managers and urban authorities. Residential, retail and leisure, and related business functions are often managed by third parties.

According to this part of the conceptual framework, universities can benefit from the network of functions available in cities and supplied by university partners across places. The extent to which universities can benefit depends on the physical distance between campus and city functions. City–campus benefits depend on both functional and physical aspects, a circumstance that highlights the importance of connecting both dimensions in a conceptual framework.

Combining Physical and Functional Campus–City Relations

Den Heijer (2011) combined the physical and functional campus–city typologies as illustrated in Fig. 13.3. She identified the most common city–campus relations practiced at that time, including one without academic functions (called "business community," which was found in in three different physical, urban settings). In that study attention was given to six types of communities defined by types of functional mix: academic community, residential community, sociocultural community, business and science community, campus community, and business community. The academic and business communities are two functional extremes.

Academic communities refer to learning environments (Worthington, 2009), which are described as holistic, loosely coupled, on and off campus, formal and informal, and virtual and physical. In terms of programmatic requirements, learning environments should provide (a) spaces that are less specialized than traditional ones, where boundaries blur and operating hours approach 24 hours a day, 7 days a week; (b) types of space designed primarily around patterns of human interaction rather than specific needs of individual departments, disciplines, or technologies; and (c) new models of space that enhance the quality of life as much as support the learning experience.

Business communities are environments with a concentration of companies, a nexus that does not necessarily exclude a collaborative relationship with universities. In these settings research is a more representative activity on campus than teaching and learning are for academic communities. Existing research refers to these environments as high-tech campuses or corporate campuses (Hoeger & Christiaanse, 2007). Usually, governments acknowledge the potential economic power of these environments for cities and regions (Buck, 2012). Tenant diversity, shared research facilities, high-quality buildings, and spaces that facilitate networking are examples of programmatic requirements on these types of campuses (Buck, 2016).

Besides defining types of communities, den Heijer (2011) defined more detailed categories when crossing these functional types with the physical campus–city

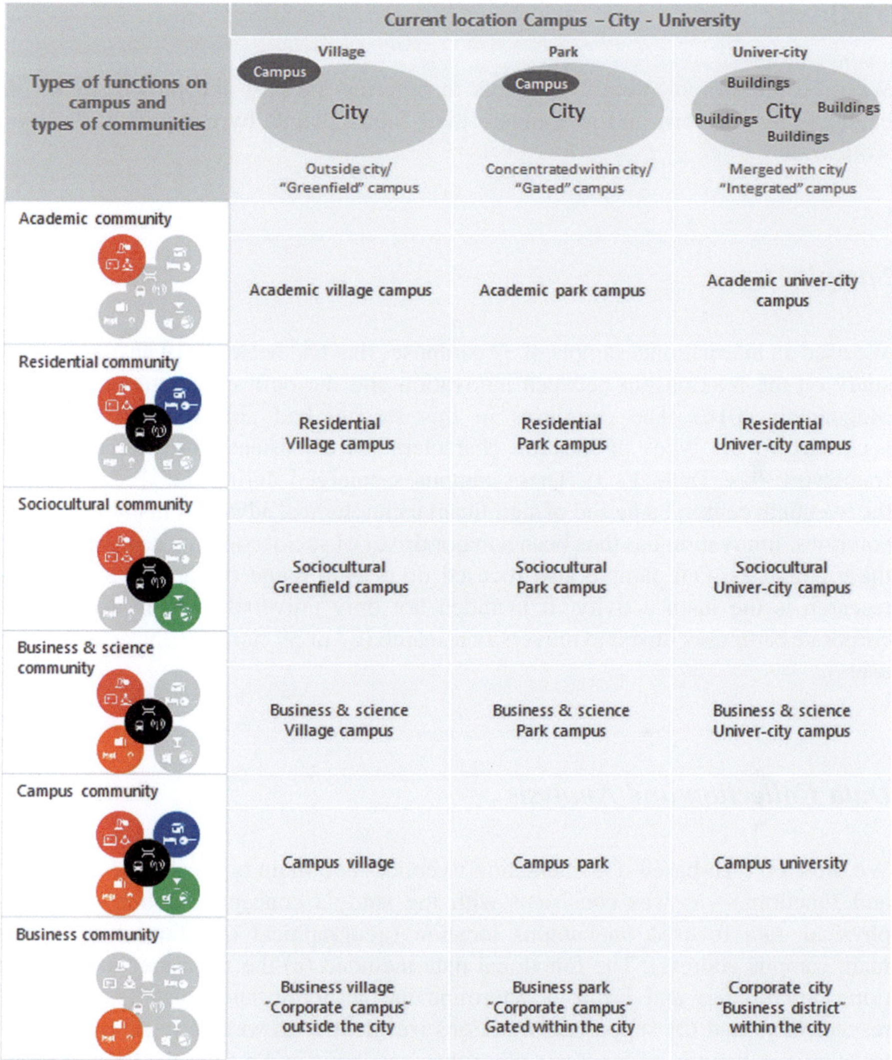

Fig. 13.3 Physical and functional typologies combined to describe the most common campus–city relations. Source: Den Heijer (2011, p. 185). Adapted and reprinted with permission.

relations. These combinations provide a conceptual framework for the analysis of past, present, and future campus–city trends presented in this chapter.

Methods

We used the aforementioned concepts to describe the past and present trends in campus–city relations and to estimate their future course by comparing 39 cases worldwide.

Sample

We used an international sample of 39 campuses that had been part of an exploratory study on the relationship between innovation and the built environment (Curvelo Magdaniel, 2016). The campuses in this sample had already been classified according to the types of location characteristics consistent with the conceptual framework (see Table 13.3). These campuses emerged during the second half of the twentieth century, a period of significant technological advances in industrialized countries. Innovation has thus been a major driver of socioeconomic development in these countries. Our sample also focused on a wide range of campuses at which research is the main activity. It included not only university campuses but also corporate campuses linked to university research (27 of 39 campuses have university users).

Data Collection and Analysis

We drew on web-based documentation to collect two main types of data—physical and functional—as was consistent with the study's conceptual framework. The physical data focused on campus location (geographical coordinates using the main campus address). The functional data included (a) the main user's organizations (universities and firms to determine the academic and business functions, respectively) and (b) supporting functions (residential as well as retail and leisure to determine the mix of functions other than academic and business). This information was collected in 2013 as part of a wide exploratory study (Curvelo Magdaniel, 2016). The collected data was publicly available through various sources. Primary data sources included official websites of the campuses, institutional documents and reports, and open map software. Secondary sources included existing empirical research documenting the cases selected. Table 13.4 presents a summary of the data.

The combination of physical and functional campus–city relations was used to categorize and analyze the data for (past, present, and future trends. Tables featuring the different types of locations and mix of functions were developed to position the campuses within the existing categories based on the two types of data collected. Mapping the campus location enabled us to describe and interpret physical campus–city relations over time. For instance, historical imagery provided by Google Earth

Table 13.3 Sample of 39 campuses exhibiting the physical campus–city relationships identified by Curvelo Magdaniel (2016)

No.	Campus	City, state, and country	Campus–city relationship
1	Stanford Research Park	Palo Alto, California, United States	Touches/ Overlaps
2	Cornell Business & Technology park	Ithaca, New York, United States	Touches
3	TU/e Science Park	Eindhoven, The Netherlands	Contains
4	Akademgorodok Academic Town	Novosibirsk, Russia	Disjoints*
5	Research Campus Garching—Technical University of Munich	Garching/Munich, Germany	Disjoints
6	Research Triangle Park	Durham, Raleigh, and Chapel Hill, North Carolina, United States	Disjoints
7	ETH Hönggerberg Science City	Zürich, Switzerland	Touches
8	MIT Campus & University Park at MIT	Cambridge, Massachusetts, United States	Overlaps
9	Drienerlo Campus University of Twente & The Innovation Campus Kennispark Twente	Enschede, The Netherlands	Touches
10	TU Delft District & Technopolis and Innovation Campus Delft	Delft, The Netherlands	Touches/ Overlaps
11	Tsukuba Science City	Tsukuba, Japan	Disjoints*
12	Cambridge Science Park	Cambridge, United Kingdom	Touches
13	Sophia-Antipolis Park	Côte d'Azur Region, France	Disjoints
14	Taedok Science Town & Daedeok Innopolis	Daejeon, South Korea	Disjoints*
15	Hsinchu Science and Industrial Park	Hsinchu City, Taiwan	Touches/ Overlaps
16	Singapore Science Park	Singapore City-State, Singapore	Contains
17	Leiden Bio Science Park	Leiden, The Netherlands	Contains/ Overlaps
18	Surrey Research Park	Guildford, United Kingdom	Touches
19	Western Australia Technology Park	Perth, Australia	Contains
20	Otaniemi Science Park & Otaniemi Technology Hub	Espoo, Finland	Contains
21	Sendai Technopolis & Izumi Park Town Industrial Park	Sendai city, Japan	Disjoints*
22	Kansai Science City	Kansai, Japan	Disjoints
23	Zhong Guan Cun Science Park	Beijing, China	Overlaps
24	Technology Park Bremen & University of Bremen	Bremen, Germany	Touches
25	Brandenburg Technical University Campus	Cottbus, Germany	Contains
26	Zhangjiang Hi-Tech Park	Shanghai, China	Touches
27	Taguspark	Lisbon, Portugal	Disjoints
28	Berlin Adlershof Humboldt University	Berlin, Germany	Touches/ Contains
29	Shenzhen Hi-Tech Industrial Park	Shenzhen, China	Touches/ Contains

(continued)

Table 13.3 (continued)

No.	Campus	City, state, and country	Campus–city relationship
30	Tainan Science Park	Tainan City, Taiwan	Disjoints
31	High-Tech Campus Eindhoven	Eindhoven, The Netherlands	Touches
32	Science Park Amsterdam	Amsterdam, The Netherlands	Touches/ Contains
33	Biopolis	Singapore City-State, Singapore	Touches/ Contains
34	Taichung Science Park	Taichung, Taiwan	Disjoints
35	Biocant Park	Cantanhede, Portugal	Disjoints
36	Chemelot Campus	Sittard-Geleen, The Netherlands	Touches
37	Barcelona City of Knowledge	Barcelona, Spain	Contains
38	GIANT Innovation Campus [Grenoble Innovation for Advanced New Technologies]	Grenoble, Isère, France	Touches/ Contains
39	RWTH Aachen University—Research Campus Metalen	Aachen, Germany	Disjoints

*Previously categorized as Equals.
Source: Design by authors.

Table 13.4 Data collected on features of 39 campuses throughout the world

Content	Evidence	Sources
	Physical	
Main address Geographic coordinates	Campus's location characteristics in the city	Campus's official websites iTouchMap (open access online software) and Google Earth
	Functional	
Main user's organizations Supporting functions	Academic and business functions on campus Residential and leisure/retail functions on campus	Campus's institutional reports and existing empirical research Campus's institutional reports and existing empirical research Google Earth

Source: Design by authors.

made it possible to determine past physical relations and to prepare estimates of future spatial developments based on observations over the years since the emergence of the campuses. The estimation of dynamic functional trends on campus was more limited. The past and the future functional trends are based on institutional documents or statements in which the representatives of organizations serving campus users explicitly described how they have changed or intended to change functionally (e.g., by increasing or decreasing the number of functions on campus). However, this information was not available for all the campuses. In such cases the existing functional situation remained unchanged in the analysis.

Results

As shown by the overview of campus–city relations in 39 international cases in Fig. 13.4, each campus proved to occupy different categories, depending on the current physical and functional data pertaining to it. Figure 13.4 also illustrates our estimations of each campus's past and future position. Physical data from all 39 cases revealed that campuses exhibited detailed location characteristics already categorized (Curvelo Magdaniel, 2016; den Heijer, 2011). The combination of these physical relations is symbolized at the top of the figure. Functional data on 37 valid cases[4] showed that campuses allowed for a variety of functional combinations. Empirical data on the 39 cases expanded den Heijer's (2011) classification scheme from 6 to 11 types of community based on the different mix of five functional categories.

Trends in Physical Campus–City Relations

The physical data showed an enduring shift in campus development from peripheral to inner-city locations. Most of the campuses in the sample (36 of 39 cases) used to be on the periphery. Of the 39 campuses we studied, 22 remain outside the cities today, 10 are currently in the city, and another 9 campuses are in transition. These nine campuses exhibit the following dual relationships: Touches/Contains (five cases), Touches/Overlaps (three cases), and Overlaps/Contains (one case). The former two dual relationships illustrate the transition from peripheral to inner-city locations. The number of inner-city locations is expected to increase (i.e., 20 of 39 cases estimated). This trend substantiates the transformation of the current campuses into "integrated campuses," which are physically merged with the city.

Trends in Functional Campus–City Relations

The functional data showed an enduring shift from monofunctional to multifunctional campuses regardless of their distinct locations in relation to the city. This shift was illustrated by both extremes of the functional categories (i.e., a change from solely academic or business communities toward campus communities). Indeed, the number of monofunctional campuses had decreased from 13 of 39 cases to 3 of 39 cases. We also estimated that monofunctional campuses would decrease to 1 of 39 cases. Accordingly, the number of campuses that have all the

[4]Functional data on two of the thirty-nine cases was not found. Most campuses highlighted their functional mix in their institutional documents. It was safe to assume that these campuses were monofunctional and to categorize them as such.

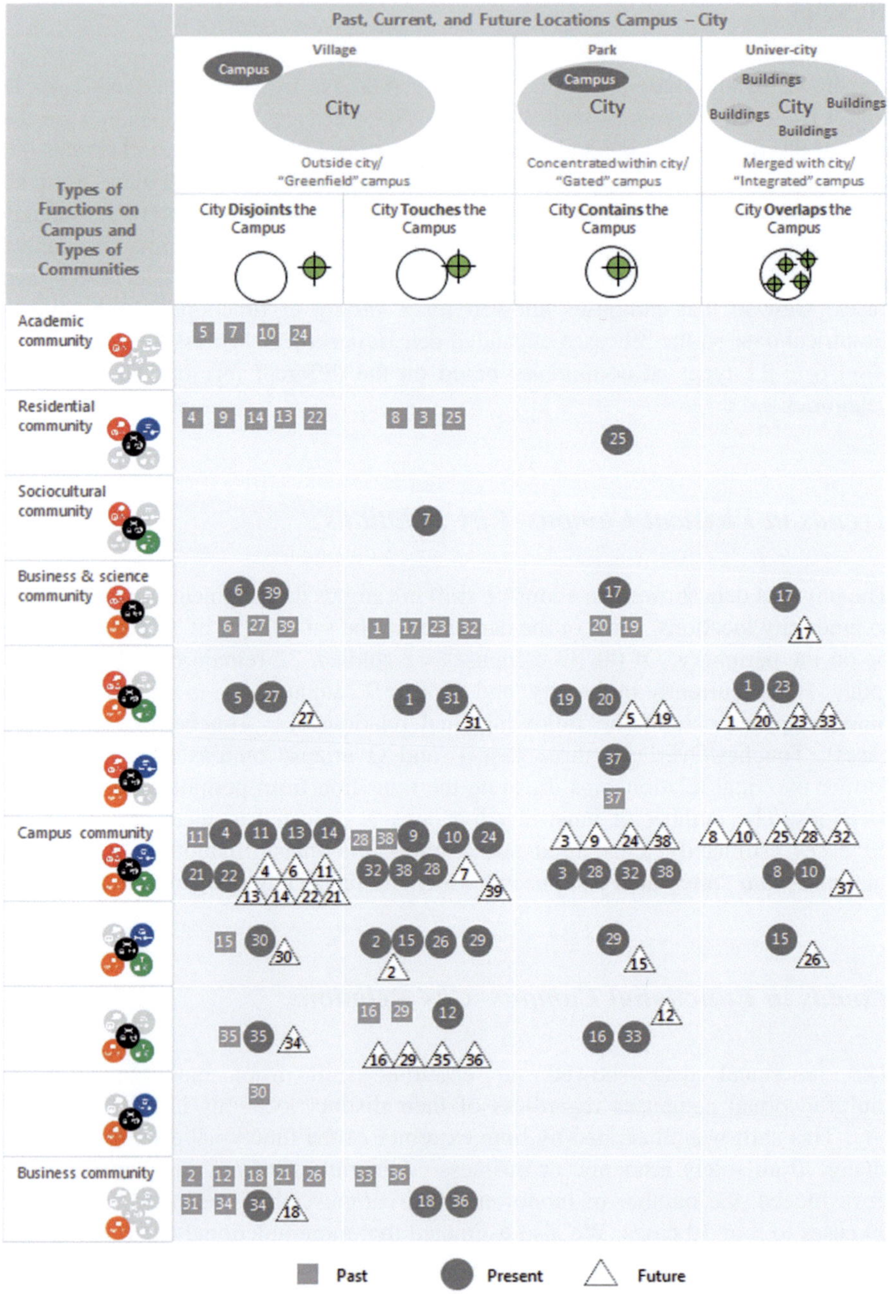

Fig. 13.4 Past, current, and future campus–city relations of 39 cases studied. Source: Design by authors.

required functions had increased over the years—from 3 of 39 cases to 16 of 39 cases. The estimations based on our research suggested that the number of multifunctional campuses was likely to rise to 19 of 39 cases. Overall, the combination of two or three functions was predominant (23 of 39 cases) and still is (20 of 39 cases), and we expect it to decrease slightly to 19 of 39 cases. This trend substantiated the transformation of the current campuses into campus communities that provide all the required functions.

Discussion

The results confirm research that has outlined the urban shift in the accommodation of universities and other technology-driven organizations in the knowledge-based economy (Aasen & Haugen, 2015; Carvalho, 2013; den Heijer, 2011; Katz & Wagner, 2014; van Winden & Carvalho, 2016). Our findings on the campus's shift from peripheral to inner-city locations confirms the work of den Heijer (2011), who documented the physical signs of universities transitions in the Dutch context. In changing from small and exclusive institutions to large institutions open to the masses, Dutch universities have built their campuses in three stages (see Fig. 13.5). First, universities in the early 1900s were small institutions physically integrated into the urban fabric. As they grew, their campuses expanded to the edges of the city (1950s through the 1970s), and some universities left their inner-city buildings to intensify the use of their newly built campuses. With the rapid urban growth of Dutch cities in recent decades, the university campuses are again becoming part of the city. The sample used in this study provides evidence of the latter two developmental stages.

In practice, the idea of the city as an ideal environment supporting innovative activities (Jacobs, 1961; Katz & Wagner, 2014) may be influencing location decisions of universities and other organizations. There are already extraordinary examples of universities in Europe and the United States organizing their move from suburban to urban locations rather than just waiting for the physical expansion of the city to happen (Aasen & Haugen, 2015; Lange, 2012).

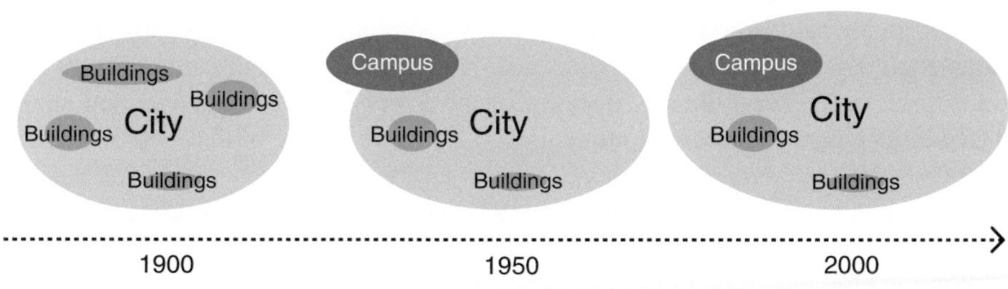

Fig. 13.5 Example of development stages that built some of the current Dutch campuses. Source: Den Heijer (2011, p. 61). Adapted and reprinted with permission.

The shift from monofunctional to multifunctional campuses supports the findings of Carvalho (2013), who observed similar trends while studying knowledge locations. Several campuses, science parks, and technology parks are being "urbanized" because new functions such as housing, amenities, and cultural facilities have been added to these places. Van Winden and Carvalho (2016) have argued that many places are being transformed from monofunctional business and research-oriented into diverse, open, and urban environments. The empirical data illustrating the past, present, and future functional trends away from exclusively academic or business communities and toward mixed campus communities strengthen these positions (Fig. 13.4).

These shifts in campus–city relations make evident that universities and other organizations on campuses increasingly share physical and functional resources with cities. These resources could be efficiently used and managed to attain shared goals (e.g., stimulating innovation and increasing sustainability, which are already on the agenda of universities and local governments alike). As illustrated in Figure 13.6, universities can benefit from the urban network of functions that is supplied by university partners across the city (den Heijer, 2011). Cities may benefit from the presence of university communities of students and knowledge workers, adding to the vitality of areas neighboring the campuses.

The current and future trends outlined in this chapter provide an opportunity for campus decision-makers to work together and mutually benefit from closer campus–city relations. However, there are two important conditions for such an endeavor's success: (a) the size of the city and (b) the commitment of the stakeholders. First, the potential success of campus–city relations depends on the distance between the city and the campus, which is influenced by the degree of accessibility afforded by public transport. This dependence is critical for campuses located on the periphery. Depending on the size of the city and its available transportation modes, this distance ranges from 12 minutes (e.g., for Technology Park Bremen) to 2 hours (for the Sophia Antipolis Park). The size of the city makes a significant difference in the opportunities to share physical and functional resources.

Second, successful relations between a campus and a city require the commitment and active participation of the campuses' and the city's decision-makers, who may be willing to assume different roles to achieve their mutual goals. Just as the borders of the inner-city campus are unclear, so are the boundaries defining the types of stakeholders who have to be involved in managing it (e.g., owners, users, policy-makers, and beneficiaries). Although such ambiguity may come across as a threat to campus governance, many universities and other organizations of campus users still want to be in (or near) the city because of the liveability, convenience, and abundance of its places for breeding innovation. Taking advantage of these opportunities requires coordinated management of campus–city relations.

In this context it is worth pointing out those campuses that are currently changing physically and functionally. Campuses in transition can be used as living labs to exploit the aforementioned opportunities and test the ways stakeholders may act on these trends. We have identified several of the campuses whose urban transition is physically evident. Some of the greenfield campuses, however, leave us uncertain

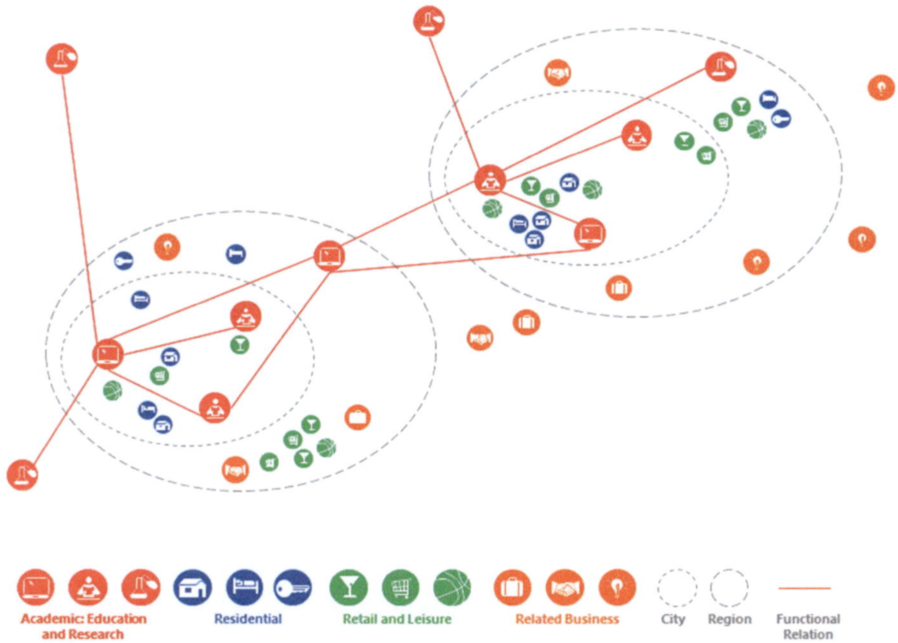

Fig. 13.6 Required university functions in a fictitious example, supplied by a network of university partners in neighboring cities or regions.
Source: Den Heijer (2011, p. 183). Adapted and reprinted with permission.

about their physical and functional integration into cities. In these instances, there is a need to be aware that efficient transportation infrastructure and the provision of multiple functions may prevent these campuses from becoming functionally self-contained. (Curvelo Magdaniel, den Heijer, & de Jonge, 2018). Nevertheless, the decision to be the city or part of the city depends on the campus decision-makers, the type of organization they support, the size of the city, and the campus-city distance as measured in travel time.

In summary, there are two extreme campus settings, each with its planning advantages and disadvantages. The functionally self-contained greenfield campus may encourage universities and other organizations on campus to remain relatively autonomous in creating their future campuses. However, campus development becomes relatively expensive for these organizations because they will underuse—if not neglect—the associated advantages of the campus–city relations. This model is not always voluntarily chosen; it can also be an unintentional or imposed strategy if the physically isolated setting is the sole option. The greater the campus–city distance is, the more complex the relation to the city becomes, both physically and functionally. By contrast, the campus integrated into the city has the advantage that there is plenty of opportunity to collaborate with the municipality and third parties. Disadvantages of this model are that the space for campus expansion is limited and that the university's identity can be diffuse, a characteristic that might affect the sense of community.

Overall, the possibility of sharing functions or having other parties supply them does not mean that the university should take it. Some universities choose to manage and own most university functions themselves in order to keep control over functions that are crucial for achieving the university's goals. However, exclusive use, ownership, and management also come at a high price for the university, competing with primary resources for education and research. The actors responsible for universities should consider these advantages and disadvantages when selecting future campus models. In terms of management information, they are looking for references and experiences of other universities on which to base their choices. Further case studies may help universities make these challenging campus-related decisions.

Conclusions

What are the past, present, and future trends in the physical settings and functional mix of campuses? How can universities and cities respond to these trends? We have answered the first question by providing a descriptive overview of the changing physical and functional relations between the campus and the city in 39 international cases. Two main shifts were identified through observation of the developments between the past and the current situations and through estimation of the future changes. The first shift is a physical one from peripheral to inner-city locations that exhibit the dynamics of urban growth affecting the accommodation of universities and other organizations on campuses. The second is a change from monofunctional to multifunctional campuses in academic and business communities alike, which manifests opportunities and hazards of collaborating and competing in campus–city planning. Discussing these results to answer the second question, we raise awareness of the changing and diverse campus–city relations exemplified by two extreme models: the functionally self-contained greenfield campus and the campus integrated into the city. These models illustrate profound transformations from autonomous campus development to coordinated campus–city development. We have explored the advantages and disadvantages of each model to help stakeholders in universities and cities make better decisions that support their mutual organizational goals.

The empirical findings of this study have improved the existing conceptual framework on both the physical settings and the functional mix of campuses. First, it has added detail to the picture of campus–city physical relations by combining insights from den Heijer (2011) and Curvelo Magdaniel (2016). Second, by increasing the empirical data base, we have expanded the types of functional communities to include the combination of functions (den Heijer, 2011). For instance, the different types of campuses in the sample suggest that several campus developments are driven by business communities without universities (e.g., R&D parks). This knowledge augments the literature on campus planning and may be helpful to other researchers investigating campus–city relations.

Our study also has limitations. Although the sample used may be representative of the variety of existing campuses, a larger sample may be more appropriate for interpreting these developments more accurately as trends with global impact. Moreover, this study's assessment of change from current to future relations between cities and campuses is based mainly on interpretations of statements in institutional documents (functional relations) and formal analysis (physical relations). Further research on this topic will require the use of more appropriate methods to estimate the changing physical and functional relations in the future. Case studies may have a part in contributing to an understanding of the dynamics explaining such changes and, ultimately, may help universities, cities, and other organizations act on these trends.

References

Aasen, T. M., & Haugen, T. I. (2015, December). *Campus alive: Transformation and integration of university work and campus space*. Paper submitted to the APROS/EGOS conference on Spaces, Constraints, Creativities: Organization and Disorganization, Sydney, Australia.

Audretsch, D. B., & Feldman, M. P. (1996). R&D spillovers and the geography of innovation and production. *American Economic Review, 86,* 630–640. Retrieved from https://www.jstor.org/stable/2118216?seq=1#page_scan_tab_contents

Audretsch, D. B., & Feldman, M. P. (2004). Knowledge spillovers and the geography of innovation. In J. V. Henderson & J.-F. Thisse (Eds.), *Handbook of regional and urban economics: Vol. 4* (pp. 2713–2739). Amsterdam: Elsevier.

Baltzopoulos, A., & Broström, A. (2013). Attractors of entrepreneurial activity: Universities, regions and alumni entrepreneurs. *Regional Studies, 47*, 934–949. doi:https://doi.org/10.1080/00343404.2011.602335

Beaudry, C., & Schiffauerova, A. (2009). Who's right, Marshall or Jacobs? The localization versus urbanization debate. *Research Policy, 38*, 318–337. doi:https://doi.org/10.1016/j.respol.2008.11.010

Buck, R. (2012). Actueel beeld campussen in Nederland [Current image of campuses in the Netherlands]. Retrieved from http://www.bciglobal.com/data/file/Actueel%20beeld%20campussen%20in%20Nederland(1).pdf

Buck, R. (2016, October). *Science parks, a promising new asset class*. Paper presented at Expo Real: The Future of Science Parks, Munich. Retrieved November 12, 2017, from https://investment-briefings.propertyeu.info/media/science-parks-a-promising-new-asset-class/

Carvalho, L. (2013). *Knowledge locations in cities: Emergence and development dynamics* (No. EPS-2013-274-S&E). *ERIM Ph.D. Series Research in Management*. Erasmus Research Institute of Management. Retrieved from http://hdl.handle.net/1765/38449

Chapman, M. P. (2006). *American places: In search of the twenty-first century campus*. American Council on Education. Series on Higher Education. Westport: Praeger.

Curvelo Magdaniel, F. T. J. (2016). *Technology campuses and cities: A study on the relation between innovation and the built environment at the urban area level* (Doctoral dissertation). doi:https://doi.org/10.7480/abe.2016.12

Curvelo Magdaniel, F. T. J. (2017). *Campuses, Cities and Innovation: 39 international cases accommodating tech-based research* (A. C. den Heijer, M. Arkesteijn, & H. de Jonge, Eds.). Delft: TU Delft, Faculty of Architecture, Department of Management in the Built Environment.

Retrieved from https://pure.tudelft.nl/portal/files/19480097/Campuses_Cities_and_Innovation_16062017_online_version_BW.pdf
Curvelo Magdaniel, F., den Heijer, A., & de Jonge, H. (2018). The locations of innovation described through thirty-nine tech-campuses. *Competitiveness Review, 28*(1), 58–74. doi:https://doi.org/10.1108/CR-01-2017-0014
den Heijer, A. C. (2011). *Managing the university campus: Information to support real estate decisions*. Delft: Eburon Academic Publishers.
den Heijer, A. C., & Curvelo Magdaniel, F. T. J. (2012). The university campus as a knowledge city: Exploring models and strategic choices. *International Journal of Knowledge-Based Development, 3*, 283−304. doi:https://doi.org/10.1504/IJKBD.2012.048392
den Heijer, A. C., & de Vries, J. C. (2007, October 25–26). *Bouwen aan de Kennisstad, verslag expertmeeting* [Building knowledge cities: Report on an expert meeting], "Bouwen aan de Kennisstad," expert meeting hosted by Inholland University of Applied Sciences, Rotterdam.
Drucker, J., & Goldstein, H. (2007). Assessing the regional economic development impacts of universities: A review of current approaches. *International Regional Science Review, 30*, 20−46. doi:https://doi.org/10.1177/0160017606296731
Fernández-Maldonado, A. M., & Romein, A. (2008). A knowledge-based urban paradox: The case of Delft. In T. Yigitcanlar, K. Velibeyoglu, & S. Baum (Eds.), *Knowledge-based urban development: Planning and applications in the information era* (pp. 221–239). Hershe: Information Science Reference.
Florida, R. L. (2002). *The rise of the creative class: And how it's transforming work, leisure, community and everyday life*. New York: Basic Books.
Florida, R. L. (2008). *Who's your city? How the creative economy is making where to live the most important decision of your life*. New York: Basic Books.
Florida, R. L. (2010). *The great reset: How new ways of living and working drive post-crash prosperity*. New York: HarperCollins.
Florida, R. L. (2014). *Startup City: The urban shift in venture capital and high technology*. Retrieved from http://martinprosperity.org/media/Startup-City.pdf
Glaeser, E. L., Kallal, H. D., Scheinkman, J. A., & Shleifer, A. (1992). Growth in cities. *Journal of Political Economy, 100*, 1126–1152. doi:https://doi.org/10.1086/261856
Hoeger, K., & Christiaanse, K. (Eds.). (2007). *Campus and the city: Urban design for the knowledge society*. Zurich: gta.
Jacobs, J. (1961). *The death and life of great American cities*. New York: Random House.
Katz, B., & Wagner, J. (2014). *The rise of innovation districts: A new geography of innovation in America*. Retrieved from https://www.brookings.edu/wp-content/uploads/2016/07/InnovationDistricts1.pdf
Lange, A. (2012, October 15). Silicon Island. *The New Yorker*. Retrieved October 9, 2017, from https://www.newyorker.com/culture/culture-desk/silicon-island
Laursen, K., Reichstein, T., & Salter, A. (2011). Exploring the effect of geographical proximity and university quality on university-industry collaboration in the United Kingdom. *Regional Studies, 45*, 507–523. doi:https://doi.org/10.1080/00343400903401618
OECD. (1996). *The knowledge-based economy*. Retrieved from https://www.oecd.org/sti/sci-tech/1913021.pdf
O'Mara, M. (1999). Strategic drivers of location decisions for information-age companies. *Journal of Real Estate Research, 17*, 365–386. Retrieved from https://www.jstor.org/stable/24886929?seq=1#page_scan_tab_contents
Porter, M. E. (2008). *On competition: Updated and expanded edition*. Boston: Harvard Business Review.
Simmie, J. (2005). Innovation and space: A critical review of the literature. *Regional Studies, 39*, 789–804. doi:https://doi.org/10.1080/0034340050021367l
TU Delft. (2016). Campus NL—Investeren in de toekomst [Campus NL—Investing in the future]. Delft: Technische Universiteit. Retrieved from http://www.vsnu.nl/files/documents/Campus%20NL%20digital%20version%20(for%20websites).pdf

van den Berg, L., Pol, P. M. J., van Winden, W., & Woets, P. (2005). *European cities in the knowledge economy: The cases of Amsterdam, Dortmund, Eindhoven, Helsinki, Manchester, Munich, Münster, Rotterdam and Zaragoza*. EURICUR European Institute for Comparative Urban Research. Aldershot: Ashgate.

van Oort, F. G., & Lambooy, J. G. (2014). Cities, knowledge, and innovation. In M. M. Fischer & P. Nijkamp (Eds.), *Handbook of regional science*: *Vol. 1* (pp. 475–488). Berlin: Springer.

van Winden, W. (2008). Urban governance in the knowledge-based economy: Challenges for different city types. *Innovation: Organization and Management, 10*, 197–210. doi:https://doi.org/10.5172/impp.453.10.2-3.197

van Winden, W., & Carvalho, L. (2016). Urbanize or perish? Assessing the urbanization of knowledge locations in Europe. *Journal of Urban Technology, 23*(1), 53–70. doi:https://doi.org/10.1080/10630732.2015.1090194

Wissema, J. G. (2009). *Towards the third generation university: Managing the university in transition*. Cheltenham: Edward Elgar.

Worthington, J. (2009, June). Univer-cities: Ivory tower or landscape for learning. Keynote address at the Conference on Campus of the Future, organized by HOI (Campus Managers Dutch Universities). Amsterdam.

Chapter 14
Coevolution of Town and Gown: The Heidelberg International Building Exhibition in Search of a Knowledge-based Urbanism for the Twenty-first Century

Carl Zillich

The evolution of cities has always been closely linked to political, social, and economic ideas of the time. Often, forces reflecting the resulting paradigms, such as the "car-friendly city" of the second half of the twentieth century, shaped the spatial configuration of urbanization. The Heidelberg International Building Exhibition (IBA) has set out to present solutions that mirror and foster what is called the knowledge-based society through processes and projects of city planning, urban design, and architecture by 2022.[1] The focus on knowledge and space is consistent with the coevolution of the city and its university and of its other research institutions thus far. For more comprehensive, integrated, and proactive solutions than every-day governance can deliver, the municipality chose the strategic process of an IBA, relying on a German tradition of experimental case studies in and with the built environment (Stadt Heidelberg, 2012).

[1]Whereas the evolution of cities has always been based on knowledge, the knowledge-based society has induced a more inclusive urban design on multiple scales. In acknowledgment of both the separation of functions in the modern city and the ever-increasing disciplinary fragmentation of knowledge, a new paradigm for the spatial production of our city is on our agenda, one that inspires exchange between, innovation in-between, and integration of formerly separated disciplines, milieus, and spaces. See http://www.iba.heidelberg.de.

C. Zillich (✉)
International Building Exhibition, IBA Heidelberg GmbH, Heidelberg, Germany
e-mail: c.zillich@iba.heidelberg.de

P. Meusburger et al. (eds.), *Geographies of the University*, Knowledge and Space 12,
https://doi.org/10.1007/978-3-319-75593-9_14

Fig. 14.1 Avant-garde housing by Ludwig Mies van der Rohe in the 1927 Weissenhof Siedlung in Stuttgart was used as an iconic background for the Mercedes-Benz limousine type 8/38 PS in 1931. The challenge of conceptualizing spatial opportunities for things to come seems more difficult a century later, with science and technology supersede human physical reality.
Source: Picture Archive Daimler AG. Copyright: Daimler AG. Reprinted with permission.

Germany's IBAs

100 Years of IBA Experience

From the outset the strategies of the IBAs have been adapted to the societal system in which the IBAs were based, but they have always been about implementing or building the next spatial practice over a period of years, often a decade. For example, the detailed history of IBAs in Germany by the architecture historian Werner Durth (2010b) has identified the Mathildenhöhe in Darmstadt as the first comprehensive approach to integrating design ambitions with a "life reform movement." In 1901 Grand Duke Ernst Ludwig (1868–1937) and his master architect, Joseph Maria Olbrich (1867–1908), achieved the integration of art, craft, and life with an art nouveau campus and buildings that provided spaces for exhibitions, studios, and housing.

Whereas this endeavor ran counter to the industrial revolution that changed society at large, subsequent IBAs embraced those forces. Under the urban design and artistic guidance of Ludwig Mies van der Rohe (1886–1969), avant-garde architects from all over Europe were invited to design the Weissenhof Siedlung in the outskirts of Stuttgart. There, in 1927, a testimony of "New Building" delivered new typologies in housing with building technologies and aesthetic coherence revolutionary in scope (Fig. 14.1). Berlin followed up by using the IBA strategy twice to deal with very different challenges under different circumstances. In the 1950s both East and West Berlin adopted large-scale urban renewal schemes to showcase their capacities to envision cities for different futures. East Berlin unveiled "palaces for the people" on Stalinallee in 1952, and West Berlin presented what was called the "Interbau" in the Hansaviertel, where planners and architects from across the world fulfilled a modernist dream of living in an urban landscape of dwellings with innovative layouts.

By 1987 the opposite paradigm was implemented when Berlin's old structures in Kreuzberg were either repaired, often with the engagement of the inhabitants, or rebuilt by young architects, mostly from abroad, who experimented with contemporary concepts of design. The social dimension of architecture and urbanism was further developed in a larger context in the deindustrialized context of the Ruhr District. In 1999 a regional network of strategic brownfield developments with housing and service industry was presented. It marked a turnaround for a declining agglomeration and transformed industrial ruins to monuments. This incomplete overview of the IBA ends with the city of Hamburg, which used a holistic approach far beyond spatial terms to guide the transformation of Wilhelmsburg, an island in the Mitte district at the mouth of the Elbe. Addressing education, infrastructure, energy supply, and other issues, the project worked on different scales with a balance of top-down and bottom-up tools for planning a city fit for the future.

State of the Art: The IBA Today

Although an IBA is not certified or licensed or guaranteed any funding, around 2010 a federal commission developed criteria for ensuring a level of discourse and quality (Durth, 2010a). An informal network of past, current, and future IBAs now aims to take standards of excellence derived from the past and strike a balance between them and necessary evolutions in processes and implementation that respond adequately to challenges of today. Projects should be about

1. linking society's evolution and spatial development;
2. addressing not just innovation in architecture but also new concepts of urban space;
3. deriving the agenda from local or regional necessities;
4. developing prototypical solutions to address spatial, economic, ecological, and social aspects;
5. linking excellence in built projects to adequate processes and procedures;
6. maintaining an international dimension from the outset through projects on-site and relevance abroad;
7. establishing exceptional conditions for the duration of an IBA by providing a cross-disciplinary laboratory and by pooling resources;
8. bringing all participants to agree that the IBA, as an experiment based in reality, requires all involved to take risks and be courageous; and
9. finding an appropriate structure for imagination and challenging established procedures.

These rather soft criteria show acceptance that architecture and the society it reflects have changed radically over the last century. Although developers and politicians desire certification in architectural production, it does not help spearhead innovation. The intention of the federal commission is therefore to give guidelines for the excellence of an IBA without requiring it to reinvent itself anew in every locality and context. An IBA today is about the urban realm and its underlying governance as much as about architecture and its aspects of function and representation.

The Question Concerning Heidelberg

Discursive and Specific

For the first time in the post-World-War II era, an attempt is underway to make an IBA proactive. With the whole city as its testing ground and no urgent problem crying out for urban renewal, the quest for an intensified knowledge-based urbanism has begun. The discourse on the service economy and its shift from dynamic capital and predefined labor to dynamic knowledge and self-organization has been heard for some time, but the resulting potentials or prerequisites for the production of space in terms of architecture and urban design have received little attention. Responding to new working environments that are evolving around the digital economy and networks and to the growing interest in built space as the third teacher or educator[2] (Hubber & Ramseger, 2017, p. 58; see also Malaguzzi & Cagliari, 2016), a discussion has begun in which Heidelberg offers the testing grounds to focus, connect, and offer new solutions on different scales. As a medium-sized city with excellent universities, Heidelberg represents a "knowledge pearl" (Van Winden, Van den Berg, & Pol, 2007, pp. 540–542) and is thus predestined for reevaluation of the relationship between a city and the production and distribution of knowledge.

As the site of Germany's oldest and most prestigious university as well as numerous international scientific research clusters and institutions (e.g., the European Molecular Biology Lab, four different Max-Planck-Institutes, the German Cancer Research Center, the Heidelberg Institute for Theoretical Studies and others), Heidelberg is embedded in a region of strong global players (e.g., BASF, Freudenberg, and SAP). The question is how the city of romanticism can also become a contemporary city of knowledge, not just substantively but also spatially. Such a transformation could help its population build on today's traditions and strengths to be fit for a future of ever-growing competition, especially with knowledge hubs like Berlin, London, Paris, New York, and Silicon Valley.

[2] Adults and children are the first two teachers.

Fig. 14.2 Large administrative and faculty buildings of Heidelberg University, such as the Old University (1735), the New University (1931), and the Neues Kollegiengebäude (1965), alter the urban fabric of the Old Town. Debate continues about the direction in which the next leap forward is to go (e.g., the urbanization of monofunctional campuses?).
Source and copyright: Carl Zillich.

Coevolution of Town and Gown

Heidelberg manifests a clear sequence in the coevolution of the city and its university (Matthiesen, 2015). Each sequence has a different typology of how university and city are intertwined, also reflecting the town-and-gown relationship from which its spatial configuration resulted. The first spatial refuge of the evolving sciences was in bourgeois mansions, which spread westward below the castle. In the stage thereafter, with its Old University (a building from 18th century) and then with increasingly specialized buildings within the boundaries of the Old Town (Fig. 14.2), the university became a driving force behind claims on new lands beyond the already urbanized territory. Technical imperatives and the drive for specialized buildings found expression in a succession of architectural eras, beginning with the hospitals in Bergheim and eventually the accommodation of the natural sciences across the Neckar river in Neuenheimer Feld. This sequence has continued into the present, with independent research facilities being located on the hillside and recent attempts being undertaken to establish a research agglomeration on brownfield developments south of the central railway station as part of the new *Bahnstadt*. Rearrangement across different campuses of Heidelberg University since 2009 has had social sciences replace clinical functions in Bergheim, which are being concentrated north of the river, with great potential for both neighborhoods.

For the IBA this coevolution of university and city exemplifies the claim in knowledge-based urbanism that each phase of development brings about a different identity, often through ambitious architecture. Yet a closer look reveals that the quality of public spaces and the possibilities for interdisciplinary and even nonacademic encounters have been gradually diminishing. Mixed-use developments, spaces for interaction and identification, integrated planning of open spaces, and integrated planning of traffic infrastructure are four layers for possible IBA engagement in this realm.

Agents of Change

Based in reality as a living lab fostering research by means of design, the Heidelberg IBA is a forum for initiating processes and inviting agents of change to join in the search for a next practice in governance of spatial production. In addition to the need to prepare for upcoming requirements in research and development and in education for an inclusive society, two particular interests have emerged over the first few years of the Heidelberg IBA. One is the infrastructure of everyday life. How do people connect the different knowledge hubs in and beyond Heidelberg, and what potential lies in public spaces, old and new, for an open and interactive society? The concepts range from express-bike paths to experimental design of the urban landscape. Other issues of today's city metabolism, or urban modes of exchange, range from biodiversity to urban agriculture and their potential for education, health, and leisure.

The Heidelberg Case and Its Strategies

Communication and Inspiration

Working on the invisible and built aspects of education, research, and development, the IBA is a platform, think tank, and start-up agency at once. Its work is therefore based on discourse and communication within different formats for different clienteles established in the first two years.

As part of a double strategy for gaining in-depth knowledge about Heidelberg and conveying the message by looking at each neighborhood anew, events known as IBA_LOCAL attracted 80 to 120 people on various Saturday afternoons. In eight dialogical walks inspired by what Burckhardt (2015) called "strollology," spatial experts foreign to Heidelberg talked with local stakeholders and IBA officials about what sites of knowledge exist in this city already, how they work, and what spaces are felt to be missing (see Figs. 14.3 & 14.4).

Fig. 14.3 IBA_LOCAL, No. 7—A dialogical stroll through the neighborhoods of Rohrbach (picture) and Kirchheim on November 7, 2014, to address the identified qualities and future potential for spaces of and for knowledge. Official partners of this particular public discourse were Ursula Baus (architecture critic, Stuttgart), Rolf Stroux (architect and member of the IBA's board of trustees), Matthis Bacht (artist, Heidelberg), and Hans-Jürgen Fuchs and Jörn Fuchs (chairs of the two community boards involved).
Source: Picture Archive IBA Heidelberg. Copyright: IBA Heidelberg, Valentina Meuren. Reprinted with permission.

The reverse strategy is implemented with the IBA_LABs, where examples of European best practice are presented to the public in Heidelberg and discussed with local users, stakeholders, and decision-makers as experts. In that forum issues such as innovative campus design, hybrids of working and housing, and potential in facilities for future research and development initiate knowledge transfers in many directions for administrators, educators, and researchers.

The IBA_ACADEMY takes a more distant, but also inspiring, look at Heidelberg. It enables students and faculty from international schools of architecture and planning (e.g., from Norway, Mexico, and Switzerland) to make Heidelberg their testing ground for ideas. Be it an entire neighborhood or certain institutions designed for innovation, a fresh view from the outside helps to increase receptiveness among different interest groups.

Last but not least, the IBA_SUMMIT aims for an international exchange among knowledge pearls. In this part of the IBA, mayors and university presidents (e.g., from Cambridge, England; Cambridge, Massachusetts in the United States; Leuven, Belgium; Lund, Sweden; and Stanford University in California) present their

Fig. 14.4 IBA_LAB, No. 3—An interdisciplinary and international debate between planners and scientists about a next practice for buildings for academia, research, and development on October 7, 2015. Pictured on the podium are (from left to right) architect Dietmar Eberle (ETH, Zürich); moderator Carl Zillich (IBA, Heidelberg); linguist and vice president of Teaching Beatrix Busse (Heidelberg University); and (standing) Andreas Schleicher, director of the PISA Report, the Programme for International Student Assessment (OECD, Paris).
Source: Picture Archive IBA Heidelberg. Copyright: IBA Heidelberg, Iman Mohammadi. Reprinted with permission.

experience of "town-and-gown" relationships as a challenging, but necessary, basis for cocreating a future vision of a built environment beneficial to both aspects.

Structure and Process

For all that these discourses can do to help build a foundation for a long-term project such as an IBA, the crucial elements are the innovative projects and the processes leading to them. In this respect the IBA faces the same challenge as its sisters of recent years did: that of securing funding for the platform only, not the construction to result from it. Nevertheless, Heidelberg has set up a structure for eliciting engagement at different public levels and by private businesses and foundations.

The IBA office is organized as a limited liability corporation of the city. The executive board has a majority consisting of members of Heidelberg's city council and three external delegates. The board of trustees is exceptional in its engagement. It includes acclaimed professionals with international reputations in architecture,

Fig. 14.5 The international and interdisciplinary board of trustees (without Iain M. Mattaj, Director General of European Molecular Biology Laboratory, Heidelberg) in 2015. From left to right: Bernd Müller (Public Real Estate, Germany), Ernst Hubeli (Urban Design, Switzerland), Angelika Fitz (Cultural Practice, Austria), Johan Deburchgrave (Economics, Belgium), Michael Braum (Director, IBA), Volker Staab (Architecture, Germany), Silke Steets (Sociology, Germany), Stephen Craig (Arts, Ireland), Undine Giseke (Chair, Landscape, Germany), Andreas Schleicher (Education Policy, France), Ulf Matthiesen (Anthropology, Germany), Carl Zillich (Curatorial Director, IBA), Siegfried Zedler (Teaching, Germany), and Annette Friedrich (Urban Planning, Germany). Source: Picture Archive IBA Heidelberg. Copyright: IBA Heidelberg, Christian Buck. Reprinted with permission.

planning, sociology, life sciences, business, and other fields (see Fig. 14.5). This interdisciplinary group advises the IBA office in its procedures and selection of projects. In the first year five criteria for projects worthy of the IBA were developed on the basis of the standards set forth by the federal commission around 2010:

1. Public Relevance: No self-referential building projects can be accepted.
2. Extra Competence: Processes and discourses should include the brightest minds available.
3. Prototype: The resulting project should include an experimental part.
4. Structural Impact: The project should contribute to its immediate surrounding and neighborhood.
5. Polyvalence: A degree of heterogeneity or mix of functions should be incorporated in order to foster the transfer of knowledge and to be open to future adaptations.

Candidates and Projects

The IBA office is now also initiating projects with selected partners, but the basis for ideas remains an open call for projects, which was first announced in October 2013. Different formats for qualification and strategies for funding and realization are developed and applied by academic institutions, established associations, and engaged individuals. So far 26 candidates have benefited from an international network of advisors and strategic partners who help them clarify their contribution to the desired aspects of innovation, which are individually defined as well. Whereas some concepts focus on new spatial arrangements of functions, others diversify their program to expand the number of people or disciplines they can reach. Not all ideas will succeed on this path to a building project, for it is full of obstacles, especially for newcomers to the field of real-estate development. As of spring 2017, 16 candidates have met the IBA criteria (Fig. 14.6).

In 2015 four concepts were recommended by the board of trustees and designated by the board of administrators as official projects of the Heidelberg IBA. After treatment in workshops, elaboration of their concept, and, in some cases, creation of a reliable organizational structure, these former candidates were selected as being especially promising. One is an independent educational facility focusing on craftmanship as a neglected aspect of knowledge. Another project is about building a self-governed and collaborative student residence hall that offers academic guidance internally and externally through preparatory training and formats open to the general public. The municipality also has an IBA project through which a new building will house a kindergarten, an elementary school, and a community center and will foster the exchange between different generations and milieus. The fourth project is the reinvention of a museum for outsider art (*art brut*, or art by people with psychiatric experience), which offers the university hospital's department of psychiatry a unique interface with the public and makes an exhibition space and a discursive laboratory available to art researchers. Additional projects are under consideration for eligibility as IBA projects (Fig. 14.7), and recruitment of academic institutions and real-estate development is in progress to meet the challenge of a diversified approach to a knowledge-based urbanism. The first projects are to be completed in time for the interim exhibition in 2018, so most of the projects must be well advanced by then and completed by 2022.

Challenges for Heidelberg and IBAs in the Twenty-first Century

Not every idea for a building project that is to be completed by 2022 has been assigned a location. The search for these sites ties into the ambitions to have the project concepts qualify for the Heidelberg IBA and secure funding for them. Parallel processes have resulted with each project, challenging established modes

Fig. 14.6 Map of Heidelberg showing the anticipated location of some projects as of 2016 (orange: ideas; green: real projects under development with IBA support). The aim is to achieve spatial dispersal across Heidelberg and thereby bring catalysts of the knowledge-based society to each neighborhood and to different milieus.
Source and Copyright: IBA Heidelberg. Reprinted with permission.

Fig. 14.7 A design concept for an as-yet unfunded IBA candidate project, "Center for Biodiversity," on the Neckar river alongside the campus Im Neuenheimer Feld (Filke, 2015). Illustrating the generation of hubs for transdisciplinary discourses in Heidelberg, the programming and spatial solutions offered by this design benefit scientist and nonscientist alike.
Source: Picture Archive IBA Heidelberg. Copyright: Thomas Harry Filke, TU Braunschweig. Reprinted with permission.

of cooperation between many stakeholders. Innovation in governance is therefore a crucial part of the IBA in Heidelberg. In the reality-based laboratory of the IBA, municipality and state and all other established stakeholders need to establish open models of cooperation.

In May 2016 the City of Heidelberg commissioned an experimental development process based on this agenda. The IBA conceived and directed four scenarios and one vision for development of the Patrick Henry Village (PHV)—the 100-hectare site of the former U.S. military housing compound—for up to 15,000 people to live and work. For one year the IBA office conducted a "planning-phase zero." Based on conceptual workshops (at SAP AppHaus, called "design thinking") with local and external advisors and on a subsequent collaborative planning process headed by architect Kees Christiaanse (Fig. 14.8), proposals for a future "knowledge city" now combine everyday life with facilities for research and development, innovative digital solutions, sustainable mobility, and experimental resource management.

Time will tell whether it is possible to improve the link between integrative strategic program development and the spatial practice of urban development by bringing together diverse actors from the corporate, scientific, and administrative worlds and society at large. Through ambitious planning processes and architecture,

Fig. 14.8 IBA Heidelberg and Kees Christiaanse Architects & Planners were responsible for the collaborative process involving many experts to shape the vision of a new Patrick Henry Village. Among them (front row): Kees Christiaanse (Rotterdam and Zürich, focus on overall urban design), Herbert Dreiseitl (Überlingen and Singapore, focus on urban metabolism), Carlo Ratti (Turin, and Cambridge, Massachusetts, focus on digitalization and mobility), and Winy Maas (Rotterdam and Delft, focus on the nexus of science and the corporate world). The group is gathered around a working model built to discuss the vision with the general public, stakeholders, and potential investors.
Source: Picture Archive IBA Heidelberg. Copyright: IBA Heidelberg, Christian Buck. Reprinted with permission.

the IBA serves as an "intermediary agent" (Selle, 2017, pp. 117–118) acting beyond existing paradigms, habits, and different point of views. Questions of densities and typologies also point out the need to engage with logics of real-estate development, both in the private and public sector, where a wide variety of regulations are still based on the modernistic paradigm of the separation of functions. Together with the need for cocreation in an existing city and society, the task has expanded to include the updating of procedures as well as of projects.

Ultimately, particular networks and windows of opportunity will determine the success of the Heidelberg IBA. Although many arrangements can be planned and executed, the lack of money for the necessary experiments calls for resilient innovation. Will there be public–private partnerships to benefit both ends in terms of content, not just profit? Is the middle ground between specialization and the commons a fruitful terrain for next practices in and for a new knowledge-based society? What the image and shape of Heidelberg will be in all its variety is an open question.

With a process stretching over more than 10 years, questions, not solutions, are on the table. This spirit of uncertainty alone is worth the endeavor. It is based on reflexive practice and destined to show how a knowledge pearl can change to meet the demands of, and deliver the integration for, a twenty-first century in which people want to work and live.

References

Burckhardt, L. (2015). Strollology: A minor subject. In conversation with Hans Ulrich Obrist. In M. von Ritter & M. Schmitz (Eds.), *Why is landscape beautiful? The science of strollology* (pp. 7–16). Berlin: Birkhäuser, de Gruyter.

Durth, W. (2010a). Ein Memorandum zur Zukunft Internationaler Bauausstellungen: Zehn Empfehlungen zur Durchführung einer Internationalen Bauausstellung [A declaration on the future of international building exhibitions: Ten recommendations for putting on an international building exhibition]. In Internationale Bauausstellung IBA Hamburg (Ed.), *Netzwerk IBA meets IBA: Zur Zukunft Internationaler Bauausstellungen* (pp. 64–73). Berlin: Jovis.

Durth, W. (2010b). Eine Zeitreise durch die Baukultur: Zur Geschichte der Internationalen Bauausstellungen [A time travel through the Baukultur: On the history of international building exhibitions]. In Internationale Bauausstellung IBA Hamburg (Ed.), *Netzwerk IBA meets IBA: Zur Zukunft Internationaler Bauausstellungen* (pp. 16–31). Berlin: Jovis.

Filke, T. H. (2015). *Das Bio-Center for Biodiversity* Heidelberg [The Biocenter for Diversity, Heidelberg] (Unpublished master's thesis). Institute of Architectural Design (IAD), Technical University Braunschweig.

Hubber, P., & Ramseger, J. (2017). Physical learning environments for science education: An ethnographic field study of primary classrooms in Australia, Germany and Taiwan. In M. Hackling, J. Ramseger, & H.-L. S. Chen (Eds.), *Quality teaching in primary science education: Cross-cultural perspectives* (pp. 51–77). Cham: Springer.

Malaguzzi, L., & Cagliari, P. (2016). *Loris Malaguzzi and the schools of Reggio Emilia: A selection of his writings and speeches, 1945–1993.* London: Routledge.

Matthiesen, U. (2015). Wissensgesellschaft und soziale Wissensstadt: Skizzen zu Kerndynamiken und neuen Arbeiten/Wohnen-Kopplungen in der Wissensstadt [The knowledge-based society and the social city of knowledge: Sketches of underlying dynamics and new links of working and living in the city of knowledge]. In Wüstenrot Stiftung (Ed.), *ZukunftsWerkstattWohnbauen 2014: Polyvalente Konversion—Wohnen und Arbeiten in der Wissensstadt Heidelberg* (pp. 12–29). Ludwigsburg: Wüstenrot Stiftung.

Selle, K. (2017). *Making of . . . Die ersten Jahre einer Internationalen Bauausstellung* [The Making of . . . The initial years of an international building exhibition]. In IBA Heidelberg (Ed.), Die Wissensstadt von morgen: Reflexionen (pp. 109–123). Zurich: Park Books.

Stadt Heidelberg. (2012). *Wissen schafft Stadt—Stadt schafft Wissen: Memorandum IBA Heidelberg. Perspektiven der Europäischen Stadt in der Wissensgesellschaft* [Knowledge-based urbanism and urban-based knowledge: A memorandum by the Heidelberg IBA. Perspectives for the European city in the knowledge-based society]. Retrieved from http://www.heidelberg.de/site/Heidelberg_ROOT/get/documents_E1837061966/heidelberg/Objektdatenbank/61/PDF/61_pdf_IBA_Wissenschafftstadt_Memorandum.pdf

Van Winden, W., Van den Berg, L., & Pol, P. M. J. (2007). European cities in the knowledge economy: Towards a typology. *Urban Studies*, 44, 525–549. doi: https://doi.org/10.1080/00420980601131886

Part IV
The University and the Region

Chapter 15
The Economic Impact of the Universities in the State of Baden-Württemberg

Johannes Glückler, Robert Panitz, and Christian Wuttke

A geographical understanding of the role of universities in society includes an appraisal of the impact they have on their regional economies. Over the past twenty years, academic and public interest in the economic effects of higher education institutions has risen dramatically. Two reasons for this change are the increasing globalization of and the mounting economic competition for knowledge and innovation. Because economic development depends ever more on creativity and innovation, academia and the public have sought to improve the understanding of the role that universities play in economic growth. This search is manifested in conceptual approaches such as national and regional innovation systems (Cooke, Heidenreich, & Braczyk, 2004; Lundvall, 1992), varieties of capitalism (Hall & Soskice, 2001), and the triple-helix model (Etzkowitz, 2008; see chapter by Etzkowitz in this volume). Further reasons for the elevated interest in the impact of universities are recent retrenchment policies in the aftermath of the financial and economic crises in North America and Europe since 2008. Budget constraints have intensified the competition for public funding, calling for more detailed assessments of the benefits of higher education as compared to alternative uses of funding (e.g., Hamm & Wenke, 2002). The competition is not only for direct funding but also for land use in urban planning, with universities pressing the argument that their positive economic impact on the city trumps alternative uses.

It is utterly impossible to quantify all the diverse economic impacts that universities have on the economy, especially over long periods and from geographical perspectives. A research activity, for example, may promote the production of new knowledge as well as the qualification of students and, many years later, may generate spin-offs or royalties from patent-related contracts. The sheer breadth of the university's activities and contributions to society and the economy is

J. Glückler (✉) · R. Panitz · C. Wuttke
Department of Geography, Heidelberg University, Heidelberg, Germany
e-mail: glueckler@uni-heidelberg.de; panitz@uni-heidelberg.de; wuttke@uni-heidelberg.de

P. Meusburger et al. (eds.), *Geographies of the University*, Knowledge and Space 12,
https://doi.org/10.1007/978-3-319-75593-9_15

extraordinary (Bathelt & Schamp, 2002; Drucker & Goldstein, 2007; Goldstein, Maier, & Luger, 1995; Goldstein & Renault, 2004; Salter & Martin, 2001). Universities generate new knowledge and qualify experts who, being mobile, can foster regional innovative strength (Breschi & Lissoni, 2009). Universities are engaged in knowledge transfer partly through contract research on the regional and interregional economy (Ponds, van Oort, & Frenken, 2010), and they are sources of technical innovations that spur the economy through patents or spin-offs (Vincett, 2010). Universities also often constitute the core of regional knowledge infrastructures by facilitating knowledge exchanges permeable enough for private business and the public sector (Owen-Smith & Powell, 2004). Moreover, universities are often committed to their regional community: They actively participate in civil society in general through political processes of decision-making and policy formation (Glückler & Ries, 2012; Glückler & Suarsana, 2014; Goddard, Hazelkorn, Kempton, & Vallance, 2016; Goddard & Vallance, 2013; see also chapter by Goddard in this volume) and influence the regional milieu through intellectual, cultural, artistic, and moral stimulation that dynamizes social and economic life. All these dimensions of qualification, innovation, social responsibility, and empowerment are so diverse, often overlapping, and fundamentally long term in their effects on a society and its economy that they are indisputably still too complex to be expressed monetarily.

In addition to these qualitative long-term impacts, universities have quantitative short-term monetary effects. What the university and its members spend on regional goods and services increases autonomous demand and leverages production and employment within a region and beyond. Although this quantifiable periodic impact of universities is just one of their many effects, we make it the sole focus of this chapter in the context of Baden-Württemberg, a large federal state of Germany. It is one of sixteen states in the country's relatively decentralized politicoadministrative system, which grants each state considerable autonomy in matters of educational policy and budgeting. Studies on the impact attributable to organizations of higher education have a long tradition (Eisen, 1948) and have taken place on various geographical scales. In Germany, however, such studies have been conducted mostly at the geographic level of local catchment areas of university regions. In Baden-Württemberg these assessments are documented for the universities of Freiburg (Drude, 1995), Heidelberg (Glückler & König, 2012; Gormsen, 1981; Grabitz, 1990), Constance (Fürst, 1979; Oser & Schroeder, 1995), and Stuttgart (Becker, Heinemann-Knoch, & Weeber, 1976), and for the Karlsruhe Institute of Technology (Kowalski & Schaffer, 2012). With rare exceptions, such as Berlin (DIW econ, 2013) and Rhineland-Palatinate (Spehl et al., 2005), the impact assessments focus on local catchment areas rather than an entire university landscape in large economic regions.

If Baden-Württemberg were a sovereign state, it would rank as the world's 18th largest national economy, immediately before Belgium.[1] It is home to 80 universities

[1]This calculation is our own for 2012 and is based on OECD and Macroeconomic Accounts of the Federal States (Volkswirtschafliche Gesamtrechnung der Länder, VGRdL).

and 55 nonuniversity research centers, with 49 of the latter facilities being located next to one of the universities. Of these 80 universities, nine are public higher education institutions whose interests are collectively represented in the Baden-Württemberg State Rectors' Conference, which has called for an independent analysis of the economic impacts of its nine comprehensive universities.[2] The significance of the state universities in Baden-Württemberg's university landscape is remarkable. Although they account for only 11% of the state's institutions of higher learning (9 out of 80), they train more than half (160,000 students in 2012) of the state's future generation of academics (Statistisches Landesamt Baden-Württemberg, 2012). These nine universities raised over 90% of Baden-Württemberg's third-party funding (or *Drittmittel*, hereafter called external funding) for research and teaching (Tanzmann, 2015), another sign of the disproportionate significance of the state universities.

In this chapter we first outline our research strategy for surpassing the validity and precision of conventional impact analyses of regional effects. Second, we describe our original data collection, methodology, and several key economic indicators with which to measure the monetary impact of the university landscape in Baden-Württemberg. The validity of the method rests on detailed, regionalized primary data on the expenditures of all the universities, enabling us to provide improved estimates of the regional multiplier effects by simultaneously considering rises in production and employment. Third, we present a differential incidence analysis to assess the "impact surplus" of state universities relative to alternative uses of the same public funding. Overall, we argue that the university landscape's monetary economic impact on the regional economy of Baden-Württemberg is hardly attainable by alternative uses and that its true compound social and economic benefit is still vastly underestimated.

Research Strategy and Methods

The pecuniary quantitative assessment of the economic impact of university spending is subject to three quality risks that our investigation overcomes with especially valid primary data and several methodological adjustments. The first challenge lies in correctly determining the payment flows of university spending, especially their regionalization, for the share of regional expenditures will be decisive for computing overall regional impact. This study is based on a unique database that sharply distinguishes between university spending and other expenditures at the state level.

[2]This chapter elaborates on an impact assessment of Baden-Württemberg's nine state universities (Glückler, Panitz, & Wuttke, 2015) and a study of Heidelberg University's impact on its local catchment area (Glückler & König, 2012). The analysis encompasses the universities of Freiburg, Heidelberg, Hohenheim, Constance, Mannheim, Stuttgart, Tübingen, Ulm, and the Karlsruhe Institute of Technology (including the medical schools of the universities of Freiburg, Heidelberg, Tübingen, and Ulm, but not their hospitals).

The second challenge lies in assessing the multiplier effects of direct regional demand, inasmuch as additional demand for goods in the region leads to simultaneous increases in production and employment in the other upstream and downstream sectors of the economy. Theoretically, assessment of this multiplier effect requires detailed knowledge of the intersectorial division of labor in the regional economic structure and of the regional population's income-dependent consumption profiles. Little of this information is available at the regional level, however, so analysts must estimate these multipliers on the basis of assumptions. In conventional impact analyses the regional economic effect is ascertained either by means of regional supplier interdependencies (Giarratani, 1976) or through an estimate of the regional increases in income (Bathelt & Schamp, 2002; Glückler & König, 2012). Either approach alone underestimates overall regional impact. The procedure in this chapter combines the two multipliers, improving the quality of the results (Pischner & Stäglin, 1976).

The third challenge lies in the erroneous assumption that the achieved regional effects would be absent if their source did not exist (Blume & Fromm, 1999). Although this assumption seems defensible for small areas, it is not tenable in large regions (Stoetzer & Krähmer, 2007) such as Baden-Württemberg. Particularly in large regions one must assume that funds saved in one place can (and generally will) be spent elsewhere within the limits set by the public budget. A few impact analyses pertaining to a federal state (DIW econ, 2013; Spehl et al., 2005) have ascertained only the absolute incidence. The study presented in this chapter is special in that it characterizes the differential impact of an entire university landscape in relation to alternative uses.

Computation of the Direct Economic Effect

The overall effect of the regional economic impact analysis is the sum of direct, indirect, and induced effects. The analysis starts with the gross spending of the universities and their members. This demand comprises three expenditure flows (Blume & Fromm, 1999): (a) university expenditures for investments and for goods and services, (b) university expenditures for the wages and salaries of their employees, and (c) student living expenses. Proceeding from this gross spending, analysts must first determine its direct effect, that is, what part of it translates into demand in the region. To do so, they must calculate two things: the gross spending's impact on consumption, and that consumption's impact within the region.

First, there is the impact on demand—the direct consumption-related demand as expressed by the university's total spending on personnel, investments, and materials and by student expenditures after deduction of taxes, social security premiums, and other levies. This computation is easy for investments and spending on goods because those amounts have direct impact as payments to suppliers. It is harder to determine the impacts that wages, salaries, and student budgets have on consumption. As in previous studies, wage and salary taxes and social security premiums are deducted from wages and salaries[3] because the taxes and premiums are not directly available for consumption. University wages and salaries are reduced by an additional amount presumed to be saved. One arrives at that sum by multiplying wages and salaries by the average savings ratio of the population in Baden-Württemberg. Because almost all students have a low income, their consumption ratio is assumed to be 100%, minus administrative contributions, which are transferred in full to the state administration.

Second, regionalization determines the share of expenditures that are made within the region under study and that have a direct effect on demand. For lack of detailed data, researchers normally make numerous assumptions to estimate the regional distribution of expenditures (Friedrich & Rahmig, 2013; Kowalski & Schaffer, 2012). However, it is the geographic differentiation of expenditure flows that influences the overall regional effect most. Downstream specifications of the impact model depend significantly on the quality of these primary expenditure flows. In late January 2013 we joined with controlling experts of the nine state universities of Baden-Württemberg to standardize the requirements for the relevant data. Hence, all expenditures accounted for in the present study are itemized according to their type of origin and the geographic distribution of the payment flows (Table 15.1). This highly selective procedure has enabled us to track the flows of university spending in unprecedented geographical detail.

Nevertheless, analysts must still make numerous assumptions when computing effects (Table 15.2). To take account of the mobility of university personnel, we assume, as a correction based on Blume & Fromm (1999), that 80% of the short-term outlays by university personnel residing in Baden-Württemberg occur at their place of residence and 10% at their place of work. These figures sum to a regional quota of 90%. Conversely, we assume that 10% of the short-term outlays by university personnel residing outside Baden-Württemberg occur at their place of work. Student spending within and outside Baden-Württemberg is regionalized analogously. Students residing in Baden-Württemberg thus have a regional quota of 90%; those outside, 10% (Blume & Fromm, 1999). The direct effect of university demand is defined as the funding that is computed above as having an impact on demand and as being spent in the region.

Computation of the Indirect and Induced Economic Effects

Direct spending constitutes an autonomous rise in regional demand, which triggers a corresponding rise in production and associated supplies in related sectors of the economy. This rise in production is called the indirect effect of university demand and is ascertained during the analysis of production multipliers (Leontief, 1936). The

[3]In macroeconomic accounting the civil servant allowance is declared as a private household expenditure, so it is disregarded in the calculation.

Table 15.1 Funding of the core universities (without their medical schools), by origin and use (in euros) in the federal state of Baden-Württemberg (BW)

		Basic funding		External funding (*Drittmittel*)		
Gross spending (including VAT)	Origin	University budget[a]	Additional state funding[b]	From BW	From elsewhere	Unknown
Goods and investments	In BW[c]	254,250,743	42,761,298	15,433,086	102,051,219	18,840,726
	Unknown[d]	8,860,949	27,795,166	3,055,067	22,883,612	1,326,544
	Total	369,626,362	103,705,117	35,447,063	235,439,801	9,824,777
Personnel (civil servants such as professors]	In BW	269,759,715	6,308,788	245,121	4,085,617	1,427,669
	Unknown	70,209	11,732	15,997	45,914	0
	Total	307,986,282	7,884,146	279,535	4,547,629	1,569,857
Personnel (employees)	In BW	540,402,225	71,882,158	25,627,738	320,778,087	43,712,054
	Unknown	266,917	141,943	37,018	1,023,309	197,664
	Total	597,178,804	81,525,826	29,673,708	361,961,145	48,855,292
Contracted Students	Total	29,860,603	14,667,178	1,813,971	22,344,918	3,598,241

Source: Design by authors.

[a]Basic financing assured by the federal state of BW. [b]Exceptional grants from the federal state (e.g., due to higher numbers of students than expected). [c]The abbreviation for Baden-Württemberg, as agreed at the EU level pursuant to ISO 3166-2. [d]Exact invoices and residential addresses were not available at the university level for all suppliers and individuals employed. Hence, unknown expenditure flows are assigned proportionately to expenditures in and outside BW.

Table 15.2 Assumptions and bases of calculating the regionalization of university spending flows in Baden-Württemberg

Variable	Assumed value	Source
	Employer contributions	
Mandatory insurance	19.58%	Inquiry at *Deutsche Rentenversicherung* [German pension insurance authority]
including health insurance	7.3%	
including nursing care insurance	0.975%	
including pension insurance	9.8%	
including unemployment insurance	1.5%	
	Employee contributions	
Mandatory insurance	20.48%	Inquiry at *Deutsche Rentenversicherung* [German pension insurance authority]
including health insurance	8.2%	
including nursing care insurance	0.975%	
including pension insurance	9.8%	
including unemployment insurance	1.5%	
	Regionalization of expenditures	
Students	10% at place of studies	(Blume & Fromm, 1999)
80% at place of residence		
University personnel (including civil servants such as professors)	10% at place of work	(Blume & Fromm, 1999)
80% at place of residence		
Employees in Baden-Württemberg	90% in Baden-Württemberg	
Median income tax rate	20.7%	(Vöhringer, 2012)
	VAT rate	
Expenditures: goods and investments	12.0% (universities without medical schools) 12.4% (with medical schools)	Authors' computation based on information from the state universities
Personnel	15.6%	Authors' computation based on typical shopping carts and mean VAT rates per sector (Destatis, 2013d; Statistisches Landesamt Baden-Württemberg, 2013b)
Students	15.9%	
Average for Baden-Württemberg	15.5%	
	Consumption ratio	
University personnel	79.6%	Authors' computation based on typical shopping cart (Destatis, 2010)
Students	100%	
Inhabitants of Baden-Württemberg	75.7%	

Source: Design by authors.

basic idea is that demand for particular goods within a region predicates prerequisite labor output inside or outside a particular region. Demand for prerequisite work generates further prerequisite work in additional upstream sectors of the economy, and so on. Eventually, this effect approaches zero asymptotically. The interdependencies of performance between the individual sectors of the economy are modeled in input-output calculations as part of macroeconomic accounting. Because input-output tables for Baden-Württemberg were regularly compiled by the State Statistical Office only until 1993 (Vogt, 2011), we had to compile our own input-output matrix for Baden-Württemberg. It is modeled on the national matrix of performance interdependencies and scaled to the regional, sector-specific circumstances, as in the Flegg location quotient (*FLQ*) procedure (Flegg, Webber, & Elliott, 1995; Lindberg, 2010).[4]

This indirectly triggered rise in production also causes a rise in employment in the upstream sectors, which leads to additional income. These additional earnings by employees further increase demand for goods and operate as an induced effect on the overall regional effect of demand (Pischner & Stäglin, 1976). The Keynesian income multiplier describes this induced rise of demand for goods by means of the underlying circulation of income, that is, by the spending of additional income, and depends decisively on the willingness of the households to consume. However, neither the production nor the income multiplier takes account of the effects of the other multiplier. We therefore use an integrative approach, bringing a Keynesian element into the regional input-output model by computing a combined multiplier (Kowalski & Schaffer, 2012; Pischner & Stäglin, 1976). Lastly, the direct effect of demand at production prices[5] was multiplied by the foregoing combined multiplier of indirect and induced effects to determine the overall effect of regional demand at production prices. We then added in the value-added tax (VAT) for the direct and induced effects to obtain the overall effect of demand at market prices (Fig. 15.1).

The Regional Economic Impact of Baden-Württemberg's State Universities

Impact on Demand

The impact analysis starts with the gross expenditures of the universities in 2012, which amounted to €3.049 billion for investments, goods, and personnel and €1.568 billion for consumption by students. These two figures sum to a total gross demand of €4.6 billion, which constitutes the point of departure for computing the direct,

[4]This procedure is described in the technical appendix.

[5]Assumed shopping carts of university personnel and students and the expenditures of the education sector served as the basis for ascertaining each share of value-added tax, which was then deducted from demand at market prices.

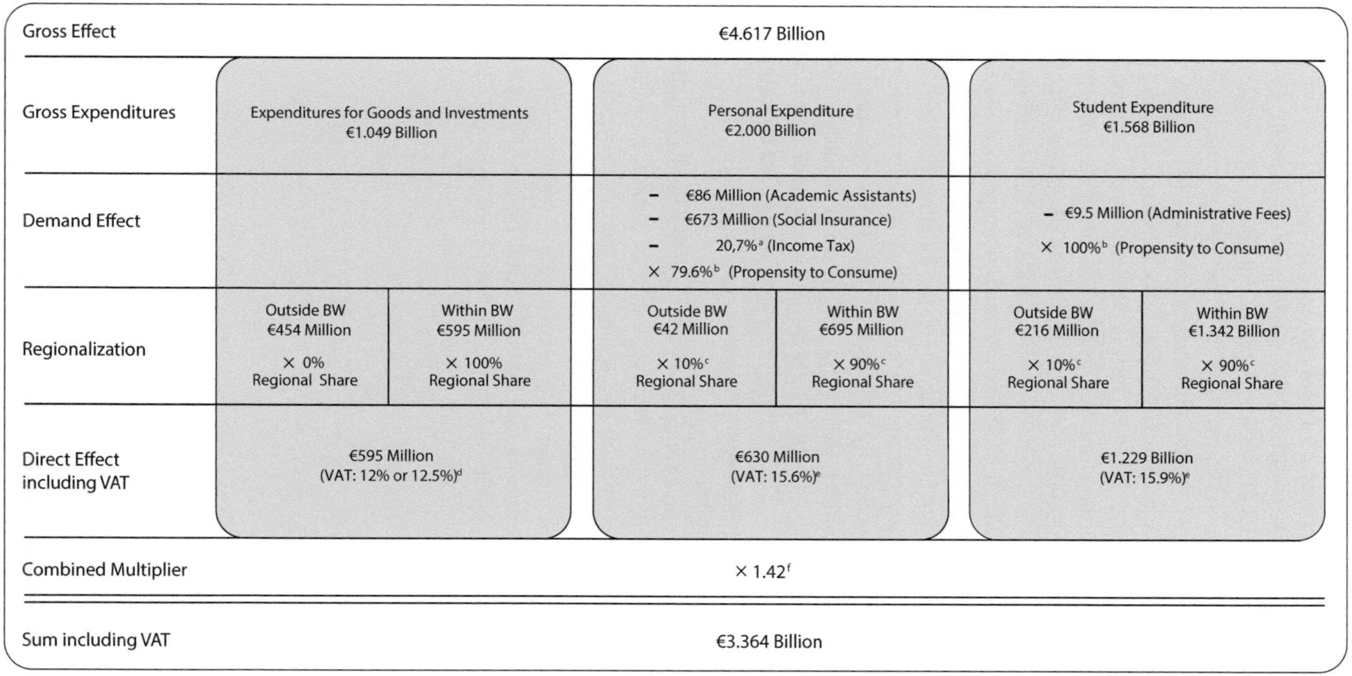

[a] Vöhringer (2012).
[b] Destatis (2010) EVS 2008 income class (*Haushaltsklasse*) €2,600–3,600; calculation: private consumption ÷ disposable income = €2,486 ÷ €3,121 = 79.6%; income class < €1,300; private consumption > disposable income leading to a marginal propensity to consume of 100%.
[c] Blume and Fromm (1999).
[d] We calculated different rates of VAT for the medical faculties (12.5%) and the core universities excluding these faculties (12%).
[e] Calculation based on Statistisches Landesamt Baden-Württemberg(2013a, 2013b) and Destatis (2013d).
[f] The multiplier is multiplied by the direct effect excluding VAT. Adding the VAT of the direct and indirect effect to this product yields the total effect, including VAT.

Fig. 15.1 Determination of the overall effect of the demand of state universities in Baden-Württemberg (BW), 2012. Source: Design by authors.

indirect, and induced effects. The state universities, except for their medical schools, spent €754 million for goods and investments, of which €470 million had an impact on demand in Baden-Württemberg (Table 15.1). Gross expenditures for personnel came to €1.441 billion, of which €451 million had an impact on demand in the region. The consumption by the 163,427 students enrolled at these universities in 2012 also increased demand for goods and services in Baden-Württemberg. With average annual receipts of €799.50 per month and student (HIS GmbH, 2010), the corresponding overall budget for students totaled €1.568 billion. After deducting the related administrative fees and taking account of the students' semester postal addresses, we found that €1.229 billion had an impact on demand directly in Baden-Württemberg. This sum corresponds to €2.150 billion as the direct effect of overall demand. In addition, the medical schools spent a total of €295 million on goods and investments in 2012, of which €125 million had an impact on demand in Baden-Württemberg. Of the €559 million in total personnel expenditures, €514 million went to persons living in Baden-Württemberg; the remaining €45 million, to persons living elsewhere. The expenditures of the medical schools accounted for €304 million of direct effect exerted by demand in Baden-Württemberg. In brief, the expenditures of the universities (including the medical faculties) came to €1.049 billion for goods and investments, €2 billion for personnel costs, and €1.568 billion in student expenditures. The resulting regional impact led to a direct effect of €595 million for goods and investment, €630 million for personnel spending, and €1.229 billion in student expenditures.

Lastly, the impact that the core universities and their medical schools had on demand in Baden-Württemberg totaled a direct effect of €2.454 billion. Because of the interdependencies of prerequisite work in the economic sectors to which this demand was directed, there were also indirect and induced effects, which were computed by means of a combined multiplier. We estimate their overall impact on demand at market prices to have been moderate, about €3.364 billion (Fig. 15.1). The overall impact of university expenditures on Baden-Württemberg's economy are also ascertainable for four other impact parameters: value creation, income, employment, and taxes (Fig. 15.2).

Impact on Gross Value Added

The impact on value added is the sum of the personnel expenditures of the state universities and the increase of gross value creation triggered by university demand in the remaining sectors. In macroeconomic accounting, gross value creation is the production value of an economic sector minus that proportion of prerequisite work which is drawn on by other economic sectors. Consequently, value creation refers only to the part of the value of goods that is added within the given sector of the economy (in the region). Because universities in Germany are governed under public law and thus not allowed to earn profits, the total personnel expenditures of €2 billion in 2012 constitute a direct impact on value creation and, hence, an expression

	Demand	Gross Value Added	Income	Employment	Taxes
Gross Effect	Gross expenses	Personal expenses of the universities	Gross income of employees and students	Number of university employees	Federal taxes paid by universities and its members
Direct Effect	Gross expenses in Baden-Württemberg	Personal expenses of the universities	Gross income of employees and students in Baden-Württemberg	Number of university employees from Baden-Württemberg	State share of paid federal taxes induced by direct demand and income effects
Indirect Effect	Increase in demand through intermediates in other sectors	Increase in value added through intermediates in other sectors	Increase in income through intermediates in other sectors	Increase in employment through intermediates in other sectors	State share of paid federal taxes induced by indirect demand and income effects
Induced Effect	Increase in demand through additional income in other sectors	Increase in value added through the demand induced by additional income in other sectors	Increase in income through the demand induced by additional income in other sectors	Increase in employment through the demand induced by additional income in other sectors	State share of paid federal taxes generated by induced demand and income effects
Total Effect	= Sum direct + indirect + induced effects				

Fig. 15.2 Types and composition of economic impacts on a region.
Source: Design by authors.

of the knowledge work performed in the universities (Spehl et al., 2005). Moreover, the expenditures for goods and investments in the state universities contributed €315 million of gross value creation to the economy of Baden-Württemberg; the consumption expenditures of their personnel and the students, another €886 million. The direct effect on demand thus led to indirect gross value creation of €1.201 billion. Additional income induced by direct demand increased gross value creation by €472 million, bringing the overall effect that the state universities had on gross value creation in Baden-Württemberg to €3.673 billion in 2012 (Table 15.3). This figure amounted to approximately 1% of gross value creation in Baden-Württemberg that year.

Impact on Income

The impact on income is the sum of the university personnel's direct gross income and the additional gross income from labor that the demand of the universities gives rise to in the other sectors of the economy (Rosner & Weimann, 2003). From the macroeconomic perspective, company profits due to effect on demand ought to be

Table 15.3 Regional economic impacts of Baden-Württemberg's state universities, 2012 (in billions of euros)

Type of effect	Demand	Value-Added	Income	Employment[a]	Taxes[b]
Gross	4.617	2.000	3.127	40,836[c]	849
Regional					
Direct	2.454	2.000	2.762	36,191	364 (156)
Indirect	630	1.201	506	19,558	399 (174)
Induced	280	472	52	7,564	46 (20)
Overall	3.364	3.673	3.320	63,313	809 (350)

Source: Design by authors.
[a]Refers to the number of jobs. [b]The parenthetic figures indicate the amount of tax revenue passed on to the state of Baden-Württemberg. [c]Corresponds to 32,918 full-time jobs.

included as well (Destatis, 2013b). However, it is difficult, if not impossible, to quantify these profits reliably, so the income effect is based solely on the university and regionally induced gross income from dependent labor. The overall effect comprises three factors. First, the direct incomes are the €1.412 billion of gross salaries paid by the state universities, and the incomes of their 140,667 students with semester addresses in Baden-Württemberg (€1.350 billion). Second, the demand of the universities and their members for goods and services in Baden-Württemberg create or secure further jobs, which generate €506 million of indirect incomes. Third, these indirect incomes likewise have impacts on demand. Because of the production interdependencies between the other sectors, they generate an additional €52 billion of induced incomes. The overall effect of direct, indirect, and induced incomes came to €3.320 billion in 2012 (Table 15.3).

Impact on Employment

An alternative to representing the impact of income monetarily is to use an equivalence computation to determine the impact on the employment stemming from university demand. The total number of employees at the state universities in 2012 was 40,836 (equivalent to 32,918 full-time employees). The number of these employees residing in Baden-Württemberg was 36,191 (equivalent to 27,965 full-time employees). The indirect impact on employment results from the increases in regional production value that are due to university expenditures and the consumption of university personnel and students. One can ascertain the number of indirectly created jobs by multiplying sectorial rise in demand with sector-specific labor coefficients. According to this calculation, the expenditures of the state universities seeded an additional 19,558 jobs in Baden-Württemberg, inducing extra demand and thereby creating an additional 7,564 jobs. Conservative computation implies that the overall effect of impact on employment was at least 63,313 jobs in Baden-Württemberg alone. This figure comes to approximately 1.2% of all persons who earn a living in Baden-Württemberg. The computation still does not include the many scientists, administrators, and technicians who are gainfully employed at the other universities and at research facilities outside the universities, most of whom have purposely settled near the universities. These jobs are especially valuable to Baden-Württemberg because they are relatively secure when crises hit and fairly insensitive to cyclical economic downturns.

Impact on Taxes

Baden-Württemberg receives tax revenues from university spending (VAT) and the incomes it generates (wages and income taxes). These two types of taxes account for 86% of Baden-Württemberg's tax revenues (Statistisches Landesamt Baden-Württemberg, 2013a). Altogether, total revenues of approximately €809 million within Baden-Württemberg in 2012 are attributable to the demand of the state universities and their members: approximately €401 million in VAT and €408 million in wage taxes. The other taxes are disregarded in our impact model because they are relatively unimportant. Since VAT and income taxes are federal taxes, they are shared between the federation, the states, and the municipalities through compensatory payments between the states and the federation pursuant to the vertical distribution of taxes (Bundesministerium der Finanzen, 2013). The tax revenues from Baden-Württemberg before these compensatory payments came to €177 million in VAT revenues and €173 million in wage tax revenues—a combined sum of €350 million in tax revenues from Baden-Württemberg. Adding in the €9 million of students' administrative fees that likewise accrue to the benefit of Baden-Württemberg brought Baden-Württemberg's total receipts to €359 million in 2012.

Expanding the Regional Impact Model to Include Statutory Health Insurance

Quite beyond the university expenditures, which are usually taken into account, there are further short-term effects that, in theory at least, would be measurable in monetary terms if it were not for the scarcity of precise empirical regional data. Their absence made it impossible for the regional impact model to include measurements of the indirect and induced effects of demand that reflect the company profits arising from increased production (exact operating surpluses) and the resulting additional impacts on income. Likewise, we could not take account of the returns on higher education, which have impacts on the income in a region along with subsequent impacts on the regional economy (Schubert et al., 2012).

Another unconsidered dimension in impact studies is the effect that mandatory insurance premiums have on demand for the system of statutory social insurance. Statutory social insurance in Germany consists of health insurance, nursing-care insurance, old-age pension insurance, and unemployment insurance. Put simply, these insurance schemes are cost-covering, pay-as-you-go systems in which the premiums received from the community of the insured are also paid out yearly as insurance benefits. With old-age, nursing-care, and unemployment insurance, however, the premiums are hardly ever paid out to the same employees who paid them in. Yet the benefits of statutory health insurance can certainly be modeled regionally because the premiums paid in the pay-as-you-go system have their impact on demand in the health-care sector during the period in which they are paid in.

We take only the university employees who have a residence in Baden-Württemberg and for whom social insurance is mandatory into account when modeling the regional economic impact of health-insurance premiums. Their insurance premiums, including the employer's contributions, amounted to €189 million in 2012. In keeping with the solidarity principle, these premiums flowed into the federal health fund and to transregional health insurers. Given the solidarity principle of equal premiums regulated by law and because wage levels differ from one region in Germany to the next, there are compensatory payments between the states. Baden-Württemberg, being a net payer, must transfer up to 5% of its collected premiums to other states (Wasem, Buchner, Lux, Manouguian, & Schillo, 2007), so it follows that only 95% of the health-insurance premiums paid in Baden-Württemberg are actually used by insured persons in Baden-Württemberg and thus have an impact on demand in the health-care sector (Destatis, 2011a). Because part of this demand also goes to providers outside Baden-Württemberg, there is a need for a regional quota. In 2011 roughly 4.6% of all hospital patients with a residence in Baden-Württemberg were treated in hospitals beyond its borders (Destatis, 2013c). For lack of available data, our model rests on the assumption that 95% of all health-care services rendered to persons employed by the universities also have their impact in Baden-Württemberg.

We therefore had to compute a direct effect of the €171 million impact on demand in Baden-Württemberg's health-care sector. This effect is to be attributed solely to the expenditures of the state universities for statutory health insurance. This direct effect triggers a multiplier impact on indirect and induced rises in production and income in other sectors, adding up to €248 million as an overall effect of demand. This sum corresponded to a gross value creation of €124 million, an impact on employment of 1,753 jobs, and gross income of €36 million. The result was €9 million in VAT revenues and €8 million in income tax revenues, from all of which Baden-Württemberg received €7 million from the federal government.

Significance of the State Universities' Impact on Baden-Württemberg's Economy

According to the absolute-incidence analysis, the nine state universities secured more than 63,000 jobs, €3.7 billion (or 1.2% of value creation), and annual tax revenues of some €350 million in Baden-Württemberg alone in 2012. But how are these impacts to be assessed? The following section offers two comparative perspectives for assessing them: the overall regional economic effect of the impact on value creation in relation to the state funds used, and a differential incidence analysis comparing alternative uses of these state resources.

Assessment of the Impact Relative to the Basic Funding Spent

Baden-Württemberg provided its nine state universities with €2.045 billion of basic funding for 2012. These universities also raised additional funds and attracted the consumption expenditures of their enrolled students, achieving a total value creation of €3.673 billion. This impact is 1.8 times the basic funding of the state universities. Taking into account Baden-Württemberg's parallel tax revenues, which result from the impact that its state universities have on demand and incomes, one finds that this state's actual net use of funds declines to €1.686 billion. The overall effect of the university impact on value creation in Baden-Württemberg is thereby 2.2 times this net use of funds.

Differential Incidence

The second possibility for assessing the degree of the impact that Baden-Württemberg's nine state universities have on that state's regional economy consists of comparing alternative uses, that is, of taking the current allocation of basic funding and comparing it to allocations of those resources to other public uses. The difference between the absolute incidences of their original use and alternative uses yields the differential incidence (Stoetzer & Krähmer, 2007) of the state universities on Baden-Württemberg's economy. So far, empirical assessments of the differential incidence are available only for impact analyses on a small scale (Assenmacher, Leßmann, & Wehrt, 2004; DIW econ, 2008). Because numerous conceivable options exist and the analyses of specific alternatives would entail great effort, the following procedure has proven helpful. It distinguishes between three fundamental alternative uses of basic funding (DIW econ, 2008): (a) for state universities, that is, funding them from the state budget, without additional effects of external funding and student expenditures; (b) for goods and investments only; and (c) for personnel costs only, without raising additional means.

If the state universities achieve a greater absolute impact than the foregoing alternatives would, it will indicate that public funding of education and research has a surplus of short-term impact on the regional economy. This interpretation is based on the assumptions that alternative public uses of the basic funding (e.g., for administrative facilities) do not bring in any additional (external) funding and do not attract stakeholders (students) to move permanently to Baden-Württemberg. The economic impact of the state universities must therefore be compared with the multiplier effect of an alternative use by which only the state funds with regional impact on demand enter the regional circulation of the economy and induce additional demand through interdependencies of intersectorial prerequisite work. Scenarios for Baden-Württemberg's economy that result from an increase or decrease of state funding for the state universities can then be worked out and simultaneously weighed against the potential effects of reallocating state funding to other areas (Fig. 15.3).

In 2012 the use of one euro of state funding corresponded to a direct value-creation impact of €1.80 in Baden-Württemberg. Cutting funds would forfeit value creation and jobs that alternative uses of public funding could not completely recoup. The quantitative difference between the impact of the state universities and that of alternative uses of the funding depends on how those resources are used. Sole reliance on basic funding of the state universities would result in a value-creation effect of €1.937 billion (Fig. 15.3, scenario 1). Using state funding exclusively for personnel (Fig. 15.3, scenario 2) would imply greater regional impact on value creation than using it only to purchase goods (Fig. 15.3, scenario 3) but would still fall short of the impact that the state universities have. The scenario shows a net difference of at least €1.173 billion in value creation even with these reallocations of the previous funding, which did not result in any savings for the state itself.

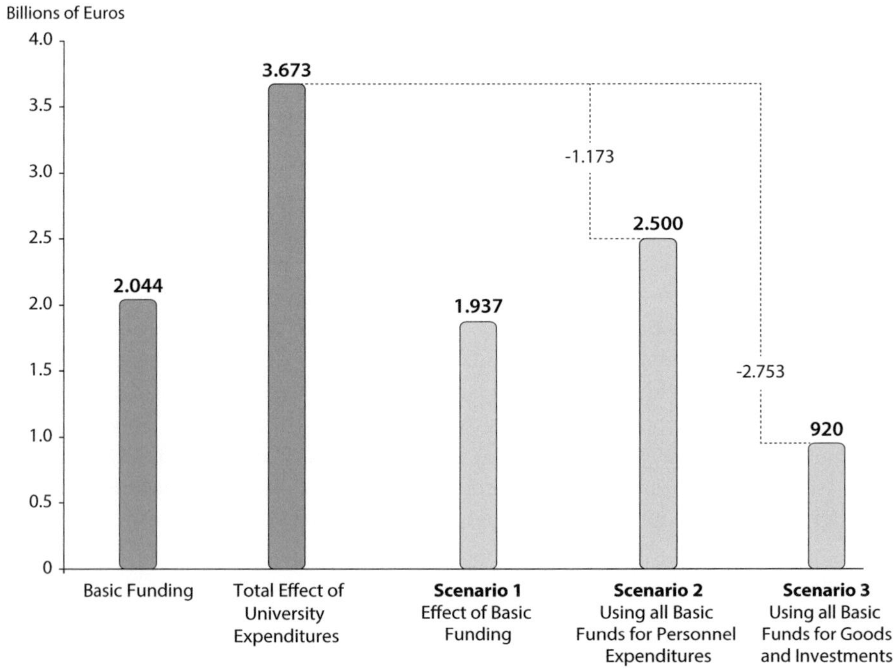

Fig. 15.3 Differential incidence of the impact that Baden-Württemberg's state universities have on gross value added.
Source: Design by authors.

Three Factors Leveraging the Relative Economic Impact Surplus of Universities

The degree of impact that university spending has on a regional economy depends on three basic parameters: attraction of additional income due to the migration of students, acquisition of external funding, and the region's specific production and consumption profiles. The differential incidence analysis has shown that these factors are precisely what give the state universities greater leverage to strengthen regional demand than other public facilities have. Table 15.4 shows that the universities with high shares of both spending and out-of-state students are especially the ones on Baden-Württemberg's periphery. This point underlines the extraregional attractiveness of the nine state universities, which are associated with a regional flow of capital out of Baden-Württemberg as well as a flow of capital into it. Unlike the police, fire brigades, and many other public services with a clear regional jurisdiction, the universities have an impact beyond regional borders.

An exceptional case in this respect is Heidelberg University. About 55% of its students come from outside Baden-Württemberg. In the winter term 2012–2013, approximately 17% hailed from other countries; an additional 38%, from other

Table 15.4 Share of expenditures, students, and external funding in Baden-Württemberg (BW)

State universities	Expenditures outside BW (share of all expenses), in €	Number of students from other states and countries (share)	External funding from outside BW (share of all funding), in €
	Universities in the center of BW		
University of Hohenheim	13,235,229 (11%)	2,117 (25%)	25,240,666 (20%)
University of Tübingen	19,993,965 (9%)	7,370 (28%)*	56,120,312 (23%)
University of Stuttgart	79,911,115 (18%)	6,910 (29%)	156,905,252 (35%)
	Universities close to BW borders		
University of Konstanz	39,190,240 (25%)	3,545 (32%)	46,205,997 (29%)
University of Freiburg	56,904,647 (18%)	8,998 (43%)	90,536,573 (29%)
Heidelberg University	80,023,267 (22%)	15,846 (55%)	91,069,964 (28%)
Karlsruhe Institute of Technology	77,744,543 (20%)	8,546 (37%)	120,469,916 (30%)
University of Ulm	36,714,467 (29%)	4,135 (44%)	22,808,000 (17%)
University of Mannheim	36,206,797 (36%)	6,268 (53%)	14,936,813 (14%)

Source: Design by authors.

*For lack of data, this figure is estimated as being similar to that for Stuttgart and Hohenheim.

federal states within Germany. Some 22% of the university's expenses went for the consumption of goods and services and for the salaries of employees outside the region.[6] Figure 15.4, which shows the geography of spending for staff, goods, and investments in 2008, indicates the dispersed character of expenses for material goods and investments, whereas employees are concentrated largely within the universities' core region.

Overall, the attractiveness of the universities for students is an initial, important factor leveraging the impact of the regional multiplier. It was in Baden-Württemberg that 59% of all the students at the universities of Baden-Württemberg earned their university entrance qualification. Hence, 41% of all the students in Baden-Württemberg moved from other states of Germany or another country to Baden-Württemberg in order to study there. The difference between the students whose semester address was in Baden-Württemberg and those who earned their university entrance certificate in Baden-Württemberg yields a net influx of 27% of all persons enrolled at the state universities, or 44,437 students. On average in Baden-Württemberg, each student has an annual income of €9,594 (HIS GmbH, 2010), so this net influx of students alone constitutes a gross effect of €426 million of annual demand.

A second factor leveraging the great impact that the state universities have on the regional economic of Baden-Württemberg is their acquisition of external funding. Empirically, for all the states of Germany, there is a linear relationship between the level of public funding for universities and the level of external funding acquired competitively. German universities and other institutions of higher learning acquire on average an additional €0.31 of external funding for each euro received in state funding. Amounts exceeding the expected values attest to above-average acquisition of external funding, as in Berlin, Saxony, and Baden-Württemberg (Fig. 15.5). The external funding that Baden-Württemberg attracted in 2012 (€1.137 billion) surpassed that of the two other geographically large states (North Rhine-Westphalia and Bavaria). The sum lay about €109 million, or 10%, above the empirically determined expected value. Baden-Württemberg's attraction of external funding has thus been an abiding strength since 1995 (Destatis, 2013a, 2015) and is attributable mostly to the state universities. In 2012 these universities acquired an additional €1.005 billion of external funding (based on our own data collection) or roughly one half of their basic funding. In fact, Table 15.4 shows that most universities achieve a positive balance between expenditures and funding from outside Baden-Württemberg.

A third factor of the regional multiplier impact is the specific production and consumption structure of the regional context. To begin with, the portion of disposable income, which has an impact on demand as a direct effect, varies with the consumption profiles of a region's population. The higher a population's savings ratio (i.e., the percentage of savings from income not spent on consumption), the less impact income has on demand. In Baden-Württemberg the population's average

[6] Glückler and König (2012) offer a more detailed geographical overview of the expenses of Heidelberg University.

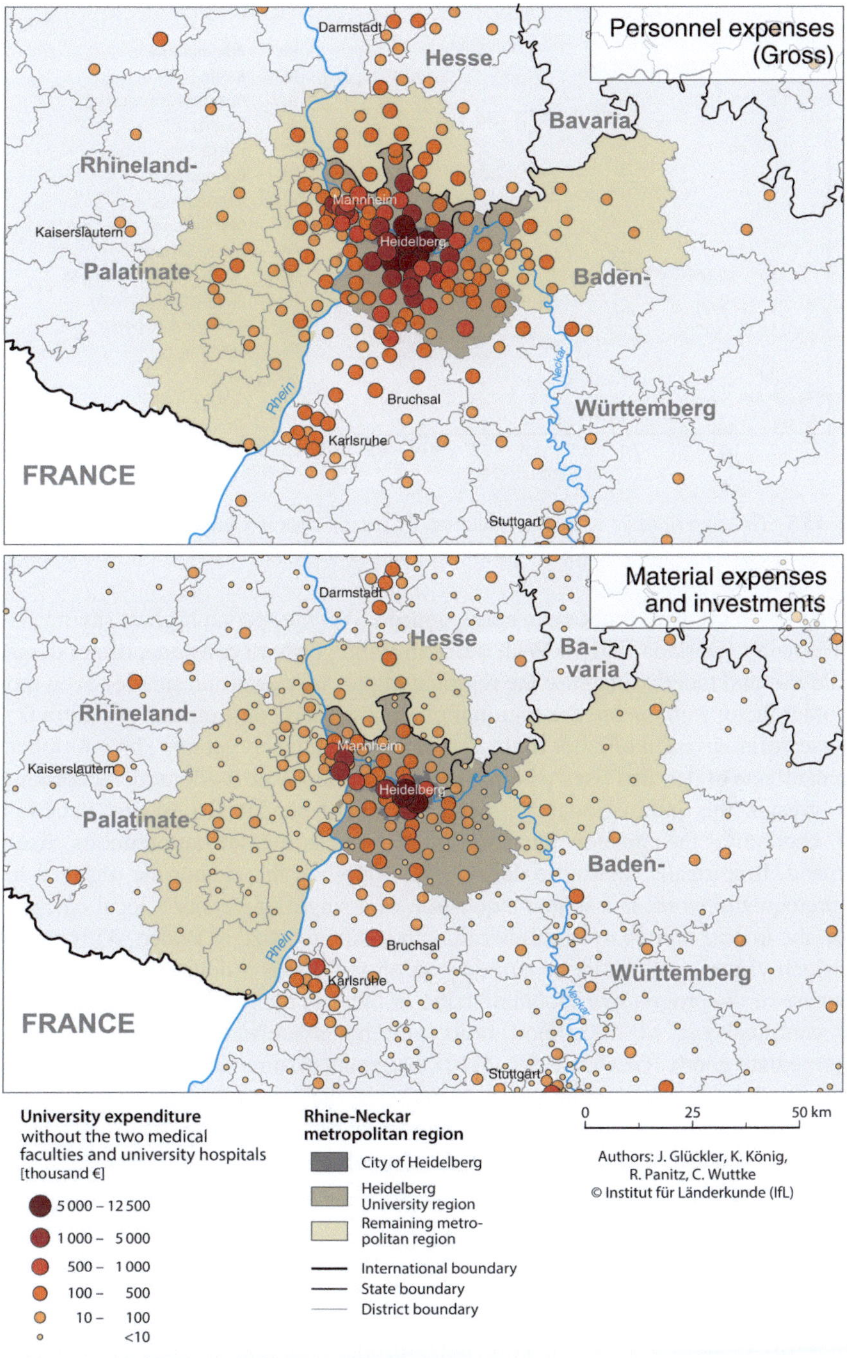

Fig. 15.4 Geographical distribution of Heidelberg University's personnel and expenditures on material (not counting the faculties of medicine), 2008.
Source: Adapted according to Glückler & König (2012, p. 347). Copyright: Heidelberg University and IfL. Reprinted with permission.

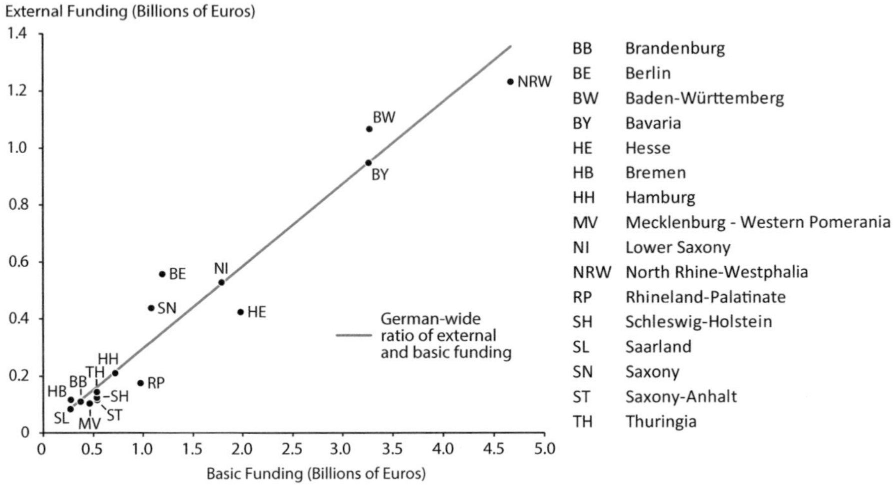

Fig. 15.5 The ratio of basic to external funding, by federal state in Germany, 2012. Source: Design by authors, data retrieved from Destatis (2015, pp. 16–18).

savings ratio (12.6%) is above average; indeed, it is the second highest savings ratio in Germany (Destatis, 2010). With a rise in willingness to consume, direct demand could rise and thereby increase the regional impact of wages and salaries even more. Another factor influencing the regional production and consumption structure is the intersectorial division of labor in the regional production structure. The multiplier of direct effects of demand rises with the regional prerequisite work ratio. Accordingly, the greater the intraregional performance interdependencies of the sectors of the economy, the greater the multiplier impacts of an autonomous rise in demand. Empirically, the more delimited a region is, the greater the regional ratio of prerequisite work. If it is just a question of a single university's local catchment area, the import ratio is much higher than in a state the size of Baden-Württemberg, in which virtually the entire economic portfolio of goods production is represented. Because of the progressive global division of labor in many sectors of the economy, the continual rise of the export ratio, and the associated increase of trade in intermediate goods (OECD, 1999, 2010), one must assume that the regional prerequisite work ratio is rather likely to decrease in the long term. Be that as it may, universities can attract students and raise external funding but are themselves largely unable to influence these regional production and consumption profiles.

Conclusion

Universities are key actors in knowledge-driven economies. They are centers of knowledge creation and of the training of highly qualified knowledge workers who shape societies and economies in the long run. Simultaneously, universities are part of the short-term economic system of regional and national economies because they spend money and attract students, employees, and external funding from other regions. By assessing the short-term, periodic economic impacts that an entire university landscape has on one of the largest federal economies in Germany, we have offered an insight into the magnitude of the economic impact of universities on a large economy. Our approach has several original aspects: (a) a unique database whose information on the spatial distribution of spending has been offered with unprecedented precision by university accounting departments, (b) advanced differential impact analysis, (c) the geographical scale of a large federal state in the German economy, and (d) the inclusion of statutory health-insurance premiums.

According to our analysis, the nine state universities in Baden-Württemberg stimulate additional demand in production, value creation, and employment that almost doubles the initial basic funding that these institutions of higher education receive from the federal government. For each euro of public funding, the universities generate an impact on gross value added of at least €1.8 in Baden-Württemberg alone. At the same time, the regional demand of the universities secures more than 63,000 jobs in Baden-Württemberg and generates immediate tax revenues of €350 million (€359 million including student fees) just for Baden-Württemberg's state administration. The aggregate impact even expands—to €1.9 per euro of basic funding—if the health-care expenditures are included. Discounting the tax revenues that Baden-Württemberg earns directly and indirectly by spending the basic funding within the same year it is received, one finds that the actual net impact of the universities is 2.3 times greater than the initial net public funding (Fig. 15.6).

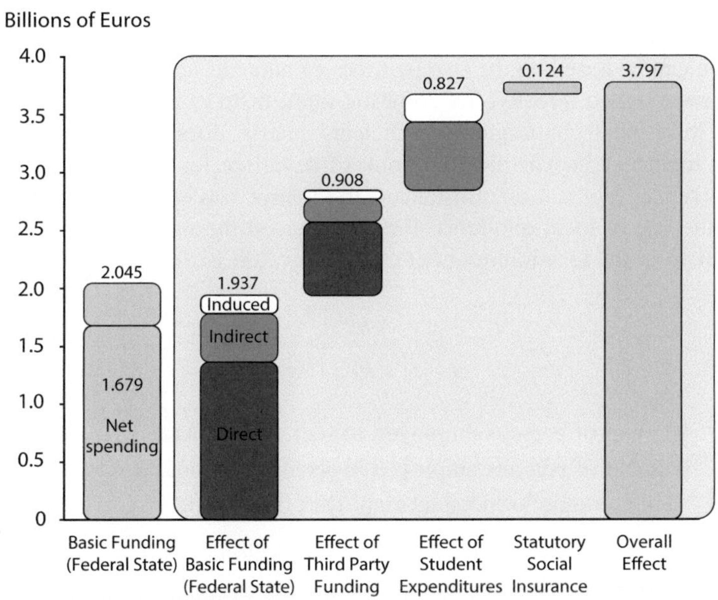

Fig. 15.6 The overall effect of value added in Germany's federal state of Baden-Württemberg, including statutory health insurance, 2012.
Source: Design by authors.

Because the universities attract many students with purchasing power from other regions and countries and raise external funding for research and employment, it is difficult to conceive of alternative uses of public funding that would exceed these regional economic impacts.

Lastly, a disadvantage of impact analyses is that they capture only existing supplies with upstream sectors of the economy and, hence, only the static effects of the current regional economic system and its division of labor (Drucker & Goldstein, 2007). But the decisive character of university activities lies precisely in the long-term, dynamic changes in these economic interdependencies (e.g., new technologies and sectors of the economy) and the conditions of production (e.g., increases in productivity). Experience shows that there are scarcely any important cross-sectorial technologies for which public promotion of research has not been a pivotal factor. Clearly, assessing only short-term economic impacts greatly underestimates the long-term overall economic impacts that universities have on their region and federal state.

Technical Appendix

To compute the production multiplier, we used the input coefficient matrix from a computation of Baden-Württemberg's input and output matrix. This kind of matrix is not provided by the state offices of statistics, so it must be constructed. There are two ways to construct a regional coefficients matrix: either with data based on surveys or, as in this chapter, with an estimation. The underlying model for such estimation can be based on the two matrices of national input coefficients. Because imports are assigned directly, the domestic input matrix makes for a more realistic estimate than the technological coefficients matrix does (Lindberg, 2010). The domestic input matrix was thus the point of departure for the remaining computations. To reflect regional circumstances, the matrix was scaled to a regional level through the use of local quotients. First, we scaled the national input coefficients with the help of the location quotient *(LQ)* (Kowalski & Schaffer, 2012).

$$LQ_i = \frac{b_i^R}{b_i^N} * \frac{b_{\sum}^N}{b_{\sum}^R} \tag{1}$$

where b_i^R: Number of persons employed in sector i (regional).

b_i^N: Number of persons employed in sector i (national).

$b_{\sum}^R$: Total number of persons employed (regional).

$b_{\sum}^N$: Total number of persons employed (national).

Each LQ_i shows the relative significance of regional economic sector i in comparison to its national counterpart as measured by sectorial employment. If a quotient lies in the range of $LQ_i \geq 1$, one assumes that the sector is sufficiently specialized in the region to satisfy the demand of the other production areas and of the final consumers. In this case each regional input coefficient equals the corresponding national coefficient ($a_{ij}^R = a_{ij}^N$). With a value of $LQ_i \leq 1$, one assumes that regional production lies below the national average. Consequently, additional prerequisite work must be imported to satisfy regional demand. The national input coefficients (a_{ij}^N)

$$a_{ij}^R = a_{ij}^N * LQ_i \tag{2}$$

must therefore be corrected through use of the LQ_i in order to avoid an overestimate for the region.

This method can be refined by using the cross-industry location quotient (*CILQ*). The quotient compares the regional share of the persons employed in the production sector with the national value and relates it to the sector to be supplied. Thus, both the size and structure of the production sector and the sector supplied are taken into account (Kowalski & Schaffer, 2012).

$$CILQ_{ij} = \frac{b_i^R}{b_i^N} * \frac{b_j^N}{b_j^R} = \frac{SLQ_i}{SLQ_j} \tag{3}$$

A special case of adjusting the regional input coefficients that use the $CILQ_{ij}$ arises with the diagonals of the matrix ($i = j$). The value is 1 by definition. Smith and Morrison (1974) addressed this problem and proposed that the LQ_i rather than the $CILQ_{ij}$ be used for the cells affected.

Another problem arises because the relative sizes of the regions are not taken into account. Flegg et al. (1995) tackled this problem and developed a new procedure based on the $CILQ_{ij}$, which they adjusted in later works by introducing the parameter λ (Flegg & Tohmo, 2013; Flegg & Webber, 2000).

$$FLQ_{ij} = CILQ * \lambda \tag{4}$$

$$FLQ_{ij} = \frac{b_i^R}{b_i^N} * \frac{b_j^N}{b_j^R} * \lambda \tag{5}$$

$$\text{where } \lambda = \left(\log_2\left(1 + b_{\sum}^R / b_{\sum}^N\right)\right)^{\delta} \tag{6}$$

The *FLQ*, like the *CILQ*, includes the producing and supplied sectors of the economy in the computation. However, by means of the parameter λ ($0 \leq \lambda < 1$), the *FLQ* also take account of the relative size of a region (as measured by employment). The value of λ approaches 1 with increasing region size. The exponent δ can be adjusted to influence the convexity of the function λ. This choice, however, is based on empirical work.

Recent studies on Baden-Württemberg use an exponent δ at the interval $0.11 \leq \delta \leq 0.17$ because of the diversified industrial structure (Kowalewksi, 2015). In comparison to other studies, however, and in the absence of further verification, this value seems to be too low. For this study we therefore adopted the value of the empirical studies and used the *FLQ* formula with $\delta = 0.3$ (Kowalski & Schaffer, 2012; Lindberg, 2010; Schaffer & Siegele, 2008) when generating the regional input-output table.

Application of *FLQ* leads simultaneously to the procedure for deriving other localization quotients.

$$a_{ij}^{R} = a_{ij}^{N} * FLQ_{ij} \tag{7}$$

In the cases of $FLQ_{ij} \geq 1$, one has $a_{ij}^{R} = a_{ij}^{N}$.

First, the input coefficients at the federal level were scaled down according to economic sectors to the federal state level with the *FLQ* procedure through the use of the figures for employment for which social security is mandatory (Bundesagentur für Arbeit, 2012). The resulting intersectorial interdependencies table *(A)* for the region of Baden-Württemberg covered all 71 economic sectors pursuant to the industrial classification of 2008 and thus constituted the point of departure for the subsequent steps of the analysis.

To compute the increase of regional production caused by an increase in demand in a given sector, one must first compute the inverse prerequisite work matrix (*X*):

$$X = (\mathrm{I} - \mathrm{A})^{-1} \tag{8}$$

The prerequisite work matrix *(A)* was subtracted from the identity matrix *(I)*,[7] and the result was inverted. This step yielded the inverse coefficients of the intersectorial interdependencies table, also known as Leontief inverses. In the final step, we prepared an input-output table based on the regional input coefficients matrix by including the gross value creation (Destatis, 2011b) of the economic sectors in Baden-Württemberg. In adjusting gross value creation to the 71 sectors, we assumed that the relative regional productivity differences of the sectors can be derived from the relative differences at the federal level. The input-output multiplier computed in this way describes the indirect effect of rendering prerequisite work that is brought about by the direct expenditures (ΔY) of the universities and their employees.

$$\Delta Y = \begin{bmatrix} 1 \\ 0 \\ 0 \end{bmatrix} \tag{9}$$

To determine this value more precisely, the vector of additional demand (ΔY) is multiplied by the inverse coefficients of the intersectorial interdependencies table. We can express this procedure in the following manner, using (8).

$$X = (\mathrm{I} - \mathrm{A})^{-1} * \Delta Y \tag{10}$$

Computation of the income multiplier (Keynesian multiplier)

The expansion of intersectorial production leads to an induced effect on income. The Keynesian multiplier theory describes this effect as a function of the willingness of households to consume. It states the number of units by which the income of employees grows because of marginal increases in state expenditures, consumption,

[7] An identity matrix is a square matrix in which the elements a_{ii} of the diagonal are 1 and the other elements 0.

investments, the tax rate, and exports. This multiplier rises with each euro of additional income that the households spend. The model consists principally of two elementary equations: the balance equation and the consumption function. The balance equation defines GDP (Y) in the Keynesian model as the total demand of the final consumers:

$$Y = C + N_0, \tag{11}$$

where $Y =$ gross domestic product (GDP).

$C =$ expenditures of the private households.
$N_O =$ autonomous consumption (demand of the state, or some other entity).

The consumption function

$$C = \mathrm{C(Y)} \tag{12}$$

implies the relationship between GDP and the amount of consumption by private households. The consumption function also satisfies

$$0 < \frac{dC}{dY} < 1 \tag{13}$$

This equation results in the following relationship for the marginal consumption ratio c:

$$c(Y) = \frac{dC}{dY} \tag{14}$$

which also lies between 0 and 1.When the marginal consumption ratio is constant, the GDP (Y) can be expressed as

$$Y = \mathrm{cY} + \mathrm{N_0} \Leftrightarrow Y = \left(\frac{1}{1-c}\right) N_0 \tag{15}$$

In this equation and under these circumstances, the GDP can be explained solely by the consumption ratio c and autonomous consumption N_O.

Computation of the combined multiplier

The combined multiplier is based on the assumption that the marginal consumption ratio stays constant over the cycles and lies in the range $\omega_2 < 1$. This value causes the effect to weaken over the cycles and eventually come to a standstill. Mathematically, we can draw on equations (8) and (16) to quantify this effect on demand as follows (Kowalski & Schaffer, 2012; Pischner & Stäglin, 1976; Schaffer & Siegele, 2008):

$$\Delta Y_1 = \omega_1 * \omega_2 * A^P * (I - A)^{-1} * \Delta Y_0 \tag{16}$$

where ΔY_1: vector of additional final demand in period 1.

ω_1: marginal consumption structure of the private households.[8]
ω_2: marginal consumption ratio of the private households.[9]
A^P: primary input coefficients matrix (quadrant III).
ΔY_0: demand trigger in period 0.

As already described, the additionally generated final demand (ΔY_1) is the trigger for the next round. Accordingly, equation 17 can be generalized as

$$\Delta Y_t = \omega_1 * \omega_2 * A^P * (I - A)^{-1} * \Delta Y_{t-1} \tag{17}$$

Lastly, we can derive the combined total effect by including the final demands (ΔY_t) that were generated in the rounds.

References

Assenmacher, M., Leßmann, G., & Wehrt, K. (2004). Regionale Entwicklungsimpulse von Hochschulen: Einkommens-, Beschäftigungs- und Kapazitätseffekte der Hochschulen Anhalt und Harz (FH) [Regional development spurred by universities: Effects of the Anhalt and Harz universities of applied sciences on income, employment, and capacity]. *Harzer Hochschultexte: Vol. 7*. Wernigerode: Hochschule Harz.

Bathelt, H., & Schamp, E. W. (Eds.). (2002). *Die Universität in der Region: Ökonomische Wirkungen der Johann Wolfgang Goethe-Universität in der Rhein-Main-Region* [The university in the region: Economic impacts of the University of Frankfurt on the Rhein-Main region]. Frankfurter wirtschafts- und sozialgeographische Schriften: Vol. 71. Frankfurt am Main: Institut für Wirtschafts- und Sozialgeographie.

Becker, R., Heinemann-Knoch, M., & Weeber, R. (1976). *Zur Rolle einer Universität in Stadt und Region: Universität als regionaler Wirtschaftsfaktor, Mobilität von Hochschulabsolventen, Verhalten von Hochschulangehörigen im Reproduktionsbereich* [On the university's role in the city and region: The university as a regional economic factor, the mobility of university graduates, and the behavior of university members in economic reproducion]. *Texte und Daten zur Hochschulplanung*. Berlin: De Gruyter.

Blume, L., & Fromm, O. (1999). Regionale Ausgabeneffekte von Hochschulen: Methodische Anmerkungen am Beispiel der Universität Gesamthochschule Kassel [Regional effects of university spending: Methodological comments on the University of Kassel]. *Raumforschung und Raumordnung*, 57, 418–431. doi: https://doi.org/10.1007/BF03184510

Breschi, S., & Lissoni, F. (2009). Mobility of skilled workers and co-invention networks: An anatomy of localized knowledge flows. *Journal of Economic Geography*, 9, 439–468. doi: https://doi.org/10.1093/jeg/lbp008

Bundesagentur für Arbeit (Federal Employment Agency). (2012). Sozialversicherungspflichtig Beschäftigte nach Wirtschaftszweigen (WZ 2008) für Baden-Württemberg [Employees for which social security is mandatory by industry sectors (WZ 2008) in the federal state of

[8]Destatis (2013d): Macroeconomic accounting: Personal consumption spending and available income with the following assumptions: employed persons (household type: €2,600–€3,600 of monthly net income) and students (household type: < €1,300 of monthly net income)

[9]Destatis (2010): Average German household with disposable monthly income of €2,965 and consumption expenditures of €2,245.

Baden-Württemberg]. Retrieved from https://statistik.arbeitsagentur.de/Navigation/Statistik/Statistik-nach-Themen/Statistik-nach-Themen-Nav.html

Bundesministerium der Finanzen. (2013). Der bundesstaatliche Finanzausgleich [The federal system of fiscal transfer]. Retrieved from http://www.bundesfinanzministerium.de/Content/DE/Standardartikel/Themen/Oeffentliche_Finanzen/Foederale_Finanzbeziehungen/Laenderfinanzausgleich/BFAG.pdf

Cooke, P., Heidenreich, M., & Braczyk, H.-J. (2004). *Regional innovation systems: The role of governance in a globalized world* (2nd ed.). London: Routledge.

Destatis. (2010). Wirtschaftsrechnungen: Einkommens- und Verbrauchsstichprobe. Einnahmen und Ausgaben privater Haushalte 2008 [System of national accounts: Sample Survey of Household Income and Expenditure [EVS], 2008]. Fachserie 15, Heft 4. Retrieved from https://www.destatis.de/DE/Publikationen/Thematisch/EinkommenKonsumLebensbedingungen/EinkommenVerbrauch/EVS_EinnahmenAusgabenprivaterHaushalte2152604089004.pdf?__blob=publicationFile

Destatis. (2011a). Gesundheitsausgabenrechnung: Methoden und Grundlagen 2008 [National account of health spending: Methods and principles]. Retrieved from https://www.destatis.de/DE/Methoden/Methodenpapiere/Download/Gesundheitsausgabenrechnung.pdf?__blob=publicationFile

Destatis. (2011b). VGR der Länder—Reihe 2, Kreisergebnisse Band 1 [GDP of the federal states—Series 2, District results Vol. 1]. Retrieved from http://www.statistik-portal.de/VGRdL/tbls/RV2014/R2B1.zip

Destatis. (2013a). Bildung und Kultur: Finanzen der Hochschulen 2011 [Education and culture: Financial accounts of higher education, 2011]. Fachserie 11, Reihe 4.5. Retrieved from https://www.destatis.de/GPStatistik/servlets/MCRFileNodeServlet/DEHeft_derivate_00010199/2110450117004.pdf

Destatis. (2013b). Glossar: Volkseinkommen [Glossary: National income]. Retrieved from https://www.destatis.de/DE/ZahlenFakten/GesamtwirtschaftUmwelt/VGR/_Doorpage/Glossar_VGR.html

Destatis. (2013c). Krankenhauspatienten: Bundesländer, Jahre, Geschlecht, Wohnort des Patienten [Hospital patients: Federal states, years, gender, residence of patients]. Retrieved from https://www-genesis.destatis.de/genesis/online/data;jsessionid=8C2E6A75CA542F2131729D8BD3E8A113.tomcat_GO_2_2?operation=abruftabelleBearbeiten&levelindex=1&levelid=1495199774592&auswahloperation=abruftabelleAuspraegungAuswaehlen&auswahlverzeichnis=ordnungsstruktur&auswahlziel=werteabruf&selectionname=23131-0012&auswahltext=&nummer=2&variable=1&name=GES025&werteabruf=Werteabruf

Destatis. (2013d). Wirtschaftsrechnungen: Laufende Wirtschaftsrechnungen. Einnahmen und Ausgaben privater Haushalte 2011 [System of national accounts: Current private household income and spending, 2011]. Retrieved from https://www.destatis.de/DE/Publikationen/Thematisch/EinkommenKonsumLebensbedingungen/EinkommenVerbrauch/EinnahmenAusgabenprivaterHaushalte2150100117004.pdf?__blob=publicationFile

Destatis. (2015). Bildung und Kultur. Finanzen der Hochschulen 2013 [Education and culture: Financial accounts of higher education, 2013]. Fachserie 11, Reihe 4.5. Retrieved from https://www.destatis.de/DE/Publikationen/Thematisch/BildungForschungKultur/BildungKulturFinanzen/FinanzenHochschulen2110450137004.pdf?__blob=publicationFile

DIW econ. (2008). Wirtschaftsfaktor TU Berlin: Welchen Einfluss hat die TU Berlin auf die Berliner Wirtschaft? [The Technical University of Berlin as an economic factor: What is the impact of the Technical University of Berlin on the economy of Berlin?]. Retrieved from https://diw-econ.de/wp-content/uploads/2014/01/20081113_Report_TU_Berlin_v1.0.pdf

DIW econ. (2013). Berliner Universitäten als Wirtschaftsfaktor: Die regionalökonomischen Effekte der Berliner Universitäten [The universities of Berlin as an economic factor: The regional economic effects of Berlin's universities]. Retrieved from https://diw-econ.de/wp-content/uploads/2014/11/Wirtschaftsfaktor-Berliner-Universitaeten-Endbericht-v1.0.pdf

Drucker, J., & Goldstein, H. (2007). Assessing the regional economic development impacts of universities: A review of current approaches. *International Regional Science Review*, 30, 20–46. doi: https://doi.org/10.1177/0160017606296731

Drude, M. (1995). *Die Albert-Ludwigs-Universität Freiburg als Wirtschaftsfaktor für die Region Freiburg/Br* [The University of Freiburg as an economic factor in its region]. Freiburg: Universität Freiburg.

Eisen, E. E. (1948). *Educational land use in Lake County, Ohio* (University of Chicago Department of Geography Research Paper: No. 2). Chicago: Department of Geography, University of Chicago.

Etzkowitz, H. (2008). *The Triple Helix: University-industry-government innovation in action*. New York: Routledge.

Flegg, A. T., & Tohmo, T. (2013). Regional input-output tables and the FLQ formula: A case study of Finland. *Regional Studies, 47*, 703–721. doi: https://doi.org/10.1080/00343404.2011.592138

Flegg, A. T., & Webber, C. D. (2000). Regional size, regional specialization and the FLQ formula. *Regional Studies, 34*, 563–569. doi: https://doi.org/10.1080/00343400050085675

Flegg, A. T., Webber, C. D., & Elliott, M. V. (1995). On the appropriate use of location quotients in generating regional input-output tables. Regional Studies, 29, 547–561. doi: https://doi.org/10.1080/00343409512331349173

Friedrich, K., & Rahmig, A. (2013). *Die Universität Halle als hochrangiger regionaler Wirtschaftsfaktor: Analyse der Nachfrageeffekte der Hochschule* [The University of Halle as a prominent regional economic factor: An Analysis of the demand effects of the university]. Retrieved from http://digital.bibliothek.uni-halle.de/pe/content/titleinfo/2416470

Fürst, D. (1979). *Die Universität als Wirtschaftsfaktor einer Region: Die wirtschaftliche Bedeutung der Universität Konstanz für die Bodenseeregion* [The university as a regional economic factor: The economic significance of the University of Konstanz in the Lake Constance region]. *Konstanzer Blätter für Hochschulfragen, 16*(61), 50–59.

Giarratani, F. (1976). Application of an interindustry supply model to energy issues. *Environment and Planning A, 8*, 447–454. doi: https://doi.org/10.1068/a080447

Glückler, J., & König, K. (2012). The impact of Heidelberg University on the regional economy. In P. Meusburger & T. Schuch (Eds.), *Wissenschaftsatlas of Heidelberg University: Spatio-temporal relations of academic knowledge production* (pp. 344–347). Knittlingen: Bibliotheca Palatina.

Glückler, J., Panitz, R., & Wuttke, C. (2015). Die wirtschaftliche Wirkung der Universitäten im Land Baden-Württemberg [The economic impact of the state universities in the federal state of Baden-Württemberg]. *Raumforschung und Raumordnung, 73*, 327–342. doi: https://doi.org/10.1007/s13147-015-0360-9

Glückler, J., & Ries, M. (2012). Why being there is not enough: Organized proximity in place-based philanthropy. *Service Industries Journal, 32*, 515–529. doi: https://doi.org/10.1080/02642069.2011.596534

Glückler, J., & Suarsana, L. (2014). Unternehmerisches Engagement im philanthropischen Feld: Das Beispiel Heilbronn-Franken [Corporate social responsibility in the philanthropic field: The case of Heilbronn-Franconia]. *Berichte: Geographie und Landeskunde, 88*, 203–221.

Goddard, J., Hazelkorn E., Kempton, L., & Vallance, P. (2016). (Eds.). *The civic university: The policy and leadership challenges*. Cheltenham: Edward Elgar.

Goddard, J., & Vallance, P. (2013). *The university and the city*. Regions and Cities: Vol. 63. Abingdon: Routledge.

Goldstein, H. A., Maier, G., & Luger, M. (1995). The university as an instrument for economic and business development: U.S. and European comparisons. In D. D. Dill & B. Sporn (Eds.), *Emerging patterns of social demand and university reform: Through a glass darkly* (pp. 105–133). Oxford, UK: Pergamon.

Goldstein, H. A., & Renault, C. S. (2004). Contributions of universities to regional economic development: A quasi-experimental approach. *Regional Studies, 38*, 733–746. doi: https://doi.org/10.1080/003434042000265232

Gormsen, E. (1981). Stadt und Universität in Heidelberg [City and university in Heidelberg]. In W. Fricke & E. Gormsen (Eds.), *Heidelberg und der Rhein-Neckar-Raum: Sammlung sozial-*

und stadtgeographischer Studien (pp. 111–154). Heidelberger Geographische Arbeiten: Vol. 46. Heidelberg: Selbstverlag des Geographischen Instituts der Universität Heidelberg.

Grabitz, J. (1990). Die wirtschaftliche Bedeutung der Universität Heidelberg [The economic significance of Heidelberg University]. In P. Meusburger & J. Schmude (Eds.), *Bildungsgeographische Studien über Baden-Württemberg* (pp. 155–186). Heidelberger Geographische Arbeiten: Vol. 88. Heidelberg: Selbstverlag des Geographischen Instituts der Universität Heidelberg.

Hall, P. A., & Soskice, D. W. (Eds.). (2001). *Varieties of capitalism: The institutional foundations of comparative advantage*. Oxford, UK: Oxford University Press.

Hamm, R., & Wenke, M. (2002). Die Bedeutung von Fachhochschulen für die regionale Wirtschaftsentwicklung: Eine Darstellung am Beispiel der Fachhochschule Niederrhein [The role of polytechnical universities in regional economic development: The case of the University of Applied Sciences Niederrhein]. *Raumforschung und Raumordnung, 60*, 28–36. doi: https://doi.org/10.1007/BF03185691

HIS GmbH. (2010). Grundauszählung zur 19. Sozialerhebung. Auszählungsreihe: Länder: Baden-Württemberg [Basic calculation for the Nineteenth Social Survey. Level: Federal States: Baden-Württemberg.]. Retrieved from http://www.sozialerhebung.de/download/19/Soz19_Ga_Baden-Wuerttemberg.pdf

Kowalewksi, J. (2015). Regionalization of national input-output tables: Empirical evidence on the use of the FLQ formula. *Regional Studies, 49*, 240–250. doi: https://doi.org/10.1080/00343404.2013.766318

Kowalski, J, & Schaffer, A. (Eds.). (2012). *Das Karlsruher Institut für Technologie—Impulsgeber für Karlsruhe und die TechnologieRegion* [The Karlsruhe Institute of Technology—Catalyst for the city of Karlsruhe and the technology region] (KIT Scientific Reports: Vol. 7630). Karlsruhe: KIT Scientific Publishing.

Leontief, W. W. (1936). Quantitative input and output relations in the economic systems of the United States. *The Review of Economics and Statistics, 18*, 105–125. doi: https://doi.org/10.2307/1927837

Lindberg, G. (2010). *On the appropriate use of (input-output) coefficients to generate non-survey regional input-output tables: Implications for the determination of output multipliers*. Retrieved from https://www.econstor.eu/bitstream/10419/119005/1/ERSA2010_0800.pdf

Lundvall, B.-Å. (Ed.). (1992). *National systems of innovation: Towards a theory of innovation and interactive learning*. London: Pinter.

OECD. (1999). OECD Science, Technology and Industry Scoreboard, 1999: Benchmarking knowledge-based economies. Retrieved from http://www.oecd-ilibrary.org/docserver/download/9299071e.pdf?expires=1495290117&id=id&accname=guest&checksum=56B8FB003244002B9BB99117099F4D74

OECD. (2010). *Measuring globalisation: OECD economic globalisation indicators 2010*. Paris: OECD Publishing.

Oser, U., & Schroeder, E. (1995). Die Universität Konstanz als Wirtschaftsfaktor für die Region [The University of Konstanz as an economic factor in the region]. Retrieved from http://nbn-resolving.de/urn:nbn:de:bsz:352-opus-731

Owen-Smith, J., & Powell, W. W. (2004). Knowledge networks as channels and conduits: The effects of spillovers in the Boston biotechnology community. *Organization Science*, 15, 5–21. doi: https://doi.org/10.1287/orsc.1030.0054

Pischner, R., & Stäglin, R. (1976). Darstellung des um den Keynes'schen Multiplikator erweiterten offenen statischen Input-Output-Modells [Description of the open static input–output model extended by the Keyensian multiplier]. *Mitteilungen aus der Arbeitsmarkt- und Berufsforschung, 9,* 345–349. Retrieved from http://doku.iab.de/mittab/1976/1976_3_MittAB_Pischner_Staeglin.pdf

Ponds, R., van Oort, F., & Frenken, K. (2010). Innovation, spillovers and university-industry collaboration: An extended knowledge production function approach. *Journal of Economic Geography, 10*, 231–255. doi: https://doi.org/10.1093/jeg/lbp036

Rosner, U., & Weimann, J. (2003). Die ökonomischen Effekte der Hochschulausgaben des Landes Sachsen-Anhalt [The economic effects of university spending by the federal state of Saxony-Anhalt]. Retrieved from http://ids.hof.uni-halle.de/documents/t655.pdf

Salter, A. J., & Martin, B. R. (2001). The economic benefits of publicly funded basic research: A critical review. *Research Policy, 30*, 509–532. doi: https://doi.org/10.1016/S0048-7333(00)00091-3

Schaffer, A., & Siegele, J. (2008, November). *Regional impacts of modern transport infrastructure—Analysis of the southern German infrastructure project Baden-Württemberg 21*. Paper presented at the 55th Annual North American Meeting of the Regional Science Association International, New York.

Schubert, T., Baier, E., Hufnagl, M., Meyer, N., Schricke, E., & Stahlecker, T. (2012). Endbericht zur Metastudie Wirtschaftsfaktor Hochschule [Final report on the meta study entitled *The university as an economic* factor]. Retrieved from http://www.wissenschaftsmanagement-online.de/sites/www.wissenschaftsmanagement-online.de/files/migrated_wimoarticle/stifterverband_wirtschaftsfaktor_hochschule.pdf

Smith, P., & Morrison, W. I. (1974). *Simulating the urban economy: Experiments with input-output techniques*. London: Pion.

Spehl, H., Sauerborn, K., Sauer, M., Benson, L., Feser, H.-D., v. Malottki, C., Schulze, P. M., & Flohr, M. (2005). Regionalwirtschaftliche Wirkungen der Hochschulen und Fo rschungseinrichtungen in Rheinland-Pfalz: Wertschöpfungs-, Einkommens- und Beschäftigungseffekte durch Bau und Betrieb der Einrichtungen. Langfassung [The regional economic impact of higher education facilities in Rhineland-Palatinate: Added value, income, and employment effects of constructing and operating the facilities, extended version]. Retrieved from https://mwwk.rlp.de/fileadmin/mbwwk/Publikationen/Wissenschaft/regionalwirtschaftliche_wirkung_HS_lang.pdf

Statistisches Landesamt Baden-Württemberg. (2012). Studierende an baden-württembergischen Hochschulen im Wintersemester 2011/12 [Students enrolled at universities in Baden-Württemberg in the 2011–2012 winter term]. Retrieved from https://www.destatis.de/GPStatistik/servlets/MCRFileNodeServlet/BWHeft_derivate_00000797/3234_12001.pdf

Statistisches Landesamt Baden-Württemberg. (2013a). Staatliches Steueraufkommen in Baden-Württemberg seit 1952 nach Steuerarten [Tax revenues in Baden-Württemberg]. Retrieved June 25, 2013, from http://www.statistik-bw.de/FinSteuern/Steuern/LRt1808.jsp

Statistisches Landesamt Baden-Württemberg. (2013b). Umsätze und ihre Besteuerung 2011 [-Revenues and taxation 2011]. Retrieved from https://www.destatis.de/GPStatistik/servlets/MCRFileNodeServlet/BWHeft_derivate_00003343/3941_11001.pdf;jsessionid=E3FD4B00B7$32#952737F01E3D4F7866C9D3

Stoetzer, M.-W., & Krähmer, C. (2007). Regionale Nachfrageeffekte der Hochschulen—Methodische Probleme und Ergebnisse empirischer Untersuchungen für die Bundesrepublik Deutschland [Regional effects of universities on demand— Methodological challenges and findings of empirical research in Germany]. *Jenaer Beiträge zur Wirtschaftsforschung 06/2007*. Retrieved from http://hdl.handle.net/10419/43652

Tanzmann, L. (2015). Die Finanzen der Hochschulen in Baden-Württemberg—ein Überblick [The finances of the universities in Baden-Württemberg—An overview]. *Statistisches Monatsheft Baden-Württemberg, 1,* 3–8. Retrieved from http://www.statistik-bw.de/Service/Veroeff/Monatshefte/PDF/Beitrag15_01_01.pdf

Vincett, P. S. (2010). The economic impacts of academic spin-off companies, and their implications for public policy. *Research Policy, 39*, 736–747. doi: https://doi.org/10.1016/j.respol.2010.02.001

Vogt, V. (2011). Schätzung regionaler Exporte und Importe als Vorarbeit zu einer Input-Output-Tabelle für Baden-Württemberg [Estimation of regional exports and imports as groundwork for an input–output table for Baden-Württemberg]. *Statistisches Monatsheft Baden-Württemberg,*

2, 30–34. Retrieved from https://www.statistik-bw.de/Service/Veroeff/Monatshefte/PDF/Beitrag11_02_08.pdf

Vöhringer, S. (2012). Einkommensmillionäre in Baden-Württemberg [Income millionaires in Baden-Württemberg]. *Statistisches Monatsheft Baden-Württemberg, 9,* 37–40. Retrieved from http://nbn-resolving.de/urn:nbn:de:0168-ssoar-405577

Wasem, J., Buchner, F., Lux, G., Manouguian, M.-S., & Schillo, S. (2007). Die Regionaldimension in der Gesetzlichen Krankenversicherung vor dem Hintergrund des GKV-WSG [The regional dimension of statutory health insurance] (Diskussionabeitrag aus dem Fachbereich Wirtschaftswissenschaften Universität Duisburg-Essen Nr. 153). Retrieved from https://www.tdr.wiwi.uni-due.de/fileadmin/fileupload/BWL-MEDMAN/Forschung/DiskussionspapierGesu$32#ndheitsreformNr153.pdf

Chapter 16
The Nonmetropolitan University's Regional Engagement in the African Context: The Case of Cameroon

Eike W. Schamp

Claiming that the university should be responsible for the social and economic development of the region in which it is located is a fairly recent idea. Since the 1960s policy-makers in the Global North saw the location of new universities in remote areas as a means of unlocking potential for higher education and as a powerful regional demand-driven economic stimulus generated by the university. Later, the university came to be considered "a warehouse of precious goods" (van der Wusten, 1998, p. 1), the bearer and developer of various kinds of knowledge that, once having spilled over after its intentional transfer, contributes to regional development. The literature on the university's mode of functioning in this respect is now abundant. The vast bulk of it, however, focuses on the Global North. Very little work on the topic is devoted to Africa.

The past half century has also been one of challenging debate about the role of the university in society and for societal development. Suffice it to recall the dispute about nineteenth-century models such as the European Humboldtian university and the traditional American engaged university (often found at land-grant universities in the United States) as opposed to more contemporary models such as the entrepreneurial (or triple-helix) university (see chapter by Etzkowitz in this volume) and the revived developmental university[1] (Goldstein, 2010; Göransson & Brundenius, 2011). With globalization, this dispute figured prominently throughout the world. A World Bank study (Altbach, Reisberg, & Rumbley, 2009) posited a true "academic revolution" in the societal repositioning of the university. Exceptional pressures on university systems in developing countries in general, especially in Africa,

[1] A developmental role had already been assigned to new universities in early postcolonial Africa (see Teferra, 2014, for example). However, they have failed to perform it in the persistent crises since the 1970s.

E. W. Schamp (✉)
Department of Human Geography, Goethe University Frankfurt, Frankfurt am Main, Germany
e-mail: schamp@em.uni-frankfurt.de

P. Meusburger et al. (eds.), *Geographies of the University*, Knowledge and Space 12,
https://doi.org/10.1007/978-3-319-75593-9_16

came from the shift to the entrepreneurial model in the United States and the launch of the Bologna process[2] in Europe. According to scholars from Africa (Munene, 2009; Obasi & Olutayo, 2009) and elsewhere (Brock-Utne, 2003; Robertson, 2009; Singh, 2010), Africa has become a battlefield of conflicting notions of the university's responsibilities.

Closely connected to this dispute has been the addition of a third mission, outreach, to complement the two traditional university missions, education and research. The outreach mission positions regional engagement by the university as a management task of the university staff. Knowledge creation, knowledge transfer, and spillovers have become the major challenge for the university's role in society and a core issue in economic geography and regional policy (Goldstein, 2009, 2010; see also chapter by Glückler, Panitz and Wuttke in this volume). A prerequisite for knowledge to spill over from the university to its proximate environment is communication, a complex process shaped by many factors, such as the capabilities of actors, the availability of media, and the opportunities and barriers created by institutional settings (see Meusburger, 2013, 2017). The type of university may also be among the factors determining the kind and degree of knowledge spillovers that may occur as universities evolve according to their disciplinary scope (e.g., full-fledged university, technical university, or applied university), organizer (public or private), size, age, and location (metropolitan, urban, or rural).

In this chapter I argue that regional knowledge spillovers in Africa emerge differently from those on other continents, not least because of dissimilar educational systems and societal contexts conditioned by the burdensome legacies of colonial rule by various European powers (United Kingdom, France, Portugal, Spain).[3] African universities are not well researched. Literature that does exist on the subject refers chiefly to the principal "flagship" universities, many of which were created during colonial times or shortly thereafter—such as Cheikh Anta Diop University (Dakar), Makerere (Kampala), University of Ghana (Legon), and the Nigerian universities of Ibadan and Nsukka. This chapter focuses on a rather neglected university type: the young, postcolonial, ordinary public university in a nonmetropolitan context, often meaning small urban centers in rural areas. This type of university is peripheral in terms of its resource endowments and academic

[2]The Bologna Process is a European initiative, launched in 1999 in Bologna by European ministers of education. It aims at creating a common European space for higher education to foster intra-European mobility of students, teachers, and graduates as a response to changes in the systems of higher education in the United States and Asia. A major element is the introduction of the bachelor-master-doctorate (B-M-D) system across Europe. Harmonization of the European higher education systems is a continuing and sometimes contested process. The Bologna Process became globalized when Australia wished to join in 2001 and the EU Commission's cooperation policies with third countries incorporated it (Charlier & Croché, 2009, Obasi & Olutayo, 2009).

[3]Although the legacy of colonial rule is indisputable in the broad sense, the debate continues about the degree to which colonial heritage currently matters in higher education. For instance, Nyamnjoh (2012) pointed to "Africa [as a] victim of a resilient colonial and colonizing epistemology" (p. 129), whereas Teferra (2016), taking a more balanced view, has claimed that "contemporary higher education in Africa . . . is a new phenomenon" (p. 80).

performance within the university system, its power relationships to the political center, and its region's economic development and absorptive capacities. These conditions constitute a common sociopolitical and regional environment for many young public universities in Africa. It differs from urban agglomerations or metropolises. Hence, the main universities observed in this chapter will be called "nonmetropolitan."[4]

The current state of nescience on the forms and extent of regional engagement by universities in Africa calls for a case study that uncovers regional relationships and communication. For this reason three nonmetropolitan universities in Cameroon were studied at a particular historical moment, the early 2000s, when the Cameroonian government implemented policies conceived to increase the universities' societal responsibility through "professionalization," a French term implying also the commercialization and privatization of higher education. This research has shown that the country's unique history under German, French, and British colonial power still matters in society and politics. For example, a leading staff member of Cameroon's only Anglophone university, Buea, contended that the Anglo-Saxon traditions were being followed as much as possible (Vice Rector for R&D, University of Buea, personal communication, December 12, 2008). Yet the current structure and policies in higher education—blamed for "excessive centralization, authoritarian management style and political control" (Konings, 2009, p. 213)—were largely dominated by a clone of the strict and inflexible French implementation of the Bologna system (Croché & Charlier, 2012; Djouda Feudjio, 2009).

In the following sections I first point out the specifics of the sub-Saharan African context, briefly discuss the methodology generally used in academic research and political debate to analyze regional knowledge spillovers, and ask to what extent it is regarded as inappropriate in this chapter (see chapter by Meusburger in this volume). I then turn to the case of Cameroon and its nonmetropolitan universities. The cardinal assertion is that knowledge spillovers measured by codified means of knowledge communication (e.g., licenses, joint and contract research, or collaborative publications, which are often directed to formal organizations such as companies) essentially provide no framework or vehicle for communicating tacit knowledge to informal societal groups and nonprofit organizations that may be crucial to regional development in rural areas of developing countries. I try to uncover those less visible, even invisible, forms of regional knowledge spillovers by examining three fairly young, small, ill-equipped universities. Particularly in the African context, which is often noted for its weak institutions, heavy financial restrictions, and lack of the kind of university leadership called for by Muriisa

[4]This term is intended to take into account the proximity that a university in a small or medium-sized African town has to its rural neighborhood, the foremost activities of which are agriculture and animal husbandry. Implicitly, these universities are characterized by poor endowment and a focus on undergraduate studies and human sciences. The term is ambiguous, however, because it does not subsume institutions such as technical universities specialized in mining, although they, too, may be located in "rural" areas. I am grateful to Michel Simeu Kamdem for pointing out the difficulties and vagueness of the terminology on types of universities in this context.

(2014), this approach is no assertion that invisible forms of knowledge spillovers sufficiently promote regional development. But given the dearth of codified knowledge that can be commercialized, the analysis can nevertheless reveal the university's unexpected regional engagement with various local stakeholders.

The African Context

The African university faces a sociopolitical environment much different than that in other parts of the developing world. That fact codetermines the university's limited abilities to engage in regional development. The roots of these constraints stem largely from weak African economic development in the 1980s and 1990s—the "lost decades"—during which most countries on the continent, still dependent on their natural resources, descended into crisis (Lawrence, 2010). State budgets shrank, and governments were unable to close the sizable gap between the levels of education in Africa and those on other continents (see Sawyerr, 2004, for a masterly summary of the long-term development and crises of the university across many African countries).

A host of problems have contributed to this malaise. Demographic growth on the continent has remained the highest globally. A persistently high rate of illiteracy has coincided with an explosion in the young population seeking improved education in Africa. The share of the Africans who have graduated from an institution of higher education—less than 1% of the population older than 15 years (Teal, 2011)—is the lowest in the world. Yet the young generation's quickly rising willingness to study, combined with a population explosion, has caused the "massification" of higher education everywhere.[5] Despite the economic crisis affecting many African countries, they have had to scale up tertiary education in response to the young generation's unrest. This decision conflicts with the World Bank's recommendations in favor of primary and secondary education—which, paradoxically, has fueled the demand for higher education. Unsurprisingly, most authors stress the overall weakness of the African university (Ajakaiye & Kimenyi, 2011; Munene, 2009; World Bank, 2009) and the science taught there (UNESCO Science Report, 2010). The shortcomings mirror those of the state, whose governance of the educational system ranges from aspired state centrism in a neopatrimonial political system (as is prevalent in Francophone Africa) to neoliberalism imposed by international organizations such as the World Bank and UNESCO (see Robertson, 2009; St. George, 2006).

Another macroeconomic explanation of weaknesses in the system of higher education in Africa calls into question the political strategies pursued by African

[5]Massification is the term usually used in the pertinent literature. Cloete, Maassen, & Bailey (2015) argued that university systems in Africa are not "massified" but are "overcrowded elite systems" (p. 5), for most of the young generation still has little or no access to higher education. This assessment is confirmed by Sawyerr (2004), who actually sees "novel forms of social exclusion" (p. 22) in student enrollment.

governments. African governments in the 1980s and 1990s were not able to codevelop primary, secondary, and tertiary education. They did not see investment in higher education and industrialization as a parallel process, whereas policies on higher education and industrialization went hand in hand in successful Asian economies (Cloete, Bailey, Pillay, Bunting, & Maassen, 2011, p. 3; St. George, 2006). According to the World Bank (2009), spending on higher education per capita of enrolled students slid from U.S. $6,800 in 1980 to an average of U.S. $981 some 30 years later in 33 low-income countries of sub-Saharan Africa. Many African policy-makers, however, have meanwhile come to see the university as many academics do: as an engine that drives development. This new perspective has raised a continuing controversy between the advocates of applied research and those of basic research and between supporters of a service orientation as part of the entrepreneurial model and those who endorse "engagement" as part of the developmental model of the university (Cloete, Maassen, & Bailey, 2015).

In summary, the African university's weakness has extended to all fields of education and research for a number of reasons. Equipment and infrastructure have been scarce, especially with regard to libraries and Internet access (Willinsky, Jonas, Shafack, & Wirsiy, 2005). Hygienic facilities (toilets) and classrooms are in short supply as well (Djouda Feudjio, 2009). Economic crises have triggered a massive brain drain of university staff to wealthier countries in the Global North and partly within Africa (at first to Nigeria, then to South Africa after the fall of the apartheid state). The use of colonial languages (English and French), too, has accelerated the brain drain inasmuch as it was seen to be a "grave problem at the university" because students have trouble mastering the foreign languages (Brock-Utne, 2003, p. 44). Time and again, curricula have been assessed as inadequate for the private sector. In addition, leading figures at the university prefer kinship relationships when it comes to the appointment of doctoral students and assistant lecturers. In short, "tribalism"[6] has figured as a ubiquitous mechanism of social relationships in the recruitment of new staff, a fact certainly not limited to Cameroon (Affa'a & Des Lierres, 2002; Kemayou, 2012).[7] The university's local public infrastructure—transport, the power supply, and student housing—is deficient as well. A complementary problem is the academic weakness of young and nonmetropolitan universities, which used to focus on undergraduate studies only. At many universities growth in the humanities, arts, and social sciences,

[6]In traditional anthropology the term *tribe* refers to an ethnic group that adheres to its traditional rules and identities. Tribalism has been common in Cameroonian politics as a form of behavior that favors reciprocal relations with people from the same place of origin. It is a specific form of clientelistic relationship embedded in ethnic politics in Cameroon (see Bayart, 1979; Ngeve & Orock, 2012; Nyamnjoh, 1999). In the academic world, it denotes behavior expressing antimodernism as opposed to basic meritocratic principles in modern academia and society at large.

[7]A sociopolitical system that is explicitly manifested in a nation's institutional settings is indisputably a powerful determinant of the quality of a university and its effectiveness as a leader of development. Researchers in political science and political economics have repeatedly pointed to this link. I am sympathetic to the approach by Acemoglu and Robinson (2012), who see the necessity of inclusive institutions and recognize that extractive political institutions inhibit a nation's long-term development.

which are less capital intensive than the technical and natural sciences, far outpaced that in the latter two fields.

African universities are thus among the world's weakest in terms of performance, much as Ajayi, Goma, and Johnson (1996, p. 229) concluded in the 1990s: The African university is marginalized "from the intellectual and informational mainstreams that shape development possibilities in the rest of the world" (p. 229). Aside from the considerable number of "transnational partnerships" arranged through Africa's development cooperation with the Global North and with emerging Asian economies (Koehn & Obamba, 2014; Schamp & Schmid, 2008; Teferra, 2014), that statement still applies. It is also substantiated by Jöns and Hoyler's (2013) evaluation of world university rankings. Writing on Uganda, Muriisa (2014) highlighted what he saw as an even worse situation. Of course, there used to be exceptions (e.g., Makerere University and Ibadan University), and some South African universities do stand out. But aside from South Africa, no sub-Saharan African university, flagship universities included, has ranked among the world's top 1,000 as of 2008 (Charlier & Croché, 2009; Jöns & Hoyler, 2013).[8] The number of research publications from universities in that part of the continent is extremely low, only about 27,000 papers annually from more than 30 nations (including the two regional "giants," Nigeria and Kenya). The total scarcely exceeds that of The Netherlands, a small country of the Global North (Adams, King, & Hook, 2010; Mugabushaka, 2008). The share of students going abroad is larger in sub-Saharan Africa than anywhere else on the globe (5.6% of worldwide student mobility in 2005; Charlier & Croché, 2009). Massive emigration of sub-Saharan academics to OECD countries persists. Predictably, these enfeebled universities have been unable to cope with new government policies promoting the commodification and commercialization of teaching and research since the early 2000s.

Lastly, the socioeconomic environment in sub-Saharan Africa is far less favorable to regional knowledge spillovers than is generally assumed in the literature on the Global North's university-industry linkages, an issue that is more closely examined in the following section. There are few efforts to pursue innovation policies, as indicated by a dearth of patent registrations, the weak innovation capacities of state enterprises and the private sector, and the fact that multinationals tap into innovation from their home country rather than from Africa. A paucity of organizational structures for improved communication between academia and the business world exacerbates the situation. The share of manufacturing remains negligible and has even declined in some parts of Africa. Small local industries, sometimes clustered at certain locations, appear to be less keen on gaining access to knowledge from universities than their counterparts from the Global North are (Zeng, 2008). Moreover, most of the sub-Saharan African population still lives in the urban informal sector and in the rural noncommercial economy. What kind of knowledge spillovers for what kind of regional development can one expect under these conditions?

---

[8]This indicator of the performance of universities is currently in wide use but may be seriously criticized for its onesidedness. As Jöns and Hoyler (2013) put it, "world university rankings represent best those investment-intensive areas of the technosciences that facilitated American hegemony in the second half of the 20th century and that China is now trying to emulate" (p. 55).

Stylized Facts about the University—Regional Development Nexus

Over the last few decades, mounting attention to innovation as a basis for economic prosperity in times of global competition has led to concepts of national and regional innovation systems that emphasize the links between the university as a knowledge creator and companies as codevelopers and users of innovation (Mowery & Sampat, 2005; Reddy, 2011). Issues such as the entrepreneurial behavior of universities in regional engagement and university-industry relationships have prevailed (Perkman et al., 2013; Rothaermel, Agung, & Jiang, 2007). They emerged partly when the intellectual property rights to knowledge that is generated at universities were shifted from the state to the researchers and university administration[9] and when the demand for codifying such knowledge in commodifiable patents and licenses intensified. Knowledge commodification and commercialization, the turn from public to private knowledge production at universities, and other forms of an increasing "academic capitalism" have spread around the world since then (Slaughter & Rhoades, 2004). Formal mechanisms of knowledge spillovers specific to the technology transferred became critically important (e.g., licensing of patented research findings; contract research in cooperation with industry; and university spin-offs or company start-ups by university teachers and/or researchers). Much of the knowledge transferred is codified by those means. They develop between formal organizations—between the university and a specific company, private research lab, or public organization, for example. From this innovation-oriented perspective, research-intensive universities perform well if they have well-developed channels and institutions facilitating communication with private companies and if there are private companies interested in and capable of innovation because they face strong competition. However, this view of innovation is narrow, for patents, licensing, copublications, and spin-offs refer chiefly to technologies originating in the natural sciences, engineering, and life sciences (Benneworth & Jongbloed, 2010). These fields tend to be commercialized to companies, a decidedly stimulating prospect for policy-makers in both the Global North and Africa, not least because these mechanisms can be quantified.

These facts are obviously stylized and may hold for the highly industrialized countries of the Global North and for large emerging economies such as Brazil, India, China, and South Africa. Even in those societies, such mechanisms seldom appear in the humanities, arts, and social sciences, where consultancy, transfer-based research, information services, training, qualification, further-education services, and network-forming services are more common mechanisms of knowledge spillovers (Froese et al., 2014). The burgeoning literature on university engagement in regional development refers principally to codified knowledge outputs that can be commercialized by codified spillover mechanisms, be they copublications by, or contracts between, university staff and nonuniversity actors. These mechanisms may occasionally come into play at some nonmetropolitan universities in Africa as well, but in my

[9]The U.S.'s Bayh–Dole Act (1980) is generally regarded as the starting point of a neoliberal turn in university strategies.

estimation tacit knowledge spillovers and informal mechanisms are more important in engagement with other social partners, such as peasants, nomads, nonprofit organizations, and local communities.[10] It follows that most of the current approaches to studying the linkages between the university and economic development hardly apply to the African context, especially to rural Africa, except for some agricultural research and related extension activities. Hence, neither the concept of regional "engagement" and its addressees nor the methodology for surveying and analyzing such engagement really seem appropriate.

The concept of stakeholders as opposed to a "shareholder" view on university–industry relationships has therefore arisen (Benneworth & Jongbloed, 2010; OECD, 1999). For the purposes of this chapter, I do not take into account the university's internal stakeholders, such as students, teachers, and other members of the institution. Instead, the focus is on external stakeholders—companies, authorities, social groups, nonprofit organizations, even individuals in the regions concerned. Knowledge spillovers to these stakeholders are often not formalized, not codified, and, hence, not easily discerned. Furthermore, codified knowledge from published research findings is scarce to come by. The strength and weakness of the university's engagement is thus less visible than it is with codified knowledge and may even be invisible, so revealing regional engagement requires a different methodology. Evidence will be based less on quantitative data than on anecdotes and interpretations.

Similarly, certain notions of the capacity of the regional society and its stakeholders to "absorb" knowledge from the university are common in economics and economic geography, but they do not strictly apply. Casper (2013) called attention to the "pulling" power of regional industries for promoting spillovers from the university. However, the concept clearly focuses on companies, in line with the Penrosian theory of the resource-based firm (Lane, Koka, & Pathak, 2006). According to Nooteboom (2000), the concept of absorptive capacity keys on the cognitive distance between sender and receiver organizations in knowledge spillover and is thus basically a relational concept. Scholars have frequently pointed out the weakness of links between the predominating, small-scale companies and the university in Africa and to difficulties of establishing those links (Benneh, Awumbila, & Effah, 2004; Oyelaran-Oyeyinka, 2006; Zeng, 2008). It seems overtly clear that the cognitive distance between, say, an illiterate peasant and a university teacher would be even greater and that only enormous effort would bridge it for regional development.

[10] Analyzing the transfer of knowledge from the academic world to industry in Mozambique, Zavale and Macamo (2016) supported this view when they ascertained that informal exchange of "embodied" (not codified) knowledge dominated. Notably, some literature on the Global North, too, calls for broadening the view on the university's regional engagement and for applying an analytical methodology different from mere commercialization (Breznitz & Feldman, 2012; Perkman et al., 2013).

The Hardships of Regional Engagement at Nonmetropolitan Universities in Cameroon

The nonmetropolitan university's regional engagement is seldom an issue in the Global North and, to the best of my knowledge, it is even more rarely so in Africa. The following sections present an exploratory inquiry into the links between three rather young universities and regional stakeholders in the institutional setting of Cameroon.[11] The gathering of data from the universities' staff was designed to gain insight into their visible and invisible interactions with local stakeholders external to the universities in terms of their three missions: teaching, research, and outreach. Because these universities concentrated on teaching and because qualified graduates were regarded as the university's most important contribution to regional development, a team of Cameroonian doctoral students conducted tracer studies on the whereabouts of the graduates. Tracer studies, too, are urgently needed in Africa. The nonmetropolitan universities researched in Cameroon were the University of Buea in the Anglophone part of the country; the University of Dschang in a border region between the Anglophone and the larger Francophone part of Cameroon; and the University of Ngaoundéré, which serves the sizable Francophone part of Cameroon's north and two neighboring countries: Chad and the Central African Republic.

The Cameroonian Socioeconomic Context of Higher Education

Cameroon's universities are a result of the country's unique colonial history, which began with German colonial power, followed by division under French and British colonial power after World War I and a partial reunification in independence after 1964. Literature brims with studies on conflictual issues in Cameroon, such as the

[11]The following section draws widely on a research project on Cameroonian universities in 2006 to 2010 (DFG Scha 237/14-1/-2), which I had supervised. Through collaboration with Cameroonian professors Alexander Asong, Martin Kuete, Michel Simeu Kamdem, and Michel Tchotsoua, it brought together a team of doctoral students consisting of one German and several Cameroonians. On the German side the study encompassed a quantitative survey on current research projects at the universities, which was combined with qualitative information from 70 in-depth interviews with research-project leaders, university officers, and officials from the Ministry of Higher Education and the Ministry of Research and Innovation. The Cameroonian members of the research team then complemented this input with tracer studies conducted at each of the three universities in order to gather information on how and where their graduates find a job. For detailed results see Mediebou and Tchotsoua (2012), Schamp and Zajontz (2010), Simeu Kamdem and Schamp (2014), and, in particular, the doctoral dissertation by Zajontz (2010). Mediebou (2011) and Gondié (2014) from the University of Ngaoundéré and Tchomga (2016) from the University of Dschang have meanwhile successfully defended their doctoral dissertations.

We acknowledge the funding received for this work from the DFG (German Research Foundation), the German Federal Ministry of Cooperation, and the Association of the Friends of Goethe University.

political order, official language policies, and "culture," which are further aggravated by two decades of severe economic crisis. These matters still affect the university staff's self-understanding, the organization of relationships with local stakeholders, and the behavior of local authorities.

For example, Cameroon is officially a bilingual country with French and English as languages in the formal sector because of its colonial past and the absence of a common vernacular "local" language. Of the 279 living indigenous languages, a few serve as a lingua franca in daily life in particular regions. Language thus becomes a substantial problem at the university. University teachers have been obliged to move between the minor Anglophone and major Francophone part of the country in the government's pursuit of an official language policy favoring French. However, many young students do not sufficiently master both official languages (as noted for African universities in general, Brock-Utne, 2003). This inability holds especially true for English in the Francophone part of the country. As an English teacher at Ngaoundéré university wrote in an article about the "hostile environment" for English: "The over-dominance of [the] mother tongue is such that even French[,] which is presumably the medium of instruction in our university[,] is rarely spoken outside the classrooms" (Zogang, 2007, p. 158). But English is also a foreign language for students in the Anglophone part of the country, who use Pidgin English in daily life. Ironically, globalization has triggered a current rush for school teaching in English in urban milieus of southern Cameroon (Fonyuy, 2010), working in favor of Buea as an educational hub. A further consequence of incomplete bilingualism is that university staff sometimes cannot communicate in the language of local stakeholders.

The colonial legacy is still manifest in the prevailing university models, the visions that university staff members have for the university's societal role and their global communication. Schamp and Zajontz (2008) had documented such path dependence in the still dominant patterns of international academic communication. Academics from the Anglophone University of Buea had a relatively global pattern—with the United States as a main travel destination—whereas academics from the predominantly Francophone University of Dschang preferred communication with France and French-speaking West Africa.[12]

Favoritism of ethnic groups is another burden in the current political system of Cameroon (Ngeve & Orock, 2012). The Cameroonian political system has been widely characterized as neopatrimonial, with the head of state presiding over an extended patronage system and playing the ethnicity card (e.g., Bayart, 1979; Nyamnjoh, 1999; Orock & Mbuagbo, 2012). Some regulations facilitate neopatrimonialism. For example, the head of state appoints university rectors or

[12]This pattern of communication is arguably due to a kind of institutional "fix" stemming from the place where teachers had earned a doctorate and where transnational partnerships have been created. However, the burgeoning literature on policies and instruments of research collaboration between universities of the Global North and South predominantly focuses on the U.S.- and U.K.-Africa nexus, essentially neglecting the France-Africa nexus (Koehn & Obamba, 2014; Teferra, 2014). For the Germany-Africa nexus see Schamp and Schmid (2008).

vice chancellors (the heads of the universities in the Anglophone part of Cameroon), and the Secretary of Higher Education appoints the deans and directors of institutes. This arrangement has two inimical effects on university life: corruption and tribalism. Indeed, Cameroon has been repeatedly accused of being "the most corrupt country in the world" (Nyamnjoh, 1999; Orock & Mbuagbo, 2012), with the vice pervading academia as well. Tribalism that shows little regard for merit has been attested several times and causes a *misère intellectuelle* at universities (Nyamnjoh, 1999, p. 107; see also Affa'a & Des Lierres, 2002; Kemayou, 2012).

The system of higher education in Cameroon also suffered from persistent economic crisis in the 1980s and 1990s (Ombga, 2011). A 50% devaluation of the currency in 1994 induced deep cuts in the state budget, the closure of many state enterprises, subsequent lay-offs of company and government employees, a 60% decline in public-sector salaries (at universities, too), additional reduction of purchasing power, and escalating political unrest. Teachers continued to escape by going abroad or looking to earn money in other ways (e.g., various forms of corruption) to support their families (Gaillard & Khelfaoui, 2000; Orock & Mbuagbo, 2012). The economic situation improved when the Cameroonian government achieved a partial debt write-off through the Heavily Indebted Poor Countries (HIPC) Initiative launched by the International Monetary Fund and the World Bank. The government used the arrangement to set up a permanent fund for higher education in 2009. Salaries of university staff rose swiftly, sometimes threefold, from a very low level. Hundreds of lecturers and researchers were hired. The government also invested in new university buildings and new public universities. New departments, faculties, and schools specialized in technological disciplines such as veterinary medicine, pharmaceutics, geology, and fishery were opened ("Cameroon puts," 2010; Minesup, 2014; UNESCO, 2010, p. 302). Establishment of a new satellite link rapidly improved access to the Internet in many towns (Willinsky et al., 2005, p. 13). Nevertheless, these decades of economic crisis have remained powerful memories among academics.

Pressures from a relentless rise in unemployment among university graduates prompted several governmental initiatives fostering professionalization of higher education to enhance employability or, because of weak growth in Cameroon's formal private sector, to spur the entrepreneurship of graduates and strengthen the commitment of the universities to regional development (for details see Zajontz, 2010, pp. 166–170; critically also Chatue, 2016). Since a government decree of December 2007, for example, universities have been "called upon to tailor the competences which they deliver to the students to fit with what enterprises, cooperations, and companies need" (Vice Rector for R&D, University of Buea, personal communication, December 5, 2008). As contradictory as it may sound, a new point of departure seems to be emerging in Cameroon for the university's engagement in development at large, regional development included.[13]

Young Nonmetropolitan Universities in a Young University System

The country's first university, today called the "mother university," was founded in the capital Yaoundé with the backing of France and UNESCO (Ajayi et al., 1996, p. 134) shortly before Cameroon's independence in 1962 and is modeled on the Humboldtian university. It is divided into faculties to which entrance is unrestricted and into schools having entrance examinations. In 1977 the government responded to the soaring demand for higher education in applied sciences by establishing schools in provincial centers as branches of the mother university. These new institutions were located in Buea, the Anglophone part of Cameroon, for language and interpreting; in Douala, the country's leading economic center, for business studies and teacher training in technologies; in Dschang, the densely populated agrarian region of West Cameroon, for agriculture; and in Ngaoundéré, the sparsely populated mid-Cameroonian region, for food science and food technology. Ever greater demand for higher education, combined with periodic student unrest, led to a university reform in the midst of economic crisis in the early 1990s. In addition to laying the ground for the privatization of higher education (an issue that did not become relevant until the 2000s), the public university reform of 1993 transformed these provincial schools into full-fledged universities with soaring numbers of students (Ngwe & Pokam, 2016; Njeuma, 2003; Zajontz, 2010). Student unrest fueled by unsatisfactory work and living conditions and political pressures continued, however (for Buea see Fokwang, 2009, and Konings, 2009; for Ngaoundéré see Woudamike, 2008). Private universities have emerged rapidly since 2004, especially in two urban agglomerations, Yaoundé and Douala (Tsafack Nanfosso, 2006). Two other public universities have opened in remote provinces, Maroua in the Francophone Extreme North Province (2010), and Bamenda in the Anglophone North-West Province (2011) (see Fig. 16.1).

The three universities examined in this chapter focused principally on undergraduate studies in faculties, not schools, and labored under the disadvantages of student massification, a deficient university system, inadequate urban infrastructure, and young, rather inexperienced university teachers intent mostly on advancing their academic careers (for details see Schamp & Zajontz, 2010; Zajontz, 2010). The universities were located in somewhat small urban centers, with Buea having about

[13]There have been several initiatives, notably at universities in Cameroon's large urban centers. The faculty of arts in Yaoundé I, for instance, introduced new curricula on hotel management and tourism, and the Universities of Buea and Douala organized stakeholder meetings to establish partnerships with companies. However, our data was collected between 2006 and 2008, just as reorientation of the universities started, so they do not capture recent developments.

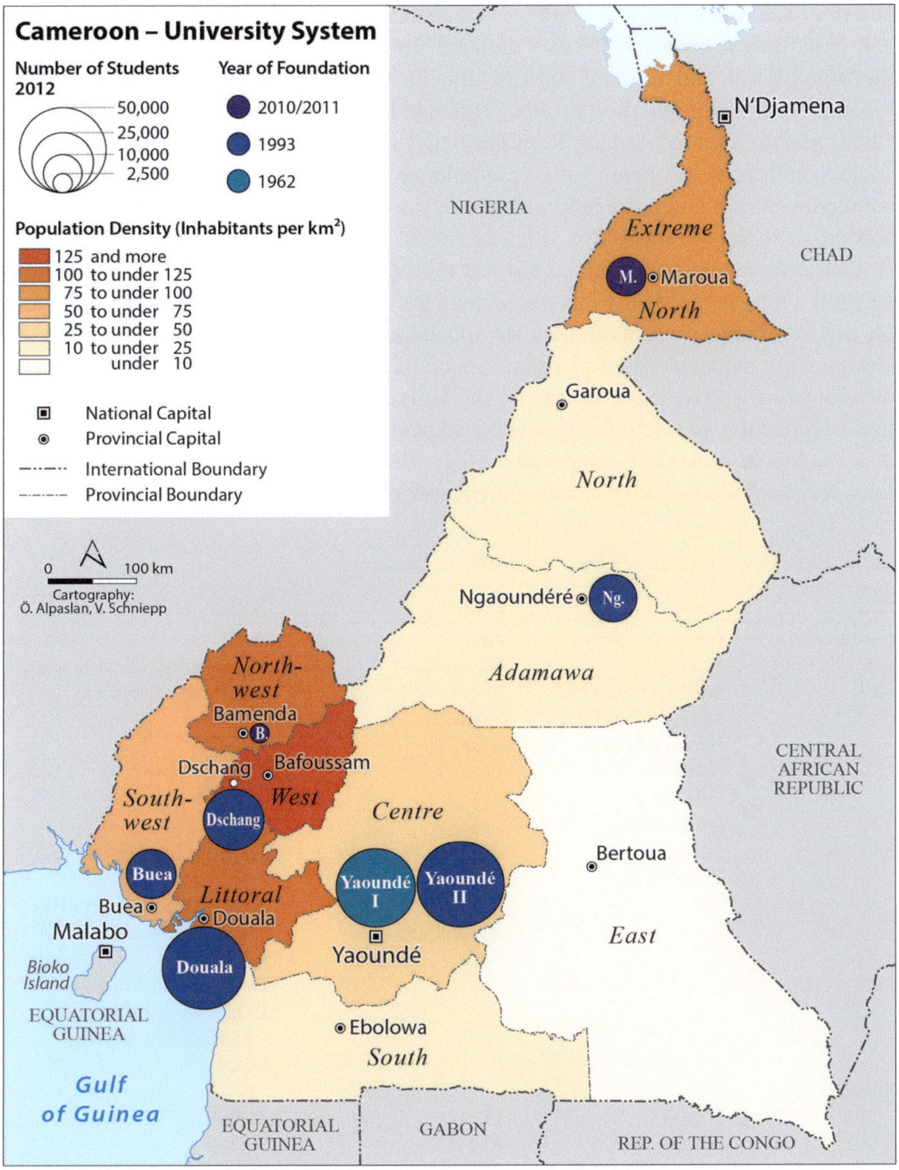

Fig. 16.1 The growth of the public university system in Cameroon, 1962–2012.
Sources: Njeuma (2003), Minesup (2014).

90,000 inhabitants; Dschang, 64,000; and Ngaoundéré, 152,000 (population figures of 2005 as reported by BUCREP, 2010). These settlements represent different rural environments—the plantation economy of South West Cameroon (Buea); the mountainous, small-scale commercial agriculture of the Ouest province (Dschang); and the cattle-raising plateaus of Adamaoua, the northern plains of cash crops (cotton and rice), and subsistence agriculture (Ngaoundéré). As dissimilar as these locations are, they have in common the low demand for university graduates and research in the agricultural and mineral sectors, the burden of widespread adult illiteracy, violence between population groups, violence from neighboring countries (Nigeria, Chad, and the Central African Republic), and violence from authorities. This entire socioeconomic world bears little resemblance to Yaoundé and Douala—although Kemayou (2012) noted the ambivalence of the relationships between the university and urban society in those two cities as well.

Curricula in most disciplines did not take the particular local environment into account. Only very recently has a new state policy on higher education put emphasis on differentiating the universities by specialization. Officials at the University of Dschang are reconsidering its original specialization in agriculture and rural societal development, and decision-makers at the University of Ngaoundéré are strengthening its capacities in scientific and technological disciplines, returning to an old idea of a "technical university" (see Table 16.1). Both institutions, however, still have a long way to go before they achieve their new objectives.

Table 16.1 Basic facts about three universities in Cameroon

Structure	Buea	Dschang	Ngaoundéré
Faculties[a]	Arts Science Social Science and Management Health Sciences (1997–1998) Education (1998–1999)	Arts Economics Legal and Political Sciences Sciences	Arts Legal and Political Sciences Economics Sciences
Schools[b]	Language and Translation	Applied Technology (IUT)[c] "Faculty" of Agriculture	Food Technology (ENSAI) Applied Technology (IUT)[c] Veterinary Medicine (2007) (ESMV) Geology and Mineral Exploitation (2011) Chemical Engineering (2014)
Students			
2006–2007	10,300	11,600	11,900
2012	16,500	25,500	15,300
Professorships			
2006–2007	13	14	7
2012	13	11	13
Teachers			
2006	242	329	183
2012	308	379	264
Student-teacher ratio			
2006–2007	43	35	65
2012	54	67	58

Note. Minesup (2007, 2013, 2014).
[a]Unlimited enrollment, except at the University of Buea. [b]Limited enrollment. [c]Undergraduate studies.

Patterns of Interactions with Local Stakeholders

It is evident from the previous section that these young, small, remote universities were usually incapable of contributing to regional development in the visible ways that are commonly assumed to have a major impact on development. Their dominant mix of nontechnological disciplines and the focus on undergraduates left them with few patents and few opportunities to become seedbeds for university spin-offs. There were almost no copublications by academics and authors from the private industrial sector. Neither collaborative research with industry nor researcher mobility from the university to industry or vice versa was widely practiced. Some minor consultancy relationships were oriented to the political and economic centers, as in the field of accounting. Moreover, the strong hold that the neopatrimonial central state had on staff management[14] lessened the likelihood that day-to-day activities would bring the university and leading regional administrators to network with the few "formal" organizations that exist.

Consequently, the university–regional development nexus seemed all but nonexistent, at least quantitatively invisible. Yet it seemed nearly inevitable that university staff would come into contact with regional stakeholders in the course of teaching and research, if not by contributing to the third mission, outreach, as marginal as it may be. All these interactions may be more adapted, but less apparent, to the local rural societies than quantitatively visible ones are. To discern the kind and scale of regional engagement by the three universities, we members of the research team surveyed each of them for its external ego relationships according to the missions of the university. This work rested on the assumption that regional spillovers occur when both personal and organizational social networks emerge between a university and the region in which it is located.

First Mission: Teaching

Depending on the level and quality of teaching at a university, the teaching mission offered several occasions to establish contact with regional stakeholders and to facilitate intentional or unintentional spillovers of knowledge. The overriding purpose of these three universities was to teach undergraduates, who in 2007 accounted for 95% of the student body at Buea, 87% at Dschang, and 91% at Ngaoundéré. The overwhelming majority of the students were enrolled in faculties (e.g., 93% at the University of Ngaoundéré). Schools, with their entrance examinations and limited enrollment, had comparatively few students. At Ngaoundéré, for instance, the school of food technology accounted for only 3% of the student body; the school of applied technology, for 4% (Zajontz, 2010, pp. 149, 151). Human resources in teaching were weak, as indicated by a high student-to-teacher ratio and a low rate of professorships (Table 16.1). Some graduate courses have been established since the early 2000s, a

[14]By contrast, local elites in the Cameroonian provinces had long lobbied for a public university to be founded there as a sign of their political power (Sikombe, 2016, p. 151).

process interfering with the introduction of the bachelor-master system (Bologna) in Cameroon. The universities awarded their first doctorates in 2001 (Buea), 2004 (Ngaoundéré), and 2005 (Dschang).

Contact and cross-fertilization in knowledge may occur, first, through teaching, chiefly by means of teacher exchange and student internships. Teacher exchange had occasionally been reported only within the public sector because of a regulation that university teachers attached to the Ministry of Education in Yaoundé had to teach part-time. Internships were obligatory in some disciplines, primarily in technical fields where bachelor and master theses, too, were written in liaison with companies. This requirement applies to most universities in the world. However, these contacts were poorly managed at the three universities covered in this chapter, so partners in the regional private sector were rarely found. Internships came about principally in the two distant metropolises and essentially depended on the willingness and motivation of the students and university teachers involved. The university administrators undertook only a few initiatives, such as pursuing formal cooperation agreements between the faculty of health at Buea and regional hospitals and, at the school of food technology at Ngaoundéré, administering a regular data bank on organizations that offer internships.

Supplying the region with a qualified labor force was a fundamental target of the university's teaching mission. Aside from the low absorptive capacity of the regions in question, the universities generally had no basic knowledge about the whereabouts of their graduates. Placement of graduates was fairly rare and was based on the personal relationships of certain teachers only. Tracer studies of university graduates were as few in Cameroon (Fohopa, Garro, & Mortelette, 2006) as elsewhere in sub-Saharan Africa (Mugabushaka, Schomburg, & Teichler, 2007), a paucity even more conspicuous at the regional level. Detailed tracer surveys at the three universities plausibly revealed that academic labor markets were nonregional, fragmented, and difficult to access and that the supply of "human capital" from the universities was therefore fairly marginal (Mediebou & Tchotsoua, 2012; Simeu Kamdem & Schamp, 2014). This statement applied especially to disciplines such as the humanities, whose ministry-approved curricula seemed inappropriate ("traditional" or "unfeasible," according to several interviewees), at least for the private sector. Those fields of study may be more in tune with the few available public-sector jobs for which most graduates still strive (education, administration). In fact, unemployment of university graduates was high, yet graduates showed little aspiration for self-employment or entrepreneurship. Graduates seeking employment preferred to migrate to Yaoundé and Douala, and it seemed that only the less successful graduates tended to return to their home region to eke out a living (Tchomga, 2016).

Second Mission: Research

Although the research mission of universities is usually the source of a key argument for their engagement in regional development, there are several reasons why it is unlikely to be so at the three universities in the study presented in this chapter. Research is less a mission for them than it is elsewhere, and it may be less technological, given the dominance of their faculties.[15] Can comparatively few research projects become an important font of knowledge spillovers to regional stakeholders? Can those projects become that wellspring if they are at least of "applied" character and if they focus on local empirical evidence? In this section I tackle these questions in a search for less visible and sometimes less purposive interactions with local stakeholders.

The limited knowledge that university administrators have about current research, particularly when projects are self-funded by academic staff, is not the only obstacle in this quest. It is also difficult to formulate a clear definition of "research" and "project." Self-funded projects can stretch over a long unspecified period, and projects include different activities such as writing a book, establishing a data bank, consulting, supplying services to laboratories, drawing on disciplinary knowledge about different local environments (applied research), and conducting basic research. In the Francophone universities research is partly self-organized and partly organized in what are called research laboratories (with physical equipment only in the sciences and technical sciences), which must be approved officially by the university. These research labs are sometimes umbrella organizations that perform no work. To shed light on the local reach of the research mission, Zajontz (2010) surveyed 104 research projects that came to attention at the three universities. She found that they encompassed a striking diversity of disciplines and topics (Table 16.2).[16]

The preconditions of research at these three institutions of higher education are even worse than those discussed generally for African universities (Sawyerr, 2004). Basic infrastructures are insufficient (library, Internet access, laboratories), and funding is marginal. The low number of professorships leads most of the teachers to pursue research in order to advance their own academic careers. Questioned about the sources of funding, 30% of the 87 researchers who responded in the 2006–2008 survey had no funds (Zajontz, 2010). When funded by their own university (another 30%), most researchers at the universities of Buea, Dschang, and Ngaoundéré had only one thousand to three thousand Euros per research project at their disposal. Allocation of university funds was regarded as nontransparent at Francophone universities, where more transparent regulation was said to exist at the Anglophone university. Another 38% of the projects were partly or totally funded externally, often by international funding organizations, partly by governments (development aid), partly by foundations and research councils (Table 16.3). For this 38% of the projects, support averaging several tens of thousands of Euros was available to the project leaders (Zajontz, 2010, p. 217). Whereas these sources of funding were sometimes seen critically as a sign of postcolonial dependence (and of academic

[15] A report on research policy in Cameroon (Gaillard, van Lill, Nyasse, & Wakata, 2014) stated that academic research still suffers from "a lack of organisation and management, scattered topics and multiplication of microresearch groups unable to reach the critical mass for productive research, and lack of financial resources" in general (pp. 9–10).

[16] The response rates were 35% for project leaders and 80% for leaders of laboratories. The survey was supplemented by 21 in-depth interviews with research and laboratory leaders. We are not able to present all the research efforts that took place at each university at a given time, though.

Table 16.2 Pattern of research projects at Cameroonian universities surveyed between 2000 and 2008, by discipline and type of university structure

Structure	Buea	Dschang	Ngaoundéré	Total
Faculties				
Humanities	13	13	15	41
Science	10	20	14	44
Schools	2	2	15	19
Total	25	35	44	104

Note. Zajontz (2010); E. W. Schamp's own calculation reconsidering raw data of Dschang and Ngaoundéré and, hence, slightly differing from Zajontz (2010), Schamp and Zajontz (2010), and Simeu Kamdem and Schamp (2014).

Table 16.3 Surveyed research projects at three Cameroonian universities, by geographical and thematic scope

Type of project	Buea	Dschang	Ngaoundéré	Total
Academic projects on local issues	11	6	15	32
Academic projects on nonlocal issues	14	23	27	64
Local services by research	...	6	2	8
Total	25	35	44	104
Projects funded or cofunded externally[a]	6	15	17[b]	38

Note. Author's own calculation.
[a]The projects addressed local and nonlocal issues, were generally in sciences and technology, and were funded by international organizations and foundations. [b]Two projects were cofunded by a Swiss pharmaceutical company and one by a Cameroonian firm.

services migrating abroad), they may, conversely, also indicate a capacity of researchers to approach international donors for project funding.

Broadly speaking, cultural differences pose a further barrier to researchers' interaction with local stakeholders. Many African regions are highly diversified in their societal statuses, number of ethnicities, local languages, and ways of life (e.g., urban, rural commercial, rural subsistence, agrarian, and nomadic). In Cameroon most of the university teachers are not natives of the region. (There are fewer out-of-region teachers at the universities of Buea and Dschang than at Ngaoundéré.) How can and do they communicate with regional rural societies involved in their research projects. What do teachers do to disseminate their results to local stakeholders? In some social science research projects a practical solution to this problem was to have local students participate as translators and interpreters.

To assess a project's possible relatedness to the local environment, we also distinguish between the research's geographical scope (location of university) and type (academic research or research services). Regional knowledge spillovers from research may take different forms, such as an outcome of academic research proper or a service that uses the university's technical equipment or provides consulting for public authorities and the private sector. Research may be defined as "local" if it addresses local stakeholders such as peasants, local administrators, or local health-care service providers. It is "nonlocal" if it grapples with national issues or issues of basic research. Individually and university-funded research was viewed as "local" when the topics had to do with the domestic natural and social environment, and it sometimes treated questions of applied research. Service projects relating to public authorities were found mainly in the agricultural sciences (Dschang) and geomatics (Ngaoundéré, Dschang).[17] Many more nonlocal research projects than local ones have been totally or collaboratively funded externally—generally by international research organizations, public donor organizations for the Francophone world, and international foundations. In some of these cases, it was not clear whether the researchers were able to apply for funding or whether they acted as local service providers to international research programs. In exceptional instances researchers found support from foreign companies. We therefore found little evidence of patents that had emerged from academic research (three or four projects only), and foreign firms seemed to be the patent holders in those cases. Apparently, most of the acquired new knowledge was unpatentable because of the research disciplines involved, unsatisfactory national patent legislation, or lack of assistance from university and government in the patenting process.[18]

There is, however, much overlap between and ambiguity in the categories of Table 16.3. In terms of local development, the table therefore serves only as an initial approach to a typology of research. The research team was unable to make a quantitative assessment of the impact that the research projects had on local development. Only the number and kind of projects were evaluated, and the method of data collection did not enable us to gauge the data's representativeness and reliability.

Nonetheless, 32% of the academic research and service projects included in our case study had a local character. Moreover, 61% of the projects that were categorized as treatment of a nonlocal topic had a local dimension as well. For example, medicinal plants were seen as an important focus of university research in Cameroon (Gaillard & Khelfaoui, 2000), and research on local medicinal plants was conducted at each university, though in different disciplines and with different approaches. The researcher may cooperate with international organizations and companies abroad *and* be in contact with local traditional doctors (*tradipraticiens*), either to gather information on plant choice or to provide information on properties of plants in order to improve treatment. Of the 15 "local" projects at Ngaoundéré, five were oriented to farmers (pest control, nutrition from different plants, beekeeping); another five, to traditional doctors (analysis of medicinal plants); and an additional two, to small agroindustries (photovoltaic energy supply, establishment of a food-value chain). In Buea, local topics encompassed a wide range of social problems in semiurban and rural settings (e.g., the identity crisis of migrant workers, the education of girls in a nomadic society, local languages, political autonomy, health, and environmental

[17]Projects concerned soil and water pollution tests, for instance, or the use of geographical information systems for the improvement of power supply configurations that have emerged historically in a chaotic way.

[18]Cameroon is faulted for its weak legislation and institutions governing intellectual property rights, as exemplified by the implementation of an access and benefit-sharing policy on biodiversity (Rosendal, 2010).

management). At the University of Dschang, we could not sufficiently discriminate between local research and services, especially with projects in the faculty of agriculture, where, according to information from the university, researchers working on integrated agriculture seemed to be collaborating with farmers and agricultural services in the region.

Hence, certain topics existed across these universities, with each institution addressing them according to its particular capabilities. Common research subjects in science and agriculture were medicinal plants, food and food storage, and water and waste management. In the social sciences (including arts, economics, management, sociology, and education), local projects helped to document cultural artifacts (local heritage), to study local crises (identity of labor migrants, the coffee economy, environmental damages), and to cope in general with manifold societal crises at the local and regional level.

These findings appear to show that university research can have an important meaning in local development. Indeed, many researchers claimed to be conducting applied research and wished to contribute to regional development. Their viewpoint was consistent with that reported by Kruss, Visser, Aphane, and Haupt (2012), who found a great deal of commitment and responsiveness to social development among teachers and researchers at a South African rural university. Table 16.4, though, reveals that communication and cooperation with local stakeholders is the exception rather than the rule at the nonmetropolitan Cameroonian universities studied in this chapter.

Obviously, research projects rarely provide for knowledge dissemination to local stakeholders, except public authorities. Although research leaders casually expressed their wish to communicate to local stakeholders, they were unable to conceptualize communication channels, let alone use them. Because there were few, if any, project structures and real initiatives therein, possible knowledge spillovers were not discernable. We members of the research team nevertheless uncovered outstanding examples of less visible communication from projects conducted in cooperation with traditional doctors, peasants, nomads, nurses in rural areas, and

Table 16.4 Number, kind, and percentage of partnerships in "local" Cameroonian research projects

	University		
Partnerships	Buea	Dschang	Ngaoundéré
Local academic projects	11	6	15
Total of cooperative partnerships[a]	74	21	37
Percentage of			
academic partnerships[b]	58	57	65
partnerships with local administration	19	14	13.5
partnership with groups from the informal sector and peasantry	5	. . .	8
other partnerships[c]	18	29	13.5

Note. Authors' own calculation.

[a]Partnerships in the survey were ranked from 1 (*very important*) to 5 (*unimportant*). The total number of cooperative partnerships is the sum of the rankings 1 through 3. [b]With other disciplines at the same university (majority), at other universities in Cameroon (primarily the "mother university," Yaoundé I), and at universities abroad. [c]Mostly foreign organizations.

other stakeholders (for details on the University of Ngaoundéré, see Schamp & Zajontz, 2010). In-depth interviews revealed interesting ways in which academic, often nontechnical research had of communicating with and having impacts on the local society. In one project local social science students had established communication with rural communities by serving as translators. In another project researchers were working on medicinal plants and communicating with professional groups (traditional doctors). A third case was a university-initiated project on geographical information systems (GIS) that had become a municipal service and consultancy. In yet another instance networking effects had emerged in a project on nomads and had forged links between nomadic society and development-aid NGOs. A fifth project had resulted in regional further education. Other outstanding examples were Ngaoundéré projects on analysis of the enabling environment and on the introduction of beekeeping where awareness, interest, initiative, and ideas generated over the radio, creation of a beekeeper's association, and training of beekeepers had contributed to the spread of beekeeping as an additional source of income for subsistence farmers. There were thus many different anecdotal examples of communication between university research and local civil society. Cultural and ideological barriers to communication exist, too, as demonstrated by policy-makers and health-care officials who, sticking to Western medicine, refused to test and accept knowledge from research on medicinal plants—a stance quite the opposite of that in Asia (project leader Ngaoundéré, personal communication, February 2, 2008).

Many research projects from faculties such as the arts, economics, and legal and political sciences tackled current problems of cultural and economic life in Cameroon. But whether local or not, their local reach was limited for want of communication channels. The projects tended to generate "background publications" that are available to policy-makers and administrators only. It seems plain that many of the projects in the sciences and technical sciences (e.g., engineering and agriculture) helped improve technical knowledge—but principally in the major urban centers, not the rural areas.

Summing up, there is sufficient anecdotal evidence that academic researchers are willing to deal with problems of the local society and that interactions with local stakeholders in rural areas are sometimes unexpected and overwhelming. These interactions remain largely invisible, however, and become apparent only through in-depth interviews. Further socioanthropological analysis is required for full assessment of their impact on the development of local societies.

Third Mission: Outreach—Building Local Intermediaries of Knowledge Spillovers

Teaching and research per se do not make the university available for engagement in regional development. Both missions brought about rather disordered, invisible, and ineffective forms of engagement. This assessment also holds for what can be seen as the high road of university engagement in regional development, namely, the creation of research-based firms (university spin-offs) by university staff and graduates. However, another consequence of the focus on undergraduate-level teaching and of the weak research landscapes at the three universities singled out in this chapter is that spin-offs as defined in the literature are relatively unlikely. A detailed search for enterprises created by graduates of the University of Ngaoundéré between 1993 and 2006 identified 17 small firms, mostly in the School of Food Technology and the School of Applied Technology. Nine of these persons were available for an interview. The others had disappeared because the firms had either closed or relocated. None of the firms was related to any research project but drew on standard knowledge from the relevant disciplines (food technology, computer sciences, maintenance). Needing urban clientele, firms in food-processing, beverage production, and maintenance services have all been created in Yaoundé and Douala, far from the small town of Ngaoundéré and its rural neighborhood. Unsurprisingly, the graduate entrepreneurs have kept little contact with their alma mater.

Unlike teachers and graduates of engineering, those of science and the humanities in general were able to create their own private enterprises at the two other university locations, Dschang and especially Buea. University staff from the social sciences and humanities founded a range of private educational institutions—a secondary, a polytechnical, and an evening school—and have meanwhile hired several graduates of those universities as teachers. Staff from the faculty of science and health at Buea established a private high school for vocational training, mainly in health care, and a private hospital, whose trained nurses then created drug shops in residential areas. On one hand, these activities respond to an increasing inability of the public education system to keep pace with the rising demand at all levels of education and with the privatization policies introduced into the educational system in the early 2000s. A challenging entanglement of public-private linkages in the education system has emerged because these private schools must be officially sanctioned by the Ministry of Education and supervised by the public university—requirements that often result in the hiring of part-time teachers from the nearby public university. Several shortages were thereby addressed at the same time: those in the provision of public services such as health care and education; in the labor market for graduates; and, occasionally in personal income at the university. On the other hand, Buea and, to a more limited extent, Dschang offer advantages deriving from an environment that is more commercialized and closer to urban centers than Ngaoundéré. Buea in particular, an Anglophone setting, has become an educational hub for ordinary and advanced education in West Cameroon, for the city benefits from both its proximity to Douala's Francophone metropolitan area and that society's rising demand for English as a medium of instruction in the educational system (Fonyuy, 2010).

These private initiatives have emerged in a spontaneous and disordered way, as did the occasional university services for public and private organizations. This pattern characterized the previously mentioned teaching in GIS at the University of Ngaoundéré, which evolved into diverse services provided to local authorities and the Cameroonian utilities company and which led to the establishment of a local commercial training unit. It typified services that the faculty of agriculture at Dschang extended to large plantations held by major companies. It also occurred with workshops that the department of journalism in Buea offered to professional journalists. These services go hand in hand with policies on the professionalization of higher education, which were relaunched in the mid-2000s when the

Cameroonian government obliged universities to expand the university's budgetary "autonomy." Nonmetropolitan universities were less able to follow metropolitan universities in commercializing master's degree courses at fees up to 15 times higher than the similar public ones. However, they did attempt to strengthen their outreach, or "third mission." Based on visible communication, formal network-building, and formal contracts between the university and external stakeholders, outreach has become the cornerstone of the entrepreneurial and developmental university models.

Accomplishing the third mission as discussed in the Global North means creating new intermediary institutions that connect university functions to stakeholders' needs systematically. The University of Dschang established the *Groupement d'intérêt économique* in 2006 as Cameroon's first university center for knowledge transfer, forming a coalition consisting of the university, the Friends of the University's Personnel, the University Teacher Association, and a local bank (Zajontz, 2010, pp. 252–253). In 2007 the University of Buea launched a series of conferences with key external stakeholders, even companies from nearby Douala, to improve contacts, establish "partnerships," and formalize procedures for internships. Additional initiatives may have followed, but it seems far too early to expect visible results, especially because they remain embedded in a neopatrimonial system.

Obviously, the nonmetropolitan universities still have little capacity for organizing the outreach function. Linkages and knowledge spillovers are usually undocumented and nontransparent, and university administrators are generally uninformed about them. However, most of the linkages seem to be nonlocal and oriented to the metropolises, as exemplified by the university's staff consultancy in disciplines such as accountancy, business studies, and legal sciences.

Conclusions

As noted at the beginning of this chapter, the university's regional engagement in Africa differs from that on other continents because of Africa's particular historical, societal, economic, and political context. I have sought to confirm this observation with a case study on three nonmetropolitan universities in Cameroon. It appears that the empirical analysis of these African examples yields few visible testimonials on regional engagement as usually studied in the literature in the Global North and in Asia's emerging economies. Such publications focus on private companies, which hardly exist in these African cases.

This statement needs qualification, though. First, interpretation of the results can differ depending on what the term *engagement* is taken to mean. Usually, the university's engagement is seen "as knowledge-related collaboration by academic researchers with non-academic organisations" (Perkman et al. 2013, p. 424), including profit-oriented companies and nonprofit-oriented societal groups and individuals. This understanding is generally based on direct, purposive communication between university staff and nonuniversity stakeholders. If communication is formalized, it becomes visible and measurable for empirical analysis. Because of numerous shortcomings—such as weak management capacities of the universities and overall weakness of the institutional context—the Universities of Buea, Dschang, and Nagoundéré had not yet seriously developed either strategies or organizational structures for regional engagement. There was no feasible intermediary for communication between the university and regional stakeholders. It would be misleading, however, to conclude that these universities were not engaged in their region at all and that personal communication between university staff and the region did not exist. If the term *engagement* is used broadly to mean any outcome of knowledge that reaches local societies—whether planned or unplanned, direct or indirect, and hence "tacit" and undocumented, "invisible"—then the university is indeed "engaged" in regional processes.

Given the universities' orientation to the mission of teaching undergraduates and given the scarcity of resources for research, some results nevertheless seem surprising. Regional engagement came about less through teaching (e.g., organizing internships and providing graduates) than through research. University staff members widely believed they should pursue "applied research" that responds to the needs of the region's "ordinary people"—a link long said to be "really missing" in Africa (Brock-Utne, 2003, p. 46). Unfortunately, neither personal strategies nor the universities' organizational framework has provided the means of communication necessary to achieve this goal. The university's "service" role has been criticized for preventing innovation-oriented basic research at universities and even for restricting academic freedom and identity (Chachage, 2006; Niang, 2005). This discomfort may characterize research-led universities, the flagships (Cloete et al., 2015), more than nonmetropolitan universities, a crucial distinction in my argumentation. Scholars have occasionally faulted Africa's university systems for their lack of diversification (Brundenius, Lundvall, & Sutz, 2009), a view that basically holds also for the current literature on the African university. The tension between the service orientation and research orientation of universities, between the supply of higher education in general and the demand for special technological disciplines (such as veterinary medicine, biogenetics, and geology) certainly calls for increased diversification of the African university system.

Second, the likelihood that visible forms of regional engagement by the university can be unraveled also depends on the region's endowments and societal structures. The manufacturing sector is negligible in the university regions considered in this chapter, urban services are underdeveloped in the rather small towns where nonmetropolitan universities are located, and neither public administrators nor the few major agricultural companies that exist are interested enough in communicating with the university to engage in regional development.

In brief, regional capacity to absorb knowledge spillovers is slight. Kruss et al. (2012), whose South African case study traced collaboration with partners other than firms, emphasized the strong regional commitment that a rural university's staff showed to community and welfare organizations, government, and civic society in general. Such variety and availability of social partners may not yet exist in Cameroon, for that country's civil society still suffers from many weaknesses (Nkwi, 2006). Nevertheless, Kruss et al.'s findings support my call for an analysis of less visible, "soft" mechanisms of regional engagement on behalf of noncommercial stakeholders. Anecdotal evidence of such mechanisms surfaced at each of the three universities examined in this chapter.

The insight that the African context is peculiar for university engagement in regional development is not new. It has been stressed repeatedly in comparisons between African and Asian models of educational systems and development (Lawrence, 2010, p. 27; St. George, 2006). Catchwords for the Asian model are a high literacy rate, rapid expansion of public investment in education, simultaneous development of education and industry, the common Confucian heritage with its high priority on meritocratic principles, and interventions by a strong developmental state (St. George, 2006). None of these characteristics applies to Africa, where the literacy rate is low, the state weak, state budgets are restricted, and growth (if any exists) is based on raw materials. Last but not least, the World Bank has long recommended that higher education in Africa be neglected in favor of primary education.

Worse still, African universities, notably those in Cameroon, have been subject to government initiatives that obviously transform the university's underlying developmental model into an entrepreneurial one. This change may respond to many pressures from globalization in higher education. However, the relevant policies are fragmented and contradictory and ultimately favor commercialization of university output and privatization of higher education in large urban areas. The nonmetropolitan universities analyzed in this chapter are ill equipped to respond as developmental universities in practice, but they have not really been able to become entrepreneurial in their region either, partly because of their region's character. They have been unable to foster engagement by networking, communicating with local stakeholders and by providing relevant institutions for local knowledge spillovers. Hence, they have been unable to turn the invisible forms of engagement into visible ones and to improve the ways they function. Unfortunately, our findings may stand for hundreds of nonmetropolitan universities that have been established in Africa. To put it more generally, the developmental university model described by Göransson and Brundenius (2011) seems more a dream than reality.

Acknowledgements I thank Peter Meusburger, Andreas Stamm (German Development Institute, Bonn), and anonymous referees for their helpful comments on an earlier draft of this chapter. I am particularly indebted to Michel Simeu Kamdem (Institut de Cartographie du Cameroun, Yaoundé), Mediebou Chindji (University of Yaoundé I), and Philippe Tchomga (Dschang) for their suggested improvements in the text.

References

Acemoglu, D., & Robinson, J. A. (2012). *Why nations fail: The origin of power, prosperity, and poverty*. New York: Crown Business.

Adams, J., King, C., & Hook, D. (2010). *Global research report: Africa*. Leeds: Thomson Reuters. Retrieved from http://researchanalytics.thomsonreuters.com/m/pdfs/globalresearchreport-africa.pdf

Affa'a, F.-M., & Des Lierres, T. (2002). *L'Afrique Noire face à sa laborieuse appropriation de l'université: Les cas du Sénégal et du Cameroun* [Black Africa in view of its painful appropriation of the university: The cases of Senegal and Cameroon]. Paris: L'Harmattan.

Ajakaiye, O., & Kimenyi, M. S. (2011). Higher education and economic development in Africa: Introduction and overview. *Journal of African Economies, 20* (AERC Supplement, ii3–ii13). doi:https://doi.org/10.1093/jae/ejr027

Ajayi, J. F. A., Goma, L. K. H., & Johnson, G. A. (1996). *The African experience with higher education*. Athens: Ohio University Press.
Altbach, P., Reisberg, L., & Rumbley, L. E. (2009). *Trends in global higher education: Tracking an academic revolution.* Paris: UNESCO.
Bayart, J.-F. (1979). *L'Etat au Cameroun* [The state in Cameroon]. Paris: Presse de la Fondation Nationale des Sciences Politiques.
Benneh, G., Awumbila, M., & Effah, P. (Eds.). (2004). *African universities, the private sector and civil society: Forging partnerships for development.* Accra: Ghana University Press.
Benneworth, P., & Jongbloed, B. W. (2010). Who matters to universities? A stakeholder perspective on humanities, arts and social sciences valorisation. *Higher Education, 59*, 567–588. doi: https://doi.org/10.1007/s10734-009-9265-2
Breznitz, S. M., & Feldman, M. P. (2012). The engaged university. *Journal of Technology Transfer, 37*, 139–157. doi:https://doi.org/10.1007/s10961-010-9183-6
Brock-Utne, B. (2003). Formulating higher education policies in Africa: The pressure from external forces and the neoliberal agenda. *Journal of Higher Education in Africa, 1*, 24–56. Retrieved from http://www.codesria.org/spip.php?article561&lang=en and http://www.jstor.org.proxy.ub.unifrankfurt.de/stable/i24484365
Brundenius, C., Lundvall, B.-A, & Sutz, J. (2009). The role of universities in innovation systems in developing countries: Developmental university systems—Empirical, analytical and normative perspectives. In B. A. Lundvall, K. J. Joseph, C. Chaminade, & J. Vang (Eds.), *Handbook of innovation systems and developing countries: Building domestic capabilities in a global setting* (pp. 311–333). Cheltenham: Edward Elgar.
BUCREP (Bureau Centrale des Recensements et des Etudes de Population, Cameroun). (2010). *Rapport de présentation des résultats définitifs* [Presentation of final results of the 2005 census]. Yaoundé: BUCREP.
Cameroon puts brain drain into reverse. (2010, January 5). Retrieved April 25, 2017, from http://www.scidev.net/global/migration/news/cameroonputs-brain-drain-into-reverse.html
Casper, S. (2013). The spill-over theory reversed: The impact of regional economies on the commercialization of university science. *Research Policy, 42,* 1313–1324. doi:https://doi.org/10.1016/j.respol.2013.04.005
Chachage, C. (2006). The university as a site of knowledge: The role of basic research. *Journal of Higher Education in Africa, 4*(2), 47–67. Retrieved from http://www.codesria.org/spip.php?article557&lang=en
Charlier, J. E., & Croché, S. (2009). Can the Bologna process make the move faster towards the development of an international space for higher education where Africa would find its place? *Journal of Higher Education in Africa, 7*(1–2), 39–59. Retrieved from http://www.codesria.org/spip.php?article168&lang=en
Chatue, J. (2016). Epistemologie et auto-constitution minimum d'un système universitaire: remarques sur la philosophie de la réforme universitaire au Cameroun [Epistemology and minimum self-constitution of an academic system: Remarks on the philosophy of university reform in Cameroon]. In L. Ngwé & Pokam, H. de P. (Eds.), *L'enseignement supérieur au Cameroun depuis la réforme de 1993: Dynamiques et perspectives* (pp. 81–104). Dakar: Codesria.
Cloete, N., Bailey, T., Pillay, P., Bunting, I., & Maassen, P. (2011). *Universities and economic development in Africa.* Cape Town: CHET (Centre for Higher Education Transformation).
Cloete, N., Maassen, P., & Bailey, T. (Eds.). (2015). *Knowledge production and contradictory functions in African higher education*. Cape Town: African Minds Publishers.
Croché, S., & Charlier, J. E. (2012). Normative influence of the Bologna process on French-speaking African universities. *Globalisation, Societies and Education, 10,* 457–472.
Djouda Feudjio, Y. B. (2009). L'adoption du « système LMD » par les universités du Cameroun: Enjeux, contraintes et perspectives [The adoption of the BMD system by the universities of Cameroon: Challenges, constraints, and perspectives]. *Journal of Higher Education in Africa, 7 (*1–2), 141–157. Retrieved from http://www.codesria.org/spip.php?article168&lang=en
Fohopa, R., Garro, O., & Mortelette, J.-P. (2006). *L'emploi et la formation au Cameroun. L'enquête génération 2000* [Employment and education in Cameroon: The generation 2000 survey]. Yaoundé: Imprimerie Saint Paul.

Fokwang, J. (2009). Student activism, violence and the politics of higher education in Cameroon: A case study of the University of Buea (1993–2003). In D. Chimanikire (Ed.), *Youth and higher education in Africa: The cases of Cameroon, South Africa, Eritrea and Zimbabwe* (pp. 9–33). Dakar: Codesria.

Fonyuy, K. E. (2010). The rush for English education in urban Cameroon: Sociolinguistic implications and prospects. *English Today, 26*, 34–42. doi:https://doi.org/10.1017/S0266078409990289

Froese, A., Mevissen, N., Böttcher, J., Simon, D., Lentz, S., & Knie, A. (2014). *Wissenschaftliche Güte und gesellschaftliche Relevanz der Sozial- und Raumwissenschaften: Ein spannungsreiches Verhältnis* [Epistemic robustness and social relevance in the social and spatial sciences: A relationship of tensions]. WZB discussion paper (SP III 2014-602). Retrieved from https://bibliothek.wzb.eu/pdf/2014/iii14-602.pdf

Gaillard, J., & Khelfaoui, H. (2000). *La science en Afrique à l'aube du 21ème siècle: La science au Cameroun* [Science in Africa at the dawn of the twenty-first century: Science in Cameroon]. Aide No. ERBIC 18 CT 98 9164 Commission Européenne, DG XII, Rapport Final. Paris: IRD. Retrieved from http://horizon.documentation.ird.fr/exl-doc/pleins_textes/divers07-09/010033546.pdf

Gaillard, J., van Lill, M., Nyasse, B., & Wakata, A. (2014). *Science granting councils in Sub-Saharan Africa: Country report Cameroon.* Stellenbosch: Centre for Research on Evaluation, Science and Technology.

Goldstein, H. A. (2009). What we know and what we don't know about the regional economic impacts of universities. In A. Vargas (Ed.), *Universities, knowledge transfer and regional development* (pp. 11–35). Cheltenham: Edward Elgar.

Goldstein, H. A. (2010). The 'entrepreneurial turn' and regional economic development mission of universities. *The Annals of Regional Science, 44*, 83–109. doi:https://doi.org/10.1007/s00168-008-0241-z

Gondié, H. (2014). *Diplômés des parcours sciences sociales et humaines de l'université de Ngaoundéré et développement socio-économique du Nord-Cameroun* [Graduates from social and human sciences of the University of Ngaoundéré and the socioeconomic development of North Cameroon] (Unpublished doctoral dissertation). Faculty of Arts, University of Ngaoundéré, Cameroon.

Göransson, B., & Brundenius, C. (Eds.). (2011). *Universities in transition: The changing role and challenges for academic institutions*. New York: Springer.

Jöns, H., & Hoyler, M. (2013). Global geographies of higher education: The perspective of world university rankings. *Geoforum, 44*, 45–59. doi:https://doi.org/10.1016/j.geoforum.2012.12.014

Kemayou, L. R. (2012). L'Université de Douala, entre ancrage et désenchantement [The University of Douala between anchorage and disappointment]. *Journal of Higher Education in Africa, 10* (2), 95–117. Retrieved from http://www.codesria.org/spip.php?article1792&lang=en

Koehn, P.H., & Obamba, M.O. (2014). *The transnationally partnered university: Insights from research and sustainable development collaborations in Africa.* New York: Palmgrave Macmillan.

Konings, P. (2009). University crisis and student strikes in Africa: The case of the University of Buea (Cameroon). *Journal of Higher Education in Africa, 7*(1–2), 211–233. Retrieved from http://www.codesria.org/spip.php?article168&lang=en

Kruss, G., Visser, M., Aphane, M., & Haupt, G. (2012). *Academic interaction with social partners: Investigating the contribution of universities to economic and social development.* Cape Town: HSRC.

Lane, P. J., Koka, B. R., & Pathak, S. (2006). The reification of absorptive capacity: A critical review and rejuvenation of the construct. *Academy of Management Review, 31*, 833–863. doi: https://doi.org/10.5465/amr.2006.22527456

Lawrence, P. (2010). The African tragedy: International and national roots. In V. Padayachee (Ed.), *The political economy of Africa* (pp. 19–39). London: Routledge.

Mediebou, C. (2011). *Les diplômés des parcours sciences et techniques de l'université de Ngaoundéré et le développement du Nord-Cameroun* [Graduates from sciences and technological sciences of the University of Ngaoundéré and the development of North Cameroon] (Unpublished doctoral dissertation). Faculty of Arts, University of Ngaoundéré, Cameroon.

Mediebou, C., & Tchotsoua, M. (2012). Analyse géographique du parcours sciences et techniques. Cas de l'université de Ngaoundéré [Geographical analysis of the occupational career [of graduates] in science and technology: The case of the University of Ngaoundéré]. *Journal of*

Higher Education in Africa, 10(2), 119–138. Retrieved from http://www.codesria.org/spip.php?article1792&lang=en

Meusburger, P. (2013). Relations between knowledge and economic development: Some methodological considerations. In P. Meusburger, J. Glückler, & M. El Meskioui, (Eds.), *Knowledge and the economy* (pp. 15–42). Knowledge and Space: Vol. 5. Dordrecht: Springer. doi:https://doi.org/10.1007/978-94-007-6131-5_2

Meusburger, P. (2017). Spatial mobility of knowledge: Communicating different categories of knowledge. In H. Jöns, P. Meusburger, & M. Heffernan (Eds.), *Mobilities of knowledge* (pp. 23–50). Knowledge and Space: Vol. 10. Dordrecht: Springer. doi:https://doi.org/10.1007/978-3-319-44654-7_2

Minesup (Ministère de l'Enseignement Supérieure, République du Cameroun). (2007). *Annuaire statistique du ministère de l'enseignement supérieur* [Statistical Yearbook of the Ministry of Higher Education]. Yaoundé: Minesup.

Minesup (Ministère de l'Enseignement Supérieure, République du Cameroun). (2013). *Annuaire statistique du ministère de l'enseignement supérieur* [Statistical Yearbook of the Ministry of Higher Education]. Yaoundé: Minesup.

Minesup (Ministère de l'Enseignement Supérieure, République du Cameroun). (2014). *L'enseignement supérieur au Cameroun: Cartographie 2014 des institutions* [Higher education in Cameroon: Mapping of institutions 2014]. Yaoundé: Minesup.

Mowery, D. C., & Sampat, B. N. (2005). Universities in national innovation systems. In I. Fagerberg, D. C. Mowery, & R. R. Nelson (Eds.), *The Oxford handbook of innovation* (pp. 209–239). Oxford, UK: Oxford University Press. doi:https://doi.org/10.1093/oxfordhb/9780199286805.003.0008

Mugabushaka, A.-M. (2008). Great disparities: A bibliometric analysis of patterns and trends of African scientific publications, 1990–2003. In E. W. Schamp & S. Schmid (Eds.), *Academic cooperation with Africa: Lessons for partnership in higher education* (pp. 37–54). Münster: Lit.

Mugabushaka, A.-M., Schomburg, H., & Teichler, U. (Eds.). (2007). *Higher education and work in Africa: A comparative empirical study in selected countries.* Werkstattberichte: Vol. 69. Kassel: INCHER.

Munene, I. I. (Ed.). (2009). *Transforming the academia: Exploring African universities in a comparative context.* New York: Nova Science Publishers.

Muriisa, R. K. (2014). Rethinking the role of universities in Africa: Leadership as a missing link in explaining university performance in Uganda. *Journal of Higher Education in Africa, 12*(1), 69–91. Retrieved from http://www.codesria.org/spip.php?article2307&lang=en

Ngeve, R. E., & Orock, T. T. E. (2012). In the name of "development": Ethnic politics and multicultural public policy in Cameroon. *International Journal of Sociology and Social Policy, 32*, 214–232. doi:https://doi.org/10.1108/01443331211214776

Ngwé, L., & Pokam, H. de P. (Eds.). (2016). *L'enseignement supérieur au Cameroun depuis la réforme de 1993*: Dynamiques et perspectives [Attendance in higher education in Cameroon since the 1993 reform: Dynamics and perspectives]. Dakar: Codesria.

Niang, A. (2005). La recherche et ses facteurs de blocage dans les universités francophones d'Afrique: l'expérience des universités sénégalaises [Factors blocking research at the universities of Francophone Africa: The experience of Senegalese universities]. *Journal of Higher Education in Africa, 3*(1), 77–100. Retrieved from http://www.jstor.org/stable/24486243

Njeuma, D. L. (2003). Cameroon. In D. Teferra & P. G. Altbach (Eds.), *African higher education: An international reference handbook* (pp. 215–223). Bloomington: Indiana University Press.

Nkwi, W. G. (2006). The dilemma of civil society in Cameroon since 1990: Which way forward? *African Journal of International Affairs, 9*(1–2), 91–106. doi:https://doi.org/10.4314/ajia.v9i1-2.57246

Nooteboom, B. (2000). *Learning and innovation in organizations and economies.* Oxford, UK: Oxford University Press.

Nyamnjoh, F. (1999). Cameroon: A country united by ethnic ambition and difference. *African Affairs, 98*, 101–118. doi:https://doi.org/10.1093/oxfordjournals.afraf.a007989

Nyamnjoh, F. (2012). 'Potted plants in greenhouses': A critical reflection on the resilience of colonial education in Africa. *Journal of Asian and African Studies, 47*, 129–154. doi:https://doi.org/10.1177/0021909611417240

Obasi, I. N., & Olutayo, A. O. (2009). Globalization, the Bologna process and African universities: Limits and contradictions of market-oriented higher education reforms. *Journal of Higher Education in Africa, 7*(1–2), 159–187. Retrieved from https://www.codesria.org/IMG/pdf/8-ObasiOlutayo.pdf?257/

OECD. (1999). *The response of HEI to regional needs*. Paris: OECD.

Ombga, L. D. (2011). Oil wealth and non-oil sector performance in a developing country: Evidence from Cameroon. *Oxford Development Studies, 39*, 487–503. doi:https://doi.org/10.1080/13600818.2011.620088

Orock, T. T. E., & Mbuagbo, O. T. (2012). 'Why government should not collect taxes': Grand corruption in government and citizen's views on taxation in Cameroon. *Review of African Political Economy, 39*, 479–499. doi:https://doi.org/10.1080/03056244.2012.710837

Oyelaran-Oyeyinka, B. (2006). *Learning to compete in African industry: Institutions and technology in development.* Aldershot: Ashgate.

Perkman, M., Tatari, V., McKelvey, M., Autio, E., Broström, A., D'Este, P., . . . & Sobrero, M. (2013). Academic engagement and commercialisation: A review of the literature on university-industry relations. *Research Policy, 42*, 423–442. doi:https://doi.org/10.1016/j.respol.2012.09.007

Reddy, P. (2011). The evolving role of universities in economic development: The case of university-industry linkages. In B. Göransson & C. Brundenius (Eds.), *Universities in transition: The changing role and challenges for academic institutions* (pp. 25–49). New York: Springer

Robertson, S. (2009). Market multilateralism, the World Bank Group, and the asymmetries of globalizing higher education: Toward a critical political economy analysis. In R. Basett & A. Maldonaldo (Eds.), *International organizations and higher education policy: Thinking globally, acting locally* (pp. 113–131). New York: Routledge.

Rosendal, K. (2010). *Access to and benefit sharing of genetic resources in Cameroon: Legal and institutional developments and challenges*. FNI Report 8/2010. Lysaker, Norway: Fridtjof-Nansen-Institute.

Rothaermel, F., Agung, S. D., & Jiang, L. (2007). University entrepreneurship: A taxonomy of the literature. *Industrial and Corporate Change, 16*, 691–791. doi:https://doi.org/10.1093/icc/dtm023

Sawyerr, A. (2004). Challenges facing African universities: Selected issues. *African Studies Review, 47*, 1–59. Retrieved from http://www.jstor.org/stable/1514797

Schamp, E. W., & Schmid, S. (Eds.). (2008). *Academic cooperation with Africa: Lessons for partnership in higher education*. Münster: Lit-Verlag.

Schamp, E. W., & Zajontz, Y. (2008). Patterns of international academic communication at African universities: Cameroon, a case study of a bilingual country. In E. W. Schamp & S. Schmid (Eds.), *Academic cooperation with Africa: Lessons for partnership in higher education* (pp. 55–75). Münster: Lit-Verlag.

Schamp, E. W., & Zajontz, Y. (2010). Wissensorganisation Universität und regionale Entwicklung im afrikanischen Kontext, am Beispiel der Universität Ngaoundéré, Nordkamerun [The university as knowledge organization and regional development in an African context: The University of Ngaoundéré, North Cameroon]. *Geographische Zeitschrift, 98*, 133–154. Retrieved from http://www.jstor.org/stable/23031949

Sikombe, A. K. (2016). La dynamique des institutions universitaires dans la politique de développement urbain au Cameroun [Dynamics of university institutions in the politics of urban development in Cameroon]. In L. Ngwé & H. de P. Pokam (Eds.), *L'enseignement supérieur au Cameroun depuis la réforme de 1993: Dynamiques et perspectives* (pp. 138–166). Dakar: Codesria.

Simeu Kamdem, M., & Schamp, E. W. (Eds.). (2014). *L'université africaine et sa contribution au développement local: L'exemple du Cameroun* [The African university and its contribution to local development: The case of Cameroon]. Paris: Karthala.

Singh, M. (2010). Re-orienting internationalization in African higher education. *Globalisation, Societies and Education, 8*, 269–282. doi:https://doi.org/10.1080/14767721003780439

Slaughter, S., & Rhoades, G. (2004). *Academic capitalism and the new economy: Markets, state, and higher education*. Baltimore: The John Hopkins University Press.

St. George, E. (2006). Positioning higher education for the knowledge based economy. *Higher Education, 52*, 589–610. doi:https://doi.org/10.1007/s10734-005-0955-0

Tchomga, P. (2016). *University decentralization and graduates' migration in Cameroon. Failure and hindrance for rebalancing disadvantaged regions* (Unpublished doctoral dissertation). Jiangsu University, China.
Teal, F. (2011). Higher education and economic development in Africa: A review of channels and interactions. *Journal of African Economics, 20* [AERC Supplement, iii50–iii79]. doi:https://doi.org/10.1093/jae/ejr019
Teferra, D. (2014). The shifting landscape of development cooperation: Repercussions for African higher education. *Journal of Higher Education in Africa, 12*(2), 1–28. Retrieved from http://www.codesria.org/spip.php?article2531&lang=en
Teferra, D. (2016). African flagship universities: Their neglected contributions. *Higher Education, 72*, 79–99. doi:https://doi.org/10.1007/s10734-005-0955-0
Tsafack Nanfosso, R. (2006). La dynamique de l'enseignement supérieur privé au Cameroun [The dynamics of private higher education in Cameroon]. *Journal of Higher Education in Africa, 4* (2), 99–122. Retrieved from http://www.codesria.org/spip.php?article557&lang=en
UNESCO. (2010). *UNESCO science report, 2010*. Paris: UNESCO.
van der Wusten, H. (1998). A warehouse of precious goods: The university in its urban context. In H. van der Wusten (Ed.), *The urban university and its identity: Roots, location, role* (pp. 1–13). Dordrecht: Kluwer Academic Publishers.
Willinsky, J., Jonas, R., Shafack, R., & Wirsiy, K. C. (2005). Access to research in Cameroonian universities. *The Electronic Journal on Information Systems in Developing Countries, 21*, 1–15. Retrieved from http://www.ejisdc.org/ojs2/index.php/ejisdc/article/view/137/133
World Bank. (2009). *Accelerating catch-up: Tertiary education for growth in sub-Saharan Africa*. Washington: World Bank.
Woudamike, J. (2008). Contestations étudiantes à Ngaoundéré répressions et insécurité des personnes et des biens sur la route nationale no 1 [Student protests in Ngaoundéré, repression, and insecurity of persons and freight on National Highway No. 1]. *Journal of Higher Education in Africa, 6*(2–3), 135–156. Retrieved from http://www.codesria.org/spip.php?article2531&lang=en
Zajontz, Y. (2010). Afrikanische Universitäten als Motoren regionaler Entwicklung? Eine Analyse von Wissenskanälen an drei peripheren Universitäten in Kamerun [African universities as motors of regional development? An analysis of knowledge channels at three peripheral universities in Cameroon]. Hamburg: Dr. Kovac.
Zavale, N. C. & Macamo, E. (2016). How and what knowledge do universities and academics transfer to industry in African low-income countries? Evidence from the stage of university-industry linkages in Mozambique. *International Journal of Educational Development, 49*, 247–261. doi:https://doi.org/10.1016/j.ijedudev.2016.04.001
Zeng, D. Z. (Ed.). (2008). *Knowledge, technology, and cluster-based growth in Africa*. Washington, DC: World Bank.
Zogang, L. (2007). Case study: Teaching English as a second language (ESL) to adult beginners in a hostile environment. *Annales de la Faculté des Arts, Lettres et Sciences Humaines, Université de Ngaoundéré, IX*, 151–166.

Chapter 17
African Universities as Employers of Returning Graduates from Germany: The Example of Ghana and Cameroon

Julia Boger

The Potential of Universities for Development in Sub-Saharan Africa

According to the Human Development Report of the United Nations Development Program (UNDP), economic underdevelopment continues to be a severe problem for most Sub-Saharan African countries (UNDP, 2015, p. 58). Policy makers have pinpointed universities in Sub-Saharan Africa as a potential tool for combating this underdevelopment. Universities with their three mandates of capacity building, research, and community have the potential to significantly stimulate development, especially in the periphery, by driving technological catchup (Bloom, Canning, & Chan, 2006; World Bank, 2000, 2009, 2010). They seem to "play an important role as society's knowledge hubs, where concentrations of highly qualified senior faculty at the PhD level can engage in innovative research that contributes to national development" (World Bank, 2010, p. 26).

These high expectations have to be scrutinized (Schamp, 2014; see also chapter of Schamp in this volume) because most universities in Sub-Saharan Africa face great challenges, which makes it difficult to fulfill this potential. The higher education sector lacks financial resources, infrastructure, and—moreover—qualified personnel (Bloom, Canning, & Chan, 2006, p. 6; World Bank, 2000, p. 23; World Bank, 2010, p. 22). Sub-Saharan African governments have made efforts to improve this situation and to increase the number of universities since the 1970s (Chien & Kot, 2012, p. 1), but they are still unable to meet the growing demand for higher education. More capacity is needed to successfully educate the growing young population in Sub-Sahara Africa (UNDP, 2015, p. 157). An indicator of this need

J. Boger (✉)
World University Service, Deutsches Komitee e.V., Wiesbaden, Germany

P. Meusburger et al. (eds.), *Geographies of the University*, Knowledge and Space 12,
https://doi.org/10.1007/978-3-319-75593-9_17

is the region's low gross enrollment ratio (GER)[1]. Even though there has been a rapid increase in enrollment in Sub-Saharan Africa, which stands at more than 5.2 million students, the GER for Sub-Saharan Africa is only 38% (UNESCO Institute for Statistics [UIS], 2012, p. 128), which is the lowest worldwide. It is likely that a lack of universities is responsible for this low enrollment ratio. Despite a rapid growth of universities, their number is still too low to build academic capacities on a large scale. Teferra and Altbach (2004) refer to "no more than 300 institutions that fit the definition of a university" (p. 22) for the whole of the African continent. This number seems to have increased little since then. The homepage of the Association of African Universities (AAU) lists 359 member universities (AAU, 2015), not counting technical colleges and private universities. Another problem is the lack of academic personnel: The student-teacher ratios are very high. Lecturers have to deal with quite large class sizes and are thus only able to provide basic theoretical teaching. In consequence, the quality of tertiary education has deteriorated (World Bank, 2010, p. 23). Migration is the only way to meet the expectations that highly skilled graduates can improve development processes in Sub-Saharan African countries. However, educational migration cuts both ways. From the individual's perspective, educational migration is the best way to improve career opportunities, whereas from the labor market perspective it is seen as the cause of a brain drain of qualified personnel. This brain drain, as I point out in this chapter, creates a demand that could be filled by returnee graduates.

Educational migration has been common in most African countries since the 1960s. Large numbers of students have in the meantime gone abroad to pursue their higher education. Those were the early days after independence, when the higher education sector was in its beginnings in many countries, meaning that they lacked higher education institutions. Even though the situation of higher educational institutions has improved in many African countries, educational migration is still typical for Sub-Saharan African youth. Educational migration can even be regarded as a "cultural event,"[2] as, for instance, Mbah describes it in her research on Nigerian university graduates (2017, p. 249).

In fact, statistics show that Sub-Saharan African students are the most mobile students worldwide (Chien & Kot, 2012, p. 15). About 4.5 million students are internationally mobile. Of them, about 350,000 tertiary students from Sub-Saharan Africa are enrolled abroad and pursuing their tertiary education outside their own country (Organization for Economic Co-operation and Development [OECD], 2014, p. 361). Thus, 4.9% of Sub-Saharan African students are studying abroad, which is three times the global average of 1.9% (UIS, 2010, p. 4). About two-thirds of these

[1]The GER presents "the number of pupils or students enrolled in a given level of education, regardless of age, expressed as a percentage of the population in the theoretical age group for the same level of education. For the tertiary level, the population used is the 5-year age group starting from the official secondary school graduation age" (UIS, 2012, p. 66).

[2]For more on an African perspective on "cultures of migration" see Hahn and Klute (2007).

mobile students enroll in tertiary institutions in the Global North, preferably in North America and Western Europe.

Germany plays an important role as a host country for these students from the Global South, a situation that also has to do with increasing competition for the best talent among European universities (Kuptsch, 2006). These students are subject to diverging policy approaches. On the one hand, actions are taken to retain the best of these international students in the face of an expected demographic change that will supposedly result in a shortage of young, skilled labor. On the other hand, since the 1970s, the German government has been acting to facilitate the return migration of these graduates through, for instance, organizations such as the Centre for International Migration and Development (CIM)[3] or by offering specific reintegration preparation seminars by the German Academic Exchange Service. The German government supports such reintegration schemes because highly skilled migrants are perceived to be "change agents" (Faist, 2008) who have the potential to stimulate development processes through knowledge transfers upon their return (Skeldon, 2005). However, to become such a change agent, returning graduates need career opportunities allowing them to apply their acquired knowledge. Recent studies (Morris-Lange & Brands, 2015, p. 20) indicate that 44% of international university students who successfully graduate in Germany remain there, while the other 66% migrate to a third country or return home.[4] Is this also the case for Sub-Saharan African graduates from German universities?

However, the proportion of potential change agents from Sub-Saharan African countries is quite moderate in Germany. About 10,000 students from Sub-Saharan Africa are currently registered in Germany, a figure that has not changed much since the start of the millennium. This amounts to around 10% of all students from countries in the Global South studying in Germany (German Federal Statistical Office, 2015a).[5] Even though these figures are not high, this comparably small group of migrants could have a great impact on development. Many empirical studies suggest that Sub-Saharan African educational migrants wish to contribute to their home countries' development and that they enter the higher education sector upon returning there. This assumption has been strongly supported by Mbah's research on educational migrants from Nigeria (2014) as well as by a multicountry

[3]These reintegration subsidy schemes include transport and travel tickets, salary subsidies, and workplace equipment. Institutions that administer these subsidies are, for instance, the Migration for Development Program run by CIM and funded by the Federal Ministry for Economic Cooperation and Development (BMZ). The CIM partnered in the returning expert subsidy scheme with the German committee of World University Service (WUS), a non-government organization (NGO) that earlier designed the reintegration subsidy scheme of the 1980s. Another program, similar in its function but much smaller in terms of participants, is run by the German Protestant Relief organization *Brot für die Welt* (Bread for the world). (Further information about the German government's reintegration subsidies can be found in Schmidt-Fink, 2009).

[4]The study also offers an overview on the different estimations about return and retention rate of foreign students which the latter differs between 22%–56% (Morris-Lange & Brands, 2015, p. 20).

[5]Currently, the majority of foreign students are from China and Russia, followed by India and Cameroon (German Federal Statistical Office, 2015b).

investigation carried out by SOFRECO (2015, p. 72). This SOFRECO study examined the contribution made by students from thirteen different Sub-Saharan African countries who had studied abroad to capacity building in the field of higher education in Africa. But in my opinion, these studies limit their focus to the individuals' motivations, while ignoring the key issue of how they find jobs. This information gap concerning the job-finding phase is quite surprising, because it would be very interesting for policy makers in higher education as well as in migration management to know if and how these students from Sub-Saharan Africa are actually able to apply professionally what they learned in Germany.

I therefore offer insight in this chapter into the job-search process of Ghanaian and Cameroonian graduates returning from Germany. I compare these two Sub-Saharan African countries because both suffer from similarly high numbers of outgoing educational migrants. This exodus, as I argued previously, causes a shortage of qualified personnel in the economy—and in the countries' universities. One could, therefore, assume that returning academics have excellent job opportunities in this section of the labor market. But the experiences of returning graduates in both countries reveal that despite this foreseeable demand for their expertise, they still find it quite difficult to obtain employment in the higher education and, moreover, to sustain it. I show as well that even though the challenges of reintegration are high, an impact on development can be surmised at an institutional level, at least for the country case of Ghana.

In this chapter I present the theoretical background of the migration and development nexus, and sketch the theoretical outlines on return migration. I then use statistical material to describe the educational migration from Ghana and Cameroon to Germany and the return rate of graduates. My empirical material consists of 14 selected interviews from my PhD survey's sample ($n=50$) conducted during fieldwork in Ghana and Cameroon (2008–2010). These interviewed returning graduates had started their professional careers in the higher educational sector. Eight of them had returned to Ghana and six of them to Cameroon. The interviewees were selected from the STEM disciplines (science, technology, engineering, and mathematics) because of these subjects' popularity among foreign students in Germany.

I conclude by asserting that graduates who start as lecturers in Cameroon seem to have mastered higher hurdles than their Ghanaian counterparts. Thus, it appears that Ghanaian universities have been more welcoming of their returning graduates than those in Cameroon. Hence, the problems upon return are similar in both countries but the weighting of the challenges varies according to the country context. This particular finding is discussed against the background of the debate surrounding the migration and development nexus.

The Migration and Development Nexus

The interconnectedness of migration and development processes can be explained in two ways. Firstly, migration can be regarded as a response to development and underdevelopment. From this perspective, lacking economic development pushes people from the poorer countries in the Global South to the more developed countries in the Global North. Better income and living prospects attract people and motivate them to leave their homes. Secondly, migration itself can create underdevelopment and development. The latter perspective has become a comprehensive interdisciplinary debate labeled as the migration and development nexus (Nyberg-Sørensen, Van Hear, & Engberg-Pedersen, 2002).

Economists investigate the extent to which the financial transfers of migrants, the remittances, influence development processes; sociologists and social anthropologists explore the impact of those living abroad (today often are referred to as *diaspora* communities)[6] on cooperation projects, and last but not least, geographers scrutinize to what degree migration causes a brain drain, brain gain or a brain waste. The migration and development nexus debate started in the 1960s and studies particularly on return migration were conducted by different disciplines, often yielding very divergent results. In some cases, return migration was seen as a success, in others as a total failure. In fact, the findings seemed to swing back and forth between optimism and pessimism like a pendulum (de Haas, 2008).[7] However, today's paradigm of transnationalism understands return migration as a part of the whole migration story—which allows investigating this form of migration without judging it.

What impact migration has on development depends on the notion of development being applied. Against the backdrop of an economically oriented development terminology, a negative impact was diagnosed in discussions about the negative effects of the brain drain in regard to the 1960s south-north migration processes. By then, most countries in the Global South, including the Sub-Saharan African countries, had declared their independence from the colonial powers in Europe. In consequence, many young citizens were encouraged to leave their native countries for the Global North to increase their knowledge and to return with this additional know-how. These highly skilled returning migrants were supposed to build the administrations of these young nations that were in need of highly qualified

[6]Today's discourse in migration studies relates diaspora to many groups of persons, mostly ethnic or religious minority groups living outside their countries of origin (e.g., on the African Diaspora, see Zips, 2003). Despite its widespread use, the term diaspora has become increasingly controversial, especially among authors of Jewish Studies who point out the religious and symbolic aspects of the term. Originally, the term diaspora (Greek, for scatteredness) was related to the situation of the Jews outside their Promised Land. Instead of a general definition, Krings proposes a more sensitive and refined treatment of the concept of diaspora as an analytical concept for the designation of a "specific transnational community form" (Krings, 2003, p. 151).

[7]For more details on how the migration and development debate shifted from a positive to a negative frame, and back again, see de Haas (2012).

specialists. But very soon the students did not return after graduation. They started to settle in their host countries. There they found a better lifestyle, better income prospects, and political stability, as well as better working and research conditions. Brain drain, the permanent loss of highly skilled persons, had started. This brain drain intensified with the economic decline of many African countries in the 1970s. Moreover, austerity measures in the countries' public sectors due to structural adjustment programs made most countries' economies rather unattractive for well-educated graduates who had been exposed to a more developed economic environment, with the result that even fewer academics returned from abroad. This permanent outmigration from the Global South to the richer North is still ongoing and continues to cause great problems for the labor markets in the Global South. It has severe negative effects on sensitive sectors in the source countries, such as the health and education sectors, because of a shortage of physicians, health staff, and university lecturers (Manuh, Asante, & Djangmah, 2005; Nyonator & Dovlo, 2005).

Return migration of highly skilled migrants has been strongly regarded as an effective countermeasure against brain drain since the 1960s. Thus, many policy makers vigorously promoted return migration and even launched reintegration programs. They assumed that migrants could contribute to the development processes in their native countries by transferring their gained knowledge, while also disseminating democratic ideas. In this positive notion of return migration, it was assumed that returnees would generally be able to stimulate development processes upon their return. However, this was and is not always the case, with the actual impact of returning migrants depending on the individuals' situation and their motivation for the return. Cerase (1974) developed of a typology of returning migrants' motivation. He distinguished between a return of failure, in cases when migrants do not achieve their goals abroad; a return of conservatism, when migrants consider their savings to be economically better invested in their home countries; a return of retirement; and finally a return of innovation, which might be the most promising type in connection with development processes. The effectiveness of a return of innovation has been questioned, for not all returnees will turn out to be successful innovators. The outcome depends on the country context and the individuals' resources.

Studies suggest that return migration to Asian countries can be quite successful. This is especially the case for China, India, and, in general, countries with emerging markets. Their economies depend on innovation and they therefore urgently need highly skilled employees (Saxenian, 2005). High demand can lead countries such as China to actively encourage their academics to return from overseas. The Chinese government even sends out delegations to recruit Chinese students while they are still studying abroad (Zweig, 2006, p. 195). Sub-Saharan African countries are generally less economically developed and politically stable. Moreover, many African governments have recruited foreign-born development staff. These three factors greatly reduce employment opportunities for those highly skilled graduates who return (Thomas, 2008). And employment opportunities are one of the three greatest motivations for graduates to return from abroad, as a recent study (SOFRECO, 2015, p. 64) shows.

Apart from this country context, the chances of success depend to a great degree on the individual returning migrant's resources and degree of preparedness (Cassarino, 2004). This preparedness includes—besides the willingness to return—tangible resources (financial reserves, educational certificates) and intangible ones (social contacts, knowledge of the country's labor market) that have to fit into the country context. However, these former approaches and theoretical concepts are all based on the assumption that migrants either return or remain abroad, whereas today's transnational paradigm assumes that migrants increasingly have circular mobility patterns. Vertovec (2007, p. 5) even suggests that such transnational migrants are likely to continue to migrate because each of their migrations and experiences abroad provides them with more knowledge about the country context, making them better able to choose their place of working and living. Unhindered transnational migration seems to describe only a minority (Portes, 2003). The majority of migrants still settle rather permanently either in the host or the source country. One major factor among others that can influence whether a person settles permanently or leaves again after returning to their home country is socioeconomic welfare. Returnees need to find a secure income generating activity (Thomas, 2008, p. 654), but finding a job is not easy, particularly if the jobs are in an environment with intransparent labor markets and comparably small formal job market sectors, as is often the case in the Global South. Thus, the SOFRECO study identifies possible difficulties finding a job as one of the greatest fears of Sub-Saharan African students abroad. In their view governments are negatively involved in employment decisions regarding academics, with recruitment not being based solely on merit, but on ethnic affiliation and social background. This entanglement was claimed by the respondents of the study in Cameroon, Ethiopia, and Sudan: "decisions for appointment are not taken on the basis of skills, talents and possible contributions but on political views and affinity" (SOFRECO, 2015, p. 78; see also Mbah, 2017 and chapter of Schamp in this volume). In the following I address the question of whether this rather opaque job placement practice also affects Ghanaian and Cameroonian graduates returning from Germany to work in their countries' higher education sectors.

Returning to Ghana and Cameroon as Academics

Ghana and Cameroon are Sub-Saharan African countries that are severely affected by a vast outmigration of their citizens to the Global North. An estimated 1.5 to 3 million Ghanaians (Quartey, 2009, p. 13) and 2.5 to 4 million Cameroonians (Owono, 2011; Sapouma, 2011) are living abroad. Compared to this total stock of emigration, educational migration flows are small in volume. In 2010, 20,093 Cameroonians went abroad to study (UIS, 2012, p. 136), which according to the Migration Fact Book (World Bank, 2011, p. 85) amounted to only 7% of the total emigrationstock of 279,200 in that same year. For Ghanaians, the proportion was even smaller: Of the total emigration stock of 824,900 (World Bank, 2011, p. 124), only 7,845 (1%) were students (UIS, 2012, p. 137). This percentage distribution is reflected in the figures for Germany, where Cameroonian students far outnumber Ghanaian students. In 2015, about 579 Ghanaians pursued their tertiary education in Germany (German Federal Statistical Office, 2015a). This is only about 2% of the total number of 29,590 Ghanaians registered in Germany (Ausländerzentralregister, 2016, p. 39). In contrast, the number of Cameroonian students is currently 6,672 (German Federal Statistical Office, 2015a), which amounts to almost 34% (Ausländerzentralregister, 2016, p. 39) of the total number of 19,800 Cameroonians living in Germany in 2015. Statistics show that the majority of students from both countries are enrolled in technical subjects. Cameroonian students tend to prefer electrical engineering, while Ghanaian students seem to have a strong preference for master's programs in natural resource management (Boger, 2013, pp. 97–99).

Of these highly skilled migrants, only a small proportion return to their home countries. Estimates based on the internal statistical data of the reintegration subsidy programs administered by CIM suggest that 323 Ghanaian graduates returned from Germany during the decade 2000–2010, whereas only 288 Cameroonian graduates returned during the same period (Boger, 2014, pp. 137–139).

Despite its comparably small volume, the return of these educational migrants is my main interest. The question is not only whether they return, but also whether those who do return will be able to find adequate employment that also has a positive impact on the situation of their countries work in order to change their home countries for the better. How do they find their jobs and could they be better informed about the transition they face upon return?

The tertiary sector with its services is the main employer of university graduates in many Sub-Saharan African countries. A large proportion of these graduates find work in the field of education (Asong & Awama, 2014, pp. 159, 164; Mugabushaka, Teichler, & Schomburg, 2003). The institution of the university enjoys a very popular image as an employing structure within this field for several reasons. Being a professor at a university is a comparably secure job that also offers financial and social advancement, with a lecturer position in many countries supplemented by nonmonetary benefits, such as a car, housing, and health insurance. Moreover, working as a professor at a university creates a rising social status. Higher social status may open doors to a career in politics or as a consultant for international donor organizations (Neubert, 2008, p. 100). Coming back to this paper's empirical cases of returning Ghanaian and Cameroonian graduates, the situation in their native countries is similar: There is a great demand for highly qualified lecturers in both countries. In Ghana, there are currently 10 public universities, 10 technical colleges, and 63 private universities (Ghanaweb, 2015; National Accreditation Board, 2015). According to the Cameroon Ministry of Higher Education (2014) that country currently has 8 public universities,[8] 6 vocational training centers (p. 31), and 163 private institutions (p. 40).

[8]The eighth university, the University of Bamenda in the Anglophone region of the country, opened in 2011, after the data collection took place.

As was mentioned previously, the demand for qualified university staff is high because enrollment in the tertiary sector is increasing in many Sub-Saharan African countries, which means that university capacities also have to rise. Moreover, the demand for lecturers with practical expertise is growing because of expanding implementation of what is known as the third mission of tertiary institutions, as Zajontz (2010) described for the case of Cameroon.

This third mission goes beyond teaching and research (second mission) to include practical entrepreneurial skills. These additional and less theoretical skills should prepare the graduates for starting their own spinoffs in a labor market in which formal employment is scarce. To fulfill this task, universities need highly qualified staff members with knowledge about modern techniques and practices. Returning graduates who have been exposed to more proactive educational environments and modern curricula abroad could offer such practical skills and would therefore seem the ideal choice for such academic positions. It has been observed that returning graduates have indeed found employment at their home countries' higher education institutions, such as universities, as well as at research centers and think tanks attached to universities (SOFRECO, 2015, p. 7; for the case of Ghana see Martin, 2005, p. 206; for the case of Cameroon see, Schamp & Zajontz, 2008, p. 65).

This positive description of the university as a potential employer ignores the fact that workplaces at most Sub-Saharan African universities are not well equipped, which is regarded by African scholars abroad as a major constraint to their becoming more involved in African higher education (SOFRECO, 2015, p. 7). These deficits cause teaching quality to deteriorate further. Martin (2005, p. 209) and Manuh, Asante and Djangmah (2005) described this for the case of Ghana:

> For many academics, frustration in many cases is not due only to low salaries, but also to the conditions under which teaching and learning occurs; the shortage of space for lecturers and tutorials; the lack of adequate materials including books and chemicals; increasing enrolments leading to inadequate counselling for guidance for each student; production of low grade/half-baked intellectuals. Job dissatisfaction, disillusionment and similar factors lead to withdrawal and stress of faculty, and contribute to the push factors for brain drain. (p. 265)

This detailed status description applies not only for Ghanaian universities, but to similar situations at most Sub-Saharan African universities, including those in Cameroon. Similarly, Zajontz (2010, p. 267) noted that respondents criticized university recruitment practices in Cameroon, claiming that universities would only "officially" follow clear hiring guidelines, such as age, number of publications, and educational experience, but in reality their recruitment procedures were rather opaque and incomprehensible to outsiders.

Summing up, although the tertiary sector in Ghana and Cameroon seems to offer a career opportunity for those graduates who return, entering the academic sector and working under the conditions there seems problematic. The people interviewed for this study spoke about the issues involved with seeking employment and working in the educational field.

Entering the Academic Labor Market in Ghana and Cameroon

Graduates who study abroad and return to their native country undergo a double transition (see also Mbah, 2017). They not only have to reintegrate into their native countries' social environment, find a place to live, and readjust to the native's way of living, they must also enter their home countries' labor markets. This can, of course, be somewhat difficult because they usually left their home countries with little or no knowledge about the labor market there, and any knowledge they originally had has become outdated during the several years they were abroad studying. The job search is difficult for graduates in Sub-Saharan African countries in general, as Mugabushaka, Teichler and Schomburg (2003) pointed out in their comparative case study on Ghana, Nigeria, Kenya, Tanzania, Uganda, and Malawi:

> Most African countries have public job centers, which allocate jobs to graduates, but in times of anticipated difficult transition, graduates cannot rely on those agencies alone. They are often forced to find more resources and pro-active solutions to the problem of finding employment. (p. 65)

This finding also applies to my case study on Ghana and Cameroon. The interviewees' narratives show that although they job-searched proactively they still faced difficulties identifying a vacant position, overcoming bureaucratic hurdles, and eventually remaining employed and improving their employment status. Apart from these generally similar findings, there was a significant difference. Ghanaian graduates seemed to fear the process of establishing a career in the university sector less than the Cameroonian graduates, who described their entry into the academic sector as comparably more dependent on personal contacts, which for some was a cause of great uncertainty. It also appeared that universities offered less career opportunities than their Ghanaian counterparts.

Identifying a Vacant Position at the Universities

In general, finding a job greatly depends on the flow of information about open vacancies. Today, information about jobs is often channeled through what is known as the *hidden* jobs market. Positions are not openly published, but announced via personal contacts and social networks. But how can returning graduates activate their social networks in the home country while they are still abroad?

In both countries, most interviewed returning graduates revealed that they had first started to search for a job upon returning. They reported that it was initially difficult for them to find an open position at a university. Many interviewees described the process of finding a job as based on luck and said that they had coincidentally met the right person, who had given them valid background information about potentially vacant positions.

Dr. E.[9], a Ghanaian graduate who had studied a quite specific subject, aquatic science, at the University of Bremen, knew that "to get a permanent job like this [at the university], it takes time" (Ghana interview #22, 2009). Looking back at the time of his job search in 2001, he recalled that it was a personal contact, a former professor, who told him about a vacancy in a new department that would be opening soon:

> So when I came back and I was looking for job, one day I came to the university. I think I came to look for school in Cape Coast. I came to the university and I saw one of my old professors. When I told him he said, oh, they have started new department fisheries and aquatic science .... That I can apply to it. When I applied, I think it took about six months. Then, they wrote to me that I should come for interview. So they called me for interview .... Then all the questions, I tried my best to address them. So after two weeks or so, they wrote to me that I have been appointed as assistant lecturer on one year probation. (Ghana interview #22, 2009)

What Dr. E. clearly outlines in his statement is that it was a coincidence that he was able to find a vacancy. He had been looking actively for a position for a while and it was only when he met a former professor by chance that he learned about a position opening. He even frankly declared that without this particular information from his professor, he would not have gotten the job: "So, I think Prof., he did well because if he hadn't told me, I wouldn't have even applied" (Ghana interview #22, 2009).

It seems crucial to have good contacts who can provide information, as the following case of Mrs. L.[10] from Ghana (case #20) also shows. Upon returning, a friend advised her where to search for a job:

> I think, I knew a friend who was an assistant lecturer here. So, I think I gave him a phone call and he told me that I just have to go to the head of department. Find out from the head of department if they need somebody. If they need somebody, they will let me apply. So he said if the head of department assures me that there is a vacancy and that they need my services, then it won't be a problem. So I think I asked and I was shown to the agricultural engineering department. When I went there, I met the immediate past head of department. I told him my intentions, he said ok. He said ok because they need people. They need people ... even still one person teaches about four to five courses. So if I come, it will be of help. (Ghana interview #20, 2009)

Interestingly, Mrs. L. had been advised by her friend to inquire about valid information directly by the head of department. This, from a northern European standpoint, could seem quite surprising. In the Global North it is general policy to

[9]Dr. E. studied aquatic sciences in Germany and returned to Ghana on September 1, 2001. Upon return he had difficulty finding a job. Eventually, after a cumbersome job search, he secured a position as an assistant lecturer at the University of Cape Coast in the department of Fisheries. Because he could not advance in his career, he decided to continue his education and to earn a doctorate in Germany. The second time he returned was in May 2008. The interview took place one year later on September 4, 2009, in his office in Cape Coast.

[10]Together with her husband, Mrs. L. studied resources engineering in Germany and they both returned to Ghana on December 1, 2006 with a master's degree. She secured a job as a lecturer five months later at the University of Ghana. The interview took place in her office almost three years later, on September 1, 2009.

advertise a vacancy more transparently in media such as newspapers, radio, job placement centers, or the Internet. The friend's advice shows that this is not the case in Ghana and that personal contacts serve as better information channels.

As in the previous cases from Ghana, Dr. L. from Cameroon (case #05)[11] described that he secured information about a prospect vacancy from a personal contact. He coincidentally met a Cameroonian scientist during his years at the Technical University in Berlin. The scientist was a guest lecturer from a newly opened university in Cameroon and thus had access to information that the university needed qualified lecturers:

> So I met him [the Cameroonian guest lecturer] and we discussed and he told me because I was really interested in knowing how the market job is in the country and he told me. He said that university lecturers, they have a chance to be recruited and particularly in the University of Buea . . . it was a new university and . . . he told me that they were looking for lecturers particularly in my field . . . . So he told me that I had good chances. So I said ok, I will not miss that opportunity. So he came here and he met the head of department. So the head of department welcomed me in 2002. I came here for a visit in 2002 to get in touch with them. I discussed with the head of department and he was very happy to hear that I want to come and join them here. (Cameroon interview #05, 2008)

The demand for qualified lecturers at Buea, which by then was a new university was great, as shows the following case of Dr. G.[12] from Cameroon (case #04). In contrast to the previous example, she had no personal contact who gave her information prior to her application phase. She investigated the situation in Cameroon during a visit before she graduated in Germany. During this visit, she applied at various universities:

> I have been sending my applications to four universities here but it is only in University of Buea where they said "ok here is a free place". In other universities there was no open position . . . .They [the universities] normally don't reply. They reply if they want you to come . . . . If you don't get any reply it means there is no position . . . yes the university of Buea replied and we settled everything before I come back everything was already done, yeah [laughs]. (Cameroon interview #04, 2008)

In contrast to the previous narrative, she did not mention any personal contact. Thus, I became curious during the interview and inquired whether she had known someone in Buea. She negated this:

> No, not really, to tell you the truth, when I came in February 2005, I was still working in Germany and I asked for permission, one week permission, to come for the interview. It was my first time in Buea . . . and I met the head of department the one I was talking of, Dr. A., yes, and he was very nice. He welcomed me and said "Ok, since you are already in

[11]Initially, Dr. L. studied geology in Russia and returned for the first time to Cameroon in 1991. There, he soon realized that the country was in deep recession, at which point he decided to earn a PhD in Berlin. He eventually returned to Cameroon on May 1, 2002, and secured a position as lecturer at the University of Buea. The interview took place on October 2, 2008, at his office in Buea.

[12]Dr. G. earned her PhD in inorganic chemistry at the University of Leipzig and returned on February 1, 2005, to Cameroon. Upon return she secured her position as lecturer at the University of Buea. The interview took place in her office in Buea on October 2, 2008.

> Cameroon we will inform the other members of the faculty so that we can make it for the interview". And then I went back and then he called me again and said the interview was set on one day .... I did the interview and then two days after, I left waiting for the flight. (Cameroon interview #04, 2008)

The quote suggests that Dr. G. did not have a personal contact who gave her information about a vacancy and that she simply applied. But this was only partly true. She mentioned the head of department, Professor A. who invited her for the appointment and was very flexible regarding her situation. Professor A., I found out during my research, was a returnee himself. He thus knew of the struggles of returning graduates. He was known to be very helpful to returning graduates searching for jobs. Hence, in this case again, it was indirectly a contact person that helped Dr. G. to find a vacancy and obtain a job.

Summing up the experiences recounted above, it seems to be difficult to identify the employment demand at universities in both countries. Returnees in Ghana as well as in Cameroon reported that they received valid information about vacancies through their personal contacts rather than through the official advertisement channels of the universities. Luckily, these returning graduates seem to have had well-embedded personal contacts and were eventually able to locate positions as lecturers. However, this job search period upon return creates frictional unemployment, which causes problems for the returning graduates for two reasons. It delays earning an income and it also creates a devaluation of the degrees they acquired in Germany. The longer they have to wait to apply what they learned abroad, the more outdated their knowledge will become.

Overcoming Bureaucratic Hurdles

The image of being a lecturer at a public university used to be that of having a secure position for lifetime. However, the position as lecturer at a university no longer offers this security. The interviewed returning graduates reported that even if they had identified a position, successfully mastered the job interview, and finally even received a job offer, they often still had to persevere to finally get the employment contract. Then, they mostly only received a fixed-term contract of as assistant lecturers instead of as full-time lecturers. Many of their statements reveal that the process of getting a full-time contract is lengthy and full of bureaucratic hurdles.

Many returnees are surprised by the fact that it takes so long to simply receive their appointment letter once they do have a contract. Those who take a sabbatical for studying abroad on unpaid study leave may assume that they will easily reenter their fixed appointment upon return and immediately receive their salary again. [13] This is

[13]For further information, see SOFRECO (2015, p. 64).

often not the case, as Dr. V.[14] from Ghana (case #12) remembered. She had to be very patient until she could work as a lecturer again:

> When I was leaving I told them I would come back. So when I finished I wrote to them that I had finished and would like to come back. They [the university officials] also gave me the assurance that they were waiting for me. I started [the process] very early before I finished my defense. They said when the certificate is ready I could apply. So when it was ready, I applied. They also played some game on me. They could not transport me back from Germany, so they refused to give me the appointment letter. That is the university condition . . . so they did not want to give me my appointment letter. So although they assured me they did not give me the appointment letter. So fortunately CIM [reintegration subsidy scheme] bought my ticket. When I came [back] the first week, I went to tell them that I am back. So I got my appointment letter and contract. I was also trying other places like Kumasi. But one of my professors didn't want me to leave. He wanted me to stay here . . . . He was one of my lecturers. He is now at the head of Central University. So it was not that much of a problem, not that difficult. But it is just because of the way they do things. And you have to have patience. (Ghana interview #12, 2008)

Even though Dr. V. stated that for her the situation was "not that much of a problem, not that difficult," the bureaucratic hurdles severely delayed her starting work. If she had not been able to get external funding for her return flight from one of the reintegration subsidies and if her professor had not insisted on keeping her as a lecturer, she probably would have experienced a much more difficult transition upon her return. Although this scenario is speculative, it should be kept in mind.

Bureaucracy can create tremendous delays as shown by the case of Dr. Y.[15] from Ghana (case #14). It took him half a year to receive a contract and payment at a public university in Ghana and another two years to successfully conclude his probation period and became a permanent staff member:

> Six months after arrival, that was when I formally got employed with university of Ghana. But before then you had to go through some processes: getting to the department, fundraising yourself, going to an interview giving a presentation [pause]. I went through all that and after six month I got a job as a lecturer here in this department. And once you are employed as a lecturer you are on probation for two years. They want to see whether after those two years your output is good before they recommend you that "yes," now you can stay on full [emphasizes]. So I am three years old here now. So from 2004 to now, three years. So I have gone through my probation period and now I am a full lecturer at the department and it has been a big challenge [emphasizes] because of the huge pressure. (Ghana interview #14, 2008)

---

[14]Dr. V. earned her doctorate in natural sciences at the University of Bremen. Her return to Ghana was on September 9, 2005. She returned to her employment in the University of Cape Coast, Department of Organic Chemistry as a lecturer. The interview took place at her workplace in Cape Coast, on July 7, 2008.

[15]Dr. Y. earned his doctorate in water management at the Center for Development Studies in Bonn and returned in August 1 2004. After a cumbersome bureaucratic act he eventually got a job as a lecturer in agricultural economics at the University of Ghana. The interview was at his office in Accra on July 9, 2008.

What Dr. Y. clearly points out in this section corresponds to Dr. V's experience: That securing a university position takes time, during which returnees lack income. They have to develop a strategy like Dr. Y. for "fundraising yourself."

> So when I came down to Ghana of course there was no job .... There was no job but I was quite prepared, knowing the way our system is .... So when I came down in August I knew there wouldn't be an immediate job for me but fortunately I had some contacts around. So I was able to survive by getting some small jobs to do even with the Ministry of Agriculture and other consultancies ... I was actually surviving making some money by using my expertise that I had gathered and especially the PhD it is good for them. So they made use of my talents and I made some money. So that was keeping me going. That was the way when I arrived and six months after arrival that was when I formally got employed with university of Ghana but before then you had to go through some processes. (Ghana interview #14, 2008)

Unlike the returnees mentioned previously, Dr. Y. seemed very aware of the fact that he would not get a job appointment immediately after his return and therefore sought consultant jobs on a project basis. He was able to earn money and even supplement his CV with working experience at the same time.

Whereas Dr. Y. in Ghana was proactive, Dr. L. in Cameroon (case #05) was not looking ahead as much and thus became the victim of a poorly run university administration. It took him one year to understand the bureaucratic process, which was very personalized:

> I came back but it was not so easy, I mean the concept was clear that I will come and teach here but it was not easy because of one or two things. I mean so when I came, so I started teaching as part time .... So I was not satisfied because I cannot leave Germany to come and teach here as part time .... So it was like a self revolution, a rebellion when I went there and met the vice rector in his office in charge of teaching. So he assured me that it was a mistake in terms of the budget. So I was recruited. It was a mistake. It was really a mistake and I had to be recruited immediately. (Cameroon interview #05, 2008)

The difference between the previously presented cases is obvious. The Ghanaian interviewees found bureaucratic systems that were more transparent—ones they were able to master by planning ahead—than the ones encountered by their Cameroonian counterparts.

These graduates' experiences show that the transitional phase does not end with identifying a vacant position and filling it. Bureaucracy, more precisely the bureaucratic formalities regulating the contract procedure and the promotion path within the university, often significantly delay the process of earning an income for the returning graduates. In consequence, returning graduates must be patient, have financial resources to bridge this interim phase, and, last but not least, the confidence to stay and continue their job search. Those who return willing to start their careers in academia have to believe that they will eventually be able to secure and maintain their positions as full-time lecturers at universities in their native countries.

Obtaining and Maintaining a Position

Not all returning graduates are convinced that the university is going to be their permanent, full-time employer. This is apparent in the following statement by an interviewee from Ghana, Mrs. N.-C. (case #23)[16]. She found out about a vacant position at her former university in Ghana—admittedly, thanks to the support from her former professor. Despite this personal contact, she, too, had to be patient while waiting to become a permanent member of staff. This, she said, made her very uncertain and convinced her to develop an alternative income generating activity:

> I have just been employed—that was during the last season. I need to get confirmation. Normally when you are employed it is just probation. Two years and after the two years you will be confirmed. And even when you are confirmed your appointment will be on contract basis. Four year contract until you attain the position of associate professor that your appointment will be automatically extended until you reach retirement. So it is not so secure the position and that is what has kept the idea of going to the farming on hold because I believe we could have started on a smaller scale then see it grows. But for my job I have to consolidate it for now. I have to sit down and be writing because that is what will get me going. (Ghana interview #23, 2009)

Dr. N.-C. explained that she had a plan B, in case her initial plan of becoming a university professor failed, which was going into farming as she frankly admitted. At the same time, however, she clearly knew what she had to do for her career at the university, which was writing and publishing texts, and she believed that she could succeed in the field of academia through hard work. Today, she is a full-time associate professor, having obtained her PhD in the interim, without which her position would have been out of reach.

Job insecurity seemed to be more severe in Cameroon according to the professional trajectories of the interviewees at Cameroonian universities. One of the greatest differences observed between the cases from Ghana and those from Cameroon was that most Cameroonian lecturers were teaching part-time simultaneously at several universities to earn a living. This, of course, is an indication of the high degree of employment insecurity in higher education sector.

A good example of this is the career path of Dr. N.[17] (Cameroon case #07). Despite her qualification from Germany she still only was a part time lecturer in physics at three different Cameroonian universities:

> I wish I was posted here [University of Yaoundé I] completely, because now in my field at the University of Yaoundé I they are going to open the program for medical physics and I have the medical physicist certificate. And there it would be good for me to be at the

[16]Mrs. N.-C. earned her MBA in small and medium-sized enterprise development at the University of Leipzig, returned to Ghana on August 13, 2007, and became a lecturer in management studies at the University of Cape Coast. I interviewed her at a restaurant in Cape Coast on September 4, 2009. The contact continued on a more private level via email and phone.

[17]Dr. N. earned her doctorate in Berlin in the field of medical physics. She returned in September 1999, and at the time of the interview, almost ten years later in 2008, was still not employed full-time. We conducted the interview in a university room at Yaoundé I on October 4, 2008.

> University of Yaounde I and not at the University of Dschang .... I would teach people up to the master's level directly in my field. Currently, I teach only second year in Dschang and third year and I'm only in my field in Douala where I teach the master and doctorate students. So if Yaoundé selects me, I will be more in medical physics and will be available for the medical school of Yaoundé I .... Myself, I would agree. If it works that I am a lecturer, the assignment will be done quickly. But it's something that will happen in one year maybe. (Cameroon interview #07, 2008)[18]

It is not too much to assume that there was not much competition in her unique field of medical physics and it is thus surprising that Dr. N. still did not have a full-time position at her favorite university, despite her doctorate and additional working experience. Perhaps it was an individual case, but it seems to have been a structural problem instead, because the next case shows a similar pattern. It is again that of Dr. L. (Cameroon case #05). He, like Dr. N., was teaching at universities in three different cities at the same time—in Buea, Douala, and Yaoundé:

> I also teach at Yaoundé I university since 2004, 2005 and I am teaching also in Douala, where you have an engineering school which is a branch of a university in Central Africa, a branch .... It is French, a corporation. So the program is a typical French program for French Cameroonians and they train engineers, I mean mechanical engineers. But there I am not teaching geology. There I am teaching but what we call quality management. That is quality security and quality safety. (Cameroon interview #05, 2008)

What is surprisingly similar to the previous case from Cameroon is that Dr. L. was only teaching his main subject in one of his three lecturer positions. Of course this could have indicated Dr. L's broad expertise in other fields, but instead it revealed the low demand for his expertise (geological mining) at local universities.

Another, quite outstanding example from Cameroon is that of Dr. H. (case #02),[19] a returnee PhD graduate. It shows that some returning academics will even go as far as to found their own institution of higher education: a polytechnic institute. He and his wife returned in 2001, after having lived and studied in Leipzig for 14 years. He recalls that his plan to establish such an institution came up when he realized he would not be able to find adequate work in his field of applied physics because of the lack of equipment and funding at the local universities:

> The plan for the creation of his institute was fundamentally motivated by the fact that I chose to do the experimental physics ... when I decided on that topic for my PhD thesis my comrades said I would not be prepared if I wanted to return some day, because those from the fundamental mathematical and physical sciences who want to work in their countries as lecturers realize that they do not have a laboratory to work in because we have no equipment, but we have to be able to carry out experiments to explain applied physics .... I submitted my application dossier and checked for open positions at universities in .... I even sent my application to the teaching ministry, but got no reply ... and I then realized that I would have

[18]This interview quote has been translated from French to English by the author of this paper.

[19]Together with his wife, Dr. H. studied physics and geography, after which he pursued his PhD at the University of Leipzig. They both returned on August 27, 2001, to Cameroon and started their private educational institute. The interview took place on October 2, 2008, at their institute's office in Douala.

to build a working laboratory space that met my needs ... that was one of my motivations. (Cameroon interview #02/03, 2008)[20]

Of course, the example of Dr. H. is not a common one, because few returning graduates have enough financial capital for such a large scale investment. It also needs to be mentioned that it took Dr. H seven years to redeem his investment in the private institution. But despite the fact that it was a difficult path he chose—maintaining his career in academia by founding an educational institute—it is a very good example of how a long-term investment can ultimately convince returnees to settle permanently in their home country.

Finally, the cases from Ghana and Cameroon discussed in this section reveal that returning academics from both countries face difficulties retaining their employment in academia. However, the cases from Cameroon show that the level of insecurity is higher than in Ghana.

Contribution to Development

In this final section I go beyond the question of how to find and maintain employment and discuss whether educational migrants achieve structural development effects through their return migration as professionals. Here, the notion of development is one of sustainable development, which includes social, political, environmental, and economic aspects. Analysis of data on the impact of returning graduates on development processes in higher education suggests that an impact on development, especially concerning the economic aspects, was possible. Most returning graduates improved their individual lives and that of their close family members through their limited financial resources. Nearly all returned graduates of this study talked about their presents and gifts, mainly consumer goods and electronic devices, which they brought from abroad for their families and kin. However, this support on an individual level can, of course, only contribute to punctual development. But the lecturers took on the mission of stimulating development within a broader context. This was achieved, interestingly, in the field of natural resource management—and surprisingly was more relevant in the interviews with Ghanaian lecturers.

The notion that Ghanaian lecturers have more impact on development processes could have to do with the fact that many Ghanaians from the sample studied in the field of renewable resources and resource management, which is developmentally oriented per se. That Cameroonian graduates have the same potential for developmental change was made evident by the case of Mr. I. from Cameroon (case #11)[21].

[20] This interview quote has been translated from French to English by the author of this paper.

[21] After studying environmental resource management at Brandenburgische Technische Universität in Cottbus, Germany, Mr. I. returned on November 1, 2007, to Cameroon. After a job search lasting five months he found employment as an environmental manager at a local NGO in Yaoundé. We met for the interview at his workplace on October 9, 2008.

He was a graduate of an English language master's degree program in natural resource management in Germany. He pointed out that he went to study in Germany because Cameroonian universities lacked expertise in the field of natural resource management:

> Cameroon needs many environmental experts. There are no experts. So I said, "Oh, so if there are no experts, if I can study this thing and come back at least I will be an expert". That's when the idea started. And this is [what I said] when they asked me at the German embassy [why I wanted] to go and study. That's when the motivation actually started. And that's when I saw that it's actually true. There is no college; there is no university in Cameroon training people on environmental stuff. (Cameroon interview #11, 2008)

Mr. I. saw an increasing need for environmental experts in Cameroon arising because of the country's rich natural resources. They represent the country's primary export commodity, but Mr. L. thought that Cameroon lacked the necessary accountability to exploit these resources responsibly:

> Our problem in Cameroon is resource management and accountability. If our president is struggling to install accountability, we need people to start teaching resource management. How can we use [our resources]? I think if we will do this, I think Americans will come to Cameroon to work. Yes! Because we may not know how to go about it harnessing diamonds. Diamonds have been discovered in Cameroon, ha? There is bauxite, there is iron that has been discovered and iron ore another, another there is enough gas, natural gas that has been discovered and other, oil, eh, rubber has been discovered in Kumba, still going to be tapped. What will we need to be doing in Europe? Making beds or washing toilets? In fact, it is so upsetting to see a PhD student cleaning toilets and we are singing, "Hallelujah!" over it. Singing songs over toilets! (Cameroon interview #11, 2008)

Mr. I. suggested in this quote that there was a need to "start teaching resource management" in Cameroon in order to "install accountability" and to make the local population capable of managing resource exploitation themselves. This, he claimed, would be a countermeasure against the increasing emigration of economic migrants who, in his view, could not access decent employment abroad and often end up in menial jobs. However, despite his expertise in this important field he only held a part-time lecturer position and generated his main income as an employee in a NGO.

Having said that sustainable development also includes a political dimension, it would now be interesting to take a closer look at the returnee graduates' impact in terms of their implementing structural changes. An example is the case of Mr. U. (Ghana case #10)[22]. Like Mr. I. from Cameroon he only holds a master's degree. But in contrast to Mr. I., he was able to obtain a higher position in the very beginning:

> Yeah the qualification I got from Stuttgart, of course it has been very helpful because to lecture in a tertiary institution, the minimum is a master's level and if you have a B. Sc. [bachelor of science] in a polytechnic you can be an instructor but your conditions of service is not good. Getting the master's degree had positioned me in a better place. In fact, there

[22]Mr. U. studied water resources engineering and management in Germany and returned on September 27, 2005, and became head of department at the polytechnic in the Volta Region at Ho. The interview took place three years later on July 4, 2008, in Ho.

> were other people, who had been in the department before I came. But because they didn't have their master's even though they had a lot of experience in terms of "on the job,", they are now at the background and I just came and I was made the head of department, it was a new, a new thing to me [smiles] .... It is because of the qualification, yes. (Ghana interview #10, 2008)

From Mr. U.'s perspective, his becoming head of department despite his lack of teaching experience was "because of the qualification" in natural resource programs. He indicated that he has even more influence in this position than a lecturer and that he designed the curriculum of the department:

> There are some of the courses [in the field of renewable energies] you can't get them here ... there is no institution in Ghana that handles waste management and K-Poly is now trying to work on it. We are developing a curricular .... So for such a course like this we need to train people at the MSc [master of science] level, there is no any fair degree, we don't have technology, we don't have it. MSc we don't have it but each course is run in Germany, water resources ... so for such courses if you don't travel outside then you can't have it and if you don't have it there is no chance, there is no way Ghana will have any education in that line. (Ghana interview #10, 2008)

Mr. U.'s impact was clearly related to capacity building in the field of environmental resource management because at that time, in 2008, there was "no institution in Ghana that handles waste management." Thanks to the specific knowledge he had gained in Germany, Mr. U. not only improved his individual situation, but also was able to bring significant change to his department by adopting the course curriculum of German universities. He contributed to institution and capacity building.

Ultimately this section about the impact of returning graduates on development processes suggests that Ghanaian returning graduates working currently as lecturers can contribute to environmental education on a larger scale. In comparison, the Cameroonian lecturers were not mentioning such development related activities in the context of their official work and in contrast to their Ghanaian counterparts had greater challenges during their entry into the university. In addition, the need to have a PhD degree was much more obvious in Cameroon. All interviewees among the group of Cameroonian lecturers had their doctorates, whereas it was possible in Ghana find employment with a master's degree, as the example of Mr. U. (case #10) showed.

Conclusions

At first glance universities in Sub-Saharan African countries would seem to be ideal employers for returning graduates who studied in the Global North. By disseminating their knowledge from abroad they could contribute to technological catchup, build capacities, and counter the ongoing massive brain drain of highly skilled workforces in the higher education sector. I reveal the contrary. The universities do not fully reap the great potential of returning graduates. My comparison of the labor market entry of graduates in the two Sub-Saharan African countries of Ghana and Cameroon shows that the job-finding process in higher education in both countries is lengthy and cumbersome and the matching process rather difficult. What differs is the weighting of the difficulties. The quotes from the interviews (which of course reflect the situation eight years ago) lead to the assumption that recruitment at Cameroonian universities at the time was even more intransparent and insecure than at those in Ghana. In addition the development impact of lecturers seemed to be clearer in the case of Ghanaian academics. Returned graduates in Ghana who worked in the higher education reported how they delivered community services, and many had a particular impact in the field of renewable energies. Ghana's comparably more transparent and open higher education system made it possible for Ghanaian return migrants to make use of their educational expertise in the field of natural resource management. Their educational migration empowered them to foster development processes in their home country of Ghana. In contrast, the system in Cameroon seemed to give returnee graduates more difficulties finding a vacant position at a university, let alone securing it, which even led some graduates to create their own job opportunities by establishing an educational institution (see the case of Dr. and Mrs. H.; Cameroon case #02). Hence, my interpretation of the empirical material leads me to suggest that Ghanaian universities used the potential of returning academics more effectively than their Cameroonian counterparts.

Does this in consequence mean that career opportunities in academia are too limited for returnee graduates and thus that graduates should not consider returning? This, despite all the obstacles shown in this chapter, is not the case. The solution to overcoming these obstacles lies in the graduates' resources, which they should acquire prior to their return. In order to prepare for their home country's professional reality they first need to build professional academic networks to help in the search for a vacant position. Second, returnees should have secure intermediary financial funding at their disposal because the job search phase may be long. And third—and probably most importantly, they need to acquire an educational profile that matches the country's specific needs. These findings also suggest that preparation schemes, such as reintegration seminars offered by the German government, are still a very important instrument increasing chances that the return of educational migrants will be a brain gain for the source countries as well as for the host country of Germany.

References

AAU (Association of African Universities). (2015). Member Universities. Retrieved from https://www.aau.org/subs/membership/

Asong, A. A., & Awama, S. G. (2014). Le suivi des diplômés de l'université de Buea: 1996–2006 [Tracking graduates of the University of Buea]. In M. S. Kamdem & E. W. Schamp (Eds.), *L'université africaine et sa contribution au développement local: L'exemple du Cameroun* (pp. 149–174). Paris: Karthala.

Ausländerzentralregister. (2016). *Bevölkerung und Erwerbstätigkeit: Ausländische Bevölkerung 2015. Ergebnisse des Ausländerzentralregisters* [Population and employment: Foreign population 2015. Results of the Central Register of Foreigners]. Wiesbaden: German Federal Statistical Office.

Bloom, D., Canning, D., & Chan, K. (2006). *Higher education and economic development in Africa* (No. 102). Retrieved from http://ent.arp.harvard.edu/AfricaHigherEducation/Reports/BloomAndCanning.pdf

Boger, J. (2013). Ghanaian 'green-degree' graduates returning from Germany and their impact on environmental development in Ghana. *Challenging Notions of Development and Change from Everyday Life in Africa*, 4, 86–122. Retrieved from https://epub.uni-bayreuth.de/113/1/BIGSASworks_4_05oct2013.pdf

Boger, J. (2014). *The job hunt: Return migration and labour market entries of Cameroonian and Ghanaian graduates from German universities*. Wiesbaden: World University Service.

Cassarino, J.-P. (2004). Theorising return migration: The conceptual approach to return migrants revisited. *International Journal on Multicultural Societies*, *6*, 253–279. doi:https://doi.org/10.1590/S1980-85852013000200003

Chien, C.-L., & Kot, F. C. (2012). *New patterns in student mobility in the southern Africa development.* UIS Information Bulletin No. 7. Retrieved from http://unesdoc.unesco.org/images/0021/002170/217095e.pdf

Cerase, F. P. (1974). Expectations and reality: A case study of return migration from the United States to Southern Italy. *International Migration Review, 8*, 245–262. doi:https://doi.org/10.2307/3002783

de Haas, H. (2008). *Migration and development: A theoretical perspective*. International Migration Institute (IMI Working Papers). Retrieved from http://www.imi.ox.ac.uk/pdfs/wp/wp-09-08.pdf

de Haas, H. (2012). The Migration and Development pendulum: A critical view on research and policy. *International Migration, 50*(3), 8–25. doi:https://doi.org/10.1111/j.1468-2435.2012.00755.x

Faist, T. (2008). Migrants as transnational development agents: An inquiry into the newest round of the migration—development nexus. *Population Space Place, 14,* 21–42. 10.1002%2fpsp.471

German Federal Statistical Office. (2015a). *Origin of foreign students, Bildungsauslaender and Bildungsinlaender students in 2015 by continents, regions and countries of origin.* Retrieved from http://www.wissenschaftweltoffen.de/daten/1/2/1?lang=en

German Federal Statistical Office. (2015b). *The number of Bildungsauslaender students from the 20 most important countries of origin in 2015—development in medium-term from 1999 to 2015.* Retrieved from http://www.wissenschaftweltoffen.de/daten/1/2/3?lang=en

Ghanaweb. (2015). *Universities in Ghana.* Retrieved from http://www.ghanaweb.com/GhanaHomePage/education/universities.html

Hahn, H.-P., & Klute, G. (2007). *Cultures of migration: African perspectives.* Beiträge zur Afrikaforschung: Vol. 32. Berlin: LIT. Retrieved from https://openaccess.leidenuniv.nl/bitstream/handle/1887/12923/ASC-140170847-131-01.pdf?sequence=2

Krings, M. (2003). Diaspora: Historische Erfahrung oder wissenschaftliches Konzept? Zur Konjunktur eines Begriffs in den Sozialwissenschaften [Historical experience or scientific concept? On a booming concept in the social sciences]. *Paideuma, 49*, 137–156. Retrieved from http://www.jstor.org/stable/40315515

Kuptsch, C. (2006). Students and talent flow—the case of Europe: From castle to harbour? In C. Kuptsch & E. F. Pang (Eds.), *Competing for global talent* (pp. 33–61). Geneva: International Institute for Labour Studies. Retrieved from http://citeseerx.ist.psu.edu/viewdoc/download;jsessionid=30B2EE7ADDA235E6CA8ED89F52374B1A?doi=10.1.1.120.2008&rep=rep1&type=pdf

Manuh, T., Asante, R., & Djangmah, J. (2005). The brain drain in the higher education sector. In T. Manuh (Ed.), *At home in the world? International migration and development in contemporary Ghana and West Africa* (pp. 250–276). Legon: Sub-Saharan.

Martin, J. (2005). *Been-To, Burger, Transmigranten? Zur Bildungsmigration von Ghanaern und ihrer Rückkehr aus der Bundesrepublik Deutschland* [Been-To, Burger, Transmigrants. Educational migration and the reintegration of Ghanaians from Germany]. Beiträge zur Afrikaforschung: Vol. 22. Berlin: LIT.

Mbah, M. (2014). *Brain Drain aus Entwicklungsländern? Migrationsmotive- und prozesse Hochqualifizierter am Beispiel von Nigeria.* [Brain drain from developing countries? Migration motives and processes of the highly skilled as exemplified by Nigeria]. Karlsruhe: University of Karlsruhe. Retrieved http://digbib.ubka.uni-karlsruhe.de/volltexte/1000042111

Mbah, M. (2017). Formal education as a facilitator of migration and integration: A case study of Nigerian university graduates. In H. Jöns, P. Meusburger, & M. Heffernan (Eds.), *Mobilities of knowledge* (pp. 247–268). Knowledge and Space: Vol. 10. Cham: Springer. doi:https://doi.org/10.1007/978-3-319-44654-7_13

Ministry of Higher Education. (2014). *L'enseignement supérieur au Cameroun: Cartographie 2014 des Institutions* [Higher education in Cameroon: 2014 atlas of the institutions]. Retrieved from http://www.minesup.gov.cm/1.1.%20cartographie%202014%20de%20l'enseignement%20sup%20au%20Cameroun1%20%20%20%20%20.pdf

Morris-Lange, S., & Brands, F. (2015). *Zugangstor Hochschule: Internationale Studierende als Fachkräfte von morgen gewinnen* [The university as entryway: How to recruit international students as tomorrow's professionals]. Retrieved from http://www.svr-migration.de/wp-content/uploads/2015/06/Studie_Zugangstor-Hochschule_SVR-FB_Web.pdf

Mugabushaka, A.-M., Teichler, U., & Schomburg, H. (2003). Failed or self-hindering prophecies? Employment experiences of African graduates in the 1990s. *Journal of Higher Education in Africa, 1*(1), 57–77. Retrieved from http://www.jstor.org/stable/24486114

National Accreditation Board. (2015). *Public universities Ghana*. Retrieved from http://www.nab.gov.gh/public-universities

Neubert, D. (2008). Academic cooperation between Germany and Africa—challenges and some lessons learnt. In E. W. Schamp & S. Schmid (Eds.), *Academic cooperation with Africa: Lessons for partnership in higher education* (pp. 93–110). Berlin: LIT.

Nyberg-Sørensen, N., Van Hear, N., & Engberg-Pedersen, P. (2002). The migration-development nexus: Evidence and policy options. *International Migration*, 40(5), 3–47. doi:https://doi.org/10.1111/1468-2435.00210

Nyonator, F., & Dovlo, D. (2005). The health of the nation and the brain drain in the health sector. In T. Manuh (Ed.), *At home in the world? International migration and development in contemporary Ghana and West Africa* (pp. 227–249). Legon: Sub-Saharan.

OECD (Organization for Economic Co-operation and Development). (2014). *Education at a Glance 2014: OECD indicators*. Paris: OECD. Retrieved from https://www.oecd.org/edu/Education-at-a-Glance-2014.pdf

Owono, J. (2011, October 13). Cameroun: La diaspora s'est abstenue de voter à l'élection présidentielle [Cameroon: Diaspora fails to participate in elections]. *Global Voices en Français*. Retrieved from http://fr.globalvoicesonline.org/2011/10/13/83633/print/

Portes, A. (2003). Conclusion: Theoretical convergencies and empirical evidence in the study of immigrant transnationalism. *International Migration Review*, 37, 874–892. doi:https://doi.org/10.1111/j.1747-7379.2003.tb00161.x

Quartey, P. (2009). *Migration in Ghana: A country profile 2009*. Geneva: International Organization for Migration. Retrieved from https://publications.iom.int/system/files/pdf/ghana_profile_2009_0.pdf

Sapouma, C. (2011, August 19). Enfin le droit de vote pour les Camerounais de l'étranger [Finally the right to vote for Cameroonians from abroad]. *Slate Afrique*. Retrieved from http://www.slateafrique.com/print/22427/elections-enfin-le-droit-de-vote-pour-les-ca

Saxenian, A. (2005). From brain drain to brain circulation: Transnational communities and regional upgrading in India and China. *Studies in Comparative International Development*, 40(2), 35–61. doi:https://doi.org/10.1007/BF02686293

Schamp, E. W., & Zajontz, Y. (2008). Patterns of international academic communication at African universities: Cameroon, a case study of a bilingual country. In E. W. Schamp & S. Schmid (Eds.), *Academic cooperation with Africa: Lessons for partnership in higher education* (pp. 55–76). Berlin: LIT.

Schamp, E. W. (2014). Université et développement régional en Afrique: Le défi de nouvelles idées et de nouvelles exigences à l'heure de la globalisation [University and regional development in Africa: The challenge of new ideas and demands in the age of globalization]. In M. S. Kamdem & E. W. Schamp (Eds.), *L'université africaine et sa contribution au développement local: L'exemple du Cameroun* [The African university and its contribution to local development: The example of Cameroon] (pp. 19–39). Paris: Karthala.

Schmidt-Fink, E. (2009). *Erfassung der Rückkehrhilfesysteme in den Bundesländern: Untersuchung im Auftrag der Landeshauptstadt München: Büro für Rückkehrhilfen Projekt Coming Home* [Compilation of assisted voluntary return schemes in the federal states: Investigation commissioned by the state capital Munich: Office for the Coming Home project] [Brochure]. Retrieved from http://www.muenchen.info/soz/pub/pdf/303_rueckkehrhilfesysteme.pdf

Skeldon, R. (2005). *Globalization, skilled migration and poverty alleviation: Brain Drains in context* (Working Paper No. T15). Retrieved from http://www.migrationdrc.org/publications/working_papers/WP-T15.pdf

SOFRECO. (2015). *Study on the contribution of the alumni and diaspora to the Joint Africa-EU Strategy: Final report* (Framework Service Contract No EAC/02/2010). Retrieved from http://ec.europa.eu/education/news/2015/documents/study-sofreco-eu-africa_en.pdf

Teferra, D., & Altbach, P. G. (2004). African higher education: Challenges for the 21st century. *Higher Education*, 47, 21–50. doi:https://doi.org/10.1023/B:HIGH.0000009822.49980.30

Thomas, K. J. A. (2008). Return migration in Africa and the relationship between educational attainment and labour market success: Evidence from Uganda. *International Migration Review*, 42, 652–674. doi:https://doi.org/10.1111/j.1747-7379.2008.00141.x

UIS (UNESCO Institute for Statistics). (2010). *Trends in tertiary education: Sub-Saharan Africa* (UIS Fact Sheets No. 10). Retrieved from http://unesdoc.unesco.org/images/0019/001926/192603e.pdf

UIS (UNESCO Institute for Statistics). (2012). *Global Education Digest 2012: Opportunities lost: The impact of grade repetition and early school leaving*. Retrieved from http://unesdoc.unesco.org/images/0021/002184/218449e.pdf

UNDP (United Nations Development Program). (2015). *Human Development Report 2015: Work for Human Development* (Publication No. E.15.III.B.1). Retrieved from http://hdr.undp.org/sites/default/files/2015_human_development_report_1.pdf

Vertovec, S. (2007). *Circular migration: The way forward in global policy?* (International Migration Institute Working Papers: No. 4). Retrieved from https://www.imi.ox.ac.uk/publications/wp-04-07

World Bank (2000). *Higher education in developing countries: Peril and promise* (Published for the Task Force on Higher Education and Society). Retrieved from http://siteresources.worldbank.org/EDUCATION/Resources/278200-1099079877269/547664-1099079956815/peril_promise_en.pdf

World Bank (2009). *Accelerating catch-up: Tertiary education for growth in Sub-Saharan Africa*. (Directions in development: Human development, No. 45275). Retrieved from http://wwwwds.worldbank.org/external/default/WDSContentServer/WDSP/IB/2009/01/27/000333038_20090127223030/Rendered/PDF/462750Revised017808213773830Revised.pdf

World Bank (2010). *Financing higher education in Africa*. (Directions in development: Human development, No. 54441). Retrieved from http://wwwwds.worldbank.org/external/default/WDSContentServer/WDSP/IB/2010/05/11/000333037_20100511005406/Rendered/PDF/544410PUB0EPI01BOX0349416B01PUBLIC1.pdf

World Bank (2011). *Migration and remittances factbook 2011* (2nd ed.). Washington DC: The World Bank.

Zajontz, Y. (2010). *Afrikanische Universitäten als Motoren regionaler Entwicklung? Eine Analyse von Wissenskanälen an drei peripheren Universitäten in Kamerun* [African universities as motors of regional development? An analysis of knowledge channels at three peripheral universities]. Lehre & Forschung—Hochschule im Fokus: Vol. 7. Hamburg: Dr. Kovac.

Zips, W. (Ed.) (2003). *Afrikanische Diaspora: Out of Africa—Into New Worlds* [African Diaspora: Out of Africa—into new worlds]. Afrika und ihre Diaspora, Vol. 1 [Africa and its diaspora], Berlin: LIT.

Zweig, D. (2006). Learning to compete: China's efforts to encourage a "reverse brain drain". In C. Kuptsch & E. F. Pang (Eds.), *Competing for global talent* (pp. 187–213). Geneva: International Institute for Labour Studies.

Chapter 18
China's Southern Borderlands and ASEAN Higher Education: A Cartography of Connectivity

Anthony Welch

The traditional Chinese aphorism "The Mountains are High and the Emperor Far Away" (*Shan gao, huangdi yuan*; 山高皇帝远) pithily illustrates distinctive qualities of regionalism in China's southern borderlands, including in higher education. The two provinces of Yunnan and Guizhou, for example, together with the Guanxi Zhuang Autonomous Region (GUAR),[1] adjoin Association of Southeast Asian Nations (ASEAN) member states, such as Viet Nam and Myanmar. Long considered fringe dwellers in more than the geographical sense by both imperial and postimperial Chinese rulers, these borderlands have recently been accorded much greater prominence and are now seen as a *qiaotoubao* (bridgehead) to the South China Sea and Southeast Asia, notably in higher education. Indeed, the fact that ASEAN now arguably constitutes China's highest regional priority (Wen, Luo, & Hu, 2014) accords a much greater prominence to China's southern borderlands, notably its universities. This greater prominence, together with several major pillars of China-ASEAN relations, many of which are of long standing but evolving, comprises thickening cross-border relations, including in higher education. Nonetheless, these same borderlands continue to display distinctive qualities, which are argued to be reflective of their peripheral geography and distance from political centers of power. In some respects, it is argued, the emperor is still far away. In the following sections I examine the context for China-ASEAN regionalism, with illustrations of regional relations in higher education relations. In light of thickening

[1]The first two are provinces, the third is designated an autonomous region, a status equivalent to other border regions of China, such as the Xinjiang Uyghur Autonomous Region, in China's Northwest.

A. Welch (✉)
Department of Education and Social Work, The University of Sydney, Sydney, Australia
e-mail: anthony.welch@sydney.edu.au

P. Meusburger et al. (eds.), *Geographies of the University*, Knowledge and Space 12,
https://doi.org/10.1007/978-3-319-75593-9_18

relations in the region, notably between universities, I then outline the overall state of China-ASEAN relations in higher education, as well as finally posing the question of how far China-ASEAN may be considered as a region, especially in higher education.

Regionalism and Borderlands

The following analysis both highlights and problematizes the conventional view of regionalism. There are a number of elements to this. Several analysts have recently argued that the more embryonic quality of ASEAN regionalism differs significantly from the more well-developed and longstanding regional architecture of the European Union (Jayasuriya, 2003, 2010; Robertson, 2008). This certainly includes higher education where, as seen below, regional mobility schemes are much less well developed and less well funded. As argued below, this more embryonic quality of higher education regionalism is in part the legacy of colonialism, which has led to reluctance by national authorities in higher education to cede authority to regional initiatives (Wen, 2016; Zhao, 2011). In part too, as other authors have pointed out, this is due to the sometimes substantial gap between expressed ASEAN aspirations and declarations; and the more limited range of actual achievements (Welch, 2012b). As part of his significant work on regulatory regionalism, Jayasuriya, for example, has pointedly criticized the triumphalism of ASEAN regionalism, pointing to its focus on "formal regional institutions . . . to the detriment of the understanding of the domestic political mainsprings of regional governance" (Jayasuriya, 2003, p. 199). This, he has argued, has limited the reach and capacity of regulatory regionalism within ASEAN. The not uncommon incidence of nationalist resistance to regional undertakings among several ASEAN member states has been seen by some as constituting a further brake on progress (Pesek, 2012), including in higher education.

As Gregory, Meusburger, and Suarsana (2015, p. 1) pointed out, Foucault underlined the link between region and (the reach and mechanisms of) political power. Foucault pointed out that regime and region both stem from the same Latin verb *regere*, to rule, (from which, for example, in English, the terms regiment, regime, and regulate all derive). In a discursive disquisition on geography and power, in which he acknowledged his spatial obsession, he argued that "Region is a fiscal, administrative, military notion" (Foucault, 1980, p. 68). He went on to point out the substantial implications for power-knowledge relations:

> Once knowledge can be analyzed in terms of region, domain, implantation, displacement, transposition, one is able to capture the process by which knowledge functions as a form of power and disseminates the effects of power. There is an administration of knowledge, a politics of knowledge, relations of power which pass via knowledge and which, if one tries to transcribe them, lead one to consider forms of domination designated by such notions as field, region and territory. And the politico-strategic term is an indication of how the military and the administration come to inscribe themselves . . . on a material soil. (p. 69)

Effectively, Foucault is highlighting here the relationship between geographic space and administrative power and control. Given this thesis, borderland regions—

by definition peripheral—can be seen as a good litmus test of the (limits of) regionalism. The notion of limit here is arguably inherent. Cogent arguments have been made that borderland regions serve dual roles: both as markers of the limit of state sovereignty—"the boundary which marks the name of the state" (Chan, 2013, p. ix)—and also as sites that test and at times defy the power and authority of the state. Borderlands regions, where quotidian realities commonly include cross-border currents of people, languages, concepts, cultures, and currencies, are sites where ways of life collide and interweave. In what follows, the southern provinces of Yunnan and Guizhou, and the GUAR, and their relations with ASEAN higher education systems and institutions, are deployed as a test case of the limits of regionalism, illustrating both the tensions and options that borderlands spaces occupy (Chan, 2013; Eimer, 2014; Scott, 2009). The porous, liminal quality of borderlands regions evokes earlier eras characterized as "territorial nonchalance in the peripheries of states" (Chan, 2013, p. 90). This porosity of relevant borders, which as seen below are routinely breached, as well as the extralegal forms of much transnational trade and commerce, challenges conventional views of regulatory regionalism, and might better be characterized as instances of irregular regionalism. The implications for understanding relations between space and knowledge are significant.

The following section outlines the character and limits of China-ASEAN regionalism, more particularly Viet Nam and Myanmar (formerly known as Burma), also alluding, where appropriate, to Malaysia, Philippines, Indonesia, and Singapore. Several examples of China-ASEAN higher education relations are provided, both historical and contemporary. An analysis is then made of regional relations in higher education, illustrating the relationship via examples from a regional university within China's southern borderlands that for the purposes of this chapter is referred to as *Borderlands University*. It is an institution of substantial significance in building and sustaining relations with higher education in the ASEAN region.

The Limits of Regionalism

China's southern borderlands, selected as an illustration of wider China-ASEAN regionalism in higher education, reveal a rich and complex tapestry of relations extending over more than two millennia, and which crucially embrace forms of higher learning and knowledge mobility. Six pillars of China-ASEAN relations are sketched below—economics, knowledge mobility, historical background, Chinese regional diaspora, regional perceptions of Chinese minorities, territorial disputes—before a specific focus is placed on China's southern borderlands region, which has long featured close relations with Viet Nam, for example. This might be considered an asymmetric relationship, yet Chan has argued that, while Viet Nam has maintained its independence, China will have to buy its way into Southeast Asia, via Viet Nam ("Asia on Alert," 2015; Chan, 2013, pp. 121–122; Kwok, 2015).

Given the intensity of current regional territorial disputes between the two, as outlined below, the ease with which this might be accomplished should not be taken for granted ("Hanoi Anger," 2016). The reestablishment of formal diplomatic relations after the Sino-Vietnamese war of 1979, and the reopening of the border in 1991, together with the increased priority accorded Southeast Asia by China, was paralleled by an intensified regionalism, including in higher education. Despite a series of obstacles outlined below, some of which are intensifying, the last three decades or so of "Reform and Opening," initiated by Deng Xiaoping in 1978, and its equivalent *Doi Moi* in Viet Nam from 1986, has, inter alia, deepened and widened cross-border flows of knowledge, students and academic staff, and educational personnel, enriching and extending the character of regionalism.

Yet it remains the case that China-ASEAN relations are conventionally viewed through the prism of economics and trade, a biased perspective that is misleading in at least two senses. Even within the trade portfolio, the emphasis is conventionally on goods, obscuring swiftly rising service-sector trade across a wide range of areas. This notably includes higher education, which now contributes billions to regional economies, including China's (Wen, 2016; Wen, Luo, & Hu, 2014). This error may well become even more important, as China deliberately moves from an emphasis on production of goods, to a more service-based economy, in which financial services, tourism, and education services assume a larger part. In 2015, for the first time, for example, China's service sector comprised more than 50% of the total economy (Wildau, 2015). The second error consists of the emphasis on trade itself, which is misleading, in that China-ASEAN relations are far richer, longer, and more varied than mere trade relations indicate.

As illustrated below, for example, the concourse in trade and ideas between China and Viet Nam is more than a millennium old. But since the end of the Sino-Vietnamese war of 1979, and more particularly since the implementation of structural economic reforms of each economy from the late 1970s in China and late 1980s in Viet Nam, and the resumption of diplomatic relations between the two "sisters in socialism" in 1991, relations have mushroomed, especially in the three abovementioned provinces and/or regions, where border crossings—literal and symbolic, legal and illegal, regulated and irregular—have become ever more routine. As illustrated below, this embraces various forms of higher education relations, including student mobility, staff mobility, language learning, and growing research collaboration.

Within China more broadly, the increased priority accorded Southeast Asia has seen a process of revalorizing China's southern borderlands. Seen until relatively recently as "backward" (*luo hou*, 落后), with intimations of danger, remoteness, and of primitiveness, the central government moved to renominate Yunnan province, for example, as a bridgehead (*qiaotoubao*, 桥头堡) to Southeast Asia of great strategic significance (Sigley, 2014). This includes higher education, where Yunnan's universities compete vigorously for ASEAN students and broader regional influence with their peers from both Guanxi and Guizhou.

From the other side of the border, Vietnamese borderlands minorities in Viet Nam's Northern Mountain Region are also still widely perceived as backward by lowlands Vietnamese (Tran, 2003). This includes universities in the Northern Mountainous Region, for example, such as Thai Nguyen, which although well-regarded domestically, is much less well known within the Asia Pacific region, and more broadly, than leading national universities such as Viet Nam National University (VNU) Hanoi and VNU Ho Chi Minh City[2]. Through much of its history, China's southern borderlands were considered marginal and problematic. In fact, Yunnan, which borders Viet Nam, and whose universities are actively engaged in cross-border collaboration, was only brought under China's dynastic system after the Mongol conquest, during the Yuan dynasty (1271–1368) and subsequent Ming dynasty (Crossley, Siu, & Sutton, 2006). Even thereafter, Yunnan remained unstable: The former Dai kingdom of Sipsongpanna[3] was only gradually incorporated into the Qing dynasty, and the last Dai king was only finally compelled to abdicate by the Chinese Communist Party, in 1953 (Atwill, 2005). Yet another instance of the emperor still being far away.

These particular qualities of regionalism have significant implications for the quality of China-ASEAN higher education relations (Welch, 2014). As detailed below, China's preferred form of transnational higher education is via partnerships; the greater priority currently accorded to relations with Southeast Asia means that universities in the borderland regions of Guangxi, Guizhou, and Yunnan assume greater importance in establishing relations with ASEAN neighbors. This is seen as an example of the Chinese principle of *yu lin wei shan, yi lin wei ban* (be good to one's neighbors, turn one's neighbors into partners) and includes specific frameworks, such as ASEAN+3 and ASEAN+1 and equivalent structures within the ASEAN University Network (AUN) (Welch, 2011a; Wen, 2016). At the same time, however, ASEAN's preferred mode of regional relations in higher education conforms to the wider ASEAN way, in which dialogue and consensus are accompanied by the principle of noninterference in the affairs of individual member states. Influenced by the historical legacy of colonialism in the region, in effect, this principle manifests itself as a reluctance to cede authority beyond the bounds of the nation state, and acts as a brake on more substantial regional initiatives in higher education (OECD 2003; Welch, 2012b; Wen, 2016; Zhao, 2011).

The Six Pillars of China-ASEAN Regionalism: Implications for Higher Education

Economic Relations

The first pillar comprises economic relations: China's dramatic economic rise over the past three decades, as well as growing China-ASEAN industrial and commercial complementarity, helped stimulate deeper relations, significantly including higher

[2]Formerly Saigon.

[3]The Chinese term *Xishuangbanna* reflects the Dai name, meaning 12 rice fields, or 12 pieces of land. It stems from Ming dynasty times.

education (Jarvis & Welch, 2011; Welch, 2011a, 2011b), although by 2015 China's slowing economy was causing significant problems for a number of ASEAN member states (Gough, 2015). But more than China's stellar GDP growth was responsible ("More Chinese," 2014; State Council, 2015). China's growing regional engagement and notable support for her ASEAN neighbors each helped strengthen regional relations. Repeated assurance of "China's peaceful rise," endorsement of the Treaty of Amity and Cooperation with ASEAN, and the Joint Declaration on Strategic Partnership for Peace and Prosperity, in 2003, cemented closer relations (Cheow, 2004; Osborne, 2006; Vaughan & Morrison, 2006; Whitney & Shambaugh, 2008). Soft loans to ASEAN member states and substantial funding for key development projects, as well as the recent One Belt, One Road (Yi Tai Yi Lu) initiative, have further strengthened regional relations ("Chinese Outbound," n.d.; Hirono, 2010; Laksmana, 2011; "Outbound Tourism," 2014).

The signing of the China-ASEAN Free Trade Area agreement in early 2010, subsequently upgraded at the November 2015 protocol meeting in Kuala Lumpur, accelerated trade in both goods and services, a trend that has been boosted by the spectacular growth of the Chinese middle class in recent years, now estimated to total between 110 and 160 million, (depending on the measure used) (Goodman, 2008; Hodge, 2016; Tomba, 2011). Latest analyses project that almost 90% of the next billion to join the middle class over the next decade will come from Asia, including 350 million Chinese and 210 million from "other Asia," especially Southeast Asia, notably Singapore, Malaysia, Thailand, and Indonesia (Kharas, 2017, p. 13).[4]

With intraregional tourism only set to expand (Tong & Chong Siew Keng, 2010, p. 7), notwithstanding some of the issues sketched below ("Guangxi Ready," 2014; State Council, 2015), the implications of a rising Asian middle class are particularly important for regional higher education. Growing regional prosperity is increasing demand for both Chinese higher education and Chinese language services in the region, as well as student flows in both directions. Indeed, Malaysia deliberately markets its universities to the Chinese middle class, while Singapore has well-established training programs for Chinese officials ("Singapore's NTU," 2009; Yi, 2015). This increased demand also includes the burgeoning number of Confucius Institutes, (the Chinese equivalent of Germany's Goethe-Institut and France's Alliance Française), now numbering more than 30 within Southeast Asia (15 in Thailand alone), despite some local resistance in Viet Nam ("Chinese Premier," 2013; Omi, 2017; Wen, 2016; Yang, 2012). But Confucius Institutes, which are always based within local universities and always involve a partnership with a Chinese university, are by no means the whole story. As detailed below, numerous analyses reveal that rising intraregional service sector trade embraces a substantial rise in international student mobility. The growth of the middle class in both China and Southeast Asia, is evident in a much greater disposable income, which is often spent on sending their children abroad for university study. Fueled also by traditionally high aspirations for

[4]As well, an added 380 million will come from India.

their children's education among Chinese and Southeast Asian parents and a willingness to incur the additional costs of an overseas degree, the growth of student flows between China and ASEAN, underpinned by supportive policies on both sides, is a clear outcome of the spectacular growth of the Asian middle class (Hodge, 2016; Illeva, Killingley, Tsiligiris, & Peak, 2017, p. 10; Kharas, 2011, 2017, pp. 2, 20).

The regional dimension of this growth is important, and sustains greater intra-regional mobility in higher education. ASEAN-China trade overtook that with Japan in 2011, to become China's third largest, after the European Union and United States (Chang, 2012), while major Chinese investment in the region proceeded, despite a history of some local "ambivalence" ("China Faces Resistance," 2017; Kwok, 2015; Laksmana, 2011; Lo & Leng, 2017; "Philippines Accuses," 2015). Trade with ASEAN is predicted to surpass that with the United States and the European Union in the next few years, "Thanks to zero tariffs, preferential trade policies, and geographic advantages." (Chang, 2012; China-ASEAN 2013). Two-way trade reached US$480 billion in 2014, and a target of US$1 trillion has already been set for 2020. Ongoing China-ASEAN discussions regarding implementation of the Regional Comprehensive Economic Partnership would further boost regional trade, including service sector trade such as higher education (Lo, 2014; "Singapore to Give, " 2012; Wu & Mealy, 2012). Most recently, the announcement of the *Yi Tai Yi Lu* (One Belt, One Road) initiative (most particularly the Maritime Silk Road), as well as the Asian Infrastructure and Investment Bank, both designed to accelerate and deepen regional connectivity and associated infrastructure, should significantly extend existing regional trade and further boost the size of the Asian middle class, upon which much transnational higher education trade depends (Asian Infrastructure Investment Bank, n.d.; Hofman, 2015; Tiezzi, 2014).[5]

But increased regional dependence on China is something of a two-edged sword for ASEAN economies. Are there too many China eggs—which contribute as much as 80% of developing East Asia's GDP—in ASEAN's basket? (Rahardja, 2012). If so, a rebalance towards service sector trade, notably including trade in higher education services, could potentially mitigate such effects.

Regional Chinese Diaspora

Regional trade and closer cultural ties are further deepened by the significant mainland Chinese diaspora spread throughout Southeast Asia, a pattern that began centuries ago. Zheng He's voyages (see Fig. 18.2) and active regional trade by

[5]The U.S. proposed Trans-Pacific Partnership, which excluded China, is an alternative regional framework, with significant implications for ASEAN member states (Boudreau, 2015; Department of Foreign Affairs and Trade, 2015; Williams, 2013). It was formally rejected by incoming U.S. President Donald Trump, in 2017.

Table 18.1 The dragon and the tiger cub

	China	Viet Nam
Population in Millions (2011)	1,346	87.9
Percentage of Chinese in Population	100*	1.5
GDP per Capita, (PPP, US$), 2011	8,400	3,300
Development Status	Developing	Developing
Status, HDI Rating, 2011	.687	.593
FDI to China,(US$ million), 2008	-	2
FDI from China, (US$ million), 2008	-	120
GDP Growth, 2011	9.24%	6.24%

Note. PPP = Purchasing Power Parity. HDI = Human Development Index. FDI = Foreign Direct Investment. *Approximately 8.5% of the population is from China's 55 designated minorities. Sources: CIA, Country Comparison: GDP per capita; Population Reference Bureau, 2011: World Population Data Sheet.

Chinese merchants, principally Fujienese and Hokkienese, laid the foundations for a contemporary regional diaspora now estimated to total between 16 and 20 million (Chang, 2008; Welch, 2012b). Although proportions of Chinese ethnicity vary considerably across ASEAN, from as little as 1.5% in Viet Nam, to 25% in Malaysia, and around 60% in Singapore (see Table 18.1), the ubiquitous presence of a Chinese community forms a bridge to China throughout the region. Research suggests that Chinese identity among its worldwide diaspora remains strong, at least among the current generation, and that increasingly highly educated overseas Chinese (*hua qiao*) communities retain a keen interest in China's development, and largely wish to contribute (Da & Welch, 2016; Wang, 2000, 2005; Wang & Wang, 1998; Welch & Cai, 2011; Welch & Hao, 2013; Welch & Zhang, 2008; Yang & Welch, 2010).

Now, China's growing importance—including in higher education (Yang & Welch, 2012)—and rising disposable incomes in Southeast Asia, as well as a growing sense of Chinese identity among its Southeast Asian diaspora, means that increasing numbers of such families choose to enroll their children in Chinese universities to deepen their knowledge of Chinese language and culture and to avail themselves of related job and business opportunities. Malaysia's decision to recognize 820 Chinese universities for degree purposes, in 2012, was also instrumental in giving greater legitimacy to study sojourns in China (Hu, Wotipka, & Wen, 2016; "Malaysia Recognises," 2012).

Despite differing proportions of Chinese within ASEAN member states, the disproportionate economic weight of ethnic Chinese minorities is relatively common (Chang, 2008; Yeung, 1999). Moreover, as the character of Chinese migration has changed, more and more overseas Chinese are highly educated, commonly with university degrees, and many with postgraduate qualifications ("China's Diaspora," 2015; Welch, 2017). This is the case in ASEAN, where it is no longer uncommon to see Chinese researchers and teachers working in universities in the more developed higher education systems, such as Singapore, Malaysia, and Thailand (Fredrickson, 2013; Krishna & Sha, 2015; Omi, 2017). As seen below, some will return to Chinese universities, lured by China's proliferating foreign talent schemes (Lu & Zhang, 2015; Welch, 2010a; Welch & Cai, 2011).

Across ASEAN, however, it is not merely proportions of ethnic Chinese that differ significantly within national populations. Of all ASEAN member states, Viet Nam, with perhaps the most minuscule proportion of ethnic Chinese, persistently reveals the most troubled relationship. This includes Vietnam's claims to islands claimed by China and some cultural resistance to learning Mandarin ("China Gets," 2012; Pak, 2012).

Epistemic Routes

But flows of ideas, also centuries old, are just as important as trade in goods. The unification of China (ca. 220 BCE), including territory now part of Viet Nam, entailed the spread of Chinese poetry, astronomy, medicine, and arithmetic through much of current northern Viet Nam. Confucian education became the major framework for higher learning in the region, (Gelber, 2007, p. 52; Welch, 2010a, pp. 197–199). The *Four Books* and *Five Classics* became the centerpiece of the temple curriculum in what is now Viet Nam as early as the Tang dynasty (618–907): "examinations based on Confucianism, as applied in China, were organized . . . for more than eight centuries (from 1075 to 1919) in Viet Nam" (Yang, 1993, p. 217).

Scholarly mobility was a notable feature, with scholars from current Viet Nam and elsewhere traveling particularly to centers of higher learning in China. Scholarly centers of Confucian higher learning were established in parts of what is now ASEAN. A prominent and venerable example was the *Quoc Tu Giam* (Imperial Academy) established in Hanoi by Emperor Ly in 1070: Open for the next 700 years, it educated not merely bureaucrats, but also scholars and nobles. A new academy was founded in Hue (seat of Nguyen Dynasty emperors and the national capital from 1802–1945) in 1802, after which the Hanoi academy fell into decline. Restored recently, it is still possible to see the names of major scholars inscribed in Chinese on the stelae within the temple garden. Confucius is traditionally revered in Viet Nam as the Teacher of Ten Thousand Sovereigns and some scholars in Viet Nam feel the Confucian doctrine of "managing state affairs and bringing peace to all under heaven" (As cited in Welch, 2011b, p. 133; see also Welch, 2010a; Welch & Cai, 2011, p. 11) is still very significant for Viet Nam:

> As for the recruitment of officials, examinations based on Confucianism as applied in China were organized . . . for more than eight centuries (from 1075 to 1919) in Viet Nam. . . . southern Viet Nam (before the reunification of the country) remained particularly faithful to his thought, especially in the domain of moral education. (Yang, 1993, p. 217)

Religion formed a further epistemic bridge. Buddhists in China and current Southeast Asia were in contact from the sixth century CE, while Islamic scholars in what is now China and ASEAN were also in contact (Chang, 1988; Gelber, 2007; Welch, 2012a, 2012b). The terrestrial and maritime Silk Roads (each of which, as seen in

Figure 18.1, has now been revived) both contributed to intellectual concourse between Muslims throughout the region (Somers Heidhues, 2001).

Longstanding Regional Relations

Clearly, both intellectual and commercial concourse sit on ancient and substantial foundations. Beginning in the third century BCE, regional trade grew during the Three Kingdoms period, and further during the Tang and later dynasties (Wang, 2000). Chinese voyages of exploration to Southeast Asia took place in the third century CE, significant contact between Buddhists in China and counterparts in Southeast Asia occurred during the fifth and sixth centuries, as also later concourse among Muslim communities in China and what is now Indonesia (Chang, 1988; Gelber, 2007; Welch, 2012a; 2012b). Growing regionalism included both Kublai Khan's expansionism during the Yuan dynasty (1297–1368), as well as increasing bilateral trade and exploration, although massacres of Chinese settlers also occurred, in Manila and Batavia (now Jakarta) during the Ming dynasty (Reid, 2008).

But Admiral Zheng He's[6] voyages to the region (1405–1433) remain the most potent example, with the clearest implications for higher education regionalism. (see Fig. 18.2; "China Beat," 2006; National Geographic, 2005). Himself a Muslim, Zheng He included a number of Muslim scholar-teachers on his ships, and although the region now comprising Indonesia and Malaysia had already been influenced by Islam the Indonesian scholar Hamka (1908–1981) has argued that "the development of Islam in Indonesia and Malaya is intimately related to a Chinese Muslim, Admiral Zheng He" ("Zheng He," n.d.a.; see also Wu, 2004). This has specific implications for higher education. It is not drawing too long a bow to argue a connection between Zheng He's deliberate inclusion of Muslim scholar-teachers among his crew, and his propagation of Islam within the region, and the existence of current higher educational institutions, such as the International Islamic University of Malaysia (IIUM), which over the years 2013–2015 increased its international enrollments from China from 136 students to 275. Although Malaysia currently attracts over 10,000 mainland Chinese students annually to its universities, IIUM's cohort of Chinese Muslims is both distinctive and an underanalyzed element of higher education regionalism.

[6]Originally known as Ma Sanbao (Ma Ho; 馬三保), Zheng He was given his more well-known name by the Yongle emperor, as a reward for his loyalty.

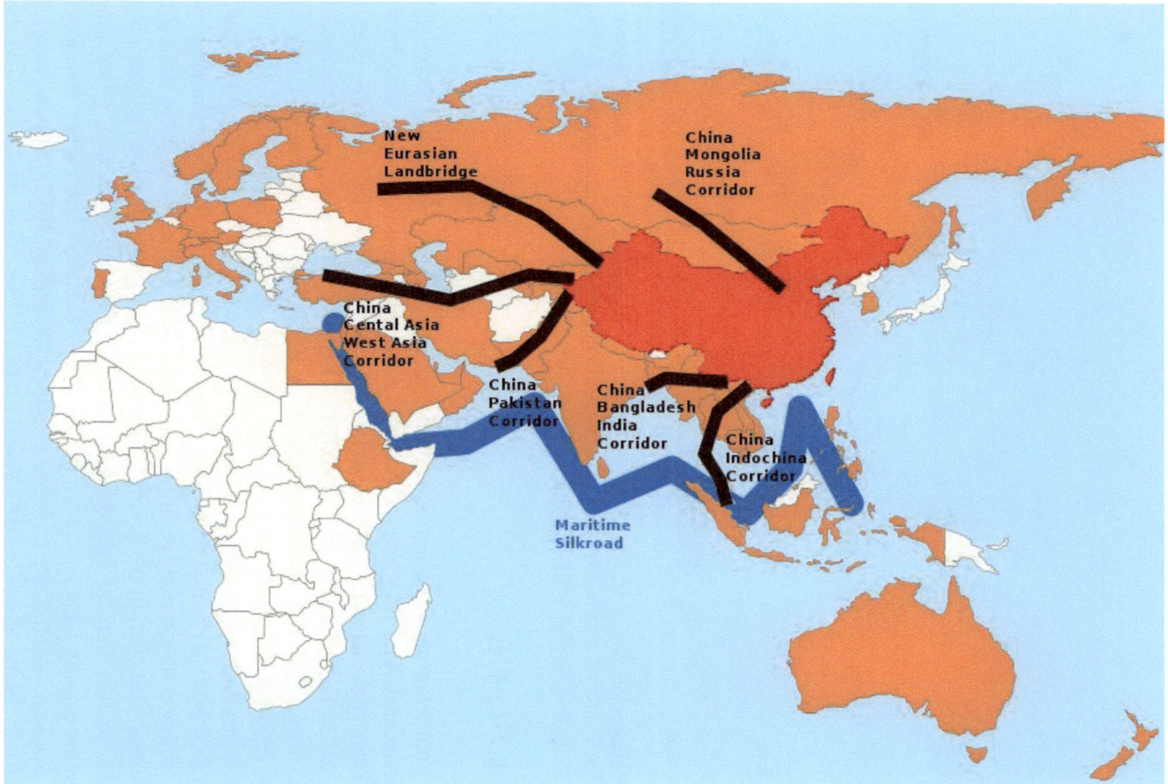

Fig. 18.1 One Belt, One Road Initiative. The Terrestrial Silk Road is marked in black.
Source: © Lommes (2017), via Wikimedia Commons. Used under Creative Commons Attribution-Share Alike 4.0 International (CC BY-SA 4.0). Retrieved from https://upload.wikimedia.org/wikipedia/commons/c/cb/One-belt-one-road.svg.

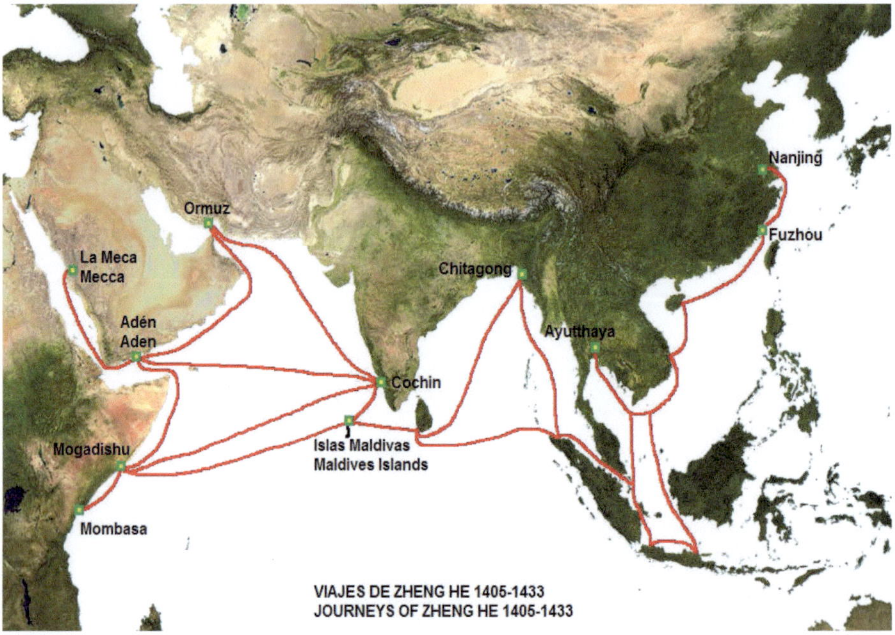

Fig. 18.2 Voyages of Zheng He, 1405–1433. Source: Continentalis (2013), via Wikimedia Commons. Used under Creative Commons Attribution-Share Alike 3.0 Unported (CC BY-SA 3.0). Retrieved from https://en.wikipedia.org/wiki/File:Zheng_He.png.

Territorial Disputes

While regional relations in higher education continue to grow, spreading territorial disputes between China and several ASEAN neighbors (maritime and terrestrial) may constrain regional relations (Bowring, 2012; Hayton, 2017; Prantl, 2012; Severino, 2012; Tiezzi, 2014, Welch, 2014). Ongoing differences over territorial ownership of minor shoals, reefs, and islands in the South China Sea have proven stubbornly resistant to resolution. As late as mid 2017, only an agreement to commence consultations on a code of conduct (begun in 2002) had been reached, a fact ascribed both to China's muscular diplomacy, and ASEAN's inability to achieve a common negotiating position (Buckley, 2015). Further disputes with the Philippines and Viet Nam in the South China Sea in 2011, that involved sovereignty over *Hoang Sa* (the Paracel Islands, known as in Chinese as *Xisha*), *Quần đảo Trường Sa* (the Spratly Islands, known in Chinese as *Nansha*), and Fiery Cross Reef (*Yongshu Jiao)* provoked sharp Chinese criticism (Bowring, 2012; "Hanoi Anger," 2016; Li, 2012; "Subs Keep," 2016; Torode, 2011). Viet Nam, Malaysia, and the Philippines all scrambled to boost their defense capabilities and ties, including with Japan, the United States, and India ("Asia on Alert," 2015; "Beijing Downs," 2016; Conelly et al., 2016"; "India Will," 2015; Nguyen, 2012; Nicholson, 2015; Parameswaran, 2015; "Philippines Security," 2016; Thayer, 2015).

Hence, despite enhanced regional economic integration, and a perception by several ASEAN states that China's rise may offset their dependence on the United States, China's assertiveness in the region has led to some hedging on the part of ASEAN neighbors (Kwok, 2015; Osborne, 2006; "Philippines Security," 2016; Taylor, 2012; Wang, 2005; Whitney & Shambaugh, 2008). Despite this, China announced in November 2014 that it "ranked relations with its neighbours higher in priority than relations with the United States and other great powers" (Beng, 2015). While still robustly asserting its territorial claims, China now emphasizes win-win diplomacy, further trade, and cultural cooperation, including in higher education (Sutter & Huang, 2012; Tiezzi, 2014).

Overall the effects are mixed: Hedging against China is balanced by concerns that the promised U.S. "pivot to Asia" is more rhetorical than real, which is leading several ASEAN states to explore closer links with the Middle Kingdom (Callick, 2016). Such strategies, together with supportive policies at national and regional levels in China, are only deepening higher education relations by universities on both sides of China's southern borders ("2016 Guizhou," 2016; "Vietnamese Students," 2016; Yi, 2015).

Anti-Chinese Sentiments

The disproportionate importance of Chinese minorities on national economies has erupted into violence against local Chinese groups at times, as seen most recently in Viet Nam ("Vietnam Stops," 2014). Earlier examples include the murder of many *Partai Komunis Indonesia* members in Indonesia in 1965, numbers of whom were killed simply because of their Chinese origin (Farram, 2010, p. 392; Pramudatama, 2012; Suryadinata, 2003; Wang, 2005). Diplomatic relations were only restored in 1990, while anti-Chinese riots broke out again in 1998, in Jakarta, Medan, and elsewhere (Conboy, 2002; Laksmana, 2011).

Such sentiments also resulted in incidents directly affecting higher education, including, for example, attacks on Res Publica University, in Jakarta, in October 1965. On the pretext that it was a headquarters for training communists, anti-communist youths invaded the campus. Res Publica students barricaded themselves inside the College of Technology, which was then burned to the ground by the mob. In response, the minister of education decreed the establishment of a new private university (Trisakti) on the site, while also banning one third of former students from reenrolling, allegedly due to their communist sympathies. While ethnic Chinese enrollments remained substantial, enrollments by ethnic Indonesians rose significantly in the years following (Welch, 2011b, pp. 33–34).

In Malaysia, too, bloody riots occurred between Chinese Malaysians and ethnic Malays (known as *Bumiputras*), in 1969. Preferential policies for Bumiputras, including university quotas that effectively forced many Chinese-Malaysians into private sector higher education, or overseas for study, were subsequently introduced to redress the comparatively weak position of Bumiputras in society, economy, and universities. The measures, which included limiting senior appointments for Chinese-Malaysian academic staff, continue to discriminate against ethnic Chinese, despite being formally abandoned in 2003.

Irregular Regionalism and the Southern Borderlands

The above sketch of the six pillars defies the conventional wisdom that views Sino-ASEAN relations simply through the prism of economics. History, culture, ethnicity, security allegiances, nationalism, and domestic political agendas, including those of the great powers, all help shape both the quantity and quality of regionalism, notably in higher education.

Regionalism is by no means uniform, however, and borderland universities represent a test case of the limits of Sino-ASEAN regionalism. The liminal quality of borderlands regions continues to resist and at times defy the regulatory architecture of both state and regional authority, including in higher education, with borderland universities, more remote from central authority, pursuing more specifically regional agendas (Chan, 2013; Eimer, 2014; Yang, 2012).

This is clearly the case in China's southern borderlands of Guangxi, Guizhou, and Yunnan, where much of cross-border flows is both irregular and illegal (Chan, 2013, pp. 89–105, 108–115; Chinese Jade Miners," 2015; Eimer, 2014, pp. 186–188, 200, 225–231; Evans, Hutton, & Eng, 2000; "Guangxi Ready," 2014;"The Plunder," 2015). In addition, for borderlands minorities such as the Dai, Wa, Kokang, and Chuang, ethnicity still transcends nationality: "Borderlanders cross state boundaries daily and conduct everyday cross-cultural interactions" (Chan, 2013, p. xi; see also Barrett, 2012, pp. 183–191; Eimer, 2014, pp. 163, 175, 188–189; Fuller, 2015; Moe, 2015; Sigley, 2014; Tiezzi, 2014).

Although locating precise data is difficult, common ethnicities influence cross-border student flows. Hence, in order to understand such flows, researchers, too, must enter that same liminal world, suspending conventional perceptions of borders and themselves becoming border-crossers engaged in transgressive, disruptive modes of thinking. Geopolitics—also complicit in blurring borders—matter, including in cross-border university links. For such universities, the emperor may still remain far away.

But Beijing's view that the local minorities on its southern borderlands are no threat also led to fewer resources for regulating cross-border flows. Accusations have been made of Han orientalism, with which locals may be at least superficially complicit—while lamenting in private the loss of their language and culture: "None of the schools teach Dai now, . . . You have to become a monk to learn how to read and write it" (Eimer, 2014, pp. 175, 181).

China-ASEAN Relations in Higher Education

How do these intertwined elements play out, in current China-ASEAN relations, in higher education? Clearly, while ASEAN can be seen as a region, including related higher education architecture such as the AUN,[7] it is internally highly diverse, rendering generalizations perilous. Equally, higher education relations embrace far more than trade in education services, however important. The following section charts both trade in higher education services, as well as wider higher education relations.

Although data is not precise, service sector trade between China and ASEAN has clearly been growing for some time; this certainly includes higher education, (United Nations Conference on Trade and Development, 2004; Welch, 2011a). Foreign direct investment (FDI) flows to service sector industries in ASEAN accounted for around half of total ASEAN FDI in 2008 (ASEAN, 2009, pp. 12−13). Global estimates of total worldwide trade in education services were around US$2.2 trillion. Growth has been faster in developing Asia, where the service sector accounts for a lower proportion of total GDP than in the OECD (Asian Development Bank, 2012; Ng & Tan, 2010). The move to list education as part of service sector trade under the Global Agreement on Trade in Services (GATS) is also enabling the more precise tracking of transnational delivery of higher education (Welch, 2011a).[8]

International branch campuses (IBCs) are one index of growth in transnational education. By June 2016, there were 232 IBCs worldwide, with a further 25 under development (Hu & Willis, 2016). A significant proportion are situated in Asia, and while transnational education has been traditionally dominated by Anglophone countries, there has been a pronounced shift to Asia in recent years. A recent Asian Development Bank study underlined that earnings from trade in higher education services still favored the wealthier and English language systems; but that much faster growth was evident in Asia. Although earnings from trade in higher education services still favored traditional exporters, such as the United States, the United Kingdom, and Australia, newer Asian competitors, such as Singapore, Malaysia, and China, were making substantial inroads (Asian Development Bank, 2012, pp. 37–38; Welch, 2011b). China alone has more than 400,000 international students enrolled in its higher education system, of whom an ever-increasing number stem from ASEAN (see Fig. 18.3).

Although China's priorities for transnational education, which is formally controlled by the ministry of education,[9] are more concerned with extending knowledge of Chinese culture and language, as compared with the more entrepreneurial

[7]As seen below, AUN includes a component that includes a number of Chinese universities, within an ASEAN+3 framework.

[8]Although this should not be taken as endorsement of education as a tradable commodity, or a denial of its importance as a public good.

[9]At least for degree level programs. Subdegree levels are monitored by provincial or municipal governments.

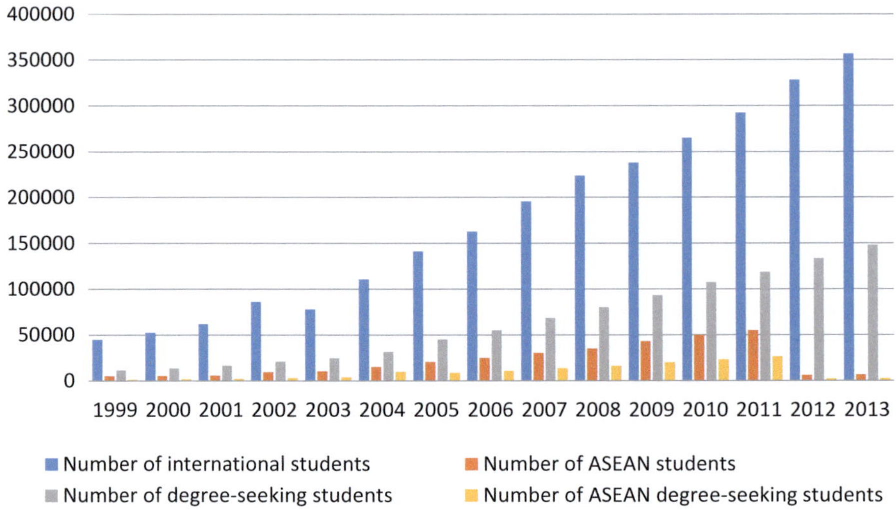

Fig. 18.3 International students, ASEAN students, and degree-seeking students in China (1999–2013). Source: Ministry of Education statistics (1999–2013), cited in Wen (2016, p. 177). Reprinted with permission.

approach of traditional Anglophone providers, such as the United Kingdom, Australia, and New Zealand, early estimates that China earned around US$2.3 billion from international students may well be conservative (Wen, 2016; Wen, Luo, & Hu, 2014). China emphasizes partnerships in domestic transnational education initiatives. Requirements include partnerships based on collaboration between foreign and Chinese higher education institutions (HEIs) or providers. This involves mutual investment from both Chinese and foreign HEIs in capital, land-use rights, intellectual property rights, institutional brand, curriculum, and administrative systems (Hu & Willis, 2016; Welch & Cai, 2011). Although China acceded to the GATS in 2001, its limited commitments to the agreement effectively prohibit foreign HEIs/providers from providing education services in China, unless partnering with a local Chinese HEI. The general principle is as follows:

> Foreign corporate, individuals, and related international organizations, in cooperation with educational institutions or other social organizations with corporate status in China, jointly establish education institutions in China, recruit Chinese citizens as major educational objectives, and undertake education and teaching activities. (McNamara, Knight, & Fernandez-Chung, 2013, p. 13)

Overall China data show that by the end of 2015, 590 Chinese HEIs were engaged in collaboration with over 400 foreign HEIs/providers, operating a total of 2,376 cooperative programs and institutions. Approximately 55,000 Chinese students were enrolled on campus and a further 1.5 million students were already graduates of such programs (Hu & Willis, 2016).

Most recently, Chinese HEIs have moved to establish both programs and IBCs, including a number within the ASEAN region. Overall, by 2015, 98 transnational education programs had been established, as well as 4 IBCs, of which 3 were based in ASEAN. Each formally approved by the Ministry of Education, the 3 ASEAN IBCs were, in chronological order of establishment:

- Suzhou University, Laos: founded in July 2011, offers undergraduate programs in international economy and trade, international finance, Chinese language, and computer science and technology
- Bangkok Business School, jointly established by Yunnan University of Finance and Economics and Thailand's Rangsit University in December 2013, offers six programs in economics and administration at both undergraduate and graduate levels.
- Xiamen University Malaysia, founded in July 2014, and the most substantial example, with a major new campus, offers 12 undergraduate programs including Chinese studies, traditional Chinese medicine, marine biotechnology, new energy science and engineering, and international business (Bevins & Phillips, 2017; Foon, 2016; Hu & Willis, 2016).

At least three points are worth making here. As Hu and Willis underline, China's transnational education initiatives tend to cluster within the ASEAN region. Secondly, at the program level, there is a concentration on traditionally prestigious disciplines, such as Chinese language, culture, traditional Chinese medicine and martial arts. Lastly, and unsurprisingly, of the three examples above, one involves a Chinese higher education initiative from one of the borderlands regions that are the particular focus of this analysis.

Frameworks for China-ASEAN Collaboration in Higher Education

Certainly, regional plans to extend China-ASEAN collaboration in higher education are ambitious. As part of the Guiyang Declaration signed in August 2010, a target was set of 100,000 Chinese and ASEAN students enrolled in each other's universities by 2020: the Double 100,000 Plan. As seen below, significant progress has been made towards attaining that goal. The imbalance indicated above, in China's favor, however, is likely to persist. China is also generous with scholarships to students from ASEAN: Its plans are to raise the number of scholarships to ASEAN students to 10,000 by 2020.

Of China's major source countries for international students, several are ASEAN member states, as the following table illustrates:

Table 18.2 shows that, in aggregate, ASEAN students form a substantial proportion of China's top 15 source countries: over 50,000 students, and almost 15% of the top 15 total enrollments. Importantly too, for this analysis, it is precisely the borderlands regions and universities where most ASEAN students are located—

Table 18.2 Top 15 source countries, China's international education market, 2013

	Country	Enrollment	Market Share (%)
1	South Korea	63,029	17.68
2	United States	25,312	7.10
3	**Thailand**	**20,106**	**5.64**
4	Japan	17,226	4.83
5	Russia	15,918	4.47
6	**Indonesia**	**13,492**	**3.78**
7	**Vietnam**	**12,799**	**3.59**
8	India	11,781	3.30
9	Kazakhstan	11,165	3.13
10	Pakistan	10,941	3.07
11	France	9,649	2.71
12	Mongolia	8,054	2.26
13	Germany	7,058	1.98
14	**Malaysia**	**6,126**	**1.72**
15	United Kingdom	5,465	1.53
	TOTAL ASEAN	52,543	14.73

Note: ASEAN countries in bold. From the Chinese Ministry of Education statistics (1999–2013), adapted from Wen (2016, p. 178). Reprinted with permission.

the southwest provinces of Yunnan, Giuzhou, and Guangxi neighboring ASEAN (Wen, 2016; Yang, 2012).

But student mobility fits within wider China-ASEAN frameworks that govern higher education relations. Broad cultural ties between China and ASEAN were inaugurated by an ASEAN + 3 Meeting of Ministers Responsible for Culture and Arts (AMCA+3) in Kuala Lumpur in October 2003 and separately endorsed by the ASEAN + China Summit that same year. The latter pledged to intensify cooperation in key areas, including education and human resource development, and the exchange of relevant personnel. In this sense, it broadly functioned within the expressed goals of the ASEAN Socio-Cultural Community: to "nurture talent and promote interaction among ASEAN scholars, writers, artists, and media practitioners to help preserve and promote ASEAN's diverse cultural heritage while fostering regional identity, as well as cultivating people's awareness of ASEAN" (ASEAN, 2003). While enthusiasm exists on both sides to strengthen relations in higher education, major accomplishments are not always simple.

Trade Agreements and Consortia

One regional trade agreement includes cross-border higher education, while three principal higher educational consortia exist that either target promoting ties with China or already include Chinese universities as members.

From its inception, the Asia Pacific Economic Cooperation (APEC) included an education component. Originally formed as the APEC Education Forum in 1988, it then became part of the APEC Human Resources Development Working Group, which includes China, Malaysia, Singapore, and Viet Nam (as well as several other countries). While APEC infrastructure to support regional initiatives in education remains modest, its ambitious aims included, notably, the attainment of free trade and investment within the Asia Pacific region for developed countries by 2010, and for developing countries by 2020. A relevant instrument consisted of an intergovernmental consultative group, the APEC Education Forum, while the Human Resource Development Working Group had wider aims (lifelong learning, capacity development, sustainable development, and labor and social protection). University Mobility in Asia and the Pacific (UMAP), which is seen by APEC as a vehicle to promote links in education, organizes one and two semester study abroad programs for undergraduates. Members include Singapore, Malaysia, and Viet Nam, while China's membership, discussed at a UMAP meeting in Japan in March 2003, was subsequently confirmed, notwithstanding difficulties arising from already having Taiwan as a member. Activities consisted of the further development of the existing UMAP Credit Transfer Scheme, while also leveraging the comparative advantage of member states, and regional language learning.

An early APEC publication focused specifically on identifying barriers to trade in cross-border educational services trade, as well as measures to promote it, included access and equity and the integrity of national systems. Several countries within the region voiced concern on this latter issue, noting the necessity to "retain the . . . sovereign right to determine . . . domestic funding and regulatory policies/measures" (OECD, 2003, p. 51), as well as the integrity of the public system and local standards.

The three regional consortia in higher education are:

1. The AUN, which in 2001 inaugurated the ASEAN-China Academic Cooperation and Exchange Programme. Members include University Sains Malaysia, University of Malaya, National University of Singapore, and Nanyang Technological University. Its activities include the ASEAN-China Rectors' Conference, the Round Table, and Joint Research and Training Grants, as well as the ASEAN-China Distinguished Professors and Lecturers Exchange Programme. One of the first actions taken was an ASEAN-China rectors' meeting in Bangkok, (2002), which was followed by a further round table in Beijing (2004) that took additional steps to advance ASEAN-China collaboration. Relevant strengths in both China and ASEAN meant marine science was selected as an initial vehicle for joint research and training grants. Chinese scientists from Qing Dao Maritime University were afforded opportunities to conduct research in tropical and equatorial water environments, while ASEAN scientists were able to take advantage of the temperate water environment offered by northeastern China. Researchers from each side received three-month grants to conduct maritime research. This framework agreement on research cooperation was confirmed at the Hanoi meeting in 2007.

   A further element of the agreement included a joint training activity, whereby some forty academics from ASEAN and China were selected for two training courses, one to be held in ASEAN and another in China. A final component was the ASEAN-China Distinguished Professors and Lecturers Exchange Programme, which aimed to strengthen relations between ASEAN and Chinese scholars via short-term academic exchanges (up to one month). Exchange activities included lectures, laboratory exercises, and demonstrations; advising research students; and collaborative development of curricular and teaching-learning materials.[10]

2. A second regional higher education consortium involving member universities in China and ASEAN is the Association of Pacific Rim Universities comprising 36 leading research universities from Singapore, China, Malaysia, Philippines, Thailand, and elsewhere. Member universities from China are among its leading research institutions: Peking University, Nanjing University, Hong Kong University of Science and Technology, Hong Kong University, Tsinghua University, and Zhejiang University.
3. The third consortium is Universitas 21, which includes, inter alia, three major Chinese universities (Fudan, Shanghai Jiaotong, and the University of Hong Kong), as well as a major research university from Singapore. Although Universitas 21 began with great fanfare—and ambitions—such aspirations soon needed to be tempered. Current initiatives include summer schools for member institutions, and student exchange programs, comprising both short-term programs and longer sojourns of one to two semesters, at partner institutions.

A further element supporting the architecture of Sino-ASEAN relations in higher education is the availability of an array of what are known in China as foreign talent schemes, which operate at both national and provincial levels. Designed to deploy overseas experts to work in China, either on a part-time or full-time basis, a wide range of such schemes exist, including the Hundred Talents, Thousand Talents, Yangtze River Scholars schemes, the Spring Light and Distinguished Overseas Scholar Programs, and the 111 Project. In practice, although several schemes are open to all overseas talent, the large majority of candidates selected are overseas Chinese. Such schemes provide a further plank in the intellectual bridge between China and ASEAN, and mean that Chinese academics working in Singapore, for example, may elect to return by taking advantage of such schemes, or may remain employed in Singapore, but travel back and forth to China, sustained by one or other of such programs. A major change in China's diaspora policy some years ago, from "returning to serve the country" (huiguo fuwu) to the more flexible "serving the country" (*weiguo fuwu*), meant that Chinese specialists based overseas could continue to serve China while remaining abroad (and visiting China often, for example, to teach, recruit Ph.D. students, or engage in collaborative research with Chinese peers). Examples exist of mainland Chinese researchers heading laboratories and departments

[10]For a list of AUN members, including the three Chinese borderlands universities of Guanxi, Guizhou, and Yunnan, see http://www.aunsec.org/membership.php.

in Singapore, while continuing to build bridges to China (Krishna & Sha, 2015, p. 407).

China Viet Nam Relations in Higher Education

While the Six Pillars sketched above represent a common framework for understanding Sino-ASEAN relations in higher education, in order to narrow the focus some selection is necessary. Given its proximity, and borderlands focus, Viet Nam was selected for further analysis. A sketch of similarities and differences in the two systems is followed by a case study of China's borderland regions' higher education relations with Viet Nam.

With a population of around 90 million, Viet Nam is still classed as a developing country, with significant economic growth rates, albeit substantially dependent on China. Almost 30% of Viet Nam's imports stem from China, while China is Viet Nam's third largest export destination (Boudreau, 2015; Tran, 2015). Its higher education system, while substantial, is much less well-developed than either China, or several other ASEAN member states. As sketched above, it has a particularly complex history of China relations: having fought a war as late as 1979, while also often following China's policies and programs, including in higher education. This includes a planned major increase in private higher education to the year 2020, as part of the Higher Education Reform Agenda (Welch, 2010a, 2010b). Although ethnic Chinese make up no more than 1.5% of the overall population, their influence on the economy significantly outweighs this modest proportion. Perhaps in part because of this, and despite the fact that Viet Nam is both a sister in socialism to China and of much smaller size and economic weight, deadly anti-Chinese riots broke out as recently as 2014. Viet Nam has long insisted on its independence (as French, Chinese, and U.S. and Allied military forces can attest). *Doi Moi*, the process of "reconstruction" instituted in 1986, is the most obvious example of Viet Nam implementing reforms broadly paralleling those of China, if somewhat later. Viet Nam's accession to World Trade Organization membership, six years after China, was another. Bilateral trade and investment were given further impetus by the China-ASEAN Free Trade Agreement of 2010. Some key major projects saw China's FDI to Viet Nam leap from 371 million U.S. dollars in 2012 to 2.3 billion U.S. dollars in 2013, according to Viet Nam's Ministry of Planning and Investment figures ("More Chinese," 2014). While bilateral trade and investment still strongly favor China, relations have been affected by the recent territorial disputes, while the anti-Chinese riots in Hanoi led to a fall of some 30% in China outbound tourism to Viet Nam in 2014, and perhaps, too, a plateauing in Vietnamese students at Chinese universities (Vu & Nguyen, 2014). Nonetheless, student flows strongly favor China: In 2011 13,549 international enrollments in Chinese universities stemmed from Viet Nam (a rise from less than 650 in 2000, and 7310 in 2006) (Clark, 2014; Welch, 2011b; Welch & Cai, 2011, p. 18; see also Fig. 18.3). By contrast, Viet Nam, with few international students enrolled at its universities, but at least 50,000 of its own enrolled overseas, remains a net importer of higher education services, including from China ("Vietnam to Send," 2013; Welch, 2010a, 2011a, 2011b).

Crossing Borders and Sino-Vietnamese Higher Education

In Asian higher education, hierarchy and stratification can never be ignored. Nonetheless, the above analysis modifies this key point significantly. While no Chinese university from the three borderlands areas of Yunnan, Guizhou, and Guangxi are listed among the ranks of the Project 985 universities (China's top tier), or in the authoritative Shanghai Jiaotong Academic Ranking of World Universities (ARWU),[11] they are critical in sustaining Sino-Vietnamese relations in higher education, as is true in some of China's other border regions (Welch & Yang, 2011). Some 11 universities in the North of Viet Nam offer Chinese language programs and take part in annual Chinese Bridge language competitions ("Vietnamese Students," 2016). Annual Chinese university expos held in Viet Nam to encourage Vietnamese students to enroll in Chinese universities often feature borderland regions and their universities. Examples included the Guizhou Education Fair in Hanoi in 2016, which featured 22 universities from Guizhou, and the Guangxi Education Fair of the same year, which traded on the extent of existing cross-border relations. Forty universities and colleges in Guangxi had already established ties with some 60 universities and colleges in Vietnam. This meant that over 3,000 Vietnamese students studied in Guangxi in 2015, while in the other direction nearly 1,000 students from Guangxi were studying in Vietnam ("China's Guangxi," 2016).

Nonetheless, hierarchy and status still differentiate the two systems. The ARWU ranking lists 35 Chinese universities among the top 500 worldwide. Viet Nam, like much of Southeast Asia, still has no HEIs listed. Its plans are to create "model universities," using loans from the World Bank and the Asian Development Bank and the expertise and resources of selected foreign partners to lift quality and act as a benchmark for domestic HEIs (Welch, 2010a). No Vietnamese university is yet a member of the Asia Pacific Research Universities (APRU) network (which includes five leading Chinese universities), or of Universitas 21, which includes two from China.[12] Viet Nam National University (VNU), Hanoi, and VNU Ho Chi Minh city, are, however, both members of the ASEAN + 3 network, which includes, of five Chinese universities, three key HEIs from the three abovementioned border provinces: Guangxi University, Guizhou University, and Yunnan University (ASEAN Universities Network, n.d).

Viet Nam's relatively less-developed higher education system and levels of infrastructure position it less well to leverage cross-border collaboration. China's

[11] See http://www.shanghairanking.com/.

[12] For APRU members, see https://apru.org/members/member-universities and for Universitas 21, see http://www.universitas21.com/member.

Table 18.3 Key forms of Sino-Vietnamese cross-border higher education services

	Viet Nam	China
Mode I	--	--
Mode II	VNU and other HEIs' language courses for Chinese students Chinese students at Vietnamese universities	*Vietnamese students at Chinese universities`* *Training of Vietnamese civil servants and teachers*
Mode III	--	--
Mode IV	*Chinese consultants training Vietnamese*	

Note. Italics indicate Chinese exports; non-italics indicate Chinese imports. VNU = Viet Nam National University. Modified from Welch (2011a, p. 105). Reprinted with permission.

greater size and weight, and much more developed higher education system, leaves it in the driver's seat, to some extent, although as indicated above, Viet Nam's history is one of stubborn independence, while, as seen above, China also needs Viet Nam. Even relative to its more well-developed fellow ASEAN member states, Viet Nam's developing country status, including a less developed higher education system, confer fewer advantages. A further inhibitor consists of intermittently difficult relations with its large and troublesome northern neighbor (although recent steps have been taken by both sides to reduce bilateral tensions).

Viet Nam's long and complex history of China relations includes key higher education elements, notably an enduring Confucianism (Welch, 2010a, 2010b; Welch & Cai, 2011). Viet Nam's peripheral, if rising, status within the global knowledge network also translates into generally dependent relations with Chinese higher education (Welch, 2011a, 2011b). Many more Vietnamese students study at Chinese universities than the reverse, a situation broadly paralleled by scholarships. This however, does not do justice to the much closer relations between borderlands institutions, the "quiet achievers" in China's GUAR and Yunnan (Yang, 2012). Training of Vietnamese civil servants and teachers occurs in both Yunnan and Guangxi, which each enrolled several thousand students by around 2008, while several thousand students from Guangxi studied in ASEAN countries, notably Viet Nam. Yunnan links include 3+1 programs with ASEAN countries and joint degree programs with Viet Nam. Hundreds of Chinese language teachers, trained at Yunnan University and Yunnan Normal University, now work in ASEAN (Yang, 2012). Using the four GATS modes, Table 18.3 summarizes Sino-Vietnamese bilateral relations in higher education, albeit not indicative of scale (World Trade Organization, n.d).

Research conducted at Borderlands University in China's South, in 2014 and again in 2015, highlighted mutual, if unequal, patterns of cross-border cooperation. Recruitment of ASEAN scholars was the major feature, but some forms of research cooperation were also evident: "we would like to accommodate researchers from ASEAN countries to do their research here, in our university. Meanwhile, we plan to send out our researchers to the ASEAN countries." (I1, 2014).

Almost 90% of the 300 or so international students stemmed from ASEAN, notably from Viet Nam: "In recent years, a large number of Vietnamese students come to study at our Business school, majoring in international trade, business administration and accounting." (I1, 2014, see also I4). Table 18.4, following, shows that while international enrolments have declined in recent years within Borderlands University's business school, (which enrolls the largest number of international students), Vietnamese enrollments still constitute the largest country of origin).

Flows to Viet Nam were smaller and only in certain areas: "in terms of student exchange, our students go to Vietnam . . . primarily for language study" (I1, 2014), although it was pointed out that this was to a major Vietnamese university: "they also went to very famous universities in Vietnam. And they sent 20 students from (Borderland) to Vietnam, for one semester" (I4, 2015). Another interviewee pointed to cooperation with Viet Nam in environmental studies. China's more developed higher education system meant that cooperation, while mutually valued, was unequal: "this is one-way cooperation, which means we help those ASEAN countries to train their talents in the field" (I1, 2014). "In terms of theory, we're relatively more advanced than Laos and Vietnam" (I5, 2014). Unsurprisingly, hierarchy was significant in accounting for student flows: "Few students from more developed members like Singapore, Malaysia and Philippine [*sic*] come to our institution for study. We don't have even a single student from Philippine, Malaysia or Singapore" (I5, 2014).[13]

Moreover, for ASEAN students in China, borderlands universities were not necessarily their first choice: "Even if they come to China for study; most of them will go to Beijing rather than (Borderland)" (I1, 2014). The highly stratified nature of Chinese higher education was cited as an obstacle for Borderlands University by I4: "What they know are only Peking University or Tsinghua," both of which, it was acknowledged, had academic strengths that Borderlands lacked (I1, 2015). An interviewee from the business school argued that Borderlands University held an interstitial position in its regional relations: "for our university, if we go to Singapore or Malaysia, we learn from them, and it we go to Laos and Vietnam, they learn from us" (I5, 2014). For another interviewee from the same school, however, the problem of hierarchy was a larger one: "We don't really know ASEAN countries, and we understand them in our respective way. But we treat them as inferior to us" (I6, 2014).

The less developed status of borderlands regions and ongoing border disputes were raised by an interviewee from an ASEAN research institute, as two barriers to further development:

> The first may be the economic and the economic development of (Borderland) is not very good, I mean our development level is lower. Second, there are many disputes between

[13]A total of 14 interviews were conducted at the university in 2014 and 2015, with both academic staff and administrative leaders.

Table 18.4 Borderlands University Business School: International student enrollment, by country.

	Year and Level									
	2010		2011		2012		2013		2014	
	Bachelor	Master	Bachelor	Master	Bachelor	Master	Bachelor	Master	Bachelor	Master
Vietnam	76	16	22	6	39	13	19		23	
Thailand	2	1		2	3	9	5		2	
Laos	2	1	1		1	2	2		1	
Cambodia	2		4		3		1		2	
Indonesia		2		6	1	10	2		1	
Myanmar		1		4	1	4				
Burundi	1									
Malawi			1							
Korea			1		1		1			
Taiwan					1					
Russia									1	
Congo-Brazzaville									1	
Central African Republic				1						
Total	104		48		88		30		31	

Source: Borderlands University data (compiled from personal communications).

> China and Vietnam, Indonesia, Philippines [*sic*], because our country wants island [*sic*], there are a lot of troubles. (I1, 2015)

Other interviewees saw it differently, arguing that China's borderlands HEIs had advantages of regional proximity that enabled them to function as a *qiaotoubao*:

> First, the geographical advantage: (Borderland University) is proximate to ASEAN countries. Second, the geographical advantage will bring policy learning. Since we are close to ASEAN countries in terms of geography, the university may pay more attention to cooperation with ASEAN countries and develop particular policies to boost the cooperation. Second, . . . we could operate employment-oriented continuing education programs. . . . programs with internship opportunities will be more attractive to students. (I2, 2014, see also I5)

Other interviewees viewed transport linkages as a regional advantage for borderland institutions, pointing to the success of colleagues in securing a substantial research grant, to work on aspects of regional relations. Both central and regional Chinese governments provide substantial support for such projects. Ethnicity was seen as a further advantage, with a specialist in minority research pointing to cross-border links: "Among the minorities in the Northern part of Viet Nam, there are many descendants of Chinese Chuang. I have been to the villages in Northern Vietnam [*sic*]; people there are very hospitable to the Chinese Chuang". (I4, 2014). A number of 2015 interviewees, both staff and student, echoed these comments, highlighting issues of geography and culture, including similarities between the Zhuang minority language, Thai, and Vietnamese, as well as a history since ancient times of "people-to-people exchanges" (I1, 2015; I2, 2015; I5, 2015). China's *Yi Tai, Yi Lu* (One Belt One Road initiative) was seen as a further opportunity for Borderlands University:

> we have access to ASEAN countries on land and also on sea, whether in terms of the strategic layout on economics, this is the great opportunity for us. (Borderlands) should take advantage. (Borderlands) is the only way to the Philippines, to Singapore, to Indonesia and Malaysia. (I1, 2015)

Territorial tensions in the relationship between China and its neighbors were acknowledged, but "economic cooperation is used as a means to conciliate those conflicts" (I1, 2014). Higher education can play its role in such conciliation: "China has to do more work in promoting Chinese culture in our ASEAN members" (I1, 2014). Possible resistance on the part of some Vietnamese students was acknowledged: Unwillingness to integrate with local people was common and some students "will disseminate their negative attitudes towards China" when they return home (I4, 2014). Better communication was seen as a solution to differences between systems: "I think we don't know each other adequately. A regular communication mechanism has not yet been established" (I6). Overall there was still considerable optimism that disputes in the South China Sea could be insulated from student exchanges:

> The South China sea (disputes) does [*sic*] not influence our overseas communication because before we have very good relationship, but only the outside, America, Japan, they have something to provoke Vietnam to do something against China. Even (if) there is tension between China and ASEAN, it didn't influence our overseas students. During those days, they still played basketball with the Vietnamese. There is very good relationship between people and people. (I2, 2015)

Ideological affinities provided another axis of cross-border cooperation, as one interviewee highlighted: "the majority of international students come from Socialist countries such as Vietnam and Laos" (I6, 2014). Research collaboration, too, sometimes hinged around socialist ideology, with another interviewee pointing to a bilateral Borderlands program on Marxism in Viet Nam and China that was being conducted with selected Vietnamese universities.

The greater status now accorded current Vietnamese students was linked to Chinese efforts to enhance its regional influence:

> Before, in the 1990s, universities in China didn't want Vietnamese students since usually they were poor and couldn't bring financial benefits to our universities. But now things changed. They are welcomed now because we want to boost China's influence in those countries. (I4, 2014)

Higher Education, Irregular and Regulatory Regionalism

The above sketch of borderlands relations in higher education problematizes conventional conceptions of both region and regionalism. While ASEAN may generally be accepted as a region, the spread and increasing depth of cross-border flows across China's southern border, notably in higher education, raises the prospect of China-ASEAN as a region. At the same time, as illustrated above, the fact that so much of cross-border flows are both irregular and illegal problematizes conventional views of regulatory regionalism.

The changing, if still somewhat peripheral, status of both China's southern borderlands, and of Viet Nam, in the global knowledge network highlights the impressive range and depth of regional relations in higher education (Graham, Hale, Stephens, & Mayer-Schönberger, 2011; Yang, 2012). While more research is needed into borderlands regional relations in general, there is clearly considerable growth potential. Notwithstanding the substantial obstacles articulated above, common interests and joint practical priorities are able to proceed beneath great power politics and regional territorial disputes. This is particularly the case in higher education, where the diligence and determination of lesser-known borderlands universities, Yang's quiet achievers are continuing to build cross-border relations, in the face of the complex geopolitics of the region sketched above. A cross-border cartography of intellectual connectivity continues.

But the analysis above also problematizes the character and limits of regionalism. As indicated above, the more embryonic, emergent quality of ASEAN regionalism has been contrasted with the more established regional architecture of the European Union, including in higher education (Robertson, 2008). More than one author has pointed to the substantial gap between lofty ASEAN declarations and actual accomplishments (Welch, 2012b). This, it has been argued, limits the actuality of regulatory regionalism, as does persistent nationalist resistance to the actual extension of ASEAN regionalism, including ceding domestic control over university governance.

If this is the case for ASEAN, what of the case for China-ASEAN regionalism, which despite growing and thickening links, including in higher education, would not be commonly acknowledged to be a region? Even more so in the peripheries and borderlands, characterized by floating populations, mobility, and connectivity (Carney, 2009): a "cartography of connections" (Larsen & Beech, 2014, p. 207) that also embraces, in this instance, historical and contemporary, legal and illegal, fluxes and flows of ideas, people, timber, drugs and gems? In a context where significant progress has been made on the ground by borderlands quiet achievers in higher education (Yang, 2012), against a wider background of largely irregular flows of people, goods, and services (Evans et al., 2000), might this not be better characterized as irregular regionalism (Scott, 2009)?

Notwithstanding the complexities and even tensions of regional relations, the quiet determination of universities on both sides of China's southern borders, strengthened by supportive government policies on both sides, and the broader influence of regional strategies, such as One Belt, One Road, point to closer cross-border ties between Chinese borderland universities and their ASEAN peers into the future.

References

Academic Ranking of World Universities (ARWU). Retrieved from http://www.shanghairanking.com

ASEAN. (2003). *Declaration of ASEAN Concord II*. Retrieved from http://asean.org/?static_post=declaration-of-asean-concord-ii-bali-concord-ii-2

ASEAN. (2009). *ASEAN investment report 2009*. Retrieved from http://investasean.asean.org/files/upload/ASEAN%20Investment%20Report%202009.pdf

ASEAN Universities Network. (n.d.). *ASEAN+3 UNet member universities*. Retrieved from http://www.aunsec.org/membership.php

Asian Development Bank. (2012). Counting the cost: Financing Asian higher education for inclusive growth. Retrieved from https://www.adb.org/sites/default/files/publication/29679/counting-cost.pdf

Asian Infrastructure Investment Bank. (n.d.). Introduction. Retrieved from https://www.aiib.org/en/about-aiib/index.html

Asia on alert. (2015, January 17). *South China Morning Post.*

Asia Pacific Research Universities. (2013). Member universities. In *Members* (section 2). Retrieved from http://apru.org/members/member-universities

Atwill, D. G. (2005). *The Chinese sultanate: Islam, ethnicity, and the Panthay Rebellion in southwest China, 1856–1873.* Stanford: Stanford University Press.

Barrett, T. C. (2012). *The Chinese diaspora in South-East Asia: The overseas Chinese in Indochina*. Library of Modern China Studies. London: I.B. Tauris.

Beijing downs Vietnam protest. (2016, January 13). *The Australian.*

Beng, K-K. (2015, January 8). Repairing China's image of 'regional bully' a priority for Xi. *The Straits Times*. Retrieved from http://www.asiaone.com/asia/repairing-chinas-imageregional-bully-priority-xi

Bevins, V., & Phillips, T. (2017, July 7). 'Going global': China exports soft power with first large-scale university in Malaysia. *The Guardian.* Retrieved from https://www.theguardian.com/

world/2017/jul/07/going-global-china-exports-soft-power-with-first-large-scale-university-in-malaysia

Boudreau, J. (2015, October 9). The biggest winner from TPP trade deal may be Vietnam. *Bloomberg L.P.* Retrieved from http://www.bloomberg.com/news/articles/2015-10-08/more-shoes-and-shrimp-less-china-reliance-for-vietnam-in-tpp

Bowring, P. (2012, April 22). Island mentality ignores history. *South China Morning Post*. Retrieved from http://www.scmp.com/article/998912/island-mentality-ignores-history

Buckley, C. (2015, December 31). China says it is building its second aircraft carrier. *The New York Times*. Retrieved from https://mobile.nytimes.com/2016/01/01/world/asia/china-says-it-is-building-its-second-aircraft-carrier.html

Callick, R. (2016, May 4). Meet China's middle class. *The Australian*. Retrieved from http://www.theaustralian.com.au/business/opinion/rowan-callick/meet-chinas-middle-class/news-story/81454cc5b2b763f200ef4fe2aea2253b

Carney, S. (2009). Negotiating policy in an age of globalization: Exploring educational "policyscapes" in Denmark, Nepal, and China. *Comparative Education Review, 53,* 63–88. Retrieved from http://www.journals.uchicago.edu/doi/abs/10.1086/593152?journalCode=cer

Chang, B. (2012, April 20). ASEAN, China to become top trading partners. *China Daily*. Retrieved from http://www.chinadaily.com.cn/cndy/2012-04/20/content_15094898.htm

Chang, H. Y. (1988). The Ming empire: Patron of Islam in China and Southeast-West Asia. *Journal of the Malaysian Branch of the Royal Asiatic Society, 61*(2), 1–44. Retrieved from https://www.jstor.org/stable/41493101?seq=1#page_scan_tab_contents

Chang, S.-D. (2008). The distribution and occupations of overseas Chinese. In A. Reid (Ed.), *The Chinese diaspora in the Pacific* (pp. 33–51). The Pacific World: Lands, Peoples and History of the Pacific, 1500–1900: Vol. 16. Aldershot: Ashgate.

Chan, Y. W. (2013). *Vietnamese-Chinese relations at the borderlands: Trade, tourism and cultural politics*. Routledge Contemporary Asia Series: Vol. 42. Abingdon: Routledge.

Cheow, E. T. C. (2004, May 3). China's rising soft power in Southeast Asia. *PacNet, 19a*. Retrieved from https://csis-prod.s3.amazonaws.com/s3fs-public/legacy_files/files/media/csis/pubs/pac0419a.pdf

China-ASEAN trade to bloom on new target (2013, October 10). *China Daily*.

China beat Columbus to it, perhaps. (2006, January 12). *The Economist*. Retrieved from http://www.economist.com/node/5381851

China faces resistance to a cherished theme of its foreign policy. (2017, May 4). *The Economist*. Retrieved from https://www.economist.com/news/china/21721678-silk-routes-are-not-always-appealing-they-sound-china-faces-resistance-cherished-theme

China gets tough as Viet Nam claims disputed islands. (2012). *CambodiaWatch*. Retrieved from http://camwatchblogs.blogspot.com.au/2012/06/china-gets-tough-as-vietnam-claims.html

China's diaspora brings it home. (2015, November 12). *Nature*, 527, 68–71. https://doi.org/10.1038/527S68a

China's Guangxi educational offerings a huge lure for Vietnamese students. (2016, July 22). *Xinhua*. Retrieved from http://news.xinhuanet.com/english/2016-07/22/c_135532703.htm

Chinese jade miners in overdrive in Myanmar. (2015, December 19). *Sydney Morning Herald*. Retrieved from http://www.smh.com.au/world/chinese-jade-miners-in-overdrive-before-institution-of-myanmars-new-government-20151217-glqk9d.html

Chinese outbound tourism in 2014. (n.d.). Retrieved from https://www.travelchinaguide.com/tourism/2014statistics/outbound.htm

Chinese Premier Li Keqiang attends the signing ceremony of the Confucius Institute at Hanoi University, Vietnam. (2013, October 18). *Hanban News*. Retrieved from http://english.hanban.org/article/2013-10/18/content_512308.htmCIA

The CIA (n.d.). *The world factbook: Country comparison*. Retrieved from https://www.cia.gov/library/publications/the-world-factbook/rankorder/2004rank.html

Clark, N. (2014, May 5). Higher education in Vietnam. *World Education News and Reviews*. Retrieved from https://wenr.wes.org/2014/05/higher-education-in-vietnam

Conboy, K. (2002). Troubled courtship: Military ties between China and Indonesia. *China Brief, 2* (20). Retrieved from https://jamestown.org/program/troubled-courtship-militaryties-between-china-and-indonesia/

Conelly, A. L., Liow. J. C., Laksmana, E. A., Heydarian, R. J., Thu, H. L., Win, C., & Noor, E. (2016). Southeast Asian perspectives on US-China competition. *Lowy Institute.* Retrieved from https://www.lowyinstitute.org/publications/southeast-asian-perspectives-us-china-competition

Crossley, P. K., Siu, H. F., & Sutton, D. S. (Eds.). (2006). *Empire at the margins: Culture, ethnicity, and frontier in early modern China.* Berkeley: University of California Press.

Da, W.-W., & Welch, A. (2016). Educative and child-rearing practices among recent Chinese migrants in Australia: Continuity, change, hybridtiy. In C. P. Chou & J. Spangler (Eds.), *Chinese education models in a global age* (pp. 231–245). Education in the Asia-Pacific Region: Issues, Concerns and Prospects: Vol. 31. Singapore: Springer.

Department of Foreign Affairs and Trade. (2015). Trans-Pacific Partnership Agreement. Retrieved from http://dfat.gov.au/trade/agreements/tpp/pages/trans-pacific-partnership-agreement-tpp.aspx

Eimer, D. (2014). *The emperor far away: Travels at the edge of China.* London: Bloomsbury.

Evans, G., Hutton, C., & Eng, K. K. (Eds.). (2000). *Where China meets Southeast Asia: Social and cultural change in the border regions.* Singapore: Palgrave Macmillan.

Farram, S. (2010). The PKI in West Timor and Nusa Tenggara Timur: 1965 and beyond. *Bijdragen tot de Taal-, Land- en Volkenkunde, 166,* 381–403. Retrieved from https://www.jstor.org/stable/41000134

Foon, H. W. (2016, January 17). Latecomer Xiamen U a big draw. *The Star.* Retrieved from http://www.thestar.com.my/news/nation/2016/01/17/latecomer-xiamen-u-a-big-draw-with-support-from-public-and-private-sectors-as-well-as-the-malaysian

Foucault, M. (1980). Questions on geography. In C. Gordon (Ed.), *Power/knowledge: Selected interviews and other writings, 1972–1977* (pp. 63–77). New York: Pantheon.

Fredrickson, T. (2013, October 8). Teaching Chinese in Thailand. *Bangkok Post.* Retrieved from http://www.bangkokpost.com/learning/learning-news/373584/teaching-chinese-in-thailand

Fuller, T. (2015, February 13). Rebels kill dozens of soldiers in Myanmar in fighting near Chinese border. *The New York Times.* Retrieved from https://www.nytimes.com/2015/02/14/world/asia/kokang-rebels-kill-dozens-of-soldiers-in-myanmar-near-border-with-china.html

Gelber, H. G. (2007). *The dragon and the foreign devils: China and the world, 1100 BC to the present.* London: Bloomsbury Books.

Goodman, D. S. G. (Ed.). (2008). *The new rich in China: Future rulers, present lives.* London: Routledge.

Gough, N. (2015, December 22). Higher interest rates lower expectations in emerging economies. *The New York Times.* Retrieved from https://www.nytimes.com/2015/12/22/business/international/indonesia-economy-interest-rates.html

Graham, M., Hale, S. A., Stephens, M., & Mayer-Schönberger, V. (2011). *Geographies of the world's knowledge.* Retrieved from https://www.oii.ox.ac.uk/archive/downloads/publications/convoco_geographies_en.pdf

Gregory, D., Meusburger, P., & Suarsana, L. (2015). Power, knowledge, and space: A geographical introduction. In P. Meusburger, D. Gregory, & L. Suarsana (Eds.), *Geographies of knowledge and power.* Knowledge and Space: Vol. 7 (pp. 1–18). Dordrecht: Springer.

Guangxi ready to crackdown on drugs. (2014, November 3). *China Daily.* Retrieved from http://english.gov.cn/news/top_news/2014/11/03/content_281475004895412.htm

Hanoi anger at 'sovereignty breach.' (2016, January 4). *Sydney Morning Herald.*

Hayton, B. (2017, July 24). South China Sea: Vietnam halts drilling after 'China threats.' *BBC News.* Retrieved from http://www.bbc.com/news/world-asia-40701121

Hirono, M. (2010, June 26). How to play a "responsible great power" role: China's post-tsunami assistance to Aceh. *East Asia Forum.* Retrieved from http://www.eastasiaforum.org/2010/06/26/how-to-play-a-responsible-great-power-role-chinas-post-tsunami-assistance-to-aceh

Hodge, A. (2016, November 12). Malaysian Philippines leaders belt out warmer relations. *The Australian*. Retrieved from http://www.theaustralian.com.au/news/world/malaysianphilippine-leaders-belt-out-warmer-relations/news-story/21b768f0ffb5bc5487887fa5fdb40f3c

Hofman, B. (2015, April 12). China's One Belt One Road initiative: What we know thus far. Retrieved from http://blogs.worldbank.org/eastasiapacific/china-one-belt-one-road-initiative-what-we-know-thus-far

Hu, C., Wotipka, C. M., & Wen, W. (2016). International students in Chinese higher education: Choices, expectations, and experiences by region of origin. In K. Bista & C. Foster (Eds.), *Global perspectives and local challenges surrounding international student mobility* (pp. 153–178). Advances in Higher Education and Professional Development. Hershey: Information Science Reference.

Hu, M., & Willis, L.-D. (2016). Towards a common transnational education framework: Peculiarities in China matter. *Higher Education Policy, 30,* 245–261. doi:https://doi.org/10.1057/s41307-016-0021-9

Illeva, J., Killingley, P., Tsiligiris, V., & Peak, M. (2017). The shape of global higher education: International mobility of students, research and education provision: Vol. 2. Retrieved from https://www.britishcouncil.org/sites/default/files/h002_transnational_education_tne_ihe_report_final_web_2.pdf

In China, few real victories in 'people's war' on drugs. (2015, January 24). *The New York Times.* Retrieved from https://www.nytimes.com/2015/01/25/world/despite-a-crackdown-use-of-illegal-drugs-in-china-continues-unabated.html

India will seek global role to secure its traditional local focus. (2015, November 11). *The Australian.* Retrieved from http://www.theaustralian.com.au/news/world/india-will-seek-global-role-to-secure-its-focus-on-subcontinent/news-story/b1d3b9ebbe0ac9d7c25bb296755ad6b7

Jarvis, D. S. L., & Welch, A. (2011). Introduction: Structural changes, economic challenges and political accommodation: Managing the rise of China in southeast Asia. In D. S. L. Jarvis & A. Welch (Eds.), *ASEAN industries and the challenge from China* (pp. 1–17). Basingstoke: Palgrave Macmillan.

Jayasuriya, K. (2003). Introduction: Governing the Asia Pacific—beyond the "new regionalism". *Third World Quarterly, 24*, 199–215. doi:https://doi.org/10.1080/0143659032000074556

Jayasuriya, K. (2010). Learning by the market: Regulatory regionalism, Bologna, and accountability communities. *Globalisation, Societies and Education, 8*, 7–22. doi:https://doi.org/10.1080/14767720903574009

Kharas, H. (2011, June 12). Can the Asian middle class come of age? *East Asia Forum*. Retrieved from http://www.eastasiaforum.org/2011/06/12/can-the-asian-middle-class-come-of-age/

Kharas, H. (2017). *The unprecedented expansion of the global middle class: An update*. Global Economy and Development. (Working Paper 100). Retrieved from https://www.brookings.edu/wp-content/uploads/2017/02/global_20170228_global-middle-class.pdf

Krishna, V. V., & Sha, S. P. (2015). Building science community by attracting global talents: The case of Singapore Biopolis. *Science, Technology and Society, 20*, 389–413. doi:https://doi.org/10.1177/0971721815597152

Kwok, K. (2015, January 17). Can Vietnam have its cake and eat it? *South China Morning Post*. Retrieved from http://www.scmp.com/news/asia/article/1680943/can-vietnam-have-its-cake-and-eat-it

Laksmana, E. A. (2011). Variations on a theme: Dimensions of ambivalence in Indonesia-China relations. *Harvard Asia Quarterly, 13*(1), 24–31. Retrieved from https://ssrn.com/abstract=1979863

Larsen, M. A., & Beech, J. (2014). Spatial theorizing in comparative and international education research. *Comparative Education Review, 58*, 191–214. doi:https://doi.org/10.1086/675499

Li, M. J. (2012, June 14). China's non-confrontational assertiveness in the South China Sea. *East Asia Forum.* Retrieved from http://www.eastasiaforum.org/2012/06/14/china-s-non-confrontational-assertiveness-in-the-south-china-sea

Lo, J. (2014, December 12). Forging ahead with shared vision. *China Daily*. Retrieved from www.chinadailyasia.com/asiaweekly/2014-12/12/content_15202531.html

Lo, K., & Leng, S. (2017, February 24). Chinese factory in Myanmar ransacked by hundreds of angry workers. *South China Morning Post*. Retrieved from http://www.scmp.com/news/china/diplomacy-defence/article/2073775/chinese-factory-myanmar-ransacked-hundreds-angry

Lu, X., & Zhang, W. (2015). The reversed brain drain: A mixed-method study of the reversed migration of Chinese overseas scientists. *Science, Technology and Society, 20,* 279–299. doi:https://doi.org/10.1177/0971721815597127

Malaysia recognises 820 higher learning institutions from China. (2012, June 29). *Borneo Post*. Retrieved from http://www.theborneopost.com/2012/06/29/malaysia-recognises-820-higher-learning-institutions-from-china/

McNamara, J., Knight, J., & Fernandez-Chung, R. M. (2013). The shape of things to come: The evolution of transnational education: Data, definitions, opportunities and impacts analysis. Retrieved from https://www.britishcouncil.org/sites/default/files/the_shape_of_things_to_come_2.pdf

Moe, W. (2015, May 6). Why Kokang rebels are giving fits to Burma's military. *Foreign Policy*. Retrieved from http://foreignpolicy.com/2015/05/06/why-kokang-rebels-are-giving-fits-to-burmas-military-myanmar/

More Chinese investment boosts China-Viet Nam trade ties. (2014). *Xinhua*. Retrieved from http://europe.chinadaily.com.cn/business/2014-01/16/content_17240017.htm

National Geographic. (2005, July). China's great armada: Voyages of Zheng He, 1405–1433. Retrieved from http://ngm.nationalgeographic.com/ngm/0507/feature2/map.html

Ng, P. T., & Tan, C. (2010). The Singapore global schoolhouse: An analysis of the development of the tertiary education landscape in Singapore. *International Journal of Educational Management, 24*, 178–188. doi:https://doi.org/10.1108/09513541011031556

Nguyen, H. (2012, May 11). Internet stirs activism in Viet Nam. Retrieved from http://yaleglobal.yale.edu/content/internet-stirs-activism-vietnam

Nicholson, B. (2015, December 19). Security at risk from China's 'great wall of sand.' *The Australian*. Retrieved from http://www.theaustralian.com.au/opinion/security-at-risk-from-chinas-great-walls-of-sand-in-the-sea/news-story/50ebaa04625186280a2147d2d27d4033

OECD. (2003). *Migration and the labour market in Asia 2002: Recent trends and policies*. Paris: OECD. doi:https://doi.org/10.1787/migr_labour-2002-en

Omi, S. (2017, June 29). Inside Beijing's project teaching Thai students Chinese. *The News Lens*. Retrieved from https://international.thenewslens.com/article/71997

One Belt One Road initiative. (n.d.). In Wikipedia. Retrieved August 2015 from https://en.wikipedia.org/wiki/One_Belt_One_Road_Initiative#/media/File:One-belt-one-road.svg

Osborne, M. (2006). *The paramount power: China and the countries of Southeast Asia* (Lowy Institute Paper 11). Retrieved from https://www.lowyinstitute.org/publications/paramount-power

Outbound tourism flies high with 100m visits. (2014, December 12–18). *China Daily*.

Pak, J. (2012, February 22). Is English or Mandarin the language of the future? *BBC News*. Retrieved from http://www.bbc.co.uk/news/magazine-17105569

Parameswaran, P. (2015, January 22). Philippines blasts China on South China Sea reclamation work. *The Diplomat*. Retrieved from http://thediplomat.com/2015/01/philippines-blasts-china-on-south-china-sea-reclamation-work/

Pesek, W. (2012, June 23). Millions stagnate in poverty as Indonesia indulges in economic nationalism. *The Sydney Morning Herald*. Retrieved from http://www.smh.com.au/business/millions-stagnate-in-poverty-as-indonesia-indulges-in-economic-nationalism-20120622-20tgm.html

Philippines accuses China of widening reclamation works in disputed South China Sea waters. (2015, January 22). *South China Morning Post*. Retrieved from http://www.scmp.com/news/china/article/1688651/philippines-accuses-china-widening-reclamation-works-disputed-south-china

Philippines security pact gets go-ahead. (2016, January 17). *The Australian.*

The Plunder of Myanmar (2015, January 23). [Editorial]. *The New York Times*. Retrieved from http://www.nytimes.com/2015/01/24/opinion/the-plunder-of-myanmar.html?_r=1

Pramudatama, R. (2012, May 13). May tragedy remembered. *The Jakarta Post*. Retrieved from http://www.thejakartapost.com/news/2012/05/13/may-tragedy-remembered.html

Prantl, J. (2012, June 7). Five principles for a new security order in the Asia Pacific. *East Asia Forum.* Retrieved from http://www.eastasiaforum.org/index.php?s=%2C+Five+principles+for+a+new+security+order+in+the+Asia+Pacific

Rahardja, S. (2012, August 9). What China's economic prospects mean for Indonesia. *East Asia Forum.* Retrieved from http://www.eastasiaforum.org/2012/08/09/what-china-s-economic-prospects-mean-for-indonesia/

Reid, A. (2008). Introduction. In A. Reid (Ed.), *The Chinese diaspora in the Pacific* (pp. xv–xxviii). Aldershot: Ashgate.

Robertson, S. (2008). 'Europe/Asia' regionalism, higher education and the production of world order. *Policy Futures in Education, 6*, 718–729. doi:https://doi.org/10.2304/pfie.2008.6.6.718

Scott, J. C. (2009). *The art of not being governed: An anarchist history of uplands Southeast Asia.* Yale Agrarian Studies Series. New Haven: Yale University Press.

Severino, R. (2012, August 17). A code of conduct for the South China Sea? *PacNet, 45A.* Retrieved from https://csis-prod.s3.amazonaws.com/s3fs-public/legacy_files/files/publication/Pac1245A.pdf

Sigley, G. (2014, April 16). Yunnan: China's bridgehead to Southeast Asia and beyond. *The Diplomat*. Retrieved from http://thediplomat.com/2014/04/yunnan-chinas-bridgehead-to-southeast-asia-and-beyond/

Singapore's NTU sets up center for Chinese officials. (2009, December 14). *Crieenglish.com.* Retrieved from http://english.cri.cn/6909/2009/12/14/1461s535442.htm

Singapore to give full license to 2 Chinese banks. (2012, July 6). *Reuters*. Retrieved from http://in.reuters.com/article/2012/07/06/singapore-china-banksidINL3E8I624T20120706

Somers Heidhues, M. F. (2001). *Southeast Asia: A concise history.* New York: Thames and Hudson.

State Council. (2015). China's GDP Growth in Past 20 Years. In *English.goc.cn. Policies. Infographics* (section 5). Retrieved from http://english.gov.cn/policies/infographics/2015/01/20/content_281475042218063.htm

Subs keep an eye on China's dream. (2016, January 8). *Sydney Morning Herald.*

Suryadinata, L. (2003). Patterns of political participation in four ASEAN states: A comparative study. In L. L.-C. Wang & G. Wang (Eds.), *The Chinese diaspora: Selected essays: Vol. 1* (pp. 64–83). Singapore: Eastern Universities Press.

Sutter, R., & Huang, C.-H. (2012). China-Southeast Asia relations: China muscles opponents on the South China Sea. *Comparative Connections*. Retrieved from http://csis.org/files/publication/1202qchina_seasia.pdf

Taylor, B. (2012, May 11). Storm in a teacup over South China Sea. *The Australian.*

Thayer, C. (2015, January 23). The Philippines, Malaysia and Vietnam race to South China Sea defense modernization: Malaysia, the Philippines, and Vietnam are all striving to bolster their maritime capabilities. *The Diplomat*. Retrieved from http://thediplomat.com/2015/01/the-philippines-malaysia-and-Vietnam-race-to-south-china-sea-defensemodernization

Tiezzi, S. (2014, February 13). The maritime silk road vs. the string of pearls. *The Diplomat.* Retrieved from thediplomat.com/2014/02/the-maritime-silk-road-vs-the-string-of-pearls

Tomba, L. (2011, August 25). Who's afraid of China's middle class? *East Asia Forum.* Retrieved from http://www.eastasiaforum.org/2011/08/25/whos-afraid-of-chinas-middle-class/

Tong, S. Y., & Chong Siew Keng, C. (2010). China-ASEAN free trade area in 2010: A regional perspective (EAI Background Brief No. 519). Retrieved from http://www.eai.nus.edu.sg/publications/files/BB519.pdf

Torode, G. (2011, June 21). Rocky patch. *South China Morning Post*. Retrieved from http://www.scmp.com/article/971158/rocky-patch

Tran, D. V. (2003). Culture, environment, and farming systems in Vietnam's northern mountain region. *Southeast Asian Studies, 41,* 180–205. Retrieved from http://hdl.handle.net/2433/53759
Tran, K. K. (2015). *How will Vietnam's economic relationship and dependency on China affects its response to China's increasing threat to its sovereignty?* (Master's thesis). Retrieved from http://hdl.handle.net/10945/45953
2016 Guizhou Education Fair held in Hanoi for Vietnamese students. (2016, May 27). *Xinhua.* Retrieved from http://news.xinhuanet.com/english/photo/2016-04/27/c_135317768.htm
United Nations Conference on Trade and Development. (2004). World investment report, 2004: The shift towards services (United Nations Publication No. E.04.II.D.33). Retrieved from http://unctad.org/en/Docs/wir2004_en.pdf
Universitas 21, Members. Retrieved from http://www.universitas21.com/member
Vaughan, B., & Morrison, W. M. (2006). China-Southeast Asia relations: Trends, issues, and implications for the United States. Retrieved from https://fas.org/sgp/crs/row/RL32688.pdf
Vietnam stops anti-China protests after riots, China evacuates workers. (2014, May 18) *Reuters.* Retrieved from http://www.reuters.com/article/us-china-vietnam-idUSBREA4H00C20140518
Vietnamese students dream of being bridges between Vietnam and China. (2016, May 26). *Xinhua.* Retrieved from http://news.xinhuanet.com/english/2016-05/26/c_135389166.htm
Vietnam to send 3,000 teachers abroad for higher education in 2014–2015. (2013, December 19). www.asean-china-center.org/english/2013-12/20/c_132984280.htm
Vu, T. K., & Nguyen, A. T. (2014, September 9). Oil rig standoff with China won't hurt FDI, Vietnam says. *The Wall Street Journal.* Retrieved from http://www.wsj.com/articles/oil-rig-standoff-with-china-wont-hurt-fdi-Vietnam-says-1410254314
Wang, G. (2000). *The Chinese overseas: From earthbound China to the quest for autonomy.* Cambridge, MA: Harvard University Press.
Wang, G. (2005). China and Southeast Asia: The context of a new beginning. In D. Shambaugh (Ed.), *Power shift: China and Asia's new dynamics* (pp. 187–204). Berkeley: University of California Press.
Wang, L. L.-C., & Wang, G. (1998). *The Chinese diaspora: Selected essays*: *Vol. 1.* Singapore: Eastern Universities Press.
Welch, A. (2010a). Internationalisation of Vietnamese higher education: Retrospect and prospect. In G. Harman, M. Hayden, & T. Nghi Pham (Eds.), *Reforming higher education in Vietnam: Challenges and priorities* (pp. 197–213). Higher Education Dynamics: Vol. 29 . Dordrecht: Springer.
Welch, A. (2010b). Vietnam, Malaysia, and the global knowledge system. In V. D. Rust, L. M. Portnoi, & S. S. Bagley (Eds.), *Higher education, policy and the global competition phenomenon* (pp. 143–160). New York: Palgrave.
Welch, A. (2011a). The dragon, the tiger cubs and higher education: Competitive and cooperative China-ASEAN relations in the GATS era. In D. S. L. Jarvis & A. Welch (Eds.), *ASEAN industries and the challenge from China* (pp. 39–122). London: Palgrave Macmillan.
Welch, A. (2011b). *Higher education in Southeast Asia: Blurring borders, changing balance.* Routledge Research on Public and Social Policy in Asia: Vol. 5. Abingdon: Routledge.
Welch, A. (2012a). Contributing to the Southeast Asian knowledge economy? Australian offshore campuses in Malaysia and Viet Nam. In A. R. Nelson & I. P. Wei (Eds.), *The global university: Past, present, and future perspectives* (pp. 55–81). New York: Palgrave Macmillan.
Welch, A. (2012b). Locating Indonesia within the emergent regionalism of southeast Asian higher education. In J. Hawkins, K. Mok, & D. Neubauer (Eds.), *Higher education regionalization in Asia Pacific: Implications for governance, citizenship and university transformation* (pp. 91–115). London: Palgrave Macmillan.
Welch, A. (2014). Richer relations? Four decades of ASEAN-Australia relations in higher education. In S. P. Wood & B. He (Eds.), *The Australia-ASEAN dialogue: Tracing 40 years of partnership* (pp. 145–166). New York: Palgrave Macmillan.
Welch, A. (2017). Culture and identity. In A. Welch (Ed.), *Education, change and society* (4th ed., pp. 139–186). Melbourne: Oxford University Press.

Welch, A., & Cai, H. (2011). Enter the dragon: The internationalisation of Chinese higher education. In J. Ryan (Ed.), *China's higher education and internationalisation* (pp. 9–33). Routledge Contemporary China Series: Vol. 57. Abingdon: Routledge.

Welch, A., & Hao, J. (2013). Returnees and diaspora: Twin sources of innovation in Chinese higher education. *Frontiers of Education in China*, *8*, 214–238. doi:https://doi.org/10.3868/s110-002-013-0016-7

Welch, A., & Yang, R. (2011). A pearl on the Silk Road? Internationalizing a regional Chinese university. In J. D. Palmer, A. Roberts, Y. H. Cho, & G. S. Ching (Eds.), *The internationalization of East Asian higher education: Globalization's impact* (pp. 63–89). New York: Palgrave Macmillan.

Welch, A. & Zhang, Z. (2008). The Chinese knowledge diaspora: Communication networks among overseas Chinese intellectuals. In D. Epstein, R. Boden, R. Deem, F. Rizvi, & S. Wright (Eds.), *Geographies of knowledge, geometries of power: Framing the future of education* (pp. 338–354). World Yearbook of Education: Vol. 14. London: Routledge.

Wen, W. (2016). China's approach towards HE regional cooperation with ASEAN. In C. S. Collins, M. N. N. Lee, J. N. Hawkins, & D. E. Neubauer (Eds.), *The Palgrave handbook of Asia Pacific higher education* (pp. 173–182). New York: Palgrave Macmillan.

Wen, W., Luo, Y., & Hu, D. (2014). Bridging the gap between Western and Chinese evidence in the international education market. *China and World Economy*, 22(6), 87–102. doi:https://doi.org/10.1111/cwe.12093

Whitney, C. B., & Shambaugh, D. (2008). Soft power in Asia: Results of a 2008 multinational survey of public opinion. Retrieved from https://www.brookings.edu/wp-content/uploads/2012/04/0617_east_asia_report.pdf

Wildau, G. (2015, December 6). China services sector key to growth. *Financial Times*. Retrieved from https://www.ft.com/content/0f6f0018-9817-11e5-bdda-9f13f99fa654

Williams, B. R. (2013, June 19). Trans-Pacific Partnership (TPP) countries: Comparative trade and economic analysis. Retrieved from https://fas.org/sgp/crs/row/R42344.pdf

World Trade Organization. (n.d.). The general agreement on trade in services (GATS): Objectives, coverage and disciplines. In *Trade Topics* (section 4). Retrieved from http://www.wto.org/english/tratop_e/serv_e/gatsqa_e.htm

Wu, D., & Mealy, M. (2012, June 26). Explaining "competing" visions: ASEAN-RCEP, TPP, FTAAP. *PacNet, 40*. Retrieved from https://csis-prod.s3.amazonaws.com/s3fs-public/legacy_files/files/publication/Pac1240.pdf

Wu, J. (2004). Zheng He's voyages of discovery. Retrieved from http://www.international.ucla.edu/china/article/10387

Xin, D. (2012, April 13). More travelers from mainland going overseas. *China Daily*. Retrieved from https://www.chinadaily.com.cn/travel/2012-04/13/content_15042405.htm

Yang, H.-Y. (1993). Confucius. *Prospects: Quarterly Review of Comparative Education*, *23*, 211–219. Retrieved from http://unesdoc.unesco.org/images/0009/000985/098500eo.pdf

Yang, R. (2012). Internationalisation, regionalization, and soft power: China's relations with ASEAN member countries in higher education. *Frontiers of Education in China*, *7*, 486–507. Retrieved from https://www.researchgate.net/publication/289917999_Internationalization_regionalization_and_soft_power_China's_relations_with_ASEAN_Member_countries_in_higher_education

Yang, R., & Welch, A. (2010). Globalisation, transnational academic mobility and the Chinese knowledge diaspora: An Australian case study. *Discourse: Studies in the Cultural Politics of Education, 31*, 593–607. doi:https://doi.org/10.1080/01596306.2010.516940

Yang, R. & Welch, A. (2012) A world-class university in China? The case of Tsinghua, *Higher Education, 63*, 645–666. doi:https://doi.org/10.1007/s10734-011-9465-4

Yeung, H. W.-C. (1999). The internationalization of ethnic Chinese business firms from Southeast Asia: Strategies, processes and competitive advantage. *International Journal of Urban and Regional Research, 23*, 88–102. doi:https://doi.org/10.1111/1468-2427.00181
Yi, T. X. (2015, October 9). Attracting China's middle class. *The Star*. Retrieved from http://www.thestar.com.my/opinion/columnists/check-in-china/2015/10/09/attracting-chinas-middle-class/
Zhao, S. (2011). China's approaches toward regional cooperation in East Asia: Motivations and calculations. *Journal of Contemporary China, 20*, 53–67. doi:https://doi.org/10.1080/10670564.2011.520846
Zheng He. (n.d.-a). In *New World Encyclopedia*. Retrieved from http://www.newworldencyclopedia.org/entry/Zheng_He
Zheng He. (n.d.-b). In *Wikipedia*. Retrieved July 2017 from https://en.wikipedia.org/wiki/Zheng_He

Part V
The International University

Chapter 19
Placing the University: Thinking in and Beyond Globalization

Allan Cochrane

For many of those who work in them, universities simply are. We (for I am one of those who does) take them for granted as our workplaces, often to the extent that, as Calhoun (2006) sardonically comments, "Professors tend to think universities exist naturally, or as a gift of history, in order to employ them" (p. 34). But, of course, it is increasingly difficult to maintain such a position. There was always a tension between a vision of the academy (as a more or less global thought community) and the university (as an institution in place). And even the most insulated of academics must be aware of the changes that have been shaping and reshaping higher education in the last few decades with the shift from elite to mass higher education, the rise of a more market-based approach to study, the increase of competition globally and locally, the growth of the expectation that universities will contribute to economic and social development, and so on (see, e.g., Goddard & Vallance, 2013, for a thoughtful engagement with the wider debates about the relationship between universities and urban development).

Changing Spatial Imaginaries

These changes have also been associated with changes in the ways in which the geographies of universities have been imagined. The contested and overlapping conceptualizations of the geographies of higher education provide a powerful way into thinking about the contemporary university. These can be briefly summarized. The first (traditional) version imagines academic life in terms that resist any sense of spatial fixity or embeddedness. It celebrates knowledge for its own sake as expressed

A. Cochrane (✉)
Faculty of Arts & Social Sciences, The Open University, Milton Keynes, UK
e-mail: allan.cochrane@open.ac.uk

P. Meusburger et al. (eds.), *Geographies of the University*, Knowledge and Space 12,
https://doi.org/10.1007/978-3-319-75593-9_19

through networks of academia that stretch beyond, into national and global networks. From this perspective, place does not matter—where a university is located is barely relevant (except sometimes as an irritant) to the life of those connected into those networks. Elite universities may be located *in* place, but that does not mean that they understand themselves as being *of* the place in which they find themselves. The second (globalized) version seems equally reluctant to accept the restrictions of territorial identity, but in this case the emphasis is on the global markets expressed in student mobility and the transferability of qualifications, the possibilities of overseas and campuses, as well as the promise of virtual higher education (through Massive Open Online Courses, or MOOCs). It is also accompanied by an obsession among elite or "wannabe" elite institutions with their position in global rankings of one sort and another (as expressed, for example, in the extensive publications associated with the Observatory on Borderless Higher Education, n.d.). The third (regionalized) version also starts from within the vision of a globalized knowledge economy but in this case one in which universities themselves are placed as development nodes and transmission belts and as active partners in communities (Harding, Scott, Laske, & Burtscher, 2007; Organization for Economic Cooperation and Development [OECD], 2007). It connects to a longer tradition with an emphasis on the potential of civic universities and (in the U.S. context) service learning (see, Goddard, 2009).

Of course this summary is not an exhaustive one. Although there has undoubtedly been a shift toward global narratives, identified as a "revolution" by Altbach, Reisberg, and Rumbley (2009), there is a parallel political world dominated by national policy initiatives, which are themselves shaped by the attempts of national governments to make their higher education systems deliver on a range of priorities, some of which are often expressed in a language of global competitiveness (as reflected, for instance, in the claims made by Universities UK, 2015). So, in England, for example, it is currently impossible for any university (however hard it tries and whatever its global ambitions) to ignore the implications of a hybrid national funding regime apparently sustained through the payment of student fees, but in practice underpinned by state-backed loans (McGettigan, 2013).

It would also be a mistake to see these different versions as somehow alternatives, in the sense that any one of them captures the reality of contemporary higher education. On the contrary these are active geographies, mobilized in different ways at different times, often by the same players. And what matters is how these imaginaries are mobilized to justify particular institutional strategies and to incorporate a range of different higher education actors—not just senior managers (or leaders as they increasingly call themselves), but also academics, students, and even a wider range of those who are from time to time identified as stakeholders, including local and regional government agencies, as well as national governments and funding agencies, both private and public.

In this context, it is helpful to reflect on some of the myths that swirl around universities. The old myth of the ivory tower, dedicated to research and study is of course, as Calhoun (2006) suggests, highly attractive to many academic staff. Perhaps no one has ever actually worked in such an institution and none probably ever existed, but it is comforting to know that it might have. This myth is frequently presented from the outside as one that positions academics as an elite disconnected from the problems and possibilities of the "real" world, with the new myths seeming to offer a very different take, even if the elite status of the academy remains central to them. The university is constructed as a special place of expertise and knowledge production, but now the task is to find ways in which our knowledge can be transferred from us to "them" in a language of impact, knowledge transfer, and engagement. At its most inspiring this is translated into a language of missions—with the missionary complex of universities hard to miss in their official statements seeking to define their roles—and contribution to the public good, as well as to economic transformation. The mission of the University of Cambridge "is to contribute to society through the pursuit of education, learning and research at the highest international levels of excellence" (University of Cambridge, n.d.); that of the Open University is "to be open to people, places, methods and ideas" and to "promote educational opportunity and social justice by providing high-quality university education to all who wish to realise their ambitions and fulfil their potential" (The Open University, n.d.).

Significant claims are increasingly being made about the importance of universities for their regional economies and societies. It is believed they are central to building competitiveness (e.g., through their role in the knowledge society and knowledge exchange), as well as contributing to the transformation of local populations through upskilling and reskilling (see chapters of Etzkowitz; Glückler, Panitz & Wuttke; Goddard in this volume) They are understood to be powerful agents of cultural change, as well as potential sources of initiatives to challenge social disadvantage (as no more than an indicative and far from comprehensive list see, for example, Addie, Keil, & Olds, 2015; Benneworth, 2012; Benneworth & Hospers, 2007; Goddard & Vallance, 2013; Harding et al., 2007; Kitigawa, 2004; OECD, 2007; Pinheiro, Benneworth, & Jones, 2012; Pinheiro, Langa, & Pausits, 2015; Rohe, 2011; Sir Witty, 2013; Stachowiak, Pinheiro, Sedini, & Vaattovaara, 2013; Yusuf & Nabeshima, 2007).

Universities as Institutions

All of this in its different ways starts from the assumption that somehow universities have a special role, whether in their ivory tower or in enabling processes of change. Much of the writing emanates from universities, of course, and where it does not, it is clearly targeted at celebrating their role (see, e.g., Etzkowitz, 2008; Etzkowitz & Leydesdorff, 1997).

This chapter has a rather different starting point. Instead of assuming a special status for universities (and their staff), it seeks to understand them through their institutional and discursive practices—as far as possible looking at them from the outside in the same way as academic researchers do when exploring how other organizations operate. That is to say, it is framed by the nagging question, "what if we are not so special after all?" The aim is to think differently, moving beyond academic and institutional self-regard to reflect on the ways in which universities operate as more or less successful businesses, positioning themselves within and attempting to influence the changing geographies of knowledge production and knowledge exchange.

The reflections that follow are informed by and build on work initially undertaken with colleagues at the Open University and the University of East Anglia from 2008 to 2010 (Michael Amoah, Alice Bennion, John Brennan, Yann Lebeau, and Ruth Williams) as part of an Economic and Social Research Council funded project on Higher Education and Social and Regional Transformation (ES/E017894/1). This project has also led to the publication of a book (Brennan et al., 2018). However, the conclusions that I draw from this work are very much my own, and I would not want any of my colleagues to face criticism for anything that is said here—they may well disagree! The research was conducted on four universities located in different regional contexts across the United Kingdom, each with its own particular institutional mission. It involved the review of a wide range of documents (strategic plans, mission statements, and more) as well as extensive interviews with university staff, local business and community stakeholders (including other universities in the region), as well as hybrid or partnership organizations working with or around universities, ranging from agencies seeking to foster widening participation by people from socially disadvantaged groups to those associated with the promotion of science or technology parks.

It is impossible to capture the full variety of universities in ways that fully reflect the significance of differences between them. But the four cases explored in the research make it possible to consider some of the ways in which the universities, each in its own way, sought to engage with and actively define the spaces of higher education. A range of options was being mobilized and managed by universities in the face of the challenges generated by the changing political economy of higher education. The experiences the research team identified are far away from some of the more visionary expressions of the role of universities. The extent to which these institutions were operating in worlds not of their own making and needing to negotiate with others was apparent. It was also clear that they sought to draw on a range of resources at their disposal to bring in funding as well as other resources from elsewhere—student fees (domestic and international), funding bodies, and other private and public sources. Institutional survival and expansion were significant drivers, even if they were often expressed in apparently well-meaning statements of intent about community, economic contribution, and cultural transformation. Of course, as with all institutions, it is important to acknowledge that their identity is more complex and less unitary than this implies—the acts of individual academics or their collective actions may sometimes fit uneasily with stated and implicit corporate priorities. But it is nevertheless possible to identify institutional framings within which even the most troublesome of individual players find themselves positioned (for some of the tensions see, Rolfe, 2012).

Universities in Their Regions

The universities and regions on which the project focused may not deliver an exhaustive set of potential cases, but they do provide some key illustrative evidence. It was apparent that the specifics of each case were significant, even as it was possible to draw out wider conclusions. The experience of the four institutions can be summarized along the following lines:

Case 1: Wannabe global city regionalism In this case, the university had a clear strategy of positioning itself as a global institution (and was highly ranked in international tables of university reputation), while at the same time the city was attempting to define itself as a global city, or at the heart of an emergent global city region. In both instances the extent to which each actor had the capacity to take on these global roles may have been questionable, but in each it translated into a powerful positioning within the national context. The university attracted students from across England, as well as from overseas, although a significant proportion were drawn from the wider region in which it was located. The university was centrally positioned as a change agent and was supported in its development ambitions by local and regional agencies. This was reflected in a continuing process of property development, making up an urban campus in the heart of the city and dramatically transforming one of the main transport arteries as it does so. The university's strategic plan described the success of the city region as being vital to the university's own chances of realizing its ambitions and also declared the institution's commitment to working with city authorities to enhance the standing of the city as a dynamic node in the global knowledge economy.

Case 2: Local of necessity The university was located within a declining industrial area with a poor external image, with a student body largely drawn from that area. It was understood by stakeholders to be the *local* university within its urban subregion, but there was a division of labor with another (more elite-oriented) university that had a branch campus in the same region. It presented itself as a business university, oriented toward the needs of local businesses, although not providing the elite labor force required to support the specialist work undertaken within some of the local industries. The main campus of the university was located in one of the subregion's urban centers, and mayor, council, and university actively engaged with each other to challenge the region's negative external image— making it attractive to potential students as well as investors are seen as complementary processes. The investment in and development of the central campus had helped to shape a modest university quarter and had begun to open up consumption based services around the university. The university had drawn on local and regional support to develop new campuses and new areas of curriculum and research, identifying and building on particular institutional strengths, such as digital media, in the global market place. Such initiatives both repositioned the university as more than local and were intended to shift perceptions of the city from a home of old industry to a center for emerging industries.

Case 3: Inventing a (sub)region Here the uncertainty of the university's position was particularly striking. It operated within a complex (mega)city region, with many institutions and blurred boundaries. Its focus was on recruiting students with relatively low educational qualifications, often from black and minority ethnic populations. This was a quintessentially urban institution whose identity was defined by the broadly defined subregion in which its campuses had historically been located, although it had sought to concentrate its activities on a more limited number of sites. The urban area in which the university was based had been the subject of major regeneration initiatives over the last 25 years, as well as being the site of at least one mega project and one significant private-sector-led development project. This had also created opportunities for the university as partner. It had learned to play the regeneration game, often responding to the possibilities created by the operation of development agencies, moving into new spaces and taking advantage of development opportunities as it had created its new campuses and sought to collaborate with other universities to create shared spaces. It had attracted some criticism from those left behind when old campuses have closed, but had sought to maintain its imagined place-based identity. The process of identity formation was fragile, but often confident, as academics also engaged in the process of place definition in their writing and promotion of workshops rooted in the local experience.

Case 4: Geographically embedded The fourth case was of a university that operated within a narrowly defined local geography, in a city that had faced major problems of economic restructuring over the past few decades. Even more than the other institutions this university saw itself as responding to national (in this case Scottish) policy, but its students were predominantly local. Like Case 2, it had sought more effectively to concentrate its activity, in this case to a more central campus within the city. There was a clear division of labor between this university and the city's other (more elite and much larger) university, which attracted students from a wider catchment area within Scotland. In this context the university identified a carefully focused set of ambitions that emphasized work in the areas of environmental sciences, policing, forensics, and criminal justice, as well as in inclusive technologies for sustainability, well-being, and security. There was an implicit partnership between the universities in terms of the way in which they came together to find ways of reimagining the city, enabling it to escape from past associations with decline by creating a cultural quarter, incorporating the campus, and seeking to transform place. In this, the university was part of a wider public-private partnership aiming to reposition the city as a place for the future, although some of the academics interviewed raised concerns about the extent to which this approach effectively excluded those more disadvantaged people outside the transformed center who had previously been served by the institution.

In all four cases, it was possible to identify ways in which the places of the university were being actively reimagined. There was a complex dance in which institutional self-image and the presentation of place were brought together (Cochrane & Williams, 2013). At first glance there is a surprising alignment of regional and local priorities with university priorities despite the different drivers to which each seemed to be responding. So in one case there was an alignment between the local ambitions to be a world city and university ambitions to be a global university; in another there was a coming together around the vision of developing a digital city and the ambition to be a center of digital media; in a third the two were focused on the prospects of large scale urban regeneration; while in the fourth, there was a promise of synergy around the university and city in fostering a creative city and creating a cultural center.

The Everyday Practices of Universities in Place

But it is also necessary to reflect on some of the mundane stories of transformation associated with universities and their development—without having any particular strategic focus on locality or region, what universities do may nevertheless have a dramatic impact on place. And the day-to-day activity of most universities will of necessity involve them in formal and informal negotiation as major employers, major landowners, and major developers of property, often in high-value, central areas and sometimes on the edge of cities. It is clear (and presumably not controversial to note) that the business strategies of universities are driven by their own priorities—student recruitment, income generation, and survival, even if, as McGettigan (2013) notes, the marketized rules of the game have changed the ways in which those priorities are pursued. Of course, this is sometimes veiled by a language of social mission or commitment to disinterested knowledge production, and the ways in which many staff espouse such ambitions is sincere enough. But wider institutional logics are stubbornly apparent, however much some members of academic and other staff seek to position themselves outside or beyond them.

One of these relates to the extent to which universities are adept at playing the grant-getting game, playing by the rules in ways that benefit them as institutions and ways that position them as regional players. So, for example, without necessarily changing the main direction of activity, all of the institutions visited responded directly to initiatives launched by university funding councils—for example, in the New Labour era forming partnerships in the context of Aimhigher (targeted at widening participation) or around the Beacons initiative, which was specifically targeted at developing forms of public engagement. Once the funding was withdrawn, most of the specific activities ceased unless they could be incorporated (costlessly) into the mainstream activities of the universities. All of them also participated in local and regional partnerships, in formal ways (for example through the representation of senior staff on partnership boards of various sorts) but becoming more actively engaged when such bodies were identified as potential sources of funds—for example in enabling the development of new campuses in the context of regeneration or in sponsoring institutional mergers that strengthened their position nationally as well as locally.

The institutional thrill of property development was frequently apparent. All of the universities were engaged with other local agencies in a series of property-related initiatives. The development of new campuses and the expansion and rationalization of existing ones was a feature in all four cases. In one the scale was such that it was the focus of a development partnership involving the city council as well as other partner agencies (such as the local hospital trust) and offered the prospect of transforming a significant area of the city. The university's institutional priorities were incorporated into the city's wider growth ambitions, as its civic leaders sought to move from an older industrial base to a service-based and possibly even knowledge-based or creative economy. In the other instances, although the scale was less ambitious, similar trends were apparent—as in one case the campus was consolidated; in a second a new campus was developed in the context of wider urban regeneration strategies; while in a third stress was placed on creating a cultural area around an urban campus, in sometimes uneasy collaboration with another local university.

But perhaps the most significant impact some universities have on the areas in which they are located has little to do with any straightforward regional or place-based institutional strategy. Instead what may matter more are the practices of higher education as a business, both in the ways they operate and in the consequences of their operation. Universities have a significant impact as businesses in their own right, both through their property strategies and as big employers of technical and professional staff (including but not only knowledge professionals). In many cities universities are the largest single employer, rivaled only by the National Health Service and (until recently at least) local government. In the context of significant regional inequality across England, they are unusual in offering comparable rates of pay as part of a national (and even global) higher education system. In this context, although less evenly, they may also act as nodes for the transmission of globalized competitive agendas emphasizing the significance of the knowledge economy and fostering or providing environments within which the language of digitization and creativity is increasingly taken for granted.

The unintended consequences of university activity for places are as important as the planned or intended ones. That is the case with property development initiatives that transform particular areas, introducing new uses and moving older ones (and sometimes older populations) away in the context of strategies whose core purpose is to underpin the university's own further development, even as it is sometimes framed in partnership terms through the rhetoric of wider ambitions to urban regeneration. Meanwhile, the significance of studentification has been increasingly widely acknowledged (see, e.g., Smith, 2008, Smith & Hubbard, 2014). As full-time student numbers in England have increased inexorably since the early 1990s, new consumption patterns have developed in many of England's older cities, and housing tenure patterns have changed, too, as some areas have effectively become dominated by student rental housing, whether houses in multiple occupation or in the form of purpose-built student accommodation.

This was a particular issue at one of the researched institutions, with interviews eliciting ambiguous responses from residents and local political actors: A residents' association highlighted the changing demographics of their area, noting that houses were often empty for long periods in vacation time and complaining that some traditional shops had gone to be replaced by others more focused on the student market; in the central area of the city, the local authority was more positive, highlighting the extent to which a more exciting (24 hour) economy was being generated, which made the city more attractive to visitors and potential campus investors. Instead of a reputation for industrial decline it was possible to call on a series of images from a more vibrant youth culture (for a more critical view, see Chatterton, 1999). Matters were less clear-cut in other cases where the student body was more local and often part-time, living at home. Even here, however, there was some evidence in two of the cases that new consumption activities were being enabled around the university.

During the period of research, one of the most actively pursued policies linking higher education to a wider social mission, which also had an implicit and sometimes explicit regional focus related to widening participation, and in their different ways all of the institutions had their own widening participation strategies. These strategies varied significantly between institutions. For the university most clearly identifiable as an "elite" institution (in the wannabe global region), widening participation was seen in terms of a responsibility to raise aspirations in local schools, in a sense inspiring pupils by fostering an engagement with education, showing what was possible. There was no particular expectation that the students inspired in this way would attend the university, since the emphasis was on recruiting across the United Kingdom and was focused on applicants with high entry qualifications. A similar approach was taken by another elite university that had a campus in the same subregion as the institution identified as local of necessity. It too saw its job as being to bring ideas from outside into the local context, without any expectation that local young people would necessarily become its students, although they might be inspired to pursue routes into higher education.

The other three universities were far more oriented toward the recruitment of students previously at risk of exclusion from higher education, with partnerships with local schools being oriented toward that recruitment. In a sense recruitment of students with lower educational qualifications was a central aspect of their business plans as much as it was a social mission. Within the project's four subregions, widening participation activities tended to provide social mobility opportunities for the few without necessarily altering patterns of inequality that affect the many, as one of our respondents commented. And in the subregions with several higher education providers, the implicit stratification of institutions mapped onto and reinforced wider patterns of inequality. One of our (private sector) respondents commented that by playing the "local" card in developing its curriculum, the university identified as local of necessity was effectively positioning its students in ways that left them disadvantaged, even as others commented that those with qualifications tended to use them to find employment outside the region.

Conclusions: The Many Spatialities of Higher Education

From different perspectives, although sometimes in the same strategic plans, universities can be understood to be everywhere, nowhere, and somewhere. The language of globalized higher education tends to focus on their ubiquity—they can be accessed virtually through MOOCs or otherwise; or by more traditional face-to-face means through an increasingly wide range of distributed campuses and partnerships associated with home institutions mainly based in the United Kingdom, the United States, or Australia; and, of course, the transnational mobility of students means that institutions stretch their connections far beyond any territorial or national borders. Alongside this form of globalized language, an older tradition survives that emphasizes the existence of a global academy, in which it is ideas and key thinkers or thinking that straddle the globe. In principle, institutional location is secondary to the networks of globalized academics meeting in conferences and workshops, writing for international journals, and sometimes even enjoying global reputations of their own.

And yet, universities are generally (even if not always) more or less fixed in place. This means that they always have a relationship with their regions, even if that is not what frames university strategies. Their day-to-day activities often have significant local and localized impacts. Their images and brands can help to define the places within which they are located, just as those locations can help to define the universities and the student markets on which they are able to draw. In other words, universities shape and are shaped by the places within which they find themselves. So, even as they are always necessarily connected to elsewhere, stretching out for students, funding, ideas, and academic networks, they are also necessarily located in place by the practices of everyday institutional life, often including significant sunk investment in infrastructure, in addition to any formal civic commitment they may espouse. In that sense, not only are they both global and local, but their local presence is formed through their global ambitions or global positioning, while the versions of globalization they help to construct are also formed through practices of local emplacement.

References

Addie, J.-P. D., Keil, R., & Olds, K. (2015). Beyond town and gown: Universities, territoriality and the mobilization of new urban structures in Canada. *Territory, Politics, Governance, 3*, 27–50. doi:https://doi.org/10.1080/21622671.2014.924875

Altbach, P. G., Reisberg, L., & Rumbley, L. E. (2009). *Trends in global higher education: Tracking an academic revolution*. Paris: UNESCO.

Benneworth, P. (Ed.) (2012). *University engagement with socially excluded communities*. Dordrecht: Springer.

Benneworth, P., & Hospers, G.-J. (2007). Urban competitiveness in the knowledge economy: Universities as new planning animateurs. *Progress in Planning, 67*, 105–197. doi:https://doi.org/10.1016/j.progress.2007.02.003

Brennan, J., Cochrane, A., Lebeau, Y., & Williams, R. (2018). *The university in its place: Social and cultural perspectives on the regional role of universities*. Heidelberg: Springer.

Calhoun, C. (2006). The university and the public good. *Thesis Eleven, 84*(1), 7–43. doi:https://doi.org/10.1177/0725513606060516

Chatterton, P. (1999). University students and city centres—the formation of exclusive geographies: The case of Bristol UK. *Geoforum, 30*, 117–133. doi:https://doi.org/10.1016/S0016-7185(98)00028-1

Cochrane, A., & Williams, R. (2013). Putting higher education in its place: The socio-political geographies of English universities. *Policy and Politics, 41*, 43–58. doi:https://doi.org/10.1332/030557312X645775

Etzkowitz, H. (2008). *The Triple Helix: University-industry-government innovation in action*. Abingdon: Routledge.

Etzkowitz, H., & Leydesdorff, L. (Eds.). (1997). *Universities in the global knowledge economy: A triple helix of university-industry-government relations*. London: Continuum.

Goddard, J. (2009). *Re-inventing the civic university*. London: National Endowment for Science, Technology and the Arts.

Goddard, J., & Vallance, P. (2013). *The university and the city*. Abingdon: Routledge.

Harding, A., Scott, A., Laske, S., & Burtscher, C. (Eds.). (2007). *Bright satanic mills: Universities, regional development and the knowledge economy*. Aldershot: Ashgate.

Kitigawa, F. (2004). Universities and regional advantage: Higher education and innovation policies in English regions. *European Planning Studies, 12*, 835–852. doi:https://doi.org/10.1080/0965431042000251882

McGettigan, A. (2013). *The great university gamble: Money, markets and the future of higher education*. London: Pluto.

OECD (Organization for Economic Cooperation and Development). (2007). *Higher education and regions: Globally competitive, locally engaged*. Paris: OECD.

Observatory on Borderless Higher Education. (n.d.). Retrieved from http://www.obhe.ac.uk/what_we_do/news_articles_reports/news_articles_reports

Pinheiro, R., Benneworth, P., & Jones, G. A. (Eds.). (2012). *Universities and regional development: A critical assessment of tensions and contradictions*. Abingdon: Routledge.

Pinheiro, R., Langa, P. V., & Pausits, A. (2015). One and two equals three? The third mission of higher education institutions. *European Journal of Higher Education,* 5, 233–249. doi:https://doi.org/10.1080/21568235.2015.1044552

Rohe, W. M. (2011). *The research triangle: From tobacco road to global prominence*. Philadelphia: University of Pennsylvania Press.

Rolfe, G. (2012). *The university in dissent: Scholarship in the corporate university*. Abingdon: Routledge.

Sir Witty, A. (2013). *Encouraging a British innovation revolution: Sir Andrew Witty's review of universities and growth*. London: Department of Business, Innovation and Skills.

Smith, D. P. (2008). The politics of studentification and '(un)balanced' urban populations: Lessons for gentrification and sustainable communities? *Urban Studies, 45*, 2541–2564. doi:https://doi.org/10.1177/0042098008097108

Smith, D. P., & Hubbard, P. (2014). The segregation of educated youth and dynamic geographies of studentification. *Area, 46*, 92–100. doi:https://doi.org/10.1111/area.12054

Stachowiak, K., Pinheiro, R., Sedini, S., & Vaattovaara, M. (2013). Policies aimed at strengthening ties between universities and cities. In S. Musterd & Z. Kovács (Eds.), *Place-making and policies for competitive cities* (pp. 263–291). Chichester: Wiley-Blackwell. doi:https://doi.org/10.1002/9781118554579.ch16

The Open University. (n.d.). *Mission*. Retrieved from http://www.open.ac.uk/about/main/mission
University of Cambridge. (n.d.). *The university's mission and core values*. Retrieved from https://www.cam.ac.uk/about-the-university/how-theuniversity-and-colleges-work/the-universitys-mission-and-core-values
Universities UK. (2015). *The funding environment for universities 2015: The economic role of UK universities*. London: Universities UK. Retrieved from http://www.universitiesuk.ac.uk/policy-and-analysis/reports/Documents/2015/the-economic-role-of-uk-universities.pdf
Yusuf, S., & Nabeshima, K. (Eds.). (2007). *How universities promote economic growth*. Washington: World Bank.

Chapter 20
The University Unbound: How Roots and Routes Intersect

Jane Kenway

Around the world considerable numbers of students are on the move to study undergraduate and postgraduate courses in universities outside their home country. According to the Organisation for Economic Co-operation and Development (OECD), 4.5 million students studied abroad in 2012, in contrast to 4.1 million in 2010, 2.01 million in 2000, and 1.03 million in 1990 (OECD, 2014). Clearly the numbers have increased steadily over the last three decades. My main purpose in this chapter is to consider some consequences of such student mobility and the university practices associated with it. I argue that these have implications for both conventional geographical understandings of the university and for contemporary geographies of knowledge. Meusburger (2015) observes that the geography of higher education recognizes that:

> The generation and diffusion of scientific knowledge is influenced by local milieus and spatial relations; that scientific practices vary from place to place; that social environments have an impact not only on generating, evaluating, and legitimizing new scientific knowledge, but also on the reading and interpretation of texts; that universities are not simply locations but social spaces, epistemic venues, and knots of scientific networks. (p. 169)

This chapter speaks to these themes and points to the complex politics that exist at the microscopic and telescopic nexus of knowledge, place, student mobility, and academic territoriality—and indeed the books we read.

Thank you to Hongzhi Zhang for producing the tables, to Hongzhi and Philip Wing Keung Chan for inviting me to be so centrally involved in the *Asia as Method in Education Studies* project and to all the other participants. The process was deeply educative for me. I also acknowledge that the inspiration for this chapter arose from this project and that I draw from different parts of our book, Zhang, Wing Keung Chan, and Kenway (2015). Permissions have been granted.

J. Kenway (✉)
Faculty of Education, Monash University, Victoria, Australia
e-mail: jane.kenway@monash.edu

P. Meusburger et al. (eds.), *Geographies of the University*, Knowledge and Space 12,
https://doi.org/10.1007/978-3-319-75593-9_20

The Reconfigured University

Global student mobilities, and the technologies that facilitate them, have changed the configuration of the university itself. In many ways it has become *unbound*. It can no longer be understood as unambiguously bounded with an obvious in and outside, with self-evident insiders and outsiders, or as having a clear-cut community. As a place it is being remade, with the links between place and institutional identity and identification being reworked.

The notion that place is socially and politically produced and contested invites a consideration of the manner in which the university is "made" or "performed" in the contemporary asymmetrical global context. The university is now a place where various expressions of globalization intersect and conflict. Indeed, as Goddard (this volume) points out, the *civic university* navigates globalization as it is expressed in the city. His analysis and mine here suggest that contemporary universities are involved in various new place- and scale-making projects. This involves them instigating global connections and imaginations while also attempting to remain connected to place. In geographical terms, then, the university, as an institution, now needs to be understood not just as territorial, as a place of institutional, national, and subnational *roots*, but also as a place of global, regional, and transnational *routes*.

I offer an illustration of what this reconfigured university of "routes and roots" (Clifford, 1997) looks like up close for particular sets of students and for the types of knowledge they are exposed to and seek. I show some ways that roots and routes intersect. In so doing I point to certain related "power geometries" (Massey, 1993, p. 59) and "emotional geographies" (Bondi, Davidson, & Smith, 2005) on different scales. A myriad of microscopic realignments occur as intertwined institutional, subnational, and national cultures adjust.

Such complexity needs to be made comprehensible, and Williams (1977) is helpful here. He talks about culture involving dominant, residual, and emergent elements. Dominant elements are those that are sanctified. They need to be understood in terms of the unequal power relationships that underpin them. Emergent elements are those that are in the process of being developed. They are coming into view; they may start from the margins but are establishing a presence. They are essentially different from the dominant and arise out of new sets of social and economic conditions. They may or may not become dominant. Residual elements arise from earlier times and social formations and usually represent earlier stages in a spatial entity's biography. They tend to be subordinate. Indeed, residual elements may persist even when the conditions that initially made them powerful have passed. Dominant elements may incorporate, reinterpret, or dilute both emergent and residual elements. They may also marginalize them, particularly if they are uncomfortable and/or oppositional. When linked to roots and routes these notions speak to the dynamic and contradictory quality of today's university unbound.

Patterns of Student Mobility

Let me offer a brief indication of the numbers associated with student mobility and draw attention to the related asymmetrical patterns of global power. The international agencies UNESCO and the OECD and the U.S.-funded Project Atlas, among others, statistically document mobile students' movements over time, place, and institution. The involved researchers show how and where the numbers rise and fall and speculate about why. The data records such things as the total numbers globally, regionally, and nationally. For example, UNESCO documents the main countries that students travel to (see Table 20.1) and depart from (see Table 20.2). Invariably national comparisons are made. For example, in Australia and the United Kingdom international students comprised 20% of total higher education enrollment in 2014. In comparison, the proportion in France and Canada was 12%, in Germany 11%, in the United States and Japan 4%, and in China 1% (Project Atlas, 2014). In comparison, for France and Canada the proportion was 12%, for Germany 11%, for the United States and Japan 4%, and for China 1% (Project Atlas, 2014).

Regional comparisons are also drawn. UNESCO (2015) shows that in 2012 the regions hosting the most mobile students were "North America and Western Europe (57% of total mobile students), East Asia and the Pacific (20%), Central and Eastern Europe (10%)." UNESCO also points out, however, that increasingly the rise of regional hubs has led to a shift in such region flows. The agency says, for example, "Australia and Japan, traditional destinations in East Asia and the Pacific, are rivaled by newcomers China, Malaysia, the Republic of Korea, Singapore and New Zealand, which hosted 6% of the global share of mobile students in 2012" (UNESCO, 2015). The global, regional, and national picture of flows is both patterned and fluid. Exit and entry numbers go up and down, locations fluctuate somewhat in popularity, new patterns emerge, and older patterns adjust, but, to date, remain dominant.

National reporting agencies, Australian in the examples used here, focus on such things as total numbers and income. The number of overseas students in Australia was 249,990 in 2014, 225,720 in 2009, and 164,044 in 2004. For 2014–2015 income generated by their presence in Australia was AU$18.1 billion (Universities Australia, 2014). Such agencies also focus on student load by university (see Table 20.3), overseas students by country of birth and gender (see Table 20.4), and, of course, a wide array of other matters.

Such matters include what is often called transnational education. This involves offshore programs delivered in partnership with an international institution or through offshore national university campuses. As Universities Australia (2014) points out it also involves "formal institution-to-institution agreements between Australian universities and overseas higher education institutions. Agreements include cooperation facilitating student exchanges, study abroad arrangements, staff exchanges and academic/research collaboration" (p. 21).

Regions, countries, and institutions have their own distinctive histories of flows. For instance, student mobility in Europe has been significantly boosted as result of

Table 20.1 Top 10 countries for number of inbound internationally mobile tertiary students studying abroad from 2004–2013 (2013 ranking)

Country	2004	2005	2006	2007	2008	2009	2010	2011	2012	2013
United States	572,509	590,158	584,719	595,874	624,474	660,581	684,807	709,565	740,482	784,427
United Kingdom	300,056	318,399	330,078	351,470	341,791	368,968	389,958	419,946	427,686	416,693
Australia	166,954	177,034	184,710	211,526	230,635	257,637	271,231	262,597	249,588	249,868
France	237,587	236,518	247,510	246,612	243,436	249,143	259,935	268,212	271,399	239,344
Germany	260,314	..	..	..	..	..	..	207,771	206,986	196,619
Russian Federation	75,786	90,450	77,438	60,288	136,791	129,690	..	165,910	173,627	138,496
China	..	..	36,386	42,138	51,038	61,211	71,673	79,638	88,979	96,409
Austria	33,707	..	39,329	43,572	53,396	59,705	68,619	70,558	58,056	70,852
Netherlands	26,154	26,387	27,037	27,449	30,052	23,674	27,968	38,367	57,506	68,943
Saudi Arabia	12,199	12,999	13,687	17,716	18,725	19,906	26,871	34,922	46,566	62,143

Note. Adapted from UNESCO (2015). Copyright 2016 by UNESCO.

Table 20.2 Top 10 source countries for students going abroad for tertiary education, 2004–2013 (2013 ranking)

Country	2004	2005	2006	2007	2008	2009	2010	2011	2012	2013
China	365,916	403,527	407,280	430,355	459,892	517,335	568,578	650,632	693,972	712,157
India	133,849	146,033	145,539	161,492	183,646	203,217	208,723	204,246	188,791	181,872
Germany	57,564	64,213	70,750	78,222	83,023	92,753	104,853	113,795	117,691	119,123
Republic of Korea	96,937	100,800	104,763	109,872	117,929	127,054	126,822	127,832	121,023	116,942
France	46,573	49,154	53,352	54,611	46,076	52,946	57,174	60,135	63,713	84,059
Saudi Arabia	11,704	12,398	13,765	20,123	24,828	31,366	42,651	51,679	63,833	73,548
United States of America	50,629	52,699	54,419	57,455	55,294	57,017	57,506	59,759	60,297	60,292
Malaysia	48,294	47,395	49,000	52,654	56,285	59,892	59,539	59,855	58,485	56,260
Viet Nam	17,030	20,801	23,330	28,012	36,514	43,945	47,268	52,029	53,004	53,546
Nigeria	23,476	26,906	27,969	30,185	34,105	42,535	45,062	49,492	49,568	52,066

Note. Adapted from UNESCO (2015). Copyright 2016 by UNESCO.

Table 20.3 Equivalent full-time student load (EFSL) for commencing and all overseas students by higher education institution and gender, 2014

	Commencing Enrollment			Total Enrollment		
	Males	Females	EFSL	Males	Females	EFSL
RMIT University	3,926	3,702	7,629	10,270	9,772	20,043
Monash University	3,587	4,272	7,859	9,355	10,278	19,634
University of Melbourne	2,533	3,197	5,729	5,835	7,365	13,200
Curtin University of Technology	2,275	2,137	4,412	6,041	5,505	11,546
University of Sydney	2,120	2,980	5,100	4,646	6,332	10,978
University of New South Wales	2,384	2,293	4,677	5,771	4,834	10,605
University of Queensland	1,779	2,359	4,138	4,417	5,386	9,802
Non-University Higher Education Institutions	3,507	2,163	5,670	5,977	3,619	9,596
University of Wollongong	2,048	1,470	3,519	5,343	3,867	9,210
Non-University Higher Education Institutions	2,785	2,424	5,210	4,878	4,203	9,081

Note. Adapted from Australian Government (2014). Copyright 2016 by Department of Education and Training. EFSL = Equivalent full-time student load (EFSL) is a measure of a student's study load. An EFSL of 1.0 is the standard annual study load of a student undertaking a program on a full-time basis. Equivalent full-time student load (EFSL) is a measure of a student's study load. An EFSL of 1.0 is the standard annual study load of a student undertaking a program on a full-time basis.

Table 20.4 Overseas students by country of birth and gender, 2014

Country of Birth	Male	Female	Total
China (excludes SARs and Taiwan)	45,825	54,175	100,000
Malaysia	14,525	14,543	29,068
Singapore	13,073	15,194	28,267
India	18,143	7,875	26,018
Viet Nam	8,809	10,052	18,861
Hong Kong (SAR of China)	6,727	5,754	12,481
Indonesia	5,992	5,762	11,754
Nepal	5,272	3,310	8,582
Republic of South Korea	3,255	3,253	6,508
United States	2,528	3,926	6,454
Total	180,602	166,958	347,560
Total 2013	168,089	160,313	328,402
% change from 2013	7.4	4.1	5.8

Note. SAR = special administrative area. Adapted from Australian Government (2014). Copyright 2016 by Department of Education and Training.

the Erasmus Program and the Bologna Process in higher education. The European Commission regards such mobility as one way to integrate the European region. Australia is distinctively tied to Asia, but the Asian Region has no imperatives similar to Europe.

Although certain policy bodies and institutes mention the quality of the education offered and the advantages to the parties involved, the primary concern is invariably the trade in education services, competition, and market share. Typically the OECD points to the individual, social, and cultural benefits but appears to give greater priority to economics. It says: "Studying abroad has considerable economic impact. For host countries enrolling international students can not only help raise revenues for higher education, but also can be part of broader strategy to recruit highly skilled immigrants" (OECD, 2014, p. 32). The following remark from Universities Australia (2014) is also typical.

> An international education contributes to the development of global citizens, regional leaders and strong global relationships that are so critical for facilitating diplomacy, business and trade. This is the case whether it is overseas students studying in Australia or Australian students completing at least part of their degree overseas. There is great potential to further expand the international reach of the Australian university sector. With global demand for higher education set to grow significantly in coming decades, even maintaining or marginally increasing Australia's market share could deliver major export growth.[1]

Such policy-oriented comments about the benefits rarely involve a close-up examination of what actually happens to students on the ground. I will return to this matter. But despite such intense coverage, insufficient attention is drawn to the asymmetrical patterns of power and wealth connected with such student mobility. In contrast, UNESCO's 2012 Global Flow of Tertiary-Level Students interactive map enables such an understanding of student flows in relation to the rich and poor nations of the global South and North, West and East. For example, it points out that

> the Arab states of Egypt and the United Arab Emirates (Dubai) are popular destinations for high-level studies for Arab students, and South Africa received 17% of mobile students from Sub-Saharan Africa in 2010.
>
> France remains the top destination for Arab and Sub-Saharan African students, receiving 29% and 19% of these students respectively. Germany and Russia are the top destinations for students from Central and Eastern Europe, and Central Asia, receiving 16% and 46% of these regions' students respectively.
>
> The US is the top destination for East Asia and the Pacific, South and West Asia, and Latin America and the Caribbean, receiving 28%, 38% and 33% respectively of their mobile students. And North America and Western Europe's top destination is the UK at 23%. (UNESCO, 2015)

Such figures indicate that these flows of students are primarily from the "third" world to the "first" or, what is better referred to as the *majority* and *minority* worlds. But interestingly and most importantly for this chapter, these student flows are often either from formerly colonized countries to former colonizing countries, or to the globally imperialist power of the United States with its Cold War history in Asia, and to these countries' satellites. As Table 20.4 shows for Australia, for example, nine of the top ten sending countries are from different parts of Asia.

As some governments in the magnet countries reduce their financial support for university education, the global market in tertiary education has stepped into the

[1] https://www.universitiesaustralia.edu.au/global-engagement.

vacuum. Mobile students' fees now increasingly underwrite the survival of their university sector and thus students from poorer countries help to underwrite university education systems in the richer countries. This is a contemporary echo of colonialism.

This echo resonates even more strongly when the flows of knowledge are taken into account. These too are usually from the global North to the global South (e.g., Chakrabarty, 2000; Connell, 2007). Plainly, the current global university system has its own "geographies of centrality and marginality" (Sassen, 1998, p. XXIV), its own integrating and fragmenting tendencies.

Transnational education complicates such geographies. Offshore campuses allow students to study "abroad" while remaining in their home countries. Mixed modes of study, which combine online and face-to-face teaching, may involve no, or little, physical presence on the campus of origin. This virtual and embodied mobility of people (students, staff) and knowledge (curriculum and research) is accompanied by other mobility—the mobility of ideologies, images, and imaginations; of finance, feelings, and fantasies. Such mobilities have been facilitated by the space-and-time-altering technologies of cheap transport and instantaneous information communication technologies. Hence university relationships that would otherwise have been stretched out over extended literal space have been both compressed and reconfigured with great consequence for the geography of the university.

The Unequal Geographies of Traveling Ideas

The following case is but one example of the complicated intersections of roots and routes in the contemporary university. It shows how they are entangled with dominant, residual, and emergent elements in relation to various spatial entities.

I begin with a book as it traveled to the hands, hearts, and minds of a group of doctoral students at Monash University, Australia. As Table 20.3 indicates, Monash had the second highest number of international students of any Australian university, with enrollment of 19,634 in 2014. All the doctoral students involved in the project were from Asia and as a result of this book sought to decolonize knowledge and themselves. In telling their story I foreground the dominant, residual, and emergent power, knowledge, and emotional geographies involved. While the notion of emotional geographies is widely evoked in geographical studies, the emotional geographies of education, particularly of universities, have attracted much less attention (Kenway & Youdell, 2011a, b)[2].

The book is called *Asia as Method: Toward Deimperialization* (Chen, 2010). It was written by Kuan-Hsing Chen, a well recognized cultural studies scholar. I found it in Singapore, and given that Australia generally, and Monash specifically, draws

[2]Our edited collection (Kenway & Youdell, 2011) is dedicated to the emotional geographies of various educational settings.

most international students from Asia, I took it back to Monash's Faculty of Education for a graduate student reading group. It provoked a great deal of interest from the students. This resulted in a multicountry reading, discussion and reflection, writing, and public presentations project that lasted several years (2010–2013) and eventually led to the edited book *Asia as Method in Education Studies* (Zhang, Wing Keung Chan, & Kenway, 2015), in which the students wrote chapters and, after each, shared their reflections on their experiences of trying to decolonize knowledge.

In 2010, when the project started, in total 34 people were involved; 7 academics and 20 PhD students. Also there were 10 observers who did not join the project but attended some seminars and workshops. Those involved were from 13 countries—Australia, Bangladesh, China, Indonesia, Iran, Malaysia, Mongolia, Pakistan, New Zealand, Saudi Arabs, Sri Lanka, Thailand, and Vietnam. Many different religions and native languages were represented in the group, with the students coming from different social class backgrounds. Twenty-two students were involved in the final book. They came from Australia, Bangladesh, China (including mainland and Hong Kong), Indonesia, Japan, and Vietnam.

Collectively the project involved people connected to 13 other universities. Many were staff on leave undertaking their PhDs who planned to return to their universities once they had completed their doctorates. These universities were the University of Hong Kong, Monash University, Swinburne University of Technology, Australia, RMIT University Vietnam, Universidad de Valparaíso Chile, Sanata Dharma University Indonesia, Indonesia University of Education Indonesia, State Polytechnic of Malang Indonesia, Hanoi National University of Education Vietnam, Hanoi University of Languages and International Studies Vietnam, University of Dhaka Bangladesh, BRAC University Bangladesh, and the University of Western Sydney, Australia.

These details alone indicate some ways Monash, at least, is a place of flows; one at which different people's Asian routes intersect. But more interestingly it is also a place where geographies of knowledge, power, and emotion are entangled with regard to Asia. Chen's book starts to explain the geopolitics involved. The West, he says, has understood and portrayed itself, to itself and to "the rest," as "the best." It has presented itself as the most modern, mature, developed, progressive, scientific, rational, enlightened, and civilized. Such views traveled to colonized countries and peoples and impacted how they felt about themselves as well as about their colonizers. Consequently, knowledge production in Asia, Chen (2010) argues, is constantly either positively or negatively referencing the West. It has been an opposing entity, a system of reference, an object from which to learn, a point of measurement, a goal to catch up with, an intimate enemy, and sometimes an alibi for serious discussion and action (p. 216). He makes similar arguments about the conceptual frameworks that arose out of the Cold War and are connected to North American imperialism in Asia.

Previous and current generations of postcolonial scholars have, of course, been critical of this. Chen identifies various types of postcolonial response. For example, one body of such scholarship identifies and challenges Eurocentric and/or North America-centric thinking, by, among other things, highlighting the binary and hierarchical logics deployed when the West constructs its Others. These logics, the argument runs, create global hierarchies of knowledge and modernity. They also involve implicit claims that "Western" knowledge is universal, when in fact it is provincial or regional. Such postcolonial critiques ask, rhetorically, whether the analytical frameworks that have been developed in Europe or North America are suitable for understanding Asia. And they are highly critical of the ready absorption of such frameworks into research about, on, and in Asia.

According to Chen, new scholarship in Asia needs to be in a dialogue with such postcolonial responses to the West. He also insists, however, that from an "Asia as Method" perspective, certain postcolonial studies should stop the West as their primary referent (2010, p. 222). Otherwise, he says, they may contribute to the use of new unhelpful binary reversals, the "East is best" for example. Problems also occur around three other approaches to knowledge that may arise in newly independent nations in their attempts to shake off the shackles of the colonizer. Chen names these as nationalism, nativism (notions of ancient wisdoms, for example), and civilizationism (competing claims about whose civilization is the more civilized).

Overall, Chen pleads for habitual practices of thought to change. He argues for "alternative mappings" that move "the point of reference" (2010, p. 212). An essential ingredient is to shift this point of reference toward Asia and other third world locations—to those places with greater resemblances to each other, to those that share and address similar problems. This point leads him to emphasize Asian studies in Asia, as opposed to Asian studies through European and North American theorizing about Asia. Such theories do not usually, he claims, help people in Asia to properly understand or address local conditions or issues.

He proposes, instead, "international localism" (p. 223). This is local, but also transborder and regional. The main point though is that it involves "inter-referencing" (p. 223). Here Asian countries become each other's reference point; they provide each other with new opportunities for comparison other than the constant comparison with the West. The purpose of the inter-referencing approach is to avoid judging any country, region, or culture as superior or inferior to any other and to tease out historical transformations within the *base entity* of each country so that the differences can be properly explained (p. 250). New intellectual alliances and solidarities need to be built, Chen claims.

In terms of emotional geographies and their implications for knowledge, Chen focuses on three things: structures of sentiment, the psyche, and subjectivity (p. 74). He draws from postcolonial psychoanalytic theory to help explain various paradoxes. First, Chen explains the psychology of colonization from the perspective of the colonizer. For instance, this way of thinking includes the practice of regarding people in the colonies as at an immature stage of development and, thus, as needing guidance from the colonizing country to enable them to reach maturity (p. 74). Secondly, against such deficit thinking alternative postcolonial analytics arose and existing psychoanalytic theory was rewritten in support of decolonizing movements. These analyses were regarded as a "weapon in the anti-colonial struggle" (p. 74). Such postcolonial theorists sought to analyze the psychology of the colonized and pointed to such things as colonial identification, the unconscious yearning (desire) for the colonizer's recognition, and the accompanying psychic suffering and resentment (p. 78). They focused, too, on the manner in which colonizer and colonized constitute each other's subjectivity. And, they examined the psychology of decolonization and national independence, pointing to the unfortunate continuation of earlier subjectivities (pp. 78–79).

One methodological implication of this line of analysis is that the researcher's subjectivity and identifications must come under scrutiny. Not least, the research community's "structural flow of desire" (Chen, 2010, p. 225) towards the United States and Europe needs to be challenged. Chen calls for a ruthless self-questioning, a process of self-critique, self-negation, and self-rediscovery (p. 3). Alternative cultural practices of identification are required. In the first instance, this involves acknowledging that the psyche may still be framed in such deleterious ways and that moving beyond the limits of such framings is crucial.

Dominant and Emergent Institutional and National Roots

Chen's ideas traveled to Australia though the work of the students. I explain how and also deploy his ideas to analyze the students' responses, and in so doing ask if, and how, his work on the geopolitics of knowledge spoke to the students and their situation. But also, as I have indicated, in the contemporary university global routes invariably meet institutional and national roots. So I also explore the character of the dominant, residual, and emergent roots the students encountered and their implications for these mobile PhD students and moving ideas.

First, to a consideration of some dominant roots: Universities potentially constrain international students' engagements in the geopolitics of knowledge through time, money, and language. These students paid exorbitant fees, and felt an intense pressure to complete on time. Additional time pressure was often due to scholarship and visa requirements. For those seeking work and/or permanent resident status in Australia more intensity arose. The requirement to write and talk in the English language was very demanding and required high levels of linguistic and cultural translation. While these did not prevent students' engagement with Chen's ideas, they did impinge on it.

The manner in which these students' identities and research imaginations were constructed at Monash, during their time there, was deeply rooted in a residual colonial imagination. Practices of infantalization and deficit models were subtly present. First, the knowledge the students brought with them was often downplayed or ignored, thus involving a denial of their full selfhood. Secondly, they were not usually encouraged to use research sources from their own countries or region, especially those not written in English. Thirdly, they were usually steered toward the knowledge and theory that their supervisors were most familiar with. This was almost inevitably knowledge from what might be called empires of English-language-knowledge—particularly the United States or the United Kingdom. Such knowledge has long been sanctified in this education faculty.

The students brought with them national attachments that, in some instances, intensified during their overseas studies as they struggled for self-differentiation in the Faculty (Chen, 2010, p. 87). They did not necessarily have an unconscious yearning (desire) for the *knowledge colonizer's* recognition as might their forebears. But in some ways the academic culture into which they came required of them some forms of self-negation and mimicry. To adopt this pose become a strategic choice of survival and progress for some.

Postcolonial psychoanalytic theory makes it clear that imperial centers and colonies were mutually constituted. Each shaped the other. For the colony to be constituted as inferior, the center has to constitute itself as superior. Such sentiments linger. Plainly, the students and the institution constituted each other's subjectivity in asymmetrical ways.

Chen draws attention to the importance of *deimperializing knowledge* and this call is directed toward those at the centers of empires of knowledge. These are the peoples who most benefit from Eurocentric and North America-centric epistemologies and ontologies. Thus an important ingredient of Asia as Method is the deimperialization of knowledge by those who are or have been imperial powers in different parts of Asia. This includes England, France, the Netherlands, Portugal, Spain, and the United States. It also includes those countries within Asia that have been colonizers as well as colonized. Chen (2010) says that deimperializing research starts with rethinking the wrongs and pains of past imperialist interventions.

> This must be performed by the colonizer first and then on the colonizers' relation with its former colonies. The task is for the colonizing or imperializing population to examine the conduct, motives, desires and consequences of imperialist history that has formed its own subjectivity. (p. 4)

The students were mindful of Chen's arguments about the importance of deimperializing knowledge. And, in terms of education studies they found succor in Lin's (2012) assertion that "Knowledge production (e.g., textbook/curriculum production, university research and publication, teacher preparation) and knowledge circulation (e.g. schooling, curriculum, and pedagogy) constitute the major sites in which imperialism operates and exercises its power" (p. 164).

Clearly, according to this logic, those living and working at the centers of various empires of knowledge must move beyond what Hall calls "the West as Method" (1992). This meant that the students' non-Asian academic colleagues in the education faculty needed to engage with the sorts of issues they were raising and attend to the emergent knowledge they were working with. The students did not subscribe to the hard-to-justify essentialism that leads to the view that only Asians can do Asia as Method and that white, non-Asian scholars should be excluded. Their view was that such scholars should be seriously involved in order to deimperialize their own supervision practices as well as their own scholarly enquiries. They thus invited others in the faculty to join this project.

But within the wider faculty little such deimperializaton occurred. The dominant roots mentioned above were too deeply entrenched. Few members of academic staff became involved and explored, with students, the implications of Asia as Method for curriculum and educational inquiry and practice, let alone for their post graduate pedagogies. More widely, at conferences and seminars it was mostly Asian staff and students who attended students' papers and symposia. The education main stream seemed to remain pretty much untouched and unchanged.

That said, emergent potentially decolonizing knowledges were available. Some PhD supervisors encouraged their students to engage with postcolonial theory, with debates about the geopolitics of globalization and the implications for education. Some also encouraged them to draw on the research and theorizing of scholars from their home country and region. However that was relatively rare.

Of course all international students are subject to their host country's policies and practices with regard to them. Australia has been in the "industry" since the 1980s, so these policy roots are well established and overriding. Australia's higher education system is dependent on international students from Asia, so it has high policy priority. But also emergent elements in relation to Asia came from Australia's national polity during the time of the project, ones that might have been expected to elevate the students' institutional status.

These involved the Australian national government's attempts to reimagine its relationships with Asia. At the time, Asia was strongly on the agenda in Australian political circles and Australia was thinking hard about its place in the Asian region. Australia is geographically located *in* Asia but it is not *of* Asia. It is a wealthy, "first" world country surrounded by many countries that are considered "third world." It has strong historical links with Britain and the Commonwealth of Nations but its political and military alignments are, primarily, with the United States and Europe. That said, Australia is, as Takayama (2016, p. 70) points out, a "rich peripheral country" plagued by a sense of ambivalence about its own identity internally and on the world and regional stages (see, further, Kenway & Fahey, 2011).

Australia has strong trade links with South Korea, Japan, and, particularly, China and India. And, it is because of the increasing dependency of Australia on these trade links that it has become much more conscious of its need to better attend to its Asian neighbors. This impulse is signaled in the policy document *Australia in the Asian Century* (Australian Government, 2012) that says:

> The Asian century is an Australian opportunity. As the global centre of gravity shifts to our region, the tyranny of distance is being replaced by the prospects of proximity. Australia is located in the right place at the right time—in the Asian region in the Asian century. (p. 1)

And, in terms of school education, the Australian Curriculum named "Asia and Australia's engagement with Asia" as one of only three cross-curriculum priorities (Australian Curriculum Assessment and Reporting Authority, 2014). It seeks to ensure that Australian students learn about and recognize the diversity within and among the countries in the Asian region. It, also, aims to develop students' knowledge and understanding of Asian societies, cultures, beliefs, and environments and the connections between the peoples of Asia, Australia, and the rest of the world.

Chen notes that capital's globalization has led to "economic and cultural regionalization" and along with this has come "the rise of Asia as a pervasive structure of sentiment" (2010, p. 214). It might be added however, as Rizvi (2012) notes, the not-too-subtle subtext to such policy mantra is usually instrumental and economically opportunistic. Understanding Asia in the manner Chen proposed, as a shifting historical, political, cultural, as well as economic entity, was not what this was about.

But even though Asia was strongly on the agenda in Australian political and national curriculum circles at the time and even though the time seemed ripe for deimperializing knowledge in the faculty, the institutional roots attached to empires of English-language-knowledge were too deep.

Routes that Settle and Unsettle

The graduate students traveled to Australia with much highly contradictory, even split, emotional baggage. This was not just about the "flow of desire" to the West, according to Asia as Method; although as one of my former graduate students made clear in her doctoral thesis (Nguyen, 2013), students in Asia are sold "Western dreams" via the intense marketing practices of Australian universities in Asia.

Most of the graduate students from Asia tended to focus their research on educational issues in their home country, yet were usually encouraged to draw on Western perspectives to study their own country's educational practices and systems. They often, therefore, utilized Western concepts and theories to interpret the issues involved. The emotional geographies involved were complicated.

Some appeared to readily accept this situation and were eager for knowledge from the West. Others were uneasy. Of the uneasy some saw this as under-valuing Asian intellectual traditions and practices and as unsuitable for Asian circumstances. Their typical counter-response was to turn to their local knowledge to interpret local educational issues in their home country. My conversations with them concerned the dilemmas they faced about the relationships between knowledge and location, universalism and specificity. Yet others found it quite difficult to specify why they felt uncomfortable—they just did.

The students' Asian-ness was affirmed through the project. International students who come from "other/othered" countries often develop affinities across nation states that they may not otherwise feel. And the project contributed to such new communities of affect, to new solidarities. The following remarks from our group's final reflective session illustrate the point.

> For me, the way I perceive myself in a different context is slightly different. I don't perceive myself as Vietnamese, but I perceive myself in the broader spectrum with other Asians. When I was in the United States, I didn't think of myself as Vietnamese, I thought of myself as Asian. So when I see a Chinese, I feel close to the Chinese person. When I'm in Vietnam, if a Chinese person came and visited, I would think, "He's so different."
>
> When you're somewhere else, you tend to draw on who you're not. So for example, back home, I never think of myself as Indonesian-Muslim. That's something you take for granted. But here, it's like the religion is being foregrounded and my skin colour as well.
>
> As an international student away from home . . . you may not find a lot of international students from the same country in the same subject, in the same faculty, so you just think, "It's a student from an Asian country, we can form a community so we support each other.

But the project also exemplified some of the knowledge problems that Chen identifies. Early on, the students organized themselves into small subgroups to develop papers for a half-day faculty seminar. The groups were all nation-state based. They later presented papers at national peer-refereed conferences. The feedback required them to deepen their understandings of Chen and they then began to undertake the ruthless self-questioning, and process of self-critique, self-negation, and self-rediscovery that Chen calls for. This was necessary but difficult intellectual and psychic work. The reasons were manifold.

As time went on, there was some worry around the politics of the project. Various concerns surfaced about the publication and what it might mean politically when students returned home. It is not uncommon for students who become politicized in certain ways in the West to face the prospect that such politics may not be welcome at home. But these anxieties were not just about the possible problems of return.

They were also concerned about disciplinary territories and borders, about, for example, whether ideas from cultural studies should travel into education studies—were they too political, were they useful for teachers, they asked. For such students the project unsettled firmly installed and held disciplinary identities. They sought to reconfigure their own intellectual horizons and to try to reimagine the discipline of education and its methodologies. They came to acknowledge, for instance, that the discipline of education is not simply apolitical and practical. But they also raised for discussion the many problems of translation across disciplines and political standpoints.

Of knowledge and decolonization the project did not do as much as Chen would have liked. The papers for the book did not do the necessary "inter-referencing" (p. 223) that Chen (2010) calls for. But some spatial unsettling occurred. At the project's start, the students were quite ignorant about each other's countries. But they developed a curiosity where little was there before. They wanted to know more. This shifted their points of reference to some extent and, in this sense, the project was a space of emergence wherein routes challenged roots.

The project raised question about the geopolitics of students own knowledge horizons. It led them to ask why they knew so much about some countries in Asia and so little about others or why was it that some issues had prominence and others did not? Why, for example, was religion always in the forefront with regard to Indonesia? This led some to the view that Chen did not actually go far enough with regard to certain issues, such as the power of religion and its implications for knowledge. As Roshid, Siddique, Sarkar, Mojumder, and Begum (2015) said in the reflective section of their chapter:

> Chen emphasizes freeing our minds from the influence of colonization, imperialization, and Cold War as they are deeply rooted in many Asian contexts, eventually posing challenges for Asian scholars when they move forward in knowledge generation. In the context of Bangladesh, we would argue for considering other aspects to liberate our minds and move forward. In particular, the influence of religious ideology plays a vital part in shaping people's mindset. For example, a considerable part of the population neither accepts Western knowledge nor tries to generate local knowledge. Instead, it relies on Islamic knowledge and continues a solely Islam-based education system in Bangladesh, adding another dimension to Chen's critique. (2010, pp. 141–142)

All that said, throughout the students held strong, spatially organized knowledge disposition and in many ways continued to think through the prism of the nation state. Chen claims that "if critiques remain within the limits of the nationalist framework, it will not be possible to work towards regional reconciliation" (2010, p. 159). As indicated, he points to the importance of affect (feeling) when it comes to the national, the colonial, and the postcolonial and to the fact that the affective investments and attachments of all three need to be acknowledged.

To understand the moods suffusing a nation-state, it is necessary to have a comprehension of its global situated-ness currently and over time. For instance, being positioned as tangential in global power formations has the potential to negatively impact the national self-image and mood. So too can the historical conditions of a nation-state, particularly if it has been on the receiving end of colonization and imperialism. These links became manifest in the collective national character and consciousness and in the national unconscious and can result in what is sometimes called a "colonial mentality," which may also involve a *dependant culture*. The emotional archive of the nation-state then, includes a collection of feelings and susceptibilities stored over time in both the body politic and the national culture. It includes the nation-state's underlying desires and defenses.

There were many barriers to undertaking the type of transnational imagining proposed by Chen. But nationalism was a major barrier. Nationalism and its affective registers remained an issue as our reflection sessions made clear. As one student explained, when inviting us to talk about

> the nation as the basic defining unit. How come we know that it's problematic, but we always come back to it? How do we escape from this idea? We know it's very problematic and lots of stereotyping and prejudice stems from it, essentialism and all such negative connotations are attached to it. But it's just hard to escape from the notion of taking national entity as the basic defining unit. How do we deal with this?

Making one of their aspirations clear, it was proclaimed, "We need to change this. We can't move on if we don't, if we keep doing things with the same mindset as we had yesterday". But changing was difficult. The politics of knowledge is transnational, regional, national, and subnational. These politics intersect and are hard to untangle.

Students recounted stories, not just of Western colonialism and American imperialism, but also of wars and international tensions within Asia itself. Some arose from a history of war and intra-Asian colonialism. Within Asia, the major powers of China and Japan have been involved in various colonizing projects. Even less powerful countries have had their own imperial fantasies, seeking to move from being colonized to become neo-colonialists. Such national tensions led to some strains within the group. Raw emotions were exposed. For example, we discussed the tense relationships within Indonesia in relation to West Papua. Chen's idea that we could rise above such rifts though reflections, conversations, and dialogues was seen as, simply, too utopian. Even so, there were glimmers here of what Chen calls "Asian studies in Asia" (2010, p. 2).

We also pondered those rivalries between Asians that do not seem to be connected to the matters Chen raised. "Why is it that Asians felt more threatened by other Asians, than by Western influences?" one person wondered. Another put is more colorfully, "Why are we so allergic to other Asians?" saying "When I think back to when I was a student, as a Vietnamese scholar, I didn't quote Chinese scholars. And Chinese scholars, in the same class, didn't quote Japanese. We all quoted Western people". For example, the question was raised: Why are mainland Chinese regarded by Singaporeans as more of a danger than the West?

> I think we are threatened by fellow Asians more than by the West. I just came back from Singapore and that sentiment is everywhere in Singapore. . . . European culture pervades every aspect of culture there, theatre, architecture, everyday conversation, Starbucks and everything, and people just don't care. All of a sudden, a few Chinese brands came in, "What is this brand? I don't like that.

But the tensions can also be on a much more personal scale, as the following comment indicates:

> Suppose I work with you for a long time and I listen to you a lot, and I have no problem. . . . And one day Philip is better than me, he's published more, he gets a rise in salary. I'm threatened. Why? . . . I feel threatened because he's so similar to me, he's yellow, he's male, he's studying in the West, he's so similar. Suddenly, one day he's being valued by the system and I feel threatened.

Interestingly it was the emotional geographies involved that led to these various practices of "inter-referencing" and "international localism" in Chen's terms (2010, p. 223).

Concluding Thoughts

Now back to the contemporary university as a place of roots and routes. Admittedly this chapter has a particularity to it, as it's only one small project in one university, in a rich country that is a British and U.S. satellite. But it offers an example of the manner in which the flows of people, knowledge, and emotion intersect. It shows how such flows can come up against the dominant roots of the grounded university itself and those nation-bound roots that exist in the students' psyches. And it illustrates how power and knowledge relationships can be implicated.

More abstractly, I have suggested some benefits of bringing Chen's theorizing to studies of global flows in the contemporary university in Australia in relation to its mobile students from Asia. And I surmise that his approach could be usefully deployed in other locations in Asia and in the universities around the world to which Asian students travel. Chen not only highlights the links between colonial, postcolonial, and imperialist history and geography, power, and knowledge, but also claims that that these links *live on* in certain ways. I have offered an example of how they live on not just in the directions of student movement but also in the direction of knowledge flows.

Chen's framework, with the assistance of Williams's (1977) cultural categories, provide a way of interpreting the knowledge politics of the territorial university into which Asian students travel and within which they temporarily settle. It helps to explain why they are implicitly constructed as deficit by a "Western" university. I have intimated that this is probably because a residual element in such universities involves colonial and racist views of students from Asia. And, in turn, this helps to explain at least some of the barriers that arose in relation to the emergent knowledges, which the students sought to engage. Along with the time, money, and language issues they and many international students face, such elements make the deimperialization of knowledge difficult.

Notions about the mutual constitution of colonizer and colonized are also helpful in unpacking the relationships between students from former colonies in countries that have a history of colonization, even if by association, as in the case of Australia, a former colony of the British Empire, but one with an enduring sense of its British connections. Through this notion it is possible to see the ways in which the mobile students' "desire" for a Western education and recognition potentially confirms the knowledge rooted in the dominant power relations of the university and the global university system. Equally, the argument that certain versions of postcolonialism may inadvertently confirm the colonial script provides a further explanation of the knowledge problems students might face. This happens if they try to challenge the knowledge politics of their host university by the most obvious route of binary reversal—the "East" is best.

But this study also points to some limitations of Chen's mode of analysis. One example is its restricted reading of the emotional geographies of the encounter, along the resentment or desire spectrum, and in relation to recognition though mimicry. As these students unpacked their emotional baggage it became clear that their emotional investments were more nuanced and also arose from such things as disciplinary territoriality, a rather defensive nationalism, and an emergent comparative national, not so much international, sensibility. This defensive nationalism was aimed less against the so-called West than at other forces in Asia itself. Such nationalism is a force for stasis in this space of flows, one that makes the forging of new intellectual alliances and solidarities through an Asia as Method approach improbable, but not impossible.

Finally, to suggest, as I have, that the contemporary university is unbound is not to suggest it is without roots. As indicated, in this one university at least, dominant and residual knowledge can be so deeply rooted in the institution that it is inhospitable to the emergent knowledges that flow in—particularly those knowledges that are subversive and oppositional. The existing *power geometries* can actually harden even when the university and the country are involved in a range of activities designed to attract and cater to international students. In short, the unbound university may also, in certain ways, be rather root-bound.

References

Australian Curriculum Assessment and Reporting Authority. (2014). *The Australian curriculum.* (ver. 7.5). Retrieved from www.australiancurriculum.edu.au/

Australian Government. (2012). *Australia in the Asian century: White paper*. Canberra: Australian Government Publishing Service.

Australian Government. (2014). Selected higher education statistics—2014 student data. In *Australian government: Department of education and training*. Retrieved from https://education.gov.au/selected-higher-education-statistics-2014-student-data

Australian Government Department of Education and Training. (n.d.). *International student data.* In Data and Research (section 5). Retrieved from https://internationaleducation.gov.au/research/International-Student-Data/Pages/default.aspx

Bondi, L., Davidson, J., & Smith, M. (2005). Introduction: Geography's 'emotional turn'. In J. Davidson, M. Smith, & L. Bondi (Eds.), *Emotional geographies* (pp. 1–19). Ashgate: Aldershot.

Chakrabarty, D. (2000). *Provincializing Europe: Postcolonial thought and historical difference.* Princeton: Princeton University Press.

Chen, K.-H. (2010). *Asia as method: Toward deimperialization*. Durham: Duke University Press.

Clifford, J. (1997). *Routes: Travel and translation in the late twentieth century*. Cambridge, MA: Harvard University Press.

Connell, R. W. (2007). *Southern theory: The global dynamics of knowledge in the social sciences.* Sydney: Allen and Unwin.

Hall, S. (1992). The West and the rest: Discourse and power. In S. Hall & B. Gieben (Eds.), *Formations of modernity* (pp. 275–332). Cambridge, UK: Polity.

Kenway, J., & Fahey, J. (2011). Getting emotional about 'brain mobility'. *Emotion, Space and Society, 4*, 187–194. doi:https://doi.org/10.1016/j.emospa.2010.07.003

Kenway, J., & Youdell, D. (2011). The emotional geographies of education: Beginning a conversation. Emotional geographies of education [special issue]. *Emotion, space and society, 4*, 131–136. doi:https://doi.org/10.1016/j.emospa.2011.07.001

Kenway, J., & Youdell, D. (Eds.) (2011b). The emotional geographies of education [Special issue]. *Emotion, space and society, 4.*

Lin, A. M. Y. (2012). Towards transformation of knowledge and subjectivity in curriculum inquiry: Insights from Chen Kuan-Hsing's "Asia as Method". *Curriculum Inquiry, 42*, 153–178. doi: https://doi.org/10.1111/j.1467-873X.2011.00571.x

Massey, D. (1993). Power-geometry and a progressive sense of place. In J. Bird, B. Curtis, G. Robertson, & L. Tricker (Eds.), *Mapping the futures: Local culture, global change* (pp. 59–69). London: Routledge.

Meusburger, P. (2015). Education, geography of. In J. D. Wright (Ed.), *International encyclopedia of the social & behavioral sciences* (2nd ed., pp. 165–171). Amsterdam: Elsevier.

Nguyen, T. (2013). *Selling western dreams: Australian transnational education in Vietnam and the formation of student identities*. Australia: Monash University.

OECD (Organisation for Economic Co-operation and Development) (2014). *Education at a glance 2014: Highlights*. Paris: OECD. doi:10.1787/eag_highlights-2014-en

Project Atlas: Trends and Global Data 2014. Retrieved from https://www.iie.org/Research-and-Insights/Project-Atlas.

Rizvi, F. (2012). Engaging with the Asian century. *ACCESS: Critical Perspectives on Communication, Cultural & Policy Studies, 31*, 73–79.

Roshid, M. M., Siddique, M. N. A., Sarkar, M., Mojumder, F. A., & Begum, H. A. (2015). Doing educational research in Bangladesh: Challenges in applying Western research methodology. In H. Zhang, P. Wing Keung Chan, & J. Kenway (Eds.), *Asia as method in education studies: A defiant research imagination* (pp. 129–143). Abingdon: Routledge.

Sassen, S. (1998). *Globalization and its discontents: Essays on the new mobility of people and money*. New York: The New Press.

Takayama, K. (2016). Deploying the post-colonial predicaments of researching on/with 'Asia' in education: A standpoint from a rich peripheral country. *Discourse: Studies in the Cultural Politics of Education, 37,* 70–88.

UNESCO, Institute of Statistics (2015). *Global flow of tertiary-level students* (interactive map). Retrieved from http://www.uis.unesco.org/EDUCATION/Pages/international-student-flow-viz.aspx

Universities Australia (2014, June 16). International links data. In *Global engagement*. Retrieved from Offshore Programs and International Links https://www.universitiesaustralia.edu.au/global-engagement

Williams, R. (1977). *Marxism and literature*. Oxford, UK: Oxford University Press.

Zhang, H., Wing Keung Chan, P., & Kenway, J. (Eds.). (2015). *Asia as method in education studies: Towards a defiant research imagination*. Abingdon: Routledge.

Chapter 21
International Education Hubs

Jane Knight

The forces and opportunities of globalization have dramatically impacted higher education, especially cross-border education. The term *cross-border education* is often used interchangeably with the terms *transnational education, borderless education,* and *international academic mobility*. Cross-border education is the preferred term for this chapter and refers to the mobility of people, programs, providers, projects, and policies between and among countries. Studies of higher education shows that international academic mobility is fundamental to the mission of universities (Altbach, 2013). Scholars and knowledge have been moving around the world for centuries. The fact that the notion of universe is the root concept for *university* is strong evidence of the internationality of higher education.

There is no question that the landscape of cross-border higher education has changed significantly in the past three decades. It is no longer just students and scholars who are moving to other countries for education opportunities. Academic programs, education institutions, and new providers are moving across borders to deliver education and training programs in foreign countries. New actors, new international partnerships, new binational universities, and new modes of program delivery characterize the dynamic and expanding area of cross-border higher education.

International education hubs are the latest development. Labeled the third generation of cross-border education, they build on the first generation of student mobility and the second generation of program and provider mobility. Education hubs can be at the country, zone, or city level and involve a critical mass of and collaboration between international-local universities, students, research institutes, and private

J. Knight (✉)
Department of Leadership, Higher and Adult Education, OISE, University of Toronto, Toronto, Ontario, Canada

Ali Mazrui Centre for Higher Education Studies, University of Johannesburg, Johannesburg, South Africa

P. Meusburger et al. (eds.), *Geographies of the University*, Knowledge and Space 12,
https://doi.org/10.1007/978-3-319-75593-9_21

industry. This chapter will focus on the country level hubs, of which there are six in the world—Botswana, Hong Kong, Malaysia, Singapore, Qatar, and United Arab Emirates (Knight, 2014b).

Important to recognize is that an education hub reflects a country's plan and priority to serve and be recognized as a center of education expertise, excellence, and economic activity in the region and beyond. Education hub countries have different objectives and characteristics, but in general the term education hub is used by countries seeking to position themselves as centers for student recruitment, education and training, research, and innovation. A variety of factors are driving these efforts and include modernizing the domestic tertiary education system, generating income, creating a skilled work force, remaining or becoming economically competitive, developing a regional profile, promoting knowledge diplomacy, and transitioning to a knowledge- and service-based economy (Knight, 2011).

The purpose of this chapter is to examine the phenomenon of education hubs within the context of three generations of cross-border education[1]. Framing education hubs as the third generation makes it possible to analyze the evolution of education hubs in respect to both the growing numbers of students moving to another country for their academic studies and the more recent boom in the number and types of programs and in provider mobility, including twinning programs, international joint-, double-, or multiple-degree programs, branch campuses, internationally cofounded institutions, and franchise universities (Knight, 2015a).

This chapter has four objectives. The first is to position and analyze education hubs within the frame of three generations of cross-border higher education. The second objective is to conceptually analyze the phenomenon by proposing a definition and a typology of the three major types of education hubs—student, talent, and knowledge–innovation. The third is to identify highlights of six current education hub countries in Southeast Asia, Africa, and the Gulf states by using the proposed typology to categorize them. Finally, the fourth is to examine how education hubs relate to the two previous generations of cross-border education activities in terms of geographic outreach and impact.

Three Generations of Cross-border Education

Numerous studies of higher education show that international academic mobility has been happening for a very long time and has evolved in diverse ways. To provide a brief overview of the evolution of cross-border education Table 21.1 summarizes the highlights of each of the three generations. Worth noting is that these generations are not mutually exclusive. In fact, education hubs build on and extend first and second generation activities.

[1]Sections of this chapter have been taken from Knight (2011, 2014a).

Table 21.1 Three generations of cross-border education

Primary Focus	Description
First Generation	
People Mobility Students, faculty, and scholars move to a foreign country for education and research purposes.	*Students* move to engage in full degree or short-term study, research, field work, internships, exchange programs. *Faculty* moves to teach, engage in professional development, and pursue research. *Scholars* move to strengthen international research collaboration and networks.
Second Generation	
Program and Provider Mobility Programs, institutions, or companies move across jurisdictional borders to deliver education and training.	*Programs* Types include twinning and franchise, joint-, double-, or multiple-degree, online, or distance programs. *Providers* Types include branch campuses, franchise universities, codeveloped universities, independent institutions.
Third Generation	
Education Hubs Countries, cities, or special zones attract foreign students, researchers, employees, programs, providers, and research and development (R&D) companies for purposes of education, training, knowledge production, and innovation.	*Student Hub*: Students, programs, and providers move to a foreign country for education purposes. *Talent Hub*: Students and workers move to a foreign country for education and training and stay for employment purposes *Knowledge* and *Innovation Hub*: Education researchers, scholars, higher education institutions, and research and development centers move to a foreign country to produce knowledge and innovation.

Note. Adapted from Knight (2014c, pp. 43–58). Reprinted with permission of Symposium Books.

In the following section, each generation is examined in more detail to illustrate the differences and similarities among them and to raise related issues and challenges.

The First Generation: Student Mobility

Nobody could have predicted the meteoric rise in all forms of student mobility in the last 50 years. The increase in mobile students from about 238,000 in the 1960s (Chen & Barnett, 2000) to 4 million in 2012 (UNESCO, 2014) is staggering. If forecasts are correct, this number will almost double in another 10 to 15 years. In the past four decades, the numbers of students, the types of mobility experiences, the driving rationales, and the destination countries have changed substantially.

When the term student mobility is used in a comprehensive sense, it usually refers to students who are taking a full degree in a foreign country or students who are participating in a semester or year abroad program as part of their academic studies at their home university. More recently, it also involves students enrolled in collaborative degree programs, such as joint, double, or multiple degree programs or franchise and twinning arrangements. Student mobility involves more than enrolling in academic courses abroad; it can include research or field work, internships or practicums. Given the importance of understanding foreign cultures and languages, students, especially those who cannot afford the time or costs of semester abroad, are participating in short-term cultural workshops, tours, and activities.

New forms of virtual mobility are emerging. Virtual mobility involves the use of information communication technologies to encourage cross-border collaboration for teaching and learning and eliminates the necessity of international travel. Working together virtually with counterpart teachers and students helps to enrich the learning experience and enhance intercultural understanding and the exchange of knowledge. Virtual mobility should not be confused with online or distance education, for it involves direct collaboration and exchange in a virtual learning environment and not merely access to learning opportunities or programs through electronic means. Virtual classrooms are yet another form of extending the internationality and geographic reach of universities and merit further attention and research.

Table 21.2 identifies the top 20 destination countries for students wishing to take their degree in a country different than their residence. According to the 2012 data from the UNESCO Institute for Statistics, five destination countries hosted nearly one-half of the total mobile student population: the United States (hosting 18%), United Kingdom (11%), France (7%), Australia (6%), and Germany (5%) (UNESCO, 2014). But the top five also saw their share of international enrollment decline from 55% in 2000 to 47% in 2012. There are new serious players in the field of international student recruitment, especially those from Asia.

In East Asia and Pacific region the traditional destinations of Australia and Japan are rivaled by newcomers China, Malaysia, the Republic of Korea, Singapore, and New Zealand, which collectively hosted 6% of the global share of mobile students in 2012. Among the Arab States, Egypt, Saudi Arabia, and United Arab Emirates are making efforts to recruit students from abroad. These three countries hosted 4% of the global share of mobile students (UNESCO, 2014). Although these numbers may be small, it is anticipated that they will increase substantially in the coming years.

Just as no one anticipated the growth in student mobility, no one predicted that international student recruitment campaigns would be linked to national innovation, science, and technology strategies or to trade and immigration policies, all in the quest for human talent to serve the knowledge economy. The brain train, or circulation concept, is the current term used to describe the trek of students and young professionals from country to country for study and employment reasons. But the notion of circulation masks the reality that there is net brain drain for some countries, usually smaller developing countries, and net brain gain for more economically advanced countries (Knight, 2008). By 2025, it is estimated that 7.8 million students will be enrolled in foreign countries for their tertiary education

Table 21.2 Top destination countries for international students in 2012

Rank	Destination Country	Percentage of Total International Students	Number of International Students
1	United States	18	740,482
2	United Kingdom	11	427,686
3	France	7	271,399
4	Australia	6	249,588
5	Germany	5	206,986
6	Russian Federation	4	173,627
7	Japan	4	150,617
8	Canada	3	120,960
9	China	2	88,979
10	Italy	2	77,732
11	South Africa	1	70,428
12	Malaysia	>1	63,625
13	South Korea	>1	59,472
14	Austria	>1	58,056
15	Netherlands	>1	57,509
16	Spain	>1	55,759
17	United Arab Emirates	>1	54,162
18	Singapore	>1	52,959
19	Egypt	>1	49,011
20	Saudi Arabia	>1	46,566

Note. Adapted from UNESCO (2014). Copyright 2014 by the UNESCO Institute for Statistics. Reprinted with permission. These statistics refer to students who crossed a national border to study or were enrolled in a distance-learning program abroad. These students were not residents or citizens of the country where they studied. Part-time, full-time, undergraduate, and postgraduate students are included. Exchange students are not included.

(Boehm, Davis, Meares, & Pearce, 2002), indicating that the first-generation cross-border education activities will continue to expand in scope and scale. The rationales and impact of student mobility will change as countries look to attract and retain students to fulfill their need for knowledge workers and skilled labor.

The Second Generation: Program and Provider Mobility

In the early 1990s the movement of programs and providers across borders began to increase substantially, raising the number of students able to access foreign higher education programs and qualifications without leaving home. Examples of cross-border program mobility include twinning and franchise programs as well as ones offering joint, double, and multiple degrees, with branch campuses, internationally cofounded institutions, franchise universities, and virtual universities being examples of cross-border institution and provider mobility (Knight, 2015b). Both program and provider mobility have become more popular, creating opportunities for large numbers of students wanting a foreign academic program and qualification.

Unfortunately, there is no comprehensive and reliable database on program and provider mobility. Many countries do not collect this data at the national level. More challenging is the reality that countries do not use the same definition or set of criteria to identify the different modes of program and provider mobility. Although the definitional issue has existed for collecting international student statistics, it is even more problematic to capture reliable data on program mobility (McNamara & Knight, 2015).

Provider mobility, in the form of branch campuses, presents a different scenario. Universities have been setting up campuses in foreign countries for decades, albeit in very small numbers and often without accreditation or licensing from the host country. Factors driving this growth include the increased demand for tertiary education arising from larger secondary school cohorts and the knowledge economy's need for a skilled labor force. Many countries found it more attractive to host branch campuses of foreign public and private universities than to invest in the physical and human infrastructure needed for an expanded domestic higher education sector (Verbik & Merkley, 2006). At the same time, regional and world trade agreements now include education as a tradable service, leading private and public education providers to explore new commercial opportunities in cross-border education. Clearly, large numbers of students have found it more attractive and economical to study at home at international branch campuses or internationally cofounded universities than to go abroad.

An international branch campus is defined as a "satellite operation of a recognized higher education institution or provider which offers academic programs and credentials in a different country than the home institution" (Knight, 2008, p. 181). According to the Observatory on Borderless Higher Education (Lawton & Katsomitros, 2012) there were just 24 branch campuses in 2002. But 13 years later there were 249 operating in all regions of the world. It is revealing to see the distribution and growth of these new initiatives by region. Table 21.3 shows that as of 2015, Asia was home to 83 of the 249 branch campuses around the world. This represented the largest number in a single region, with the forecast indicating continual growth. Of particular interest is that the number of receiving or host countries of branch campuses doubled from 36 in 2006 to 76 in 2015. In that period there were also some branch campus closings: 5 from 2006 to 2009, 12 between 2009 and 2011, and 15 from 2012 to 2015.

Overall, this unanticipated increase in branch campuses highlighted the second generation of cross-border education and strongly influenced the emergence of the third generation.

The Third Generation: Education Hubs

Education hubs are the most recent development and constitute the third wave of cross-border education initiatives. Education hubs build on and can include first and second generation cross-border activities, but they represent a wider and more

Table 21.3 Increase in the number of branch campuses, 2002–2015

	2002	2006	2009	2011	2015
Total number of branch campuses	24	82	162	200	249
Number of source countries		17	22	24	33
Number of host countries		36	51	67	76
Number of branch campuses hosted by region					
Africa			5	18	19
Asia Pacific			44	69	83
Europe			32	48	74
Latin America			18	10	9
Middle East			55	55	51
North America			8	10	12
Branch closures		6	5	12	15

Note. Data from Garrett, Kinser, Lane, & Merola (2016, pp. 51–52) and Lawton & Katsomitros (2012). Copyright by Observatory on Borderless Education and C-BERT.

strategic configuration of actors and activities. An education hub is a concerted and planned effort by a country, zone, or city to assemble a critical mass of local and international actors to support its efforts to build the higher education sector, expand the talent pool, or contribute to the knowledge economy.

There are only a handful of countries around the world seriously attempting to develop themselves as education hubs. These include Hong Kong, Singapore, Malaysia, United Arab Emirates, Qatar, and Botswana (Knight, 2014b). Others still in initial or perhaps "stalled" stages are Bahrain, Mauritius, Korea, and Sri Lanka. The economic crisis of 2008–2010 impacted plans by Botswana and Hong Kong to invest in hub development and slowed their progress considerably. And there are other countries that seem to be using the term education hub only as a branding label to attract more international students and providers.

In addition, there are cities around the world, for instance Panama City, Bangalore in India, and Monterey in Mexico, that have been seeking to brand themselves as education or knowledge cities. Some city-level initiatives, Panama being a prime example, are trying to be international in scale, while others are local level initiatives. The diversity of approaches to and motives for developing an education hub raises the questions of what, exactly, an education hub is and what it involves.

There is no single model of an international education hub or any one-size-fits-all approach to establishing one. Each country or jurisdiction has its own set of drivers, strategies, and expected outcomes. A new feature of the third generation of cross-border education is the emphasis on knowledge production and innovation. Education and training initiatives have been traditionally associated with the first two generations of cross-border education, so the addition of knowledge generation and application is a noteworthy development and feature of education hubs.

The Definition and Types of International Education Hubs

With countries, scholars, and policy makers defining education hubs differently, it is important to have a common understanding of the term. The purpose of this section is to examine the proposed working definition and typology of three different types of education hubs.

Key Concepts of an Education Hub[2]

Given the diversity of education hub models, plus the lack of any systematic study of the phenomenon to date, an analysis of the common characteristics of education hubs is warranted. On the assumption that the number and types of education hubs will increase, any working definition needs to be generic enough to apply to all levels of education hubs as well as to their scope of engagement and impact. A proposed definition, regardless of the level of the hub (country, zone, or city) or the region of the world where it is located, is, "an education hub is a planned effort to build a critical mass of local and international actors strategically engaged in cross-border education, training, knowledge production and innovation initiatives" (Knight, 2011, p. 227). The identification of driving rationales, expected outcomes, sponsors, major actors, and specific types of activities is intentionally omitted to allow the definition to apply to the emerging diversity of hubs. To fully understand the meaning and dimensions of the proposed definition, it is helpful to examine each core concept.

The concept of a *planned effort* indicates that a hub is an intentional or deliberate project that normally involves a strategy, policy framework, and some public and private investment. In other words, a hub is more than a coincidental interaction or colocation of actors working in the education and knowledge sectors. The notion of being planned helps to decrease the chances that it is merely a fad, a branding exercise, or a serendipitous set of temporary interactions among key players.

The notion of *critical mass* suggests that there is more than one actor and set of activities involved. This means that a single branch campus, franchise program, science and technology park, or internationally engaged institution does not constitute a hub. A hub is different from individual first and second generation cross-border activities because it brings these kinds of initiatives together in some kind of planned or coordinated project. The concept of critical mass intentionally goes beyond a random collection of cross-border activities by denoting the presence of a key combination of actors. The term *colocation* was considered and deliberately excluded from the definition even though it is significant to the meaning of a hub. Colocation can mean different things at the city, zone, or national levels. Actors can be colocated in a single location or multiple ones because of complementarities of

[2]The definition discussion is based on Knight (2011).

services, but this does not imply that all actors must be colocated in one designated area. Larger countries such as Malaysia and United Arab Emirates are good examples of multiple activities and multiple colocation sites, while Hong Kong and Singapore are small enough that the notion of one colocation site can apply.

The mention of *local and international actors* indicates that an international education hub involves both domestic and foreign players. These can include local, regional, and international students, scholars, institutions, companies, organizations, research centers, and knowledge industries. The term *actor* is used in an inclusive manner to cover providers, producers, and users of the education, training, and knowledge services and products. The diversity of actors will vary from hub to hub, depending on the rationales and functions of the hub, so different types of actors are intentionally not specified in the definition.

The idea of *engaging strategically* is central to the definition because it emphasizes a deliberate sense of interaction or relationship among the actors. While the nature of the engagement will differ from hub to hub, a fundamental principle is that there is added value when the actors connect, collaborate, or share common facilities and resources. This does not deny that there will be competition among actors offering similar services or products, but the pros of being part of a strategic and interactive initiative appear to outweigh the cons. The nature and number of interactions is unlimited, given the diversity of local and international actors and users. In addition, an education hub normally involves a master plan or overall strategy that is augmented by aligned policies and regulations, a match-up that enhances the chances of success and sustainability, as well as substantiating the importance of a strategic approach laid out in the definition.

Cross-border education, training, and knowledge and innovation initiatives depict the broad categories of activities and outputs of hubs. There is a wide selection of initiatives and services available, depending on the type of hub, priorities of the individual actors, and strategic plan of the sponsor.

Worth noting is that the definition does not refer to the level (zone, city, or country) of the hub because that quality is determined by the hub sponsors, as are the reach or engagement of its actors and the spread of its impact and influence. For example, a zone-, city-, or country-level education hub can aim to attract actors from its immediate vicinity or beyond, with its impact being local, national, regional, or global. The level and scope of a hub's activities are, therefore, not specified in the generic definition, although these elements would normally be part of the description of a specific education hub.

Finally, an education hub has not been defined in physical or spatial terms, for instance, as a designated area, because this could be too limiting. Rather, the central concept involves connectedness or a network of interactions among engaged local and international actors who undertake cross-border education activities to achieve their individual objectives as well as the collective goals and outcomes of the sponsoring body, whether it is a city, zone, or country.

The Types of Education Hubs: Student, Talent, and Knowledge-Innovation

As indicated, different rationales, actors, and activities characterize education hubs. Some countries see hubs as a means to build a critical mass of foreign students and providers to generate income as well as modernize and internationalize their domestic higher education. Other sponsors want to become hubs in order to train foreign and local students and employees to become part of a skilled labor force. And other countries focus on attracting foreign students and labor, institutions, and companies to build a vibrant research, knowledge, and innovation sector to lead them into the knowledge economy.

In order to capture the differences among various hub approaches and allow for a more nuanced understanding and exploration of education hubs, a typology of three categories of hubs is suggested (Knight, 2011). The typology is based on the rationales driving hub development, not on the location or level of hubs.

The *Student Hub* is the most focused type of education hub. Its key activity is educating and training local, expatriate, and international students. In addition to recruiting students, this model also seeks to increase access for all types of students by attracting foreign higher education institutions offering franchised and twinning programs or establishing branch campuses. The primary objectives of student hubs are a) to provide increased access to higher education for local students, b) to generate revenue from international student fees, c) to expand the capacity of local higher education institutions (HEIs), d) to internationalize the domestic higher education system, and e) to enhance the profile, branding, and ranking of local HEIs and the host country.

In the student hub scenario, both local HEIs and foreign providers recruit local and international students to their programs and campuses. A student hub often gives priority to foreign student enrollment even though there is an interest in providing wider access for local students. A student hub may intend to attract students from all parts of the world, but in many cases the majority of students come from neighboring countries. In a student hub model, foreign students are recruited to complete their studies in the host country and then return home or move to a third county. Generally, they are not encouraged or provided incentives to stay in the host country.

The *Talent Hub* (*Skilled Workforce*) model focuses on student education and training but differs from the student hub because its overarching goal is human resource development for a skilled work force. Foreign students are therefore encouraged to remain in the host country for employment purposes. Retention of foreign students (and workers) is central to the talent hub model. International HEIs, as well as private training and education companies, are encouraged to offer academic programs and professional development opportunities aimed at international, expatriate, and national students, as well as local employees. The overall goal is human resource development. The driving objectives are to a) expand talent pool of skilled workers, b) build a service or knowledge based economy, c) increase economic competiveness and influence in region and beyond, and d) strengthen the quality and relevance of labor. Education and training institutions and providers are often colocated in a common zone to facilitate the use of shared facilities and promote collaboration among them and with industry. In order to develop a critical mass there can be more than one colocation site in a country.

The *Knowledge–Innovation Hub* broadens its mandate beyond education and training to include the production and distribution of knowledge and innovation. Foreign actors, including universities, research institutes, companies with major research and development activities are persuaded through favorable business incentives to establish a base in the country and to collaborate with local partners in developing applied research, knowledge, and innovation. The model's primary objectives are to a) to build a knowledge- and innovation-based economy, b) to attract foreign direct investment, c) to expand the capacity of local research and development centers, d) to increase competitiveness in specialized fields, and e) to enhance soft power. Collaboration among the key players—foreign and local education institutions, industries, research centers, and companies—is a key factor in establishing a knowledge and innovation hub and providing added value for the major actors.

Highlights of Six International Education Hubs

As of 2015, the six international country-level education hubs of Hong Kong, Singapore, Malaysia, United Arab Emirates, Qatar, and Botswana were in different stages of implementation. They are located in three different regions of the world—the Middle East, South East Asia, and Africa. While the six countries or jurisdictions are very different, all are relatively small and committed to moving their economies from a dependence on national resources or manufacturing to being based on knowledge and service industries.

The term education hub is a subjective and self-ascribed label. There is no exclusive set of indicators or official body that determines whether a country meets stated requirements to be called an education hub. As the popularity and *branding value* of the concept increases, so does the number of countries seeking to become education hubs.

Singapore is one of the more serious and successful hubs. It has moved over the last 15 years from its *Global Schoolhouse* project, which concentrated on recruiting foreign students and prestigious universities, to its current strategy emphasizing investing in major research initiatives and facilities to establish sustainable international research partnerships (Sidhu, Ho, & Yeoh, 2014). Its current focus is on research, knowledge production, and innovation, with the Singapore government's investment of considerable financial, human, and structural resources in the project underlining its belief that knowledge and innovation are the cornerstones of its shift to the knowledge economy. Based on the previously described typology, Singapore can be categorized a Knowledge–Innovation education hub and is, interestingly, the only one of the six country level education hubs in this category.

Qatar has taken another approach by developing itself as a Student-Talent education hub with aspirations of becoming a Knowledge-Innovation education hub. Its centerpiece is Education City, a site housing 10 prestigious universities from the United States and United Kingdom invited there and generously supported by the Qatar Foundation, which oversees the strategy and development of the country's hub plans. Another core element is a science and technology park that Qatar established in a free zone, which is special jurisdiction offering tax and financial incentives to attract international branch campuses. Forming international research partnerships, building research facilities, developing a research culture, training researchers, and providing major research grant programs are all key components of the Qatar approach (Ibnouf, Dou, & Knight, 2014).

United Arab Emirates, a neighbor of Qatar, can also be labeled a Student–Talent education hub, although it has chosen a very different model. Four of the country's seven emirates have recruited international branch campuses (IBCs) to provide increased access for expatriate and domestic students. The Knowledge Village in Dubai and the Dubai International Academic City are the best known free zones in United Arab Emirates and together host about 25 of its 37 IBCs. Using a different approach, Abu Dhabi, the wealthiest emirate, has invited and generously supported elite universities from the United States and France and has invested in research partnerships with foreign universities to develop centers of excellence, such as Masdar City. Abu Dhabi's approach of investing in and supporting foreign branch campuses differs markedly from the commercial approach used by the other emirates, especially Dubai. There is no overall country-level strategy for developing United Arab Emirates as an education hub. This has led to the diversity of approaches, which seem to have been successfully used to date (Fox & Al Shamisi, 2014).

Malaysia is a country with a long history of international education. It has developed a comprehensive but diversified approach to positioning itself as a Student education hub with long-term aspirations of becoming a Knowledge education hub. Over the last decade, seven international branch campuses have been established throughout the country and there are more in the pipeline for approval. Malaysia has doubled its number of international students, using its attractiveness to Muslim students as a key feature. Iskandar, an ambitious new Malaysian free zone abutting Singapore, is also under development and already home to several branch campuses of major international universities. Other policies and programs have also been established that aim to increase Malaysia's attractiveness and competitiveness as an education hub. The country's efforts to date have focused on education and training rather than research (Aziz & Abdullah, 2014).

Hong Kong's intentions to position itself as an education hub and its related policy statements have been clear, but plans to move forward are less visible (Mok & Bodycott, 2014). Troubled by the economic downturn in 2008, Hong Kong's efforts have focused on offering scholarships and recruiting more "non-local students"—a term used to describe students from the region and Mainland China who officially cannot be classified as foreign or international students (Cheng, Cheung, & Yeun, 2011). Given its priority of recruiting students, Hong Kong sees itself as a Student education hub, but seeks to become more of a Talent education hub as immigration policies change and more students stay and work in the city. Aspirations are high but progress has been slow, so Hong Kong is probably best described as an education hub in the making.

Botswana has taken a rather innovative approach to planning its development as a Talent education hub. To broaden its economic base, Botswana identified and prioritized five different industrial hub sectors and areas for investment. With all of them requiring competent and trained professionals, the country envisions the education hub as serving to educate, train, and supply the required labor. While Botswana has taken steps to attract more foreign students and international branch campuses, progress has been moderated by financial challenges. The extensive consultation and planning process has provided a firm foundation for achieving its goal of becoming a Talent education hub, but finding resources to implement the plan is taking longer than the government anticipated and Botswana's progress fulfilling its plan has been limited (John, Wilmoth, & Mokopakgosi, 2014).

The Relationship of Education Hubs to First and Second Generation Cross-border Education Activities[3]

As I have discussed, education hubs have built on first and second generation cross-border education activities. It is useful, therefore, to explore whether there is any correlation between education hub locations and the most popular international student destinations and countries hosting international branch campuses. A geographic lens is used to examine these potential relationships.

Destination Countries for International Students

Table 21.2 lists the top 20 destination countries for international students (UNESCO, 2014). Interestingly, none of the six education hub countries rank among the top 10 destination countries for international students, with Malaysia in 12th place, followed by United Arab Emirates and Singapore in the 17th and 18th positions, respectively. This raises the question as to why education hub countries are not preferred destinations for larger numbers of international students.

Worth noting is that the six hub countries have reasonably well developed higher education systems but are relatively small in size, which makes them unable to host large numbers of international students. Most hubs, especially the student and talent hubs, aim to increase the number of education providers and programs, and hence the number of students. In contrast, the giants of international student recruitment

[3]This section is based on the discussion in Knight (2014a).

like the United States, the United Kingdom, and Australia are already popular destination countries and not moving toward establishing themselves as education hubs. The international education actors and activities in these countries are so diverse and numerous that trying to develop an education hub master plan would be a daunting challenge.

Economic development plans, international education engagement, and size are all factors at play in determining the desirability and potential of becoming an education hub. It appears that if an education hub is to be anything more than a branding label or status symbol, it will probably continue to be the smaller and more developed countries that are able to strategically invest considerable effort and funding in planning and developing a critical mass of local and international actors working collaboratively on cross-border teaching, training, and research activities. Thus, it will likely not be the large countries now leading international student recruitment that transition into education hubs, but these smaller countries. This may be counterintuitive but the reality of the cited student data supports the conclusion.

Zone- and City-Level Hubs

A possible alternative scenario involves these large countries so successful in international student recruitment establishing education hubs at the city or zone level. These kinds of education hubs are characterized by colocation of key actors in a specific geographic area. Boston is often referred to as an education hub because of its concentration of universities and research institutes. Interestingly, Boston did not start out with a master plan to develop itself as an education hub, but it may be the best example of a city-level education hub in the world (Crabtree, 2006). India has announced its plan to establish 12 city-level education hubs, but close examination of those plans reveals an aim to foster closer links between local HEIs and private industry. Thus, the country does not plan, at least at this stage, to make the cities a center of local and foreign actors working collaboratively on cross-border education activities. Monterrey in Mexico is an example of a city that actually did work on a strategic plan to develop and market itself as a knowledge city (Engardio, 2009). Plans, investments, and actors were on board but Monterrey's early progress stalled because of political and economic problems associated with the growing drug cartels in the region.

Panama City is another interesting example of an urban center attempting to become an education hub. Over the last ten or fifteen years, Panama's City of Knowledge has undertaken several bold initiatives (Vonortas, 2002), including the establishment of a "Techno Park" that provides infrastructure and services to research and technology companies, houses many regional offices of international government organizations, manages international cooperation projects, and hosts foreign universities' international programs and one branch campus. Although the city has yet to achieve its goal of being a preferred destination for international

Table 21.4 Top branch campus source and host countries, 2015

Source Countries		Host Countries	
United States	78	United Arab Emirates	31
Australia	15	China	32
United Kingdom	39	Singapore	12
France	28	Qatar	11
Russia	21	Malaysia	12

Note. Data from Garrett et al. (2016, pp. 48–51).

students and foreign branch campuses, it has developed an interesting model catering to the needs of the country and the demands of the market.

Silicon Valley in California is a well-known example of a successful zone-level research and innovation hub. Other countries are trying to emulate this successful initiative, with, for example, Bangalore in India often being referred to as the Silicon Valley of India (Collato, 2010). Korea is another interesting case. Given its strategy of developing two special education zones, the Songdo Global University Campus and Jeju Global Education Campus, it is still unclear whether Korea will become a comprehensive country-level education hub, integrating the two zones and other international education projects, such as Brain Korea 21, or a nation with two independent, zone-level education hubs.

International Branch Campus Host Countries

It is equally interesting to discover if there is any relationship between the location of international branch campuses around the world and the location of education hubs. As of 2011 there were about 200 international branch campuses operational in more than 67 countries around the world (Lawton & Katsomitros, 2012). Unlike the lack of correspondence seen in regard to destination countries for international students, there seems to be a direct correlation between international branch campuses and education hubs. The evidence is clear and convincing, with four of the five top host countries of branch campuses being country-level education hubs. As presented in Table 21.4, the top receiving countries in 2011 were United Arab Emirates, Singapore, China, Qatar, and Malaysia. Counting the six established country-level education hubs and four emerging ones (Korea, Mauritius, Bahrain, Sri Lanka), education hub countries hosted 40% of the total branch campuses in 2015 and were also home to the highest concentration of education hubs in individual countries (except for China).

In conclusion, it appears that the education hub countries are not necessarily the most popular destination for international students, although they do host the largest concentration of international branch campuses. There are many factors that influence a country's decision to position itself as a hub, and size appears to be one them. Smaller countries that are relatively politically and economically stable have the capacity to make and attract public and private investment, support a reasonably adequate tertiary education system, and undertake the planning and policy preparation necessary to develop themselves as education hubs. It is the countries importing international branch campuses that are hubs, not the large countries exporting them. As of 2015, the majority of hub countries were focused on attracting students and education providers for economic reasons or for developing a skilled work force. Only Singapore can be described as a knowledge and innovation type education hub building on its reputation for excellence in its higher education system and its long history of strategic international engagement with top universities around the world.

Regional Engagement of Education Hubs

The regionalization of higher education is an increasingly important trend, not only in Europe, but also in Asia, Africa, and Latin America (Aphijanyathan, 2010; de Prado Yepes, 2006; Knight, 2013). It is therefore interesting to look at the issue of regionality with respect to the geographical reach and engagement of education hubs. Reach and engagement refer to the extent to which an education hub reaches out to other countries to attract and involve key actors and participants in cross-border education activities. Given that international branch campuses are key players in hubs, the location of their home or source institutions is relevant to the question of regionality. Table 21.5 summarizes the total number of international branch campuses per hub country and indicates how many are linked to universities within the same region and how many are sourced from countries outside the region. Only three countries—United Arab Emirates, Malaysia, and Singapore—have a handful of branch campuses from universities located within the region, representing about 17% of the total. This demonstrates that the reach and engagement of education hubs in terms of branch campuses is primarily beyond the region. For some, this is an unexpected finding, for education hubs are often promoted as being regionally based.

Table 21.5 Regional source of international branch campuses (IBCs) in education hub countries

	Number of IBCs in Country	IBCs External to Region	IBCs Internal to Region
Botswana	1	1	0
Hong Kong	5	5	0
Malaysia	12	5	7
Qatar	11	11	0
Singapore	12	7	4
United Arab Emirates	31	29	2
Totals	72	58	13

Note. Data from Garrett et al. (2016, pp. 48–51).

Unfortunately, reliable data is not available on whether the international students enrolled at branch campuses or local HEIs originate from countries internal or external to the region. A best guess is that overall more students come from countries outside the region, but this is not true for all countries, with Hong Kong and Qatar being two examples where this does not apply. Furthermore, United Arab Emirates is a difficult situation to assess because many of the students at branch campuses there are the children of expatriates and though born and brought up in United Arab Emirates hold the citizenship of their parents' home country. In this case, what is their country of origin—the one where they have lived all of their lives or their country of citizenship? The situation is murky.

Looking to the Future of Education Hubs

It is fascinating and at times bewildering to imagine the evolution of education hubs over the next decade. If the pace of change continues at the rate seen in the last ten years, the education hub phenomenon is bound to experience fundamental transformations. Will one model of education hub dominant? Will the demand for international education remain strong enough to support student-type hubs dependent on physical mobility or will student hubs become virtual entities? Is the term education hub anything more than a branding exercise or marketing strategy? Will talent hubs become a powerful and sustainable strategy for brain gain? Will education hubs evolve into strong and effective agents of knowledge diplomacy? Education hubs are not static entities; they evolve in reactive and proactive ways to external exigencies, unintended consequences, and new opportunities. Are binational or multinational hubs on the horizon? Will education cities become more popular and ubiquitous? Will education hubs morph into a fourth generation or be replaced by a new cross-border education development? These kinds of questions and speculations have no immediate answers, but they do serve to stimulate more lateral thinking about the future of cross-border education and how HEIs work collaboratively and internationally with other domestic and foreign actors.

References

Altbach, P. G. (2013). *The international imperative in higher education.* Rotterdam: Sense.

Aphijanyathan, R. (2010). *A research report on East Asian internationalization of higher education: A key to regional integration* (Program Report 24). Bangkok: Regional Centre for Higher Education and Development, South East Ministers of Education Organization.

Aziz, M. I. A., & Abdullah, D. (2014). Malaysia: Becoming an education hub to serve national development. In J. Knight (Ed.), *International education hubs: Student, talent, knowledge-innovation models* (pp. 101–119). Dordrecht: Springer.

Boehm, A., Davis, D., Meares, D., & Pearce, D. (2002). *The global student mobility 2025 report: Forecasts of the global demand for international education.* Canberra: IDP Education.

Chen, T., & Barnett, G. (2000). Research on international student flows from a macro perspective: A network analysis of 1985, 1989 and 1995. *Higher Education, 39*, 435–453. doi:https://doi.org/10.1023/A:1003961327009

Cheng, Y. C., Cheung, A. C. K., & Yeun, T. W. W. (2011). Development of a regional education hub: The case of Hong Kong. *International Journal of Educational Management, 25*, 474–493. doi:https://doi.org/10.1108/09513541111146378

Collato, F. (2010). Is Bangalore the Silicon Valley of Asia? Analysis of the evolution and the structure of this Indian local economy organization. *Journal of Indian Business Research, 2,* 52–65. doi: https://doi.org/10.1108/17554191011032947

Crabtree, J. (2006). *Ideopolis: Knowledge city region: Boston case study*. Retrieved from http://www.theworkfoundation.com/downloadpublication/report/128_128_ideo_boston.pdf

de Prado Yepes, C. (2006). World regionalization of higher education: Policy proposals for international organizations. *Higher Education Policy, 19*, 111–128. doi:https://doi.org/10.1057/palgrave.hep.8300113

Engardio, P. (2009, June 1). A Mexican technology park in Monterrey. *Bloomberg*. Retrieved from http://www.bloomberg.com/news/articles/2009-06-01/a-mexican-technology-park-in-monterrey

Fox, W. H., & Al Shamisi, S. (2014). United Arab Emirates' education hub: A decade of development. In J. Knight (Ed.), *International education hubs: Student, talent, knowledge-innovation models* (pp. 63–80). Dordrecht: Springer.

Garrett, R., Kinser, K , Lane, J., & Merola, R. (2016). *International branch campuses: Trends and developments*. The Observatory of Borderless Higher Education and C-BERT. Retrieved from www.obhe.ac.uk/documents/download?id=1049

Ibnouf, A., Dou, L., & Knight, J. (2014). The evolution of Qatar as an education hub: Moving to a knowledge-based economy. In J. Knight (Ed.), *International education hubs: Student, talent, knowledge-innovation models* (pp. 43–61). Dordrecht: Springer.

John, B. P., Wilmoth, J. D., & Mokopakgosi, B. (2014). Botswana country hub: Africa's first education hub. In J. Knight (Ed.), *International education hubs: Student, talent, knowledge-innovation models* (pp. 145–164). Dordrecht: Springer.

Knight, J. (2008). *Higher education in turmoil: The changing world of internationalization*. Rotterdam: Sense.

Knight, J. (2011). Education hubs: A fad, a brand, an innovation? *Journal of Studies in International Education, 15*, 221–240.doi:https://doi.org/10.1177/1028315311398046

Knight, J. (2013). Towards African higher education regionalization and harmonization: Functional, organizational and political approaches. In A. Wiseman & C. Wolhuter (Eds.), *The development of higher education in Africa: Prospects and challenges* (pp. 347–373). International Perspectives on Education and Society Series: Vol. 21. Bingley: Emerald.

Knight, J. (2014a). Education hubs: Issues, indicators, and reflections. In J. Knight (Ed.), *International education hubs: Student, talent and knowledge-innovation models* (pp. 207–228). Dordrecht: Springer.

Knight, J. (Ed.). (2014b). *International education hubs: Student, talent, knowledge-innovation models*. Dordrecht: Springer.

Knight, J. (2014c). Three generations of crossborder higher education: New developments, issues and challenges. In B. Streitwieser (Ed.), *Internationalization of higher education and global mobility* (pp. 45–58). Oxford Studies in Comparative Education. Oxford, UK: Symposium Books.

Knight, J. (2015a). International universities: Misunderstandings and emerging models. *Journal of Studies in International Education, 19*, 107–121. doi:https://doi.org/10.1177/1028315315572899

Knight, J. (2015b). Transnational education remodeled: Towards a common TNE framework and definitions. *Journal of Studies in International Education, 20*, 34–47. doi:https://doi.org/10.1177/1028315315602927

Lawton, W., & Katsomitros, A. (2012). *International branch campuses: Data and developments*. London: Observatory on Borderless Higher Education.

McNamara, J., & Knight, J. (2015). *Transnational education data collection systems: Awareness, analysis, action*. London: British Council and German Academic Exchange Service.
Mok, K. H., & Bodycott, P. (2014). Hong Kong: The quest for regional education hub status. In J. Knight (Ed.), *International education hubs: Student, talent, knowledge-innovation models* (pp. 81–99). Dordrecht: Springer.
Sidhu, R., Ho, K.-C., & Yeoh, B. S. A. (2014). Singapore: Building a knowledge and education hub. In J. Knight (Ed.), *International education hubs: Student, talent, knowledge-innovation models* (pp. 121–143). Dordrecht: Springer.
UNESCO Institute of Statistics. (2014). *Global flow of tertiary-level students*. Retrieved from http://www.uis.unesco.org/Education/Pages/international-student-flow-viz.aspxN
Vonortas, N. S. (2002). Building competitive firms: Technology policy initiatives in Latin America. *Technology in Society, 24*, 433–459. doi:https://doi.org/10.1016/S0160-791X(02)00034-9
Verbik, L., & Merkley, C. (2006). *The international branch campus-models and trends*. London: Observatory on Borderless Higher Education.

The Klaus Tschira Stiftung

The Klaus Tschira Stiftung was created in 1995 by the physicist Klaus Tschira (1940–2015). It is one of Europe's largest privately funded non-profit foundations. The Foundation promotes the advancement of natural sciences, mathematics, and computer science and strives to raise appreciation of these fields. The focal points of the foundation are "Natural Sciences – Right from the Beginning," "Research," and "Science Communication." The involvement of the Klaus Tschira Stiftung begins in kindergartens and continues in primary and secondary schools, universities, and research facilities. The foundation champions new methods in the transfer of scientific knowledge, and supports both the development and intelligible presentation of research findings. The Klaus Tschira Stiftung pursues its objectives by conducting projects of its own but also awards subsidies after approval of applications. To foster and sustain work on selected topics, the Stiftung has also founded its own affiliates. Klaus Tschira's commitment to this objective was honored in 1999 with the "Deutscher Stifterpreis," the award conferred by the National Association of German Foundations.

The Klaus Tschira Stiftung is located in Heidelberg and has its head office in the Villa Bosch (Fig. 2), once the residence of Carl Bosch, a Nobel laureate in chemistry.

www.klaus-tschira-stiftung.de

P. Meusburger et al. (eds.), *Geographies of the University*, Knowledge and Space 12,
https://doi.org/10.1007/978-3-319-75593-9

Fig. 1 Participants of the symposium "Geographies of the University" at the Studio Villa Bosch in Heidelberg, Germany. © Peter Meusburger, Heidelberg.

Fig. 2 Villa Bosch, the head office of the Klaus Tschira Stiftung, Heidelberg, Germany. © Peter Meusburger, Heidelberg.

Index

P. Meusburger et al. (eds.), *Geographies of the University*, Knowledge and Space 12,
https://doi.org/10.1007/978-3-319-75593-9

D

S

MIX
Papier aus verantwortungsvollen Quellen
Paper from responsible sources
FSC® C105338

If you have any concerns about our products,
you can contact us on
ProductSafety@springernature.com

In case Publisher is established outside the EU,
the EU authorized representative is:
Springer Nature Customer Service Center GmbH
Europaplatz 3, 69115 Heidelberg, Germany

Printed by Libri Plureos GmbH
in Hamburg, Germany